D1543524

Statistical Thinking for Behavioral Scientists

Statistical Thinking for Behavioral Scientists

David K. Hildebrand

The University of Pennsylvania

Duxbury Press

Boston

PWS PUBLISHERS

Prindle, Weber & Schmidt • Duxbury Press • PWS Engineering • Breton Publishers •
20 Park Plaza • Boston, Massachusetts 02116

Copyright © 1986 by PWS Publishers

All rights reserved. No part of this book may be reproduced or transmitted in any form or by any means, electronic or mechanical, including photocopying, recording, or any information storage and retrieval system, without permission, in writing, from the publisher.

PWS Publishers is a division of Wadsworth, Inc.

Printed in the United States of America
86 87 88 89 90 — 10 9 8 7 6 5 4 3 2 1

Library of Congress Cataloging-in-Publication Data

Hildebrand, David K.
Statistical thinking for behavioral scientists.

Bibliography: p.
Includes index.
1. Statistics. I. Title.
QA276.12.H557 1986 519.5 85-25377
ISBN 0-87150-949-0

ISBN 0-87150-949-0

Sponsoring Editor: Michael Payne
Editorial Assistants: Deanne Hart and Traci Sobocinski
Composition: Edwards Brothers
Printing and Binding: R. R. Donnelley
Cover Printing: New England Book Components
Production: Lifland et al., Bookmakers
Production Coordinator: Helen Walden
Text Design: Julia Gecha
Cover Art: Copyright © 1985 by Michael Lasuchin

To Mrs. Dr. H. from Mr. Dr. H.
(with many thanks)

Preface

The author of another statistics text must either justify his actions or plead not-so-temporary insanity. After teaching both undergraduate and graduate psychology and education students for years, I have felt a large and growing need for a text that responds adequately to the availability of modern computing power. No such book has become available.

Three objectives for a statistics text in this era seem obvious:

1. Given that calculators or computer programs will perform most of the routine arithmetic of statistics, the focus of the book must turn away from the arithmetic and toward thoughtful planning of statistical studies and intelligent interpretation of the results.

2. Given that computers can draw plots of data very quickly, a textbook should incorporate careful examination of pictures of the data as a routine part of any statistical analysis.

3. Given that computers can simulate sampling from virtually any sort of population, a textbook should use simulation extensively, to demonstrate critical concepts, illustrate important theorems, and investigate the effect of violation of assumptions.

In this book, aimed at advanced undergraduate and beginning graduate students in psychology and education, I have incorporated the power of computing within a presentation of classical statistical methods. Along with discussion of the essential ideas of probability and statistics, there are a substantial number of computer outputs and many computer simulation results. The outputs are from the standard, nationally available packages. It is not absolutely necessary, though of course it would be helpful, for students to have access to these packages; the outputs stand on their own.

Though the concepts and methods discussed in this book are the same as those discussed in most introductory textbooks, there are some important differences in style and approach.

1. The emphasis is on critical understanding rather than on calculation. Almost every exercise that asks the student to perform a certain statistical procedure is followed by another that asks the student to judge whether the procedure is appropriate for the data or to judge whether a particular conclusion is justified.

2. The assumptions underlying each procedure are spelled out, and methods for checking the validity of those assumptions are given.

3. Computer simulations are used repeatedly to illustrate the effect of particular violations of assumptions. Sample sizes are varied in many of these simulations, so students can see which assumptions become less crucial with increasing sample sizes and which do not.

4. The basic ideas of elementary probability theory are closely integrated with the essential ideas of statistical inference, rather than being lumped in a "throwaway" chapter.

5. Nonparametric competitors to parametric procedures are introduced along with the parametric methods, rather than being segregated in another throwaway chapter.

6. Confusing technical phrases have been replaced by more meaningful alternatives. In particular, "statistically detectable" is used instead of "statistically significant," and "theoretical distribution" is used instead of "sampling distribution." The conventional language is introduced, along with an argument for the use of the alternative language.

7. The close relationship between confidence intervals and hypothesis tests is stressed. Wherever practical, the confidence interval and hypothesis test forms of a particular procedure are stated in the same section. Students should be able to use whichever form of inference is most revealing.

8. Methods for assessing association in cross-tabulated data, which are not always covered in elementary texts, are discussed in Chapter 3. Thus a student faced with a cross-tab will not have to rely solely on a chi-squared test.

9. This book contains uncommonly extensive discussions of the power of various tests. A student designing a study will be able to calculate power without having to consult hard-to-find references. Typically, the power discussion is in a separate section, so that it is available but not intrusive.

Almost all of the book should be accessible to students with only high-school mathematics background. In a few places, some calculus methods are used where they are natural. These sections and the associated exercises may be omitted without major loss. The primary requirement is that students be able to use mathematical ideas in critical thinking. Many times, students will be asked to judge whether a particular method is appropriate for a particular set of data, and also whether a particular conclusion is justified based on that test. This is not only a how-to book; it is also a why-to, when-to, when-not-to book, as well as a what's-it-mean book.

Many people have contributed much to this book. Dr. Patricia J. Hildebrand did almost all of the computer work. Amy Kohn researched the many articles from the literature referred to in the text. The students and faculty of the department of psychology at Penn provided many of the ideas and some of the data used in examples and exercises. The students in my undergraduate and graduate classes at Penn have encouraged me in this effort. Drafts of the book have been reviewed by many professors from other institutions. I have appreciated their suggestions, even when I have not been able to incorporate them. Kimberly Kist and Michael Payne have nurtured the book in their role as editors. Helen Walden and especially Sally Lifland have labored mightily to convert my prose into English.

Most of all, I thank my family for their patience as I've growled, groaned, and labored on the book.

David K. Hildebrand

Contents

5 Random Variables and Probability Distributions 135

6 Special Types of Probability Distributions 179

7 Random Sampling, Statistics, and Probability Theory 223

8 Statistical Inference: Estimation 273

13 Methods for Qualitative Data 487

14 Design and Analysis of Experiments 539

15 Regression and Correlation Methods 635

References 743

Tables 747

1 Making Sense of Data

Statistical thinking is a major part of scientific thinking, particularly in psychology, education, and all the social sciences. Many, if not most, of the scientific conclusions in psychology journals are based on statistical evidence. A basic understanding of the principles of statistical reasoning is a necessity for anyone who wants to study a social science. This text is about statistical reasoning. It is an introduction to the main problem of statistics: making sense of data.

data There are several issues in the business of making sense of data. For statistical purposes, **data** means a collection of *numerical* information, together with an explicit or implicit statement of how the numbers were obtained. Purely verbal descriptions of events, although important parts of science, are not data in the statistical sense. The first problem is how to gather the data. What information should be obtained? From what entities? Are the data to be collected in an experimental laboratory setting or observed in a more naturalistic setting? Are there standard scales to use in measuring the data? The answers to all these questions and more must be determined when the study is being planned.

population Ideas on how to gather data will occur throughout this book. Social science data will almost always be a sample from a much larger population. The word **population** is used in a slightly unusual way in statistics; it means the set of all the measurements that, in principle, one would wish to have in answering a scientific question. Suppose we wanted to study the level of computer literacy of students currently enrolled in social statistics courses in the United States. The relevant population would be computer-literacy measurements for the tens of thousands of students currently enrolled in such courses. Note the slightly peculiar usage: the population consists of the numbers, not the people.

sample

random sampling

In principle, we would like to know every number in this population; in practice, we will have to settle for much less—a **sample,** which is a much smaller subset of scores. One problem in gathering data is how to make the sample at least roughly similar to the population. Measuring students only at M.I.T. would not do, nor would measuring students only at a single community college. Ideally some form of **random sampling** should be used, to minimize potential bias in the sample. Random sampling will be discussed in Chapter 7.

Once the data have been gathered, they must be summarized. A list of 625 computer literacy scores is incomprehensible to most people; the scores must be summarized to bring out basic patterns in the data. What is a typical score? Is there a lot of variability around that typical score? Are the scores spread out smoothly or are there several clumps? Are there any *outliers*—wild values far away from the rest of the data? The most basic ways of summarizing the data for a single variable appear in Chapter 2.

A critical statistical idea is the notion of statistical relationship (or association or correlation) discussed in Chapter 3. Let us consider an example. Gallistel (1983) performed a study in which electric charges of various durations and intensities were administered to a certain portion of the brain of several laboratory rats, and the running speeds of the rats obtained. For rat S-61, the data in Table 1.1 were obtained:

Duration = 0.4		Duration = 4.0	
Intensity	Speed	Intensity	Speed
0.90	34	4.0	21
1.00	40	4.6	29
1.10	31	5.1	32
1.26	68	5.8	36
1.44	59	6.4	42
1.59	73	7.2	57
1.81	61	8.0	60
2.00	96	9.1	70
		10.1	87

Table 1.1

There appears to be a relation between intensity and speed; generally, as intensity increases, speed increases. But what kind of a relation is there? Is it a straight-line or a curved relation? How strong is the relation? Does the relation at duration 0.4 appear similar to the relation at duration 4.0? The methods for answering these questions are discussed in Chapter 15.

The Gallistel data were obtained in an experimental laboratory setting. Much of the use of statistical reasoning, in psychology particularly, is in designing experiments and in analyzing the resulting data. There are many issues. How many subjects should be used? Should each subject be measured under only one experimental condition or under all experimental conditions? Are several different experimental factors involved? How much of an effect does each experimental factor seem to have on the average response of subjects? Do certain combinations of the factors produce unusual effects? The basic methods for designing experiments and analyzing experimental data are discussed in Chapter 14.

The summary figures obtained from a sample will in general not be the same as the corresponding summary figures for the underlying population. *There is no such thing as a (perfectly) representative sample.* Almost inevitably, there will be some degree of error in estimating a summary quantity for the underlying population. Presumably, that error should be smaller for a large sample than for a small one, and for a well-chosen sample than for a sloppily chosen one. How can we assess the probable degree of error? This is one of the issues of **inference.** What can we say about the full data set (the population) with knowledge only of the partial data set (the sample)? Inference is the central problem of statistical thinking. The basic concepts are introduced in Chapters 8 and 9; these concepts are used over and over again throughout the rest of the book.

inference

Inference is always uncertain to some degree; it is never exact. Given limited information, we can only come to limited, somewhat uncertain conclusions. In making an inference, it is vital to know and to communicate the degree of uncertainty involved. The basic language of uncertainty is probability theory. Probability is at the heart of the theory of inference, as well as being an important language for stating scientific theories. The basic concepts of probability theory that we will need are developed in Chapters 4 through 7, and specific probability facts are stated as needed throughout the book.

These issues—data gathering, data summarization, statistical relation, experimental design and analysis of experimental data, probability theory, and inference—are the core of statistical thinking. Once we have an understanding of this basic core, we can develop the most commonly used statistical methods. Roughly the second half of the book is devoted to standard statistical techniques. There are some basic ideas underlying many of the methods; in particular, the so-called General Linear Model covers many of the standard methods for analyzing both observational and experimental data. By focusing on the underlying ideas and treating the various methods as special cases, we can organize the methods into a coherent system of ideas for making sense of data.

Summarizing Data

2

The first task in statistical analysis of data is summarization. Most studies in psychology, education, and other social sciences involve the collection of numerical data; the initial result may be a notebook or computer tape with a great blob of numbers. Without some careful summarization, the blob will most likely be indigestible. In this chapter and the next, we will consider some basic methods for summarizing data and making it comprehensible. This chapter is devoted to summarizing data for a single variable; the next chapter is concerned with summarizing relationships among variables.

We will focus mostly on three topics in this chapter: the general shape of the data, the middle of the data, and the variability of the data. Section 2.1 begins with a distinction between qualitative and quantitative variables, and then discusses the fine and enormously important art of drawing pictures about data. Section 2.2 contains a discussion of ways to measure a typical, average value of a variable. Section 2.3 describes some ways to measure variability around the middling value. Section 2.4 discusses some ways to measure other relevant properties of the data, particularly skewness (asymmetry) and "heavy-tailness." Section 2.5 is about the special problems that arise with *ordinal* data—data that reflect ordering, but not necessarily magnitude. Finally, Section 2.6 contains some useful mathematical properties of the mean and median of a variable.

2.1 Plot the Data!

In a discussion of methods for summarizing data, it seems useful to have some data to summarize. A statistics professor at a certain university collected information on students in an introductory class; the variables included were sex (female = 1, male = 0), major (biology = 1, psychology = 2, other social science = 3, all other = 4), self-reported degree of "math phobia" (ranging from "terrified of anything mathematical" = 1, through "not scared of it, not very good at it" = 3, to "enjoy it and good at it" = 5), age (in years), score on first midterm exam (out of 100), score on second midterm exam (likewise), and score on final exam (out of 200). Suppose that the data for 46 students were as shown in Table 2.1.

Sex	Major	Phobia	Age	M1	M2	F	Sex	Major	Phobia	Age	M1	M2	F
1	1	3	21	78	83	171	0	2	4	19	80	80	175
0	3	1	24	59	71	166	0	1	3	21	82	87	185
0	2	4	18	91	78	181	1	1	1	26	53	81	190
1	2	5	20	99	93	187	0	2	1	20	47	55	120
1	4	2	31	67	82	188	0	1	4	19	88	91	175
0	2	3	20	73	75	161	1	3	5	21	97	95	168
1	1	3	22	77	59	148	0	2	3	20	65	75	160
1	2	5	19	93	96	187	1	4	3	19	98	99	197
0	4	2	27	63	71	149	1	1	4	22	87	83	177
1	2	1	21	37	55	140	0	2	2	20	60	77	180
0	2	4	18	66	62	138	1	3	3	19	75	82	168
1	1	3	20	83	85	173	0	2	2	21	66	71	132
0	1	1	20	63	87	162	1	2	1	28	31	57	165
0	2	2	19	81	67	170	1	1	4	44	82	80	177
1	2	1	18	33	45	78	0	1	2	19	63	67	135
1	2	5	20	87	94	192	1	2	3	20	77	77	153
1	1	3	19	81	84	168	0	2	4	17	75	82	144
0	4	5	39	92	99	197	1	4	3	20	83	67	179
0	2	3	20	77	55	166	1	1	5	19	96	95	192
1	1	4	19	84	89	177	1	2	1	24	60	70	155
1	2	5	18	96	81	147	1	2	4	20	55	70	152
0	2	4	21	83	89	171	1	1	3	22	77	87	194
1	2	3	20	77	83	170	1	2	1	36	48	67	184

Table 2.1

As it is, this list is of little use; scanning the various columns, one can't get a clear idea of what's going on with any one variable, let alone any idea of the relations among the variables. The first step, and the most important one, in understanding data is to draw a picture of it—to plot it.

In this chapter, we will be looking at variables one at a time, as opposed to studying relations among two or more variables. Therefore, we want to look first at how to get a picture of the pattern of data on a single variable.

Variables such as sex, major, and (possibly) math phobia should be treated differently from variables such as age or exam score. Variables such as sex and major are **qualitative variables;** the numbers assigned to the various possible values are merely convenient codes, and the usual properties of numbers (such as magnitude and ordering) aren't relevant. Variables such as age and exam score are **quantitative variables;** the numbers are assigned according to actual measurement as opposed to an arbitrary code. Variables such as math phobia represent an intermediate case, because the numbers do reflect an ordering, but do not reflect any measured magnitude. Such variables are called **ordinal variables,** to indicate that their values only indicate an ordering. How to treat ordinal variables has been a subject of debate; we will consider some of the issues in Section 2.5. For now, we will usually treat ordinal variables as merely qualitative.

qualitative variables

quantitative variables

ordinal variables

Example 2.1

Strang (1981) reports a reaction time study in which challenging goal instructions were manipulated. Two variables of interest were trial number (baseline = 0, 1, and 2) and reaction time. Should these variables be considered as qualitative (including ordinal) or quantitative?

Solution

Reaction time is certainly a measured variable and should be treated as quantitative. Trial number could, perhaps, be treated as quantitative. But it isn't clear that the unit change from baseline to trial 1 should be treated the same way as the unit change from trial 1 to trial 2. Therefore, it would be safer to treat trial number as ordinal. □

Common methods for plotting data for a single variable are **bar charts** for qualitative variables and **histograms** for quantitative variables. A bar chart for the qualitative variable "major" of our illustrative data is shown in Figure 2.1.

bar charts

histograms

In a bar chart, the number of occurrences of a value in the data, called the **frequency** of that value, is represented by the height of the corresponding bar; if one thinks of the width of the bar as equaling 1 unit, then the area of the bar also represents the frequency. A relative-frequency bar chart is one in which the **relative frequency,** the frequency divided by the total number of observations, is represented by the height or the area of each rectangle. With a bar chart, it's easy to see the most common value (such as major 2 = psychology), as well as the rough proportions in the data of each value of the variable.

frequency

relative frequency

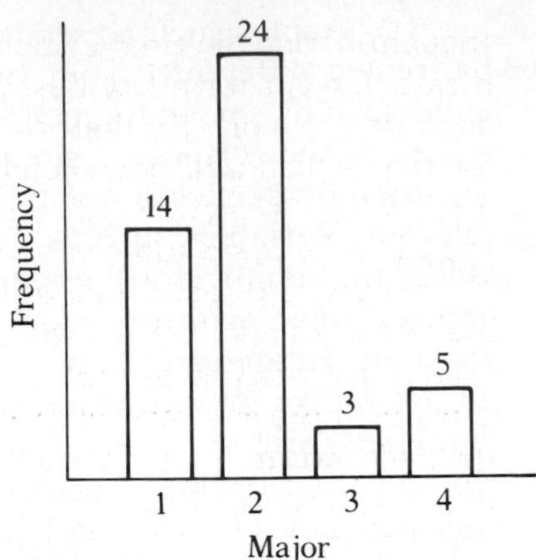

Figure 2.1 Bar chart—Major

Example 2.2 In a study of 40 children, 31 of them were found to be right-handed, 7 left-handed, and 2 ambidextrous. Construct a relative-frequency bar chart.

Solution The bar chart is shown in Figure 2.2. Note that the relative frequency of right-handedness is $31/40 = .775$. □

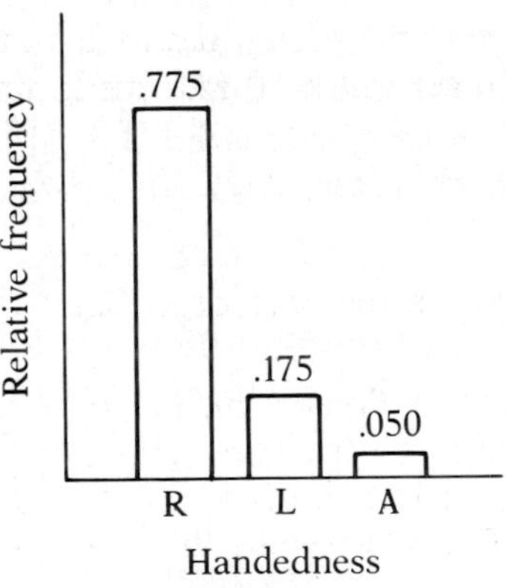

Figure 2.2 Relative-frequency bar chart for Example 2.2

One might be tempted to draw a histogram for a quantitative variable in the same way, with rectangles corresponding to each possible value of the variable. The temptation should be resisted, as a little thought about the final exam grades of our illustrative data set will indicate. Except for a few ties, there will be as many different rectangles as there are scores, and all the rectangles will have a height of 1 (except for ties, which will have height 2, or maybe even height 3). Converting 46 numbers into about 40 rectangles doesn't help much in summarizing the data. To get a more revealing histogram, we must **group** the data.

group

Grouping data amounts to rounding the data off, and there are many ways to do it. For example, with the final exam scores of our data set, we may define groups with a **width** of 5, 10, 20, or any other convenient number of points. If we have selected a width of 10 points, for instance, we may also achieve that width in many different ways; we might put scores from 161 to 170 in one group, or scores from 160 to 169, or perhaps scores from 155 to 164. Necessarily, the grouping process is somewhat arbitrary. There are, however, sensible rules of thumb to guide the grouping.

width

1. Try to have somewhere between 5 and 20 groups; having too few groups obliterates much useful information, whereas having too many groups obscures the broad patterns one is looking for.

2. All groups should have the same width. The only possible exception would be a group at one end or the other of the scale, such as an "age over 35" group for college students. Such **open-ended intervals** should only be used to include a small number of values that are widely spread out on one side of the data.

open-ended intervals

midpoints

3. Widths and **midpoints** of intervals should be chosen to be convenient, easily grasped numbers. The intervals 160–169, 170–179, etc., have width 10 and midpoints 165, 175, etc., and are easier to work with than the intervals 160–166, 167–173, etc. Widths that are multiples of 2, 5, or 10 seem to be convenient.

4. To avoid excessive detail and "jiggly" pictures, there should be an average of at least three observations per group.

For example, the final exam scores of our data set could reasonably be rounded off to groups of width 10, with endpoints such as 160 and 169; rounding to widths of 20 would yield a rather coarse summary, whereas rounding to widths of 5 would yield a large number of classes.

Older textbooks give very detailed instructions about the finer points of grouping, pointing out such minor facts as that the midpoint of the interval 160–169 is really 164.5, not 165, and that ages, unlike most variables, are rounded down rather than rounded off (a person aged 20 years and 364 days is recorded as age 20, not age 21, but a time of 20.99 seconds would be rounded off to 21 seconds). For purposes of getting a useful picture of the data, such issues aren't crucial.

Example 2.3 Assume that under certain conditions 48 reaction times were obtained from the reaction time study in Example 2.1. The times are in hundredths of a second, and are ordered from lowest to highest:

25	26	26	27	27	27	27	27	27	27	27	28
29	29	29	29	29	29	30	30	30	30	31	31
32	32	32	32	32	32	33	33	33	33	33	34
34	35	35	36	37	39	41	44	44	47	49	56

Select a reasonable set of intervals for a histogram.

Solution Grouping by 10's (like 20–29, 30–39, etc.) isn't very good; only four groups would result. Grouping by 5's (maybe 25–29, 30–34, etc.) seems about right. There will be about eight groups, so that there will be an average of 48/8 = 6 observations per group. The 25–29, 30–34, etc., grouping has the disadvantage that the midpoints are 27, 32, etc.; a marginally better grouping would be 23–27, 28–32, 33–37, etc., with midpoints 25, 30, 35, etc. □

Once a grouping method has been chosen, a histogram can be constructed by counting the frequencies in all groups and letting the areas of the rectangles be proportional to the frequencies (or to the relative frequencies—the picture will be the same). If an open-ended interval is used, the corresponding rectangle should stretch to cover the farthest-away data point, and the height should be reduced to achieve the proper area. The resulting histogram will look rather odd; therefore, open-ended intervals should be avoided. A histogram of the final exam score data is shown in Figure 2.3.

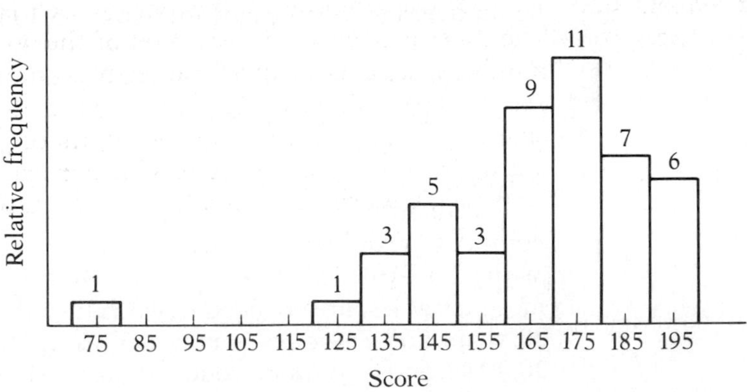

Figure 2.3 Histogram of final exam scores

Note that the rectangles in a histogram are placed side by side, rather than being separated by spaces as are the rectangles of a bar chart. Thus numerical values are shown spread out along a line in a histogram, whereas qualitative variables are shown in separate "bins" in a bar chart.

Example 2.4 Construct a frequency histogram of the data in Example 2.3, grouped as 23–27, 28–32, 33–37, etc.

Solution The frequency histogram is shown in Figure 2.4. A relative frequency histogram would have the same shape. □

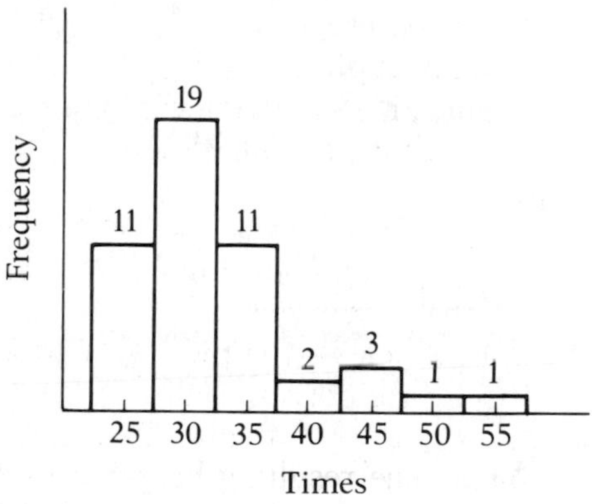

Figure 2.4 Histogram for Example 2.4

stem and leaf diagram

A variation on the histogram idea is the **stem and leaf diagram.** The most convenient illustration of the stem and leaf is with two-digit numbers, such as the first midterm exam scores of our illustrative data set. The possible first digits are placed in a "stem" column on the left side of the diagram; the second digits of the actual data are stretched out along "leaf" lines to the right. For example, the first five first-midterm exam scores are 78, 59, 91, 99, and 67. In the stem and leaf plot, they are represented as

```
5 | 9
6 | 7
7 | 8
8 |
9 | 1 9
```

Note that the 91 and 99 scores, which have the same initial digit, are shown on the same "leaf" line. The complete stem and leaf diagram for the first-midterm data is shown in Figure 2.5. The stem and leaf plot is effectively a histogram turned sideways. It has the additional virtue that the actual data are not lost in the grouping process. Note that we could improve the stem and leaf diagram a bit by sorting the data from low to high within a particular "leaf."

```
3 | 7 3 1
4 | 7 8
5 | 9 3 5
6 | 7 3 6 3 5 0 6 3 0
7 | 8 3 7 7 7 5 7 5 7
8 | 3 1 7 1 4 3 0 2 8 7 2 3
9 | 1 9 3 2 6 7 8 6
```

Figure 2.5 Stem and leaf display of exam grades

As described, the stem and leaf display corresponds to grouping the data with group widths of 10 units. Other groupings, particularly with a width of 2 or 5, are equally workable; for example, we could separate low-60s grades (60–64) from high-60s grades (65–69), etc. Other widths are awkward at best; try to do a stem and leaf display with group widths of 3, and you'll run into problems labeling the stem.

Example 2.5 Construct a stem and leaf plot of the reaction time data in Example 2.3.

Solution We must select a width for the "leaves." In Example 2.4, we argued that a width of 5 was about right. It's much more convenient to take intervals 25–29, 30–34, etc., for a stem and leaf display, despite the slightly awkward midpoints, because the initial digit for any particular leaf should be a specified number. The result is shown in Figure 2.6.

Looking sideways at the display, we see that the bulk of the data are in the high 20s and low 30s, with a "tail" of longer times to the right. Fancier stem and leaf plots are sometimes presented, with frequencies for each leaf, labels to distinguish (for instance) low 30s from high 30s, and so on. Most of these embellishments are easily understood when seen. □

```
2 | 5 6 6 7 7 7 7 7 7 7 7 8 9 9 9 9 9
3 | 0 0 0 0 1 1 2 2 2 2 2 2 3 3 3 3 3 4 4
3 | 5 5 6 7 9
4 | 1 4 4
4 | 7 9
5 |
5 | 6
```

Figure 2.6. Stem and leaf display of reaction time data

Much data analysis and summarization is now done by "packages" of computer programs. Any good package will allow you to request histograms; some, such as SAS or Minitab, will construct stem and leaf displays as well. Because the details of using the packages vary from one computer and package to another, we will not attempt to teach you how to use any one system. Instead, we will reproduce key parts of the output from several different packages. The output from most computer packages will be extensive, and will contain some more-or-less incomprehensible items. Find what you need in a computer output. Don't try to interpret every single item.

Example 2.6 A Minitab stem and leaf display for the reaction time data of Example 2.3 is shown in Figure 2.7. What features of the output aren't clear to you?

```
LEAF UNIT = 1.00
1 2 REPRESENTS 12

 18    2* | 5 6 6 7 7 7 7 7 7 7 7 8 9 9 9 9 9
(19)   3. | 0 0 0 0 1 1 2 2 2 2 2 2 3 3 3 3 3 4 4
 11    3* | 5 5 6 7 9
  6    4. | 1

       HI  44  44  47  49  56
```

Figure 2.7 Minitab stem and leaf display of reaction time data

Solution The basic stem and leaf is clear. However, the largest five numbers are not included in the stem and leaf, but are, for some reason, labeled as HI. (Later we will see that these are candidates for outlier values.) Also, some numbers appear to the left of the display, and the symbols . and * appear. We have no particular use for these features at this time, so we ignore them. □

The reason for plotting the data is to get a broad indication of the shape of the data. We want to look at the distribution of the variable—how the data are "distributed" over the possible values of the variable. There are four key questions:

1. Where is the middle of the data?
2. How spread out are the data?
3. How nearly symmetric are the data around the middle value?
4. Are there "outliers" (values far removed from the bulk of the data) or "stragglers" (a series of values that trail out a long way from the bulk of the data)?

Although we will develop numerical measures for middle, for variability, for asymmetry, and for outlier/straggler-proneness in this chapter, a picture of the data distribution is *always* useful in answering these questions. In particular, a picture will indicate the general pattern of symmetry or **skewness** (asymmetry) of the data. A histogram of the age data is shown in Figure 2.8. Note that the ages are not symmetric, but instead have a long "tail" to the larger values on the right; such a distribution is said to be **right-skewed.**

skewness

right-skewed

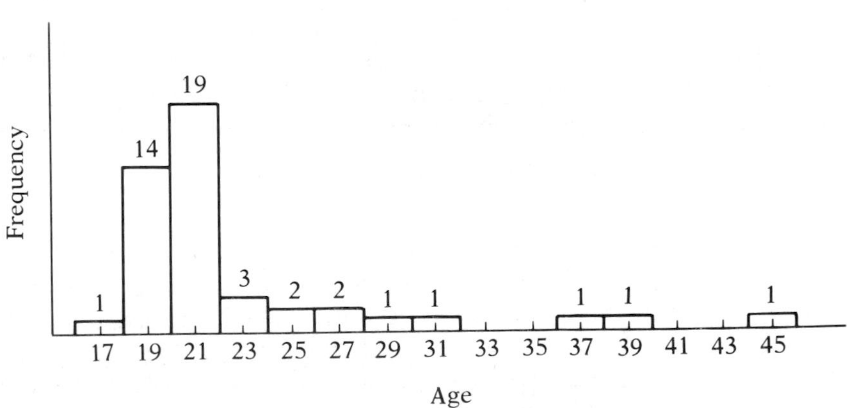

Figure 2.8 Histogram of age data

In contrast, the stem and leaf display of first-midterm grades in Figure 2.5 shows a long tail toward the lower values (which would be on the left of a histogram); such a distribution is left-skewed.

Example 2.7 Is skewness evident in the stem and leaf displays in Figures 2.6 and 2.7, which show the same reaction time data?

Solution Yes, indeed. Particularly in Figure 2.6, we can see a long tail to the right. The reaction times are right-skewed. □

Outliers and/or stragglers cause data pictures to appear heavy-tailed. Figure 2.9 shows an idealized histogram corresponding to the so-called normal distribution, which will be discussed in Chapter 6 and will be assumed in many statistical procedures later on. Also shown in Figure 2.9 is a heavy-tailed histogram, which would be much more prone to yield outliers or stragglers.

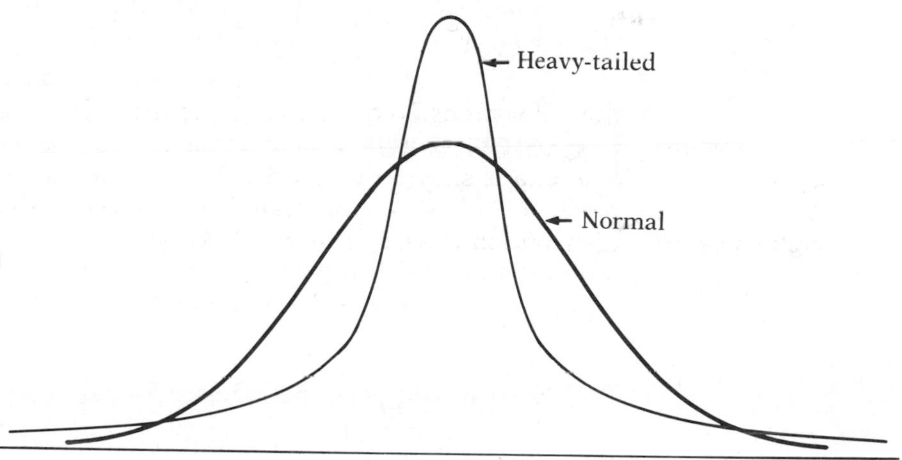

Figure 2.9 *Normal and heavy-tailed distributions*

bimodal

In addition, pictures can suggest such phenomena as the **bimodal** (two-peaked) distribution of Figure 2.10, as well as other anomalies.

Pictures of data should be looked at a bit skeptically. They can be affected both by the arbitrary grouping of the data and by random variation in the data (from one data set to another). Minor bumps and wiggles in the data aren't important; only strong, obvious patterns should be taken seriously.

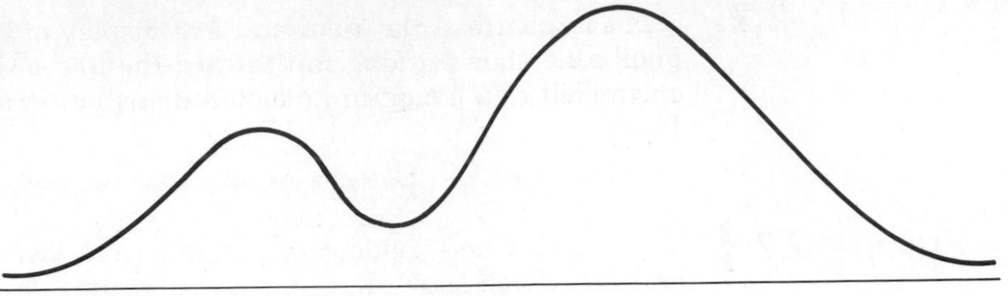

Figure 2.10 Bimodal distribution

Example 2.8 Close inspection of the data of Example 2.3 reveals that there are eight values equal to 27, only one equal to 28, and six equal to 29. Is this a reliable indication of bimodality?

Solution Not really. The broad pattern of the data is clearly unimodal, although right-skewed. If three of the six 29 values had been 28s instead, we'd never have noticed the problem. A stronger pattern should be present before we claim bimodality. □

Exercises for Section 2.1

2.1. Morelli et al. (1982) report on a study in which extrovert and introvert groups were measured on an anxiety scale. Assume that the data were as follows:

Extroverts:	4.7	5.2	5.3	5.5	5.8	6.0	6.1	6.1	6.4	6.6
	6.6	6.7	6.9	7.0	7.0	7.3	7.4	7.4	7.4	7.6
	7.9	8.2	8.4	8.8	9.1	9.8				
Introverts:	5.4	6.7	6.9	7.0	7.1	7.6	8.0	8.2	8.2	8.5
	8.7	8.9	8.9	8.9	9.0	9.0	9.0	9.1	9.2	9.3
	9.3	9.5	9.5	9.6	9.7	9.8				

a. Draw histograms of each set of anxiety scores. Use between five and eight intervals.

b. Describe the basic shape of each data set. Make a rough estimate of what a typical anxiety score is for each group.

c. Draw a stem and leaf display of each group's anxiety score. Does the stem and leaf display indicate the same basic shapes?

2.2. Raaheim, Kaufmann, and Bengtsson (1980) reported on a study in which good and poor problem solvers were evaluated on a scale measuring their ability to categorize effectively. Assume that the data were as follows:

Good:	25	28	31	33	35	38	39	40	41	42	42	43
	45	48	53	58	59							
Poor:	20	23	25	25	26	27	28	29	29	30	30	30
	31	31	32	34	34	35	35	37	37	38	38	39
	39	39	40	40	40	40	40	40	40	40	41	41
	42	44	44	44	45	47	47	48	48	48	49	50
	50	51	53	55	57	57	59	62				

a. Draw a histogram for the anxiety scores of the good problem solvers. Use about six or seven classes.

b. Draw a stem and leaf plot of the data on the good problem solvers. The stem should consist of the numbers 2, 3, 4, and 5.

c. Redraw the stem and leaf display, breaking the intervals into 20–24, 25–29, 30–34, 35–39, 40–44, etc. Does the display look much different?

2.3. Refer to the problem solvers data of Exercise 2.2.

a. Draw a histogram or a stem and leaf display of the data on the poor problem solvers.

b. Do the data on the good and the poor problem solvers have roughly the same shape?

c. Can you tell from the pictures which group has the higher average categorizing score?

2.4. Waxman (1984) obtained data on the ability of children to perform a categorizing task properly. The children were divided into three types, according to the instructions given them. The data were as follows:

Type A:	0.00	0.00	0.33	0.67	0.67	1.00	1.33	2.67
Type B:	0.33	1.00	1.33	1.33	1.67	2.00	2.33	3.00
Type C:	1.00	2.00	2.33	2.33	2.67	3.00	3.00	3.00

a. Draw a histogram for the combined data. What can you say about the shape of the data?

b. Draw histograms for each type separately. Do they all have the same general shape?

c. Can you tell which type seems to have the highest average score?

2.5. Schumm et al. (1981) performed a study that related (among other variables) a marital satisfaction scale to a marital conventionalization scale. One part of the study dealt with urban-dwelling wives. Assume that the data were as shown in Table 2.2:

ID number	Satisfaction	Conventionalization
1	37	64
2	58	46
3	77	43
4	55	51
5	29	76
6	48	68
7	84	44
8	61	54
9	55	43
10	33	72
11	55	51
12	47	49
13	56	44
14	61	42
15	20	70
16	81	36
17	52	55
18	49	41
19	58	46
20	62	42
21	49	56
22	36	67
23	77	40
24	44	57
25	40	62
26	49	53
27	72	36
28	60	49
29	27	68
30	46	54

Table 2.2

a. Draw a stem and leaf display for the conventionalization data. What is its general shape?

b. Draw a histogram for the marital satisfaction scale. Does it have the same general shape?

2.2
The Middle of the Data

The single most important summary value that can be computed about a variable in a scientific study is its "typical" value. For better or worse, the average, typical value of any variable is, in fact, taken as *the* representative value of the variable. Thus, it is critically important to examine various definitions of the "average" value, as we shall do in this section. It is at least equally important to understand the limitations of these definitions.

There are three common definitions of the "average" value of a variable:

mode 1. The **mode** is the single most common value.

median 2. The **median** is the middle value, when the data are arranged in numerical order, from low to high.

mean 3. The **mean** is the arithmetic average of the data values.

In this section we will consider these three measures, as well as percentiles—natural extensions of the median.

The mode is the most useful average in dealing with qualitative variables, such as major in our illustrative data set. The only consideration in defining the mode is the number of observations in each category; no numerical properties at all need be assigned to the variable. With quantitative variables, the mode has relatively little use. If the data are not grouped, it is likely that each data value will differ, so there may be as many modes as there are data points; even with coincidental ties, there may well be many "modes." One could group the data and look for the most frequently represented group, but grouping is inevitably an arbitrary process; arbitrary choices can have a major effect on which group is called the mode. Hence the mode as a measure of the typical value is used largely with purely qualitative variables.

Use of the median requires that an order underlie the assignment of numbers to the values of the variable, because the median, by definition, is the middle number when the data are arranged in order from lowest to highest. However, no concept of magnitude is needed, so the median can be used with ordinal variables.

If the data set contains three observations, the median is the second value; for five observations, it is the third observation; for seven, it's the fourth. In general, if the data set contains n observations, the median is the $[(n + 1)/2]$th observation, when the data are ordered from low to high. If n, the number of observations, is an even number, there is a slight ambiguity; for our illustrative data set, where $n = 46$, the $(n + 1)/2$ formula would say to take the 23.5th observation. The natural interpretation is to take the median as the average of the 23rd and 24th observations.

> **Median of a Variable**
>
> If n, the number of observations on a variable in a data set, is an odd number, the median of the variable is the $[(n + 1)/2]$th smallest observation. If n is an even number, the median is the average of the two values closest to the $[(n + 1)/2]$ th.

Example 2.9 Find the median for the data in Example 2.3.

Solution There are $n = 48$ observations on the variable. The median is the $[(48 + 1)/2]$th $= 24.5$th observation—that is, the average of the 24th and the 25th scores. The data are arranged in order already. The 24th observation is 31; the 25th is 32. The median is $(31 + 32)/2 = 31.5$. □

In cases where the data are only available in grouped form, about the only reasonable assumption is that the data are spread out uniformly within each group. An interpolation argument leads to a formula for estimating the median from grouped data; of course, the formula should only be used when the actual data aren't available.

$$\text{median} = L + (w/f_m)(.5n - c_b)$$

where

$L =$ lower limit of the interval containing the median
$w =$ width of that interval
$f_m =$ frequency in that interval
$n =$ total number of observations in the data
$c_b =$ cumulative frequency (total count) below that interval

Example 2.10 Assume that the frequency data of Example 2.3 had been presented in grouped form as shown in Table 2.3.

Interval	Midpoint	Frequency	Cumulative frequency
25–29	27	18	18
30–34	32	19	37
35–39	37	5	42
40–44	42	3	45
45–49	47	2	47
55–59	57	1	48

Table 2.3

Approximate the median. How well does the approximate value correspond to the actual value found in Example 2.9?

Solution The median is, once again, the average of the 24th and 25th values. It must lie in the 30–34 interval. The lower limit of the interval L appears to be 30, but that is because the data were rounded off to the nearest unit; the real lower limit is 29.5. The width of the interval is $w = 5$; the frequency in that interval is $f_m = 19$. The cumulative frequency below that interval is $c_b = 18$, and $.5n$ is half the sample size, or 24. Thus, the median is approximately

$$29.5 + (5/19)(24 - 18) = 31.08$$

which is fairly close to the actual median, 31.5. □

Occasionally, this grouped-data formula is used to calculate a median of very "lumpy" data. For instance, what is the median number of children born to a woman through her child-bearing years? No doubt that number is quite near 2, but it makes quite a bit of difference demographically whether that number is a bit less than 2 or a bit greater than 2. Even though it's obviously impossible to have 1.6 or 2.4 children (the data are "lumped" at 0, 1, 2, 3, ...), it's sometimes useful to pretend that the data are uniformly spread through the lumps, and therefore to use the grouped-data approximation.

Example 2.11 A survey of 1520 elementary school teachers included a question about a particular form of merit pay for outstanding teachers. The responses were coded 1 = strongly opposed, 2 = opposed, 3 = indifferent, 4 = in favor, 5 = strongly in favor. The following data were obtained:

Value	1	2	3	4	5
Frequency	737	401	213	42	127
Cumulative frequency	737	1138	1351	1393	1520

What is the median response?

Solution Literally speaking, the median is the average of the 760th and 761st ordered responses, namely 2. With these ordinal data, it might be better to use the approximation. Here, L should be taken as halfway between the stated values 1 and 2; $L = 1.5$. The width is $w = 1$, the frequency in the median interval is $f_m = 401$, the cumulative frequency below the median interval is $c_b = 737$, and the total number of values is $n = 1520$. Report the median as

$$1.5 + (1/401)[.5(1520) - 737] = 1.56$$

□

The ideas involved in calculating a median can be extended to finding other percentiles. The [100p]th percentile is the score such that $100p$ percent of the data fall below it. Thus if $p = .80$, the $100(.80) =$

80th percentile is the score such that 80% of the data fall below it. The median is the 50th percentile, so ideas about the median generalize immediately to other percentiles. We shall extend the method for grouped data to percentiles other than the median; one can always regard ungrouped data as n groups of 1 data point each. The interpolation argument used for the median leads to a similar expression for other percentiles:

$$100p \text{ percentile} = L + (w/f_p)(pn - c_b)$$

where

L = real lower limit of the interval containing the percentile (the lower limit of the interval where cumulative frequency goes from below pn to above pn)
w = width of that interval
f_p = frequency in that interval
c_b = cumulative frequency below that interval
n = total number of observations in the whole data set

Example 2.12 Find the 80th percentile for the data in Example 2.10.

Solution We know that $n = 48$, so $.80n = 38.4$. In the interval labeled 35–39, the cumulative frequency goes from 37 to 42, so the 80th percentile must be in that interval. The percentile should be closer to 35 than to 39, because 38.4 is closer to 37 than to 42. The true lower limit of the interval is 34.5 hundredths of a second, the width is 5, the frequency in the interval is $42 - 37 = 5$, and $c_b = 37$. The 80th percentile is approximated as

$$34.5 + (5/5)[.80(48) - 37] = 35.9$$ □

The same interpolation method is often used when there are discrete data with many individuals having the same score. Even if the data logically can have only distinct, separate values—such as integer values—we can usefully regard the data as grouped.

Example 2.13 Find the 80th percentile for the data in Example 2.11.

Solution We know that $n = 1520$, so $.80n = 1216$. The cumulative frequency goes from 1138 to 1351 at the value 3. Regard the value 3 as the midpoint of the interval 2.5–3.5. Then $L = 2.5$, $w = 1$, $f_p = 213$, $c_b = 1138$, and $n = 1520$, so the 80th percentile is

$$2.5 + (1/213)[.80(1520) - 1138] = 2.87$$

Note that the 80th percentile is closer to 2.5 than to 3.5, because $.80n = 1216$ is closer to 1138 than to 1351. □

Some percentile problems must be worked in the opposite direction. Rather than having the percentage and needing to find the score, we might have the score and need to find the percentage. The percentile rank of a particular score is the percentage below that score. The interpolation argument for medians and other percentiles can be reversed to give an expression for the percentile rank; the expression can be used either for grouped data or for discrete data with many ties.

$$p = \frac{(y - L)(f_p/w) + c_b}{n}$$

where

y = specified score
L = true lower limit of the interval containing the score

and the other quantities are the same as for percentiles.

Example 2.14 Referring to Example 2.10, find the percentile rank of a score of 38.

Solution We have $y = 38$ in the interval 35–39. Thus $L = 34.5$, $f_p = 5$, $c_b = 37$, $n = 48$, and $w = 5$.

$$p = \frac{(38 - 34.5)(5/5) + 37}{48} = .84375$$

Note that if we had asked for the 84.375th percentile, we would have obtained

$$y = 34.5 + (5/5)[.84375(48) - 37] = 38$$

□

The most widely used measure of the average, typical value of a variable is the **mean,** which is the arithmetic average of the numerical values in the data. Because the mean takes magnitude into account, it is not usable with qualitative data; in our math phobia data set, the mean major would not be meaningful.

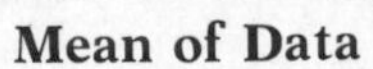

Mean of Data

Let the n observations on a variable y in a data set be denoted by $y_1, y_2, \ldots, y_n$. Then the (sample) mean $\bar{y}$ is

$$\bar{y} = \frac{y_1 + y_2 + \cdots + y_n}{n}$$

or the total sum of the individual observations divided by the number of observations. In summation notation, with the values of the variable denoted as y_i,

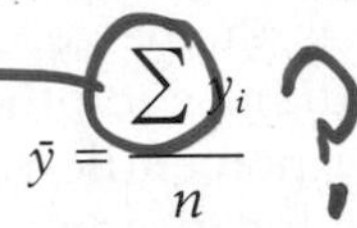

$$\bar{y} = \frac{\sum y_i}{n}$$

If the data are regarded as the entire population of relevance, the number of observations is denoted by N and the population mean by μ, the Greek letter mu. The population mean is also the sum of the population of observations divided by the number of observations:

$$\mu = \frac{\sum y_i}{N}$$

Example 2.15 Find the mean for the data in Example 2.3.

Solution Add up all the scores and divide by the number of scores, which in this case is 48. The data should be regarded as a sample, so we'll denote the mean as $\bar{y}$.

$$\bar{y} = \frac{25 + 26 + 26 + \cdots}{48} = 32.58$$

□

The mean measures the middle of the data in the sense that it is the "balancing point" of the data. Imagine equal-sized weights placed along a line, one at every data point; the weights to the left of the mean would exactly balance those to the right. For instance, the data values 11, 12, 13, 14, and 30 have a mean of 16. The one extreme value, 30, has so much leverage that it balances the four smaller values (see Figure 2.11).

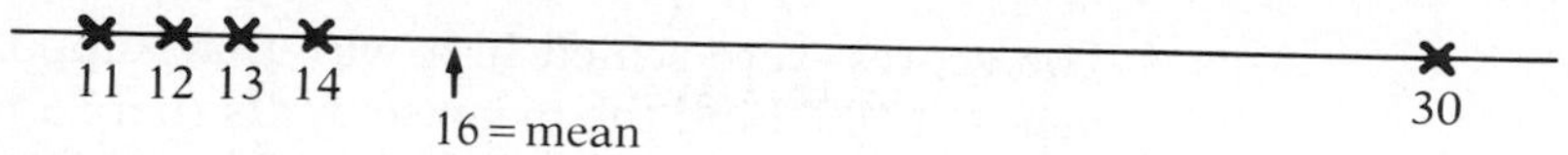

Figure 2.11 Mean as the balancing point of data

Deviations

Deviations from the mean are the differences between data values and the mean. For data values 11, 12, 13, 14, and 30 with mean 16, the deviations are $11 - 16 = -5$, $12 - 16 = -4$, $13 - 16 = -3$, $14 - 16 = -2$, and $30 - 16 = 14$. The sum of these deviations is always 0; the negative and positive deviations always cancel, so the mean is indeed the balancing point of the data.

Because of this balancing property, the mean will reflect any skewness in the data. Figure 2.12 shows an idealized right-skewed histogram. The mode (peak) of the histogram is toward the left side, the median is in the middle, and the mean is pulled to the right by the data values out in the tail. Of course, the opposite pattern would be seen in left-skewed data. Some variables used in social science research, such as reaction times and latencies (times taken to observe a certain result, such as time to next feeding), tend to be severely skewed. For such variables, a mean may be a misleading value for the "middle" of the data. This is one reason why data plots are so important in summarizing data. Another reason is that outliers have a strong effect on a mean. A single monstrously large or small value in the data can have a major effect on the mean.

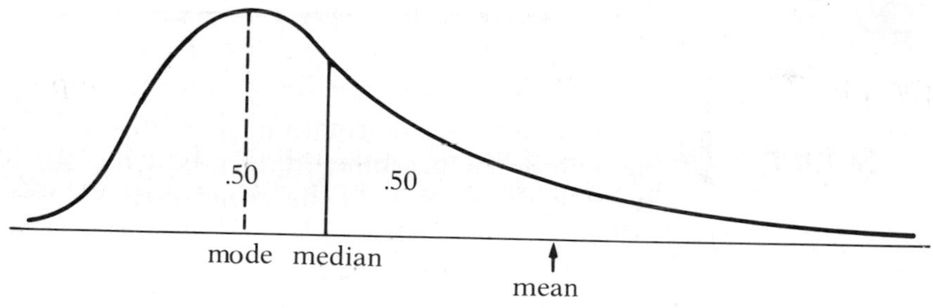

Figure 2.12 Effect of skewness on the mean, median, and mode

Example 2.16

Scores of a sample of 12 subjects on the Beck depression scale were as follows:

88 91 91 95 96 99 102 104 104 108 113 151

The scores were scaled in a way that supposedly yielded a population mean of 100. Find the mean score. Is there a reason why it deviates from 100?

Solution The mean is

$$\bar{y} = \frac{88 + 91 + \cdots + 151}{12} = 103.5$$

The highest value, 151, is far from the rest of the data. If it had been, say, 120, the mean would have been 100.92, very close to the nominal value. The 151 value seems suspicious. □

Despite its sensitivity to skewness and outliers, the mean is by far the most widely used measure of the "location," or middle, of data. Mathematically, it turns out to be far more convenient to use the mean than any other measure. In particular, means can be combined easily, whereas medians and modes can't. For example, suppose that the data consisted of three groups and that the summary figures in Table 2.4 had been obtained.

	Group		
	A	B	C
Mode	32	37	33
Median	36	42	39
Mean	40	50	45
Sample size	10	10	10

Table 2.4

From this information, it is impossible to tell what the mode or median of the combined data might be, but it follows that the overall mean is the average of the group means,

$$\frac{40 + 50 + 45}{3} = 45$$

weighted average

Even if the sample sizes are not the same, it's easy to compute the overall mean; it is a **weighted average** of the group means. A weighted average of values, call them v_j, with weights w_j, is

$$\text{weighted average} = \frac{\sum v_j w_j}{\sum w_j}$$

or the sum of values times weights, divided by the total of the weights. For combining means, the values are the group means and the weights are the respective sample sizes.

Combining Means

If there are subsample means $\bar{y}_1, \ldots, \bar{y}_K$, based on respective sample sizes $n_1, \ldots, n_K$, then the overall sample mean is

$$\bar{y} = \frac{\sum n_j \bar{y}_j}{\sum n_j}$$

Example 2.17

Suppose that a sample contained 32 three-year-olds, 48 four-year-olds, and 20 five-year-olds. Mean scores on a Piagetian conservation task were 22.4 for three-year-olds, 29.0 for four-year-olds, and 42.8 for five-year-olds. Find the sample mean.

Solution

$$\bar{y} = \frac{32(22.4) + 48(29.0) + 20(42.8)}{32 + 48 + 20}$$

$$= 29.648$$

Exercises for Section 2.2

2.6. Refer to the data of Exercise 2.1.

a. Find the medians of the extrovert scores and of the introvert scores. Which group has the higher "typical" value as measured by the median? By how much?

b. Find and compare the means for the two groups. Do the means differ by the same amount as do the medians?

c. Refer to the data histograms of Exercise 2.1. What feature of the histograms explains why the means are closer together than are the medians?

2.7. Again refer to Exercise 2.1.

a. Group the extroverts data into intervals 4.5–5.4, 5.5–6.4, 6.5–7.4, etc., and obtain frequencies.

b. Approximate the mean score based on the grouped data.

c. Approximate the median score based on the grouped data.

d. Repeat the calculations, using groups 4.0–4.9, 5.0–5.9, 6.0–6.9, etc. Do you get just about the same values?

2.8. Refer to the intervals and frequencies found in Exercise 2.7, part a.

a. Find the 30th percentile.

b. Find the percentile rank of a score of 5.8.

2.9. Refer to the categorizing-ability data of Exercise 2.2.

a. Calculate the means and medians for the good and the poor problem solvers separately.

b. In Exercise 2.3, you were asked if you could tell which group had the higher average score. Did you guess correctly?

2.10. **a.** Use the results of Exercise 2.9 to calculate the mean for the combined data.

b. Calculate the median of the combined data. Are the results of Exercise 2.8 useful in this calculation?

2.11. Refer to the Waxman data of Exercise 2.4.

a. Find the mean score for each of the three types.

b. Find the median score for each of the three types.

c. The mean scores are somewhat closer together than are the median scores. Use the histograms drawn in Exercise 2.4 to explain why.

2.12. Refer to the Waxman data of Exercise 2.4 again.

a. Find the mean score of the combined data directly.

b. Use the three type means found in Exercise 2.11 to find the combined mean.

2.13. Refer to the histogram of the marital satisfaction data in Exercise 2.5.

a. Do you expect that the mean and the median will differ from each other by much? Why?

b. Approximate the mean and median using the histogram intervals and frequencies.

c. Go back to the data and calculate the correct sample mean and median.

2.14. Refer again to the histogram obtained in Exercise 2.5.

a. Find the 20th and 80th percentiles.

b. Find the percentile rank for a score of 72.5.

2.3
Measuring Variability

As is demonstrated by the well-known story of the fellow with his head in the oven and feet in the freezer, who was comfortable on the average, it is important to measure not only the average value, but also the extent of variability around that average. Variability in aptitudes is a key issue in the study of intelligence. Variability is crucial in clinical studies; after all, variability is the manic-depressive's fundamental problem. Variability is also crucial in statistical inference. How accurately one can infer from a limited sample to a broader population depends not only on the sample size, but also on how variable the data are. In this section we will consider ways of measuring variability.

The easiest way to measure variability is one of the worst ways—namely, to assess the range of data. In ordinary English, range is specified by stating the smallest and largest values in the data, as in "The range of exam scores was from 23 to 99." In statistical usage, the range is defined as the difference between the largest and smallest values, as in "The range of exam scores was 76 points." There are two problems with this measure. First, it is extremely sensitive to outliers; second, as the number of observations increases, the range can only increase, regardless of the apparent variability of the data.

Example 2.18

Consider two sets of scores on a certain scale:

Set I: 31 47 48 49 49 50 50 50 50
51 51 52 53 69

Set II: 32 35 38 41 45 48 50 50 52
55 59 62 65 68

Which set of scores should be regarded as more variable? Which set has a larger range?

Solution

All but two of the scores in set I are within three points of the mean (50); the scores in set II are spread out nearly evenly over a much wider interval. In any reasonable sense, set I should be regarded as less variable. Yet it has a larger range (38 vs. 36) because of the outlying values at 31 and 69. □

Example 2.19

A pilot study of IQ scores yielded a sample of only four scores, 70, 85, 110, and 130. A follow-up study yielded ten more scores, 65, 90, 95, 100, 100, 100, 105, 105, 115, and 140. What is the range of the pilot study scores, and what happens to the range when the two sets of data are combined?

Solution

The range for the pilot study scores is 130 − 70 = 60. When the data are combined, the range increases to 140 − 65 = 75. Note that the bulk of the follow-up scores are close to 100, indicating that the additional scores have relatively little variability. Even so, the range of scores actually increases. □

Interquartile range

A modification of the range idea yields a more useful measure of variability, the **interquartile range.**

> **Interquartile Range**
>
> The **quartiles** of a data set are the 25th and 75th percentiles of the data. Thus, 25% of the data should lie below the 25th percentile, and 75% below the 75th percentile. The interquartile range (IQR) is
>
> $$\text{IQR} = \text{75th percentile} - \text{25th percentile}$$

The IQR is the range over which the "middle half" of the data stretches. There is a clever way to compute the IQR based on the fact that the 25th percentile must be the median of the lower half of the data and the 75th percentile must be the median of the upper half: Take the (ordered) data and divide it in half at the median; if the sample size, n, is odd, arbitrarily include the median in both halves of the data. Then find the 25th and 75th percentiles of the data by finding the median of each half of the data. (The idea is due to Tukey (1977), who called the quartiles "hinges.")

Example 2.20

Find and interpret the IQR for each of the samples in Example 2.18.

Solution

In both samples $n = 14$, so the median is the 7.5th score—that is, the average of the 7th and 8th scores. In both samples, the median is

50. The 25th percentile is the median of the bottom seven scores, that is, the 4th score in the sample. In set I, it is 49; in set II, it's 41. Similarly, the 75th percentile for set I is 51, and in set II it's 59. The IQR (the range containing the middle half of the data) is 2 for set I and 18 for set II. □

The IQR provides a useful test for potential outliers, also due to Tukey (1977). Any data value that falls more than 1.5 times the IQR below the 25th percentile, or more than 1.5 times the IQR above the 75th percentile, is a potential mild outlier. Any value that falls more than 3 times the IQR below the 25th percentile, or more than 3 times the IQR above the 75th percentile, is a serious outlier. See Figure 2.13. (The numbers 1.5 and 3 are somewhat arbitrary, but seem to work fairly well.) Some computer programs perform this calculation automatically and report potential "low" and "high" outliers. Often the limits at 1.5 and 3 IQRs away from the 25th and 75th percentiles are called the **inner fences** and **outer fences**, respectively.

inner fences
outer fences

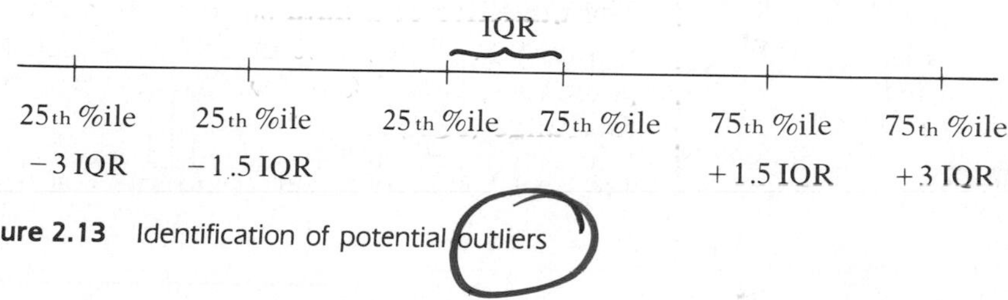

Figure 2.13 Identification of potential outliers

Example 2.21 Find all potential outliers in the samples of Example 2.18.

Solution In sample I, the IQR is 2 and the 25th percentile is 49. Any data value below 49 − 1.5(2) = 46 is a mild low outlier, and any data value below 49 − 3(2) = 43 is a serious low outlier. Note that 31 is an obvious outlier and is indeed far below 46 and 43. Similarly, the value 69 is far above 51 + 1.5(2) = 54, and even above 51 + 3(2) = 57. In sample II, there are no candidate outliers; the "inner fences"—1.5 IQRs away from the quartiles—are 41 − 1.5(18) = 14 and 59 + 1.5(18) = 86, and no data value is even close to them. □

box plot

The quartiles are also used in still another Tukey (1977) invention, the **box plot,** sometimes called the box-and-whiskers plot. See Figure 2.14. The ends of the box are at the quartiles, the line inside the box indicates the median, and the whiskers stretch out to the smallest and largest values. Sometimes potential outliers are shown as separate points, with the whiskers extending only to the largest and smallest non-outliers.

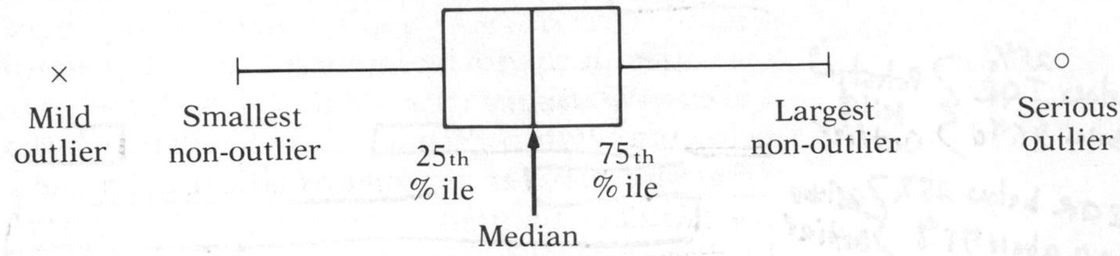

Figure 2.14 Box plot

Example 2.22 Construct separate box plots for the samples of Example 2.18.

Solution See Figure 2.15. □

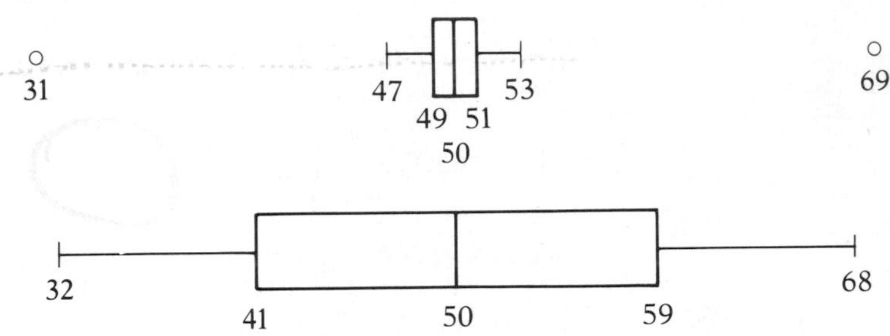

Figure 2.15 Box plots for Example 2.22 □

Box plots, because they show only a limited amount of information for each sample, are very useful when a large number of samples are to be compared. In that case, too much detail in the data summaries could be distracting. The box plot shows only some of the most important features of the data, making quick comparisons among many samples easier.

deviations from the mean

The IQR, although useful, is not so commonly reported as it might be. By far the most widely used measures of variability are the variance and the standard deviation. Both these statistics are based on **deviations from the mean**—that is, differences between individual values and the mean. If the ith individual value in a sample is denoted by y_i and the sample mean is denoted by $\bar{y}$, the deviations are $y_i - \bar{y}$. If all the deviations in a sample are small in magnitude (whether positive or negative), there is little variability in the data; if at least some of the deviations are large in magnitude, there is more variability. The average of these deviations is not a good measure, because it is always 0; the positive and negative deviations exactly cancel. To overcome the cancellation problem, the **variance** is defined as the average *squared* deviation. If the original data are in units of, say, hours, then the variance is measured in squared hours, not the easiest units to understand. To get back to the original units such as hours, the square root of the variance is calculated; this value is called the **standard deviation.** There is a slight difference between the mathematical definitions of variance and standard deviation for population data and for sample data. The reason will be discussed after the definitions.

variance

standard deviation

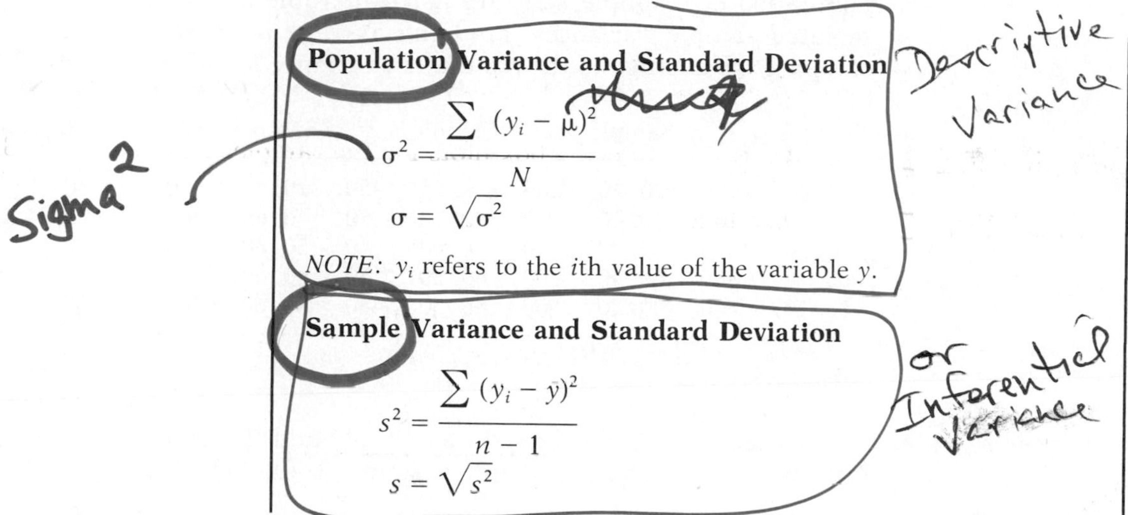

Population Variance and Standard Deviation

$$\sigma^2 = \frac{\sum (y_i - \mu)^2}{N}$$

$$\sigma = \sqrt{\sigma^2}$$

NOTE: y_i refers to the ith value of the variable y.

Sample Variance and Standard Deviation

$$s^2 = \frac{\sum (y_i - \bar{y})^2}{n - 1}$$

$$s = \sqrt{s^2}$$

Example 2.23 A (tiny) population consists of the values 20, 40, 50, 60, and 80. Find the mean, variance, and standard deviation.

Solution The mean is easily found to be 50, either by considering the symmetry of the data or by doing the arithmetic. We must compute the average of the squared deviations.

$$\sigma^2 = \frac{(20 - 50)^2 + (40 - 50)^2 + (50 - 50)^2 + (60 - 50)^2 + (80 - 50)^2}{5}$$

$$= 400$$

Therefore $\sigma = \sqrt{400} = 20$. □

The choice of $n - 1$ as the denominator of the sample variance is not arbitrary. It can be proved that if all possible samples of a specified size are taken (sampling with replacement) from a population, the average sample variance equals the population variance. Thus, the sample variance will estimate the population variance "correctly on the average." If n had been used in the denominator instead of $n - 1$, the average sample variance would have been smaller than the population variance.

Example 2.24 All possible samples of size 2, taken with replacement from the population of Example 2.23, are listed in Table 2.5, along with the associated sample variances. Find the average of these variances, and compare the average to the population variance found in Example 2.23.

Sample	s^2	Sample	s^2	Sample	s^2	Sample	s^2	Sample	s^2
20, 20	0	20, 40	200	20, 50	450	20, 60	800	20, 80	1800
40, 20	200	40, 40	0	40, 50	50	40, 60	200	40, 80	800
50, 20	450	50, 40	50	50, 50	0	50, 60	50	50, 80	450
60, 20	800	60, 40	200	60, 50	50	60, 60	0	60, 80	200
80, 20	1800	80, 40	800	80, 50	450	80, 60	200	80, 80	0

Table 2.5

Solution There are 25 s^2 values, so we simply add them up and divide the total by 25. The average sample variance is 400, which equals σ^2. □

Most computer programs use the $n - 1$ definition in calculating standard deviations. Some hand calculators allow you to call for either the $n - 1$ or the n version. The manual for the calculator or computer program should specify which definition is used. Normally, the available data will be regarded as a sample, and the $n - 1$ definition should be used. Only in the case (which is usually conceptual rather than actual) in which the data can be considered as the entire population should the population variance definition be used.

The standard deviation and variance measure variability in the data. The numerical value of the standard deviation can best be interpreted by the so-called Empirical Rule.

Empirical Rule

In "well-behaved" sample data with mean $\bar{y}$ and standard deviation s:

About 68% of the data fall in the range $\bar{y} - s$ to $\bar{y} + s$.

> About 95% of the data fall in the range $\bar{y} - 2s$ to $\bar{y} + 2s$.
>
> Almost all the data fall in the range $\bar{y} - 3s$ to $\bar{y} + 3s$.
>
> A "well-behaved" sample is not severely skewed, is not bimodal or multimodal, and contains no severe outliers.

Example 2.25 Calculate the mean and standard deviation for the sample data in Example 2.3. What fraction of the data fall within 1 standard deviation of the mean? Within 2 standard deviations? Are these fractions close to the Empirical Rule fractions?

Solution The sample mean is $\bar{y} = 32.58$ and the sample standard deviation is $s = 6.47$. The 1 standard deviation interval is $32.58 - 6.47 = 26.11$ to $32.58 + 6.47 = 39.05$; it includes 37 out of 46 values, or 80.4%. The 2 standard deviation interval is $32.58 - 12.94 = 19.64$ to $32.58 + 12.94 = 45.52$; it includes 43 of 46 values, or 93.5%. The Empirical Rule approximation is not good for the 1 standard deviation rule. The data are severely right-skewed and not well-behaved. □

Alternatively, there is a mathematically guaranteed inequality called Chebyshev's (also transliterated as Tchebisheff, Tshebyshef, and probably other ways) Inequality. This inequality applies to all samples, whether well-behaved or not, but is usually very conservative.

> **Chebyshev's Inequality**
>
> For all samples:
>
> At least 75% of the data fall in the range $\bar{y} - 2s$ to $\bar{y} + 2s$.
>
> At least 88.9% of the data fall in the range $\bar{y} - 3s$ to $\bar{y} + 3s$.
>
> For any number $k > 1$, at least $100(1 - 1/k^2)\%$ of the data fall in the range $\bar{y} - ks$ to $\bar{y} + ks$.

Example 2.26 How close an approximation is Chebyshev's Inequality for the data in Example 2.25?

Solution

In Example 2.25, we found that 93.5% of the data fell in the 2 standard deviation interval, well above the 75% bound. The 3 standard deviation interval is 32.58 − 19.41 = 13.17 to 32.58 + 19.41 = 51.99, which includes 45/46, or 97.8%, of the data, again well above the 88.9% bound. □

With a little practice, the Empirical Rule and Chebyshev's Inequality become fairly easy to use. For example, suppose someone claimed that the standard deviation of grade averages of college seniors (on the usual A = 4 to F = 0 scale) was 0.1, and that the mean was 3.0. The claimed standard deviation seems very small. According to the Empirical Rule, about 68% of the averages would be between 3.0 − 0.1 = 2.9 and 3.0 + 0.1 = 3.1 and about 95% of the averages would be between 2.8 and 3.2. These seem like very narrow ranges; I would expect the standard deviation to be much larger, perhaps 0.4.

Example 2.27

Would you accept a claim that the standard deviation of heights of adult American women is 8 inches?

Solution

Let's guess that the mean height of adult American women is something like 65 inches. If the claim were true, only 68% of women would have heights between 57 inches (4 feet, 9 inches) and 73 inches (6 feet, 1 inch). In fact, only a few percent of women have heights outside this range. The claimed standard deviation is too large. □

***z*-score**
raw score

Data are often converted into *z*-scores. A ***z*-score** is computed from the original **raw score** by subtracting the mean and dividing by the standard deviation. Thus if the original value is 65, the mean is 50, and the standard deviation is 10,

$$z = \frac{\text{score} - \text{mean}}{\text{st. dev.}} = \frac{65 - 50}{10} = 1.50$$

The *z*-score is useful in indicating where a particular data point is relative to the rest of the data. It indicates how many standard deviations the point is away from the mean. Positive *z*-scores indicate points above the mean, negative ones indicate points below the mean. The *z*-score of 1.50 that we calculated indicated that 65 is 1.50 standard deviations above the mean. Suppose that a student scores 43 ($z = -3.60$) on an examination. One might reasonably say that the student, with a score 3.60 standard deviations below the class mean, is in trouble.

The variance and therefore the standard deviation are based on *squared* deviations from a mean. As a consequence, these statistics are extremely sensitive to outliers. An outlier, by definition, has a very large deviation from the mean. Because the square of a very large number is a downright huge number, an outlier can lead to a grossly inflated standard deviation. The best strategy is to look at the data before computing. Outliers that might make the standard deviation misleading will be fairly evident, and where they appear the calculated standard deviation should be viewed a bit skeptically.

If you have no calculator or computer program available to do the arithmetic, a short-cut method can be used.

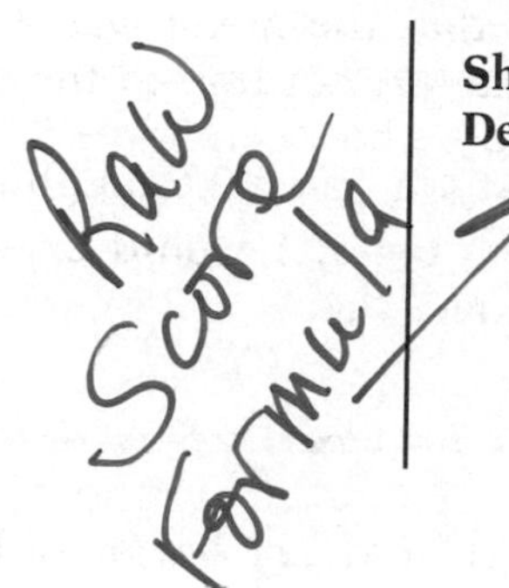

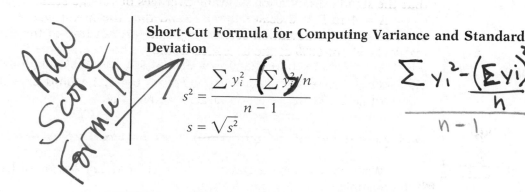

Short-Cut Formula for Computing Variance and Standard Deviation

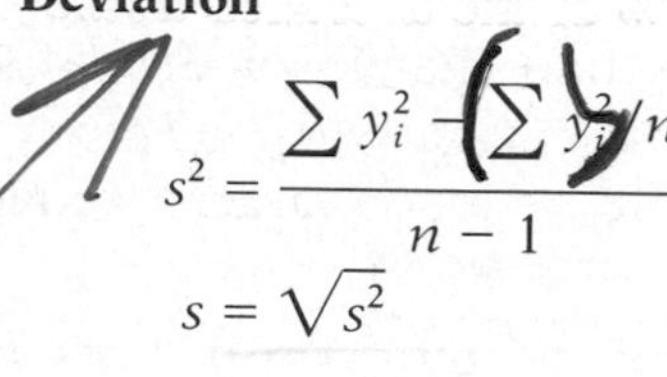

$$s^2 = \frac{\sum y_i^2 - \left(\sum y_i\right)^2/n}{n-1}$$

$$s = \sqrt{s^2}$$

Example 2.28 Compute the standard deviation of the following sample data by hand, using both the definition and the short-cut formula: 49, 77, 93, 81, 71, 62, 75.

Solution To use the definition, we first calculate $\bar{y} = 62.57$. Then

$$s^2 = \frac{(49 - 62.57)^2 + (77 - 62.57)^2 + \cdots + (75 - 62.57)^2}{7 - 1}$$

$$= 840.62$$

and $s = \sqrt{840.62} = 28.99$.

To use the short-cut method, we calculate $\Sigma y_i = 438$ and $\Sigma y_i^2 = 32450$. Then

$$s^2 = \frac{32450 - (438)^2/7}{7 - 1} = 840.62$$

and $s = 28.99$ once again. □

If only grouped data are available, the variance may be approximated by assuming that all data points within a class fall exactly at the midpoint of the class.

Variance Approximation for Grouped Data

$$s^2 = \frac{\sum (m_j - \bar{y})^2 f_j}{n - 1} = \frac{\sum m_j^2 f_j - \left(\sum m_j f_j\right)^2 \Big/ n}{n - 1}$$

NOTE: m_j is the midpoint of the jth interval (class), and f_j is the frequency in that class.

Example 2.29 Approximate the variance for the data in Table 2.6 on 88 depression scores:

Interval	Frequency
20–29.9	5
30–39.9	12
40–49.9	27
50–59.9	23
60–69.9	14
70–79.9	6
80–89.9	1

Table 2.6

Solution The midpoints are, respectively, 25, 35, and so on. (It's hardly worth worrying about whether they should be 24.95, 34.95, etc.) By calculator, we find that $\Sigma m_j^2 f_j = 242{,}200$ and $\Sigma m_j f_j = 4470$, so

$$s^2 = \frac{242200 - (4470)^2/88}{88 - 1} = 174.073$$ □

Exercises for Section 2.3

2.15. Refer to the data of Exercise 2.2.

a. Find the range of scores for the sample of good problem solvers and for the sample of poor problem solvers.

b. Refer to the histograms drawn in Exercise 2.2. Would you say that the sample of poor problem solvers was more variable than the sample of good problem solvers, as indicated by the ranges found in part a?

c. If you do not think that the sample of poor problem solvers was more variable, explain why the range was larger in that sample.

2.16. Refer to the data of Exercise 2.2.

a. Calculate the sample standard deviations for the two samples.

b. Which standard deviation is larger? Does this result agree with your judgment in Exercise 2.15?

2.17. Again refer to the data of Exercise 2.2.

a. Calculate the IQR for each sample.

b. Are there any candidate outliers in either sample, according to the 1.5 IQR rule of thumb?

c. Draw a box plot for each sample.

2.18. Refer to the data of Exercise 2.1.

a. Calculate the IQR for each sample.

b. Are there any candidate outliers in either sample?

c. Draw box plots for the two samples. How is the skewness in the introverts sample shown in its box plot?

2.19. Again refer to the data of Exercise 2.1.

a. Calculate the standard deviation for each sample.

b. In each sample, find the fraction of scores that fall within 1 standard deviation of the mean. In which sample is the Empirical Rule approximation better?

2.20. Suppose that the data of Exercise 2.5 were (implausibly) regarded as the entire population. Calculate the variance and standard deviation of both the satisfaction and the conventionalization scores.

2.21. A computer calculation showed that the sample standard deviation of the conventionalization data in Exercise 2.5 was 11.2755. In addition, the sum of the conventionalization scores was 1579, the sum of squared conventionalization scores was 86,795, and the sample size was 30.

a. Use the short-cut formula given in the text to verify the computer calculation.

b. What fraction of the scores fall within 2 standard deviations of the mean?

c. Does this fraction agree with the Empirical Rule approximation? From the looks of the data, should it agree?

d. How close to the Chebyshev Inequality bounds are the fractions in each sample?

e. Repeat parts b through d for the fraction of scores within 3 standard deviations of the mean.

2.22. Refer to the data of Exercise 2.4.

a. Find the fraction of scores that fall within 1 standard deviation of the mean and within 2 standard deviations of the mean.

b. A histogram of the data was drawn in Exercise 2.4. Does that histogram indicate that the Empirical Rule approximation will be a close one?

c. How close is the fraction of scores found in part a to the Chebyshev Inequality bound?

d. Repeat parts a through c with the data from only the A type.

2.23. The following (extremely) artificial data have been chosen to illustrate a point.

Value	1	2	3	4	5
Frequency	998	2	6000	2	998

The (sample) mean is 3.00000, and the (sample) standard deviation is 0.99931.

a. Draw a histogram of the data. Can the Empirical Rule be expected to work well for such data?

b. What fraction of the data fall within 2 standard deviations of the mean? How close is this fraction to the Empirical Rule approximation?

c. How close is this fraction to the Chebyshev Inequality bound?

2.24. The data of Exercise 2.4 were summarized by several standard computer packages (BMDP, Minitab, SAS, and SPSS, respectively). For each output shown in Figure 2.16, locate the mean, median, and standard deviation. Ignore those parts of the output that you don't understand.

2.4 Measuring Skewness and "Heavy-Tailness"

The most critical summary statistics for sample data are measures of location (average) and variability. In this section we briefly consider two other issues: the degree of skewness in data and the extent to which the data contain "heavy tails." These issues will be particularly important in statistical inference. The best choice of an inference method depends heavily on the degree of skewness and heavy-tailness.

In Section 2.2 we noted that data skewness, in the form of a long tail in one direction, pulled the mean in that direction, as compared to

```
BMDP2D - DETAILED DATA DESCRIPTION, INCLUDING FREQUENCIES

/PROBLEM TITLE IS 'SUMMARY STATISTICS FOR WAXMAN DATA'.
/INPUT   FILE='DSKC: WAXMAN.DAT'.
         VARIABLES ARE 2.
         FORMAT IS FREE.
/VARIABLE  NAMES ARE TYPE, SCORE.
           USE IS SCORE.
/PRINT  ESTIMATES.
        STEM.
/END

PROBLEM TITLE IS
 SUMMARY STATISTICS FOR WAXMAN DATA

Q1 AND Q3 ARE THE 1ST & 3RD QUARTILES.  S- AND S+ REPRESENT THE
QUANTITIES (MEAN-ST.DEV.) AND (MEAN+ST.DEV.).

NUMBER OF CASES READ. . . . . . . . . . . . . . .     24

  ************                                                                              EACH 'H'
  * SCORE    *                          MAXIMUM      3.0000000                             REPRESENTS
  ************                          MINIMUM      0.0000000                                 1
                                        RANGE        3.0000000                        H    COUNT(S)
VARIABLE NUMBER . . . . . .      2      VARIANCE     1.0179651                  H H   H H
NUMBER OF DISTINCT VALUES .     10      ST.DEV.      1.0089425                HHHH H H HHH
NUMBER OF VALUES COUNTED. .     24      (Q3-Q1)/2    0.8325000                HHHH HHH HHH
NUMBER OF VALUES NOT COUNTED     0      MX.ST.SC.    1.36                     L-------------U
                                        MN.ST.SC.   -1.61
                                                                     EACH '-' ABOVE =      0.2500
LOCATION ESTIMATES                          ST.ERROR                              L=       0.0000
              MEAN          1.6245833      0.2059495                              U=       3.2500
              MEDIAN        1.5000000      0.3839381               CASE NO. OF MIN. VAL. =   1
              MODE          3.0000000                              CASE NO. OF MAX. VAL. =  16
```

Figure 2.16(a)

```
                                        SUMMARY STATISTICS FOR WAYMAN DATA

                                                    UNIVARIATE

VARIABLE=SCORE

                MOMENTS                                           QUANTILES(DEF=4)                              EXTREMES

N                 24   SUM WGTS        24        100% MAX          3        99%         3             LOWEST      HIGHEST
MEAN         1.62458   SUM          38.99         75% Q3       2.585        95%         3                  0         2.67
STD DEV      1.00894   VARIANCE   1.01797         50% MED        1.5        90%         3                  0            3
SKEWNESS -0.0743573    KURTOSIS   -1.3004         25% Q1      0.7525        10%     0.165               0.33            3
USS          86.7557   CSS        23.4132          0% MIN          0         5%         0               0.33            3
CV           62.1047   STD MEAN   0.20595                                    1%         0               0.67            3
T:MEAN=0     7.88826   PROB>|T|    0.0001        RANGE             3
SGN RANK       126.5   PROB>|S|    0.0001        Q3-Q1        1.8325
NUM ^= 0          22                             MODE              3
      STEM LEAF                    #        BOXPLOT                         NORMAL PROBABILITY PLOT
         3 0000                    4           |          3.25+                         **********
         2 77                      2        +-----+           |                    ***++++
         2 00333                   5        |     |           |                *****++
         1 7                       1        |  +  |       1.75+             + *++
         1 000333                  6        *-----*           |         ******
         0 77                      2        +-----+           |     ++***
         0 0033                    4           |          0.25+ *****  ***
        ----+----+----+----+                                    +----+----+----+----+----+----+----+----+----+----+
                                                                     -2        -1       +0        +1        +2
```

Figure 2.16(b)

```
MTB > EXECUTE 'WAXMAN'
MTB > READ  'WAXMAN DATA A' INTO C1 C2
     24 ROWS READ
 ROW   C1    C2

   1    1  0.00
   2    1  0.00
   3    1  0.33
   4    1  0.67
   .  .  .

MTB > HISTOGRAM OF C2

  C2

  MIDDLE OF   NUMBER OF
  INTERVAL    OBSERVATIONS
      0.0       2   **
      0.4       2   **
      0.8       2   **
      1.2       6   ******
      1.6       1   *
      2.0       2   **
      2.4       3   ***
      2.8       2   **
      3.2       4   ****

MTB > DESCRIBE C2

              C2
N             24
MEAN        1.62
MEDIAN      1.50
TMEAN       1.64
STDEV       1.01
SEMEAN      0.21
MAX         3.00
MIN         0.00
Q3          2.58
Q1          0.75
MTB > STOP
```

Figure 2.16(c)

```
FILE HANDLE KIDDATA/NAME='WAXMAN DATA A'
DATA LIST FILE=KIDDATA LIST/TYPE SCORE
FREQUENCIES   VARIABLES=SCORE/
              STATISTICS=MEDIAN MEAN STDDEV

SCORE

                                                      VALID     CUM
 VALUE LABEL          VALUE  FREQUENCY  PERCENT  PERCENT  PERCENT

                        .00          2      8.3      8.3      8.3
                        .33          2      8.3      8.3     16.7
                        .67          2      8.3      8.3     25.0
                       1.00          3     12.5     12.5     37.5
                       1.33          3     12.5     12.5     50.0
                       1.67          1      4.2      4.2     54.2
                       2.00          2      8.3      8.3     62.5
                       2.33          3     12.5     12.5     75.0
                       2.67          2      8.3      8.3     83.3
                       3.00          4     16.7     16.7    100.0
                                 -------  -------  -------
                      TOTAL         24    100.0    100.0

MEAN        1.625    MEDIAN       1.500      STD DEV       1.009
```

Figure 2.16(d)

the median. This idea is the basis for one of the standard measures of skewness. If the data are symmetric, and therefore not skewed at all, the difference between the mean and median is 0. If the data are, say, right-skewed, with a long tail toward larger values, the difference between the mean and median will be a positive number; if the data are left-skewed, the difference will be negative. One measure of skewness is the difference between the mean and the median, taken as a fraction of the standard deviation:

$$\text{skew}_1 = \frac{\text{mean} - \text{median}}{\text{standard deviation}}$$

It can be proved that this statistic ranges from −1 to +1; values near 0 indicate symmetric data, and values above .2 or below −.2 indicate rather severe skewness.

Example 2.30 The data of Example 2.3 have a mean of 32.58, a median of 31.5, and a standard deviation of 6.47. Calculate the skew_1 measure. Does the value of this measure support the judgment of skewness based on a data plot?

Solution The data were found to be substantially right-skewed in Example 2.3.

$$\text{skew}_1 = \frac{32.58 - 31.5}{6.47} = .167$$

The value is positive and fairly substantial, indicating a fairly strong right-skewness. □

An alternative measure of skewness, which will be called skew_2, is based on deviations from the mean. In, say, right-skewed data, there will be many scores just below the mean, yielding many small negative deviations, and some scores far above the mean, yielding some large positive deviations. The sum of the deviations is 0, so the positive and negative deviations cancel. The alternative skewness measure is based on the cubes (third powers) of these deviations. The cube of a small negative number is a small negative number, whereas the cube of a large positive number is a huge positive number. In right-skewed data, the average cubed deviation will be positive, because the huge positive cubes will overwhelm the many small negative cubes. Usually, the average cubed value is taken as a fraction of the cubed standard deviation:

$$\text{skew}_2 = \frac{\sum (y_i - \bar{y})^3/(n-1)}{s^3}$$

Example 2.31 The sum of cubed deviations from the mean, for the data of Example 2.3, is 21404.56. What is the numerical value of skew_2?

Solution The standard deviation is 6.47 and $n = 48$. So

$$\text{skew}_2 = \frac{21404.56/(48 - 1)}{(6.47)^3} = 1.68$$ □

The third-power skewness measure is useful in mathematical statistics and is reported by many statistical computer packages. However, it has some severe limitations as a summary statistic. Because it is based on third powers, it is terribly sensitive to outliers. Furthermore, there seems to be no natural interpretation of the number, so it is almost impossible to tell if $\text{skew}_2 = +.46$ indicates large, moderate, or small positive skewness.

Groeneveld and Meeden (1984) have shown that both of these statistics are imperfect skewness measures; either statistic can indicate that sample A is more skewed than sample B, even though by any reasonable standard the actual skewness comparison is reversed. They proposed a third skewness measure and proved that, in an appropriate sense, it always gave a proper comparison of skewness in two data sets. Their measure can be defined best in a betting context. You bet that a randomly chosen observation will be above the median. If it is, you win \$1 for every unit (inch, second, point, whatever) the observation is above the median; if the observation is below the median, you lose \$1 for every unit below. If the variable is right-skewed, you have an advantageous bet. When you win, you often will win large amounts (for values far out in the right tail), and when you lose, your dollar loss will tend to be small. The Groeneveld-Meeden skewness measure, skew_3, can be interpreted as your average winnings per dollar that changes hands. It turns out that this can be expressed as

$$\text{skew}_3 = \frac{\left(\begin{array}{c}\text{mean of upper}\\ \text{half of data}\end{array} - \text{median}\right) - \left(\text{median} - \begin{array}{c}\text{mean of lower}\\ \text{half of data}\end{array}\right)}{\text{mean of upper half} - \text{mean of lower half}}$$

When n, the sample size, is odd, there is a question about what constitutes the upper and lower halves of the data. If $n = 25$, should the upper half consist of the top 12 scores or the top 13 scores? Arbitrarily, we designate both halves of the data as including the median in this case.

If skew_3 comes out $+.10$, you would win 10 cents more than you would lose out of every dollar bet. A value of $\text{skew}_3 = +.10$ indicates moderately substantial skewness; of course, $skew_3 = 0$ when there is no skewness.

This measure is far less sensitive to outliers than either $skew_1$ (which uses the outlier-sensitive standard deviation) or $skew_2$ (which uses cubed deviations). It also has an easier interpretation than either of the other measures. Unfortunately, it is not routinely computed by any of the standard statistical programs, so the arithmetic must be done by hand or a special procedure must be written within one of the packages.

Example 2.32 For the data of Example 2.3, the mean of the bottom half of the data is 28.21, the mean of the top half is 36.96, and the median is 31.5. Find $skew_3$ and interpret the resulting number.

Solution

$$skew_3 = \frac{(36.96 - 31.5) - (31.5 - 28.21)}{36.96 - 28.21} = .248$$

For these data, you would win 24.8 cents more than you lost, out of every dollar that changed hands. This indicates rather substantial skewness, as seen in a histogram or stem and leaf display. □

Another aspect of data, besides skewness, that deserves attention is the "heavy-tailness" of the variable. We have repeatedly noticed the sensitivity of various statistics to extremely large or small values—outliers, to greater or lesser degree. To what extent do data on a variable indicate outlier-proneness?

There will always be larger and smaller values in any data set, so there will always be "tails" in the data. To define heavy-tailness, we need a reference standard: heavy-tailed as compared to what? The "normal" distribution, which will be discussed in Chapter 6, is assumed for most standard statistical procedures; therefore, it is a convenient reference standard. Figure 2.17 shows a normal histogram and a heavy-tailed histogram. Notice that the heavy-tailed histogram not only exhibits long tails, but also has a sharp peak in the middle. The sharp peak tends to make the standard deviation small, whereas the heavy tails tend to make the standard deviation large. These two effects can be made to cancel out, so it is possible to have a heavy-tailed variable with the same standard deviation as a normal variable.

The usual statistic for measuring heavy-tailness is the kurtosis, which is the average of the fourth powers of deviations, as a fraction of the fourth power of the standard deviation.

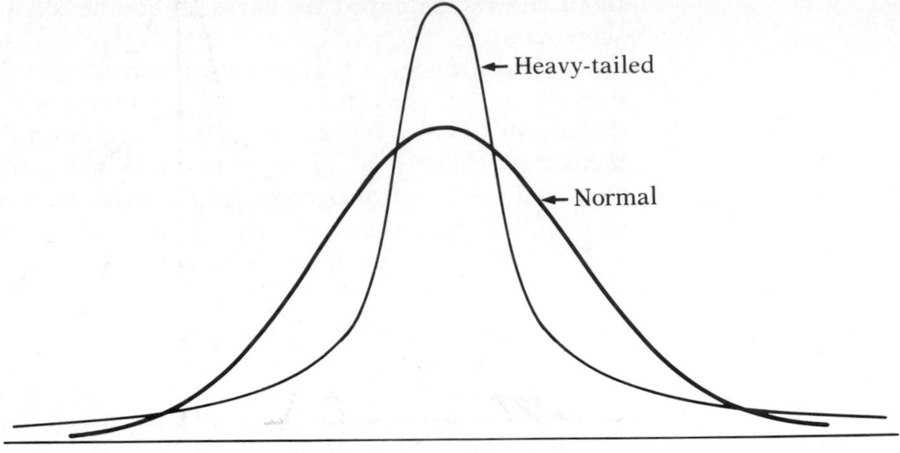

Figure 2.17 Normal and heavy-tailed distributions

$$\text{kurtosis} = \frac{\sum (y_i - \bar{y})^4/(n-1)}{s^4}$$

It turns out that for exactly normal data, the kurtosis equals +3; indeed, many texts and computer programs take our definition of kurtosis and subtract 3. When using any computer program, check the manual to see which definition was employed.

Example 2.33 For the data of Example 2.3, the sum of fourth powers of deviations from the mean is 474258.9, n is 48, and the standard deviation is 6.47. Calculate the value of kurtosis.

Solution

$$\text{kurtosis} = \frac{474258.9/47}{(6.47)^4} = 5.758$$

indicating a heavy-tailed distribution. Indeed, there are a number of very large outliers in the data. □

The reason that kurtosis measures heavy-tailness is that the fourth power of a large (positive or negative) deviation is a huge, huge number. In heavy-tailed data, the fourth powers of the extreme values will be monstrous and will lead to a large kurtosis; conversely, if there's nothing remotely close to an outlier in the data, none of the fourth powers will be large, and the kurtosis will be small.

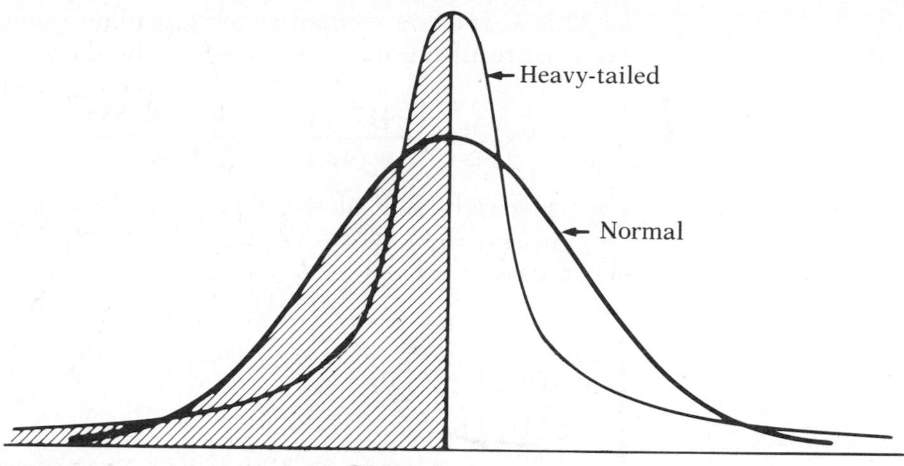

Figure 2.18 Skewness of both halves of data

Unfortunately, the kurtosis measure is terribly sensitive to just how "outlying" an outlier may be. A small change in an outlying value may yield a large change in kurtosis. Further, the numerical value of the kurtosis has no obvious meaning; a value of 3.5 is larger than the normal value (3), but is that a large discrepancy? No one has a good answer. Also, this measure combines the effect of the two tails in measuring heavy-tailness. It may be that one tail is heavy and one light (in a skewed distribution), in which case kurtosis will be "normal."

An alternative measure of heavy-tailness looks at the upper half and the lower half of the data separately. When the histogram is "normal," the right half of the data is right-skewed; but when the data are heavy-tailed, the right half of the data is even more right-skewed. See Figure 2.18. Just as in the IQR test for outliers, the data are split at the median; when the number of observations in the sample is odd, the median is arbitrarily included in both halves. The Groeneveld-Meeden skewness measure, skew_3, can be computed for both halves of the data. If the data are "normal," it can be shown that the skew_3 value for the upper half of the data will be about .27; a heavy right tail is reflected in a skew_3 value much larger than .27. Similarly, a skew_3 more negative than $-.27$ for the lower half of the data indicates a heavy left tail.

Example 2.34

A computer program took the top half of the data from Example 2.3 and calculated the mean of the top half of that (which would be the top quarter of the full data set) as 41.417, the mean of the bottom half as 32.583, and the median as 34. Calculate $skew_3$ for these data. What does the resulting number say about the data?

Solution

$$skew_3 = \frac{(41.417 - 34) - (34 - 32.583)}{(41.417 - 32.583)} = .679$$

Thus, the right half of the data is severely skewed, indicating a very long, heavy, right tail in the data. That long tail can be seen in a plot of the data. □

Exercises for Section 2.4

2.25. Refer to the data and histograms of Exercise 2.1.

a. Which group—extroverts or introverts—appears to you to be more skewed? In which direction?

b. Calculate the various skewness measures separately for each group.

c. Do the numbers you found in part b agree with your "eyeball sense" of the data, as stated in your answer to part a?

2.26. Refer to the data for good problem solvers in Exercise 2.2.

a. Do the scores appear to be highly skewed? Very heavy-tailed?

b. Calculate the $skew_1$ measure. What does the resulting number mean?

c. Calculate the $skew_3$ measure. What does the resulting number mean?

2.27. Suppose that, in the data of Exercise 2.2, the value 59 had been recorded as a 99.

a. Recalculate the two skew measures that you calculated in Exercise 2.26. How seriously are they affected?

b. Calculate the third-power skewness measure $skew_2$ for both the original data and the revised data. How seriously is the $skew_2$ measure affected by the change of one value in the data?

2.28. A Minitab program was written to measure the skewness (by the Groeneveld-Meeden $skew_3$ measure) and the heavy-tailness (by the G-M measure on each half of the data) for the satisfaction data of Exercise 2.5. The results were as follows:

ALL DATA SKEW = −0.074
LOWER HALF OF DATA SKEW = −0.458
UPPER HALF OF DATA SKEW = 0.487

a. Refer to a histogram or stem and leaf display of the data. What skewness and/or heavy-tailness do you see in the data?

b. Do the output values shown confirm your impression of the basic shape of the data?

2.29. Consider the following artificial data sets.

Set A:	26	32	37	40	42	44	46	47	48	49	49
	51	51	52	53	54	56	58	60	63	68	74
Set B:	26	26	27	28	29	31	33	35	38	43	49
	51	57	62	65	67	69	71	72	73	74	74

a. Draw a stem and leaf display of each data set; use intervals that are five points wide.

b. Describe the shape of each data set. What should the skewness value be?

c. Calculate the $skew_3$ measure for each half of data set A. How does this value compare to what should be found with normally distributed data?

d. Calculate the $skew_3$ data for each half of data set B. Explain why the left half of the data has a positive skew and the right half a negative skew.

2.5 Dealing with Ordinal Data

The fundamental summary statistics discussed in this chapter—the mean, the standard deviation, and the skewness and heavy-tailness measures—all assume that the variable is legitimately quantitative. They should not be used for qualitative variables. It would be absurd to ask for the mean religion in a sample of 12 Catholics, 19 Protestants, 4 Jews, and 3 others. **Ordinal variables** (those whose values reflect ordering but no measured magnitude) fall in between purely **nominal variables** (those which, like religion, reflect neither ordering nor magnitude) and truly quantitative variables. Ordinal variables occur quite frequently in social science, particularly in fields that attempt to deal with attitudes. Unfortunately, there is no universal agreement on the appropriate way to deal with ordinal variables.

ordinal variables
nominal variables

It is tempting to treat ordinal variables as quantitative, if only to be able to use the various summary statistics in this chapter and an enormous number of other statistical procedures. There is a real danger in doing so, because the numbers assigned to the various categories are arbitrary, except for their order. For example, one could assign 1 = strongly disagree, 2 = disagree, 3 = uncertain, 4 = agree, 5 = strongly agree, or one could assign 1 = strongly disagree, 3 = disagree, 4 = uncertain, 5 = agree, and 7 = strongly agree. The choice is arbitrary and will affect summary statistics; the second possible assignment in the previous sentence would give heavier tails than the first, and hence a larger standard deviation, for example. Thus, the danger is that apparent scientific findings, such as a substantial difference in mean attitude between two groups, may be a mere artifact of an arbitrary choice of numerical values.

Whenever possible, ordinal data should be analyzed by methods that only use order properties, not magnitudes. Many "nonparametric" methods, such as those discussed in Chapters 10, 11, and 12, involve only ranking the data from lowest to highest. Such methods do not depend on the numerical values of a variable reflecting any measured magnitudes; therefore they are suitable for ordinal variables. Unfortunately, many important statistical questions cannot be answered by ranking methods alone, so there are many cases in which a researcher with ordinal data will want to use methods that assume quantitative data.

At minimum, anyone who does that should try some alternative assignment of values to see if such arbitrary choices have an effect on the conclusion. If, for example, a 1, 2, 3, 4, 5 coding gives quite different results from a 1, 3, 4, 5, 7 coding, the results are quite probably artifacts of an arbitrary choice and should not be published. If, however, essentially the same results occur for several different assignments of values, the results should be more reliable.

Example 2.35

Two samples of 100 patients each were treated by two different therapeutic methods. After one month, the condition of each patient was assessed by an observer, who did not know which treatment had been employed on which patients. The judgment frequencies shown in Table 2.7 were obtained.

		Much Worse	Worse	No Change	Better	Much Better
Treatment	A	10	15	30	25	20
	B	5	20	40	25	10

Table 2.7

If the five categories are scored −2, −1, 0, 1, 2, which treatment has the better (higher) mean score? What if the categories are coded −3, −1, 0, 1, 3?

Solution For the first scoring system, the A mean is 0.30 and the B mean is 0.15. For the second scoring method, the A mean is 0.40 and the B mean is 0.20. In either case, the A mean is better. □

Exercises for Section 2.5

2.30. Suppose that responses on a seven-point scale (where 1 is most positive, 7 is most negative, and 4 is neutral) were obtained from two groups of students.

Group A:	1	2	2	3	3	3	3	4	4	4
	4	4	4	5	5	5	5	6	6	7
Group B:	1	1	1	1	1	3	4	4	4	5
	5	5	5	5	5	5	6	6	6	7

a. Find the median and mean scores for each group. How do they compare?

b. Suppose that the value 1 is recoded as −3, the value 2 as 0, the value 6 as 8, and the value 7 as 11. Recalculate the medians and means. Now how do they compare?

c. Which measure of "location" (that is, of the "typical" value in the data) is more sensitive to the recoding?

2.31. In Exercise 2.30, how should the recoding affect the standard deviations for the two groups? Verify your answer by calculating the standard deviations for the original and recoded data of (say) group A.

2.32. Kleinke and Nicholson (1979) asked groups of black and white children in third through fifth grade to rate their agreement with the statement "This woman could be your father's new boss," when shown a picture of a woman (race not obvious). The rating scale ran from 1 = agree very strongly to 5 = disagree very strongly. Suppose that the data were as follows:

B:	1	1	1	1	2	2	2	2	2	2	2	2	2	2	2
	2	2	2	2	3	3	3	3	3	3	3	4	4	4	5
W:	1	2	2	2	2	3	3	3	3	3	3	3	3	4	4
	4	4	4	4	4	4	4	4	4	4	5	5	5	5	5

a. Find the means and medians for the black and the white children.

b. Find the respective standard deviations.

2.33. In Exercise 2.32, replace all 1s by 0s and all 5s by 6s.

a. Recalculate the means and medians.

b. In the original data, the mean and median scores for blacks were lower than (more agreeing than) the corresponding figures for whites. Is that still true in the recoded data?

2.6 Properties of the Mean and Median

In this section we will prove some results about the mean and median as predictors of values of a variable. Both the results and the methods of proof will be useful in later chapters.

In Section 2.2 we noted that the mean was the balancing point of the data, with positive and negative deviations always canceling. Here we'll actually prove that the deviations must add to 0. Note that the proof does *not* require that the data be symmetric.

Result 2.1 $\sum (y_i - \bar{y}) = 0$

Proof

$$\begin{aligned}\sum (y_i - \bar{y}) &= (y_1 - \bar{y}) + (y_2 - \bar{y}) + \cdots + (y_n - \bar{y}) \\ &= (y_1 + y_2 + \cdots + y_n) - n\bar{y} \\ &= \sum y_i - n\left(\sum y_i/n\right) = 0\end{aligned}$$

Now suppose that we try to predict the value of a variable. Specifically, suppose that individual scores are drawn at random from a sample, and we try to predict the values. Naturally, we'll want to predict a typical value such as the mean or the median. Which typical value should be used depends on how the prediction error is evaluated. The most widely used evaluation is the total *squared* prediction error. If we predict a value p each time, the prediction error is evaluated as

$$S(p) = \sum (y_i - p)^2$$

A basic property of squared error is that the mean value becomes the best predictor, minimizing the squared error $S(p)$.

Result 2.2 The squared prediction error $S(p)$ is smallest when the predicted value $p = \bar{y}$.

Proof We give two proofs. The first uses calculus, but the second uses only algebra.

Using calculus, we take the first derivative of $S(p)$ with respect to p, and equate the derivative to 0.

$$\begin{aligned}\frac{d}{dp} S(p) &= \sum \frac{d}{dp} (y_i - p)^2 \\ &= \sum 2(y_i - p)(-1) \\ &= (-2) \sum (y_i - p)\end{aligned}$$

By Result 2.1, $\Sigma(y_i - \bar{y}) = 0$. Thus choosing p to be $\bar{y}$ makes the derivative 0. The second derivative of $S(p)$ with respect to p is $(-2)\Sigma(-1) = 2n$, which is greater than 0, so $p = \bar{y}$ in fact gives the minimum value of $S(p)$.

expanding the square

Alternatively we may use the algebraic idea of **expanding the square:** $(a + b)^2 = a^2 + 2ab + b^2$. Specifically we expand $S(p)$ "around the mean," first adding and subtracting the same $\bar{y}$ value.

$$S(p) = \sum (y_i - p)^2 = \sum (y_i - \bar{y} + \bar{y} - p)^2$$

Now expand the square with $a = y_i - \bar{y}$ and $b = \bar{y} - p$:

$$S(p) = \sum (y_i - \bar{y})^2 + 2 \sum (y_i - \bar{y})(\bar{y} - p) + \sum (\bar{y} - p)^2$$

The second term is

$$\begin{aligned}(y_1 - \bar{y})(\bar{y} - p) + (y_2 - \bar{y})(\bar{y} - p) + \cdots + (y_n - \bar{y})(\bar{y} - p) \\ = (\bar{y} - p)\left[\sum (y_i - \bar{y})\right] = 0\end{aligned}$$

by Result 2.1. The last term in the expansion is

$$(\bar{y} - p)^2 + (\bar{y} - p)^2 + \cdots + (\bar{y} - p)^2 = n(\bar{y} - p)^2$$

Therefore, we have

$$S(p) = \sum (y_i - \bar{y})^2 + n(\bar{y} - p)^2$$

The last term is a positive number unless $p = \bar{y}$, so $S(p)$ is smallest when $p = \bar{y}$.

The use of squared prediction error is mathematically convenient because of the possibility of expanding the square, but squared error is not the only reasonable evaluation of prediction error. One reasonable alternative is the total of the absolute prediction errors: $A(p) = \Sigma|y_i - p|$. The squared error prediction penalizes large prediction errors very severely, because the square of a large number is a very large number. The absolute error criterion also penalizes large errors, but less severely. When absolute error is the criterion, the median is the best predictor.

Result 2.3 $A(p)$ is minimized when $p = m$, the median of the data.

Proof Note that $|y_i - p| = y_i - p$ when $y_i \geq p$, and that $|y_i - p| = p - y_i$ when $y_i < p$. Thus

$$A(p) = \sum_{y_i<p} (p - y_i) + \sum_{y_i \geq p} (y_i - p)$$

Assume that we use a value $p < m$. Let's increase the prediction by a small amount d to $p + d$, in such a way that every $y_i > p$ is also $> p + d$. See Figure 2.19, where the subscripts are arbitrary.

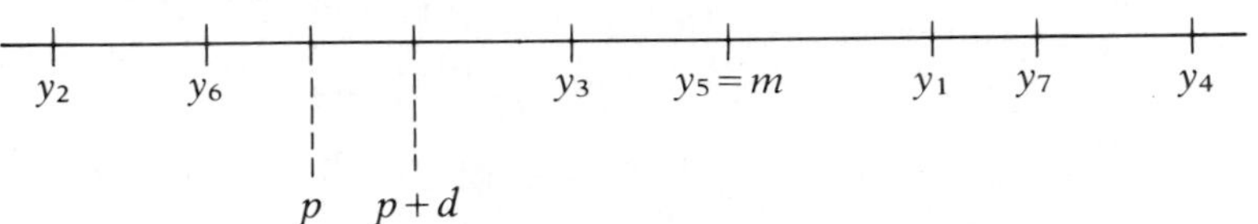

Figure 2.19 Effect of changing prediction

Then

$$A(p) - A(p + d) = \sum_{y_i<p} (p - y_i) + \sum_{y_i \geq p} (y_i - p) - \sum_{y_i<p+d} (p + d - y_i) - \sum_{y_i \geq p+d} (y_i - p - d)$$

By assumption, $y_i < p + d$ whenever $y_i < p$, and $y_i \geq p + d$ whenever $y_i \geq p$. So

$$\begin{aligned} A(p) - A(p + d) &= \sum_{y_i<p} (p - y_i - p - d + y_i) + \sum_{y_i \geq p} (y_i - p - y_i + p + d) \\ &= d\,[(\text{number of } y_i \geq p) - (\text{number of } y_i < p)] \end{aligned}$$

If p is below the median, there must be fewer y_i values below p than above it. Therefore the quantity in square brackets is a positive number. So is d. Given the assumptions, $A(p) - A(p + d) > 0$, or, equivalently, $A(p) > A(p + d)$.

We have shown that starting with a prediction p below the median, one can decrease $A(p)$ by increasing the prediction up to the next y_i value. When the prediction moves across a y_i value, one $y_i - p$ term becomes $p - y_i$, but the term is 0 anyway. Thus increasing the prediction up to the median decreases $A(p)$. A similar proof for $p > m$ shows that *decreasing* a prediction down to the median decreases $A(p)$. Therefore the smallest value for $A(p)$ occurs when $p = m$.

Example 2.36 A sample yielded values 11, 12, 13, 14, and 30. Find $S(p)$ and $A(p)$ for $p = \bar{y} = 16$ and for $p = m = 13$.

Solution

$$S(\bar{y}) = (11 - 16)^2 + (12 - 16)^2 + \cdots + (30 - 16)^2 = 250$$
$$S(m) = (11 - 13)^2 + (12 - 13)^2 + \cdots + (30 - 13)^2 = 295$$

Note that $S(p)$ is smaller when $p = \bar{y}$, as indicated by Result 2.2. $A(\bar{y})$ comes out to be 28, and $A(m)$ is 21. Result 2.3 indicated that $A(m)$ must be smaller. □

Chapter Exercises

2.34. A sample of 63 adults were measured on the Wechsler Adult Intelligence Scale, yielding the following scores:

73	96	123	88	122	91	59	102	110	100	134
98	92	107	123	119	99	91	105	118	103	115
91	97	86	109	94	94	101	111	69	89	111
111	112	97	145	100	94	124	90	104	90	101
84	109	95	85	100	122	127	86	103	98	77
105	96	70	88	103	115	85				

a. Select a reasonable set of intervals for a histogram.

b. Construct the histogram. Describe the general shape of the data.

c. IQ tests like the WAIS are supposed to yield approximately normally distributed scores. Do the data appear roughly normal?

2.35. Construct a stem and leaf display of the data of Exercise 2.34. Does the appearance of the stem and leaf indicate an approximately normal distribution of scores?

2.36. **a.** Calculate the mean and median for the data of Exercise 2.34.

b. Find the standard deviation of the data.

c. How many scores actually fall within 1 standard deviation of the mean? How does the fraction of scores in this interval compare to that given by the Empirical Rule?

2.37. Refer to the histogram constructed in Exercise 2.34.

a. Find the 95th percentile of scores.

b. Find the percentile rank of a score of 95.

c. Are the answers to parts a and b the same? Should they be? (Note that in one case 95 refers to an IQ score, in the other to a percentage.)

2.38. Refer again to the data of Exercise 2.34.

a. Calculate appropriate measures of skewness and heavy-tailness.

b. Are the values of these measures close to those for a normal distribution?

2.39. Data were collected on the time required for each of 22 laboratory rats to run a simple maze, under specified conditions. The times in seconds were as follows:

16.6	31.8	14.1	19.5	20.1	15.0	17.4	17.0
16.3	17.6	18.5	13.1	12.7	18.8	15.9	18.8
19.5	15.9	17.6	17.7	18.0	13.9		

a. Construct a histogram of the data, using about five or six classes.

b. Do the data appear reasonably symmetric? Why might one not expect such data to be symmetric?

2.40. Construct a stem and leaf display of the data of Exercise 2.39. The stem part should be 12, 13, 14, etc., and the leaf part should contain the values to the right of the decimal point.

2.41. Calculate the mean and median for the data of Exercise 2.39. Which of the two values could be expected to be larger? Why?

2.42. Calculate the skewness of the data in Exercise 2.39. Does the resulting value confirm your visual impression of the skewness in the data?

2.43. **a.** Calculate the standard deviation of the data of Exercise 2.39.

b. What fraction of the scores fall within 2 standard deviations of the mean? How well does the Empirical Rule work for these data?

2.44. A sample of 42 nine-year-old children in a certain suburb yielded the following scores on reading ability, stated in terms of norms for grades in school:

6.0	3.5	1.8	7.2	8.1	9.8	1.7	5.9	4.4	4.1	4.1
2.4	3.7	4.3	2.5	4.7	6.7	3.7	7.6	3.3	3.8	4.0
1.7	4.2	4.1	4.5	4.8	2.0	4.7	3.4	1.7	2.2	4.3
3.6	4.8	1.2	4.0	6.6	4.3	4.6	3.5	3.1		

a. Plot the data in any convenient way.

b. Do the data appear roughly symmetric? Would you call the data bell-shaped (normal) or heavy-tailed?

2.45. The nine-year-old children in Exercise 2.44 could be expected to be reading at about grade level 4.0, on the average. Is the typical reading level about 4.0? How much difference does it make which "typical" value you use?

2.46. **a.** Calculate the mean and standard deviation for the data of Exercise 2.44.

b. Show that the fraction of these scores falling within 1 standard deviation of the mean isn't very close to that predicted by the Empirical Rule. Does a picture of the data suggest a reason?

2.47. **a.** Calculate the interquartile range for the data of Exercise 2.1.

b. According to the 1.5 IQR rule of thumb, are there any candidate outliers?

2.48. **a.** What is the IQR for the data of Exercise 2.34?

b. Are there any outliers in the data?

2.49. Calculate a heavy-tailness measure for the data of Example 2.34. What does the resulting number indicate about whether or not the data are reasonably near to normally distributed?

2.50. A sample of 32 patients in a detoxification program yielded the following values for concentration of a certain chemical in blood samples:

47	47	50	51	54	58	63	65	68	69	70
77	77	79	81	91	93	94	99	103	108	113
114	138	161	175	185	194	197	199	223	241	

a. Construct a histogram of the data.

b. Summarize the shape of the data, including in your consideration issues of skewness, heavy-tailness, and bimodality.

c. Try to construct a stem and leaf display of the data. Do you find the result to be a reasonable picture of the data? Is there anything you could do to improve the presentation of the data in the display?

2.51. Refer to the data of Exercise 2.50.

a. Calculate the mean and the median of the data.

b. Mark the mean and the median on a histogram of the data.

c. What explains the discrepancy between the mean and the median?

2.52. Calculate the IQR for the data of Exercise 2.50. Are there any candidate outliers in the data? Perform the 1.5 IQR check for outliers.

2.53. **a.** Calculate the standard deviation for the sample data of Exercise 2.50.

b. Calculate the fraction of scores in that data set that are within 2 standard deviations of the mean. Is there any major discrepancy between the actual fraction and the Empirical Rule approximate fraction?

c. Is the fraction in part b close to the bound given by the Chebyshev Inequality?

2.54. Calculate the three different skewness measures defined in the text, using the data of Exercise 2.50. Do they all agree on the direction of skewness?

2.55. Data on the number of words recalled from a list of 100 were collected for 35 subjects. The values were as follows:

26	26	24	34	31	22	23	25	30	31	28	35
16	53	15	30	39	32	47	30	34	47	37	23
49	9	32	34	33	41	46	40	30	6	33	

Figure 2.20 shows a stem and leaf display of the data created using the Minitab system.

```
STEM AND LEAF DISPLAY OF C1
LEAF DIGIT UNIT = 1.0000
1 2 REPRESENTS 12.

          LO  6, 9

   4      1.  56
   8      2*  2334
  12      2.  5668
 (13)     3*  0000112233444
  10      3.  579
   7      4*  01
   5      4.  6779

          HI  53
```

Figure 2.20 Stem and leaf display for Exercise 2.55

a. From the display, what would you guess would be a typical, middling value for the data?

b. Compute the median. Note that the stem and leaf display contains the data ordered from lowest to highest.

c. Calculate the IQR and the outlier fences 1.5 IQR away from the 25th and 75th percentiles. Which scores fall outside those fences? How does the Minitab system indicate such scores?

2.56. Further Minitab output for the data of Exercise 2.55 was

```
MTAB > MEAN OF C1
       AVERAGE = 31.171
MTAB > STANDARD DEVIATION OF C1
       STANDARD DEVIATION = 10.6
```

a. How many of the scores fall within 1 standard deviation of the mean?

b. Is this fraction close to the Empirical Rule fraction? If not, what is the reason for the discrepancy?

2.57. Calculate $skew_1$ and $skew_3$ for the data of Example 2.55. What do the resulting numbers indicate about the degree of skewness of the data?

2.58. Calculate $skew_3$ for each half of the data in Example 2.55. What do these values indicate about the heavy-tailness of the data?

2.59. Subjects in a study were divided into "mood induction" and "normal" groups and a depression score was obtained for each subject. Part of the computer output (SPSS-X) that resulted is shown in Figure 2.21.

a. What do the boxplots indicate about the shape of the distributions?

b. Locate the stem and leaf display of the combined scores. Is there an obvious, severe skewness?

c. Locate the mean, median, and standard deviation, and compute $skew_1$. Does the value of $skew_1$ confirm your judgment of the degree of skewness?

2.60. The data for Exercise 2.59 were also analyzed by the SAS package. A portion of the output is shown in Figure 2.22.

a. Locate the mean, median, and standard deviation in this output.

b. Locate a skewness measure in the output. (SAS happens to print out $skew_2$.) Does it have the same sign as the $skew_1$ measure calculated in Exercise 2.59?

c. Do the pictures of the data give the same impression of shape as the pictures in Exercise 2.59?

```
FILE HANDLE DEPRDATA/NAME='FOOK22 DATA A'
TITLE 'DEPRESSION SCORES OF SUBJECTS'
DATA LIST FILE=DEPRDATA LIST/DEPRESS GROUP
VARIABLE LABELS DEPRESS 'DEPRESSION SCORE AFTER TREATMENT'/
                GROUP 'TREATMENT TYPE'
VALUE LABELS GROUP 1 'MOOD   INDUCT' 2 'NEUTRAL'
COMPUTE CONSTANT=1
MANOVA DEPRESS BY CONSTANT(1,1)/
  DESIGN=CONSTANT/
  PLOT=BOXPLOTS NORMAL STEMLEAF/
MANOVA DEPRESS BY GROUP(1,2)/
  PRINT OMEANS(VARIABLES(DEPRESS))/
  DESIGN=GROUP/
  PLOT=BOXPLOTS NORMAL STEMLEAF/
```

```
Box-Plots For variable .. DEPRESS

   31 .    0
      .    X
      .    I
      .    I       X
      .    I       I
      .  --+--     I
      .  I   I     I
      .  I * I     I
      .  I   I   --+--
      .  I   I   I   I
      .  --+--   I   I
      .    I     I * I
      .    I     I   I
      .    I     I   I
      .    X     --+--
      .            I
      .            I
      .            I
      .            I
    3 .            X
```

```
Stem-and-leaf display for variable .. DEPRESS

   0 . 34
   0 . 5669
   1 . 01122233344444
   1 . 5555667888889999999
   2 . 000001111222234
   2 . 677889
   3 . 1
```

```
          FREQUENCIES VARIABLES=DEPRESS/
                      HISTOGRAM/
                      STATISTICS=DEFAULT MEDIAN/
```

```
DEPRESS   DEPRESSION SCORE AFTER TREATMENT

COUNT   MIDPOINT    ONE SYMBOL EQUALS APPROXIMATELY   .40 OCCURRENCES

   0        2.0
   2        3.5  *****
   1        5.0  ***
   2        6.5  *****
   0        8.0
   2        9.5  *****
   2       11.0  *****
```

(continued)

Figure 2.21 Output for Exercise 2.59

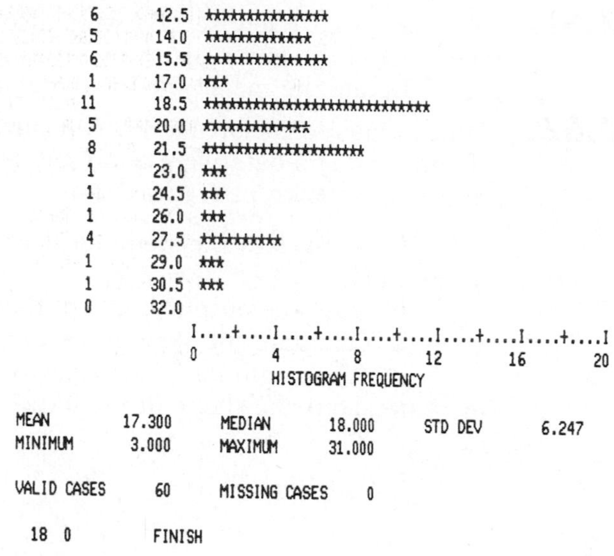

```
 6       12.5 **************
 5       14.0 ************
 6       15.5 **************
 1       17.0 ***
11       18.5 ****************************
 5       20.0 ************
 8       21.5 ******************
 1       23.0 ***
 1       24.5 ***
 1       26.0 ***
 4       27.5 *********
 1       29.0 ***
 1       30.5 ***
 0       32.0
              I....+....I....+....I....+....I....+....I....+....I
              0         4         8        12        16        20
                          HISTOGRAM FREQUENCY

MEAN        17.300     MEDIAN      18.000      STD DEV      6.247
MINIMUM      3.000     MAXIMUM     31.000

VALID CASES     60     MISSING CASES    0

 18  0          FINISH
```

Figure 2.21 *(continued)*

```
                         SUMMARY OF COMBINED DEPRESSION SCORES USING SAS

                                          UNIVARIATE

VARIABLE=DEPSCORE        DEPRESSION SCORE AFTER TREATMENT

              MOMENTS                                   QUANTILES(DEF=4)                        EXTREMES

N                 60   SUM WGTS        60     100% MAX        31      99%        31       LOWEST     HIGHEST
MEAN            17.3   SUM           1038      75% Q3         21      95%        28            3          27
STD DEV      6.24717   VARIANCE   39.0271      50% MED        18      90%      26.9            4          28
SKEWNESS   -0.148287   KURTOSIS -0.0401139     25% Q1      13.25      10%       9.1            5          28
USS            20260   CSS         2302.6       0% MIN         3       5%      5.05            6          29
CV           36.1108   STD MEAN  0.806506                              1%         3            6          31
T:MEAN=0     21.4506   PROB>|T|    0.0001     RANGE           28
SGN RANK         915   PROB>|S|    0.0001     Q3-Q1         7.75
NUM ^= 0          60                          MODE            19
      STEM LEAF                   #     BOXPLOT                   NORMAL PROBABILITY PLOT
        30 0                      1        |        31+                                               *
        28 000                    3        |          |                                          ** *
        26 000                    3        |          |                                      *** ++
        24 0                      1        |          |                                     *++
        22 00000                  5        |          |                                  ****
        20 000000000              9     +-----+       |                               *****
        18 00000000000           11     *-----*       |                           *****+
        16 000                    3     |  +  |     17+                          *+++
        14 000000000              9     |     |       |                       ****
        12 000000                 6     +-----+       |                    ****+
        10 000                    3        |          |                  *++
         8 0                      1        |          |                ++*
         6 00                     2        |          |             +***
         4 00                     2        |          |           * *
         2 0                      1        |         3+   *
           ----+----+----+----+                       +----+----+----+----+----+----+----+----+----+----+
                                                           -2        -1        +0        +1        +2
```

Figure 2.22 Output for Exercise 2.60

2.61. The data from Exercise 2.59 were also analyzed using the BMDP package. Part of the output is shown in Figure 2.23.

Locate the mean, median, and standard deviation in this output.

2.62. Times required to solve a difficult puzzle were obtained from a sample of subjects. The data were analyzed using Minitab. The commands and results are shown in Figure 2.24.

a. What does the stem and leaf display indicate about the shape of the data?

b. Does the output indicate that there are outliers?

c. Calculate the $skew_1$ measure. Does its sign make sense?

d. What summary statistic was stored in K2? What does its value indicate about the shape of the data?

```
/PROBLEM TITLE IS 'DEPRESSION SCORES OF SUBJECTS'.
/INPUT    FILE='DSKC:FOOK22.DAT'.
          VARIABLES ARE 2.
          FORMAT IS FREE.
/VARIABLE  NAMES ARE DEPRESS, GROUP.
           USE IS DEPRESS.
/PRINT    ESTIMATES.
          STEM.
/END
PROBLEM TITLE IS
DEPRESSION SCORES OF SUBJECTS
 ***********
 * DEPRESS  *                          MAXIMUM       31.0000000
 ***********                           MINIMUM        3.0000000
                                       RANGE         28.0000000
VARIABLE NUMBER . . . . . .      1     VARIANCE      39.0271190
NUMBER OF DISTINCT VALUES .     25     ST.DEV.        6.2471688                        EACH 'H'
NUMBER OF VALUES COUNTED. .     60     (Q3-Q1)/2      3.7500000             H         REPRESENTS
NUMBER OF VALUES NOT COUNTED     0     MX.ST.SC.      2.19                  H             2
                                       MN.ST.SC.     -2.29               H  HH         COUNT(S)
                                                                         HHHHHH
LOCATION ESTIMATES                      ST.ERROR                       H HHHHHH  H
               MEAN         17.3000000   0.8065060                    HH HHHHHHHHHHHH
               MEDIAN       18.0000000   1.1547011                    L-----------------U
               MODE         19.0000000                        EACH '-' ABOVE =     2.0000
                                                                          L=       2.0000
                                                                          U=      36.0000
                                                          CASE NO. OF MIN. VAL. =  43
                                                          CASE NO. OF MAX. VAL. =  26
                                                                                 Q1=  13.5000000
          SOME NEW LOCATION ESTIMATES                     VALUE      VALUE/S.E.  Q3=  21.0000000
               HAMPEL       17.3750000                                           S-=  11.0528310
               TRIM(.15)    17.3809524      SKEWNESS      -0.14        -0.45     S+=  23.5471690
               BIWEIGHT     17.4872289      KURTOSIS      -0.23        -0.36
```

Figure 2.23 Output for Exercise 2.61

```
MTB > EXECUTE 'FOOK23'
MTB > READ 'FOOK23 DATA A' INTO C2 AND C1
   117 ROWS READ
 ROW    C2     C1

   1     1    279
   2     2    239
   3     3    202
   4     4    307
   . . .

MTB > NAME C1 'TIMES'
MTB > DESCRIBE C1

            TIMES
N             117
MEAN          265
MEDIAN        168
TMEAN         206
STDEV         367
SEMEAN         34
MAX          2461
MIN             8
Q3            276
Q1             76
MTB > SORT C1 PUT BACK IN C1
MTB > STEM AND LEAF OF C1

 STEM-AND-LEAF DISPLAY OF TIMES
 LEAF DIGIT UNIT = 10.0000
 1 2 REPRESENTS 120.


   13    +0* 0112233333444
   37    +0. 555555666666666778888889
   55     1* 000011111222334444
   (9)    1. 556677789
   53     2* 000111233333344
   38     2. 556666667788889
   23     3* 00004
   18     3. 79
   16     4* 4
   15     4. 55
   13     5* 0
   12     5. 5

         HI   62,  65,  77,  78,  91,  95, 100, 104, 169, 212, 246,


MTB > PICK ROWS 1 TO 58 OF C1 PUT IN C2
MTB > PICK ROWS 58 TO 117 OF C1 PUT IN C3
MTB > NAME C2 'LOHALF' C3 'HIHALF'
MTB > MEDIAN OF C1 PUT IN K1
   MEDIAN =      168.00
MTB > LET K2=((MEAN(C3)-K1)-(K1-MEAN(C2)))/(MEAN(C3)-MEAN(C2))
MTB > PRINT K2
K2       0.520011
MTB > STOP
```

Figure 2.24 Output for Exercise 2.62

3 Summarizing Relationship and Predictability

The focus in Chapter 2 was on summarizing data for one variable at a time. Much scientific use of statistical methods involves summarizing relations between two variables, or among several variables. This chapter is devoted to some of the many statistical measures of relationship. In this chapter we will consider only qualitative and ordinal variables. Prediction and relation for quantitative variables is the topic of Chapter 15.

The key idea is that the degree of relationship can be measured by the degree of predictability. If two variables are closely related, knowing the value of one variable is highly useful in predicting the value of the other. Conversely, if the relation is weak, knowing the value of one variable will be of little value in predicting the other one. Not all measures of statistical relation are based on predictability, but the very useful ones in this chapter all have a prediction basis. In Sections 3.1 and 3.2 we consider two different measures for qualitative variables, and in Section 3.3 we discuss some measures for ordinal variables. In Section 3.4 we raise some fundamental issues that arise when one tries to control for the effects of other variables.

3.1 The λ Approach

The ideas involved in summarizing predictability are most easily seen with respect to qualitative variables (variables in which the values are only categories, possibly in some sort of order). The simplest approach to summarizing predictability is the λ (lambda) measure developed by

Guttman (1941) and by Goodman and Kruskal (1954). The ideas in the method extend to other types of variables as well.

dependent variable **independent variable**

To describe this measure, we need some definitions. In a prediction problem, the **dependent variable** is the variable that we're trying to predict, and an **independent variable** is one that we're using to make the prediction. The names are not meant to suggest any cause-and-effect relation; the choice of which variable is to be dependent and which independent can be made by the researcher according to the purposes of the investigation.

An example is helpful in describing the measure. Bennett and Grosser (1981) report data on, among other things, Bem Sex Role Inventory Rating scores on 45 women before and after a treatment. Before giving the actual data, we will consider an exaggerated version of their finding.

Assume that the frequencies in Table 3.1 had been found:

		Bem Score			
		Androgynous	Moderately Sex-typed	Sex-typed	Total
Time	Before	3	9	33	45
	After	18	21	6	45
	Total	21	30	39	

Table 3.1

How strong is the relationship between time and Bem score in these artificial data?

One way to answer the question is by looking at percentages. Before treatment 33 of the 45 women (73.3%) were rated as sex-typed, and after treatment only 6 of the 45 (13.3%). The percentages of androgynous and moderately sex-typed ratings increased substantially. Therefore it seems that there is a fairly strong relation between time and Bem rating in these data.

To get a single numerical value for the strength of relation, we take Bem score as a dependent variable and time as an independent (predictor) variable. Let us suppose that we pick a woman at random from the before group and try to predict her Bem rating. It's natural to predict "sex-typed," because there are more women in that category than in any other. If we make this prediction for all 45 women before treatment, we will make 12 errors (the 3 androgynous and 9 moderately sex-typed women). Similarly, the most common rating in the after data is moderately sex-typed, so if we were in the after ratings, we should predict "moderately sex-typed"; this prediction results in 24 errors (the 18 androgynous ratings and 6 sex-typed ratings). Overall, the predictions result in 12 + 24 = 36 errors. To see if this is good or bad prediction,

marginal frequencies

we need a comparison standard. In the lambda method, the comparison standard is based on the **marginal frequencies** for the dependent variable. In the example, these frequencies are shown at the bottom margin of the table: 21 androgynous ratings, 30 moderately sex-typed ratings, and 39 sex-typed ratings. If we do not know whether the rating was made before or after treatment (so we don't know the independent variable value), the best guess is that the rating is "sex-typed." Applying this prediction to all 90 ratings (45 before, 45 after) results in 21 + 30 = 51 errors. In this example, knowing the time allows us to make fewer errors (36) than we would not knowing the time (51 errors). The λ measure is defined as the difference of the two error rates, taken as a proportion of the error rate when the independent variable isn't known. For the example,

$$\lambda = \frac{51 - 36}{51} = .294$$

modal value

The general λ approach follows the pattern of the example. The investigator must specify which variable is to be regarded as the independent variable and which the dependent variable. First, consider the prediction with known independent variable values. For each such value, predict the **modal value** (most common value) of the dependent variable, and count up the total number of errors committed. Second, consider the prediction with *unknown* independent variable values. Predict the modal value based on the marginal frequencies of the dependent variable. Third, calculate

$$\lambda = \frac{\text{"unknown" prediction errors} - \text{"known" prediction errors}}{\text{"unknown" prediction errors}}$$

Example 3.1

Boone and Montare (1979) report an experiment in which 56 Hispanic boys, 7 and 8 years old, were measured on aggressive behavior in a certain team task. They reported the data in terms of means and standard deviations, which raises the problem of using numerical values for ordinal data. Assume that the data were as shown in Table 3.2.

		Number of Children in Family						
		1	2	3	4	5	6+	Total
Aggressiveness	H	2	5	2	1	0	0	10
	M	1	6	3	4	3	1	18
	L	0	1	11	7	7	2	28
	Total	3	12	16	12	10	3	56

Table 3.2

Calculate the λ value, treating aggressiveness as the dependent variable.

Solution For 1 child in the family, the modal class is H. Predicting H for 1 child yields 1 error. For 2 children in the family, the prediction is M, yielding 6 errors; for any number of children larger than 2, the prediction is L, yielding a total of $5 + 5 + 3 + 1 = 14$ errors. If the number of children in the family is known, the total number of prediction errors is $1 + 6 + 14 = 21$. If the number of children in the family is not known, the best prediction is L, because there are more L children overall than in any other category; this prediction results in 28 errors. So

$$\lambda = \frac{28 - 21}{28} = .25$$

□

The λ measure is a Proportionate Reduction in Error (PRE) measure. That is, it measures the reduction in prediction error resulting from knowledge of independent variable values, expressed as a fraction of the prediction error for the unknown independent variable case. In the Bem rating example, $\lambda = .294$ means that knowing whether the rating was made before or after treatment (that is, knowing the value of the independent variable time) leads to a 29.4% reduction in the prediction error below the value associated with not knowing the independent variable value. The maximum possible value is 1.000, indicating 100% error reduction, or perfect prediction. At the other end of the scale, a λ value of .000 indicates no predictive value at all. This value occurs when the two variables are statistically independent—that is, when the dependent variable percentages are the same for every value of the independent variable.

A λ value equal to .000 can also occur even when there is some degree of relationship between variables. The actual frequencies found by Bennett and Grosser (1981) are shown in Table 3.3.

		Bem Score			
		Androgynous	Moderately Sex-typed	Sex-typed	Total
Time	Before	5	14	26	45
	After	17	11	17	45
	Total	22	25	43	

Table 3.3

Note that the ratio of "sex-typed" women decreased from 26 of 45 (57.8%) before treatment to 17 of 45 (37.8%) after treatment, and the ratio of androgynous women increased sharply. Therefore, there is a statistical dependence between time and Bem score. But λ doesn't reflect this dependence. Before treatment, "sex-typed" is the mode; there is a tie in the after-treatment frequencies, so we might as well predict "sex-typed" after treatment as well. Knowing the time, we would make 5 + 14 + 17 + 11 = 57 errors; not knowing time, we would still predict "sex-typed" and make the same 57 errors. For the Bennett and Grosser data, $\lambda = (57 - 57)/57 = .000$ despite the statistical dependence.

The λ measure depends on which variable is regarded as the dependent variable. For the actual Bennett and Grosser data, if time is taken as the dependent variable, $\lambda = (45 - 36)/45 = .200$, rather than the .000 found when Bem score is dependent. Thus, the λ measure is not ideal as a measure of predictability. However, the ideas behind the measure can be adapted to several other more satisfactory measures, so λ is a useful starting point in the study of statistical predictability.

Example 3.2 Referring to the data of Example 3.1, find λ when the dependent variable is number of children in family.

Solution If H, predict 2 children (5 errors); if M, predict 2 children (15 errors); if L, predict 3 children (17 errors). Prediction error when aggressiveness is known equals 37. If aggressiveness isn't known, predict 3 children, which yields 40 errors.

$$\lambda = \frac{40 - 37}{40} = .075$$

Note that this value is quite different from the $\lambda = .25$ value found in Example 3.1. □

3.2 The ∇ Method

The λ measure defined in Section 3.1 has some drawbacks as a measure of statistical predictability. In that section, we noted that λ can equal .000 even when there is statistical dependence in the data and that the value of λ depends on which variable is declared to be the dependent variable. There are other drawbacks as well. The value of λ can be affected substantially by the way the categories of the qualitative variables are defined. In the Bennett and Grosser data of the previous section, if the androgynous and moderately sex-typed categories are combined into a less-sex-typed category, and if Bem score is taken as the dependent variable, $\lambda = .163$ rather than the .000 found in Section 3.1. Also, the prediction rule implied by λ is rigid. One must always predict a single value (the mode) of the dependent variable. If a researcher has some other prediction in mind, λ is of no use.

A different measure of predictability, the ∇ (del) measure, was developed by Hildebrand, Laing, and Rosenthal (1977). This measure allows (indeed, requires) a researcher to state a specific prediction rule; the ∇ measure then indicates how well one can predict a dependent variable value *according to that specific prediction rule.* The prediction rule may be almost anything; it may predict several values of the dependent variable, or only one, or even none in some cases. The prediction rule may be stated *a priori,* before the data are collected, or *ex post,* on the basis of the collected data.

The Bennett and Grosser data of Section 3.1 can provide a useful example of the ∇ approach. The treatment group was women enrolled in a Psychology of Men and Women course. One translation of the idea that such a course should make the students more aware of, and freer of, sex stereotyping is that before the course students would tend to be sex-typed while after the course students would tend to be androgynous *or* moderately sex-typed. Note that this prediction rule involves **set prediction;** the prediction for after the course only states that the Bem score will be somewhere in a *set* of predicted states. This prediction rule ("before" predicts "sex-typed," "after" predicts "androgynous" or "moderately sex-typed") yields 5 + 14 + 17 = 36 errors in the actual data, reproduced in Table 3.4.

set prediction

		Bem Score			
		Androgynous	Moderately Sex-typed	Sex-typed	Total
Time	Before	(5)	(14)	26	45
	After	17	11	(17)	45
	Total	22	25	43	90

Table 3.4

Example 3.3

The frequencies in Example 3.1 indicate that aggressiveness decreased as the number of children in the family increased. Suppose we predict that for 1 child in the family, aggressiveness will be H; for 2, H or M; for 3, M or L; and for 4 or more, L. How many errors are made?

Solution

The errors are (1, M) (1 case), (1, L) (no cases), (2, L) (1 case), (3, H) (2 cases), (4, H) (1 case) 4, M (4 cases), 5, H (no cases), 5, M (3 cases),

6+, H (no cases), and 6+, M (1 case). The total number of prediction errors is found by simple addition to be 13. □

A comparison to the actual number of prediction errors is needed; is 36 errors good or bad? The comparison standard of the λ measure is based on predicting the overall mode of the dependent variable; it is not appropriate for the ∇ method, which allows an investigator to predict more than one category, if desired. Instead, imagine a *randomized replication* of the prediction. In the Bennett and Grosser data "sex-typed" was predicted 45 times (the 45 "before" scores) and "androgynous or moderately sex-typed" was predicted the other 45 times. In the randomized replication, the same predictions are made the same number of times, only this time randomly, without regard to the value of the independent variable, time. Of the 90 scores, there are 47 that are not sex-typed; thus in the 45 "sex-typed" predictions we could expect to make 45(47/90) = 23.5 errors. Similarly, there are 43 scores that are not "androgynous or moderately sex-typed," so we could expect 45(43/90) = 21.5 errors for that prediction. The total expected number of errors is 45 for this randomized-replication baseline. The ∇ measure compares the actual number of errors committed by the prediction rule to this baseline, in the same proportionate reduction in error (PRE) fashion as the measure.

$$\nabla = \frac{45 - 36}{45} = .200$$

indicating that by knowing the time, and using the particular prediction rule, one gets a 20.0% reduction in prediction error.

Example 3.4 Compute the ∇ measure for the prediction stated in Example 3.3.

Solution First it is necessary to calculate the error rate for randomized prediction. The prediction H (by itself) was made 3 times; 46 of the 56 children are not H. Thus, if one randomly predicts H 3 times, one can expect 3(46/56) = 2.464 errors. For the 12 H or M predictions, we note that 28 children are neither H nor M, yielding an error rate of 12(28/56) = 6.000. Similarly, the error rate for the 16 M or L predictions is 16(10/56) = 2.857; and for the 25 L predictions, the error rate is 25(28/56) = 12.500. Adding these error rates, we get the expected error rate for randomized prediction, 2.464 + 6.000 + 2.857 + 12.500 = 23.821. The actual error total was 13. Therefore,

$$\nabla = \frac{23.821 - 13}{23.821} = .454$$

The comparison standard, randomized-replication error rate may be computed more simply by creating a table of **expected frequencies.** We begin to construct the table by entering marginal row and column totals, as well as the grand total, in the margins of the table. For the Bennett and Grosser data, we have Table 3.5.

expected frequencies

		Bem Score			
		Androgynous	Moderately Sex-typed	Sex-typed	Total
Time	Before				45
	After				45
	Total	22	25	43	90

Table 3.5

Then each entry in the body of the table may be computed by the following formula:

$$\text{expected frequency} = \frac{(\text{row total})(\text{column total})}{\text{grand total}}$$

The expected frequency for (before, androgynous) is $(45 \times 22)/90 = 11.0$; for (before, moderately sex-typed), it is $(45 \times 25)/90$; etc. The complete table is shown in Table 3.6.

		Bem Score			
		Androgynous	Moderately Sex-typed	Sex-typed	Total
Time	Before	11.0	12.5	21.5	45
	After	11.0	12.5	21.5	45
	Total	22	25	43	90

Table 3.6

The randomized-replication error rate may be calculated by summing up the entries in those "cells" of the expected frequency table which correspond to prediction errors. For the prediction rule discussed above, the errors are (before, androgynous), (before, moderately sex-typed), and (after, sex-typed). The sum of the expected entries in these cells is 11.0 + 12.5 + 21.5 = 45.0, as we found by considering the randomized replication.

In general, the ∇ statistic may be defined by specifying the **error cells** for the prediction rule, counting the observed, actual number of

error cells

prediction errors, constructing the expected frequency table, and adding up the expected errors. Then

$$\nabla = \frac{\text{expected errors} - \text{observed errors}}{\text{expected errors}}$$

Example 3.5 Compute a table of expected values for Example 3.3. Find the total expected errors.

Solution For the (H, 1) cell, the expected value is 10(3)/56 = 0.536. The rest of the table can be computed in the same way, yielding Table 3.7.

		Number of Children in Family						
		1	2	3	4	5	6+	Total
	H	0.536	2.143	2.857	2.143	1.786	0.536	10.001
Aggressiveness	M	0.964	3.857	5.143	3.857	3.214	0.964	17.999
	L	1.500	6.000	8.000	6.000	5.000	1.500	28.000
	Total	3.000	12.000	6.000	12.000	10.000	3.000	56.000

Table 3.7

The expected errors are found by adding up the entries in the expected table corresponding to prediction errors. The error cells were listed in Example 3.3. The expected errors are 0.964 + 1.500 + 6.000 + 2.857 + 2.143 + 3.857 + 1.786 + 3.214 + 0.536 + 0.964 = 23.821, as in Example 3.4. □

The value of ∇ is, once again, a PRE value. Its largest possible value is 1.000, indicating perfect prediction using the specified prediction rule. It will equal .000 if there is no relation between the variables, or if the prediction rule fails to capture what relation there is. The value of ∇ may be negative, indicating that the specified prediction rule is grossly incorrect about the nature of the relation.

The ∇ measure has some advantages over the λ measure. It allows a researcher to evaluate the success of a specific prediction. The choice of prediction rules is not rigid. The measure is related to a widely used statistical test, called the chi-squared test, which will be introduced in Chapter 13. One difficulty with the ∇ measure is a natural consequence of its generality. Because set prediction is allowed, it is possible to evaluate predictions that are too vague and imprecise to be scientifically interesting. In the Bennett and Grosser data, we might have used the prediction rule "Before predicts moderately sex-typed or sex-typed, after predicts androgynous or moderately sex-typed or sex-typed." This prediction rule isn't very precise; the only cases that violate it are the (before, androgynous) cases. Hildebrand, Laing, and Rosenthal suggest a

precision measure of the **precision** of the prediction:

$$\text{precision} = \frac{\text{expected errors}}{\text{grand total}}$$

For the original Bennett and Grosser prediction, precision = 45/90 = .500, whereas for the imprecise, vague prediction of the previous paragraph, precision = 11.0/90 = .122. Typically, a low-precision prediction will have a higher ∇ value. The "cowardly" prediction has $\nabla = (11.0 - 5)/11.0 = .545$, as compared to the ∇ value of .200 for the original, higher-precision prediction. Thus a trade-off between prediction success, as measured by ∇, and prediction boldness, as measured by precision, must be made.

Example 3.6 What is the precision for the prediction of Example 3.3?

Solution In both Examples 3.4 and 3.5, we found that the expected error rate was 23.821. Thus

$$\text{precision} = \frac{23.821}{56} = 0.425$$ □

No definitive statement can be made about what constitutes good ∇ and precision values. A rough rule of thumb is that having both precision and ∇ larger than .500 is quite good. However, such a rule of thumb doesn't allow for the fact that some situations are easily predictable with elementary measurements, whereas others are highly unpredictable, even with sophisticated instruments. Thus, "good" ∇ and precision values are very much a matter of context.

Exercises for Sections 3.1 and 3.2

3.1. Buss, Iscoe, and Buss (1979) asked parents of 355 children ages 3-12 to record any incidences of embarrassment over the past six months. The frequencies shown in Table 3.8 were reported:

		Age								
		3–4	5	6	7	8	9	10	11–12	Total
Embarrassment	Yes	9	19	32	38	37	36	26	28	225
	No	26	13	12	13	20	18	10	16	128
	Total	35	32	44	51	57	54	36	44	353

Table 3.8

Treat the age variable as qualitative, partly because the categories at both ends of the scale are lumped together.

a. Calculate the λ value for predicting whether or not a child showed embarrassment, given the child's age.

b. Carefully interpret the numerical value obtained in part a. What does it say about the ability to predict embarrassment given age?

3.2. For the data of Exercise 3.1, calculate the λ value for predicting age from embarrassment. Is it markedly different from the λ value found in Exercise 3.1?

3.3. The authors of the study in Exercise 3.1 suggested that children 5 years old and older tend to be capable of embarrassment, whereas those 3 and 4 years old tend not to be.

a. Which cells of the tabulated frequencies are errors for that prediction?

b. How many errors does the prediction yield?

c. Calculate the ∇ value for this prediction. What is the interpretation of this number?

d. What is the precision of this prediction?

3.4. An alternative prediction based on the data of Exercise 3.1 would be that children up to 5 years old will tend not to be embarrassed, whereas those 6 and older will tend to be embarrassed.

a. Calculate the ∇ and precision values for the alternative prediction.

b. Is either of the two possible prediction rules superior to the other, in the sense of resulting in higher error reduction and higher precision?

3.5. Vandewiele (1981) administered a questionnaire to 776 Senegalese secondary school students and obtained data on, among other things, sex of student and reported key figure in dreams. The data were reported as percentages, but are shown reconstructed as frequencies in Table 3.9.

		Key Figure			
		Dreamer	Other	None	Total
Sex	Female	208	92	60	320
	Male	337	96	23	456
	Total	545	188	83	776

Table 3.9

a. Calculate the λ value for predicting the sex of the dreamer given the key figure in the dream.

b. Interpret this number.

3.6. **a.** Show that, for the data in Exercise 3.5, the λ value with sex as the independent variable and key figure as the dependent variable is 0.

b. Explain why λ came out 0 in this case.

3.7. For the data in Exercise 3.5, calculate the ∇ value for the prediction that "Females will have other or none as the key figure; males will have dreamer as the key figure."

3.8. What is the precision value for the prediction in Example 3.7?

3.3 Relations of Ordinal Variables

Ordinal variables, those that reflect an ordering of the values without necessarily measuring actual magnitudes, are extremely common in research in psychology, education, and other social sciences. The Bem scale used in the examples in the two preceding sections is an example. The categories of that scale can be thought of as ordered along a continuum from androgynous to sex-typed. In fact, the before-after variable, like any other variable taking on only two possible values, can be treated as an ordinal variable by arbitrarily declaring one value, say "after," as larger than the other.

There are several ways in which social scientists measure relationships among such ordinal variables. One approach is the one used in Section 3.2. There, the ordinal nature of the variables suggested a prediction rule: in going from before-treatment to after-treatment, subjects were expected to move away from the sex-typed end of the Bem scale toward the androgynous end. The ∇ measure of the previous section can be used to evaluate the success of any such ordinal prediction rule. Alternatively, ordinal variables are often treated as legitimately numerical variables. Innumerable studies have been done that, for example, code answers to a questionnaire on a 1 = strongly disagree to 5 = strongly agree basis, then treat the coded scores as truly quantitative, measured variables. In fact, the computation of college grade point averages is a nearly universal example of treating ordinal data as quantitative. There are dangers in using quantitative-data measures of relationship on ordinal data. Ordinal-data codes are somewhat arbitrary, and it is perfectly possible that different coding schemes will give substantially different measures of the degree of relation between two variables. Furthermore, most quantitative-data measures work best when the variable has a very large number of possible values; in most ordinal data, the number of possible values is quite small, such as the three possibilities in the Bem scale.

Froman and Hubert (1980) examined the degree of understanding of two logical concepts by school children. Table 3.10 is an exaggerated version of the frequencies they found.

		Understanding of Concept A				
		1	2	3	4	Total
Understanding of Concept B	1	15	2	2	0	19
	2	8	71	4	1	84
	3	1	16	49	2	68
	4	0	1	13	15	29
	Total	24	90	68	18	200

Table 3.10

(1 is the highest level of understanding, 4 the lowest.) The high frequencies fall on or just below the main diagonal of the table, indicating that children's understanding of the two concepts is either identical or that the understanding of concept B is slightly poorer than that of concept A. The ∇ measure of Section 3.2 can be applied to the prediction that $A = 1$ predicts $B = 1$ or 2, $A = 2$ predicts $B = 2$ or 3, $A = 3$ predicts $B = 3$ or 4, and $A = 4$ predicts $B = 4$; $\nabla = .845$ with precision $= .418$, indicating rather good prediction success and fair-to-middling prediction precision.

comparing pairs

Another approach to measuring relations among ordinal variables is based on the idea of **comparing pairs** of individuals. In the Froman and Hubert example, there appears to be a rather strong positive relation between the two understanding scores, in that most children who score well on one concept score well on the other and those who score poorly on one concept tend to score poorly on the other. Suppose we compare two children. If child I scores better (lower code) on concept A than child II, it is likely that child I will also score better on concept B. If child I scores better on both variables than child II, or *worse* on both, that pair of children is **consonant** with respect to the two concepts. If child I scores better on one scale, but worse on the other, than child B, that pair is **dissonant.** The idea is that, if there is a positive relation, there should be many more consonant pairs than dissonant pairs. Several ordinal-data measures of relationship are based on this idea. The measures differ among themselves in how they handle ties—cases in which both individuals score the same on one or both of the scales.

consonant

dissonant

Computing the number of consonant, dissonant, and tied cases can be rather tedious, but is easily done by a computer program. Therefore, we will discuss the computation only briefly; a more complete discussion appears in Hildebrand, Laing, and Rosenthal (1977b). Again we'll use the Froman and Hubert frequencies for illustration. We can let any of 200 children be child I, and any of 200 children be child II. We might as well allow children to be paired with themselves; in the example, there are only 200 such pairs, so it won't make much difference. Let us first compute the number of pairs in which both A and B scores of child I are better (lower code) than those of child II. To begin, consider the 15 children who scored 1 on both scales. Any child falling below and to the right of the (1, 1) cell in the data table has poorer understanding of both concepts; there are $71 + 4 + 1 + 16 + 49 + 2 + 1 + 13 + 15 = 172$ such children, giving us $15 \times 172 = 2580$ pairs with child I scoring 1 on both concepts and child II having a poorer score on both. To that total we must next add all the pairs where child I scores 2 on A and 1 on B and child II has poorer scores. There are

$$2(4 + 1 + 49 + 2 + 13 + 15) = 168$$

such pairs, which brings the running total up to $2580 + 168 = 2748$. We continue in this way throughout the data table, multiplying the frequency in each cell by the combined frequency that is "southeast" of it in the table. The total of these products, 10,441, is the number of pairs in which child I scores better on both scales. For every (Susie, Johnny) pair, there is a (Johnny, Susie) pair, so this total is also the number of pairs in which child I scores *worse* than child II on both scales. This is the hardest step in the computation.

Example 3.7

The data from Example 3.1 are shown again in Table 3.11, but reordered so that both scales go from high to low.

		Number of Children in Family						
		6+	5	4	3	2	1	Total
Aggressiveness	H	0	0	1	2	5	2	10
	M	1	3	4	3	6	1	18
	L	2	7	7	11	1	0	28
	Total	3	10	12	16	12	3	56

Table 3.11

Compute the number of pairs of data points corresponding to cases in which subject I scores higher on both scales than does subject II.

Solution Begin in the upper left-hand corner, and work row by row. There are no (6+, H) or (5, H) cases, so we can omit these categories. For the 1(4, H) case, there are 22 cases "southeast" of it, for a total of 22 pairs. For the 2(3, H) cases, there are 8 southeast cases; the running total is 22 + 2(8) = 38. Continuing in this fashion, we get

$$1(22) + 2(8) + 5(1) + 1(26) + 3(19) + 4(12) + 3(1) + 6(0) = 177 \text{ such pairs} \quad \square$$

The rest of the computation can be done indirectly, by considering the number of tied pairs. To calculate the number of pairs in which both scores are equal, pair 15(1, 1) child I cases with the same 15 cases, to get $15^2 = 225$ pairs, plus 2^2 pairs with $A = 2$ and $B = 1$ for both, and so on; there are $15^2 + 2^2 + \cdots + 13^2 + 15^2 = 8412$ doubly tied pairs. Similarly, there are $24^2 + 90^2 + 68^2 + 18^2 = 13{,}624$ pairs tied in understanding of concept A, and $19^2 + 84^2 + 68^2 + 29^2 = 12{,}882$ pairs tied in understanding of concept B. Thus, we have computed the following parts of a new table, where ">" means "better understanding of."

	$A_I > A_{II}$	$A_I = A_{II}$	$A_I < A_{II}$	Total
$B_I > B_{II}$	10,441			
$B_I = B_{II}$		8,412		12,882
$B_I < B_{II}$			10,441	
Total		13,624		40,000

Table 3.12

The rest of the table may be filled in by using the (Susie, Johnny)/(Johnny, Susie) symmetry. The remaining cases in the Total column (40,000 − 12,882 = 27,118) must be split equally in two, so there are 13,559 $A_I > A_{II}$ pairs and 13,559 $A_I < A_{II}$ pairs. Similarly there are (40,000 − 13,624)/2 = 13,188 $B_I > B_{II}$ pairs and 13,188 $B_I < B_{II}$ pairs. The middle row and column of the table can be filled in similarly. Assign (12,882 − 8412)/2 = 2235 pairs to each unfilled entry in the middle row, and (13,624 − 8412)/2 = 2606 to each unfilled entry in the middle column. The result is Table 3.13.

	$A_I > A_{II}$	$A_I = A_{II}$	$A_I < A_{II}$	Total
$B_I > B_{II}$	10,441	2,606		13,559
$B_I = B_{II}$	2,235	8,412	2,235	12,882
$B_I < B_{II}$		2,606	10,441	13,559
Total	13,188	13,624	13,188	40,000

Table 3.13

Finally, the remaining holes may be filled in by subtraction. The $(A_I < A_{II}, B_I > B_{II})$ entry is 13,559 − 10,441 − 2606 = 512, as is the $(A_I > A_{II}, B_I < B_{II})$ entry. Table 3.14 is the final table.

	$A_I > A_{II}$	$A_I = A_{II}$	$A_I < A_{II}$	Total
$B_I > B_{II}$	10,441	2,606	512	13,559
$B_I = B_{II}$	2,235	8,412	2,235	12,882
$B_I < B_{II}$	512	2,606	10,441	13,559
Total	13,188	13,624	13,188	40,000

Table 3.14

Example 3.8 Complete the table of pairs for the data of Example 3.7.

Solution Call the row scale C for Children, and the column scale A for Aggressiveness. From Example 3.7, we have that there are 177 pairs with $A_I > A_{II}$ and $C_I > C_{II}$. The number of cases tied on both scales is the sum of the squared frequencies in the body of the table, which comes out to be 330. The number of pairs tied on C is the sum of squared column frequencies, which is 662; the number of pairs tied on A is 1208. The total number of pairs is $(56)^2 = 3136$. Knowing these values, we can complete Table 3.15 using symmetry. □

This rather tedious computation, which of course can be done by computer, is a necessary step in defining many measures of ordinal relation. These measures are based on the difference between the fraction of consonant pairs (having $A_I > A_{II}$ and $B_I > B_{II}$ or else $A_I < A_{II}$ and $B_I < B_{II}$) and the fraction of dissonant pairs (having inequalities in opposite directions). In the example, the fraction of consonant pairs is

	$C_I > C_{II}$	$C_I = C_{II}$	$C_I < C_{II}$	Total
$A_I > A_{II}$	177	166	621	964
$A_I = A_{II}$	439	330	439	1208
$A_I < A_{II}$	621	166	177	964
Total	1237	662	1237	3136

Table 3.15

(10,441 + 10,441)/40,000 = .522 and the fraction of dissonant pairs is (512 + 512)/40,000 = .026. The simplest measure is **Kendall's tau,** τ, which is the difference between these fractions. In the example,

Kendall's tau

$$\tau = .522 - .026 = .496$$

This measure ranges from +1, for a perfectly increasing relation, without ties, in the data, down to −1, for a perfectly decreasing, tieless relation. In most ordinal-data situations, there are many ties, which are not considered by Kendall's tau. Many solutions to the ties problem have been proposed. The most extreme, proposed by Goodman and Kruskal (1954), is to throw out all the tied cases and to take the difference in consonant and dissonant fractions as a fraction of their sum. The result is a measure called **gamma,** γ. In the example,

gamma

$$\gamma = \frac{.522 - .026}{.522 + .026} = .907$$

Gamma can range from −1, for a decreasing relation, to +1, for an increasing relation. The value of γ will be 0 if there is no **monotone** (increasing or decreasing) relation. There may be some other kind of relation, such as a U-shaped one.

monotone

Example 3.9

Calculate Kendall's τ and γ for Example 3.8.

Solution

From Table 3.15, the fraction of consonant pairs is (177 + 177)/3136 = .113, and the fraction of dissonant pairs is (621 + 621)/3136 = .397. Kendall's τ is .113 − .397 = −.284, and γ is (.113 − .397)/(.113 + .397) = −.557. Thus there appears to be a moderate, negative relation between number of children in the family and observed aggressive behavior. □

As the example indicates, the values of Kendall's τ and of γ may be quite far apart. There are a number of other measures, such as Somer's *d* measures, which have intermediate values; see Hildebrand,

Laing, and Rosenthal (1978). The only reason for the discrepancies is the way in which each measure treats ties. Kendall's τ includes all tied cases as relevant; γ throws them all out. Obviously, there are intermediate possibilities. A reasonable approach is to regard τ and γ as pessimistic and optimistic bounds, respectively, on the degree of ordinal relation.

All the standard ordinal measures have a limitation; they all are intended to measure the general tendency of one variable to increase as another increases. None of the measures reflects any more specific statement about the exact nature of the relation. The ∇ measure defined in Section 3.2 can be used to evaluate specific kinds of predictability, as opposed to the rather vague "tends to increase together" relationship evaluated by the various ordinal measures.

Exercises for Section 3.3

3.9. Goodwin and Michel (1981) studied infant handedness and head position preferences of 71 nineteen-week-old infants. The frequencies in Table 3.16 were reported.

		Handedness			
		Strong Right	Biased Right	Biased Left	Strong Left
Head Position	Strong Right	11	16	3	3
	Biased Right	1	8	4	3
	Biased Left	6	2	4	2
	Strong Right	1	1	3	3

Table 3.16

Calculate the ∇ value for the prediction "handedness will be exactly the same as head position."

3.10. Refer to the frequencies of Exercise 3.9.

a. Compute the number of pairs of infants having $H_{I} > H_{II}$ and $P_{I} > P_{II}$, where H refers to handedness and P refers to head position.

b. Calculate the value of Kendall's τ. What is the interpretation of the resulting number?

c. Calculate the value of γ. What is the interpretation of this number?

3.11. Inhelder and Piaget (1964) reported a study of children of various ages who were given rods to arrange in order. The children were judged on their ability to arrange the rods in series, with the possible ratings being none, small, trial and error, and systematic. The frequencies were as shown in Table 3.17.

		Ability			
		None	Small	Trial and Error	Systematic
	4	8	7	0	0
	5	6	21	4	3
Age	6	2	11	8	11
	7	0	7	5	20
	8	0	0	1	20

Table 3.17

a. Consider the prediction "4-year-olds will have no seriation ability, 5- and 6-year-olds will have small or trial-and-error ability, 7-year-olds will have trial-and-error or systematic ability, and 8-year-olds will have systematic ability." Indicate which cells of the frequency table are errors for this prediction and how many errors are committed.

b. Compute the ∇ and precision values for this prediction.

3.12. Refer to the data of Exercise 3.11, regarding both age and series ability as ordinal variables.

a. Does there appear to be a relation? How can you tell, without computation?

b. Table 3.18 is the table of pairs, with S denoting series ability and A denoting age. Find the value of Kendall's τ and the value of γ.

	$S_I > S_{II}$	$S_I = S_{II}$	$S_I < S_{II}$	Total
$A_I > A_{II}$	4638	1906	499	7043
$A_I = A_{II}$	1035	1800	1035	3870
$A_I < A_{II}$	499	1906	4638	7043
Total	6172	5612	6172	17,956

Table 3.18

c. Interpret the numbers found in part b. Do they confirm your opinion from part a?

3.4 Controlling for Other Variables

association The study of relations is the most scientifically important part of statistics. The finding of a statistical relationship, or **association,** between two variables, or among several, often suggests that a causal relation might exist. However, it is a valid cliché that "correlation does not prove causation"—that a statistical association between two variables does not imply that either one causes the other. One major reason is that both variables may be responding to still other variables. Therefore it is important to have methods for controlling for the effects of other variables. In this section we'll look at the nature of the problem and at some of the seemingly paradoxical situations that can occur.

Again, we proceed by example. The data are artificial, but the issues that arise are very typical. Suppose that, in a study of obesity, we ask if 1000 subjects eat a full breakfast and if they are seriously obese. Suppose that the frequency data are as shown in Table 3.19.

		Obese?		
		Yes	No	Total
Full	Yes	150	150	300
Breakfast?	No	250	450	700
	Total	400	600	1000

Table 3.19

There appears to be a relation. Of those who did eat a full breakfast, 50% (150/300) were obese, whereas of those who didn't, only 35.7% (250/700) were obese. This suggests, but certainly doesn't prove, that eating a full breakfast is a causal factor in obesity (or perhaps vice versa).

But now let's break the subjects into two groups, based on sedentary vs. active occupation. Suppose that the frequency data are as in Table 3.20.

Within the sedentary group, there is no relation between breakfast habits and obesity; 60% of the yes group are obese, as are 60% of the no group. Likewise, within the active group, there's no relation; 10% (6/60 and 34/340) are obese, for both the yes and the no group. In the combined data, there appears to be a statistical association, but when the subjects are separated by work habits, the association disappears!

What happened? First, there is a subconscious tendency to think

SEDENTARY

		Obese?		
		Yes	No	Total
Full Breakfast?	Yes	144	96	240
	No	216	144	360
	Total	360	240	600

ACTIVE

		Obese?		
		Yes	No	Total
Full Breakfast?	Yes	6	54	60
	No	34	306	340
	Total	40	360	400

Table 3.20

that a statistical relation must be some kind of average. In the obesity example, one might think that the relation between breakfast habits and obesity would vary by groups, with a stronger relation in some groups and a weaker one in others, but averaging out to the indicated degree of relation. The example shows that this is not true. There's a relation in the overall data, but no relation at all within either subgroup. *A statistical relation shown in an overall data set need not be an average, in any sense, of relations in subsets of the data.*

spurious relation
indirect relation

The obesity data exhibit what is often called **spurious relation.** A better name for this phenomenon might be **indirect relation,** because the association between breakfast habits and obesity is explainable in terms of associations between breakfast habits and occupation, and between occupation and obesity. Note that 40% of the sedentary workers ate a full breakfast, but only 10% of the active workers did. Of the sedentary workers, 60% were obese, but only 10% of the active workers were obese. One possible explanation for the overall association between eating a full breakfast and obesity is an indirect one; full-breakfast eaters are relatively likely to be sedentary workers, who in turn are relatively likely to be obese. This indirect association is indicated in Figure 3.1.

In the arrow diagram of Figure 3.1, the arrows between breakfast habits and occupation indicate a hypothetical direct relation, as does the arrow between occupation and obesity. The absence of an arrow directly from breakfast habits to obesity indicates that (supposedly) the only relation between these two variables is an indirect one, by way of

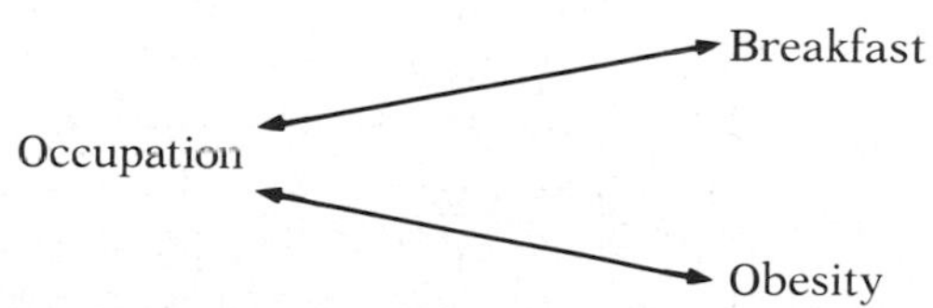

Figure 3.1 Arrow diagram of indirect correlation

intervening variable occupation. If, in fact, this is true, then occupation is an **intervening variable** which explains the indirect relation between breakfast habits and obesity.

> **Intervening Variable and Spurious Relation**
> If two variables are statistically related, but statistically independent within all levels of a third variable, then the third variable is an intervening variable, and the relation between the first two variables is called spurious or indirect.

Example 3.10

Table 3.21 shows frequencies for two ordinal variables X and Y, by sex of subject. Is there an overall relationship between X and Y? Is the relationship spurious?

FEMALE

		X: L	M	H	Total
Y	L	330	140	30	500
	M	198	84	18	300
	H	132	56	12	200
	Total	660	280	60	1000

MALE

		X: L	M	H	Total
Y	L	4	20	26	50
	M	16	80	104	200
	H	60	300	390	750
	Total	80	400	520	1000

Table 3.21

Solution

First, add the entries in the female and male tables together, cell by cell, to get the overall frequencies (Table 3.22).

COMBINED

		X: L	M	H	Total
Y	L	334	160	56	550
	M	214	164	122	500
	H	192	356	402	950
	Total	740	680	580	2000

Table 3.22

For the combined data, there's a moderate positive ordinal relation. Kendall's τ, defined in Section 3.3, is .190 for these data, and γ is

.434. Within the female table, the percentages scoring L, M, and H on the Y scale are 50%, 30%, and 20%, respectively, *for every X value.* Thus, within the female table there's no relation. There's no relation within the male table either. Thus the relation is spurious, by definition. □

Not only is it possible for a third variable to make a statistical association disappear; it is even possible for a third variable to make a relation change directions. Suppose that the obesity data had been as shown in Table 3.23.

SEDENTARY

		Obese? Yes	Obese? No	Total
Full Breakfast?	Yes	139	101	240
	No	221	139	360
	Total	360	240	600

ACTIVE

		Obese? Yes	Obese? No	Total
Full Breakfast?	Yes	1	59	60
	No	39	301	340
	Total	40	360	400

COMBINED

		Obese? Yes	Obese? No	Total
Full Breakfast?	Yes	140	160	300
	No	260	440	700
	Total	400	600	1000

Table 3.23

According to the combined data, full-breakfast eaters are relatively more likely to be obese than are non-full-breakfast eaters, 46.7% (140/300) vs. 37.1% (260/700). Yet in the sedentary group, full-breakfast eaters are relatively *less* likely to be obese than are non-full-breakfast eaters, 57.9% (139/240) vs. 61.4% (221/360). In the active group, full-breakfast eaters are also relatively less likely to be obese than are non-full-breakfast eaters, 1.7% (1/60) vs. 11.5% (39/340). In this case, the relatively weak negative relation between full-breakfast eating and obesity that is shown within each occupation is overwhelmed by stronger positive relations between full-breakfast eating and occupation, and between occupation and obesity.

Example 3.11 What relations present in Example 3.20 could explain the spurious association that was found?

Solution Women tend to score low on both scales, and men tend to score high on both. Thus, even though there is no direct relation between the X and Y scales for either women or men, low X scorers tend to be women, who in turn tend to be low Y scorers, whereas high X scorers tend to be men, who in turn tend to be high Y scorers. ☐

The preceding discussion has been about how a third variable can make an apparent relation disappear. The reverse situation is also possible. It can happen that two variables are statistically related within each level of a third variable, but are not related in the combined data. An extreme example is shown in Table 3.24.

SEDENTARY

		Obese?		
		Yes	No	Total
Full Breakfast?	Yes	192	228	420
	No	68	112	180
	Total	260	340	600

ACTIVE

		Obese?		
		Yes	No	Total
Full Breakfast?	Yes	58	22	80
	No	182	138	320
	Total	240	160	400

COMBINED

		Obese?		
		Yes	No	Total
Full Breakfast?	Yes	250	250	500
	No	250	250	500
	Total	500	500	1000

Table 3.24

In the combined group of Table 3.24, it's obvious that there's no statistical relation. Among the sedentary workers, full-breakfast eaters are relatively more likely to be obese, 45.7% (192/420) vs. 37.8% (68/180). Among the active workers, the same is true, 72.5% (58/80) vs. 56.8% (182/320). In effect, the relations between breakfast habits and obesity (within occupation), between occupation and breakfast habits, and between occupation and obesity cancel each other out, so the association between breakfast habits and obesity, within each occupation, is hidden in the combined-occupation data. Once again, the overall relation is not an average of the relation within each occupation.

The examples in this section indicate that the analysis of statistical associations is not simple. When a statistical association is found, there remains the possibility that it is indirect, explainable by a third variable. When a statistical association is *not* found, there remains the possibility that a relation exists, but is hidden by another variable. Scientific work is never final; there always remains the possibility that a different theory or hypothesis would explain the data more adequately.

The examples also indicate that statistical analysis, unaided by scientific theory, isn't very effective. Suppose, for example, that we had observed 1000 subjects over a period of time and recorded many variables for each subject; further, suppose that we had observed a statistical association between breakfast habits and obesity. To see if this association was possibly an indirect one, we would have to break down the overall data into subgroups based on every other variable that we'd observed. Many statistical computer packages will do that automatically, but the result is inevitably a mountain of computer output to be examined. Apart from the labor of climbing Mount Output, there is a major problem that sheer random variation will cause some relations to disappear and other nonrelations to appear in such a massive analysis. Statistical analysis should be guided by scientific theory, which in turn should be informed by previous statistical analyses. Just as pure theory without analysis of data makes poor science, so does pure data analysis without scientific theory.

The statistical method for examining the effects of other variables on a relation involving qualitative variables has already been indicated. The tabulation of the data is simply broken down into subgroups. Any of the qualitative-variable measures of relation that have been described in this chapter may be used for each subgroup tabulation. For a discussion of the complexities of combining the numerical results of the various subtabulations into an overall measure, see Hildebrand, Laing, and Rosenthal (1977, ch. 7).

Exercises for Section 3.4

3.13. There has been much fuss recently about the purported lack of ability of prospective school teachers, particularly in the mathematics and science areas. Assume that a sample of college students has been taken, classified by geographical region (E, MW, S, and W), by major (education or other), and by math/science aptitude (L, M, or H). We treat the data as nominal, ignoring any order relations. Assume the frequencies shown in Table 3.25.

a. Calculate a combined table by summing the frequencies over all regions.

b. Calculate the λ statistic for predicting major given aptitude, in the combined table.

REGION E

		Aptitude		
		L	M	H
Major	E	12	20	8
	O	105	42	42

REGION S

		Aptitude		
		L	M	H
Major	E	20	8	3
	O	30	67	22

REGION MW

		Aptitude		
		L	M	H
Major	E	15	5	3
	O	45	73	59

REGION W

		Aptitude		
		L	M	H
Major	E	6	5	1
	O	28	35	25

Table 3.25

c. Calculate the ∇ statistic for the prediction "L aptitude predicts E, M aptitude predicts E or O, and H aptitude predicts O," in the combined table.

3.14. Refer to the data of Exercise 3.13.

a. For each region, calculate the λ measure for predicting major given aptitude.

b. For each region, calculate the ∇ measure for the prediction given in part c of Exercise 3.13.

c. Is the relation between aptitude and major found in Exercise 3.13 spurious and explained by regional variation?

3.15. Suppose that a study is made of the effects of helplessness (inability to affect the receipt of either positive or negative rewards) on degree of depression. A research question is whether any relation between perceived helplessness and depression might be explained by the perceived stressfulness of the subject's current life. Ordinal measures are obtained from 500 subjects, measuring X_1 = perceived helplessness (L, M, or H), X_2 = perceived stress (below average, average, or above average), and Y = degree of depression (L, M, or H). Assume the following frequencies:

a. Compute a cross-tabulation of frequencies for all (X_1, Y) combinations by adding across the possible X_2 values.

b. Does there seem to be an ordinal relation between helplessness and depression? A percentage analysis may suffice to answer the question.

3.16 Refer to Exercise 3.15.

X_2 = BA		X_1 L	M	H
	L	50	15	5
Y	M	11	23	4
	H	1	2	9

X_2 = AV		X_1 L	M	H
	L	36	6	3
Y	M	13	58	23
	H	2	15	24

X_2 = AA		X_1 L	M	H
	L	22	6	2
Y	M	8	65	6
	H	2	10	79

Table 3.26

a. Consider only the X_2 = BA subjects. Is there an ordinal relation within this subgroup, in the same direction as in the table of part b of Exercise 3.15?

b. Repeat part a for the AV and AA subjects (separately).

c. Can the relation between helplessness and depression, found in Exercise 3.15, be considered spurious in that it is explained by variation in stress?

3.17. Refer to Exercise 3.15.

a. Is there an overall relation between stress and depression in these data?

b. Can it be considered spurious and explained by variation in helplessness?

Chapter Exercises

3.18. Buss, Iscoe, and Buss (1979) asked the parents of children whether the children would show signs of embarrassment. The frequency of each answer was tabulated in Table 3.27 according to the child's age.

		Age in Years 3&4	5	6	7	8	9	10	11&12	Total
Embarrassment	Yes	9	19	32	38	37	36	26	28	225
	No	26	13	12	13	20	18	10	16	128
	Total	35	32	44	51	57	54	36	44	353

Table 3.27

We will ignore the possibility of treating age as a quantitative variable, and regard it as nominal.

a. Calculate the value of the λ statistic for predicting embarrassment given age.

b. Calculate the value of the λ statistic for predicting age given embarrassment.

c. Does there appear to be any relation? Could it be called a strong relation?

3.19. Refer to Exercise 3.18. Suppose we combine all children aged 6 and above into one age group called "school age."

a. Recalculate the frequencies in a 2-row, 3-column table.

b. Calculate the λ statistic for predicting embarrassment given age. Has the value changed from the value found in Exercise 3.18?

3.20. **a.** Calculate ∇ for the frequencies in Exercise 3.18, using the prediction "3- and 4-year-olds will not show embarrassment; 5-year-olds may or may not; children 6 years old and older will." Does the result indicate a fairly strong relation between age and embarrassment?

b. Recalculate ∇ for the frequencies in Exercise 3.19. How much difference does combining the older ages make in ∇?

3.21. Goodwin and Michel (1981) performed a study of newborn infants, recording birth position (left or right) and head position preference (whether the child consistently looked right, had a bias toward looking right, had a bias toward looking left, or consistently looked left). The observed frequencies are shown in Table 3.28.

		Head Position Preference				
		Consistent Right	Biased Right	Biased Left	Consistent Left	Total
Birth Position	Left	29	8	8	4	49
	Right	18	7	12	13	50
	Total	47	15	20	17	99

Table 3.28

a. Calculate the value of λ with birth position as the dependent variable. Does it indicate that there is a relation?

b. Calculate the value of λ with birth position as the independent variable. Does it indicate that there is a relation?

3.22. Refer to Exercise 3.21. The categories for head position preference and birth position were regarded as ordinal, with "left" arbitrarily being taken to be higher and "right" lower. (No political meaning is implied.) Calculate the value of the γ statistic. Is there a strong ordinal relation between the two variables?

3.23. Suppose that a school district had three types of first-grade classrooms—type A, the traditional fixed-desks room (located largely in the poorer sections of town); type B, an open classroom (located largely in the well-to-do sections); and type C, a small-group-oriented room (located in mostly middle-class areas). Data were collected for 470 children on initial reading ability at the beginning of the year, reading ability relative to national norms at the end of the year, and type of classroom. Assume that the frequencies in Table 3.29 were found.

		Type of Classroom			
		A	B	C	Total
End-of-Year Ability	Lower than Normal	56	36	31	123
	Near Normal	57	55	80	192
	Above Normal	17	99	39	155
	Total	130	190	150	470

Table 3.29

a. Calculate λ for predicting ability given type of classroom.
b. Calculate λ for predicting type of classroom given ability.
c. Do the λ values indicate that there is some degree of relation between the two variables?

3.24. Refer to Exercise 3.23. Suppose that a researcher has predicted that classroom A students will show lower than normal ability, classroom B students will show higher than normal ability, and classroom C students will show near normal ability.

a. Calculate the ∇ statistic for the data and this prediction.
b. What is the precision of the prediction?
c. The λ statistic is not useful in evaluating the success of this prediction. Explain why.

3.25. The γ and τ measures of relationship are not particularly relevant to the data of Exercise 3.23. Explain why not.

3.26. Suppose that the data for Exercise 3.23 were divided up on the basis of the child's initial reading ability, with the results shown in Table 3.30.

a. For each category of initial reading ability, find the value of λ for predicting end-of-year ability given type of classroom. Are these values close to the one found in Exercise 3.23?
b. What does the comparison among values of λ suggest about the relation between type of classroom and end-of-year ability found in Exercise 3.23?

INITIAL READING ABILITY: NONE

		Type of Classroom			
		A	B	C	Total
End-of-Year Ability	Lower than Normal	49	16	7	72
	Near Normal	29	11	2	42
	Above Normal	2	3	1	6

INITIAL READING ABILITY: SLIGHT

		Type of Classroom			
		A	B	C	Total
End-of-Year Ability	Lower than Normal	7	9	20	36
	Near Normal	25	24	71	120
	Above Normal	8	7	29	44

INITIAL READING ABILITY: GOOD

		Type of Classroom			
		A	B	C	Total
End-of-Year Ability	Lower than Normal	0	11	4	15
	Near Normal	3	20	7	30
	Above Normal	7	89	9	105

Table 3.30

3.27. The data for Exercise 3.26 were added up to yield the cross-tabulation in Table 3.31.

a. In actuality, would you expect a monotone relation between these two variables? Should it be a positive or a negative relation?

b. Does the γ statistic indicate that there is a reasonably strong relation? There are 220,900 pairs of children, of which 91,212 are concordant and 14,256 are discordant.

		Initial Reading Ability			
		None	Slight	Good	Total
End-of-Year Ability	Lower than Normal	72	36	15	123
	Near Normal	42	120	30	192
	Above Normal	6	44	105	155
	Total	120	200	150	470

Table 3.31

3.28. The data for Exercise 3.26 also yielded the cross-tabulation in Table 3.32.

		Type of Classroom A	B	C	Total
Initial Reading Ability	None	80	30	10	120
	Slight	40	40	120	200
	Good	10	120	20	150
	Total	130	190	120	470

Table 3.32

Calculate the ∇ value for the prediction that students in classroom A will be in the none category, students in classroom B in the good category, and students in classroom C in the slight category. Does the value of ∇ indicate that there is a relation?

3.29. Combining the results of Exercises 3.23–3.28, draw an arrow diagram that summarizes the relationships found.

3.30. Lundgren (1969) collected data on successful and unsuccessful college students, including self-reported attendance at classes. The reported frequencies were as shown in Table 3.33.

	Successful	Unsuccessful
Almost Always Attend	34	19
Sometimes Absent	3	3
Often Absent	3	18

Table 3.33

a. Would you anticipate that there would be a relation? If so, what predictions would you make for the three attendance categories? (Don't peek at the data yet; consider what the variables are.)

b. Calculate the ∇ value for the prediction you stated in part a.

c. Calculate the precision of this rule.

3.31. Calculate the values of λ for predicting success given attendance and for predicting attendance given success, using the data in Exercise 3.30. Are the two values roughly the same?

3.32. Calculate the τ statistic for the data of Exercise 3.30. Is τ meaningful for the variables in that exercise?

3.33. Suppose that a study related three qualitative variables—whether or not a child exhibited prejudiced behavior, the ethnic category of the child, and the degree of racial integration of the child's school. Assume that the frequencies were as shown in Table 3.34.

LITTLE INTEGRATION

		Ethnicity		
		B	H	W
Prejudice	Yes	34	14	20
	No	7	15	7

SOME INTEGRATION

		Ethnicity		
		B	H	W
Prejudice	Yes	7	9	15
	No	12	6	12

HIGH INTEGRATION

		Ethnicity		
		B	H	W
Prejudice	Yes	10	4	15
	No	15	6	35

Table 3.34

```
                    ORDINAL ASSOCIATION
                          TABLE OF Y BY X

FREQUENCY|
 PERCENT |
 ROW PCT |
 COL PCT |VH      |H       |M       |L       |VL      |  TOTAL
---------+--------+--------+--------+--------+--------+
H        |    170 |     20 |      7 |      2 |      1 |    200
         |  34.00 |   4.00 |   1.40 |   0.40 |   0.20 |  40.00
         |  85.00 |  10.00 |   3.50 |   1.00 |   0.50 |
         |  85.00 |  40.00 |   8.75 |   3.33 |   0.91 |
---------+--------+--------+--------+--------+--------+
M        |     10 |     25 |     45 |     15 |      5 |    100
         |   2.00 |   5.00 |   9.00 |   3.00 |   1.00 |  20.00
         |  10.00 |  25.00 |  45.00 |  15.00 |   5.00 |
         |   5.00 |  50.00 |  56.25 |  25.00 |   4.55 |
---------+--------+--------+--------+--------+--------+
L        |     20 |      5 |     28 |     43 |    104 |    200
         |   4.00 |   1.00 |   5.60 |   8.60 |  20.80 |  40.00
         |  10.00 |   2.50 |  14.00 |  21.50 |  52.00 |
         |  10.00 |  10.00 |  35.00 |  71.67 |  94.55 |
---------+--------+--------+--------+--------+--------+
TOTAL         200       50       80       60      110      500
            40.00    10.00    16.00    12.00    22.00   100.00

                    ORDINAL ASSOCIATION
                  STATISTICS FOR 2-WAY TABLES
CHI-SQUARE                       448.029   DF=   8  PROB=0.0001
PHI                                0.947
CONTINGENCY COEFFICIENT            0.687
CRAMER'S V                         0.669
LIKELIHOOD RATIO CHISQUARE       472.810   DF=   8  PROB=0.0001
GAMMA                              0.876   ASE1=  0.020
KENDALL'S TAU-B                    0.721
STUART'S TAU-C                     0.745   ASE1=  0.025
SOMER'S D C|R                      0.776   ASE1=  0.026
SOMER'S D R|C                      0.669   ASE1=  0.023
PRODUCT MOMENT CORRELATION         0.782
SPEARMAN CORRELATION               0.781
LAMBDA ASYMMETRIC C|R              0.397
LAMBDA ASYMMETRIC R|C              0.623
LAMBDA SYMMETRIC                   0.510
ASE1 IS THE ASYMPTOTIC STANDARD ERROR.
R|C MEANS ROW VAR DEPENDENT ON COLUMN VAR.
```

Figure 3.2 Output for Exercise 3.35

Is there a consistent, strong relation between ethnicity and prejudice within each integration category? Use whatever summary statistic you prefer.

3.34. Add up the frequencies of Exercise 3.33 to obtain a cross-tabulation of ethnicity versus prejudice. Is there a strong relation shown in this table?

3.35. Data for two ordinal variables, X and Y, were analyzed using the SAS package. The output is shown in Figure 3.2.

a. Locate the value of the γ statistic.

b. Locate the values of λ. How can you tell which variable is independent and which variable is dependent?

c. Does there seem to be a strong relation between the variables?

3.36. Pharis and Manosevitz (1981) reported data on the preference of parents as to the sex of an about-to-be-born baby, as well as on the parents' beliefs as to the child's sex. The data were analyzed using SPSS-X. The results are shown in Figure 3.3.

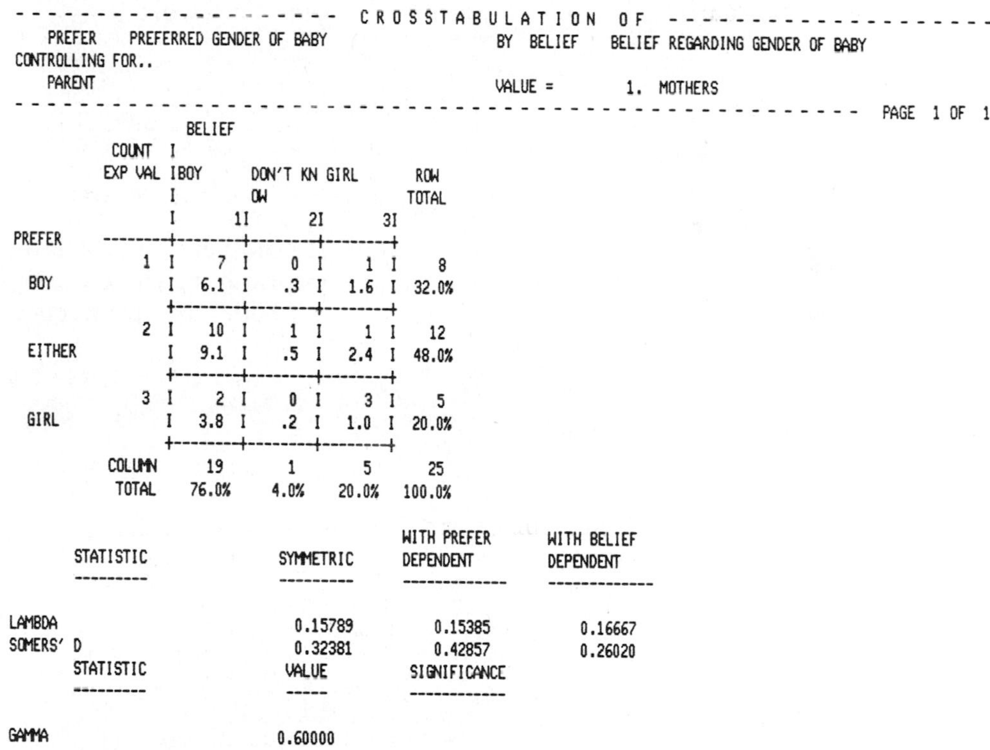

```
- - - - - - - - - - - - - - - - - - - -  C R O S S T A B U L A T I O N   O F  - - - - - - - - - - - - - - - - - -
    PREFER   PREFERRED GENDER OF BABY                            BY  BELIEF    BELIEF REGARDING GENDER OF BABY
CONTROLLING FOR..
    PARENT                                                       VALUE =         1.  MOTHERS
- - - - - - - - - - - - - - - - - - - - - - - - - - - - - - - - - - - - - - - - - - - - - - - - - -   PAGE  1 OF  1
                   BELIEF
            COUNT  I
          EXP VAL  IBOY       DON'T KN GIRL       ROW
                   I          OW                  TOTAL
                   I       1I       2I       3I
PREFER    --------+--------+--------+--------+
                1  I     7 I     0 I     1 I      8
  BOY              I   6.1 I    .3 I   1.6 I   32.0%
                   +--------+--------+--------+
                2  I    10 I     1 I     1 I     12
  EITHER           I   9.1 I    .5 I   2.4 I   48.0%
                   +--------+--------+--------+
                3  I     2 I     0 I     3 I      5
  GIRL             I   3.8 I    .2 I   1.0 I   20.0%
                   +--------+--------+--------+
           COLUMN       19        1        5      25
            TOTAL    76.0%     4.0%    20.0%  100.0%

                                                 WITH PREFER        WITH BELIEF
      STATISTIC                  SYMMETRIC       DEPENDENT          DEPENDENT
      ---------                  ---------       -------------      -------------

LAMBDA                           0.15789         0.15385            0.16667
SOMERS' D                        0.32381         0.42857            0.26020
      STATISTIC                  VALUE            SIGNIFICANCE
      ---------                  -----            ------------

GAMMA                            0.60000
```

Figure 3.3 Output for Exercise 3.36

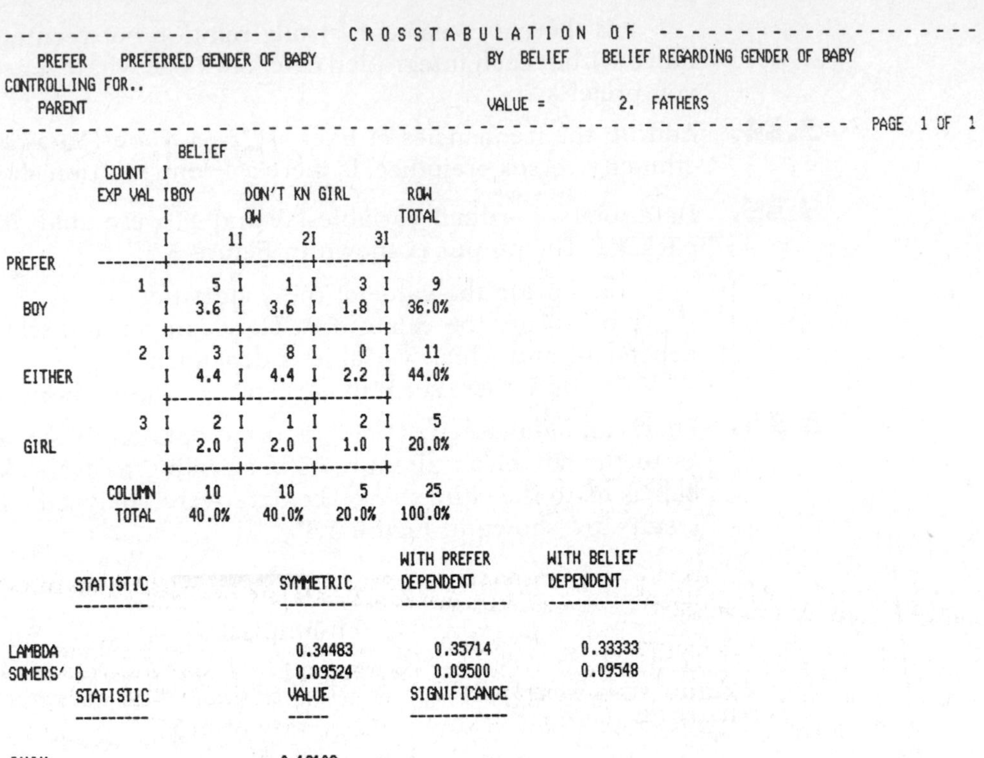

CROSSTABULATION OF

PREFER PREFERRED GENDER OF BABY BY BELIEF BELIEF REGARDING GENDER OF BABY

CONTROLLING FOR..

PARENT VALUE = 2. FATHERS

PAGE 1 OF 1

PREFER	COUNT / EXP VAL	BELIEF BOY 1	DON'T KNOW 2	GIRL 3	ROW TOTAL
BOY	1	5	1	3	9
		3.6	3.6	1.8	36.0%
EITHER	2	3	8	0	11
		4.4	4.4	2.2	44.0%
GIRL	3	2	1	2	5
		2.0	2.0	1.0	20.0%
COLUMN TOTAL		10	10	5	25
		40.0%	40.0%	20.0%	100.0%

STATISTIC	SYMMETRIC	WITH PREFER DEPENDENT	WITH BELIEF DEPENDENT
LAMBDA	0.34483	0.35714	0.33333
SOMERS' D	0.09524	0.09500	0.09548

STATISTIC	VALUE	SIGNIFICANCE
GAMMA	0.13103	

Figure 3.3 (continued)

a. Locate the λ values for mothers and for fathers.

b. Is the γ statistic appropriate for these data?

c. Is the relation between belief and preference strong for either mothers or fathers?

4 Basic Probability Ideas

This chapter is devoted to the simplest ideas of probability theory. Probability is a crucial idea in social science, both as a language for expressing scientific theories and as the basis for statistical inference. The various interpretations of the meaning of probability are discussed first, in Section 4.1. Then, in Section 4.2 the basic mathematical structure of probability is introduced, together with two of the key principles—addition and complements. In Section 4.3 we develop the notion of conditional probability and the third key principle—multiplication. Section 4.4 is devoted to the fundamental idea of statistical independence. The principles are all reasonably easy to understand; the difficulty is in using them to solve problems. Section 4.5 presents two devices, probability trees and probability tables, as aids in problem solving. Finally, in Section 4.6 we discuss conditional independence, an idea closely related to the notion of spurious relation discussed in Chapter 3.

The emphasis in this chapter is on thinking probabilistically, rather than on mechanically plugging numbers into formulas. Some aids in the thinking process are included. Most students find probability trees particularly useful.

4.1 The Role and Meaning of Probability

A wide variety of scientific theories are stated in probabilistic terms. From quantum mechanics in physics or population genetics in biology to learning models in psychology or mobility models in sociology, probability is a fundamental language for scientists. In this text, the primary use for probability will be in statistical inference; because many probability models require more mathematical sophistication than is assumed of readers of this book, we will not be able to go into great detail. However, some "feel" for probability models can be gained through examples and exercises.

The first issue in probability theory is what probability means. Three typical probability statements can be used to illustrate three possible interpretations of probability.

1. The probability of a fair coin's landing head up is .50.
2. The probability of a tack's landing point up is .72.
3. The probability that this book will sell more than 10,000 copies in its first year is .07.

These three statements can be most readily interpreted in the classical, relative frequency, and subjective senses, respectively.

classical interpretation

The **classical interpretation** of probability goes back to the beginnings of probability theory, as a study of gambling games. An experiment is assumed to consist of some number of "outcomes," all equally likely. The probability that an "event" will occur is taken to be the ratio of the number of outcomes *favorable* to the event to the number of *possible* outcomes. Symbolically,

$$P(\text{event}) = \frac{\text{number of favorable events}}{\text{number of possible events}}$$

In flipping a fair coin, there are two possible outcomes, heads and tails, which are assumed (for a fair coin) to be equally likely. Thus $P(\text{heads}) = 1/2 = .50$. Of course, the critical issue in applying the classical probability interpretation is that the outcomes be equally likely. An important application of the classical interpretation occurs when one takes a truly random sample from a population. In random sampling, each possible sample—that is, each possible outcome—is equally likely to be drawn.

relative frequency

The **relative frequency** interpretation of probability can be used even when the outcomes of an experiment cannot be considered equally likely. For example, suppose that the experiment is to drop an ordinary thumbtack onto a hard surface, so that it will wind up either point up or rolling on edge. There are only two outcomes, but the two can't be considered equally likely. The way to find the probability of the tack's landing point up is to try it! In a large number of tries, point up occurred in about 72% of the tries; therefore, it's reasonable to say that the probability of point up is (about) .72. The (long-run) relative frequency interpretation of a probability requires that one imagine a long series of trials repeating the experiment. The probability of a particular event is the ratio of the number of trials yielding the event to the total number of trials; symbolically,

$$P(\text{event}) = \frac{\text{number of trials yielding event}}{\text{total number of trials}}$$

The "long-run" idea is that there should be a large (mathematically, an infinite) number of trials to determine the probability of an event. In practice, of course, one cannot have an infinite number of trials. In Chapter 8 we'll see how to assign a plus-or-minus to a probability estimated on the basis of a limited number of trials.

Sometimes, it may be silly to imagine a long series of repetitions of an experiment. In particular, consider the question of whether this book will sell 10,000 copies in its first year. To give a long-run relative frequency interpretation of that probability, I would have to write a very large number of similar textbooks and see what fraction of them did in fact sell more than 10,000 copies in the first year. I refuse. Instead, the interpretation of the (very small) probability that the book will sell over 10,000 copies in its first year is *subjective*. It is a statement of my personal beliefs about the relative plausibility of the event. Such probabilities are often called *personal* probabilities; different people (such as textbook editors) may rationally have different probabilities for the same event. Subjective probability interpretations apply particularly to "one-shot" experiments, which cannot be repeated.

Example 4.1

State classical, relative frequency, and subjective interpretations of the following statement: "The probability that a randomly chosen student will answer exam question 1 is .70."

Solution

A classical interpretation could be made if it were known that, say, 21 of the 30 students in the class could answer question 1. Because a student is chosen randomly, each outcome (student) is equally likely; the probability of answering the question would be 21/30 = .70.

A relative frequency interpretation could be made if data were available on how many previous students had answered questions similar to question 1. Assuming that 70% of past students had answered such questions, the desired probability would be .70.

A subjective interpretation would be necessary if question 1 were a new type of question. The probability of .70 would presumably represent the instructor's personal judgment of the ability of the class and the difficulty of the question. □

4.2 Basic Concepts and Principles of Probability

Regardless of the chosen interpretation of probability, certain basic mathematical concepts apply. These concepts are a language and a set of principles. In this section we will discuss first the language, then the basic mathematical principles.

All probabilities are defined relative to a specific "experiment." The word experiment is used in a very broad sense, not restricted to laboratory work. An **experiment** is any data-gathering procedure that has a well-defined set of possible results. An experiment might be a study of the reactions of 23 8-month-old children to a certain taste; the possible results could be whether each child laughs, smiles, cries, pushes away, or has some other reaction. Or, an experiment might be conducting a survey of 21,327 married couples, with each couple indicating their weekly frequency of both arguments and sexual relations. The possible variety is endless. The critical requirement of an experiment is that, in principle, it must be possible to specify in advance all the possible results. In the child-reaction experiment, we might code the reactions of each child as L, S, C, P, O; the possible results would be all possible 23-letter sequences of these five letters. In the survey, each couple would provide two numbers, ranging from 0 up to some very large number; the possible results could be arrayed in a list with 2 columns and 21,327 rows. Each such list would be a possible result.

experiment

Probabilities are only defined within the context of a given experiment. The answer to "What is the probability that an individual will receive two identical lottery tickets?" depends very much on whether the experiment is to have a single individual buy 2 lottery tickets, to have 1,000,000 individuals buy 2 tickets, or to have 500,000 individuals buy 2 tickets, 300,000 buy 3 tickets, and 200,000 individuals buy 4 tickets.

Example 4.2

For each of the following experiments, define the possible results.

(a) Four individuals are told to indicate whether the tone heard in the left ear is louder or softer than the one heard in the right ear. A response of "equal" is not allowed.

(b) A sample of 373 students at a university take a math placement exam; the scores can range from 0 to 100.

(c) After being deprived of food, water, or both for a 24-hour period, 30 rats (10 from each of the three deprivation groups) are measured on the time required for each to learn a certain task.

Solution

(a) If we let L stand for louder and S for softer, the possible results are LLLL, LLLS, LLSL, . . ., SSSS.

(b) The possible results are all possible strings of 373 numbers, each number being a score between 0 and 100.

(c) There are several ways in which the data could be arranged. For instance, a possible result might consist of 3 rows of 10 times per row. Alternatively, we might define a possible result as a string of 30 positive numbers, where each number represents a time; the first 10 numbers would represent the times for group 1, the next 10, times for group 2, and the last 10, times for group 3. □

sample space

Corresponding to any given experiment is a **sample space,** the set, or list, of all possible outcomes of the experiment. In principle, the sample space for a given experiment could be written down in advance; in practice, it is enough to be able to specify what a typical member of the sample space (that is, a typical outcome) is. There are no absolute rules for specifying a sample space, except that the possible outcomes must be distinct and exhaust all possibilities.

Example 4.3

Describe two possible sample spaces for the experiment described in Example 4.2(a).

Solution

One possible sample space was described in Example 4.2(a); it consists of the 16 possible 4-letter sequences, where each letter is either L or S. Alternatively, if the experimenter were only interested in the number of times the left-ear tone was judged louder, the sample space could simply be the numbers 0, 1, 2, 3, 4. □

In this book, sample spaces are denoted by S and outcomes by o (or o_i to distinguish among particular outcomes).

event

An **event** is defined as a collection or set of possible outcomes. Thus, in Example 4.3, the event "as many subjects judge the left-ear sound as louder as judge it softer" consists of the outcomes LLSS, LSLS, LSSL, SLLS, SLSL, SSLL. For the alternative (0, 1, 2, 3, 4) sample space, the event consists simply of the number 2. The words "outcome" and "event" are synonymous in English, but not in probability language. An outcome is one specific possible result, whereas an event may include many possibilities.

Example 4.4 In an experiment, a single student takes two exams, and obtains two scores between 0 and 100. What outcomes are contained in the events A = total score of 197, B = average score of 99 or larger, C = average score of -1.

Solution A consists of (98, 99), (99, 98), (97, 100), and (100, 97). B consists of (99, 99), (99, 100), (100, 99), and (100, 100). C is impossible and consists of no outcomes at all. □

Generally, events will be denoted by capital letters early in the alphabet—A, B, E, etc.

Sometimes it is possible to specify probabilities for each outcome o_i in a sample space S. In particular, if the classical interpretation applies, all outcomes have equal probabilities. In such cases probabilities for events can be calculated simply by adding up the probabilities of all outcomes included in the event.

Example 4.5 In the sound-loudness experiment of Examples 4.2 and 4.3, assume that the sounds are in fact of equal loudness, so in effect each subject is guessing. What is the probability that exactly two of the four subjects will report that the left-ear sound is louder?

Solution In this situation, it is reasonable to assume that all outcomes in the sample space have equal probabilities; there are 16 outcomes, so each outcome probability should be 1/16. There are 6 outcomes in the desired event, so the probability of that event should be 6/16, or .375. □

Generally, it is not possible or convenient to assign probabilities to each outcome. In such cases, it is necessary to work with the basic mathematical principles, or axioms, of probability theory. Two of these axioms are obvious.

1. If A is any event, the probability of A is a number between 0 and 1 (both inclusive).

2. If S is the sample space, $P(S) = 1$.

The third axiom we will use requires a preliminary, but important, definition.

Mutually Exclusive Events

Events A and B are mutually exclusive if they have no outcomes in common. Synonymously, A and B are *disjoint,* or *logically incompatible.* If A and B are mutually exclusive and if A occurs, B logically cannot occur. Several events are mutually exclusive if every pair of them is mutually exclusive.

The question of whether events are mutually exclusive depends very much on the sample space being considered. The events "scoring 99 on an exam" and "scoring 93 on an exam" are mutually exclusive if the experiment consists of a single student taking a single exam; the events "student A scores 99 on an exam" and "student B scores 93 on an exam" are not mutually exclusive if the experiment consists of two students' taking an exam.

The definition of mutually exclusive events is important in the third axiom of probability theory.

3. If events A and B are mutually exclusive, then the probability that either A or B occurs is $P(A) + P(B)$.

Example 4.6

A sample space consists of 16 outcomes, numbered 1 to 16. Event A includes outcomes 1, 2, 3, and 4; event B contains outcomes 5, 6, 7, 8, and 9; and event C contains outcomes 1 and 10. The events are not all equally likely; $P(A) = .3$, $P(B) = .4$, and $P(C) = .2$. Calculate (if possible) $P(A \text{ or } B)$, $P(A \text{ or } C)$, and $P(B \text{ or } C)$.

Solution

Note that events A and B are mutually exclusive, as are B and C. However, events A and C have outcome 1 in common. It follows that $P(A \text{ or } B) = .3 + .4 = .7$ and $P(B \text{ or } C) = .4 + .2 = .6$; but we can't calculate an exact value for $P(A \text{ or } C)$. □

When events are not mutually exclusive, adding probabilities will double-count the outcomes that the events have in common. The Venn diagram in Figure 4.1 shows the events in Example 4.6. Note that each event is represented by a "blob" within the overall area. Think of probability as the area of the specified blob, relative to a total area of the square taken equal to 1. Note that events A and C overlap; the event "A and C" consists of outcome 1. In such a case, to get the correct proba-

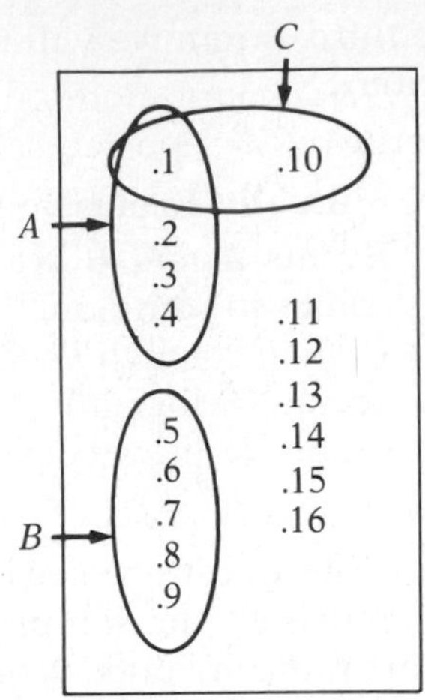

Figure 4.1 Venn diagram of overlapping events

bility of the event "A or C," we must take care of a double-counting problem.

> **General Addition Principle**
>
> For any two events, say D and E,
>
> $$P(D \text{ or } E) = P(D) + P(E) - P(D \text{ and } E)$$

Example 4.7 In Example 4.6, assume that outcome 1 has probability .05. Find $P(A \text{ or } C)$.

Solution The event "A and C" consists of outcome 1 alone and has probability .05. Therefore,

$$P(A \text{ or } C) = .3 + .2 - .05 = .45$$ □

The General Addition Priniciple may be extended to three, four, or more events. The problems with accounting for double-counts, triple-counts, and worse make the extensions hard to use. As a general rule, it is easier to try to break up an event into mutually exclusive cases, find the probability of each case, and add.

Example 4.8

In a population of depressed patients, 65% complained of loss of appetite, 60% reported sexual dysfunction, and 50% were unable to do normal work. Furthermore, 35% reported both loss of appetite and sexual dysfunction, 25% had loss of appetite and inability to work, 40% had sexual dysfunction and inability to work. A total of 20% had all three complaints. Construct a Venn diagram to find the probability that a patient chosen at random will have one or more of the complaints.

Solution

A Venn diagram is shown in Figure 4.2. The events are A = loss of appetite, B = sexual dysfunction, and C = inability to do normal work. It's given that $P(A \text{ and } B \text{ and } C) = .20$; because $P(A \text{ and } B) = .35$, it follows that $P(A \text{ and } B \text{ but not } C)$ must equal $.35 - .20$, or $.15$. The remaining probabilities can all be found by similar reasoning.

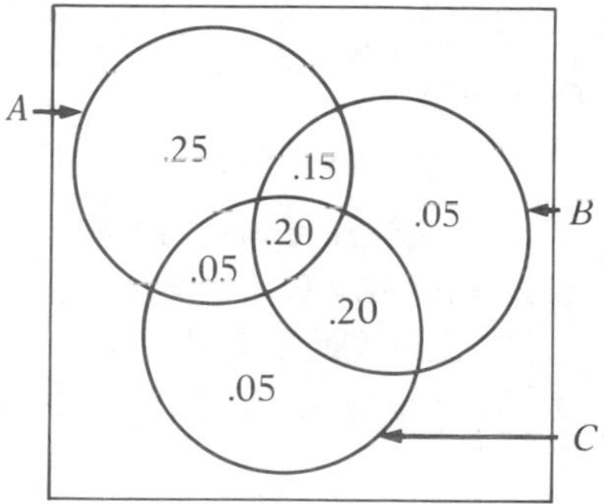

Figure 4.2 Venn diagram for Example 4.8

The probability that a randomly chosen patient will have one or more of the complaints can now be found simply by adding up the probabilities of all possible, mutually exclusive cases, as shown in the Venn diagram.

$$\begin{aligned} P(A \text{ or } B \text{ or } C) &= .25 + .15 + .05 + .05 + .20 + .20 + .05 \\ &= .95 \end{aligned}$$

Another basic probability principle is the complements principle. It is sometimes easier to find the probability that an event will *not* happen than the probability that it will. Of course, the probability that an event will happen is 1 minus the probability that it won't.

Complements Principle
For any event A,

$$P(A) = 1 - P(\text{not } A)$$

This principle follows because the events A and not-A are mutually exclusive and the event "A or not-A" is the sample space S, which has probability 1.

Example 4.9 Referring to Example 4.8, find the probability that a randomly chosen patient will have none of the three complaints.

Solution In Example 4.8, we found that

$$P(\text{1 or more of the complaints}) = .95$$

Therefore,

$$P(\text{no complaints}) = 1 - P(\text{1 or more complaints}) = .05.$$ □

Exercises for Sections 4.1 and 4.2

4.1. Should each of the following be interpreted as an objective probability, a relative frequency probability, or a personal probability? (You may be able to think of more than one interpretation in some cases.)

a. The probability that a child will be left-handed is .15.

b. The probability that a Down's syndrome baby will be severely retarded is .6.

c. The probability that a particular pre-med student will dispute at least one exam grade is .98.

d. If a child puts peanut butter on a slice of bread and then drops the bread, the probability that it will land peanut butter side down is .99.

4.2. In a pool of 100 subjects, 20 have poor enough eyesight to be useful in a particular experiment. Three different subjects are to be chosen randomly.

a. Write out a sample space for the experiment.

b. Identify the events A = exactly one of the subjects is useful, B = the first two subjects are useful, and C = all of the subjects are useful.

4.3. In Exercise 4.2, which pairs of events are mutually exclusive?

4.4. In an experiment, subjects are told that they are assessing the readability of different typefaces. They are told to find as many spelling errors in a 600-word document as possible; in fact the document is perfectly spelled. Before a subject announces how many misspellings were found, three "stooges" of the experimenter each announce that they have found eight errors. The experiment is repeated for six different subjects.

a. Describe a sample space for this experiment.

b. Describe the outcomes corresponding to the event E = "at least four of the subjects announce eight errors."

4.5. If A and B are mutually exclusive with $P(A) = .3$ and $P(B) = .4$, find

a. $P(A \text{ or } B)$
b. $P(A \text{ and } B)$
c. $P(\text{not-}A \text{ and not-}B)$

4.6. Suppose that $P(A) = .3$, $P(B) = .4$, and $P(A \text{ and } B) = .2$. Draw a Venn diagram and find each of the following:

a. $P(A \text{ or } B)$
b. $P(A \text{ occurs and } B \text{ does not})$
c. $P(\text{neither } A \text{ nor } B \text{ occurs})$

4.7. Experimental animals may display any of three reactions, labeled D, E, and F, to an electric shock. Assume that 30% display D, 40% display E, and 50% display F; 20% display D and E, 10% display D and F, and 15% display E and F; 5% display all three. An animal is chosen randomly.

a. Draw a Venn diagram for the experiment. Note that, for example, the 5% displaying all three reactions is part of the 20% displaying reactions D and E.

b. Find the probability that the animal will display reactions D and E but not F.

c. Find the probability that the animal will display none of the reactions.

4.8. In a certain hospital, 40% of the patients have been alcoholics and 30% have had severe drug dependency; 10% have had both problems. A patient is to be selected at random. What is the probability that the patient

a. has been an alcoholic, but not drug dependent?
b. has been either an alcoholic or drug dependent?
c. has been neither an alcoholic nor drug dependent?

4.3
Conditional Probability and the Multiplication Principle

Before the third basic probability principle—the multiplication principle—can be introduced, we need to define conditional probabilities.

Imagine that a certain population has the frequencies shown in Table 4.1. Each of the 1000 individuals enrolled in a required statistics course is measured on a math anxiety scale. The results for each student are recorded, along with whether the student chose to enroll in the second course of the sequence. There are 300 individuals who scored low (L) on anxiety and did take the second course, and so on. Suppose that a person is chosen at random. What is the probability that the person is Y and H? It should be clear that

$$P(Y \text{ and } H) = \frac{20}{1000} = .02$$

conditional probability

If the person is chosen randomly from only those taking the second course, what is the probability that the person is H? This probability is a **conditional probability,** because we have imposed the condition that the person be a Y rather than drawing from the entire population. It will be denoted $P(H|Y)$; the vertical bar should be read as "given," so $P(H|Y)$ is read "probability of H given Y." Because we are drawing from the 400 Y's, of which 20 are H's,

$$P(H|Y) = \frac{20}{400} = .05$$

(Note that it matters which event is the given and which random; $P(Y|H) = 20/200 = .10$.)

There is a relation between a conditional probability such as $P(H|Y)$ and a joint probability such as $P(H \text{ and } Y)$. A Venn diagram such as the one in Figure 4.3 helps to clarify the relation.

		Degree of Math Anxiety			
		L	M	H	Total
Enroll in	Y	300	80	20	400
Second Course?	N	200	220	180	600
	Total	500	300	200	1000

Table 4.1

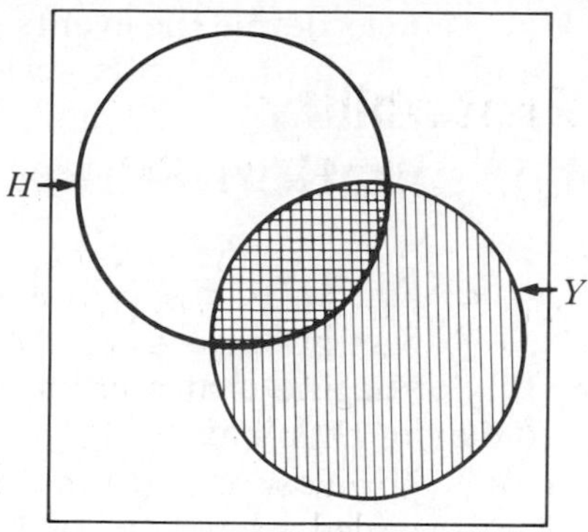

Figure 4.3 Venn diagram for conditional probability

When the "given Y" condition is specified, the sample space (set of possible outcomes) is restricted to Y. Now in order for H to occur, an outcome in both H and Y must result. Thinking of probability as area in the Venn diagram, we can see that the conditional probability must be the area of H and Y relative to (as a fraction of) the area of Y. Thus we have the following mathematical definition of conditional probability.

Conditional Probability of B Given A

$$P(B|A) = \frac{P(A \text{ and } B)}{P(A)}$$

Note that this definition works in our example. We found $P(H|Y) = 20/400 = .05$. Likewise $P(Y \text{ and } H)/P(Y) = (20/1000)/(400/1000) = .02/.40 = .05$.

Example 4.10 In a certain school district, the probability that a randomly chosen child comes from a single-parent home is .32; the probability that a child exhibits serious behavior problems is .20; and the probability that a child both comes from a single-parent home and exhibits serious behavior problems is .16.

(a) If a child comes from a single-parent home, what is the probability that he/she will exhibit serious behavior problems?

(b) If a child exhibits serious behavior problems, what is the probability that he/she comes from a single-parent home?

Solution Let's define the events A = child comes from a single parent home and B = child exhibits serious behavior problems. We are told that $P(A) = .32$, $P(B) = .20$, and $P(A \text{ and } B) = .16$.

(a) The "given" event is A.

$$P(B|A) = \frac{P(A \text{ and } B)}{P(A)} = \frac{.16}{.32} = .50$$

In words, 50% of the children from single-parent homes exhibit serious behavior problems.

(b) Now the "given" is B.

$$P(A|B) = \frac{P(A \text{ and } B)}{P(B)} = \frac{.16}{.20} = .80$$

In words, 80% of the children who exhibit serious behavior problems come from single-parent homes. □

multiplication principle

The **multiplication principle** is really just a restatement of the definition of conditional probability. In the math anxiety example at the beginning of this section, suppose we had said that 40% of all students enroll in the second semester course, so $P(Y) = .40$. Further, of the students enrolled in the second semester course, 5% show high math anxiety, so $P(H|Y) = .05$. If a student is randomly chosen from the population enrolled in the first semester course, what is the probability that the student enrolls in the second semester and is high on the anxiety scale? Because $P(H|Y) = P(Y \text{ and } H)/P(Y)$, it follows that

$$P(Y \text{ and } H) = P(Y)P(H|Y) = (.40)(.05) = .02$$

which is exactly the probability found in the original population.

General Multiplication Principle

For any events A and B,

$$\begin{aligned} P(A \text{ and } B) &= P(A)P(B|A) \\ &= P(B)P(A|B) \end{aligned}$$

Note that in this principle, the events can be taken in either order. If we had been given $P(H) = .20$ and $P(Y|H) = .10$, we could have calculated $P(H \text{ and } Y) = (.20)(.10) = .02$. Further, the multiplication principle can be extended to more than two events in the obvious way. For example,

$$P(A \text{ and } B \text{ and } C) = P(A)P(B|A)P(C|A \text{ and } B)$$

Example 4.11 Refer to Example 3.8, where we noted that 50% of all depressed patients complained of an inability to do normal work. Of these patients, 80% reported sexual dysfunction; of those reporting both inability to work and sexual dysfunction, 50% reported loss of appetite. Find the probability that a randomly chosen patient has all three complaints.

Solution Define C = inability to work, B = sexual dysfunction, and A = loss of appetite. We are given that $P(C) = .50$, $P(B|C) = .80$, and $P(A|B \text{ and } C) = .50$. Because of the way the information is specified, it's convenient to calculate

$$P(A \text{ and } B \text{ and } C) = P(C)P(B|C)P(A|B \text{ and } C)$$
$$= (.50)(.80)(.50) = .20$$

as indicated in Example 4.8. □

4.4 Statistical Independence

An idea that will be used repeatedly in later chapters is that of statistical independence. To illustrate the idea, we will once again begin with an example. A psychologist (of less-than-impeccable reputation) claims that the presence of a certain physiological trait indicates a predisposition to schizophrenia. Data were collected for a population of 5000 individuals. Frequencies are shown in Table 4.2. Presence of the trait is indicated by T, and diagnosed schizophrenia is indicated by S.

	T	not T	Total
S	48	352	400
not S	552	4048	4600
Total	600	4400	5000

Table 4.2

The unconditional probability of schizophrenia is 400/5000 = .08. The probability of schizophrenia, given that the trait is present, is 48/600 = .08. The probability of schizophrenia, given that the trait is not present, is 352/4400 = .08. It doesn't matter whether the trait is present or absent; the schizophrenia rate is 8%, regardless. The occurrence of schizophrenia is statistically independent of the trait, so the trait is utterly useless in predicting the occurrence of schizophrenia.

There are several mathematically equivalent ways of defining statistical independence. The essential idea is that the occurrence (or nonoccurrence) of one event does not change the probability of the other. Events S and T are statistically independent if the probability of S given T is the same as the probability of S given not T; equivalently, they are independent if the conditional probability of S given T is the same as the *unconditional* probability of S. A third equivalent definition follows from the multiplication principle $P(S \text{ and } T) = P(S|T)P(T)$: under statistical independence, the "given T" condition in $P(S|T)$ is irrelevant and can be dropped, leaving $P(S \text{ and } T) = P(S)P(T)$. Therefore, three equivalent definitions of independence can be stated.

Statistical Independence

Two events A and B (with nonzero probabilities) are independent if and only if

1. $P(B|A) = P(B|\text{not } A)$ or
2. $P(B|A) = P(B)$ or
3. $P(A \text{ and } B) = P(A)P(B)$

There is a very slight reason to prefer the third form of this definition. That form makes it obvious that independence is a symmetrical concept. If A is independent of B, B is independent of A. Also, the third form is valid even if $P(A) = 0$, in which case $P(B|A)$ is 0/0 and undefined. We will use the third form of the definition so many times that it deserves restatement.

Multiplication Principle for Independent Events

If A and B are independent events,

$$P(A \text{ and } B) = P(A)P(B)$$

Example 4.12

Referring to Example 4.10, are the events A = child comes from single-parent home and B = child exhibits serious behavior problems independent?

Solution

In that example, we were given that $P(A) = .32$ and $P(B) = .20$, and $P(A \text{ and } B) = .16$. $P(A)P(B) = (.32)(.20) = .064$, which is not equal to $P(A \text{ and } B) = .16$; therefore the events are not independent. Alternatively, we can compare $P(B|A) = .50$ to $P(B) = .20$, which again shows that the events are not independent. □

The idea of independence can be extended to more than two events. Unfortunately, there's a nuisance problem; the obvious extension $P(A$ and B and $C) = P(A)P(B)P(C)$ isn't sufficient. There are bizarre examples where $P(A$ and B and $C) = P(A)P(B)P(C)$, but $P(A$ and $B)$ is not equal to $P(A)P(B)$. The more useful extension is the idea of **independent processes.** For our purposes, a process can be thought of as a smaller sample space. The overall experiment (or "super sample space") is composed of some number, say k, of such processes.

independent processes

Independent Processes

Random processes $S_1, S_2, \ldots, S_k$ are independent if for any events A_1 from S_1, A_2 from S_2, ..., A_k from S_k,

$$P(A_1 \text{ and } A_2 \text{ and } \ldots \text{ and } A_k) = P(A_1)P(A_2) \ldots P(A_k)$$

Example 4.13

An examination consists of four multiple choice questions, each having five possible answers. Assume that a student randomly guesses the answer to each question. What is the probability that the student answers the first three questions correctly and the fourth question wrong?

Solution

The answers to the four questions may be assumed to be independent processes, because whether the student happened to guess right on any one question should not affect the probability of guessing right or wrong on any other question. The probability of guessing right on any given question should be 1/5, or .20; by the complements principle, the probability of guessing wrong on any given question is 1 − .20 = .80. Therefore,

$$P(R_1 \text{ and } R_2 \text{ and } R_3 \text{ and } W_4) = P(R_1)P(R_2)P(R_3)P(W_4)$$
$$= (.20)(.20)(.20)(.80) = .0064$$

We have now used the idea of independence in two ways. In the example of the physiological trait and schizophrenia, we began with a specified set of probabilities and showed that two events were independent. In Example 4.13, we began by *assuming* that events (and processes) were independent and used the assumption in calculating probabilities.

Many of the probability models we will use in this book involve an assumption of independence at some stage. In particular, when random samples are taken *with replacement*, so that an individual drawn on one trial is put back into the population before the next draw, it is very reasonable to assume that the successive draws represent independent processes. This assumption is most useful in calculating probabilities related to random sampling.

Example 4.14 Individuals in a population are classified as to whether they are susceptible (S) to peer pressure in estimating the intensity of light stimuli. Assume that 60% of the population is susceptible. A sample of three individuals is to be chosen at random from the population; assume that the population is so large that it doesn't matter whether the sample is taken with or without replacement. What is the probability that none of the three individuals drawn is S?

Solution From the information given, we may assume that the three draws are three independent processes. On each draw $P(\text{not } S) = 1 - .60 = .40$.

$$\begin{aligned} P(\text{not-}S_1 \text{ and not-}S_2 \text{ and not-}S_3) &= P(\text{not-}S_1)P(\text{not-}S_2)P(\text{not-}S_3) \\ &= (.40)(.40)(.40) = .064 \end{aligned}$$

□

Exercises for Sections 4.3 and 4.4

4.9. A population consists of 900 men, of whom 60 are colorblind, and 1100 women, of whom 30 are colorblind. An individual is drawn at random from the population.

a. What is the probability that a noncolorblind man is drawn?
b. If a man is drawn, what is the probability that he is noncolorblind?
c. What is the probability of drawing a colorblind person?

4.10. In Exercise 4.9, are the events "man" and "colorblind" independent?

4.11. A sample of three individuals is to be selected from the population of Exercise 4.9. The sampling is done without replacement.

a. What is the probability that all three will be colorblind?
b. What is the probability that all three will be either women or colorblind or both?

4.12. Answer Exercise 4.11, assuming that the sampling is done without replacement, so that the same individual could be sampled twice or even all three times.

4.13. In a helplessness experiment, a subject is rewarded for an answer to a multiple-choice question with probability .4, regardless of whether or not the answer is correct. There are five questions.

a. What is the probability that the subject is not rewarded at all?

b. What is the probability that the subject is rewarded on only the second trial?

c. What is the probability that the subject is rewarded on one of the first two trials, regardless of what happens in later trials?

4.14. Referring to Exercise 4.7, find each of the following.

a. $P(F|E)$

b. $P(E \text{ and } F|D)$

c. $P(\text{all three reactions}|\text{at least 1 reaction})$

4.15. Referring to Exercise 4.4, assume that the probability that any particular subject will report no errors is .1, the probability of reporting between 1 and 7 errors is .3, and the probability of reporting 8 or more errors is .6.

a. What is the probability that all six subjects will report no errors?

b. What is the probability that the first three subjects will report 8 or more errors, the fourth subject will report between 1 and 7 errors, and the last two subjects will report no errors?

4.16. In Exercise 4.15, what did you assume about the relation between one subject's response and another's? Can you think of experimental situations where this assumption might be reasonable? Situations where it might be dubious?

4.17. Referring to Exercise 4.15, what is the probability that a subject reports no errors, given that the subject does not report 8 or more errors?

4.18. In a certain course, 20% of the grades are A, 40% B, 20% C, 15% D, and 5% F. Also 60% of the students are women, and 40% men. Assume that sex and grade are independent. What is the probability that a randomly selected student

a. is a woman who gets an A?

b. is a woman who gets a C or better grade?

c. is either a woman or gets an A?

4.19. A random sample of three students is to be taken from the course in Exercise 4.18. Assume that the course is very large, so the issue of sampling with or without replacement is irrelevant.

a. What is that probability that all three students are women who got A's?

b. What is the probability that none of the three received A's?

c. Given that all three students are women, what is the probability that all three received A's?

4.5

Combining Principles: Probability Trees and Tables

In this chapter we have introduced the three basic principles—addition, complements, and multiplication—of probability theory, as well as the crucial idea of statistical independence. Most probability problems require the use of several of these principles to get a solution. There are many ways to get to the solution. The particular approach used is as much a matter of personal style and taste as anything. In this section we will introduce a few methods that seem generally useful in solving probability problems.

Suppose that 4% of the children in a certain school district suffer from dyslexia, a perceptual problem that causes difficulty in reading. Precise diagnosis of dyslexia is a difficult, expensive, and time-consuming process. A rough diagnostic test is available. It is known that 75% of dyslexic children yield a positive response to this test, as do 5% of nondyslexic children. Furthermore, 15% of dyslexic children and 85% of nondyslexic children yield a negative response. Finally, 10% of dyslexic children and 10% of nondyslexic children yield an ambiguous response. If a child is selected at random, what is the probability of a positive response? Given a positive response, what is the probability that the child is dyslexic?

This problem can be attacked by probability trees, by probability tables, or by direct probability reasoning. Most students seem most comfortable with probability trees, so that approach will be discussed first. To construct a probability tree, we begin by "branching" on events with known unconditional probabilities. In our example we know that the probability of dyslexia (D) is .04, and by complements the probability of not-D is .96. Therefore, we begin by drawing Figure 4.4.

Next, from the tip of each initial branch, we draw new branches for events with known conditional probabilities. In the example, we know the probabilities of positive (P), negative (N), and ambiguous (A) re-

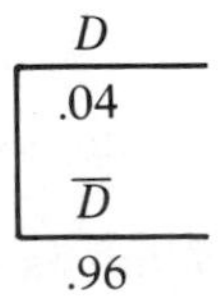

D
.04
$\overline{D}$
.96

Figure 4.4 Beginning of a probability tree

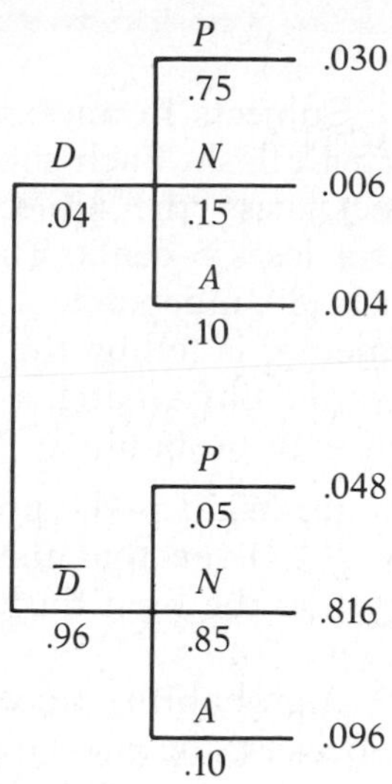

Figure 4.5 Complete probability tree

sponses to the test, given either a D or a non-D child. Therefore, we expand the tree as shown in Figure 4.5.

If necessary, we could construct further branches from the tips of the current tree. In our example, no further branching is needed. To find the probability of any particular *path*—sequence of branches—we multiply all the probabilities along that path. The path probabilities are shown on the right side of Figure 4.5. To find the unconditional probability of any particular event, we simply add the probabilities for all paths corresponding to the particular event. The probability of a positive response is the sum of the first and fourth path probabilities in Figure 4.5.

$$P(P) = .030 + .048 = .078$$

To find the conditional probability of one event given another, we use the definition of conditional probability. The probability of dyslexia given a positive response is

$$P(D|P) = \frac{P(D \text{ and } P)}{P(P)} = \frac{.030}{.078} = .3846$$

In passing, note that $P(A) = .004 + .096 = .10$ and that $P(D|A) = P(D \text{ and } A)/P(A) = .04$, which is the same as the unconditional probability $P(D)$. Because 10% of both dyslexics and nondyslexics give ambiguous (A) responses to the test, the events A and D are statistically independent.

Example 4.15 Subjects in an experiment are told that either a red or a green light will flash. Each subject is to guess which light will flash; for each correct guess, the subject wins 10 cents; for each incorrect guess, the subject loses 5 cents. The subject is told that the probability of a red flash is .80, independent of the guess. (And for once, a psychological experimenter is telling the truth to the subjects.) Assume that the subject is a probability matcher—that is, guesses red with probability .80 and green with probability .20.

(a) What is the probability that the subject guesses correctly?

(b) Given that the subject guesses correctly, what is the probability that the light flashed red?

Solution A probability tree can be drawn in a couple of ways for this problem. Because the guess occurs before the flash, it seems natural to branch first on the guess; but, because the flash and the guess are independent, we could do it the other way around. Out of sheer perversity, we branch on the flash first in Figure 4.6.

(a) The probability of a correct guess is the sum of the first and fourth path probabilities.

$$P(\text{correct}) = .64 + .04 = .68$$

(b)

$$P(\text{guess red}|\text{correct}) = \frac{P(\text{guess red and correct})}{P(\text{correct})}$$

$$= \frac{.64}{.68} = .9412$$

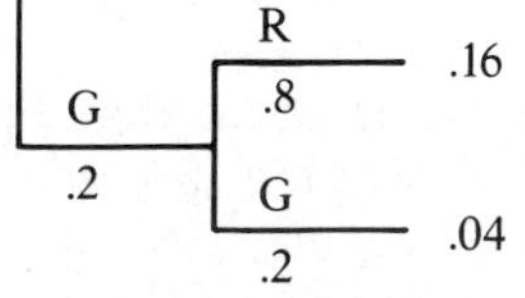

Figure 4.6 Probability tree for Example 4.14

Notice that the events "guess red" and "flash red" are independent, but the events "guess red" and "guess correct" are not. □

In constructing probability trees, remember that:

1. At any stage in branching, the branches should represent mutually exclusive events that exhaust all possibilities.

2. Probabilities at the first branching should be unconditional; later probabilities should be conditional on preceding branches. In the case of independence, it is not necessary to worry about previous conditions.

3. Path probabilities are the products of branch probabilities along the path.

4. Unconditional probabilities are found by adding path probabilities.

5. Conditional probabilities are found by applying the definition of conditional probability.

An alternative approach to solving probability problems is the use of probability tables. These tables look very much like the cross-tab tables of Chapter 3, except that they involve probabilities instead of frequencies. The bare bones of a probability table for the dyslexia example are shown as Table 4.3. Note that the events D and not-D are mutually exclusive and exhaustive, as are the events P, N, and A.

		Test Response			
		P	N	A	Total
	D	.04(.75) = .030	.006	.004	.04
Dyslexia	not D	.048	.816	.096	.96
	Total	.078	.822	.100	

Table 4.3

The table is filled in as follows:

1. Insert any known unconditional probabilities in the margins of the table. These probabilities are often called *marginal* probabilities.

2. Calculate *joint* probabilities (and-type probabilities in the body of the table) using the multiplication principle. For example, $P(D \text{ and } P) = P(D)P(P|D) = (.04)(.75) = .030$. The joint probabilities for the dyslexia example are shown in Table 4.3.

3. Calculate any other marginal probabilities by addition.

4. To calculate conditional probabilities, use the definition.

Filling in Table 4.3, we see once again that $P(P) = .078$ and $P(D|P) = (.030)/(.078) = .3846$.

Example 4.16 Use a probability table to solve Example 4.15.

Solution In this case, the marginal probabilities of the guess and of the color flashed are known. The table may be filled in using statistical independence (Table 4.4).

		Flash		
		Red	Green	Total
Guess	Red	(.8)(.8) = .64	.16	.80
	Green	.16	.04	.20
	Total	.80	.20	

Table 4.4

$P(\text{correct}) = P(\text{red, red}) + P(\text{green, green}) = .64 + .04 = .68$. Also,

$$P(\text{flash red}|\text{guess correct}) = \frac{P(\text{flash red and guess correct})}{P(\text{guess correct})} = \frac{.64}{.68}$$

as in Example 4.15. □

Finally, it is also possible to solve probability problems by direct application of the basic probability principles. For the dyslexia example,

$$P(P) = P(D \text{ and } P) + P(\text{not-}D \text{ and } P)$$

because the event P can be regarded as (D and P) or (non-D and P). Then, by the multiplication principle,

$$\begin{aligned} P(P) &= P(D)P(P|D) + P(\text{not-}D)P(P|\text{not-}D) \\ &= (.04)(.75) + (.96)(.05) = .078 \end{aligned}$$

once again.

Example 4.17 Solve Example 4.15 by direct application of probability principles.

Solution We have

$$\begin{aligned} P(\text{correct guess}) &= P(\text{red guess and red flash}) \\ &\quad + P(\text{green guess and green flash}) \\ &= P(\text{red guess})P(\text{red flash}) \\ &\quad + P(\text{green guess})P(\text{green flash}) \end{aligned}$$

by the assumed independence of the guess color and the flash color. So once again,

$$P(\text{correct guess}) = (.8)(.8) + (.2)(.2) = .68$$

□

There is no universally right way to solve basic probability problems. There are only more convenient and less convenient ways, and convenience is very much a matter of personal style. Most students seem to find trees convenient. Trees have the additional advantage that they can be extended to more than two levels of branches very easily. For more complicated problems, probability tables can be confusing. A very good strategy is to try several approaches to attack the problem. The best way to solve a problem is the way that you find clearest.

Exercises for Section 4.5

4.20. Refer to Exercise 4.19.

a. What is the probability that (exactly) two of the students received grades of C or better?

b. Given that exactly two of the students received grades of C or better, what is the probability that both received A's?

4.21. Suppose that 75% of all subjects are able to learn task A within a specified number of tries. Of those who learn task A, 60% also are able to learn task B. Of those who do not learn task A, 40% are able to learn task B. A subject is chosen at random.

a. What is the probability that the subject will learn exactly one of the two tasks?

b. Given that the subject learns exactly one of the two tasks, what is the probability that it is task A?

c. If the subject learns task B, what is the probability that the subject also learned task A?

4.22. Referring to Exercise 4.21, suppose that there is also a task C which is learned by 90% of those who learned both A and B, by 80% of those who learned A but not B, by 60% of those who learned B but not A, and by 20% of those who learned neither A nor B.

a. What is the probability of learning exactly two of the three tasks?

b. If exactly two of the three tasks are learned, what is the probability that the two tasks are A and B?

c. If exactly two of the three tasks are learned, what is the probability that one of them is A?

4.23. A population consists of 20% depressed people and 80% "normals." Of the depressed people, 60% have a certain chemical in abnormally high levels in their blood, as do 15% of the normals.

a. What is the probability that a randomly chosen person will have an abnormally high level of the chemical?

b. Given that the chemical is at an abnormally high level in a randomly chosen person, what is the probability that the person will be depressed?

4.24. Of the depressed people in Exercise 4.23, 40% will report some form of sexual dysfunction; this percentage applies regardless of the presence or absence of abnormally high levels of the chemical. Of the normal people, 10% will report some dysfunction, again regardless of the chemical level.

a. What is the probability that a randomly chosen person will both report dysfunction and have a high level of the chemical?

b. If a person reports dysfunction and has a high level of the chemical, what is the probability that the person is depressed?

c. If a person reports dysfunction or has a high level of the chemical, or both, what is the probability that the person is depressed?

4.25. A multiple-choice examination consists of four questions, each with five possible answers. A well-prepared student has a 90% chance of answering any particular question correctly, whereas a poorly prepared student has a 50% chance of answering any particular question correctly. In the class, 80% of the students are well prepared.

a. What is the probability that a randomly chosen student will answer all four questions correctly?

b. If a student answers all four questions correctly, what is the probability that the student was well prepared?

4.6 Conditional Independence

In Section 4.4 the fundamental concept of statistical independence was introduced. In this section we discuss a slightly subtler idea, that of conditional independence. Although the idea is important for understanding some of the issues of statistical thinking, it is not a critical one for understanding the rest of this book.

The idea can best be introduced by an example. Individuals in a certain population are classified by sex, by diagnosis of a certain degenerative disease, and by presence or absence of abnormally high levels of a certain chemical in the bloodstream. The proportions in each category are shown in Table 4.5.

WOMEN

		Present	Absent
Diagnosis	Yes	.012	.048
	No	.108	.432

MEN

		Present	Absent
Diagnosis	Yes	.180	.020
	No	.180	.020

Table 4.5

The joint probabilities for all combinations of diagnosis and chemical presence can be obtained by adding together the proportions for women and for men (Table 4.6).

		Present	Absent	Total
Diagnosis	Yes	.192	.068	.260
	No	.288	.452	.740
	Total	.480	.520	

Table 4.6

There appears to be an association (a statistical dependence) between a positive diagnosis and presence of the chemical. Independence would hold if the probability of yes and present had been $P(\text{yes})P(\text{present}) = (.260)(.480) = .1248$, but the actual probability is quite a bit higher, .192. This suggests that presence of the chemical is related to the disease and could even be a cause of it. Of course, statistical association alone cannot prove causality; indeed, the chemical could also be an effect of the disease.

A different indication emerges when we look at women and at men separately. Separate tables for women and for men can be calculated by dividing the probabilities shown in the first table in this section by the respective probabilities of women and of men, .6 and .4. The resulting probabilities, shown in Table 4.7, are conditional on the sex of the person.

WOMEN

		Present	Absent	Total
Diagnosis	Yes	.02	.08	.10
	No	.18	.72	.90
	Total	.20	.80	

MEN

		Present	Absent	Total
Diagnosis	Yes	.45	.05	.50
	No	.45	.05	.50
	Total	.90	.10	

Table 4.7

In each of the two conditional probability tables, diagnosis is *independent* of the presence of the chemical! For example, given a woman, the probability of yes and present is .02, which is exactly the product of the (conditional on women) marginal probabilities .10 and .20. This independence suggests that there is no causal relation between the chemical and the disease.

What happened? Somehow an apparent association has been made to disappear. This is an example of *spurious association*, as discussed in Chapter 3, in which an apparent relation disappears once a third variable is considered. The apparent association between diagnosis and presence of the chemical disappears once sex is taken into account. The reason for the apparent association is that men have a much higher incidence of a yes diagnosis than women and, quite independently, a much higher incidence of presence of the chemical. In the combined population, there appears to be an association, but taking into account the sex of the patient "explains away" the relation. This situation is very similar to one discussed in Chapter 3 in which data showed a relation between two variables that was "partialed out" by consideration of a third variable.

In probability terms, the example shows events that are conditionally independent.

Conditional Independence
Events B and C are conditionally independent given A if

$$P(B \text{ and } C|A) = P(B|A)P(C|A)$$

The formal difference between independence and conditional independence is merely that conditional probabilities are used to test conditional independence, whereas unconditional probabilities are used to test plain vanilla independence.

Example 4.18 Grade-school children were classified as having low, medium, or high difficulty with reading. Two different tests were administered. For simplicity, assume that the tests yielded only "pass" or "fail" scores. Table 4.8 gives the proportions observed in the population.

LOW		A Test P	A Test F
B Test	P	.395	.005
	F	.055	.045

MEDIUM		A Test P	A Test F
B Test	P	.162	.018
	F	.048	.072

HIGH		A Test P	A Test F
B Test	P	.036	.004
	F	.044	.016

Table 4.8

Are the events pass A and pass B conditionally independent given the event low difficulty?

Solution

$$P(\text{pass A and pass B}|\text{low}) = \frac{P(\text{pass A and pass B and low})}{P(\text{low})}$$

$$= \frac{(.395)}{(.500)} = .790$$

(Note that $P(\text{low})$ can be obtained by adding up the four probabilities in the low portion of the table.)

$$P(\text{pass A}|\text{low}) = \frac{P(\text{pass A and low})}{P(\text{low})}$$

$$= \frac{(.450)}{(.500)} = .900$$

and similarly,

$$P(\text{pass B}|\text{low}) = \frac{(.400)}{(.500)} = .800$$

The conditional probability of passing both A and B tests is .790, which is *not* the product of the conditional probabilities of passing A and of passing B, namely .900 and .800. Therefore, the events passing A and passing B are not conditionally independent given low reading difficulty. □

Just as it is often convenient to assume complete independence, we sometimes will assume conditional independence. The results of various diagnostic tests for an abnormality can often be assumed to be conditionally independent, given the presence (or given the absence) of the abnormality, at least as a good approximation. The Markov assumption used in models of how behavior evolves over time, and discussed in Chapter 6, is the assumption that future behavior is independent of past behavior, given present behavior.

Example 4.19 Three graders evaluate a sample of the written work of high school seniors. If the student is an outstanding writer, any of the graders has probability .9 of grading the sample A. Assume that 10% of the population are outstanding writers and that the evaluations of the three graders are conditionally independent given the actual ability of the student. Find the probability that a randomly chosen sample was written by an outstanding writer and receives three A grades.

Solution

$$P(\text{outstanding and 3 A's}) = P(\text{outstanding})P(\text{3 A's}|\text{outstanding})$$

$P(\text{outstanding})$ is assumed to be .1. The assumption of conditional independence allows us to calculate

$$\begin{aligned} &P(\text{3 A's}|\text{outstanding}) \\ &\quad = P(\text{first A}|\text{outst.})P(\text{second A}|\text{outst.})P(\text{third A}|\text{outst.}) \\ &\quad = (.9)(.9)(.9) = .729 \end{aligned}$$

So $P(\text{outstanding and 3 A's}) = (.1)(.729) = .0729$. □

Exercises for Section 4.6

4.26. Refer to Exercise 4.25.

a. Show that you assumed that the correctness of the answers to (say) the first two questions was conditionally independent, given the preparation of the student.

b. Does this seem to you to be a reasonable assumption? Are there conditions under which it might not be reasonable?

4.27. A course has two midterm examinations and a final. Historically, 20% of all students receive A's, 30% B's, and 50% C's or lower on the first midterm. Of those who receive A's on the first midterm, 60% score A's on the second midterm, 30% B's, and 10% C's or lower; for those who receive B's on the first midterm, the second-midterm percentages are 30%, 60%, and 10%; for those scoring C or lower on the first midterm, the percentages are 10%, 20%, and 70%. Of those scoring A on the second midterm, 70% receive A's on the final, 20% B's, 10% C's or lower; the percentages for those who score B's on the second midterm are 40%, 50%, and 10%; and those who score C's or lower on the second midterm are distributed 20%, 30%, and 50% on the final.

a. Calculate the probability that a randomly chosen student receives at least two A grades on the three tests. You may want to construct a probability tree.

b. In answering part a, what did you assume about the dependence of grades on the various exams?

c. Do you think that your assumption was a reasonable one?

4.28. Referring to the data of Exercise 3.13, pretend that the data constitute the entire relevant population.

a. Is there a statistical dependence between the events "major is E" and "aptitude is high"?

b. Are those events conditionally independent given "region is MW"?

c. Are "E" and "MW" independent given "aptitude is high"?

4.29. In a population of 500 mentally ill people, each person is classified about recent violent behavior (yes/no), perceived degree of stress (high/medium/low), and severity of illness (severe/moderate/noticeable). The following frequencies hold for the population:

SEVERE ILLNESS

		Stress		
		H	M	L
Violence	Yes	95	50	15
	No	25	10	5

MODERATE ILLNESS

		Stress		
		H	M	L
Violence	Yes	33	48	19
	No	57	102	41

NOTICEABLE ILLNESS

		Stress		
		H	M	L
Violence	Yes	3	13	34
	No	37	107	306

Table 4.9

a. Find the probability that violence is yes and stress is high.
b. Are the events yes and high statistically independent?

c. Find the probability that violence is yes and stress is high, given that illness is moderate.

d. Are yes and high conditionally independent, given moderate?

4.30. Referring to Exercise 4.29, calculate conditional probability tables given each level of illness. Does there appear to be a strong statistical dependence between stress and violence, given severity?

4.31. A study of reading abilities of elementary-school children concluded that at the end of first grade, the probability that a child would be reading above grade norm (+) was .2; the probability of reading at grade norm (0) was .5, and the probability of reading below grade norm (−) was .3. The conditional probabilities in Table 4.10 were obtained.

		End Second +	End Second 0	End Second −			End Third +	End Third 0	End Third −			End Fourth +	End Fourth 0	End Fourth −
	+	.90	.08	.02		+	.88	.10	.02		+	.86	.09	.05
1st	0	.03	.95	.02	*2nd*	0	.05	.90	.05	*3rd*	0	.08	.85	.07
	−	.01	.06	.93		−	.04	.12	.86		−	.07	.09	.84

Table 4.10

That is, given that a child is + after first grade, the probability of being + after second grade is .90; given that a child is + after second grade, the probability of being + after third grade is .88; and so on. Assume conditional independence of later years and earlier years, given the results in intermediate years.

a. Find the probability of being + after all four years covered in the study.

b. Find the probability of ending fourth grade above norm, given that the child ended first grade above norm.

4.32. What is the meaning of the conditional independence assumption in Exercise 4.31, in terms of children's learning to read? Specifically, assume that we are dealing with two children, both of whom scored at grade level in second grade. The first child was a + in grade 1, the second was a −. Which child, according to the assumption, is more likely to be a + in third grade?

Chapter Exercises

4.33. Events A, B, and C are mutually exclusive, each having probability .2. Find the following probabilities.

a. $P(A \text{ or } B \text{ or } C)$
b. $P(\text{at least one of } A, B, C)$
c. $P(A, B, \text{ and } C \text{ all do not occur})$
d. $P(A \text{ or } B \text{ occurs} | C \text{ does not occur})$

4.34. Events A, B, C, and D are from independent processes. Each event has probability .2. Find

a. $P(A \text{ and } B \text{ and } C \text{ and } D)$
b. $P(\text{none of } A, B, C, \text{ or } D \text{ occurs})$
c. $P(\text{at least 2 of the events occur})$
d. $P(\text{either } C \text{ or } D | \text{at least 2 of the events occur})$

4.35. A sample space consists of 100 outcomes, all equally likely. Event D consists of all odd-numbered outcomes. Event E consists of the outcomes numbered 51 to 100. Event F consists of the outcomes numbered 1 to 25 and 50 to 74. Find

a. $P(D \text{ or } E \text{ or } F)$
b. $P(D \text{ and } E \text{ and } F)$
c. $P(F|D)$
d. $P(\text{exactly 1 of } D, E, F)$

4.36. Which pairs of events in Exercise 4.35 are independent? Mutually exclusive?

4.37. An expert on autistic children stated that an otherwise normal autistic child has one chance in three of becoming an independently functional adult, and one chance in nine of completely outgrowing autism.

a. What is an appropriate interpretation (objective, relative frequency, or subjective) of these probabilities?

b. Assuming that the probabilities are correct, what is the probability that an autistic child will become a functional adult but not completely outgrow autism?

4.38. Only 10% of the potential subjects for a particular experiment are actually usable. Subjects are selected at random, one by one, and tested for usability. Assume that sampling is done with replacement.

a. What is the probability that the first usable subject is found at the third selection?

b. What is the probability that the second usable subject is found at the fourth selection?

c. If four potential subjects are selected, what is the probability that exactly one of them is usable?

4.39. About 20% of all students in a class receive A's on a midterm exam, 40% receive B's, 20% receive C's, 10% receive D's, and 10% receive F's. On the final exam, 30% score A's, 30% B's, 20% C's, 15% D's, and 5% F's.

a. What is the probability that a student will receive A grades on both the midterm and the final exams?

b. What did you assume (beyond the information given in the problem) in answering part a? Does this assumption seem reasonable to you?

4.40. Referring to Exercise 4.39, assume that three students are selected at random from the very large class.

a. What is the probability that all three receive B's on the midterm exam?

b. What is the probability that one of the students is graded A, one B, and one C on the midterm? (Note that which student receives which grade is not specified.)

4.41. A study of voting behavior revealed that 10% of a population were high on a SES (socioecomic status) variable, 60% were medium, and 30% were low. Conditional probabilities of voting for Democratic or Republican candidates were as shown in Table 4.11.

	SES		
	H	M	L
Democratic	.30	.50	.80
Republican	.70	.50	.20

Table 4.11

a. What is the probability that a randomly selected voter will vote Democratic?

b. If a randomly selected voter votes Democratic, what is the probability that the voter belongs to the H SES status?

4.42. Suppose that a random sample of five voters is selected from the population of Exercise 4.41. (Assume that it doesn't matter whether the sample is chosen with or without replacement.)

a. What is the probability of obtaining no H SES voters?

b. What is the probability of obtaining two H SES voters, (exactly)?

c. What is the probability that the first H SES voter is found at the fourth sample?

4.43. In a child-development study, babies were observed at ages 6, 12, and 18 months. They were rated high, medium, or low on measures of physical maturity and dependency on mother. The probabilities shown in Table 4.12 were observed (conditional on age).

6 MONTHS

		Physical Maturity		
		H	M	L
Dependency	H	.06	.10	.04
	M	.12	.20	.08
	L	.12	.20	.08

12 MONTHS

		Physical Maturity		
		H	M	L
Dependency	H	.12	.09	.09
	M	.16	.12	.12
	L	.12	.09	.09

18 MONTHS

		Physical Maturity		
		H	M	L
Dependency	H	.15	.09	.06
	M	.15	.09	.06
	L	.20	.12	.08

Table 4.12

Are physical maturity and dependency conditionally independent given age?

4.44. In Exercise 4.43, the three ages had equal probabilities.

a. Find the probability that in a randomly chosen baby, maturity and dependence were both high.

b. Are maturity and dependence unconditionally independent?

5 Random Variables and Probability Distributions

The basic language of probability theory developed in Chapter 4 applies to any kind of event, qualitative or quantitative. Many of the statistical concepts developed in Chapters 2 and 3 require quantitative measurements; it's a bit difficult to define the mean of two Lutherans, one Roman Catholic, and one Buddhist. Some additional ideas are needed to deal with experiments yielding numerical outcomes, which may reasonably be averaged. The critical concepts are those of random variable and probability distribution.

The fundamental concepts are defined in Sections 5.1 and 5.2, along with the required notation. These sections deal with discrete random variables that take on distinct, separate values. Then in Sections 5.3 and 5.4 the concepts of mean, median, and standard deviation from Chapter 2 are extended to work with the new concepts of random variable and probability distribution. In Section 5.5 some elementary calculus is used to define the properties of continuous random variables, those that take values in an entire interval of numbers. In Section 5.6 the basic probability principles of Chapter 4 are translated into the language of random variables. In Section 5.7 the key idea of correlation is defined in random variable language. Finally, in Section 5.8 there is an introduction to the basic mathematical properties of expected values, the random-variable generalization of means.

5.1 Random Variables

Virtually all social science experiments (in the wide sense of experiments) yield numerical data, whether they be simple frequency counts or complicated measurements. The data are almost always subject to some degree of randomness, because the samples are random or there is random measurement error or there are unexplainable variations in responses of individual entities in the experiment. Probability theory is the language of randomness, but we have not yet developed the language needed to deal with specifically numerical, quantitative results from randomness-plagued experiments.

random variable

The fundamental concept required is that of a **random variable.** A random variable may be as simple as the number of heads obtained in flipping a coin twice, or as complex as the result of a computer-based massage of a complex survey. For the result of an experiment to be a random variable, it must have two essential properties: The final result of the experiment must be a *number,* and that number must be subject to random variation, as indicated by the possibility that a repetition of the experiment would yield a different number. Of course, a given experiment may well yield many numbers and therefore involve many random variables. But for now, let's deal with random variables one at a time.

A sloppy, but useful, definition of random variable is based on the two essential ideas—numerical result and random variation.

Random Variable (Informal Definition)

A **random variable** is any numerical quantity derived according to specified procedures from an experiment and subject to random variation from one repetition of the experiment to the next.

Conventionally, random variables are denoted by capital letters near the end of the Roman alphabet—X, Y, Z, or maybe W.

Example 5.1

Identify one possible random variable for each of the following experiments:

(a) A student randomly guesses the answers to 12 true-false questions on a test.

(b) A rat is placed on one of ten selected points in a box, in the dark, and the time required for the rat to reach a target point (baited by food) is recorded; the procedure is repeated six times, with six different starting points.

(c) A sample of 50 New York City high school students are asked to compare relative distances between pairs of American cities; by means of a technique called multi-dimensional scaling, a set of 50 "perceptual maps" is obtained. Interest is focused on the perceived ratios of distances from New York to Chicago to distances from New York to Los Angeles.

Solution

(a) The most obvious random variable is X = the number of correct guesses made.

(b) One obvious random variable is $\bar{Y}$ = the average time required. Another is S^2 = the variance in the times. Another is R = the correlation between time and distance traveled. (Note: We said that capital letters near the end of the alphabet are usually used to denote random variables. Usually is not always.)

(c) A natural random variable of interest would be $\bar{Y}$, the sample mean of the ratios obtained. Again, the variance would be relevant. It is likely that there would be some outliers in these data, so the range might not be a very useful statistic. □

There is a problem with the informal definition of a random variable. The concept is needed to deal with randomness of numerical outcomes, but there is no connection with the probability apparatus of sample spaces and principles developed in Chapter 4. A more formal definition is needed. To understand the forthcoming definition, consider experiment (a) of Example 5.1. A sample space for the experiment could consist of ten letter strings of C (correct) and I (incorrect); the outcome CCCCCCCIII would be one such string and would correspond to $X = 7$ correct answers; so would the outcome CCICCCICIC. For any random variable, it is possible, in principle, to make each outcome of a sample space correspond to a particular numerical value of the random variable.

Random Variable (Formal Definition)

A **random variable** is a function or rule that assigns a number to each possible outcome in a sample space.

Example 5.2

A very simple experiment consists of having three subjects judge the relative weights of pairs of rather light objects. The experiment is a "forced-choice" situation, in which the subject must specify that one or the other object is heavier. Attention is centered on two particular objects that are dissimilar in shape but equal in weight. Let X = the number of trials for which object A is judged heavier than object B. Construct a sample space for the experiment, and show how to assign a number to each outcome consistent with the definition of X.

Solution The outcome in which subjects 1 and 3 judge object A to be heavier and subject 2 judges object B to be heavier could be denoted by ABA, for instance. The possible outcomes and associated X values can be listed, as follows:

Outcome	AAA	AAB	ABA	BAA	ABB	BAB	BBA	BBB
X value	3	2	2	2	1	1	1	0

□

Standard notation is to use a capital letter to denote a random variable, and the corresponding lower-case letter to denote any particular possible value. Thus for Example 5.2, X = the number of subjects judging A heavier than B, and $x = 2$ for the specific outcome ABA. The distinction between the rule (X) and the result (x) is a bit subtle, but will become clearer as we go along.

In this book we will observe an additional notational convention. The letter Y will be reserved to denote a dependent random variable—that is, a response that is to be explained by or associated with values of other, independent variables. For example, if the time required by a laboratory rat to learn a particular maze is to be explained (partially) by the length of time the rat has been deprived of food, we will let Y = learning time; some other letter, often X, will be used for an independent variable such as deprivation time.

5.2 Discrete Probability Distributions

The concept of random variable was defined in Section 5.1 to make connections with the probability apparatus of Chapter 4 in the specific case of a numerical outcome of an experiment. In this section we will take advantage of the connection to define probabilities associated with each possible value of a random variable—the probability distribution of the random variable.

The definitions of this section apply only to *discrete* random variables—those which can take on only distinct, separate values such as 0, 1, 2, ... or perhaps $6\frac{1}{2}$, $6\frac{5}{8}$, $6\frac{3}{4}$, ... Analogous definitions for *continuous* random variables—random variables such as time, which can in principle be measured along a whole continuum or range of possible values—will be given in Section 5.5.

Recall that a random variable formally assigns a number to each possible outcome in a sample space. If a probability has been assigned to each outcome in the sample space, an associated probability can be assigned to each possible value of a random variable.

Example 5.3

Assume that each outcome in the sample space in Example 5.2 has equal probability. Assign probabilities to each possible value x of the random variable X.

Solution

There are eight possible outcomes, with equal probabilities 1/8. Only one outcome (BBB) corresponds to $x = 0$, so $x = 0$ should have probability 1/8. Outcomes ABB, BAB, and BBA all yield $x = 1$, so $x = 1$ should have probability $1/8 + 1/8 + 1/8 = 3/8$. The probabilities for $x = 2$ and for $x = 3$ should be 3/8 and 1/8, respectively. □

probability distribution

The name **probability distribution** is attached to the set of probabilities of possible values of a random variable; the probability distribution of a random variable indicates how probability is distributed across its possible values. The probability distribution of a random variable, say Y, is denoted by $f_Y(y)$.

Probability Distribution of a Random Variable *Y*

$$\begin{aligned} f_Y(y) &= P(Y \text{ takes the value } y) \\ &= P(Y = y) \end{aligned}$$

The apparently redundant notation $f_Y(y)$ becomes useful when we consider more than one random variable at a time. If we merely wrote $f(x), f(y)$, etc., we would run into trouble interpreting an expression such as $f(2)$; does that mean $P(X = 2)$ or $P(Y = 2)$ or what? Writing $f_Y(2)$ resolves the ambiguity.

Example 5.4

Calculate $f_X(x)$ for Example 5.2.

Solution

In Example 5.3, we calculated the relevant probabilities under reasonable assumptions. The probability distribution is shown below.

x	0	1	2	3
$f_X(x)$	1/8	3/8	3/8	1/8

□

The actual computation of a probability distribution of a random variable, say Y, is simply another probability problem. It can be attacked by means of any of the probability principles discussed in Chapter 4. In particular, probability trees and tables are available; these techniques, in effect, reconstruct sample spaces. A random variable merely assigns numbers to all the outcomes in a sample space, so all that's needed is to assign numbers to each path of a probability tree or to each cell of a probability table. Then the probability distribution of a random variable can be calculated by simple addition.

Example 5.5 A graduate student is seeking a somewhat obscure statistical formula from textbooks in the library. There are five possible textbooks available; each has probability .4 of containing the relevant formula. Assume independence from one textbook to the next. Define X = the number of textbooks consulted until the formula is found; if the formula is in none of the five books, define X to be 5. Find the probability distribution of X.

Solution A probability tree for this problem is shown in Figure 5.1. Note that the associated x value is assigned to each path, along with a probability.

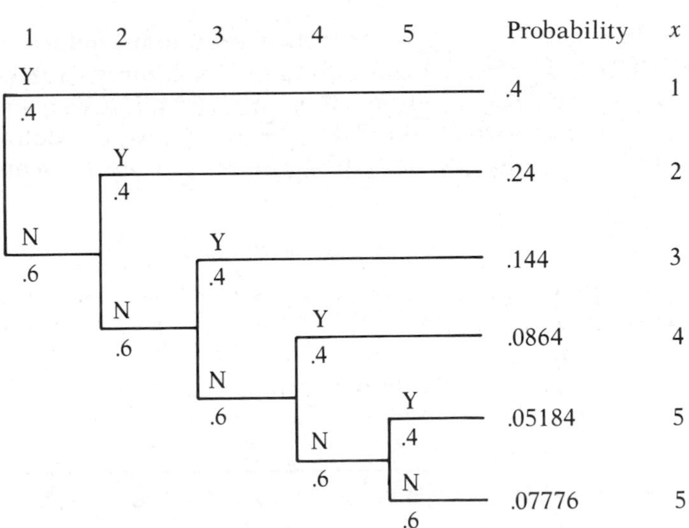

Figure 5.1 Probability tree for Example 5.5

The probability distribution is

x	1	2	3	4	5	
$f_X(x)$	.4	.24	.144	.0864	.05184 + .07776	(= .1296)

□

In principle, it is always possible (with some assumptions) to find the probability distribution of any random variable. In practice, the mathematics can get nasty. For example, finding the probability distribution for the random variable of part (c) of Example 5.1 is an unsolved and formidable problem. When the mathematics gets ugly, a technique called the Monte Carlo method can be employed. Essentially, it involves simulating an experiment repeatedly on a computer and observing the relative frequencies of the various possible results.

Example 5.6 Suppose that a computer program was written to simulate the experiment of Example 5.5 10,000 times. The possible values of X and the observed relative frequencies are shown below. Are the results reasonable, given that 10,000 is not an infinite number of observations?

Values	1	2	3	4	5
Frequencies	3961	2427	1449	849	1314

Solution The probability that $X = 1$ was calculated to be .4, so in 10,000 trials there should have been 4000 occurrences of $X = 1$; the actual frequency of 3961 is within 1% of the nominal value. Similar results hold for all other x values, so the results seem reasonable. □

cumulative distribution function

An alternative way to specify a probability distribution is by way of the **cumulative distribution function,** usually abbreviated cdf and denoted by F. As the name indicates, the individual probabilities are cumulated—added—to yield the cdf. For instance, the cdf of the random variable X in Example 5.4 is

x	0	1	2	3
$F_X(x)$	1/8	4/8	7/8	8/8

Note that $4/8 = 1/8 + 3/8$, and $7/8 = 1/8 + 3/8 + 3/8$.

Cumulative Distribution Function (cdf)

$$F_Y(y) = P(Y \leq y) = \sum_{y' \leq y} f_Y(y')$$

Example 5.7 Calculate the cdf for the random variable X of Example 5.6.

Solution The probability distribution $f_X(x)$ is shown below; the cdf $F_X(x)$ is calculated by addition.

x	1	2	3	4	5
$f_X(x)$	.4000	.2400	.1440	.0864	.1296
$F_X(x)$	.4000	.6400	.7840	.8704	1.0000

The cdf is particularly convenient for use in large tables of probabilities. If a random variable Y had possible values 0, 1, ..., 100, to find the probability that Y was between 20 and 80, both inclusive, one would have to add 61 $f_Y(y)$ values. With a cdf table, one subtraction will do the job. The value $F_Y(80)$ is the sum of probabilities for $y = 0, 1, \ldots, 80$; to get the sum of the probabilities for $y = 20, 21, \ldots, 80$, all we have to do is subtract $F_Y(19)$:

$$P(20 \leq Y \leq 80) = F_Y(80) - F_Y(19)$$

The easiest way to calculate a desired probability from a cdf table is to draw a *probability histogram*. This is just like the data histograms described in Chapter 2, except that the areas of the rectangles represent probabilities, not frequencies. One can mark the rectangles included in the desired probability. It's almost always easy to determine how to use cdf tables given the histogram.

Example 5.8 Use the cdf calculated in Example 5.7 to find

(a) $P(X \leq 2)$
(b) $P(3 \leq X \leq 4)$
(c) $P(X \geq 4)$

Solution A probability histogram is shown in Figure 5.2. The letters in the rectangles indicate the rectangles to be included in each of the three desired probabilities.

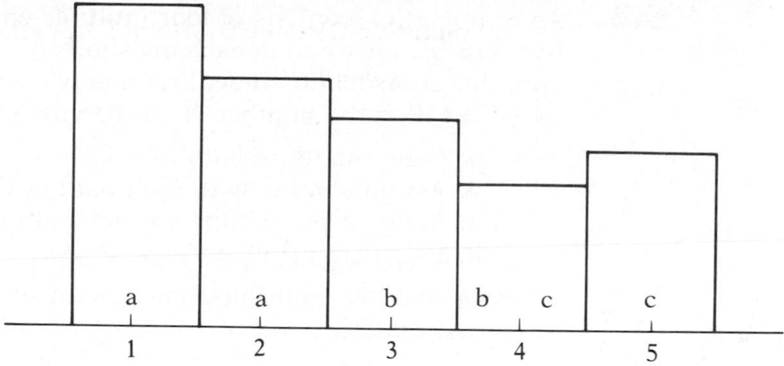

Figure 5.2 Probability histogram for Example 5.8

(a) $P(X \leq 2) = F_X(2) = .6400$
(b) $P(3 \leq X \leq 4) = F_X(4) - F_X(2) = .2304$
(c) $P(X \geq 4) = 1 - F_X(3) = .2160$ (Note that the total area of all the rectangles must be 1.)

In this example, it's also easy to calculate the same probabilities by adding up $f_X(x)$ values; the answers are, of course, the same. ☐

Exercises for Sections 5.1 and 5.2

5.1. A population consists of ten subjects, of whom six are suitable for a particular experiment. A random sample of three subjects is taken, without replacement. Let X = the number of suitable subjects in the sample.

a. Write out a sample space S.

b. Assign probabilities to each outcome. You may find it useful to construct a probability tree.

c. Assign a value x to each outcome.

d. Calculate $f_X(x)$.

5.2. Assume the same population as in Exercise 5.1. This time subjects are randomly selected, without replacement, until the first suitable subject is found. Let Y = the number of subjects sampled.

a. Write out a sample space for the experiment.

b. Assign probabilities to each outcome.

c. Assign y values to each outcome.

d. Find $f_Y(y)$.

5.3. An examination consists of four multiple-choice questions, each having five possible answers; in each question only one answer is correct. Suppose that a particular student randomly guesses answers to all the questions. Let W = the number of wrong answers.

a. Construct a probability tree.
b. Assign a value w to each path of the tree.
c. Assign a probability to each path of the tree.
d. Calculate $f_W(w)$.

5.4. Suppose that the multiple-choice exam of Exercise 5.3 consists of six questions, not four.

a. List the possible outcomes and assign a w value to each.
b. Assign probabilities to all outcomes.
c. Calculate $f_W(w)$.
d. Compare the probability distributions $f_W(w)$ found in this exercise and in Exercise 5.3. Can you see a pattern?

5.5. A rat runs a certain T maze, which is set so that if the rat turns left, it is rewarded with probability .8, and if the rat turns right, it is rewarded with probability .3. On the first try, the rat has equal probabilities of turning either left or right. On all succeeding trials, the rat will turn in the direction it did on the preceding trial with probability .9 if it was rewarded. If the previous trial was not rewarded, the rat will turn in the opposite direction with probability .6. The rat makes two trials. Let Y = the number of left turns.

a. Construct a probability tree for this experiment. The successive branches should reflect the first turn, whether or not the first turn was rewarded, the second turn, and the second reward.
b. Assign probabilities and y values to each path.
c. Find $f_Y(y)$.

5.6. If, in Exercise 5.5, X = the number of rewards received, what is $f_X(x)$?

5.7. In an experiment, a target stimulus is flashed directly in front of a subject. The stimulus is one of the digits 2 through 9; the digit is randomly selected. With probability .5, another digit is simultaneously flashed to the left of the subject, but within the subject's field of vision. The two digits are chosen independently, and both may be the same or different. Further, with probability .5 and independently of whether or not a left digit is flashed, a randomly chosen digit is flashed to the right of the subject. Let Y = the number of digits flashed.

a. Construct a probability tree for this experiment.
b. Calculate $f_Y(y)$.

5.8. Refer to Exercise 5.7. Define Z to be the number of digits flashed that are in the subject's memory set. If the digits 3 and 7 are in the subject's memory set, what is $f_Z(z)$?

5.3
Expected Value

The purpose of defining random variables was to make connections between the descriptive ideas of Chapter 2 and the probability ideas of Chapter 4. One particularly central idea in Chapter 2 was the concept of a mean of a data set. In this section we will develop the concept of the mean, or expected value, of a random variable.

Consider, for example, the random variable X of Examples 5.5 and 5.6. It has possible values 1, 2, 3, 4, and 5 with distinctly unequal probabilities. It seems reasonable that, in calculating the "average value" of X, we should take into account not only the possible values, but also their respective probabilities. Because the values 1 and 2 have relatively large probabilities and the values 4 and 5 relatively small ones, the average, or expected, value should be something smaller than 3. This can be achieved by *weighting* the possible values of the random variable by their respective probabilities. The expected value of any random variable is defined as the probability-weighted average of the values of the random variable.

Expected Value (Mean) of a Random Variable

$$E(Y) = \sum y f_Y(y)$$

In words, to find the expected value of any random variable Y, take each possible value y, multiply by the associated probability $f_Y(y)$, and sum.

Example 5.9 Find the expected value of the random variable X in Example 5.6.

Solution

$$\begin{aligned} E(X) &= \sum x f_X(x) \\ &= 1(.4000) + 2(.2400) + 3(.1440) + 4(.0864) + 5(.1296) \\ &= 2.3056 \end{aligned}$$

Note that the values 1 and 2 are relatively heavily weighted, so $E(X)$ is less than 3, the simple (unweighted) average of the possible values of X. □

The expected value of a random variable Y is often denoted by μ_Y. The Greek letter mu (μ) was also used to denote a population mean in Chapter 2. The mean (expected value) of a random variable is a generalization of the idea of population means. In the special case where the random variable Y is the result of one random draw from a population, $E(Y) = \mu_{\text{pop}}$. Of course, there are other random variables that cannot reasonably be thought of as the results of a random draw from a population, so the concept of expected value is more general than that of population mean.

Example 5.10

Assume that a certain population consists of the following values and frequencies.

Values	10	20	30	40	
Frequencies	400	300	200	100	($N = 1000$)

Define Y to be the value obtained by a random sample of size 1 from this population.

(a) Find $f_Y(y)$ and $E(Y)$.
(b) Show that $\mu_{\text{pop}} = E(Y)$.

Solution

(a) The possible values y of the random variable Y are 10, 20, 30, and 40. The respective probabilities are $f_Y(10) = 400/1000 = .4$, $f_Y(20) = .3$, $f_Y(30) = .2$, and $f_Y(40) = .1$. Therefore,

$$E(Y) = 10(.4) + 20(.3) + 30(.2) + 40(.1) = 2.0$$

(b) The population mean μ_{pop} may be calculated as

$$\mu_{\text{pop}} = \frac{10(400) + 20(300) + 30(200) + 40(100)}{1000}$$
$$= 2.0 = E(Y) = \mu_Y$$

The only difference between parts (a) and (b) is the irrelevant difference between weighting (for instance) the value 10 by .4 and by 400/1000. □

The word "mean" is used both for data and for random variables. In solving problems, it's important to ask first: Which kind of mean are we talking about?

How can one interpret the numerical value of an expected value? We already have one partial interpretation; the expected value of a random variable is a generalization of a population mean. A more general interpretation is that the expected value of a random variable is the **long-run average** of the random variable. Imagine repeating the experiment that yields a Y value many times, and averaging the observed

long-run average

values y. The long-run average Y value will be $E(Y)$, because in the long run the proportion (relative frequency) of a particular y value will be exactly its probability $f_Y(y)$.

Example 5.11 In Example 5.9, $E(X)$ was found to be 2.3056. What is the meaning of this number?

Solution If the experiment is repeated a very large number of times, the average value of X will be slightly over 2.3. Note that the "expected" value of X is not even a possible value of X. □

5.4 Variance and Standard Deviation of a Random Variable

The expected value of a random variable measures the long-run average value of the random variable. It is equally important to measure the variability of the random variable around this value. In Chapter 2 the concepts of variance and standard deviation were applied to data. In this section we will extend the concepts to random variables.

Recall that the variance of a data variable is the average squared error of the data, and that the standard deviation is the square root of the variance. Essentially the same definitions apply to random variables, with the appropriate interpretation of "average." In Section 5.3 we pointed out the necessity of weighting the possible values of the random variable by their respective probabilities in computing the mean (expected value). The same probability-weighting should be applied to the squared errors in computing the average squared error (variance).

Variance σ_Y^2 of a Random Variable Y

$$\mathrm{Var}(Y) = \sigma_Y^2 = \sum (y - \mu_Y)^2 f_Y(y)$$

Standard Deviation of a Random Variable Y

$$\text{St. dev. } (Y) = \sigma_Y = \sqrt{\mathrm{Var}(Y)}$$

Example 5.12 Calculate the variance and standard deviation of the random variable X in Example 5.9.

Solution In Example 5.9, we found that $E(X) = \mu_X = 2.3056$. Therefore,

$$\begin{aligned}\text{Var}(X) &= (1 - 2.3056)^2(.4000) + (2 - 2.3056)^2(.2400) \\ &\quad + \cdots + (5 - 2.3056)^2(.1296) \\ &= 1.9626\end{aligned}$$

Taking a square root, we find that

$$\sigma_X = \text{St. dev.}\,(X) = \sqrt{1.9626} = 1.4009$$

□

The interpretation of the standard deviation of a random variable is essentially the same as that of a data standard deviation. In Chapter 2 we used an Empirical Rule which estimated the proportion of data values falling within 1 standard deviation of the mean to be about .68, and the proportion falling within 2 standard deviations to be about .95. In addition, we had the mathematically guaranteed Chebyshev's Inequality, which said that at least 75% of the data fell within 2 standard deviations of the mean, and at least 88.9% fell within 3 standard deviations of the mean. The Empirical Rule and Chebyshev's Inequality extend to random variables.

Empirical Rule and Chebyshev's Inequality for Random Variables

If the probability histogram of a random variable is well-behaved (not very skewed, outlier-prone, or "lumpy"), then the probability that the random variable will fall within 1 standard deviation of the mean (expected value) is about .68, and the probability that it will fall within 2 standard deviations is about .95. Regardless of the shape of the histogram, it is always true that the probability that a value will fall within k standard deviations of the mean is at least $1 - 1/k^2$.

The only difference between the data Empirical Rule and the random variable Empirical Rule is that one refers to proportions and the other to probabilities. The same holds for Chebyshev's Inequality.

Example 5.13 What is the actual probability that the random variable X of Example 5.9 falls within 1 standard deviation of the mean?

Solution The mean, from Example 5.9, is 2.3056; the standard deviation, from Example 5.12, is 1.4009. A value within 1 standard deviation of the mean must be between 2.3056 − 1.4009 and 2.3056 + 1.4009; that is, between 0.9047 and 3.7065. The actual probability that X falls between those values is the probability that X will equal 1, 2, or 3, which is .4000 + .2400 + .1440 = .7840. Note that the probability histogram of X is not mound-shaped, but rather both skewed and "lumpy" (having few values). □

Computation of the variance of a random variable can be tedious if, as in Example 5.12, one has to subtract a many-decimal expected value in each term. Just as there is a short-cut formula for a data variance, there is a short-cut formula for a random variable variance.

Short-Cut Formula for a Variance

$$\mathrm{Var}(Y) = \left[\sum y^2 f_Y(y)\right] - (\mu)^2$$

Example 5.14 Use the short-cut formula to recompute Var(X), from Example 5.9.

Solution

$$\begin{aligned}\sum x^2 f_X(x) - (\mu)^2 \\ &= (1)^2(.4000) + (2)^2(.2400) + \cdots + (5)^2(.1296) - (2.3056)^2 \\ &= 1.9626\end{aligned}$$

as in Example 5.12. □

The mean (expected value) of a random variable is a generalization of the idea of a (data) population mean; similarly, the variance of a random variable is a generalization of the idea of a population variance. If the random variable Y is the result of a single random draw from a population, then Var(Y) = σ^2_{pop}.

Example 5.15 Show that for Example 5.10 both the population variance and Var(Y) are equal to 100.0.

Solution The population mean was found to be 20 in Example 5.10. Thus the population variance is

$$\sigma^2_{\text{pop}} = \frac{(10-20)^2(400) + \cdots + (40-20)^2(100)}{1000}$$
$$= 100.0$$

The variance of Y is

$$(10-20)^2(.4) + \cdots + (40-20)^2(.1) = 100.0$$

also. □

It is possible to define the expected value of a mathematical function of a random variable, such as X^2 or log (Y). For example, suppose that X has the following distribution:

x	-2	-1	0	1	2
$f_X(x)$	.1	.2	.4	.2	.1

To find the expected value of X^2, we could define the new random variable $W = X^2$. W has possible values 0 (when X is 0), 1 (when X is either -1 or 1), and 4 (when X is either -2 or 2). To find the W probabilities, we can add the appropriate X probabilities. For instance,

$$f_W(4) = P(X = -2 \text{ or } X = 2) = .1 + .1 = .2$$

by the addition principle. Similarly, $f_W(0) = .4$ and $p_W(1) = .2 + .2 = .4$. Remembering that $W = X^2$, we get

$$E(X^2) = 0(.4) + 1(.4) + 2(.2) = 0.8$$

A simpler way to do the same calculations is to define

$$E(X^2) = \sum x^2 f_X(x)$$

In words, take each possible x value, square it, and weight it by the x probability. In the example,

$$E(X^2) = (-2)^2(.1) + (-1)^2(.2) + (0)^2(.4) + (1)^2(.2) + (2)^2(.1) = 0.8$$

Of course, this is merely a different way of doing the same arithmetic, so it yields the same answer.

> **Expected Value of a Function $g(X)$**
>
> $$E(g(X)) = \sum g(x) f_X(x)$$
>
> where the sum is taken over all possible values x.
> In particular,
>
> $$\text{Var}(X) = \sum (x - \mu_X)^2 f_X(x) = E(X - \mu_X)^2$$

Example 5.16 A measure of the skewness of a random variable Y can be based on the expected value of $(Y - \mu_Y)^3$. Find this expected value if the distribution of Y is

y	1	2	3	4	5	6	7
$f_Y(y)$	.05	.10	.20	.30	.20	.10	.05

Solution The probability distribution is symmetric around $y = 4$, so $\mu_Y = 4$. Because of the symmetry, the skewness measure should equal 0 for this distribution.

$$\begin{aligned} E(Y - \mu_Y)^3 &= \sum (y - \mu_Y)^3 f_Y(y) \\ &= (1 - 4)^3(.05) + (2 - 4)^3(.10) + \cdots + (7 - 4)^3(.05) \\ &= 0.00 \end{aligned}$$

as expected. □

Exercises for Sections 5.3 and 5.4

5.9. Find the expected value and variance of X in Exercise 5.1.

5.10. Find the expected value and standard deviation of Y in Exercise 5.2.

5.11. Find the mean number of left turns and the mean number of rewarded trials for Exercise 5.4.

5.12. The distribution of grades on a ten-point quiz for a certain class (which may be regarded as a population) was as follows:

Grade	0	1	2	3	4	5	6	7	8	9	10
Percent	1	2	3	4	5	6	9	15	20	25	10

Let the random variable X be the grade of a randomly chosen student. Find the mean and standard deviation.

5.13. Assume that two students are randomly chosen from the class in Exercise 5.12, with replacement. Let T = the total of the two students' grades.

a. Use a probability table or tree to calculate $f_T(t)$.
b. Calculate the expected value and variance of T.

5.14. Compare the expected value of T in Exercise 5.13 to the expected value of X in Exercise 5.12. Do the same for the respective variances. What appears to be the relation?

5.15. Calculate the expected value and variance of the random variable in Exercise 15.8.

5.5 Continuous Random Variables

The random variables that have been considered so far in this chapter have all been *discrete* ones, involving only distinct, separate values such as 0, 1, 2, ... Many random variables can, in principle, take on an entire interval, or continuum, of possible values. For example, suppose that a laboratory rat, after having been deprived of water for a 24-hour period, is allowed to drink as much as desired, at the cost of enduring a loud noise which continues as long as it drinks. The random variable X = the amount of water drunk could be measured to many possible degrees of accuracy—cubic centimeters (cc's), or tenths of cc's, or hundredths of cc's, or whatever. There is no logical limit to the accuracy of the measurement, though there is a practical limit imposed by the precision of the measuring device used. In principle, X could come out to be any positive number, measured to infinitely many decimal places. Random variables that can take on a whole range (or interval or continuum) of nonseparated values are called *continuous* random variables. In this section we will set up the basic language needed to deal with such random variables. The discussion has to involve some basic calculus. Those who don't know calculus can skim this section without too much loss.

To see what the issues are, imagine drawing a probability histogram for X = the amount of water drunk by the rat, as measured to the nearest cc; it might look something like Figure 5.3(a). Now imagine drawing a histogram for X as measured to the nearest half a cc; it might look like Figure 5.3(b). In each figure the probability that X will be between 15 and 25 cc's is the shaded areas of all rectangles between 15 and 25. When X is measured to the nearest thousandth of a cc, there are so many super-skinny rectangles that their tops effectively form a continuous curve, as shown in Figure 5.3(c). The probability that X will be

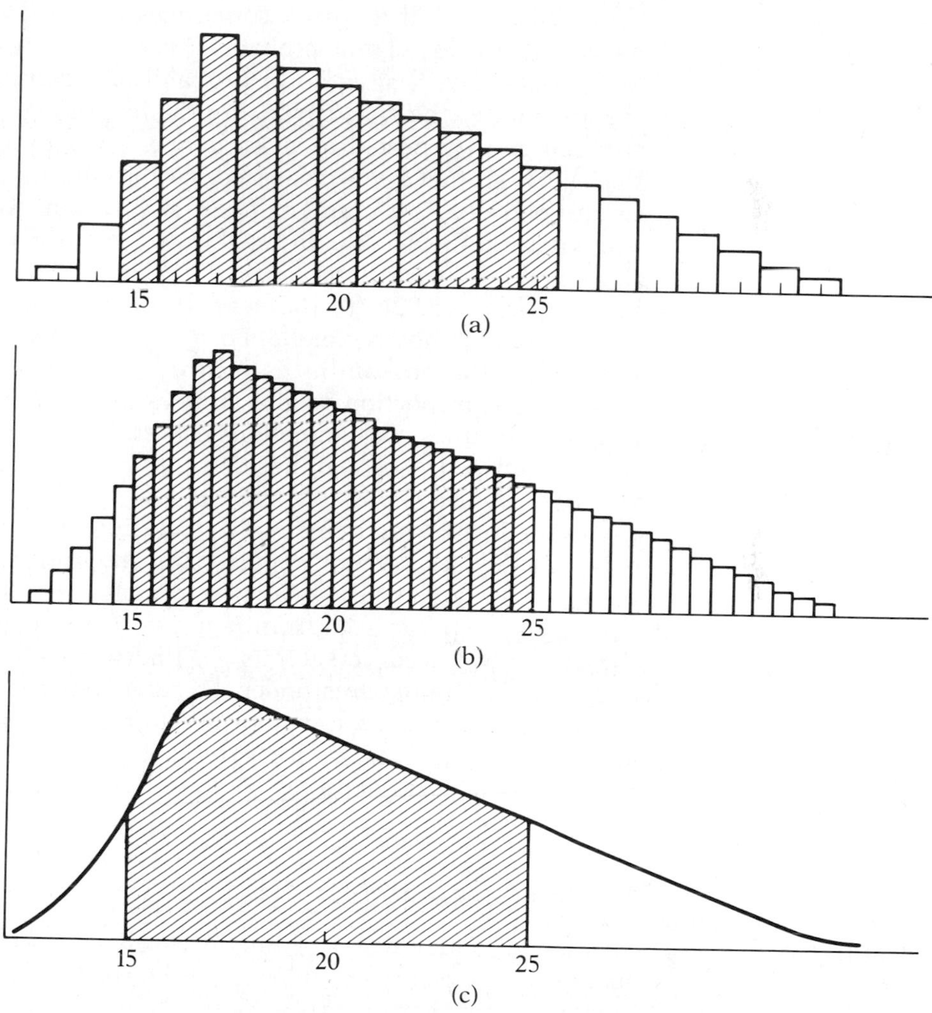

Figure 5.3 (a) X to the nearest cc; (b) X to the nearest half cc; (c) X to very high accuracy of measurement

between 15 and 25 effectively becomes the area underneath the curve. The curve is still labeled $f_X(x)$, but now, instead of calculating probabilities by plugging into the $f_X(x)$ formula, we must find areas under the $f_X(x)$ curve.

When X is a continuous random variable, we can't interpret $f_X(x)$ directly as a probability, because probability is now measured as the area under a curve. The probability that X will be 15.925731... is so small that it must be taken to be 0, but $f_X(15.925731...)$ is not 0. When dealing with a continuous random variable, how shall we interpret $f_X(x)$?

probability density function

Mathematically, $f_X(x)$ is called the **probability density function** of X; the probability is "smeared" over a whole range of possible values, sometimes thickly (high density), sometimes thinly (low density), and $f_X(x)$ measures how densely the probability is smeared at the particular x value. Mathematically, $f_X(x)$ is not a probability; probability is calculated as the area under the $f_X(x)$ curve. In practice, though, the probability density can be interpreted as being *proportional* to a probability. Assume that X has in fact been measured very accurately, to the nearest thousandth of a cc. As a mathematical idealization, we take X to be a continuous random variable. In this idealization, the probability of any particular value is 0. But in practice, the probability of obtaining a particular value such as 15.926 (to the nearest thousandth of a cc) is not 0; it is the area of a skinny rectangle. For those x values with a relatively large $f_X(x)$ value, the probability is *relatively* high; in fact the probability is basically in proportion to the $f_X(x)$ value. In dealing with continuous random variables, we can no longer literally interpret f as a probability, but in practice the probability density f still indicates the relative probabilities of various possible values.

The formal mathematical calculation of probabilities as areas under a curve uses calculus, specifically integral calculus. In fact, those readers who have studied calculus probably could see an integral sign coming as we progressed from Figure 5.3(a) to Figure 5.3(c). In calculus the integral of a curve such as $f_X(x)$ between two numbers such as 15 and 25 is exactly the area under the curve between those numbers. There are many techniques for evaluating integrals. In this book, we will evaluate almost all integrals (areas) by looking them up in tables, but more mathematical texts definitely require fluency in calculus.

Example 5.17

A professor who had taught Psych 1 for many years always gave lists of 100 nonsense syllables to all the students in the class. At the next class meeting, the students were told to write down all the syllables they could remember. (They had been forewarned that this test would be part of their grade, which guaranteed some fervent attention to the task at hand.) Define Y = the proportion of correct recalls by a randomly selected student. As an idealization, the professor treated Y as a continuous random variable, with the probability density function

$$f_Y(y) = 30y(1-y)^4, \quad \text{for } y \text{ between 0 and 1}$$

(a) What are the actually possible values of Y?

(b) Plug various values of y into $f_Y(y)$, and sketch the curve. Which values of Y are relatively highly probable?

(c) If you know calculus, use it to calculate the probability that Y will be less than .4. (Note that Y must be greater than 0.)

Solution

(a) Actually, Y can only be .00, .01, ..., 1.00.

(b) A table of a few $f_Y(y)$ values follows:

y	.1	.2	.3	.4	.5	.6
$f_Y(y)$	1.97	2.46	2.16	1.56	0.94	0.46

Note that values around .2 have relatively high probability.

(c)

$$P(Y < .4) = \int_0^{.4} 30y(1-y)^4\,dy$$

$$= \left[30\left(\frac{y^2}{2} - \frac{4y^3}{3} + \frac{6y^4}{4} - \frac{4y^5}{5} + \frac{y^6}{6}\right)\right]_{y=0}^{y=.4}$$

$$= .7667$$

In Sections 5.3 and 5.4 we used the concepts of mean (expected value) and standard deviation of a (discrete) random variable in measuring the long-run average value and the probable degree of short-run deviation around that value. These concepts can be extended to the case of continuous random variables. In general, any discrete-variable definition involving a sum has a continuous-variable analogue involving an integral. In the discrete case the expected value is calculated by multiplying values times probabilities and summing; in the continuous case it is calculated by multiplying values times probability density values and integrating. The same analogy works for variances and standard deviations, including the short-cut formula for calculating the variance.

Expected Value, Variance, and Standard Deviation for a Continuous Random Variable Y

$$E(Y) = \mu_Y = \int_{\text{all } y} y f_Y(y)\, dy$$

$$\text{Var}(Y) = \sigma_Y^2 = \int_{\text{all } y} (y - \mu_Y)^2 f_Y(y)\, dy$$

$$\text{Var}(Y) = \left[\int_{\text{all } y} y^2 f_Y(y)\, dy\right] - \mu_Y^2$$

Of course, there are technical calculus problems involved in actually finding the expected value and standard deviation of a continuous random variable. However, there are no conceptual problems at all. The expected value still measures the long-run average value of the random variable, and the standard deviation still assesses the degree of variability around that value. Rules of thumb such as the Empirical Rule work as well for continuous random variables as for discrete ones. In general, the issues involved with continuous random variables are merely technical; no new concepts are involved.

Example 5.18 Find the expected value and standard deviation of the random variable Y of Example 5.17. It can be proved using calculus that

$$\int_0^1 y^a(1 - y)^b\, dy = \frac{a!b!}{(a + b + 1)!}$$

Solution

$$\begin{aligned}
\mu_Y &= \int_0^1 y30y(1 - y)^4\, dy \\
&= 30\frac{2!4!}{7!} = .2857 \\
\sigma_Y^2 &= \int_0^1 y^2 30y(1 - y)^4\, dy - \mu_Y^2 \\
&= 30\frac{3!4!}{8!} - (.2857)^2 = .03444 \\
\sigma &= \sqrt{.03444} = .1856
\end{aligned}$$

□

cumulative distribution function The **cumulative distribution function** (cdf) of a random variable was defined in Section 5.2 as $F_Y(y) = P(Y \leq y)$. Exactly the same definition holds for continuous random variables. The cdf is very useful with continuous random variables because a cdf table can eliminate the need for doing calculus to find probabilities.

$$P(a < Y < b) = \int_a^b f_Y(y)\,dy = F_Y(b) - F_Y(a)$$

Thus, given a table of F_Y values, instead of performing an integral to find probabilities, all that we have to do is a subtraction. In fact, it's even easier to use a cdf table in the continuous case. The probability that a continuous random variable will exactly equal a particular number is 0 (as a mathematical idealization). Therefore, there's no need to worry about the distinction between $<$ and $\leq$.

Example 5.19 It can be shown that a table of cdf values for Y in Example 5.17 is

y	.00	.10	.20	.30	.40	.50	.60
$F_Y(y)$	.0000	.1143	.3446	.5798	.7667	.8906	.9590

Find the probability that Y will be less than .40.

Solution $P(Y \leq .40)$ is shown as $F_Y(.40) = .7667$. There's no need to worry about the chance that Y will exactly equal .4000..., because that probability is 0 for a continuous random variable. Thus $P(Y < .40) = .7667$. □

Exercises for Section 5.5

5.16. If a number is "drawn at random from the interval 0 to 1," the result can be any (real) number in that interval. If the number is truly drawn at random, the probability distribution is *uniform:*

$$f_U(u) = 1, \quad \text{for } 0 < u < 1$$

a. Assume that a number U is drawn at random from the interval 0 to 1. Use easy calculus to find the probability that U will be larger than .7.

b. Assume that U_1, U_2, and U_3 are independently and randomly drawn from the interval 0 to 1. Find the probability that at least two of the U_j numbers are larger than .7. (You may want to use a probability tree.)

5.17. For the random variable U defined in Exercise 5.16, calculate the expected value (mean) $E(U)$. Does your result make sense?

5.18. Find the standard deviation for the random variable U defined in Exercise 5.16.

5.19. A random variable Y is said to have a *beta* distribution if

$$f_Y(y) = \frac{(a + b - 1)!}{a!b!} y^a(1 - y)^b, \quad \text{for } 0 < y < 1$$

Assume that $a = 9$ and $b = 1$.

a. Find the probability that Y will be between .7 and .9.
b. Find the probability that Y will be less than .6.

5.20. Find the mean and standard deviation of Y in Exercise 5.19.

5.21. Refer to Exercises 5.19 and 5.20.

a. Find the probability that Y will be within 2 standard deviations of its mean.

b. How does the probability that you found in part a compare to the Empirical Rule approximation? Is it consistent with Chebyshev's Inequality?

c. The beta density with $a = 9$ and $b = 1$ is very skewed to the left. Does this skewness affect the Empirical Rule approximation severely?

5.22. When events (such as a series of stimuli) arrive "at random over time," it can be proved that the time T between successive events is a random variable with the *exponential* density

$$f_T(t) = (1/\mu)e^{-t/\mu}, \quad \text{for } t > 0$$

where $e = 2.718281828\ldots$ is the base of natural logarithms.

Suppose that stimuli arrive at random over time at a particular neuron with $\mu = .05$ second per stimulus.

a. Find the probability that the time to the next arrival of a stimulus is between .02 and .06 second. Recall from calculus that

$$\int_a^b ce^{-ct}\, dt = e^{-ca} - e^{-cb}$$

b. Find the probability that T, the time to the next stimulus, is larger than .1.

5.23. Refer to Exercise 5.22.

a. Find μ_T.
b. Find σ_T^2.

5.24. Referring to Exercises 5.22 and 5.23, find the probability that T will be within 1 standard deviation of its expected value (mean). How does this probability compare to the Empirical Rule approximation? (The exponential density is skewed to the right.)

5.25. Find the cumulative distribution function (cdf) of T in Exercise 5.22, and use it to solve the problem.

5.6 Joint Probability Distributions

Up to now in this chapter we have been considering only a single random variable, just as we considered summarizing data on a single variable in Chapter 2. Now we turn to considering probabilistic relations between variables, as we did for data in Chapter 3. In this section we will be restating many of the basic probability principles of Chapter 4 in random variable language.

The central idea in this section is the joint probability distribution of two random variables, say X and Y. For instance, suppose that we allow subjects in an experiment to select the number of practice sessions to take in learning to track a moving target on a television screen, and then measure how well they do in keeping a cursor within the target. Specifically, define X = the number of practice sessions chosen and Y = the number of seconds (rounded to the nearest 10) the cursor is within the target. Assume that the probabilities in Table 5.1 apply.

		y					
		10	20	30	40	50	60
x	1	.01	.03	.05	.06	.03	.02
	2	.02	.04	.06	.08	.06	.04
	3	.01	.01	.03	.07	.10	.08
	4	.01	.01	.02	.04	.05	.07

Table 5.1

This table gives the joint probability distribution in the sense that it specifies the probability of obtaining any particular x value *and* any particular y value for a randomly chosen subject.

Joint Probability Distribution of Two Random Variables

If X and Y are two discrete random variables, their joint probability distribution is

$$f_{XY}(x, y) = P(X = x \text{ and } Y = y)$$

and is defined for all possible combinations of values x and y.

Example 5.20

Suppose that the principal of a large elementary school records both U = the number of older siblings that a child has and V = the number of disciplinary actions taken concerning that child in a given year. Assume the joint probabilities in Table 5.2 apply.

		v		
		0	1	2
u	0	.32	.10	.08
	1	.18	.06	.04
	2	.10	.02	.01
	3	.07	.01	.01

Table 5.2

Find and interpret $f_{UV}(2, 0)$.

Solution

$f_{UV}(2, 0)$ is shown in the table as .10. If a child is chosen at random, the probability that that child both has two older siblings and has been subject to no disciplinary actions is .10. □

marginal probability distributions

Many of the basic probability principles can be applied to joint probability distributions. The addition principle can be defined to obtain **marginal probability distributions.** In the example of tracking a moving target, the obvious way to find the probabilities for Y = the number of seconds the cursor is within the target is to add the appropriate probabilities, as in Table 5.3.

The addition is justified because, for instance, the event $Y = 10$ can be broken up into the mutually exclusive cases $(X = 1, Y = 10)$, $(X = 2, Y = 10)$, $(X = 3, Y = 10)$, and $(X = 4, Y = 10)$. By the addition principle, the probability that $Y = 10$ is the sum of the probabilities of the cases.

		y					
		10	20	30	40	50	60
x	1	.01	.03	.05	.06	.03	.02
	2	.02	.04	.06	.08	.06	.04
	3	.01	.01	.03	.07	.10	.08
	4	.01	.01	.02	.04	.05	.07
	$f_Y(y)$	.05	.09	.16	.25	.24	.21

Table 5.3

Marginal Probability Distribution

If X and Y are discrete random variables,

$$f_X(x) = \sum_{\text{all } y} f_{XY}(x, y)$$

$$f_Y(y) = \sum_{\text{all } x} f_{XY}(x, y)$$

The idea of marginal probability is very similar to the idea of marginal frequency defined in Chapter 3, where we added frequencies instead of probabilities. In both cases, the name comes from where we put the "marginals"—at the margins of the table.

Example 5.21 Find the marginal probability distributions of both random variables in Example 5.20.

Solution The marginal probabilities may be found by addition, and recorded in the margins of the joint probability table (Table 5.4). □

		v			
		0	1	2	Marginal
u	0	.32	.10	.08	.50
	1	.18	.06	.04	.28
	2	.10	.02	.01	.13
	3	.07	.01	.01	.09
	Marginal	.67	.19	.14	

Table 5.4

We may also adapt the idea of conditional probability defined in Chapter 4 to obtain the conditional probability distribution of one random variable given the value of another random variable. Recall that the conditional probability of an event B given another event A is

$$P(B|A) = P(A \text{ and } B)/P(A)$$

Conditional Probability Distribution of *Y* Given X

If X and Y are discrete random variables,

$$f_{Y|X}(y|x) = \frac{f_{XY}(x, y)}{f_X(x)}$$

In the example with X = the number of practice sessions and Y = the number of seconds the cursor is within the target, we can calculate $f_{Y|X}(10|1) = f_{XY}(1, 10)/f_X(1) = .01/.20 = .05$. In fact, we can calculate an entire table of conditional probabilities, Table 5.5.

		y					
		10	20	30	40	50	60
x	1	.05	.15	.25	.30	.15	.10
	2	.0667	.1333	.2000	.2667	.2000	.1333
	3	.0333	.0333	.1000	.2333	.3333	.2667
	4	.05	.05	.10	.20	.25	.35

Table 5.5

Although this table may look like a joint probability table, it gives conditional probabilities. The probabilities in a conditional probability table add to 1 across each row (or down each column, if the "given" is the column variable), rather than adding to 1 over the whole table.

Example 5.22

Find a table giving $f_{U|V}(u|v)$for the probabilities of Example 5.20.

Solution

For instance,

$$f_{U|V}(0|1) = \frac{f_{UV}(0, 1)}{f_V(1)} = \frac{.10}{19} = .5263$$

The complete conditional probability table is shown in Table 5.6.

		v		
		0	1	2
u	0	.4776	.5263	.5714
	1	.2687	.3684	.2857
	2	.1493	.0526	.0714
	3	.1045	.0526	.0714
	Total	1.0001	.9999	.9999

Table 5.6

Note that the *column* totals add to 1, within roundoff error, because we are conditioning on the column variable V. □

An important probability principle that generalizes to random variables is statistical independence. Events A and B are statistically independent if $P(B|A) = P(B)$ or, equivalently, if $P(A \text{ and } B) = P(A)P(B)$. Again, we may take B to be the event $Y = y$ and A to be the event $X = x$. To assure complete independence of the random variables X and Y, we want the definition to hold for all possible values x and y.

Independence of Random Variables

Let X and Y be discrete random variables. They are statistically independent if and only if

1. $f_{Y|X}(y|x) = f_Y(y)$ for all x and y
2. $f_{XY}(x, y) = f_X(x)f_Y(y)$ for all x and y

As in Chapter 4, statistical independence is often a natural assumption based on the nature of the random variables. For example, suppose that 20% of all subjects obtain a score of 0 on a perception test, 30% score 1, and 50% score 2. If we take two subjects at random and define X = the score of subject 1 and Y = the score of subject 2, we may reasonably assume that X and Y are independent. There is no reason why the score of subject 1 should alter the probabilities for subject 2. Assuming independence, we may construct a table of joint probabilities by multiplying marginal probabilities, as in part 2 of the definition of independence. In this example, the table would be Table 5.7.

		y			
		0	1	2	Marginal
	0	.04	.06	.10	.20
x	1	.06	.09	.15	.30
	2	.10	.15	.25	.50
	Marginal	.20	.30	.50	

Table 5.7

Example 5.23 Are U and V independent in Examples 5.20–5.22?

Solution No. One way to see the dependence is to note that the conditional probability distributions shown in Example 5.22 are not the same as the marginal probabilities shown in Example 5.21. Alternatively, we can calculate the product of marginal probabilities and compare the result to the joint probability. For instance, $f_{UV}(0, 1) = .10$ but $f_U(0) f_V(1) = (.50)(.19) = .095$. Therefore, U and V are not independent. □

The degree of dependence can be assessed by the measures developed in Chapter 3 for frequencies. These measures apply equally well to probabilities. For example, the γ measure can be applied to assess the degree of monotone (generally increasing or generally decreasing) relation between two random variables. If the variables are independent, the measure will equal 0.

Example 5.24 The value of Kendall's τ for the probabilities in Example 5.20 is $-.0484$, and $\gamma = -.154$. What does this indicate about the degree of dependence between U and V?

Solution Both values are negative but small in magnitude. Thus there is a weak negative relation. As V increases, U tends to decrease slightly. In Example 5.22, we computed conditional probabilities of U given V. Note that there is a weak pattern in the conditional probabilities. As V increases, the U probabilities shift toward the smaller values. □

All the concepts in this section can be applied to continuous random variables as well as discrete ones. If X = the time interval between successive stimuli and Y = the reaction time to the second stimulus, both random variables are most naturally treated as continuous. With continuous random variables, $f_{XY}(x, y)$ is a **joint density function.** Probabilities are found by multiple integration:

joint density function

$$P(a < X < b \text{ and } c < Y < d) = \int_a^b \int_c^d f_{XY}(x, y)\, dy\, dx$$

Such multiple integrals are beyond the mathematical level assumed in this book, so we shall not go into detail. The ideas of marginal probability, conditional distribution, and independence all go through with sums being replaced by integrals. For details, see Mood, Graybill, and Boes (1974).

5.7 Covariance and Correlation of Random Variables

In Section 5.6 we briefly considered some of the measures of statistical relation as applied to joint probability distributions. In this section we discuss how the most important measure of statistical relation, correlation, applies to random variables and probability distributions.

For example, suppose that we measure X = the age (to the nearest year, coarsely) of our child subjects and Y = the number of correct answers to a sorting task. Suppose that the probabilities in Table 5.8 apply:

		y			
		0	1	2	Marginal
	2	.15	.10	.05	.30
x	3	.10	.20	.10	.40
	4	.05	.10	.15	.30
	Marginal	.30	.40	.30	

Table 5.8

The basic ideas of Sections 5.3 and 5.4 yield $\mu_X = 3.0$, $\mu_Y = 1.0$, and $\sigma_X^2 = \sigma_Y^2 = 0.60$. There appears to be a relation between X and Y. Younger children ($X = 2$) have a tendency to score 0 on the task, and older children ($Y = 4$) have a tendency to score 2 on the task. A measure of correlation should indicate that there is a relation in these probabilities.

covariance The fundamental correlation measure for random variables is defined in terms of an intermediate quantity, the **covariance** of two random variables.

> **Covariance of X and Y**
> For discrete random variables X and Y,
>
> $$\text{Cov}(X, Y) = \sum (x - \mu_X)(y - \mu_Y) f_{XY}(x, y)$$
>
> where the sum is taken over all possible pairs of x, y values.
>
> *NOTE:* For continuous random variables, replace the sumation by integration over all possible values.

In our example,

$$\begin{aligned}\text{Cov}(X, Y) &= (2 - 3)(0 - 1)(.15) + (2 - 3)(1 - 1)(.10) \\ &\quad + (2 - 3)(2 - 1)(.0) + (3 - 3)(0 - 1)(.10) \\ &\quad + (3 - 3)(1 - 1)(.20) + (3 - 3)(2 - 1)(.1) \\ &\quad + (4 - 3)(0 - 1)(.05) + (4 - 3)(1 - 1)(.10) \\ &\quad + (4 - 3)(2 - 1)(.0) \\ &= 0.20\end{aligned}$$

The calculation of a covariance is tedious, and can be especially unlovely if the means have many decimal places. In such cases, a short-cut formula saves a bit of labor.

> **Short-Cut Formula for Cov(X, Y)**
> For discrete random variables X and Y,
>
> $$\text{Cov}(X, Y) = \left[\sum xy\, f_{XY}(x, y)\right] - \mu_X \mu_Y$$
>
> *NOTE:* For continuous random variables, replace the summation by integration over all possible values.

In the example,

$$\begin{aligned}\text{Cov}(X, Y) &= (2)(0)(.15) + (2)(1)(.10) + (2)(2)(.05) + (3)(0)(.10) \\ &\quad + (3)(1)(.20) + (3)(2)(.10) + (4)(0)(.05) + (4)(1)(.10) \\ &\quad + (4)(2)(.15) - (3.0)(1.0) \\ &= 0.20\end{aligned}$$

Example 5.25 The joint probabilities found in Example 5.20 are shown in Table 5.9. Calculate the covariance of U and V, using both the definition and the shortcut.

		v 0	1	2	Marginal
u	0	.32	.10	.08	.50
	1	.18	.06	.04	.28
	2	.10	.02	.01	.13
	3	.07	.01	.01	.09
	Marginal	.67	.19	.14	

Table 5.9

Solution We must first find μ_U and μ_V, using the marginal probabilities.

$$\mu_U = 0(.50) + 1(.28) + 2(.13) + 3(.09) = 0.81$$

and

$$\mu_V = 0(.67) + 1(.19) + 2(.14) = 0.47$$

Using the definition, we have

$$\begin{aligned}\text{Cov}(U, V) &= (0 - 0.81)(0 - 0.47)(.32) + (0 - 0.81)(1 - 0.47)(.10) \\ &\quad + (0 - 0.81)(2 - 0.47)(.08) + (1 - 0.81)(0 - 0.47)(.18) \\ &\quad + (1 - 0.81)(1 - 0.47)(.06) + (1 - 0.81)(2 - 0.47)(.04) \\ &\quad + (2 - 0.81)(0 - 0.47)(.10) + (2 - 0.81)(1 - 0.47)(.02) \\ &\quad + (2 - 0.81)(2 - 0.47)(.01) + (3 - 0.81)(0 - 0.47)(.07) \\ &\quad + (3 - 0.81)(1 - 0.47)(.01) + (3 - 0.81)(2 - 0.47)(.01) \\ &= -0.0707\end{aligned}$$

after much arithmetic.

The short-cut method indicates that we should first compute

$$\begin{aligned}\sum uvf_{UV}(u, v) &= 0(0)(.32) + 0(1)(.10) + 0(2)(.08) + 1(0)(.18) \\ &\quad + 1(1)(.06) + 1(2)(.04) + 2(0)(.10) + 2(1)(.02) \\ &\quad + 2(2)(.01) + 3(0)(.07) + 3(1)(.01) + 3(2)(.01) \\ &= 0.31\end{aligned}$$

after less arithmetic.

$$\begin{aligned}\text{Cov}(U, V) &= \sum uvf_{UV}(u, v) - \mu_U\mu_V \\ &= 0.31 - (0.81)(0.47) = -0.0707\end{aligned}$$

once again. □

The covariance of two random variables is not a very satisfactory measure of how the variables co-vary (vary together). One problem is the units of a covariance. If X is measured in dollars and Y in seconds, the covariance is measured in dollar-seconds, not what one would call the most readily interpreted units. A more useful measure is the **correlation** of two random variables, which is an absolute number without units.

correlation

Correlation of X and Y

$$\text{Corr}(X, Y) = \rho_{XY} = \frac{\text{Cov}(X, Y)}{\sqrt{\text{Var}(X)\text{Var}(Y)}}$$

In the example, $\text{Corr}(X, Y) = 0.2/\sqrt{(0.6)(0.6)} = 1/3$.

Example 5.26

Referring to Examples 5.20 and 5.25, calculate the correlation between U and V.

Solution

In Example 5.25, we found that $\text{Cov}(U, V) = -0.0707$. We need the variances of U and V. Using the short-cut method, we find

$$\begin{aligned}\text{Var}(U) &= \sum u^2 f_U(u) - \mu_U^2 \\ &= 02(.50) + 1^2(.28) + 2^2(.13) + 3^2(.09) - (0.81)^2 = 0.9539\end{aligned}$$

and

$$\text{Var}(V) = 0^2(.67) + 1^2(.19) + 2^2(.14) - (0.47)^2 = 0.5291$$

Therefore,

$$\text{Corr}(U, V) = \frac{-0.0707}{\sqrt{0.9539(0.5291)}} = -.0995$$

□

The correlation between X and Y is 0 if X and Y are statistically independent, and also if there is no *linear* predictability of Y given X. The correlation ranges from -1 (when Y is a perfect, linear decreasing function of X) to $+1$ (when Y is a perfect, linear increasing function of X).

The numerical value of the correlation of random variables has an interpretation via the *square* of the correlation. When X is not known, the best prediction (in the squared-error sense) of Y is the mean of Y, μ_Y. When X is given to be a particular value x, the best prediction is the **conditional expected value** of Y given $X = x$, $E(Y|X = x)$. The conditional expected value merely uses conditional probabilities instead of unconditional ones.

conditional expected value

$$E(Y|X = x) = \sum y f_{Y|X}(y|x)$$

In our age/number of correct classifications example,

$$E(Y|X = 2) = 0(.15/.3) + 1(.10/.3) + 2(.05/.3) = 0.6667$$
$$E(Y|X = 3) = 0(.10/.4) + 1(.20/.4) + 2(.10/.4) = 1.0000$$
$$E(Y|X = 4) = 0(.05/.3) + 1(.10/.3) + 2(.15/.3) = 1.3333$$

In this example, but not in all examples, the conditional expected value increases linearly with x. The conditional variance of Y given $X = x$ is defined similarly, using the same conditional probabilities. In our example,

$$\begin{aligned}\text{Var}(Y|X = 2) &= (0 - 0.6667)^2(.15/.3) + (1 - 0.6667)^2(.10/.3) \\ &\quad + (2 - 0.6667)^2(.05/3) \\ &= 0.5556\end{aligned}$$

Similarly, $\text{Var}(Y|X = 3) = 0.5000$ and $\text{Var}(Y|X = 4) = 0.5556$.

When X is being used to predict Y, the best prediction is the conditional expected value, and the squared error, given that $X = x$, is the conditional variance. To find the average squared error given X, we must weight by the probabilities of the various X values. In the example, the average squared error given X is

$$(0.5556)(.3) + (0.5000)(.4) + (0.5556)(.3) = 0.5333$$

The squared error not given X is $\text{Var}(Y)$; in the example, $\text{Var}(Y) = 0.6000$. The squared correlation is, once again, the Proportionate Reduction in Error (PRE):

$$\rho_{XY}^2 = \frac{\text{Var}(Y) - \text{Avg. Var}(Y|X)}{\text{Var}(Y)}$$

In the example, $\rho_{XY}^2 = (1/3)^2 = .1111$, and the PRE is $(.6000 - .5333)/(.6000) = .1111$.

Example 5.27 Refer to Example 5.26, where we found $\text{Corr}(U, V) = -.0995$. What is the interpretation of this number?

Solution

$$\rho_{UV}^2 = (-.0995)^2 = .00990$$

Thus linear prediction reduces squared error by .00990, less than 1%. □

When the conditional mean of Y given $X = x$ is a nonlinear function of X, the squared correlation is an underestimate of the predictive value of X. Correlation measures the ability of X to predict Y in a linear fashion. If the conditional mean is nonlinear in x, the squared correlation measures the effectiveness of a least-squares approximation to the conditional expected value. Unless the nonlinearity is fairly dramatic, the squared correlation is a good approximation to the predictive value.

Example 5.28 For the probabilities underlying Example 5.26, it can be calculated that

$$E(V|U = 0) = 0.520$$
$$E(V|U = 1) = 0.500$$
$$E(V|U = 2) = 0.308$$

and

$$E(V|U = 3) = 0.333$$

Is the prediction an extremely nonlinear function of U?

Solution The prediction is certainly not a straight line. Between $U = 0$ and $U = 1$, the conditional expectation goes from 0.520 to 0.500; between $U = 1$ and $U = 2$, it decreases from 0.500 to 0.308; and between $U = 2$ and $U = 3$, it increases slightly. Thus the squared correlation 0.00990 found in Example 5.27 understates the predictive value of U somewhat. The nonlinearity doesn't seem extreme, however. □

5.8
The Mathematics of Expected Values and Variances

This section presents some of the key mathematical results involving expected values and variances of random variables. Although the results are critical to developments in later chapters, readers who are willing to take mathematical results on faith can skip this section. The results concerning expected values are fairly obvious, but the results about variances are real mathematical theorems.

First, we consider expected values of sums. For illustration, let's look at three possible joint probability distributions of the random variables X and Y (Table 5.11).

CASE A

		y		
		0	1	2
	2	.15	.10	.05
x	3	.10	.20	.10
	4	.05	.10	.15

CASE B

		y		
		0	1	2
	2	.09	.12	.09
x	3	.12	.16	.12
	4	.09	.12	.09

CASE C

		y		
		0	1	2
	2	.05	.10	.15
x	3	.10	.20	.10
	4	.15	.10	.05

Table 5.11

In each of the three cases, the marginal probability distribution of X is .3, .4, and .3 on values 2, 3, and 4. Thus the mean (expected value) of X is $2(.3) + 3(.4) + 4(.3) = 3.0$. Similarly, the expected value of Y is 1.0.

Now let's look at the total $T = X + Y$. One natural argument would say that the average of the total should equal the total of the averages—that is, that $E(X + Y)$ should equal $E(X) + E(Y)$. One way to check on this argument is to find the probability distribution of T in each of the three cases and calculate $E(T)$. The probabilities that apply to the total T can be found by simple addition: merely add up the probabilities for all (x, y) combinations that yield the desired total t. For example, in Table 5.11, case A, the probability that $T = 3$ is

$$\begin{aligned} P(T = 3) &= P(X = 3 \text{ and } Y = 0) + P(X = 2 \text{ and } Y = 1) \\ &= .10 + .10 = .20 \end{aligned}$$

A table of T probabilities for the three cases of Table 5.11 is shown in Table 5.12.

	t	2	3	4	5	6
CASE A	$f_T(t)$	.15	.20	.30	.20	.15
CASE B	$f_T(t)$	.09	.24	.34	.24	.09
CASE C	$f_T(t)$	.05	.20	.50	.20	.05

Table 5.12

In Table 5.12, each probability distribution is symmetric around the value 4. Given that $E(X) = 3$ and $E(Y) = 1$, it seems reasonable that $E(X + Y) = E(X) + E(Y)$, regardless of the dependence of X and Y.

Result 5.1 $E(X + Y) = E(X) + E(Y)$, regardless of the dependence of X and Y.

Proof As in the previous section, either we may calculate the probabilities for $t = x + y$ and then find the expected values, or we may take all possible values of x and y, form t, and weight by the joint probability of the particular x, y combination. The results will be the same. We follow the latter strategy.

$$\begin{aligned} E(X + Y) &= \sum_{x,y} (x + y) f_{XY}(x, y) \\ &= \sum_{x,y} x f_{XY}(x, y) + \sum_{x,y} y f_{XY}(x, y) \\ &= \sum_{x} x f_X(x) + \sum_{y} y f_Y(y) \\ &= E(X) + E(Y) \end{aligned}$$

because the marginal probabilities $f_X(x)$ and $f_Y(y)$ can be obtained by summing probabilities. Note that the result doesn't assume anything about the dependence or independence of X and Y.

An extension of this result indicates the effect of adding a constant c to X or multiplying X by a constant k. Result 5.1 suggests that $E(X + c) = E(X) + c$. Also, given that $E(X + X) = E(2X) = E(X) + E(X) = 2E(X)$, it seems plausible that $E(kX) = kE(X)$. Both results follow easily.

Result 5.2 $E(X + c) = E(X) + c$ and $E(kX) = kE(X)$

Proof

$$E(X + c) = \sum (x + c) f_X(x)$$
$$= \sum x f_X(x) + c \sum f_X(x) = E(X) + c$$

because $\Sigma f_X(x) = 1$, as a sum of all probabilities. Second,

$$E(kX) = \sum (kx) f_X(x) = k \sum x f_X(x) = kE(X)$$

The results for means are valid regardless of the dependence among the random variables. Results for variances, on the other hand, depend very crucially on the degree of correlation between the random variables. The correlation can be computed by the methods of the previous section as 1/3, 0, and −1/3, for cases A, B, and C, respectively. Examination of Table 5.12 indicates that the variance of $T = X + Y$ is largest when the correlation between X and Y is positive and smallest when the correlation is negative. The result can be stated in terms of covariance or correlation.

Result 5.3

$$\text{Var}(X + Y) = \text{Var}(X) + 2\,\text{Cov}(X, Y) + \text{Var}(Y)$$
$$= \text{Var}(X) + 2\rho_{XY}\sigma_X\sigma_Y + \text{Var}(Y)$$

Proof

This is another expand-the-square result. Recall that $\mu_{X+Y} = \mu_X + \mu_Y$.

$$\text{Var}(X + Y) = \sum (x + y - \mu_{X+Y})^2 f_{XY}(x, y)$$
$$= \sum (x + y - \mu_X - \mu_Y)^2 f_{XY}(x, y)$$
$$= \sum_{x,y} (x - \mu_X)^2 f_{XY}(x, y)$$
$$+ 2 \sum_{x,y} (x - \mu_X)(y - \mu_Y) f_{XY}(x, y)$$
$$+ \sum_{x,y} (y - \mu_Y)^2 f_{XY}(x, y)$$
$$= \sum_x (x - \mu_X)^2 f_X(x)$$

$$+ 2 \sum_{x,y} (x - \mu_X)(y - \mu_Y) f_{XY}(x, y)$$

$$+ \sum_{y} (y - \mu_Y)^2 f_Y(y)$$

$$= \text{Var}(X) + 2\,\text{Cov}(X, Y) + \text{Var}(Y)$$

because

$$\sum_{y} f_{XY}(x, y) = f_X(x)$$

and

$$\sum_{x} f_{XY}(x, y) = f_Y(y)$$

By definition of the correlation of random variables,

$$\rho_{XY} = \frac{\text{Cov}(X, Y)}{\sigma_X \sigma_Y}$$

Thus we may substitute $\rho_{XY}\sigma_X\sigma_Y$ for Cov(X, Y) in the expression for Var$(X + Y)$.

Our numerical example illustrates this result. In each of the three cases, Var(X) = Var(Y) = 0.6. Var$(X + Y)$ differs in the three cases, being 1.6, 1.2, and 0.8, respectively. By the computations of the previous section, Cov(X, Y) = .2, 0, and −.2, respectively. Thus, for instance, in the first case, Var$(X + Y)$ = 0.6 + 2(.2) + 0.6 = 1.6.

To complete our discussion of the basic mathematics of expected values and variances, we consider the effect on Var(X) of multiplying the random variable X by a constant, or adding a constant to X. The effect of adding a constant is to shift the distribution of X over; merely shifting the distribution over has no effect on variability. The effect of multiplying by a constant k is to change the scale by the same constant k. Because variance measures squared error, the variance is multiplied by k^2.

Result 5.4

If c and k are constants,

$$\text{Var}(X + c) = \text{Var}(X)$$
$$\text{Var}(kX) = k^2\,\text{Var}(X)$$

Proof In Result 5.2, we proved that $\mu_{X+c} = \mu_X + c$ and $\mu_{kX} = k\mu_X$.

$$\text{Var}(X + c) = \sum (x + c - \mu_{X+c})^2 f_X(x)$$
$$= \sum [x + c - (\mu_X + c)]^2 f_X(x)$$
$$= \sum (x - \mu_X)^2 f_X(x) = \text{Var}(X)$$
$$\text{Var}(kX) = \sum (kx - \mu_{kX})^2 f_X(x)$$
$$= \sum (kx - k\mu_X)^2 f_X(x)$$
$$= k^2 \sum (x - \mu_X)^2 f_X(x) = k^2\text{Var}(X)$$

Chapter Exercises

5.26. An experimenter needs at least four subjects for a particular experiment. First, a list of six potential subjects is called. From experience, the experimenter assumes that there is a 10% chance of obtaining only two subjects from the list, a 20% chance of obtaining three, a 40% chance of obtaining four, a 20% chance of obtaining five, and a 10% chance of obtaining all six. If fewer than four subjects are obtained from the first list, a second list is called. The second list can yield an additional two, three, four, or five subjects, with respective probabilities .2, .3, .3, and .2. Define X to be the number of subjects obtained from the *first* list.

a. Find $f_X(x)$ and draw a probability histogram of it.

b. From the probability histogram, what should be the expected value of X? Explain why. Verify your explanation by actually calculating $E(X)$.

c. Calculate the standard deviation of X.

d. What fraction of the probability actually falls within 1 standard deviation of the mean for this distribution?

5.27. Referring to Exercise 5.26, let Y be the total number of subjects obtained from the first list and, if necessary, the second.

a. Calculate $f_Y(y)$.

b. Find the expected number of subjects obtained.

c. Find the variance of Y.

5.28. In an experiment, subjects are asked to guess the length of a lighted bar shown on a screen, relative to a standard 5-unit bar, which is shown parallel to it but several feet away. Defining X to be the subject's estimate of the length, the probability distribution of X is as follows:

x	3	4	5	6	7
$f_X(x)$	.10	.20	.40	.20	.10

a. Find $E(X)$.

b. Draw a probability histogram. What is the indicated value of the mean of X? Why?

c. Calculate the standard deviation of X.

5.29. An experiment similar to the one in Exercise 5.28 has the reference bar perpendicular to the presented bar, rather than parallel to it. Let Y be the subject's indicated size; the probability distribution of Y is

y	3	4	5	6	7
$f_Y(y)$	.15	.20	.30	.20	.15

a. Calculate the mean and standard deviation of Y.

b. Draw a probability histogram of the Y distribution and compare it to the X histogram drawn in Exercise 5.28. Which standard deviation should be larger? Why?

5.30. A computer is programmed to choose a digit randomly from among 0, 1, 2, 3, and 4. Let U = the digit chosen. Find $E(U)$ and $\text{Var}(U)$.

5.31. The computer in Exercise 5.30 chooses a series of digits, independently from choice to choice. Let $U_1, U_2, U_3, \ldots$ be the resulting digits. Define $T_2 = U_1 + U_2$.

a. Use a probability table or tree to find $f_T(t)$.

b. Find $E(T)$ and $\text{Var}(T)$.

c. How do $E(T)$ and $\text{Var}(T)$ compare to $E(U)$ and $\text{Var}(U)$ as found in Exercise 5.30?

5.32. Referring to Exercise 5.31, let $T_3 = T_2 + U_3$ be the sum of the first three digits chosen.

a. Find $f_T(t_3)$.

b. Find $E(T_3)$ and $\text{Var}(T_3)$.

c. Again compare $E(T)$ and $\text{Var}(T)$ to $E(U)$ and $\text{Var}(U)$ from Exercise 5.30. Can you guess what the expected value and variance for a sum of 20 U's would be?

5.33. In a game, a subject guesses whether the light on the left or the light on the right will turn on. The probabilities for the two lights are equal, and the sequence is completely random. Trials continue until the subject guesses right or until a maximum of five trials have occurred. Let Y = the number of trials. Find the probability distribution, mean, and variance of Y.

5.34. Suppose that the experiment in Exercise 5.33 is changed so that the maximum number of trials is ten. Find the mean and variance of Y.

5.35. Refer to Exercises 5.33 and 5.34. Suppose that no maximum is imposed on the number of trials. Can you guess what the expected value and variance of Y would be?

5.36. A continuous random variable X has a *triangular* distribution if

$$f_X(x) = \begin{cases} x, & \text{for } 0 < x < 1 \\ 2 - x, & \text{for } 1 < x < 2 \end{cases}$$

a. Draw a picture of this density.

b. Find the probability that X will be less than .5. (Hint: To save calculus, remember that the area of a triangle is half the base times the height.)

5.37. Refer to Exercise 5.36.

a. Verify that μ_X is 1.00. Why should this be the mean of X?

b. Find the standard deviation of the random variable X.

c. Find the probability that X will be within 1 standard deviation of its mean. Would you expect this probability to be fairly close to the Empirical Rule approximation? Why?

6 Special Types of Probability Distributions

In Chapter 5, the general ideas of random variables and probability distributions were introduced. These ideas apply broadly to any experiment involving measured quantities and random variation. Certain special, but recurring, situations yield useful specific probability distributions. In this chapter, we introduce three of the most crucial special probability distributions—the binomial (Section 6.1), Poisson (Section 6.2), and normal (Section 6.4) distributions. In addition, an important class of probability models—Markov models—is briefly considered in Section 6.3.

The various formulas that will be specified are convenient, but they are not the most crucial issue in this chapter. More important are the assumptions underlying the formulas. Before applying the formulas, it is vital to consider whether the assumptions could be valid, at least approximately.

6.1 Bernoulli Trials and Binomial Probabilities

Consider the following three situations. Each of 24 5-year-old children is tested separately; each child either does or does not demonstrate mastery of a Piagetian conservation task. An experimenter makes appointments with 40 potential subjects for an experiment; each subject either does or does not appear at the appointed time. A person is taught a set of symbols in a memory set, then a series of 36 symbols is flashed on a screen; the person either does or does not specify correctly whether the flashed symbol is in the memory set. These situations, and many others, have several properties in common.

In each situation, the overall experiment can be thought of as a series of **trials,** where each trial can yield only one of two possible results; the possible results are arbitrarily called "success" and "failure." In the Piagetian experiment, each child constitutes a trial, and mastery of the conservation test by a child could be called a success. In the appointments experiment, each potential subject is a trial, and we might as well define a success to be that the subject showed up. In the memory experiment, each symbol is a trial, and a correct answer might be defined as a success. Thus, these situations all consist of a series of yes/no trials.

In each situation, it is reasonable to assume that the probability of success remains constant over trials. Assuming that the children are tested in random order or that the subjects are called randomly, there is no reason why the probability of a success should change from one person to the next. In the memory experiment, there is some possibility that learning effects could improve the probability of a correct answer or that fatigue effects could decrease the probability, but as a rough approximation we could still assume constant probability of success.

In each situation, it is reasonable to assume independence of trials. The result for any one child (or any one potential subject) should not affect the probability of success for someone else. Again, the assumption is shakier in the memory experiment. It is possible that the person may become inattentive, so a failure on one trial might signal a higher probability of failure on the next trial. If boredom and inattentiveness can be assumed to be negligible factors, however, independence of trials can be assumed.

Bernoulli trials

Together, these three assumptions define **Bernoulli trials.**

Assumptions for Bernoulli Trials

1. The experiment consists of a series of trials, with each trial yielding one of two possible outcomes—success or failure.

2. The probability of success remains constant over trials.

3. Results of the various trials are statistically independent.

Example 6.1 In each of the following situations, are the Bernoulli trials assumptions plausible?

(a) Patients in a state hospital for the mentally ill take an experimental drug. The patients are randomly selected; their diagnoses range from hopelessly ill to mildly disturbed. After 2 months on the drug, each patient is observed by a psychiatrist and rated as fit or not fit for release.

(b) A child is observed monthly by a developmental psychologist, and rated as below or above average in social skills. The process lasts for 36 months.

(c) A sample of 24 laboratory rats are placed in a stressful situation (trying to escape a painful stimulus). Each rat is graded on ability to escape (always, usually, sometimes, rarely, or never).

Solution **(a)** It seems reasonable to assume independence across patients; why should the results for one patient have any effect on other patients? The problem is nonconstant probabilities. Presumably, mildly disturbed patients are much more likely than hopelessly ill patients to be judged fit for release, however effective the drug may be.

(b) Here the problem is independence. If the child is rated below average one month, it is likely that the child will be rated below average the next month as well. Thus there is dependence in the month-to-month trials.

(c) There are 5 possible ratings, not 2. But if we focus on one particular rating (say "always" vs. "not always"), the Bernoulli assumptions seem plausible. The results for one rat should have no bearing on the results for others, so independence should hold. The rats are presumably homogeneous, so the probability of success should be essentially constant. □

Typically, the independence assumption is the most critical. The assumption that the experiment consists of a series of yes/no, success/failure assumptions either is or isn't valid on the face of the experiment. If there are many trials with more than two possible outcomes per trial, one can define any one possible outcome as a "success," lumping all other outcomes together as "failure." Alternatively, one can extend the ideas of this section to the case of several outcomes per trial. If the probability of success varies slightly from one trial to the next, the probability distributions derived under the constant-probability assumption will not be correct. In particular, if the constant-probability assumption is wrong, it is likely that the variability of the number of successes will be affected. At the extreme, suppose that the first half of the trials have probability 0 of success and the second half of the trials have probability 1 of success. Then it will be guaranteed that the number of successes will be half the number of trials, with no variability at all. Usually, the probability distribution is close to correct, unless the probabilities vary quite widely from trial to trial. If, however, there is dependence from trial to trial, the derived probability distribution may be grossly in error.

The three situations have one other aspect in common. In each situation the number of trials is fixed (at 24, 40, and 36, respectively). The random variable of interest is the number of successes occurring in the fixed number of trials, without regard to the order of successes and failures. In contrast, suppose that subjects were called one at a time, until a total of 12 subjects had shown up; the condition that the number of trials be fixed would be violated. Instead the number of successes would be fixed (at 12) and the number of trials would be random. Together with the Bernoulli trials assumptions, the condition that the number of trials be fixed and that order be irrelevant defines a **binomial experiment.**

binomial experiment

Binomial Experiment

A series of Bernoulli trials is a binomial experiment if, in addition to the three Bernoulli trials assumptions, the following assumption is met:

> The number of trials is fixed and the random variable is the number of successes, without regard to order.

Example 6.2

Refer to Example 6.1, part (c). Suppose that the experimenters want to obtain 20 rats who are always able to escape the painful stimulus. The experimenters keep testing rats until they obtain 20 "always" animals. Is this a binomial experiment?

Solution

No. The Bernoulli trials assumptions were plausible in the context of Example 6.1. But in this experiment, the number of trials is not fixed; the experimenters keep going until they get 20 successful rats. In fact, the order of successes and failures is relevant here; the last trial is guaranteed to be a success. □

There is a fairly standard notation for binomial experiments. The number of trials is denoted n. The probability of success on any trial is denoted either p or π. Because p is used to mean several other things in this book, we will use π for the probability of success. We will use Y for the random variable "number of successes in a fixed number, n, of trials."

In principle, one could find the probability distribution of Y for any values of n and π by using a probability tree. In Figure 6.1, a tree is drawn for $n = 3$ and $\pi = .2$. The resulting probability distribution is

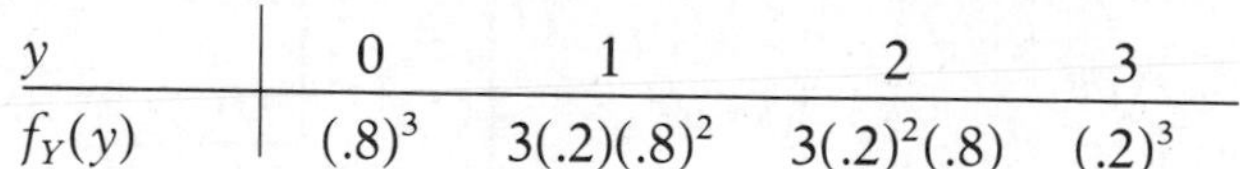

y	0	1	2	3
$f_Y(y)$	$(.8)^3$	$3(.2)(.8)^2$	$3(.2)^2(.8)$	$(.2)^3$

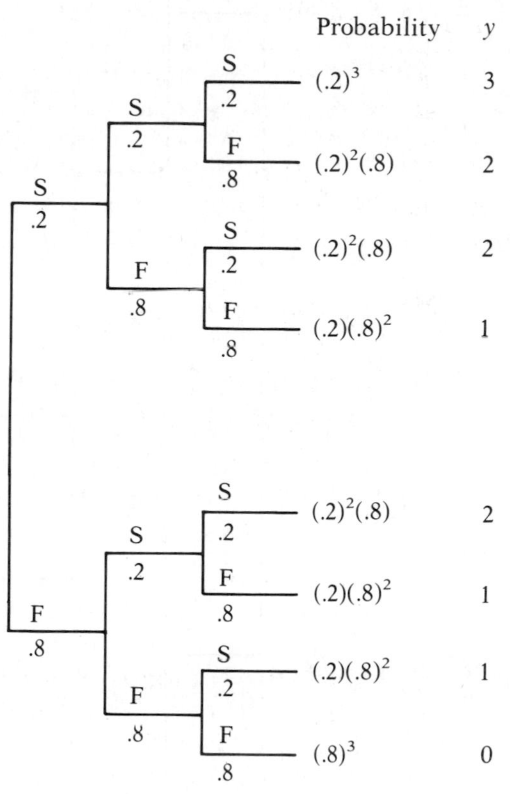

Figure 6.1 Tree for binomial experiment, $n = 3$, $\pi = .2$

Example 6.3 Construct a probability tree for a binomial experiment with $n = 4$ and $\pi = .3$.

Solution The tree is shown in Figure 6.2. ☐

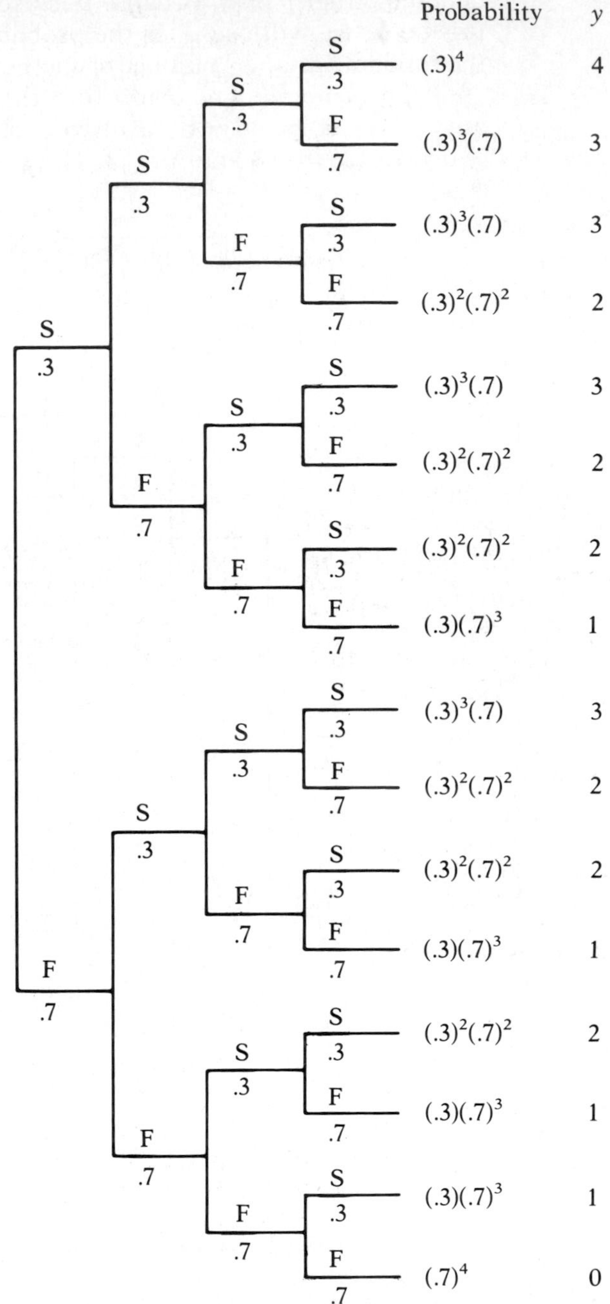

Figure 6.2 Probability tree for binomial experiment, $n = 4$, $\pi = .3$

Obviously, as the number of trials gets larger, use of a tree gets more and more impractical. For 20 trials, there would be 1,048,576 paths in the tree! Therefore, we should use the ideas in the probability tree to find a more efficient way of calculating binomial probabilities. In Figure 6.1, look at the three paths corresponding to $y = 2$. Each path by itself has probability $(.2)^2(.8)^1$, because for these paths there must be exactly 2 successes and 1 failure. In general there will be y successes and $n - y$ failures; so that if $\pi = .2$, the probability for any one path will be $(.2)^y(.8)^{n-y}$. More generally, replace .2 by π, and .8 by $1 - \pi$. The difficult part of finding binomial probabilities is counting how many distinct paths correspond to y successes in a fixed number, n, of trials. As is proved in many texts [e.g., Ott and Hildebrand (1983)], this number is

$$\frac{n!}{y!(n-y)!}$$

where $n! = n(n-1) \cdots (2)(1)$, $1! = 1$, and $0! = 1$.

To find the probability that a binomial random variable Y takes on the value y [that is, to find $f_Y(y)$], we note that all the path probabilities corresponding to y successes have the same probability, namely $(\pi)^y(1-\pi)^{n-y}$. To find $f_Y(y)$, we merely take this path probability and multiply it by the number of paths corresponding to y successes, namely $n!/(y!(n-y)!)$.

Binomial Probability Distribution

For a binomial experiment,

$$f_Y(y) = \frac{n!}{y!(n-y)!}(\pi)^y(1-\pi)^{n-y}$$

for $y = 0, 1, 2, \ldots, n$.

Example 6.4 Refer to Example 6.1. Suppose that 16 rats are selected randomly. Further suppose that for this particular strain of rats, the probability that a randomly chosen rat will always escape is .3. What is the probability that exactly 8 rats in the sample of 16 will always escape?

Solution Here $\pi = .3$, $n = 16$, and y (the desired number of successes) $= 8$. Thus

$$f_Y(8) = \frac{16!}{8!(16-8)!}(.3)^8(1-.3)^{16-8}$$

□

Some computer packages (e.g., Minitab) will compute binomial probabilities on request. Alternatively, binomial probabilities can be tabulated. Table 1 in the appendix to this book gives selected binomial probabilities. For values of π between .05 and .50, shown across the top of the table, values of y are read on the left side of the table; for values of π between .50 and .95, shown along the bottom of the table, values of y are read on the right side of the table.

Example 6.5 Find a numerical value for the desired probability in Example 6.4.

Solution We have $n = 16$, $y = 8$, and $\pi = .3$. The entry in the $n = 16$ block, $y = 8$ row, and $\pi = .3$ column of Table 1 is .0487. □

Some probability histograms of binomial probabilities are shown in Figure 6.3. Note that binomial distributions are symmetric for $\pi = .5$ and right-skewed for $\pi < .5$. Though Figure 6.3 doesn't show binomial probabilities for $\pi > .5$, it isn't surprising that these are left-skewed. Further, as the number of trials n gets larger, the skewness decreases.

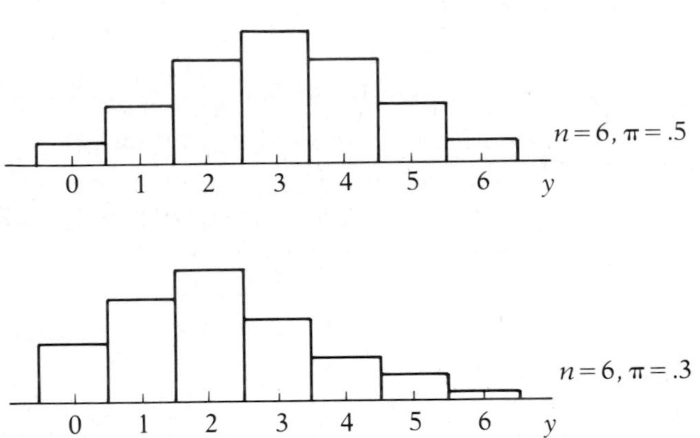

Figure 6.3 Binomial probability histograms

Expected values, variances, and standard deviations for binomial random variables can be computed using Table 1 and the definitions given in Chapter 5. For example, if $n = 3$ and $\pi = .8$, the respective probabilities for $y = 0, 1, 2, 3$ are .0080, .0960, .3840, and .5120. Therefore,

$$E(Y) = 0(.0080) + 1(.0960) + 2(.3840) + 3(.5120) = 2.4000$$

and

$$\text{Var}(Y) = (0 - 2.4)^2(.0080) + (1 - 2.4)^2(.0960) + \cdots$$
$$= 0.48$$

Once again, this process would be tedious if n were at all large. Fortunately, a short-cut formula has been derived to save arithmetic.

Expected Value and Variance of a Binomial Distribution

If Y satisfies all the assumptions for a binomial random variable,

$$E(Y) = n\pi$$
$$\text{Var}(Y) = n\pi(1 - \pi)$$

Example 6.6 Referring to Example 6.4, find the expected value and standard deviation of the random variable "number of always-escaping rats."

Solution We have $n = 16$ and $\pi = .3$. Therefore, the expected value is $16(.3) = 4.8$, and the variance is $16(.3)(.7) = 3.36$. The standard deviation is $\sqrt{3.36} = 1.833$. □

To reiterate, the critical issue with binomial random variables is whether the binomial assumptions are met. Not every yes/no, success/failure situation yields binomial probabilities. The assumptions of constant probability, independence, and fixed number of trials with order irrelevant must all be met to justify using binomial probabilities, expected value, and variance.

Example 6.7 Refer to Example 6.1, part (a). Suppose that in repeated samples of groups of 50 patients it is found that the mean number of releasable patients is 20.0, and that the variance is 7.0. Are these values compatible with binomial assumptions?

Solution The overall probability of releasability is estimated to be $20.0/50 = .40$. If the binomial assumptions hold, the variance should be $50(.40)(.60) = 12.0$. The actual variance is not nearly as high: at 7.0. Thus there is too little variability, and the binomial assumptions are highly suspect. It is known that a nonconstant probability of success leads to decreased variability. It seems likely that the probability of success (releasability) is highly variable over patients, thus decreasing the variance. □

One particular case, in which the Bernoulli trials assumptions are met but the fixed number of trials with order irrelevant assumption is not, leads to the **negative binomial distribution.** An example of this type of case is one in which the experimenter calls potential subjects at random, and keeps calling until 12 potential subjects actually appear. Now the number of *successes* is fixed, and therefore the number of trials is random.

negative binomial distribution

Negative Binomial Experiment
A series of Bernoulli trials is a negative binomial experiment if, in addition to the three Bernoulli trials assumptions, the following assumption is met:

> The desired number of successes is fixed in advance to be, say, k and the random variable of interest is Y = required number of trials.

A probability tree for a negative binomial experiment is hard to draw, because the number of trials and therefore the number of branches could, in principle, be indefinitely large. It is, however, possible to derive negative binomial probabilities. For the kth success to occur at trial y, two things must happen:

1. There must be exactly $k - 1$ successes in the first $y - 1$ trials, *in any order.*

2. Trial y must result in a success.

The probability that the first event will happen is the binomial probability of $k - 1$ successes in $y - 1$ trials, because order is irrelevant. The probability that the second event will happen is π. The two probabilities may be multiplied, because the trials are independent. This reasoning yields the negative binomial probability distribution.

Negative Binomial Distribution
If Y = number of trials required to obtain a specified number k of successes, then

1. $f_Y(y) = \dfrac{(y-1)!}{(k-1)!(y-k)!}(\pi)^k(1-\pi)^{y-k}$

2. $E(Y) = \dfrac{k}{\pi}$

3. $\text{Var}(Y) = \dfrac{k(1-\pi)}{\pi^2}$

Example 6.8 Refer to Example 6.2. Assume again that the probability that any particular rat always escapes the stimulus is .3. Let Y = the number of rats tested to obtain 20 always-escaping rats. Find the probability that $Y = 50$.

Solution

$$f_Y(y) = \frac{(50 - 1)!}{19!(50 - 20)!}(.3)^{20}(.7)^{30}$$

□

Exercises for Section 6.1

6.1. Segal (1981) reports on a study of coalition formation in groups of 4 people. Under specified group decision-making conditions, the probability that a particular type of coalition (called the BCD coalition) forms is .75. Suppose that 48 subjects are randomly allocated into 12 groups of 4 subjects each and are randomly assigned the roles of A, B, C, and D. The question is how many BCD coalitions will form.

a. Is it reasonable to assume that the 12 groups constitute 12 Bernoulli trials?

b. Is this experiment a binomial experiment?

6.2. Refer to Exercise 6.1 and assume that binomial probabilities apply.

a. Write expressions for the probabilities that exactly 8 BCD coalitions form, that exactly 9 BCD coalitions form, and that exactly 10 BCD coalitions form.

b. Use Table 1 to find numerical values for the probabilities of part a.

c. Find the probability that the number of BCD coalitions is between 8 and 10 (both included).

6.3. Refer to Exercise 6.1 and assume that binomial probabilities apply. Find the expected value and standard deviation of the number of BCD coalitions formed.

6.4. Refer to Exercise 6.1 again. Suppose that, instead of having 48 different subjects in 12 groups of 4, the experimenter has 4 subjects participate in the coalition formation 12 consecutive times (rotating the A, B, C, D roles). Are the Bernoulli and binomial assumptions questionable in this experiment?

6.5. Refer once again to Exercise 6.1. Assume that the experimenter needs to find 8 groups that do not form BCD coalitions. Successive groups of 4 subjects are tried until 8 non-BCD coalitions are obtained.

a. What probability distribution applies to this experiment? Defend your answer.

b. Find the probability that exactly 20 groups must be tested until the experimenter obtains the eighth non-BCD coalition.

c. Find the expected number of groups tested and the variance.

6.6. In a study of rat behavior, it is necessary to implant an electrode in a certain part of a laboratory rat's brain. For various reasons, only 95% of the implantations are successful. Suppose that electrodes are implanted in lab rats, and the number of successful implantations found.

a. Are there circumstances under which the Bernoulli assumptions might be suspect?

b. Assuming that the Bernoulli assumptions hold, is this a binomial experiment?

6.7. Refer to Exercise 6.6, and assume that binomial probabilities apply.

a. Find the probability that 90 or more of 100 implants will be successful.

b. Calculate the mean and the standard deviation of the number of successful implantations.

6.8. Refer to Exercise 6.6, and assume that the Bernoulli assumptions are satisfied. Suppose that electrodes are implanted until 100 successful implantations are achieved.

a. What probabilities apply?

b. Let Y = the number of attempts needed to obtain 100 successful implantations. Find $P(Y = 110)$; obviously, you don't want to carry out the arithmetic.

c. Find the expected value and variance of the random variable Y defined in part b.

6.2 Poisson Probabilities

Another typical probability situation involves events happening at random over time. Consider the arrival of new patients at a mental hospital, or the perception of flashes of light caused by environmental radiation in a darkened room. In either case, events (new patients or flashes) can reasonably be assumed to occur randomly over time. In particular, in a very short time interval, it is highly unlikely that two such events will occur essentially simultaneously. That is, events won't "clump" together. Also, the occurrence (or nonoccurrence) of an event in one time period should have no effect on the probability of occurrence in another time period. That is, events should occur independently. These two assumptions are the critical ones for Poisson probabilities to apply. A third assumption is often made in textbooks—that the expected rate of occurrence remains constant over time—but this assumption can be shown to be merely a convenience for the underlying mathematics. If the expected rate varies over the time period, the average expected rate may be used.

Poisson Distribution Assumptions

1. (Nonclumping) The probability that two or more events will occur at the same time is negligible.

2. (Independence) The occurrence or nonoccurence of an event in one time period doesn't change the probability that an event will occur in a later time period.

Example 6.9 For each of the following situations, how plausible are the Poisson assumptions?

(a) The weekly number of calls to a suicide hotline

(b) Occurrences of seizures in an epileptic

(c) Arrival of people with head injuries in a hospital emergency room

Solution **(a)** There is no obvious reason why there should be any clumping of calls, barring such farfetched possibilities as that a group of people make a suicide pact and then have second thoughts. Conceivably, there might be some dependence if callers made repeat calls at more-or-less regular intervals. However, these objections seem very minor, so the number of calls should obey Poisson probabilities. It might well be that the average rate of calls would vary, being higher (for example) around holidays, but that's not relevant to whether the number of calls is Poisson.

(b) If occurrence of one seizure increases the risk of another in the near future, the independence assumption would be violated.

(c) There's a clear possibility of multiple injuries in automobile accidents and the like. The nonclumping assumption is not reasonable. □

The expected number of events occurring per unit of time is denoted by λ, the Greek letter lambda. (The same letter was used in a completely different context in Chapter 3.) The number of time units under consideration is denoted by T. Thus, if routine arrivals at a mental hospital occur at a long-run average rate of 2.7 per week, and the concern is about the number of arrivals in a 6-week period, then $\lambda = 2.7/$week and $T = 6$ weeks. The expected number of arrivals in the specified time period is λT; the expected number of arrivals at the hospital over a 6-week period is $(2.7)(6) = 16.2$. Using this notation, we can state the Poisson probability distribution, expected value, and variance.

Poisson Distribution

Under the assumptions of nonclumping and independence, if Y = the number of events occurring in a T-unit time period,

$$f_Y(y) = \frac{e^{-\lambda T}(\lambda T)^y}{y!}, \quad \text{where } y = 0, 1, 2, \ldots$$

$$E(Y) = \lambda T$$

$$\text{Var}(Y) = \sigma^2 = \lambda T$$

Note the coincidence that the variance of a Poisson distribution happens to equal its mean.

Poisson probabilities can be calculated by hand calculator or computer program, or looked up in a table. Table 2 in the appendix gives cumulative Poisson probabilities $P(Y \leq y)$ for various values of $\mu = \lambda T$. The columns are indexed by $\mu = \lambda T$, the expected value.

Example 6.10 Refer to the suicide hotline of Example 6.9. Assume that the expected rate of calls is 1.3 per day and that the Poisson probability distribution assumptions are correct. Find the probability that there will be 3 or fewer calls in a specified 5-day period.

Solution The expected number of calls is $\mu = (1.3)(5) = 6.5$. In the 6.5 column of Table 2, the cumulative probability for $y = 3$ (which is the probability of 3 or fewer calls) is .1118. □

As indicated, if the expected rate per unit time, λ, varies, the average rate may be used. For instance, suppose that at the hospital the expected rate is 2.9 per week for the first two weeks, then drops to 2.6 per week for the next 4 weeks. The average rate is

$$\frac{(2.9 + 2.9 + 2.6 + 2.6 + 2.6 + 2.6)}{6} = 2.7$$

and the effective rate once again is 2.7.

Example 6.11 Refer to Example 6.10. How would the answer change if the expected number of calls had been assumed to be 1.2 for each of the first 4 days, then 1.7 on the fifth day?

Solution The average rate is $[4(1.2) + 1(1.7)]/5 = 1.3$, just as it was in Example 6.10. Whether the rate is constant (as in Example 6.11) or variable (as in this example) is irrelevant. The probability is exactly the same. □

A probability distribution closely related to the Poisson arises if the same nonclumping and independence assumptions (and, this time, the constant expected rate assumption) are met but we change the question. Instead of considering Y = the number of events in a fixed time period, we consider X = the time until the next event. Instead of prespecifying the time period T, we now are prespecifying the number of events (1) and allowing the time required to be random. As a time measure, X is a continuous random variable. Rather than directly specifying probabilities, we must specify a probability density and then calculate probability as area under the probability density curve. Fortunately, in this case we can steal a simple formula from calculus to get the area under the curve.

(Negative) Exponential Distribution

Under the assumptions of nonclumping and independence, if X = the time required until the next event, the probability density of X is

$$f_X(x) = \lambda e^{-\lambda x}, \quad \text{where } x > 0$$

For any numbers a and b, both ≥ 0,

$$P(a < X < b) = e^{-\lambda a} - e^{-\lambda b}$$

$$E(X) = \frac{1}{\lambda}$$

$$\text{Var}(X) = \frac{1}{\lambda^2}$$

NOTE: In this case, the assumption of a constant expected rate of occurrence *is* important, as are the nonclumping and independence assumptions.

Example 6.12 Refer to Example 6.10. Find the probability that the next call to the hotline will occur within 1.5 days.

Solution If X is the random variable "time to next call," we're trying to find $P(X < 1.5)$. We should write it as $P(0 < X < 1.5)$, because X obviously can't be negative. Recall that $\lambda = 1.3$ calls per day.

$$P(0 < X < 1.5) = e^{-1.3(0)} - e^{-1.3(1.5)}$$
$$= 1 - .1423 = .8577$$

Although we shall not do so, it is possible to find the probability distribution for the time required to get 2 arrivals, 3 arrivals, or whatever number. The resulting distribution is called the gamma (or sometimes the Erlang) distribution, and is discussed in most mathematical statistics books. See, for instance, Mood, Graybill, and Boes (1974).

A completely different use for Poisson probabilities, having nothing in particular to do with events happening randomly over time, is in approximating binomial probabilities. In some situations, there are many Bernoulli trials, but the probability of success is so low that it doesn't appear in binomial tables. In such a case, we simply use Poisson probability tables, but take the expected value as $n\pi$, the binomial expected value.

Example 6.13 It has been estimated that in criminal trials only 1% of all "innocent by reason of insanity" defenses are successful. Assuming that this is true, what is the probability that, in a sample of 200 trials in which the defense is attempted, at least 6 are successful?

Solution This book doesn't have binomial tables for $n = 200$ or for $\pi = .01$. Approximate the binomial probabilities by Poisson probabilities with $\mu = 200(.01) = 2.0$. Note that $P(Y \geq 6) = 1 - P(Y \leq 5)$. The cumulative Poisson probability for $y = 5$ and $\mu = 2$ is given in Table 2 as .9834. Thus $P(Y \geq 6) = 1 - .9834 = .0166$, approximately.

The Poisson approximation to binomial probabilities works best when n is large, say at least 20, and π is small, say such that the expected value $n\pi$ is no bigger than 5. The bigger the n and the smaller the π, the better is the approximation.

Example 6.14 Refer to Example 6.13. How good should the approximation be?

Solution It should be quite close, because n is 200 and the expected value is only 2.0. In fact, the exact binomial probability turns out to be .0160. □

Exercises for Section 6.2

6.9. Assume that under a stimulus, a neuron fires at the average rate of 4 firings per second. What must be assumed to assert that the number of firings in a fixed time period follows a Poisson probability distribution?

6.10. Refer to Exercise 6.9, and assume that Poisson probabilities apply. Consider a 2-second time interval.

a. Find the probability that there are 12 or fewer neuron firings in this interval. Use Table 2.

b. Find the probability that there are exactly 8 firings in the interval.

c. Find the probability that there are more than 10 firings.

6.11. Refer to Exercise 6.10. Find the expected value and standard deviation of the number of firings.

6.12. Refer to Exercise 6.10. Let X = time to next firing of the neuron.

a. Find the probability that X is between .1 and .4 second.

b. Find the expected value and standard deviation of X.

6.13. Refer to Exercises 6.9–6.11. Assume that the expected number of firings declines (in a straight line) from 6 per second at the onset of the stimulus to 2 per second 2 seconds after the onset of the stimulus. Which answers, if any, will be changed because of this change in assumptions? Will the answer to Exercise 6.12 be changed?

6.14. Refer to Exercise 6.7. Use a Poisson approximation to find the requested probability. How close is the approximation?

6.15. Refer to Exercise 6.6, but suppose that 97.5% of the implantations are successful and that there are 200 attempts.

a. Find the probability that 10 or fewer of the implantations are *unsuccessful*. Use a Poisson approximation.

b. How accurate should this approximation be?

6.16. In a study of the response to random rewards, a pigeon is trained to peck at a bar and rewarded with food after every peck. Once the pigeon has learned to peck, a random reward scheme is used. Randomly, the food-dispensing machine is turned on for 1-minute periods. This is achieved by a computerized device which signals a turn-on at the average rate of 1 signal per half hour. If a turn-on signal is received while the dispenser is already on, the dispenser is kept on for an additional minute.

a. What is the probability that the dispenser will be turned on exactly 4 times in 1 hour?

b. What did you assume in answering part a?

c. Find the expected number of turn-on signals in a 1-hour period, and the standard deviation.

6.17. Refer to Exercise 6.16. Assume that a time clock is automatically started every time a turn-on signal is received. What is the probability that another turn-on signal is received within 1 minute?

6.3 Markov Models

Very often, probability models are used to help understand behavior as it evolves over time. For example, suppose that a large group (sometimes called a **cohort**) of 20-month-old children is to be observed monthly for 36 months. Each child is to be rated on degree of socialization, as below normal (B), near normal (N), or above normal (A). This is a fairly typical example of a **longitudinal**, or **panel, study.** One of the questions is "To what extent does degree of socialization persist over time; do, for instance, below-normal children tend to stay in that category?" One reasonable way to attack such a question is to determine the probabilities for month-to-month behavior changes, and then to calculate the long-run consequences of these probabilities.

cohort

longitudinal, or panel, study

A formidably large number of probabilities would have to be specified in the most general case. One would need probabilities for the three ratings at the initial observation, say $P(B_0)$, $P(N_0)$, and $P(A_0)$. Then one would need conditional probabilities for month-1 observations, given the month-0 ratings, say $P(B_1|B_0)$, ..., $P(A_1|A_0)$; there are 9 such probabilities. Then one would need 27 different month-2 probabilities for all possible combinations of month-0 and month-1 ratings. Clearly it would be impossible to actually specify the enormous number of probabilities that in principle would be needed. A simplifying assumption is desperately needed.

Markov assumption

The most common, and most useful, simplifying assumption is the **Markov assumption.** The Markov assumption says, roughly, that the random process has no memory. Given the current state (e.g., sociability rating of a child), the past history of the process is irrelevant to future probabilities. In our example, the Markov assumption would mean that two children who were currently in the same socialization rating would have the same probabilities for future socialization, despite the possibility that one child had been improving and the other regressing in the recent past.

> **Markov Assumption**
>
> Assume that, over time, a random process moves through k mutually exclusive and exhaustive states, for times $t = 0, 1, \ldots$. The Markov assumption asserts that
>
> $$P(\text{state at } t+1 \mid \text{state at } t, \text{state at } t-1, \ldots, \text{state at } 0) = P(\text{state at } t+1 \mid \text{state at } t)$$
>
> In other words, future state and past states are conditionally independent, given the current state.

The Markov assumption *is* an assumption. A random process does not inevitably follow a Markov assumption. Rather, the assumption may be reasonable in some situations and unreasonable in others. In the sociability-rating example, we would be mildly skeptical of the assumption, on the grounds that a child who had been improving in the recent past might be expected to continue to improve, whereas a child who had been regressing might be expected to continue regressing. There are methods for using longitudinal data to test whether the Markov assumption is reasonable; see Bishop, Fienberg, and Holland (1975).

Example 6.15

In mastery learning situations, students attempt examinations about certain ideas until they can "demonstrate mastery" by obtaining a passing score on one examination. One evident problem is the possibility of trying to beat the system. A student might not study, but merely keep taking exams until one was eventually passed. Suppose that a Markov model is proposed, based on the assumption that the probability that a student will pass the next exam depends on whether or not the student passed the previous exam. What is the consequence of the Markov assumption here? Does the assumption seem reasonable?

Solution The Markov assumption would say that the probability that a student will pass the next exam, given (say) that the student passed the most recent past exam, doesn't depend on whether the student had passed the most recent exam on the first try or after 17 consecutive failures. The assertion is not entirely self-evident.

A much better model would focus on the number of consecutive failures leading up to the current exam. Although it seems obvious that a student with no recent failures has a better chance of passing an exam than one who has already failed several times, it isn't obvious that a student who hasn't failed any exams on the current topic or on the previous topic is a better (or worse) bet than one who hasn't failed any exams on the current topic but did fail one on the previous topic. □

Situations involving Poisson probabilities often are reasonable candidates for the Markov assumption. As an (overly simple) example, suppose that a certain form of behavior therapy requires precisely one month to complete, and that facilities are so limited that only one person may be undergoing therapy at a time. Candidates for the therapy arrive according to Poisson probabilities, at an expected rate $\lambda = 0.8$ per month. Candidates who cannot begin therapy at the beginning of the month are put on a wait list. Define the state of the system at the beginning of month t as the number of candidates on the wait list at that time. Next month's wait list will depend on next month's arrivals and this month's wait list, and this month's wait list is a function of past arrivals. But one of the basic Poisson assumptions is the independence of occurrences in different time intervals, so the number of past arrivals is irrelevant to the number of arrivals next month. Therefore, all that's needed to specify next month's wait list probabilities is the current wait list. Past history is irrelevant, and the Markov assumption holds because of the independence property of a Poisson process.

Example 6.16 Refer to the suicide hotline of Example 6.10. Assume that all persons calling the hotline during any one week are put on a treatment list for the following week. The probability that a person on the treatment list during any one week is removed from the list starting the next week is .7. Can "number of persons on the treatment list in a given week" be treated as a Markov process?

Solution Additions to the list are independent of past history, by the Poisson assumption adopted in Example 6.10. Whether this week's list depends on history further back than last week's list depends on how people are dropped from the list. If the probability that a person is dropped from the list doesn't depend on how long the person has been on the list, then past history is irrelevant and the Markov assumption is legitimate. □

transition probability matrix

The most important aspect of a Markov process is the **transition probability matrix.** Transition probabilities are conditional probabilities of future states, given current states, and the transition probability matrix is simply a square array of these probabilities. In the socialization example, suppose that children who are below normal at the start of one month have a .9 probability of being below normal at the start of the next month, a .09 probability of moving to near normal, and a .01 probability of moving to above normal. For children beginning near normal, the probabilities are .1, .8, and .1 of beginning the next month as below, near, or above normal, respectively. For children beginning above normal, the probabilities are .02, .18, and .80. All these probabilities are summarized in the transition probability matrix in Table 6.1.

		This Month		
		B	N	A
	B	.90	.09	.01
This Month	N	.10	.80	.10
	A	.02	.18	.80

Table 6.1

In general, the transition probability matrix of a Markov process contains, in a particular row and column, the conditional probability, given the row state, of being in the column state after the next time period.

stationary transition probabilities

An additional assumption is usually made in Markov models: that there are **stationary transition probabilities.** The assumption is that the transition probabilities are the same for month 0 to month 1 transitions as for month 1 to month 2 transitions as for .... This is a different assumption from the Markov assumption. In the context of the sociability of children, the Markov assumption says "no memory"; the stationary transition matrix assumption says that the month-to-month change probabilities are the same at the beginning of the study (20-month-old kids) as in the middle (say 38-month-old kids) as at the end (56-month-old-kids).

Example 6.17

Refer to the mastery learning situation of Example 6.15. Assume that "number of consecutive failures before the most recent exam" is a Markov process. Assume that the transition probabilities in Table 6.2 have been estimated.

		After Exam				
		0	1	2	3	4+
	0	.9	.1	0	0	0
	1	.7	0	.3	0	0
Before Exam	2	.6	0	0	.4	0
	3	.4	0	0	0	.6
	4	.3	0	0	0	.7

Table 6.2

(a) What is the interpretation of the 0 column of the transition probabilities?

(b) What does the assumption of stationary transition probabilities mean in this context?

Solution

(a) To say that there have been 0 failures before the most recent exam is to say that the student passed the most recent exam. The probability that the student will pass an exam, given that the student passed the previous exam, is .9, according to the table. The probabilities of passing an exam, given 1, 2, 3, or 4+ previous failures, are assumed to be .7, .6, .4, and .3, respectively.

(b) To say that these probabilities are stationary is to say that there is no trend of exam difficulty over the course. Early exams are no more or less difficult than late ones (for a student with a specified number of failures on the most recent topic). □

Just as Poisson models lead naturally to Markov models, so do they lead to stationary transition probabilities. In our discussion of the wait list Markov model in this section, we noted that the independence assumption of a Poisson process led naturally to the satisfaction of the Markov assumption. If the value of λ in the Poisson process doesn't change over time (in the example, if the expected arrivals per month stays constant), then the transition probabilities won't change.

Example 6.18

Refer to the suicide hotline of Examples 6.10 and 6.16. Assume that the rate of calls (and therefore additions to the treatment list) varies seasonally. Does the stationary transition probability assumption hold?

Solution

No. In relatively heavy calling seasons, the probabilities of an increasing list will be relatively high; in light calling seasons, these probabilities will be relatively low. □

One of the important uses of Markov models, given that the Markov and stationary transition probability assumptions are met, is the calculation of multi-period transition probabilities. If a child is below normal in sociability now, what is the probability that the child will be below normal 6 months from now? 18 months from now? No new probability principles are involved in the calculations; the assumptions merely simplify the problem of specifying so many probabilities. In particular, a probability tree can be used to calculate many-step transition probabilities. A probability tree for 3-month transitions, assuming that the child is below normal at month 0, is shown in Figure 6.4. Note that to decide what probabilities to put on particular branches, one looks only at the immediately preceding branch, not at anything that has gone before; this is a consequence of the "no-memory" Markov assumption. Also, the same transition probabilities are used in the month-2 branches as in the month-3 branches, as a consequence of the stationary transition probability assumption. The 3-month transition probabilities, starting at B, may be calculated simply by adding up the appropriate path probabilities.

Example 6.19

Refer to the (assumed stationary) transition probability matrix in Table 6.2. Find the probability of going from state 1 to state 0 in 2 steps.

Solution

The only ways to get from 1 to 0 in 2 steps are 1-2-0 and 1-0-0; transitions 1-1-0, 1-3-0, and 1-4+-0 have 0 probability. The probability of 1-0-0 is (.7)(.9) = .63, and the probability of 1-2-0 is (.3)(.6) = .18. Thus the probability of a 2-step transition from 1 to 0 is .63 + .18 = .81. □

Tree calculations are obviously going to be messy for, say, 18-month transitions. The calculations can be done in a much tidier way, using matrix algebra. Therefore, it's easy to write a computer program to do the arithmetic, and we'll not spend time on the computational prob-

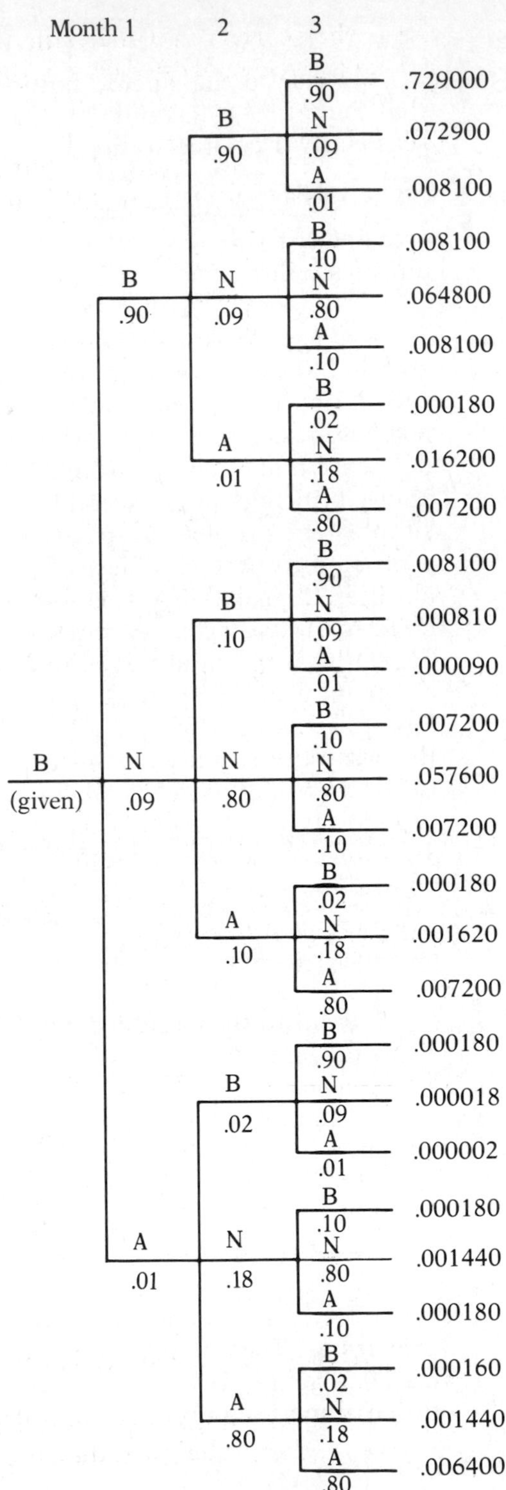

Figure 6.4 Three-month transition probabilities, by tree

lems. The point is that the same probability principles are being used. For those who know matrix notation: The 2-period transition probability matrix is found by multiplying the 1-period matrix by itself, the 3-period matrix is found by multiplying the 2-period matrix by the 1-period matrix, and so on. A "homebrew" program yielded the selected transition probabilities in Table 6.3.

TWO-MONTH

	A	B	C
A	.8192	.1548	.0260
B	.1720	.6670	.1610
C	.0520	.2898	.6582

FOUR-MONTH

	A	B	C
A	.6991	.2376	.0633
B	.2640	.5182	.2178
C	.1267	.3921	.4812

EIGHT-MONTH

	A	B	C
A	.5594	.3140	.1265
B	.3489	.4166	.2344
C	.2530	.4220	.3250

SIXTEEN-MONTH

	A	B	C
A	.4546	.3599	.1855
B	.3999	.3821	.2180
C	.3710	.3924	.2366

THIRTY-TWO-MONTH

	A	B	C
A	.4194	.3739	.2067
B	.4155	.3755	.2091
C	.4134	.3763	.2103

Table 6.3

Example 6.20 Refer to Example 6.17. The 10-step transition probability matrix in Table 6.4 was computed.

		Current State				
		0	1	2	3	4+
	0	.8587	.0859	.0258	.0103	.0192
	1	.8553	.0854	.0256	.0102	.0235
Previous State	2	.8498	.0846	.0252	.0100	.0303
	3	.8427	.0836	.0248	.0097	.0392
	4	.8400	.0832	.0246	.0096	.0426

Table 6.4

What is the interpretation of the probability in the 0, 0 cell?

Solution If an individual has passed the most recent exam (and is therefore in state 0), the probability that the individual will pass the 10th exam hereafter is .8587. Note that the probabilities for current state 0 are very similar for all previous state values, indicating that how many failures had occurred before an exam 10 tries ago has very little predictive value in assessing how one is doing on the current exam. □

A quick look at the many-month transition probabilities of the socialization example indicates another important characteristic of Markov models. The rows of the transition matrices all are tending to the same limit, namely .42, .37, .21. If all the entries of the (1-period) transition matrix are positive (and in some other cases), every row of the multi-period transition matrix will converge to the same probabilities. These are the long-run probabilities of the various states. In the socialization example, in the long run it doesn't matter where the child started. After a long time, there is a certain set of probabilities for B, N, and A (namely .42, .37, and .21), which holds regardless of the initial position. Of course, this statement only holds if the Markov and stationary transition probability assumptions are valid, and if some technical conditions are met. One way to calculate these long-run probabilities is simply to have the computer grind out the long-run transition matrices.

Example 6.21 Refer to Example 6.20. Use the 10-step transition matrix to guess the long-run probabilities.

Solution The first column entries all are about .85, which is a reasonable guess for the long-run probability of state 0. Roughly, the others seem to be about .08 (maybe .09), .02 (.03?), .01, and .02 (.03?). In fact, the long-run probabilities are .8576, .0858, .0257, .0103, and .0206. □

There are many other questions that can be asked of a Markov model. Many technical issues need to be dealt with, as well. Books such as Kemeny and Snell (1960) have been written on the subject. The key for the nonspecialist is to understand exactly what the assumptions mean, and to view skeptically any claims from a Markov model which are not accompanied by a defense of the assumptions.

Exercises for Section 6.3

6.18. Refer to Exercise 6.16. Suppose that the food dispenser is turned on or off every minute according to the following scheme: If the dispenser is on during one minute, it remains on during the next minute with probability .09; if it is off during one minute, it is turned on during the next minute with probability .03.

a. What must be assumed to make the on/off process a Markov process?

b. If the process is Markov, are the transition probabilities stationary?

6.19. Refer to Exercise 6.18. Assume that the on/off process is a Markov process with stationary transition probabilities.

a. Write out the transition probability matrix.
b. Find the 2-minute transition probabilities.

6.20. Refer to Exercise 6.19.

a. Find the 4-minute transition probabilities. (It might be easier to consider two 2-minute transitions than four 1-minute transitions.)

b. Can you guess roughly what the long-run probabilities of on and off are?

6.21. Refer to Exercise 6.20. If you have a computer program available to multiply out the transition matrices, find the 8-minute and 16-minute matrices. What are the long-run probabilities of on and off?

6.22. Laboratory rats are put through a long series of maze-learning tasks. At each task, each rat is classified as a quick (Q), average (A), or slow (S) learner. The transition matrix in Table 6.5 has been obtained from a large number of rats.

		Next Task		
		Q	A	S
This Task	Q	.9	.1	0
	A	.1	.8	.1
	S	0	.1	.9

Table 6.5

a. What is assumed when learner status is claimed to be a Markov process?

b. What is assumed when the transition probability matrix is claimed to be stationary?

6.23. Refer to Exercise 6.22. Assume that the process is Markov and that the transition probabilities are stationary.

a. If a rat is currently a slow learner, what is the probability that it will be a slow learner 2 tasks from now?

b. If a rat is currently a slow learner, what is the probability that it will be a quick learner 3 tasks from now?

6.24. Refer to Exercise 6.22. Assume that the process is a Markov process with stationary transition probabilities.

a. Calculate the 2-step, 4-step, and 8-step transition probability matrices.

b. What are the long-run probabilities of the three learning states?

6.4 The Normal Distribution

The single most important probability distribution for statistics is the normal distribution. Many naturally occurring variables have approximately normal distributions. Scores on many standardized tests are deliberately scaled to have normal distributions. Normal distributions are a formal mathematical assumption underlying many of the statistical procedures that we will discuss. And finally, normal distributions are the approximate theoretical distributions for an astonishing variety of statistics used in statistical inference. In much of the rest of this book, normal probabilities will play a leading role.

One reason for the prevalence of the normal distribution in statistics is the so-called Central Limit Theorem, which we will discuss in the next chapter. Therefore, we will defer discussion of when normal probabilities may be expected to apply until that chapter. In this section, we will discuss the calculation of normal probabilities and how these probabilities relate to the binomial and Poisson probabilities, which were introduced in Sections 6.1 and 6.2.

Mathematically, the normal distribution is a continuous distribution. A normally distributed random variable should, formally, be capable of attaining any value (integer, fraction, or decimal) between $-\infty$ and ∞. Of course, in practice, random variables are only measured to some number of decimal places, so not all numerical values can be attained. Furthermore, many random variables cannot attain negative values; although IQ scores are usually assumed to be normally distributed, there is no such thing as a negative IQ. Fortunately, these formal requirements are not critical in practice. Even when the requirements are not literally met, normal probabilities often provide a totally adequate approximation.

For continuous random variables, probabilities are calculated as areas under a curve. The normal curve depends on the assumed mean and standard deviation. It is symmetric around the mean; its "spreadoutness" is determined by the standard deviation. Some normal curves are shown in Figure 6.5.

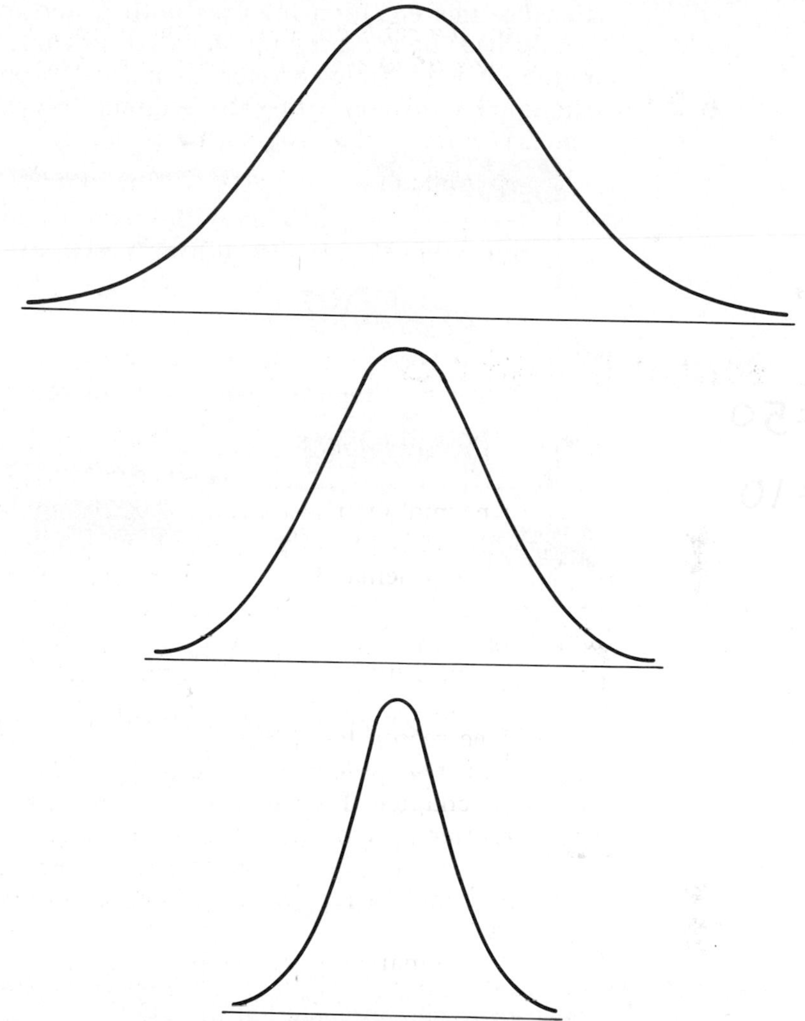

Figure 6.5 Normal probability distributions

The equation of the normal probability density is hard to handle by standard calculus methods, so normal probabilities are calculated in this book by table rather than by calculus. The equation for the normal probability density is

$$f_Y(y) = \frac{1}{\sqrt{2\pi}\,\sigma} e^{-.5(y-\mu)^2/\sigma^2}$$

(Interestingly, two of the most famous mathematical constants, $e = 2.71828\ldots$ and $\pi = 3.14159\ldots$, appear in the same defining equation.)

Because this equation involves both μ and σ, it would seem that we would need a separate table for each possible value of μ and σ. Fortunately, it is possible to refer all normal probability calculations to a single table, by converting the original "raw scores" to z-scores.

> **z-scores**
>
> For a random variable with mean μ and standard deviation σ, a z-score is calculated from a raw score y by
>
> $$z = \frac{y - \mu}{\sigma}$$
>
> Conversely, the raw score is calculated from the z-score by
>
> $$y = \mu + z\sigma$$

For example, suppose the mean is 50 and the standard deviation is 10. Corresponding to original raw scores of 65 and 35 are, respectively,

$$z = \frac{65 - 50}{10} = 1.50 \qquad \text{and} \qquad z = \frac{35 - 50}{10} = -1.50$$

Further, a z-score of -2.30 corresponds to a raw score of $50 + (-2.30)(10) = 27$.

Example 6.22 Assume a random variable with a mean of 80 and a standard deviation of 20. Find z-scores corresponding to raw scores of 100, 90, 80, 56, and 20.

Solution The raw score of 100 is 1 standard deviation above the mean, so the z-score should come out to 1.00.

$$z = \frac{100 - 80}{20} = 1.00$$

The raw score of 90 corresponds to a z-score of 0.50, indicating that it is half a standard deviation above the mean. Because 80 is right at the mean, its z-score is 0.00. The other two scores are below the mean and therefore have negative z-scores, -1.20 and -3.00, respectively. □

Example 6.23 Suppose you are told that your exam score, in z-score form, is 3.67. What does that tell you about your grade? How about a z-score of

-0.20? Assume that the mean grade is 120 and that the standard deviation is 15; calculate the corresponding raw scores.

Solution A z-score of 3.67 indicates that your score was far above average, whereas a z-score of -0.20 indicates that your score was slightly below average. A z-score of 3.67 corresponds to a raw score of $y = 120 + (3.67)15 = 175$, and a z-score of -0.20 corresponds to $y = 120 + (-0.20)15 = 117$. □

Normal tables, such as Table 3 in the appendix, are almost always stated in terms of z-scores. Therefore, the first step in calculating a normal probability is to convert to z-score form; this should become a Pavlovian reflex. Table 3 gives probabilities (areas under the normal curve) as shown in Figure 6.6. The best way to use these tables is to sketch pictures like those of Figure 6.6.

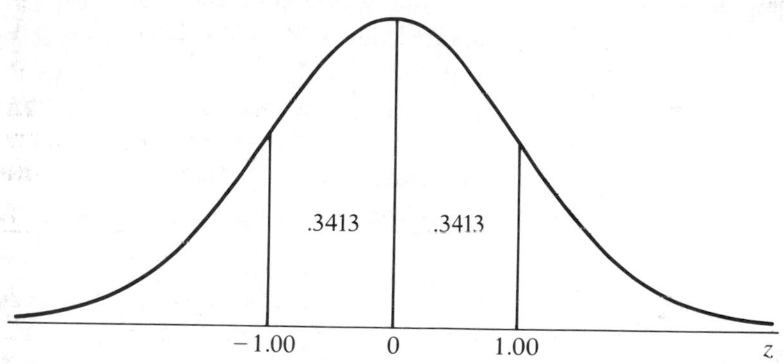

Figure 6.6 Normal table areas (probabilities)

For example, suppose that a random variable has a normal distribution with mean 50 and standard deviation 10. What is the probability that it will take a value between 50 and 65? Between 35 and 50? Above 65? Below 35? First convert to z-scores; 65 corresponds to $z = 1.50$, 50 to $z = 0.00$, and 35 to $z = -1.50$. The entry corresponding to $z = 1.50$ is .4332; this is the area (probability) for a z-score between 0.00 and 1.50, corresponding to a raw score between 50 and 65, as shown in Figure 6.7. Therefore, the probability of a value between 50 and 65 is .4332. Because the normal curve is symmetric around its mean (0 in z-score form), the area between -1.50 and 0.00 is also .4332, which is the probability of a score between 35 and 50. Again, see Figure 6.7. From this figure, it's clear that the area between 35 and 65 (z-scores between -1.50 and 1.50) is $.4332 + .4332 = .8664$, and that the area above 65 (z-score $= 1.50$) is .0668, as is the area below 35 (z-score $= -1.50$).

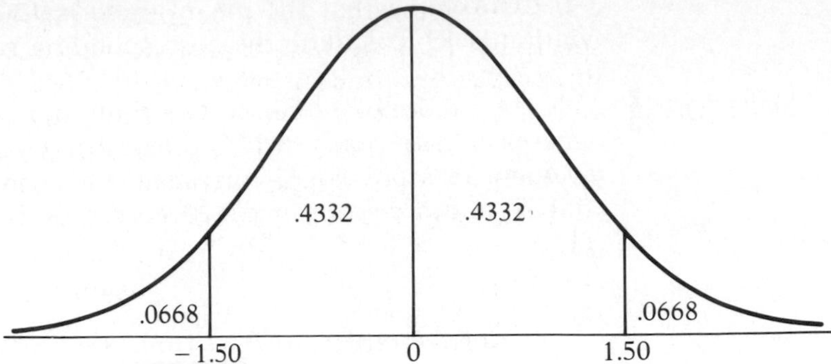

Figure 6.7 Normal curve areas

Example 6.24 Assume that exam scores are normally distributed with mean 120 and standard deviation 15. Find the probabilities of exam scores

(a) Between 105 and 135 **(b)** Between 105 and 150
(c) Between 90 and 105 **(d)** Below 90

Solution See Figure 6.8.

(a) $P(105 < Y < 135) = P(-1.00 < Z < 1.00) = .3413 + .3413 = .6826$

(b) $P(105 < Y < 150) = P(-1.00 < Z < 2.00) = .3413 + .4772 = .8185$

(c) $P(90 < Y < 105) = P(-2.00 < Z < -1.00) = .4772 - .3413 = .1359$

(d) $P(Y < 90) = P(Z < -2.00) = .5000 - .4772 = .0228$ □

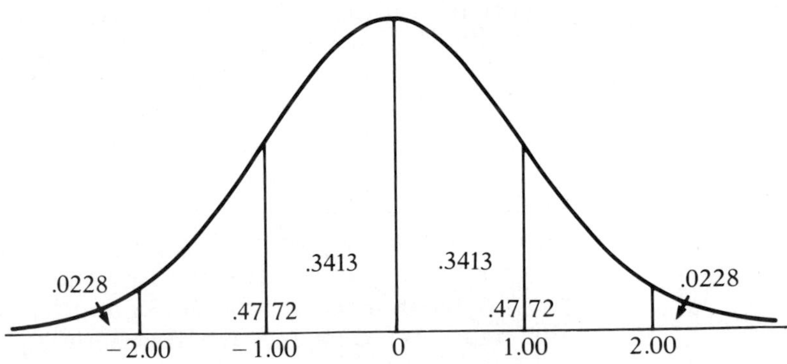

Figure 6.8 Solution for Example 6.24

The z table may also be read in the opposite direction. Instead of specifying a z-score and finding a probability, we often will specify a probability and find the associated z-score. The z-score can then be converted back to a raw score. For example, suppose that the mean of a normal distribution is 50 and that the standard deviation is 10. What is the 90th percentile of scores? According to Figure 6.9, the 90th percentile (which has 90% of the scores below it, 10% above) corresponds to a table area of .4000. The closest entry to .4000 in Table 3 corresponds to $z = 1.28$. In raw score form, the 90th percentile is $y = 50 + (1.28)(10) = 62.8$.

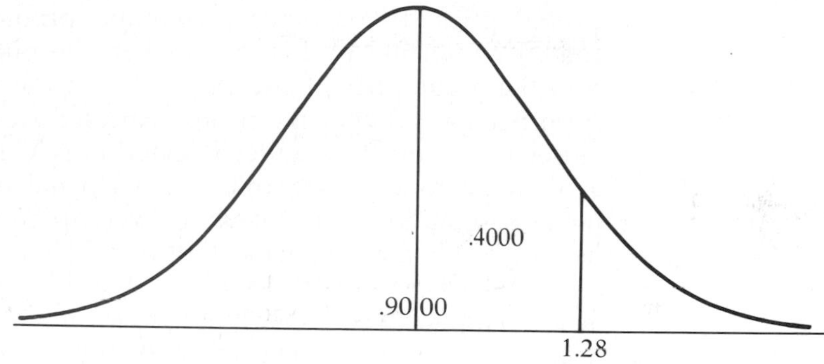

Figure 6.9 90th percentile of a normal distribution

Example 6.25 For the normal distribution of Example 6.24, find the middle 90% range of scores.

Solution See Figure 6.10. We must look for a probability of .4500 in Table 3. The closest entries are .4495 ($z = 1.64$) and .4505 ($z - 1.65$). Splitting

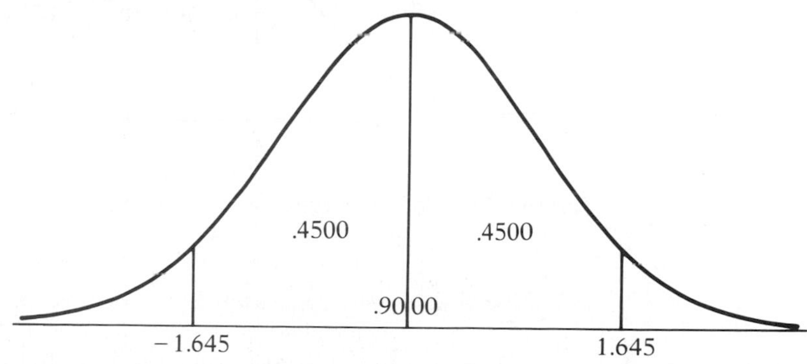

Figure 6.10 Middle 90% of normal distribution values

the difference, we take $z = 1.645$, corresponding to a raw score of $y = 120 + (1.645)(15) = 144.675$, or $y = 144.7$, rounded. The lower limit is $120 - (1.645)(15) = 95.3$. □

The Empirical Rule values introduced in Chapter 2 (68% within 1 standard deviation of the mean, 95% within 2 standard deviations of the mean) were derived on the assumption of normal probabilities. Thus they do best, as approximations, when the distribution looks roughly normal—not too skewed, not multi-modal, not heavy-tailed.

Normal probabilities may also be used as approximations to other, exact probabilities when there are no convenient tables of the exact probabilities. In particular, binomial probabilities may be approximated by normal probabilities when the number of trials, n, is large and the probability of success, π, is not too near either 0 or 1. Recall from Section 6.1 that the mean (expected value) of the number of successes is $n\pi$ and the standard deviation is $\sqrt{n\pi(1 - \pi)}$. Simply act as if a binomial random variable had a normal distribution, calculate the mean and standard deviation, convert to z-scores, and use Table 3 as usual. For instance, suppose that $n = 225$, $\pi = .20$. What is the probability that there will be between 39 and 57 successes (both inclusive)? The expected value is $225(.20) = 45$, and the standard deviation is $\sqrt{225\,(.20)(.80)} = 6.0$. The value 39 corresponds to a z-score of -1.00, and 57 corresponds to a z-score of 2.00. From Figure 6.11, we can see that the probability that there will be between 39 and 57 successes is approximately $.3413 + .4772 = .8185$.

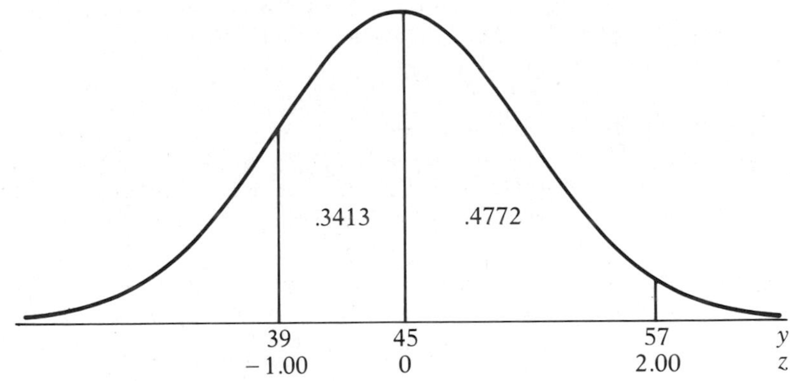

Figure 6.11 Normal approximation to a binomial

Example 6.26

A counselor at a large university believes that 10% of all seriously depressed students will make a suicide attempt (usually unsuccessful) at some time. Would it be improbable that 18 or more out of 100 seriously depressed students would make such attempts?

Solution

Binomial probabilities should apply. The probability of a suicide attempt no doubt differs somewhat across students, but probably not enough to grossly affect the probabilities. There might be some dependence effect, with one attempt triggering others, but again it seems unlikely that the dependence would be great. Assuming that binomial probabilities are a reasonable model, the expected value is $100(.10) = 10.0$ and the standard deviation is $\sqrt{100(.10)(.90)} = 3.0$. Using a normal approximation,

$$P(Y \geq 18) = P(Z \geq (18 - 10.0)/3.0) = P(Z > 2.67)$$
$$= .5000 - .4962 = .0038$$

The probability is quite small. □

The standard rule of thumb for using normal probabilities as an adequate approximation to exact binomial probabilities is that the expected number of successes, $n\pi$, and the expected number of failures, $n(1 - \pi)$, should both be at least 5.0. Generally, the larger the expected number of successes and the expected number of failures are, the better is the normal approximation. For $n = 225$ and $\pi = .20$, the expected numbers of successes and failures are, respectively, 45 and 180, so the approximation should be excellent. In fact, the exact probability that there will be between 39 and 57 successes can be calculated to be .8406, as compared to the approximation .8185.

Example 6.27

Refer to Example 6.26. Use Table 1 to calculate the exact binomial probability of 18 or more attempts. Compare it to the normal approximation value. How close are the exact and approximate values?

Solution

Enter Table 1 for $n = 100$ and $\pi = .10$; the sum of the probabilities for $y = 18, 19, \ldots$ is .0100, the exact binomial probability of 18 or more attempts. The normal approximation value was .0038. The magnitude of the error is $.0100 - .0038 = .0062$. In absolute terms this is a small error, but relative to the (small) exact probability the proportional error is large (62% in error). □

In Section 6.2 we noted that Poisson probabilities could approximate binomial probabilities when n was large and the probability of success was small. Now we know that normal probabilities can be used to approximate binomial probabilities when n is large and the probability of success is moderate. (If the probability of success is very large, reverse the definitions of success and failure, and use Poisson probabilities.) To complete the triangle, it is also possible to use normal probabilities to approximate Poisson probabilities when the Poisson mean is large (say larger than 10 or so). Recall from Section 6.2 that the standard deviation of a Poisson random variable is the square root of its mean. When the mean is large, we simply act as if a Poisson random variable had a normal distribution. For instance, if the Poisson mean is 25, what is the probability that the random variable will attain a value of 15 or less? The mean is 25 and the standard deviation is $\sqrt{25} = 5$, so the z-score corresponding to 15 is 2.00. From Figure 6.12 we see that the desired probability is approximately .0228.

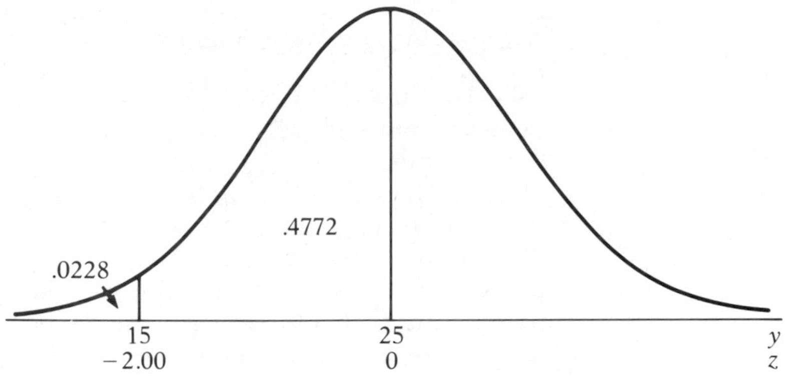

Figure 6.12 Normal approximation to a Poisson distribution

Example 6.28

Assume that Y has a Poisson probability distribution with an expected value of 9.0. Use a normal approximation to calculate the probability that Y is no larger than 12. Use Table 2 to calculate the exact probability. How close is the approximation? Note that the rule of thumb that the mean should be at least 10 is violated.

Solution

We have $\mu = 9.0$, so in the Poisson case $\sigma = \sqrt{9.0} = 3.0$. Using a normal approximation,

$$P(Y \leq 12) = P(Z \leq (12 - 9.0)/3.0) = P(Z \leq 1.00)$$
$$= .5000 + .3413 = .8413$$

The Table 2 entry for $y = 12$ and $\mu = 9.0$ is .8758, which is the exact value for $P(Y \leq 12)$. The approximation is only fair; the error is .0345. □

The quality of a normal approximation to either binomial or Poisson probabilities can be improved a bit by recognizing that we're using a continuous curve to approximate discrete probabilities. For instance, suppose that we tried to approximate $P(7 \leq Y \leq 10)$, where Y is binomial with $n = 16$ and $\pi = .50$. From Table 1 of the appendix, the correct probability is .6678, whereas the normal approximation is the area between $z = -0.50$ and $z = 1.00$, which is .5096. A picture of the situation is shown in Figure 6.13. Notice that the correct probability is the sum of the areas of the rectangles centered at 7, 8, 9, and 10, and that the bases of these rectangles extend from 6.5 to 10.5. By taking the area under the normal curve from 7 to 10, we are missing the left half of the 7 rectangle and the right half of the 10 rectangle. To improve the approximation, we use a **continuity correction**; rather than taking the area under the curve from 7 to 10, we take it from 6.5 to 10.5, thus better matching the rectangles. The revised approximation is the area under the normal curve between $z = -0.75$ and $z = 1.25$, which is .6648, a much better approximation.

continuity correction

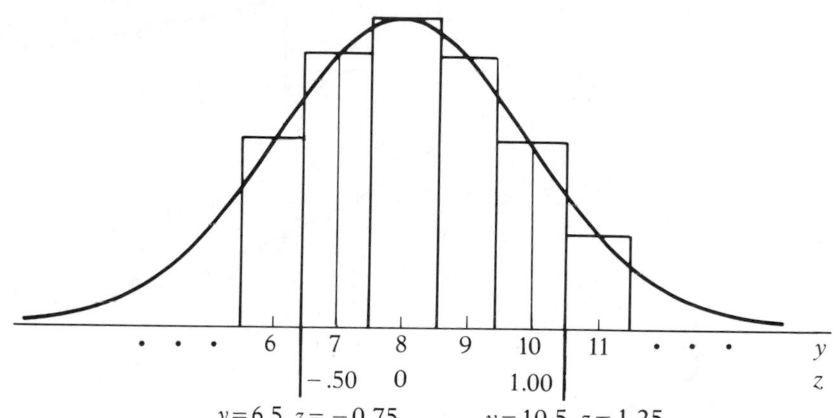

Figure 6.13 Continuity correction for approximating binomial probabilities

The best way to use the continuity correction is to sketch a picture like Figure 6.13. The picture will indicate what endpoints to use in the normal approximation.

Example 6.29

Use a continuity correction for the normal approximations of Examples 6.26 and 6.28. In each case, does the correction improve the approximation?

Solution In Example 6.26, the continuity correction replaces $P(Y \geq 18)$ by $P(Y \geq 17.5)$, as in Figure 6.14(a). The normal approximation is

$$P(Y \geq 17.5) = P(Z \geq (17.5 - 10.0)/3.0\,) = P(Z > 2.50)$$
$$= .5000 \;-\; .4938 = .0062$$

which is very close to the exact value of .0068.

In Example 6.28, correct $P(Y \leq 12)$ to $P(Y \leq 12.5)$ as in Figure 6.14(b). The normal approximation is

$$P(Y \leq 12.5) = P(Z \leq (12.5 - 9.0)/3.0) = P(Z \leq 1.17)$$
$$= .5000 + .3790 = .8790$$

which is quite close to the exact value, .8758. In both cases, the approximation is improved considerably. ☐

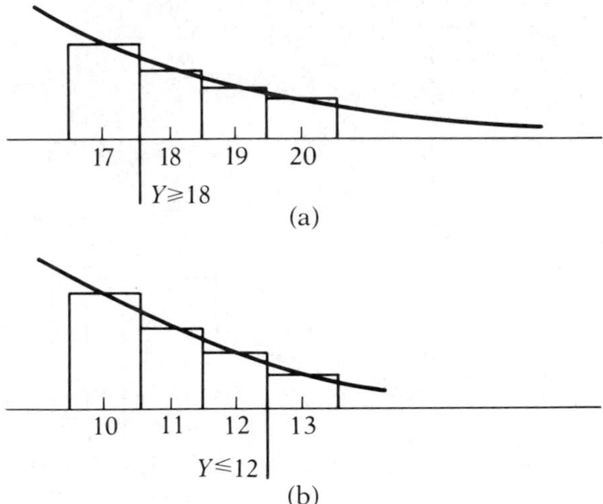

Figure 6.14 Sketch of continuity corrections

Exercises for Section 6.4

6.25. IQ scores are thought to be normally distributed with mean 100 and standard deviation 15.

a. Find z-scores corresponding to IQ scores of 121, 88, and 65.

b. Find the IQ scores corresponding to z-scores of 2.20, −0.60, and 1.33.

6.26. Refer to Exercise 6.25.

a. Find the probability of obtaining an IQ score between 80 and 120.

b. Find the probability of obtaining an IQ score above 125.

c. Find the probability of obtaining an IQ score between 75 and 85.

6.27. Refer to Exercise 6.25.

a. Find the 80th percentile of z-scores—that is, the z-score that has 80% of all z-scores below it and 20% above it.

b. What is the corresponding IQ score?

c. Find z-scores, symmetric around 0, that include the middle 90% of all z-scores.

d. What are the corresponding IQ scores?

6.28. Refer to Exercise 6.25. If an IQ score is above 130, what is the probability that it is above 140? (Use the definition of conditional probability.)

6.29. Suppose that scores on a depression scale are normally distributed with mean 80 and standard deviation 20.

a. Find z-scores corresponding to depression scores of 95, 82, 78, and 65.

b. Find depression scores corresponding to z-scores of 0.85, -1.55, and 2.68.

6.30. Refer to Exercise 6.29.

a. Find the probability that a depression score will be between 70 and 90.

b. Find the probability that a depression score will be greater than 130.

c. Find the z-score corresponding to the highest 5%, and the depression score corresponding to it.

6.31. Refer to Exercise 6.1. Assume that 80 groups of 4 subjects each are studied.

a. Find the expected value and standard deviation of the number of BCD coalitions formed.

b. Use a normal approximation without continuity correction to approximate the probability that 50 or more BCD coalitions form.

6.32. Refer to Exercise 6.31.

a. Repeat part b, using a continuity correction.

b. Is the continuity-corrected approximation likely to be close to the exact probability?

6.33. Refer to Exercise 6.9. Assume a constant expected rate of firing equal to 4 per second.

a. Determine the expected value and variance of the number of firings in a 60-second period.

b. Use an uncorrected normal approximation to estimate the probability that there will be fewer than 200 firings in such a period.

6.34. Refer to Exercise 6.33.

a. Repeat part b, using a continuity correction.

b. How important is the continuity correction in this situation? How accurate should the approximation be to the exact probability?

Chapter Exercises

6.35. Scores on standard tests are sometimes scaled to have a normal distribution with mean 50 and standard deviation 10. Assume that a certain test has these characteristics.

a. What is the probability that a randomly chosen subject will score between 38 and 62?

b. What is the probability of scoring above 68? Below 32?

c. What is the 80th percentile of the scores? The 20th percentile?

6.36. Refer to Exercise 6.35.

a. If a subject scores above 50 on the test, what is the probability that the score is below 68?

b. What is the probability that a score will be below 0 or above 100?

6.37. The probability that an infant subject will actually appear at a scheduled session is .8.

a. Suppose that 6 subjects are scheduled to appear at a session. Write an expression for the probability that exactly 4 of the subjects appear.

b. Use Table 1 in the appendix to find a numerical value for the probability indicated in part a.

6.38. Refer to Exercise 6.37. Find the expected number of subjects appearing at the session, the variance, and the standard deviation.

6.39. Assume once again that the probability that a subject will appear as scheduled is .8, but now assume that subjects are scheduled one at a time. The experimenter wants 20 subjects, in all, for a particular study.

a. What is the probability that the 20th subject appears at the 28th scheduled session?

b. What is the expected number of scheduled sessions required to obtain 20 subjects? What is the standard deviation of the number of scheduled sessions?

6.40. Refer to Exercises 6.37–6.39. What assumptions were made in answering these questions? Are there plausible conditions under which the assumptions might be grossly wrong?

6.41. Subjects who are seated in a darkened room report perceiving flashes of light randomly, at a rate of 0.75 flashes per minute. Suppose that a particular subject reports Y = the number of perceived flashes in an 8-minute period.

a. What is the probability that Y is less than or equal to 3?
b. What is the probability that Y is between 4 and 8 (both included)?
c. Determine the expected value and standard deviation of Y.

6.42. Refer to Exercise 6.41.

a. What is the probability that at least 2 minutes elapse before the first flash is reported?
b. Find the expected time to the first reported flash.

6.43. Refer to Exercises 6.41 and 6.42. What did you assume in answering the questions?

6.44. In a study of laboratory rats' behavior, a feeding machine is set so that it will respond always (A), sometimes (S), or never (N) to a learned bar-press action of a rat. The response of the machine is changed, randomly, every 5 minutes. If the machine is currently responding A, it will change to S with probability .1 and to N with probability .7. If the machine is currently responding S, it will change to A with probability .1 and to N with probability .6. If the machine is currently responding N, it will change to A with probability .1 and to S with probability .1.

a. Construct a transition probability matrix.
b. Can this matrix be assumed to be stationary?
c. Under what conditions will the current response of the machine constitute a Markov process?

6.45. Refer to Exercise 6.44, and assume that the machine is currently in the A condition. What is the probability that 3 periods (15 minutes) later it will be in the N condition?

6.46. The 4-period and 8-period transition matrices for the machine in Exercise 6.44 were calculated as shown in Table 6.6.

a. What does the 8-period matrix indicate about how well one can predict the state of the machine 8 periods hence?
b. What are the long-run probabilities of A, S, and N?

4-PERIOD

	A	S	N
A	.1112	.1248	.7640
S	.1111	.1264	.7625
N	.1111	.1248	.7641

8-PERIOD

	A	S	N
A	.111111	.125000	.763889
N	.111111	.125002	.763887
S	.111111	.125000	.763889

Table 6.6

6.47. Refer to Exercise 6.46. Assume that the long-run probability of S is .125. Would it be correct to use binomial probabilities with $n = 10$ and $\pi = .125$ to find the probability that the machine will be in S in exactly 2 of 10 consecutive periods? Why or why not?

6.48. Suppose that a multiple-choice examination has 4 possible answers to each question, and that a student guesses answers randomly. There are 20 questions.

a. What is the expected number of correct guesses?

b. What is the probability of exactly 5 correct guesses?

c. What is the probability of 5 or more correct guesses?

6.49. Refer to Exercise 6.48. Suppose that the student gains 3 points for every correct guess and loses 1 point for every incorrect guess. No gain or loss is recorded if the answer is omitted.

a. What is the expected amount of credit if a student guesses the answer to one question?

b. If the student attempts 14 questions and really is guessing, what is the probability that the student will get a positive score? (Suggestion: How many correct guesses are needed to get a positive score?)

6.50. Refer to Exercise 6.48. Now suppose that there are 100 questions.

a. Using Table 1 in the appendix, find the probability of 20 or fewer correct guesses.

b. Find the expected value and standard deviation of the number of correct guesses.

c. Use a normal approximation without continuity correction to find an approximation to the probability in part a.

6.51. Answer part c of Exercise 6.50 using a continuity correction. Which approximation is closer—the one with continuity correction or the one without?

6.52. In a mastery-learning environment, students are tested and retested until they demonstrate mastery of a particular set of skills. Assume that experience has indicated that 85% of students pass the first test for acquisition of a particular set of skills.

a. In a class of 20 students, what is the probability that 4 or more will not pass the first test?

b. What is the expected number of students who do not pass the test?

6.53. Refer to Exercise 6.52.

a. Use a normal approximation (either with or without continuity correction, as you wish) to approximate the probability in part a.

b. According to the rule of thumb for using normal approximations to binomial probabilities, should this approximation be close? Is it?

6.54. Refer to Exercise 6.52. Find the Poisson approximation to the probability in part a. How close an approximation is it?

6.55. A psychiatric social service agency receives applications for employment from qualified people at an average rate of 3.5 per year.

a. What is the probability that the agency will receive at least 5 applications from qualified people in the next year?

b. What assumptions did you make in answering part a?

6.56. Refer to Exercise 6.55. What is the probability that the next qualified applicant will apply within 1 month (1/12 of a year)?

6.57. The position of director of a state psychiatric hospital becomes vacant every so often. The probability that a current director will leave the job in any one month is .04. If the position is vacant, the probability that a new director will fill the position in the next month is .12. Define a two-state (vacant or filled) process.

a. Show that the transition matrix is

	F	V
F	.96	.04
V	.12	.88

b. What assumptions would have to be made in order to consider the process a Markov process with stationary transition probabilities?

c. Assuming that it is such a process, and assuming that the position is currently vacant, find the probability that it will be filled 4 months from now (whether or not it is filled between now and then).

6.58. Refer to Exercise 6.57. Assume that the process is Markovian, with stationary transition probabilities. The 16-month and 32-month transition probabilities can be computed as shown in Table 6.7. What fraction of the time is the director's job filled, in the long run?

6.59. A population of students consists of 80% with no learning disability, 15% with a mild disability, and 5% with a serious disability. A preliminary diagnostic test has been developed. The mean score for those with no disability is 50. For those with a mild disability the mean is 40, and for those with a serious disability the mean is 30. In each disability class,

16-MONTH

	F	V
F	.765361	.234639
V	.703918	.296082

32-MONTH

	F	V
F	.750944	.249056
V	.747169	.252831

Table 6.7

the scores are normally distributed with a standard deviation of 10. Students scoring below 35 are referred for further testing.

a. For each of the disability groups, calculate the probability of scoring below 35.

b. If a student scores below 35, what is the probability that the student has a severe learning disability? You may find a probability tree or table useful.

6.60. Refer to Exercise 6.59. A random sample of 180 students from the (very large) population is to be taken and tested. Let Y denote the number scoring below 35.

a. Find the expected value and variance of Y.

b. Find the approximate probability that at least 20, but no more than 30, students in the sample score below 35.

c. Should this approximate probability be a good approximation to the exact probability?

6.61. Refer to Exercise 6.60. Suppose that students are to be tested until 20 scores below 35 are obtained. Find the expected number that will be tested.

6.62. Scores on a certain IQ test purportedly are normally distributed with a mean of 100 and a standard deviation of 15. Individuals with IQ scores below 70 are classified as retarded, and those scoring below 55 are classified as severely retarded.

a. What is the probability that an individual is classified as retarded, but not severely so?

b. If an individual is classified as retarded, what is the probability that the individual is classified as severely retarded?

6.63. Refer to Exercise 6.62. Suppose that a random sample of 2312 individuals has been taken.

a. Is it improbable that 73 or more individuals would be classified as retarded? Use a normal approximation.

b. Should the normal approximation be accurate in this case?

6.64. Refer to Exercise 6.63. How important should a continuity correction be in this case?

7 Random Sampling, Statistics, and Probability Theory

Probability theory, as introduced in the preceding three chapters, has two roles in statistics. It is the basis of probability models for scientific pheonomena, such as binomial or Poisson models. It is also critically important in statistical inference. In this chapter, we turn to the role that probability theory has in statistical inference. Probability enters because samples are taken, at least theoretically, at random. Section 7.1 discusses the basic ideas of random sampling, how these ideas are sometimes violated in practice, and the consequences of the violations. In the same section are some ideas on stratified random sampling and some principles for determining when stratified sampling is desirable. Next, in Section 7.2 the crucial idea of experimental randomization is discussed.

Then, in Section 7.3 we turn to what is probably the most central idea of statistical inference—the theoretical, or sampling, distribution of a statistic. In Section 7.4 the concepts of expected value and standard deviation that were defined in Chapter 5 are applied to the theoretical distribution of a statistic. In Section 7.5 we consider the Central Limit Theorem for means, which indicates that the normal distribution plays a central role in the distribution of the sample mean. Section 7.6 shows that the normal distribution also applies to many, but not all, other summary statistics. Finally, in Section 7.7 some of the mathematical results of the chapter are actually proved.

7.1 Random Sampling

The information available to a researcher is almost always a small fraction of the scientifically ideal information. To determine the average time required for a laboratory rat to learn to run a complicated maze under specified conditions, ideally one would want to measure every existing rat under those conditions. To find the average age at which children in the United States learn the concept of conservation of volume, ideally one would want to observe every child in the country. These ideals are utterly impractical. Of necessity, one must almost always settle for less than complete information.

population
sample

We have defined the **population** as the ideally available set of measurements and a **sample** as an actually available set. Note that conventionally the measurements (such as the ages of conservation learning), not the individuals (such as the children), constitute the population or sample. Statistical inference depends a great deal on exactly how the sample is selected from the population—that is, on the method of sampling.

simple random sampling

The most important sampling method in statistical theory is **simple random sampling.** In this method, every possible sample has the same probability as every other possible sample. The sample is chosen "completely at random." Ideally, one has a numbered list called the **sampling frame** which identifies all the entities for which one might, in principle, obtain data; that is, ideally the sampling frame is the same as the population. Then one uses a table of random numbers, or the equivalent, to choose the sample. For example, suppose that the desired population is (the conservation-learning age of) all 8407 school children in a certain school district. Think of the children as numbered from 0001 to 8407. Select the first 4-digit random number from a random number table. If it is, say, 3266, then child number 3266 is included in the sample. Keep going (ignoring any random numbers above 8407) until the desired sample size is obtained. The sampling may be done **without replacement;** if, say, 3266 occurs twice in the random number table, child 3266 is only measured once. Or the sampling may be done **with replacement;** if 3266 occurs twice, child 3266 is measured twice. There aren't many situations where sampling with replacement is sensible. Why would one want to waste effort to obtain data that one already has? But conceptually sampling with replacement leads to easier mathematics, and the difference between with-replacement and without-replacement sampling is usually utterly negligible. When 25 children out of 8407 are being sampled, it *can't* matter much whether the sampling is done with or without replacement.

(Margin terms: **sampling frame**; **without replacement**; **with replacement**)

Example 7.1

Suppose that the desired population is (the time required for learning of a certain maze by) 200,000 rats of a specified breed. In principle, how could one select a simple random sample of size 50 from this population? How crucial would the with/without replacement distinction be?

Solution

If, for example, the rats were housed in cages identified by a 6-digit cage number, use a table of 6-digit random numbers (or a computer) to select random numbers. Because only 50 rats are to be selected out of 200,000, there is only a tiny chance that any rat will be selected twice; therefore, the with/without distinction is unimportant. □

stratified random sampling

A slightly more complex, and often very useful, sampling method is **stratified random sampling.** In this method, the population is divided into subpopulations, called strata, on the basis of some characteristic. Simple random samples of specified sizes are drawn from each stratum. For example, one might want to stratify school children by sex or by whether they had attended a nursery school. Stratified random sampling is very useful if the various strata are *homogeneous*—that is, if there is very little variability within them. Figure 7.1 shows extremely homogeneous strata. In this case, one can estimate the mean (for example) of each stratum very accurately with only a few observations, and therefore estimate the overall population mean very accurately. In contrast, if there is as much variability within each stratum as there is in the whole population, there is no advantage to stratified sampling.

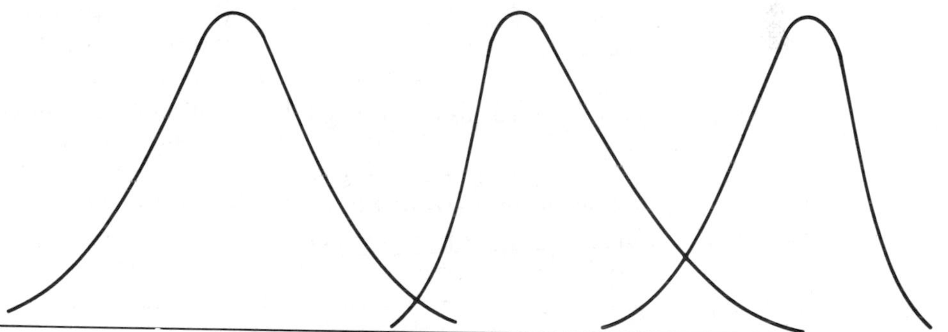

Figure 7.1 Homogeneous strata

As we will see in Chapter 14, the total variability of a set of data can be split into variability within groups (strata) and variability between strata. If there is relatively low variability (high homogeneity) within strata, there must be relatively large variability between strata.

Variability between strata is measured by differences in means. Thus, stratified sampling is advantageous when there are relatively large differences in strata means, because this implies relatively low within-strata variability.

Example 7.2 A certain new style of therapy has been proposed for alcoholics. It is planned to take a sample of psychotherapists to determine their opinion about the probable effectiveness of the therapy. Why might one stratify the sample on the basis of the type of training that the therapists have had?

Solution One reason is simply to see if there is a difference in opinion based on type of training. If there is a substantial variation in opinion between groups, according to type of training, stratification should allow for a more accurate estimate of the average opinion than would simple random sampling. □

If a stratified sample is taken, the means for various strata should be weighted to yield a reasonable estimate of the population mean. Suppose that a population is divided into 4 strata, consisting of, respectively, 10%, 20%, 50%, and 20% of the individuals in the population. To get a reasonable estimate of the population mean, we must weight the sample mean from stratum j by the proportion in stratum j. In our 4-strata population, the weights would be .1, .2, .5, and .2, respectively, corresponding to the proportions that each stratum has in the population. In general, the sample mean for stratum j is denoted by $\bar{y}_j$, and the fraction of the population represented in stratum j is denoted by N_j/N. Then the stratified sample mean is defined as shown in the box.

Stratified Sample Mean

For a stratified random sample,

$$\bar{y}_{\text{str}} = \sum_j \frac{N_j}{N} \bar{y}_j$$

where

$\bar{y}_j$ = sample mean for stratum j
N_j = size of stratum j
N = population size = ΣN_j

Example 7.3

Suppose that the therapists in Example 7.2 had been stratified into three groups based on professional training. It was known that 300 of all practicing therapists had behavioral training, 500 were classically trained, and 200 had other training. The average opinion (on a 100-point scale) by each group was as follows:

Group	B	C	O
Mean	62.4	57.2	62.9

Calculate the stratified sample mean.

Solution

$$\bar{y}_{\text{str}} = \frac{(300)62.4 + (500)57.2 + (200)62.9}{1000}$$
$$= 59.9$$

□

convenience sampling

Another, far less desirable, sampling method is **convenience sampling.** In this method, a researcher simply samples those entities that are most accessible. For example, a researcher studying the effect of different odors on the perceived taste of a food might get volunteer subjects from Psychology 1, or post a sign-up sheet in the student union. A huge problem with convenience sampling is that it may introduce biases. A **bias** is any *systematic* (as opposed to random) tendency to a distortion of the sample. A **selection bias** occurs when some members of a population have systematically higher probabilities of being sampled than do others. Convenience sampling is particularly prone to selection bias. Selecting volunteers from the Psychology 1 class rules out all nonstudents, plus those who don't study psychology. Obtaining names from a sign-up sheet for a taste/odor experiment will systematically tend to result in inclusion of students who are more interested in, and perhaps therefore more sensitive to, taste and odor. It is often possible to recognize the probable direction of a bias—that is, whether it is likely to lead to an overestimate or to an underestimate. But it is usually difficult, perhaps impossible, to know the magnitude of the bias and thus to remove the effect of the bias from data, once the bias is recognized. Therefore, the biases resulting from convenience sampling may distort the resulting data seriously.

bias

selection bias

A distressingly large percentage of the research done in psychology and education is based on convenience sampling. What percentage of the work done in psychology uses college students as subjects? Although college students are certainly convenient subjects, they differ in age, possibly in intelligence, and undoubtedly in many other ways from the general population. One has to be careful about applying the findings of research on college students to the rest of the world.

nonresponse bias

response bias

size bias

There are many other potential biases besides selection bias. Surveys are particularly prone to **nonresponse bias.** Those who choose not to answer a survey almost surely differ from those who do respond. Some data-gathering situations lead to a **response bias** by introducing a tendency to give a socially or morally desirable response. Imagine a college professor asking a sample of college students, "Have you shoplifted something from the college bookstore within the last year?" Obviously, even the stickiest-fingered student is going to respond "no." (Incidentally, note that nonresponse and response biases are not opposites, despite the names.) There is also **size bias,** in which the probability of an entity's being selected is dependent on some aspect of the size of that entity. For example, suppose that one wants to sample the mothers of children in a public school district, and selects a random sample of children. Mothers of two children are essentially twice as likely as mothers of one child to be chosen. Size bias is an unusual bias in that one can calculate the degree as well as the direction of bias. To uncover potential biases in a data-gathering plan, ask how closely the plan approximates random sampling. Do all possible measurements have equal probability (within strata, if stratification is used) of being sampled? Further, ask if there are systematic, nonrandom biases in the way the measurements are made.

Example 7.4

What biases can you detect in each of the following proposed studies?

(a) To study the childhood experiences of schizophrenics, take a sample of patients in the mental-disease wing of a nearby private hospital.

(b) To study the responses of mothers to infants, take a sample of children from day-care centers near your university, and observe them with their mothers.

(c) To study the ability of university students to reach cooperative decisions, stand at the intersection of the two most heavily travelled walkways at your university, and invite every tenth student who passes to participate.

Solution

(a) The obvious bias is that patients in a private hospital are likely to be more affluent than schizophrenics taken from the population at large.

(b) Children in day-care centers obviously are more likely to be children of working mothers. Further, they are more likely to have parents who are students, faculty, or staff members at the university. In such a sample, you would certainly underrepresent children with nonworking mothers and would probably underrepresent children with less-educated mothers.

(c) The biases here are less obvious, and perhaps less serious. You would overrepresent the gregarious student, who spends a lot of time walking around the campus, and underrepresent the shy student. Further, there's no guarantee that those who are invited to participate actually will participate. There's a distinct possibility of a nonresponse bias. □

Exercises for Section 7.1

7.1. A researcher wanted to establish norms for the ages at which U.S. children (aged 3 to 10) acquired various Piagetian conservation concepts. Ideally, what is the relevant population?

7.2. Refer to Exercise 7.1. Define an ideal sampling scheme for obtaining a simple random sample from the relevant population.

7.3. Refer to Exercise 7.1. It was proposed to take random samples of day-care centers (for kids aged 2 to 10) and to obtain conservation scores from every child in every sampled center. In what ways does this method depart from simple random sampling? In your judgment, which of these ways pose serious problems?

7.4. Recently in the news have been proposals for "merit pay" for outstanding teachers, however that may be measured. Suppose that you want to find out how current members of school boards feel about such proposals. Assume you have a questionnaire that you regard as valid. Ideally, how would you select a random sample of school board members?

7.5. Refer to Exercise 7.4. It was proposed to select entire school boards at random, in proportion to the population size served by each board. A list of school districts was available. What biases are present in such a method?

7.6. Refer to Exercises 7.4 and 7.5. Assume that a sample of school board members was selected as described in Exercise 7.5. A questionnaire was mailed to each board member; the results for all board members who responded were collected and reported. What additional biases can you see?

7.7. A research study aimed at estimating the average caseload of clinical psychologists in the United States. It was assumed that the population of practicing psychologists equaled the combined membership lists of three major professional organizations. What biases might be introduced by this assumption?

7.8. Refer to Exercise 7.7. It was proposed to stratify a random sample by type of employment (private practice, employed by state, and all other). Under what conditions would such stratification be useful?

7.9. A preliminary sample of clinical psychologists was taken as part of the study mentioned in Exercise 7.7. The means and standard deviations in Table 7.1 were obtained (based on the reported number of cases handled per week):

	Employer		
	Private	State	Other
Mean	52.6	88.2	59.9
Standard Deviation	7.9	13.9	14.6

Table 7.1

Does it appear useful to have taken a stratified sample? Explain.

7.2 Randomization in Experiments

randomization

Probability theory is used in statistical inference not only because of random sampling, but also because well-conducted experiments involve **randomization.** The idea here is that the subjects of an experiment are to be divided into two or more groups, often one or more "experimental" groups and one or more "control" groups. It is critically important in such experiments that the assignment to groups be done randomly.

confounded
placebo

For example, suppose that one wanted to test an experimental drug that promised to relieve clinical depression. A bad way to do the test would be to give the drug to a bunch of people who had been hospitalized for depression and see if they got better. This is a bad experimental design because severely depressed people might well "get better" over time for many reasons having nothing to do with the drug. In the language of experimental design, the effect of the drug would be **confounded** with the effects of all these other factors. A better experiment would be to give the drug to half the patients, and a **placebo** (a similar-appearing substance that is known to have no particular effect, such as a pill composed only of sugar) to the other half. One might allow the psychiatrists in charge to decide who should get the drug. This, too, is a bad design. Consciously or unconsciously, the psychiatrists might well give the drug to those patients who seemed to need it most, thus introducing a bias. The drug-receiving patients would look worse than they should, simply because they started off in relatively bad shape. To avoid this bias, the assignment of patients to drug should be done *randomly*.

double-blind

Experiments like this ideally should be done on a **double-blind** basis. Not only should the assignment of patients to drugs be done "blindly" (randomly), but also the person evaluating the response of each patient should not know which patients have received the drug and which the placebo. Otherwise, it is possible for the evaluator to bias the results either for or against the drug.

Example 7.5

Refer to Example 7.2. Suppose that the proposed new therapy is used on a sample of 200 alcoholics, while a competitive therapy is used with another sample of 100 alcoholics. The degree of improvement is reported by the therapists in each case. Is this a reasonable experimental design?

Solution

Ouch! As a study of the comparative effects of the two therapies, the experimental design is awful. Those who are evaluating the results of the therapies are precisely those who have an involvement in the therapies. Therefore, there are all kinds of pressures working to distort the reported results. It would be far better to have neutral observers, who didn't know which therapy had been employed, judge the improvement.

□

When experimental randomization is used, as it should be, probability theory is necessary in evaluating the results. Suppose that a drug is administered to half the depressed people (randomly chosen) in a sample, and a placebo to the other half. Further, suppose that an evaluator, who is blind to who got the drug, rates every one of the drug-receiving patients as more improved than every one of the placebo-receiving patients. That result would be mildly impressive if the total sample size were 6, but very impressive if the total sample size were 600. It is not too improbable that just by chance 3 drug patients would do better than 3 placebo patients. It is overwhelmingly improbable that all 300 drug patients would do better than all 300 placebo patients as a result of chance alone. Just how improbable the result would be must be calculated using probability theory.

Example 7.6 Suppose that in an experiment about cooperation in a committee setting, 100 subjects are tested. The subjects are divided randomly into two groups. The subjects from group A are given a set of instructions emphasizing long-term gains from cooperation; the subjects from group B are given another set of instructions emphasizing "getting what you can while you can." Then one group A member and one group B member (randomly paired) conduct a series of bargaining sessions with other committee members, who are members of neither group. At the end of the series, the experimenter records which of the two was more successful in obtaining the desired results. The same process is used for all 50 pairs of group A and group B members. What probabilities should apply?

Solution This is a binomial experiment. There are 50 trials corresponding to the 50 pairs of subjects. Each trial yields either a success (A does better, say) or a failure. The probability of success should be the same from trial to trial, because the conditions are the same. Finally, the relevant variable is the number of successes in a fixed number of trials, without regard to the order of successes and failures. These are precisely the assumptions behind a binomial random variable. Of particular relevance are binomial probabilities with $\pi = .5$. These probabilities correspond to the hypothesis that the difference between A and B groups is irrelevant and that success might as well be determined by flipping coins. If the actual result is highly unlikely given these probabilities (such as that the A group did better in 48 of the 50 trials), we would believe that something unlike mere coin-flipping is going on—that is, that the difference in instructions did yield a difference in behavior. □

It is sometimes believed that experimental randomization can make up for any bias introduced by convenience sampling. Not true. For example, compare two experiments. In the first, college students who had never smoked cigarettes were randomly divided into two groups tasting the same foods. One group tasted the food under odor-free conditions, the other while in a moderately foul-smelling environment. In the second experiment, everything was the same except that the subjects were heavy smokers aged 60 to 70. It is very reasonable to assume that the combination of age and smoking would dull the sensitivity to odor, and make the older smokers less responsive to odor than the younger nonsmokers. In both cases, the samples would most likely be convenience samples; in both cases, the assignment to groups could and should be done randomly. In neither case would the subjects be representative of the overall population and in neither case would the result reflect the average sensitivity of the overall population to odor.

Exercises for Section 7.2

7.10. An experiment about subjects' ability to distinguish red from green under varying lighting conditions was planned. The first experimental design involved 30 subjects. The first 15 were to be tested under intense light, and the remaining 15 were to be tested under dim light. Explain why this design violates the principle of randomization.

7.11. Refer to Exercise 7.10. It was proposed to have the person who assigned the color discrimination task also judge the success of each subject in discriminating red from green. What's the problem with this idea?

7.12. Refer to Exercise 7.10. After due consideration, the experimenter made a plan to take 30 undergraduate volunteers from a certain university; 15 would be assigned randomly to the intense light condition, the other 15 to dim light. A neutral observer would judge each subject's ability to discriminate between red and green. To what extent does this design resolve the problems you found in Exercises 7.10 and 7.11?

7.13. A study was planned to test the effectiveness of a new drug that was supposed to alleviate schizophrenic symptoms. One proposed design was to administer the drug to a sample of 50 patients at one state hospital, and a placebo to a sample of 50 patients at another hospital. The director of psychological services at each hospital was to judge the degree of improvement of each patient at that hospital. Criticize this experimental design.

7.14. Propose a better design for the study in Exercise 7.13.

7.15. Refer to Exercise 7.13. Under what conditions would it be useful to stratify on the basis of age?

7.3 Statistics and Their Theoretical Distributions

Both random sampling and experimental randomization require the use of probability theory. This section introduces the basic concepts and language of that use of probability theory.

The distinction between populations and samples is critical here. Recall that a population is the set of all scientifically relevant measurements that could in principle be made, and a sample is the set of

PROF. DOUGLAS GOETZ

population parameter
sample statistic

measurements that actually are made (preferably at random from the population). The critical distinction is between a **population parameter** and a **sample statistic.**

> **Parameter and Statistic**
> A statistical **parameter** is a summary number calculated from a population. A **statistic** is a summary number calculated from a sample.

The essential problem of statistical inference is to say something about a (population) parameter based on a well-chosen (sample) statistic.

If a sample is chosen randomly or if experimental randomization is used to divide the sample into subgroups, any statistic is a random variable. Recall that, informally defined, a random variable is any numerical quantity subject to random variation from one trial of an experiment to the next. A statistic is by definition numerical, and either random sampling or experimental randomization guarantees the randomness of the statistic.

theoretical distribution

Like any other random variable, a statistic has a probability distribution. In principle, the probability distribution may be derived using the tools of probability theory. In this book, we shall refer to this probability distribution as the **theoretical distribution** of the statistic.

> **Theoretical Distribution**
> The theoretical distribution of a statistic is its probability distribution—that is, the set of possible values of the statistic together with the respective probabilities of those values.

derived distribution
sampling distribution

Elsewhere, the theoretical distribution of a statistic is called its **derived distribution** or usually its **sampling distribution.** I prefer the word "theoretical" because it indicates the underlying probability theory, and also because "sampling distribution" seems to suggest "distribution of the data in the actual sample," which is something very different. The theoretical distribution of a statistic is the set of theoretical probabilities that apply to that statistic in that experiment. It could in principle be derived by repeatedly sampling from the population, calculating the statistic for each sample, and plotting the resulting values of the statistic in a histogram. The histogram of (for example) sample means chosen repeatedly from a population will be very different from the histograms of individual data points chosen in samples from the population.

The theoretical distribution of any statistic can in principle be derived by using probability theory. To illustrate, imagine we have a population in which 40% have IQ's of 100, 30% IQ's of 110, 20% IQ's of 120, and 10% IQ's of 130. (The population mean is 110, and the population variance is 100.) Suppose that we take a random sample of size 2 from this population. Figure 7.2 shows a probability tree for this situation. For each possible path, the sample mean and sample variance have been calculated. To find the probability distribution of possible values of the sample mean, we add path probabilities, as usual. The resulting probabilities are

$\bar{y}$	100	105	110	115	120	125	130
$f_{\bar{Y}}(\bar{y})$	.16	.24	.25	.20	.10	.04	.01

For the sample variance, the probabilities are

s^2	0	50	200	450
$f_{S^2}(s^2)$	.30	.40	.22	.08

Similarly, we could find the theoretical probabilities for the sample median or any other statistic. For larger sample sizes, or for a population with many different IQ values, the arithmetic would obviously be tedious, but computer programs can be written to do the dirty work.

Example 7.7

Suppose that a random sample of size 3 is taken without replacement from the following very small population:

Individual	1	2	3	4	5
Value	11	12	13	14	30

Find the theoretical probability distribution for the sample mean and the sample median.

Solution

The population is so small that it's possible to list all samples, sample means, and sample medians:

Sample	123	124	125	134	135	145	234	235	245	345
Mean	12.0	12.3	17.7	12.7	18.0	18.3	13.0	18.3	18.7	19.0
Median	12	12	12	13	13	14	13	13	14	14

For the sample mean $\bar{y}$:

$\bar{y}$	12.0	12.3	12.7	13.0	17.7	18.0	18.3	18.7	19.0
$f_{\bar{Y}}(\bar{y})$	.1	.1	.1	.1	.1	.1	.2	.1	.1

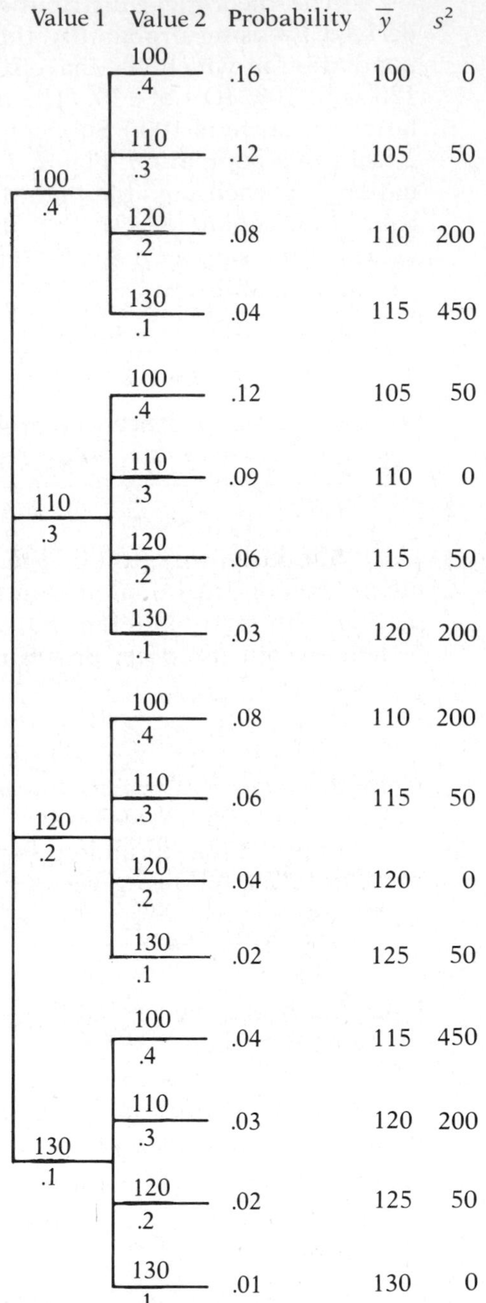

Figure 7.2 Probability tree for a random sample

For the median:

Median	12	13	14
f(median)	.3	.4	.3

□

One good interpretation of the theoretical distribution of any statistic is that it gives the long-run probabilities for the possible values of the statistic. If a specified experiment is repeated a very large number of times, then according to the long-run frequency interpretation of probability, a value of the statistic will occur in the same proportion as its theoretical probability. Thus, if we took a very large number of size-2 samples from the population of the preceding paragraph, 16% of the sample means would be equal to 100, and so on.

Example 7.8 Refer to Example 7.7. What is the interpretation of $f(\text{median} = 13) = .4$?

Solution If a very large number of samples of size 3 are taken without replacement from the population of Example 7.7, then 40% of the samples will have a median equal to 13. □

In practice, the theoretical distribution of a statistic *is* theoretical, as opposed to being empirically observed. In practice, we would *not* take a very large number of samples; we would take one sample. Therefore we would only see one value of the sample statistic, not its theoretical distribution. *A researcher will not usually be able to observe the theoretical distribution of a statistic. Rather, the theoretical distribution must be found by theoretical methods.*

Example 7.9 Each member of a very large statistics class used the Minitab computer package to draw one random sample of size 30 from a "uniform" population, which has the population distribution shown in Figure 7.3. What is the meaning of the theoretical distribution of the sample mean in this context? When could the students see the theoretical distribution?

Solution The theoretical distribution would be the probabilities (or relative frequencies) of all possible sample means obtained by an infinite number of students, each of whom drew such a sample. At the time of drawing one particular sample, no student could see the theoretical probabilities for the sample mean, because each student individually could only look at one sample mean. Only when all the results (ideally, an infinite number of them) were combined could the theoretical distribution be seen. □

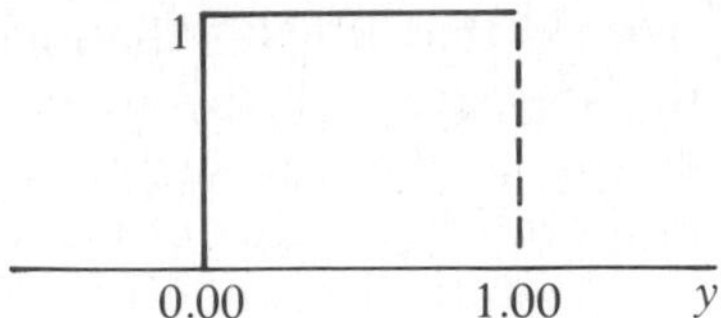

Figure 7.3 Uniform population distribution

Example 7.9 suggests an important general method for approximating the theoretical distribution of any statistic, assuming any kind of population and any sampling method. One can program a computer to take repeated samples from any specified population, using any desired sampling method. Then it's just a matter of bookkeeping, which computers do nicely, to find the proportion of samples for which any particular value of the statistic occurs. The process of having a computer imitate repeated random sampling is called the **Monte Carlo method.**

Monte Carlo method

Example 7.10

A computer program was written to draw 10,000 random samples of size 3 (without replacement) from the population of Example 7.8. The following results were obtained:

Median	12	13	14
Frequency	3019	3970	3011

Are these results close to the theoretical probabilities of Example 7.8?

Solution

The theoretical frequencies are 3000, 4000, and 3000. The Monte Carlo frequencies are quite close to these numbers. □

The computer simulation, Monte Carlo method is not the ideal way to find theoretical probability distributions. For one thing, Monte Carlo results must always be approximate; not even the fastest computer can take an infinite number of samples. Also, computer simulations must be based on extremely specific assumptions; mathematical results, when available, usually are more general. Thus computer simulations aren't a complete replacement for mathematical derivations. Simulations are very useful when no mathematical result is available.

In this book, they will also be used quite heavily to give concrete examples of somewhat abstract mathematical results.

The theoretical distribution of a statistic depends on three things:

1. What the statistic is. The theoretical probabilities for a sample mean will be different from those for a sample variance, which in turn will differ from the theoretical probabilities for a sample median.

2. How the sample was taken. Simple random sampling yields different probabilities for the sample mean than does stratified random sampling.

3. The nature of the population. If, for example, the population has a huge variance, it is reasonable that the theoretical distribution of the sample mean should have a relatively large variance. The shape of the population will affect the shape of any theoretical distribution, though the shapes will not be identical.

There is no problem in practice with the first two issues. Surely a researcher will know whether the statistic in question is a mean, a median, a standard deviation, or whatever, and the process of gathering the data determines the nature of the sampling method. But handling the third issue is more difficult. In practice, one does not know the exact characteristics of the population. If one did, there would be no need to sample. Therefore, it will be important to find statistics having theoretical distributions that don't depend heavily on the exact nature of the population. In the next section we present some results about expected values and standard deviations that are valid regardless of the specific nature of the population being sampled. Then, in Section 7.5 a critical mathematical theorem, the Central Limit Theorem, will be presented; in making inferences specifically about means, this theorem helps enormously in minimizing the effect of minor variation in the nature of the population.

Exercises for Section 7.3

7.16. A sample of 22 mentally retarded adults who were confined to state institutions were tested by Sternlicht (1979) regarding their primary fears. Of particular interest were their fears of supernatural occurrences, and the proportion of patients reporting such fears. What is the relevant parameter for this study? The relevant statistic?

7.17. Refer to Exercise 7.16. Sternlicht stratified the sample by sex. Why might this be a good idea? Under what conditions might stratification be useless?

7.18. Refer to Exercise 7.16. Assume that the theoretical probability that 11 of 22 patients will report a primary fear of supernatural occurrence is .0853. What is the interpretation of this number?

7.19. A random sample of size 16 is to be taken from a certain population, and a measure of suicide proneness is to be obtained from each person sampled. Assume that the following proportions apply to suicide-proneness scores from this population:

Score	0	1	2	3	4	5	6	7	8	9
Proportion	.900	.060	.020	.008	.004	.003	.002	.001	.001	.001

The population mean is 0.1910, and the population standard deviation is 0.7619. The theoretical probability distribution for the sample mean can be calculated from the given information. For a sample mean of 0, the theoretical probability is .1853; for a sample mean of 0.0625, the theoretical probability is .1977. Explain what these numbers mean.

7.20. A Monte Carlo simulation of the sampling of Exercise 7.19 was conducted, based on 100,000 random samples, each of size 16. The following frequencies (among others) were obtained:

Sample mean	0.0000	0.0625 ...
Frequency	18,597	19,770

Are these frequencies roughly consistent with the theoretical frequencies stated in Exercise 7.19?

7.21. Refer to Exercise 7.19. Would the same theoretical probabilities apply to the sample median statistic? Explain.

7.22. Refer to Exercise 7.19. It has been proposed to stratify the sample, so that 50% of the sample is severely depressed and the other 50% is nondepressed. Under what conditions would such sampling be a good idea? Would such sampling change the theoretical probability distribution of the sample mean?

7.23. Refer to Exercise 7.19. It was proposed to take the same kind of random sample from hospitals in a different state. Would the same theoretical probabilities apply? Why?

7.24. The λ statistic discussed in Chapter 3 is, in fact, a statistic and has a theoretical probability distribution. Under certain assumptions, the theoretical probability distribution of λ can be found. Assume that the following probabilities apply to ranges of values of the statistic, under the assumption (among others) that the true value of λ is 0.

Value	0	.001–.099	.100–.199	.200–.299	.300–.399
Probability	.21	.53	.22	.03	.01

Interpret the .53 probability.

7.25. Refer to Exercise 7.24. A Monte Carlo study involved taking 2000 random samples under the same assumptions and calculating the λ statistic for each sample. The frequencies found were

Value	0	.001–.099	.100–.199	.200–.299	.300–.399
Frequency	447	1031	427	73	22

Are these frequencies close to what would be expected if the probabilities in Exercise 7.24 are correct?

7.4
Expected Values and Standard Errors

Sample statistics are random variables and therefore have probability distributions. The general concept of the theoretical probability distribution of a statistic was discussed in Section 7.3. This section will discuss some specific aspects of these distributions, namely their expected values and standard deviations. These results depend on the statistic being considered and on the type of sampling method, but not, except where indicated, on any detailed properties of the population.

Recall that the expected value of any random variable is both its probability-weighted average and its long-run average value. For any statistic based on any sampling method drawing from any particular population, its **expected value** is its long-run average value, based on infinitely many samplings from the same population by the same method.

expected value

The expected value of a statistic can be calculated the same way that any other expected value is calculated: find the probabilities for all values, multiply values by probabilities, and sum. In Section 7.3 we calculated the probabilities for a sample mean based on a sample of size 2. The expected value is

$$E(\bar{Y}) = 100(.16) + 105(.24) + \cdots + 130(.01) = 110$$

Recall that $\mu_{\text{pop}} = 110$. We also calculated the probabilities for the sample variance. Its expected value is

$$E(S^2) = 0(.30) + 50(.40) + 200(.22) + 450(.08) = 100$$

Recall that $\sigma^2_{\text{pop}} = 100$.

Alternatively, general mathematical properties of expected values can be used to prove results for particular statistics. The easiest results are for the sample mean.

Expected Value of a Sample Mean

If simple random sampling is used either with or without replacement,

$$E(\bar{Y}) = \mu_{pop}$$

If stratified random sampling is used, and if $\bar{Y}_{str}$ is calculated as a weighted average of the means for the strata,

$$\bar{Y}_{str} = \sum \frac{N_j}{N} \bar{Y}_j$$

then

$$E(\bar{Y}_{str}) = \mu_{pop}$$

Thus, a sample mean estimates a population mean "correctly on the average." In the IQ sampling example, we calculated that $E(\bar{Y}) = 110 = \mu_{pop}$. The result in the box shows that this equality will always hold. In the long run, averaging over repeated samples, the sample mean will equal the population mean. (In particular, using weights N_j/N makes this true for $\bar{Y}_{str}$.) Of course, that does not mean that in one particular sample the sample mean will necessarily come out equal to the population mean. There will be some probability that the sample mean will be below the true population mean and some that it will be above.

Example 7.11

Population scores on an introversion-extroversion scale are assumed to be distributed as follows:

Score	1	2	3	4	5	6	7
Proportion	.05	.10	.20	.30	.20	.10	.05

The population mean is 4.0, and the population variance is 2.1. A simple random sample of 2 scores is obtained from the population; the population size is so large that we may assume sampling with replacement. The theoretical distribution of the sample mean is

$\bar{y}_2$	1.0	1.5	2.0	2.5	3.0	3.5	4.0	4.5	5.0	5.5	6.0	6.5	7.0
$f(\bar{y}_2)$	.0025	.01	.03	.07	.12	.17	.195	.17	.12	.07	.03	.01	.0025

Calculate the expected value of $\bar{Y}_2$.

Solution By definition,

$$E(\bar{Y}_2) = 1.0(.0025) + 1.5(.01) + \cdots + 7.0(.0025) = 4.00$$

From the result in the box above, we know that $E(\bar{Y}) = \mu_{pop} = 4.0$. □

Results can be proved for statistics other than the sample mean. In particular, expected values can be computed for sample variances.

Expected Value of a Sample Variance

For simple random sampling with replacement,

$$E(S^2) = \sigma^2_{pop}$$

For simple random sampling without replacement,

$$E(S^2) = \frac{N}{N-1}\sigma^2_{pop}$$

where N is the population size.

Note that when sampling is with replacement, the expected value of the sample variance equals the population variance. In the IQ example, $E(S^2) = 100 = \sigma^2_{pop}$. This is the reason why the sample variance was defined with an $n - 1$ divisor. Had we divided by n, the expected value of the sample variance would have been smaller than the population variance. When sampling is without replacement, the expected value of the sample variance is ever so slightly more than the population variance, but the discrepancy is so small that it's not worth worrying about.

Example 7.12 Refer to the introversion-extroversion scale of Example 7.11. Assume that the size of the relevant population is about 2,000,000. Calculate the expected value of the sample variance for a simple random sample of size 2, taken without replacement.

Solution

$$E(S^2) = \frac{N}{N-1}\sigma^2_{pop} = \frac{2{,}000{,}000}{1{,}999{,}999}(2.1) = 2.10000105$$

For all possible practical purposes, $E(S^2) = \sigma^2_{pop}$. □

These results only indicate the long-run average values of sample means and variances, not the variability of these statistics. To assess variability, we must consider the theoretical variance and standard deviation. Recall that the variance of a random variable is calculated as a probability-weighted average of squared deviations from the mean. In the IQ example,

$$\begin{aligned}\text{Var}(\bar{Y}) &= (100 - 110)^2(.16) + (105 - 110)^2(.20) + \cdots \\ &\quad + (130 - 110)^2(.01) \\ &= 50.0\end{aligned}$$

As always, the standard deviation is the square root of the variance; in the example, the standard deviation is 7.07.

This standard deviation is, literally, "the standard deviation of the theoretical probability distribution of the sample mean." That's a rather large mouthful, so we introduce a conventional abbreviation.

Standard Error

The standard error of any statistic is the standard deviation of the theoretical probability distribution of that statistic. It depends on the statistic being considered, on the method of sampling, and on some characteristics of the underlying population. The standard error of any particular statistic is denoted $\sigma_{\text{statistic}}$. In particular, the standard error of the sample mean is denoted $\sigma_{\bar{Y}}$.

Thus, in the IQ example, we would say that "the standard error of the sample mean is 7.07."

Example 7.13 Refer to Example 7.11. Calculate the standard error of the sample mean.

Solution In Example 7.11, we showed that $E(\bar{Y}_2) = 4.00$. Thus

$$\begin{aligned}\text{Var}(\bar{Y}_2) &= (1.0 - 4.00)^2(.0025) + (1.5 - 4.00)^2(.01) + \cdots \\ &\quad + (7.0 - 4.00)^2(.0025) \\ &= 1.05\end{aligned}$$

The standard error (theoretical standard deviation) of $\bar{Y}_2$ is the square root of this theoretical variance, $\sqrt{1.05} = 1.024695$. □

In principle, we could compute the standard error for any statistic as we did in the IQ example, by calculating the theoretical probabilities of the statistic. That would be a very tedious process, particularly if one wanted to consider many different possible populations. Fortunately, many standard errors have been derived mathematically. In particular, the standard error for a sample mean is known.

Standard Error of a Sample Mean

For simple random sampling with replacement,

$$\sigma_{\bar{Y}} = \frac{\sigma_{\text{pop}}}{\sqrt{n}}$$

For simple random sampling without replacement,

$$\sigma_{\text{pop}} = \frac{\sigma_{\text{pop}}}{\sqrt{n}} \frac{\sqrt{N-n}}{\sqrt{N-1}}$$

In the IQ example, the sample size was 2, σ_{pop} was 10, and we sampled with replacement, so the standard error of the sample mean should be $\sigma_{\bar{Y}} = 10/\sqrt{2} = 7.07$. This is exactly the value we obtained via the theoretical probability distribution.

Example 7.14 Does the standard error found in Example 7.13 agree with the formula given?

Solution The probabilities used in Example 7.13 were based on sampling with replacement, so the standard error should be, and is,

$$\frac{\sqrt{2.1}}{\sqrt{2}} = 1.024695$$

□

There are several things to note about the standard error of the sample mean. First, it depends on (but does not equal) the population standard deviation. Second, it depends on the sample size. As the sample size increases, the standard error of the sample mean decreases, so the probabilities become more and more closely concentrated around the expected value. However, the standard error does not decrease in proportion to the sample size, but rather in proportion to the square root of n. Thus, to cut the standard error in half, we must quadruple, rather than double, the sample size.

Example 7.15 Refer to Example 7.14. Recalculate the standard error of the sample mean assuming a simple random sample of size 4, then assuming a simple random sample of size 8.

Solution For $n = 4$, $\sigma_{\bar{Y}} = \dfrac{\sqrt{2.1}}{\sqrt{4}} = 0.724569.$

For $n = 8$, $\sigma_{\bar{Y}} = \dfrac{\sqrt{2.1}}{\sqrt{8}} = 0.512348.$ □

finite population correction factor

The standard error of the mean is only a bit more complicated when one samples without replacement. The extra factor $\sqrt{(N-n)}/\sqrt{(N-1)}$ is often called the **finite population correction factor.** This factor is always less than 1, so that the without-replacement standard error is always less than the with-replacement standard error. Typically, the factor is so close to 1 that its effect is negligible. For instance, suppose a simple random sample of size 50 is taken, without replacement, from a population of 1000 values. Then the finite population correction factor is $\sqrt{(1000-50)}/\sqrt{(1000-1)} = .975$. Usually, this factor is ignored in most statistical inference, leading to a slight overestimation of the standard error. Standard computer programs that report the standard error of the sample mean always use the with-replacement version.

Example 7.16 Calculate the standard error of the sample mean for both of the following situations, for both sampling with replacement and sampling without replacement. Which situation is preferable—that is, which yields the smaller standard error?

(a) $\sigma_{pop} = 20$, $n = 100$, $N = 1000$
(b) $\sigma_{pop} = 20$, $n = 1000$, $N = 1{,}000{,}000$

Solution **(a)** With replacement, the standard error is

$$\frac{20}{\sqrt{100}} = 2.000$$

Without replacement, it is

$$\frac{20}{\sqrt{100}}\frac{\sqrt{900}}{\sqrt{999}} = 1.898$$

(b) With replacement, the standard error is

$$\frac{20}{\sqrt{1000}} = 0.632$$

Without replacement, it is

$$\frac{20}{\sqrt{1000}} \frac{\sqrt{999{,}000}}{\sqrt{999{,}999}} = 0.632$$

Note that in situation (b), the effect of the correction factor is so small that it disappears when we round off to three decimal places. Furthermore, whether the sampling is done with or without replacement, situation (b) yields a much smaller standard error than does situation (a). □

Example 7.16 indicates a phenomenon that surprises many people. One might expect that situation (a) would be preferable, because in situation (a) one is sampling 10% of the population, whereas in situation (b) one is sampling only .1% of the population. Nevertheless, situation (b) yields a much smaller standard error of the sample mean. The reason is that in situation (b) the absolute sample size is larger. *In general, the absolute sample size is far more important in determining the probable accuracy of a statistic than is the fraction of the population being sampled.* Only when the sample is a large fraction (say, more than 5%) of the population does the sample fraction have an appreciable effect, via the finite population correction factor, on the standard error. Even in this unusual case, the absolute sample size is still important in determining the standard error.

The standard error expression for stratified random sampling is more complicated, and would be still more so if we had bothered with correction factors. We will have relatively little use for this expression, but you should be aware of it. Quite a few statistical studies are done by stratified sampling, only to have the investigator use the erroneous simple-random-sampling standard error. Beware!

Standard Error of a Stratified Sample Mean

Neglecting finite population correction factors,

$$\sigma_{\bar{Y}_{\text{str}}} = \sqrt{\sum \left(\frac{N_j}{N}\right)^2 \frac{\sigma_j^2}{n_j}}$$

where σ_j^2 is the population variance within the jth stratum.

Example 7.17 A population is divided into two strata, college students and similar-aged nonstudents. The population is assumed to have the characteristics listed in Table 7.2.

	Stratum	
	College	Noncollege
Size	3,000,000	2,000,000
Mean	3.60	4.60
Variance	1.60	1.80

Table 7.2

The overall population mean is 4.00, and the overall variance is 1.92. A stratified random sample of 60 college and 40 noncollege students is taken. Find the standard error of $\bar{Y}_{\text{str}}$. If the sample were incorrectly assumed to be a simple random sample, what would the calculated standard error be?

Solution Note that $N_1/N = .6$ and $N_2/N = .4$. For stratified random sampling,

$$\text{Var}(\bar{Y}_{\text{str}}) = (.6)^2\left(\frac{1.60}{60}\right) + (.4)^2\left(\frac{1.80}{40}\right)$$
$$= .0168$$

so the standard error is $\sqrt{.0168} = .1296$. For simple random sampling, the standard error would be

$$\frac{\sqrt{1.92}}{\sqrt{100}} = .1386$$

□

You should note that every result in this section depends only on population means and/or variances. The expected values and standard errors discussed here are correct regardless of the skewness, heavy-tailness, or other shape properties of the underlying population.

Exercises for Section 7.4

7.26. Refer to the population of Exercise 7.19. A simple random sample of size 16 is to be taken. Assume with-replacement sampling.

a. Find the expected value of $\bar{Y}$.
b. Find the standard error of $\bar{Y}$.

7.27. A Monte Carlo study of the sampling process of Exercise 7.26 involved generating repeated samples of size 16 from the population. The standard deviation of these means was found to be 0.1886. What is the theoretical value for this standard deviation, under the given conditions?

7.28. Refer again to Exercise 7.26. Assume now that the sample was taken without replacement from a population of size 40,000. How would your answers to Exercise 7.26 change?

7.29. Refer to Exercise 7.4. The design of the study was severely criticized because the sample was such a small fraction of the relevant population. Is this an appropriate criticism?

7.30. Data on birth weight was obtained from a stratified sample of 250 boys and 150 girls. Assume that in the population, the mean weight for boys was 6.84 pounds (standard deviation 0.92 pound) and that for girls was 6.93 pounds (standard deviation 0.96 pound). Does stratification help a great deal is estimating the overall mean weight of newborns, under these conditions?

7.31. Refer to Exercise 7.30. Assume that there are equal numbers of boys and girls in the newborn population.

a. What is the expected value of the stratified sample mean?
b. What is the standard error of that mean?

7.32. Refer to Exercise 7.26. Assume that a random sample of size 16 is to be taken from a population having a different distribution, but the same mean and standard deviation. How will the answers to that exercise change?

7.5 The Central Limit Theorem for the Sample Mean

The purpose of deriving the theoretical distribution of a statistic is to make probability statements about the accuracy of a statistical inference. For example, when we use a sample mean as an estimate of a true, population mean, we want to be able to say something like "There is a .95 probability that the error of estimate is less than 4.31 IQ points." The expected value and standard error results of the previous section do not, by themselves, allow us to make useful probability statements, although they can be of enormous help. This section discusses the Central Limit Theorem, which will usually allow us to make accurate probability statements when the relevant statistics are specifically sample means.

The content of the theorem can be illustrated in terms of the IQ sampling discussed in the last two sections. Instead of taking a sample of size 2 from the population, we could take a sample of size 4, say, or 8. The theoretical probability distribution of the sample mean can be calculated for each of these sample sizes. The resulting probability histograms are shown in Figure 7.4.

Two results of increasing the sample size are evident in Figure 7.4. First, the histograms become more concentrated (less variable) around the expected value $\mu_{pop} = 110$. The standard error formula of the previous section explains that. As n increases, the standard error (for simple random sampling, with replacement) of the sample mean, which is $\sigma_{pop}/\sqrt{n}$, decreases. Because the standard error is exactly the standard deviation of the theoretical probabilities, those probabilities must become more concentrated as the standard error gets smaller.

Second, as n increases, the theoretical distribution of the sample mean becomes more and more nearly normal. The original population distribution of Section 7.3 was quite skewed. The probability distributions of the sample mean become less and less skewed, and more and more nearly normal, as the sample size increases. This approach to normality is not a coincidence. The Central Limit Theorem says that, for sample means from essentially any population, the theoretical probability distribution will approach nearer and nearer to normality as n increases.

Central Limit Theorem

Assume that a sample is being taken from a population that has finite σ_{pop}. The theoretical probability distribution of a sample mean, whether based on simple random sampling (with or without replacement) or on stratified random sampling, is approximately a normal distribution if the sample size is large.

The fundamental consequence of this theorem is that if n is large enough, we can use normal probabilities for inferences using a sample mean, *regardless of the shape of the underlying population.* No matter what the shape of the population—symmetric, skewed, six-modal, whatever—we can use normal probabilities, provided only that the sample size is large enough. Economists assert that there is no such thing as a free lunch. This theorem is the closest thing to a free lunch that there is. With almost no assumptions about the population, we can use one probability table in calculating probabilities about a sample mean.

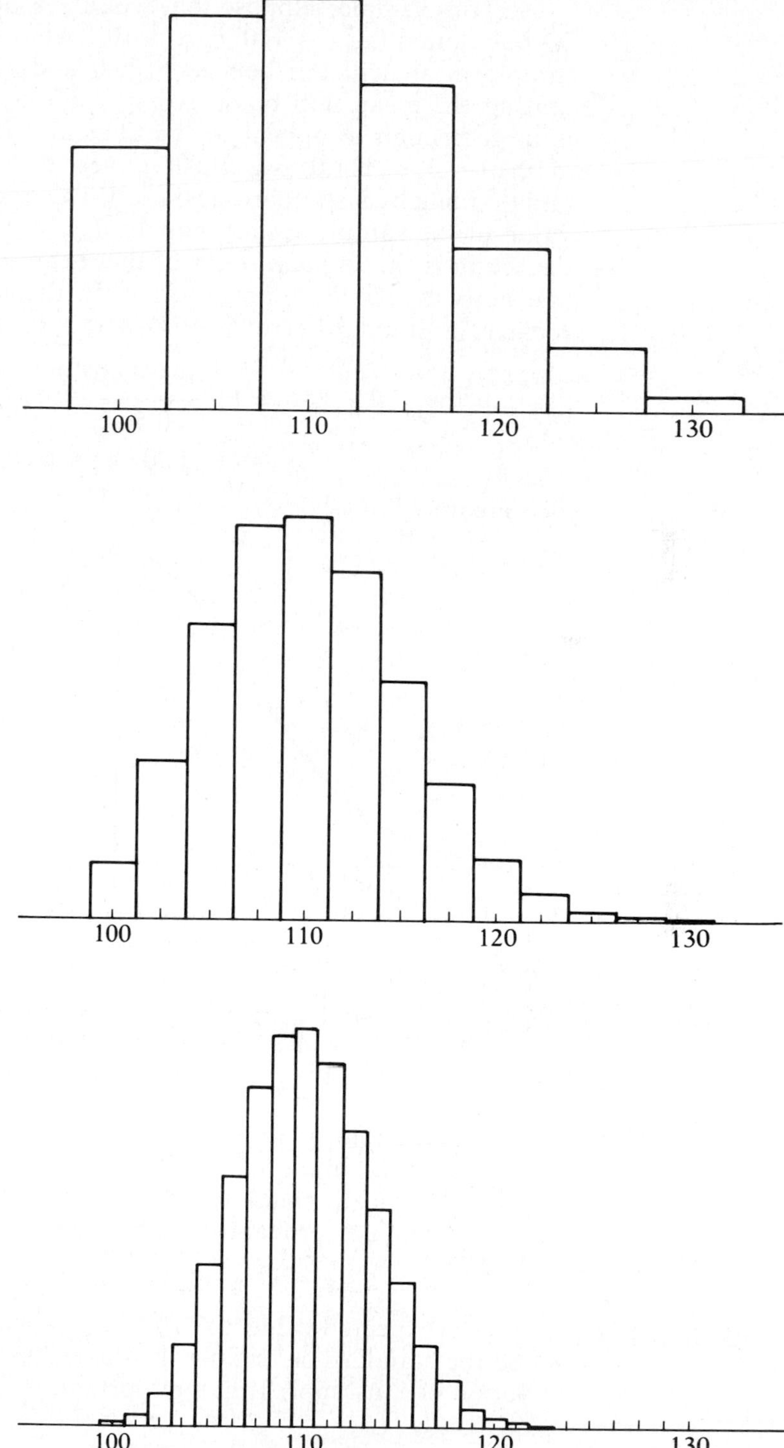

Figure 7.4 Histograms of sampling distributions, n = 2, 4, 8

For example, suppose that we take a sample of size 400 from our IQ population ($\mu_{pop} = 110$, $\sigma_{pop} = 10$). What is the probability that the sample mean will fall between 109.0 and 111.0, so our error in estimating the mean will be, at worst, 1 IQ point? Assume that $n = 400$ is large enough to appeal to the Central Limit Theorem. To calculate $P(109.0 < \bar{Y} < 111.0)$, we must convert to z form, subtracting a mean and dividing by a standard deviation. The relevant mean is the expected value of the sample mean, namely $\mu_{pop} = 110$. The relevant standard deviation is the *standard error* of the mean, because the standard error is exactly the standard deviation of the theoretical probabilities for the mean. The standard error is $10/\sqrt{400} = 0.50$. Therefore,

$$P(109 < \bar{Y} < 111) = P\left(\frac{109 - 110}{0.50} < z < \frac{111 - 110}{0.50}\right)$$
$$= P(-2.00 < z < 2.00) = .9744$$

See Figure 7.5.

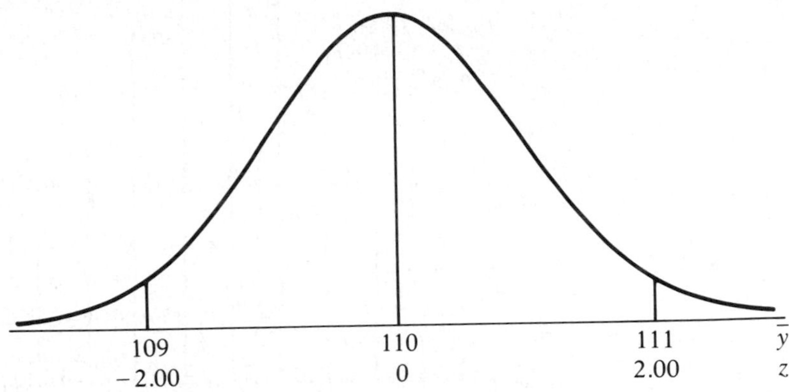

Figure 7.5 Normal distribution for $\bar{Y}$

Example 7.18 Assume that a simple random sample of size 4 is taken from the population of Example 7.11 ($\mu_{pop} = 4.0$, $\sigma^2_{pop} = 2.1$). Assume (heroically) that $n = 4$ is large enough to appeal to the Central Limit Theorem. Find the (approximate) probability that the sample mean is between 3.125 and 4.875.

Solution The expected value is $E(\bar{Y}) = \mu_{pop} = 4.0$. In Example 7.15, we computed the standard error for $n = 4$ as 0.725. Taking the theoretical distribution of $\bar{Y}$ as approximately normal,

$$P(3.125 < \bar{Y} < 4.875) = P\left(\frac{3.125 - 4.0}{0.725} < z < \frac{4.875 - 4.0}{0.725}\right)$$
$$= P(-1.21 < z < 1.21) = .7738$$

□

There is one obvious question. The Central Limit Theorem says, "if n is large"; how large is "large"? One good way to see if a sample of a specified size, drawn from a given population is large enough to make the theoretical distribution of $\bar{Y}$ approximately normal is to program a computer to draw many samples of that size from that population and to compute the resulting means. If a large number of sample means are computed, the histogram of those means will be (very close to) the theoretical distribution of means. In fact, a different picture—a **normal probability plot**—is very useful in determining whether or not a computer-generated distribution of means is approximately normal. A normal probability plot is a scatter plot of actual values versus the expected values assuming a normal distribution. If the probability distribution of means is in fact normal, then the actual and expected values should be equal except for random noise, and the plot should be a straight line. It turns out that if the actual distribution is skewed, the normal plot will be curved. If the actual distribution is heavier-tailed than the normal, the plot will be S-shaped; if the actual is lighter-tailed than the normal, it will be S-shaped in the opposite direction. Figure 7.6 shows normal probability plots for 1000 means, each drawn from a moderately skewed population and a mildly outlier-prone population.

normal probability plot

Notice that in both Figure 7.6(a) and Figure 7.6(b), the normal plots approach a straight line as the sample size increases. Once the sample size reaches 60, the plots are quite close to lines. Thus, *for these populations*, $n = 60$ appears to be large enough to invoke the Central Limit Theorem and use the normal approximation to the theoretical distribution of the sample mean.

Many textbooks use the rule of thumb that n should be at least 30 to invoke the Central Limit Theorem. Although that's not a bad rule of thumb, a better one would consider the effects of data discreteness and of population skewness. If the underlying population consists of only a few discrete values, the correct theoretical distribution will be somewhat "lumpy," in contrast to the smooth, continuous normal curve. As our IQ example and Figure 7.6 make clear, population skewness is reflected in skewness of the theoretical distribution. Increasing the sample size "de-skews" the theoretical probabilities gradually. For a symmetric population, there's no skewness problem, and relatively small sample sizes (say, n at least 15) will be good enough to appeal to the Central Limit Theorem. In fact, if the *population* itself is normal, the theoretical

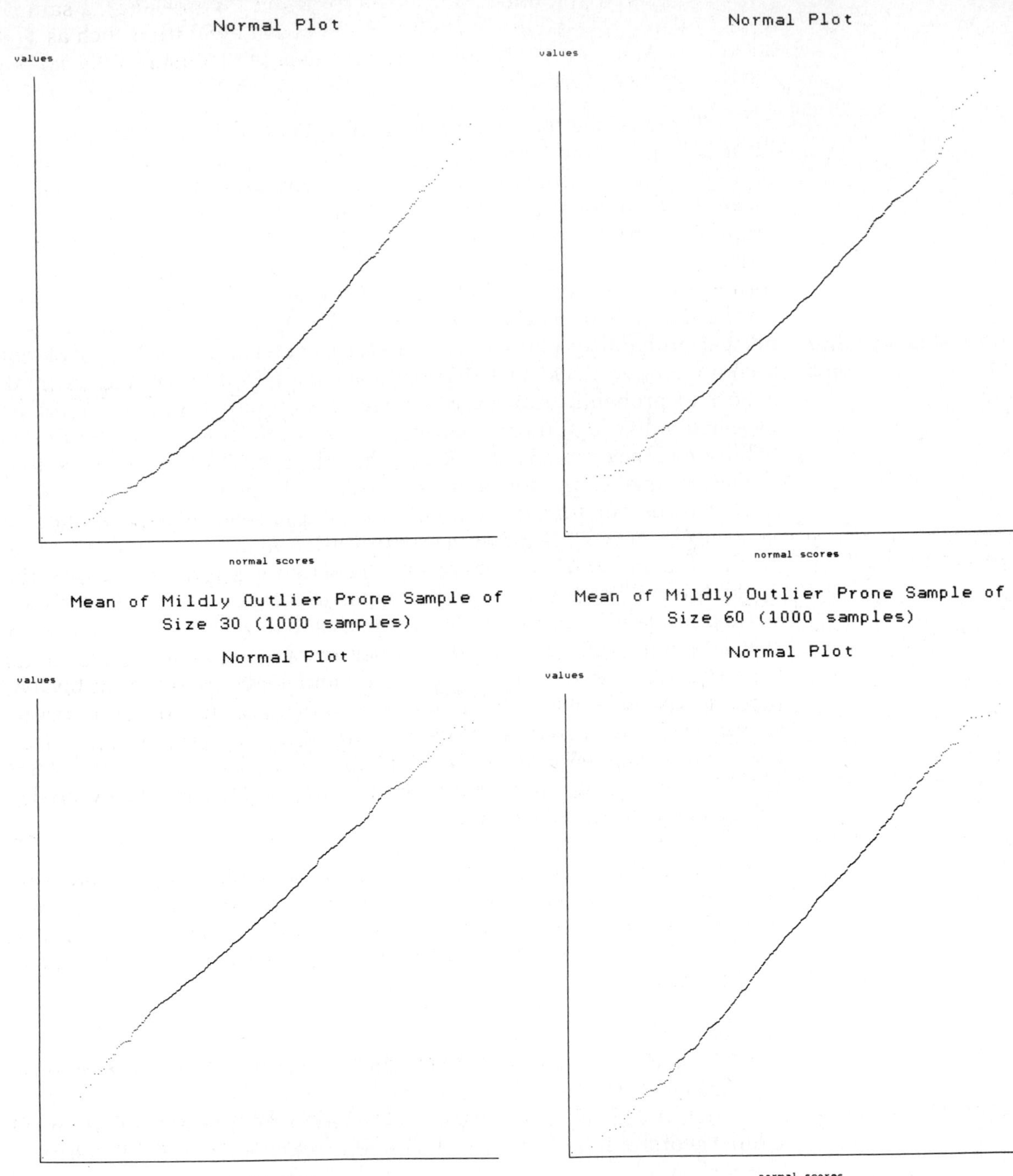

Figure 7.6 Normal probability plots: means from samples of size 30 and 60 from (a) a moderately skewed population (b) a mildly outlier-prone population

distribution of the sample mean will be normal for any sample size. For a mildly skewed population such as is shown in Figure 7.7(a), a sample size of at least 30 will serve. For a badly skewed population such as that shown in Figure 7.7(b), the sample size must be substantially larger, perhaps 100 or more.

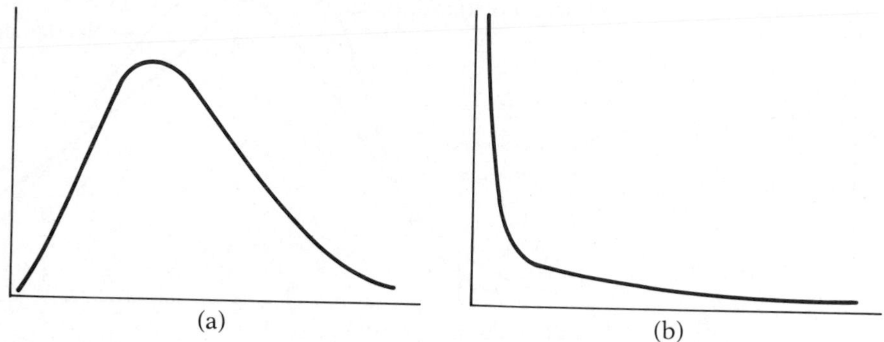

Figure 7.7 (a) Mildly skewed and (b) highly skewed populations

---

Example 7.19 How accurate can you expect the normal approximation in Example 7.18 to be?

Solution The sample size is only $n = 4$, and there are only 7 possible values (the population values are discrete), so the approximation should not be very accurate. On the other hand, the population distribution is symmetric, which indicates that the approximation should be decent. The exact probability that the sample mean falls between 3.125 and 4.875 is .7725, compared to the approximate value .7738. Partly by coincidence, the approximation is excellent. □

One other aspect of normal approximations should be noted. One-tailed probabilities are usually poorer approximations than are two-tailed probabilities. With a two-tailed probability, if the area in one tail is overestimated, the area in the other tail will usually be underestimated. See Figure 7.8.

Example 7.20 The probability computed in Example 7.19 was (the complement of) a two-tailed probability. Did this fact help to improve the approximation?

Solution No. The underlying population was symmetric, so the true theoretical probabilities for the sample mean were also symmetric. In such a case, one-tailed and two-tailed approximations are equally good. ☐

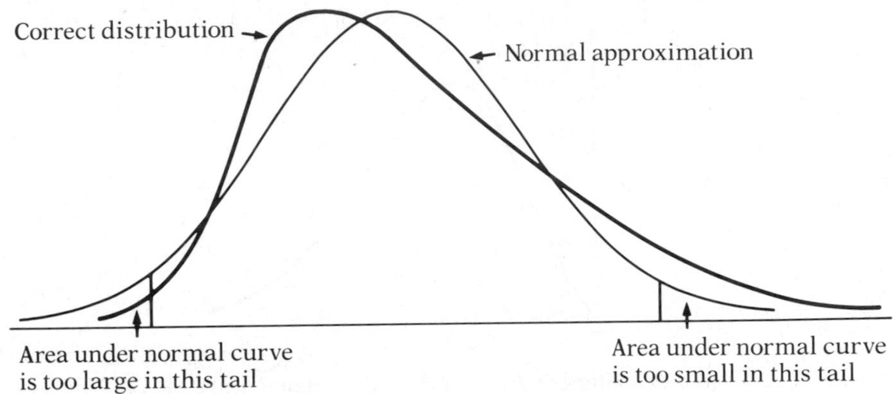

Figure 7.8 One-tailed and two-tailed probabilities

There is one technical requirement of the Central Limit Theorem for means, namely that μ_{pop} and σ_{pop} be finite. This requirement will be met for almost all real-data situations. It will certainly be met if the possible values have a finite range. Even variables that in principle could go to infinity (such as latency times) will in fact be limited to some very large but finite number. Though there is considerable mathematical interest in infinite-σ populations, the finite-σ requirement of the Central Limit Theorem for means should virtually always be met in practice.

7.6 Normal Distributions of Other Statistics

The Central Limit Theorem, discussed in Section 7.5, applies to the theoretical distribution of one particular statistic—the sample mean. Of course, there are many other statistics that we'll be considering—medians, standard deviations, correlations, and so on. Not all statistics have approximately normal theoretical distributions—a sample correlation coefficient does not have a useful approximately normal distribution—but an astonishing number of them do. Mathematical statistics contains an enormous number of theorems to the effect that "under such-and-such conditions, the so-and-so statistic has an approximately normal theoretical distribution." The sample median and indeed all sample percentiles have an approximately normal distribution. So do the λ and ∇ measures discussed in Chapter 3, as do the regression slopes of Chapter 15. The list is very long.

To use a normal approximation, one needs three things: the expected value, the standard error, and a rule of thumb about when the normal approximation is adequate. For many, not all, statistics, the expected value is the corresponding population parameter. For example, the expected value of a sample median is, to a reasonable approximation, the population median.

The standard error formula will vary from one statistic to another, and from one sampling situation to another. For example, the ∇ statistic of Chapter 3 has one (long, ugly) standard error formula for simple random sampling, and a different (longer, uglier) formula for stratified random sampling. It should be emphasized that "standard error" does not necessarily mean $\sigma_{\text{pop}}/\sqrt{n}$. That formula only works for a sample mean, under simple random sampling. Use of a different statistic or a different sampling method yields a different standard error formula.

The rule of thumb for adequacy of normal approximations also varies from one statistic to another. For the sample mean, the primary concern is the skewness of the underlying population, but other characteristics are relevant for other statistics. The very rough rule of thumb that n should be at least 30 only applies to means. For one of the kurtosis statistics discussed in Chapter 2, the rough rule of thumb is that n should be at least 1500!

Example 7.21

For the ∇ statistic discussed in Chapter 3, the rough rule of thumb for adequacy of a normal approximation is that the expected number of prediction errors should be somewhere between 5 and $n - 5$. Suppose that a certain prediction has probability .03 of being in error, and that a statistic is to be computed based on a sample of size 180. Should a normal approximation (to the true theoretical probabilities of the statistic) be a good approximation?

Solution

The expected number of errors will be $180(.03) = 5.4$. This value is just a little above the rule of thumb value, 5.0, so the approximation should be at least fair. Of course, one would have more faith in the approximation if the expected number of errors were larger. □

It's a good idea, before using a normal approximation, to find out what the rules of thumb are for a particular statistic. If one is using a standard computer program package, the user's manual should give some information or a reference.

Exercises for Sections 7.5 and 7.6

7.33. Refer to Exercise 7.26. Use the normal distribution to find the approximate probability that the sample mean will be larger than .300.

7.34. In Exercise 7.33, is there reason to believe that the approximation will be very good?

7.35. An experiment involved determining the number of trials required for a random sample of 64 laboratory rats to learn to run a certain maze. Assume that the true (population) mean is 7.50 trials, and that the true standard deviation is 2.00 trials.

a. Determine the expected value and standard error of the sample mean.

b. Find the approximate probability that the sample mean will be between 7.00 and 8.00 trials.

7.36. Refer to Exercise 7.35. Should the approximation of part b be an accurate one?

7.37. Suppose that a Monte Carlo study involved taking 10,000 samples, each of size 64, according to an assumed set of maze-learning probabilities. (The assumed probabilities yielded the given population mean and standard deviation.) Of the 10,000 samples, 9599 had sample means between 7.00 and 8.00, 166 had means below 7.00, and 235 had means above 8.00.

a. What does this study indicate about the accuracy of the approximation in Exercise 7.35?

b. What does this study indicate about the accuracy of the approximate probability that the sample mean will be above 8.00?

7.38. Suppose that given certain assumptions (in particular, a sample size of 200), the Groeneveld-Meeden skewness measure described in Chapter 2 has an expected value of 0 and a standard error of .060.

a. Explain what "standard error = .060" means.

b. Find the approximate probability that the sample skewness measure is larger than .100.

7.39. Refer to Exercise 7.38.

a. What did you assume in answering part b?

b. Can you be confident that a sample size of 200 is adequate to make the approximation close?

7.40. A computer was programmed to draw samples from a normal population with $\mu_{pop} = 50$ and $\sigma_{pop} = 10$. A total of 1000 samples were drawn, and 1000 sample means ($n = 10$) were calculated.

a. The average value of the sample means was 50.0976. If there had been an infinite number of means calculated, what should the average value have been?

b. The standard deviation of the sample means was 3.1679. What is this quantity estimating? Numerically, what should the value have been? How close is the Monte Carlo estimate?

7.41. Refer to Exercise 7.40. A normal probability plot of the means is shown in Figure 7.9. Does this plot indicate that the theoretical distribution of means is normal? Should it be?

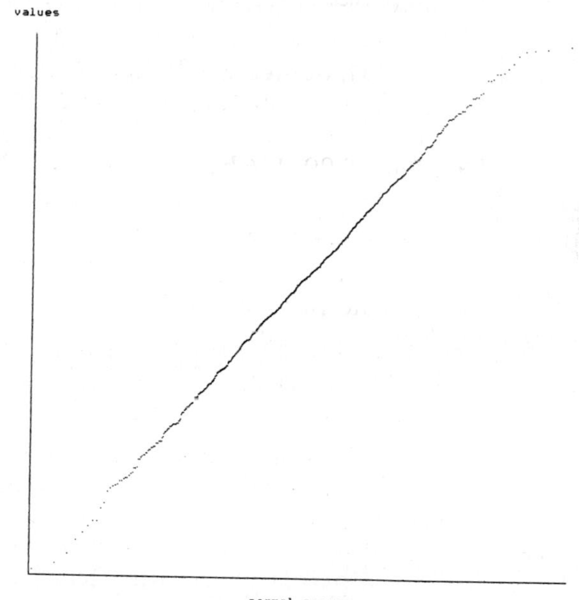

Figure 7.9 Normal probability plot for Exercise 7.41

7.7 The Mathematics of Expected Values and Standard Errors

In this section we will prove some of the basic results about theoretical distributions that were stated in Section 7.4. It should be noted that the Central Limit Theorem, stated in Section 7.5, is a more advanced result that doesn't fall within the scope of this book. See Larson and Marx (1981) or Mood, Graybill, and Boes (1974). First, let's consider the sampling distribution of the sample mean statistic. The easiest result to prove is that the expected value of a sample mean is the population mean. In Chapter 5 we stated that the expected value of an individual observation drawn randomly from a population is $E(Y) = \mu_{\text{pop}}$ and that the expected value of a sum is the sum of the expected values. These results hold regardless of independence, and hence regardless of sampling with or without replacement.

Result 7.1 For sampling either with or without replacement,

$$E(\bar{Y}) = \mu_{\text{pop}}$$

Proof

$$\begin{aligned} E(\bar{Y}) &= E\left(\sum \frac{Y_i}{n}\right) \\ &= \frac{1}{n}\sum E(Y_i) \\ &= \frac{1}{n}\sum \mu_{\text{pop}} \\ &= \frac{1}{n}(\mu_{\text{pop}} + \mu_{\text{pop}} + \cdots + \mu_{\text{pop}}) \\ &= \mu_{\text{pop}} \end{aligned}$$

When we turn to variances and standard errors of $\bar{Y}$, we need further mathematical results. Recall from Chapter 5 that if X is any random variable, $\text{Var}(X/c) = (1/c^2)\text{Var}(X)$. In our case, X will be the sum of the sample values $Y_1, \cdots, Y_n$. Further, the variance of a sum of *independent* random variables is the sum of their variances. However, as we saw in Chapter 5, when the random variables are *dependent,* we must add two covariance terms for all pairs of random variables.

Result 7.2 For sampling with replacement from a population (hence for any situation where the data $Y_1, \cdots, Y_n$ are statistically independent),

$$\text{Var}(\bar{Y}) = \frac{\sigma^2_{\text{pop}}}{n}$$

so $$\sigma_{\bar{Y}} = \frac{\sigma_{\text{pop}}}{\sqrt{n}}$$

Proof

$$\begin{aligned}\text{Var}(\bar{Y}) &= \text{Var}\left(\frac{1}{n}\sum Y_i\right)\\ &= \frac{1}{n^2}\sum \text{Var}(Y_i)\\ &= \frac{1}{n^2}(\sigma^2_{\text{pop}} + \cdots + \sigma^2_{\text{pop}})\\ &= \frac{1}{n^2}(n\sigma^2_{\text{pop}}) = \frac{\sigma^2_{\text{pop}}}{n}\end{aligned}$$

To deal with sampling without replacement, we need to consider covariance terms such as

$$\text{Cov}(Y_1, Y_2) = E(Y_1 - \mu_{\text{pop}})(Y_2 - \mu_{\text{pop}})$$

By an "expand-the-product" argument, it is possible to show that

$$\text{Cov}(Y_1, Y_2) = E(Y_1Y_2) - \mu^2_{\text{pop}}$$

where

$$E(Y_1Y_2) = \sum_{y_1}\sum_{y_2} y_1y_2f(y_1, y_2)$$

When Y_1 and Y_2 are the results of successive random draws taken without replacement from a finite population, we need to calculate $f(y_1, y_2)$. Denote the population size by N, and the number of individuals having a particular y value by $N(y)$. The probability of drawing any particular y value is thus $N(y)/N$. If a particular y value is drawn on trial 1, there are $N - 1$ values remaining, of which $N(y) - 1$ equal the drawn y value; the number of other y values remaining in the population is still $N(y)$. Thus the probability of drawing two consecutive, identical y values is

$$f(y_1, y_2) = \frac{N(y)[N(y) - 1]}{N(N - 1)}$$

whereas the probability of drawing two consecutive, different y values is

$$f(y_1, y_2) = \frac{N(y_1)N(y_2)}{N(N-1)}$$

The same probability distribution applies to any two draws y_i and $y_{i'}$. The basic reason is that the numbering of the draws from the population is totally arbitrary, so any two draws labeled i and i' may be regarded as draws 1 and 2.

We note that

$$\mu_{\text{pop}} = \mu_Y = \sum y f_Y(y) = \sum y \frac{N(y)}{N}$$

and

$$\sigma^2_{\text{pop}} = \text{Var}(Y) = \sum (y - \mu_Y)^2 \frac{N(y)}{N}$$

Result 7.3 If a random sample is drawn without replacement from a population,

$$\text{Cov}(Y_i, Y_{i'}) = \frac{-\sigma^2_{\text{pop}}}{N-1}$$

Proof As indicated, we may safely assume that i and i' are 1 and 2. Now

$$\begin{aligned}
E(Y_1Y_2) &= \sum_{y_1}\sum_{y_2} y_1y_2 f(y_1, y_2) \\
&= \sum_{y_1}\sum_{y_2 \neq y_1} y_1y_2 f(y_1, y_2) + \sum_{y_1}\sum_{y_2 = y_1} y_1y_2 f(y_1, y_2) \\
&= \sum_{y_1}\sum_{y_2 \neq y_1} y_1y_2 \frac{N(y_1)N(y_2)}{N(N-1)} + \sum_{y_1}\sum_{y_2 = y_1} y_1y_2 \frac{N(y_1)[N(y_1)-1]}{N(N-1)} \\
&= \frac{N}{N-1}\sum_{y_1}\sum_{y_2} y_1y_2 \frac{N(y_1)N(y_2)}{N \cdot N} - \frac{1}{N-1}\sum y_1^2 \frac{N(y_1)}{N} \\
&= \left(\sum y_1 \frac{N(y_1)}{N}\right)\left(\sum y_2 \frac{N(y_2)}{N}\right) - \frac{1}{N-1}\sum y_1^2 \frac{N(y_1)}{N} \\
&= \frac{N}{(N-1)}\mu^2_{\text{pop}} - \frac{1}{(N-1)}E(Y^2)
\end{aligned}$$

Thus

$$\begin{aligned}\text{Cov}(Y_1, Y_2) &= \frac{N}{(N-1)}\mu_{\text{pop}}^2 - \frac{1}{(N-1)}E(Y^2) - \mu_{\text{pop}}^2 \\ &= \frac{-1}{(N-1)}[E(Y^2) - N\mu_{\text{pop}}^2 + (N-1)\mu_{\text{pop}}^2] \\ &= \frac{-1}{(N-1)}[E(Y^2) - \mu_{\text{pop}}^2] = \frac{-1}{(N-1)}\sigma_{\text{pop}}^2\end{aligned}$$

Now, we note that there are many $\text{Cov}(Y_i, Y_{i'})$ terms in the expansion of the variance of ΣY_i. We may choose any of n terms to be represented by i, and then any of $n - 1$ to be represented by i'. But there is a double counting, because, for instance, $i = 4$ and $i' = 7$ represent the same pair as $i = 7$ and $i' = 4$. Thus there are $n(n - 1)/2$ pairs of distinct covariances to be accounted for.

Result 7.4 For sampling without replacement,

$$\text{Var}(\bar{Y}) = \frac{\sigma_{\text{pop}}^2}{n}\frac{N-n}{N-1}$$

Proof As in the proof of Result 7.2,

$$\text{Var}(\bar{Y}) = \frac{1}{n^2}\text{Var}\left(\sum Y_i\right)$$

Now

$$\begin{aligned}\text{Var}\left(\sum Y_i\right) &= \sum_i \text{Var}(Y_i) + 2\sum_i \sum_{i'} \text{Cov}(Y_i, Y_{i'}) \\ &= n\sigma_{\text{pop}}^2 + \frac{2\,n(n-1)}{2}\frac{-\sigma_{\text{pop}}^2}{N-1}\end{aligned}$$

because there are n variances and $n(n - 1)/2$ covariances in the expansion, and because the covariance is given by Result 7.3. Therefore,

$$\begin{aligned}\text{Var}(\bar{Y}) &= \frac{\sigma_{\text{pop}}^2}{n}\left(1 - \frac{n-1}{N-1}\right) \\ &= \frac{\sigma_{\text{pop}}^2}{n}\frac{N-n}{N-1}\end{aligned}$$

Now we turn to $\bar{Y}_{\text{str}}$, the mean of a stratified random sample. We note that the data from different strata are gathered completely separately, so the different strata results may be assumed to be independent. In deriving the variance of $\bar{Y}_{\text{str}}$, we will not need covariance terms. Also, the weighting by (N_j/N) that occurs in the definition of $\bar{Y}_{\text{str}}$ leads to squared terms $(N_j/N)^2$ in the variance.

Result 7.5

$$\text{Var}(\bar{Y}_{\text{str}}) = \sum \left(\frac{N_j}{N}\right)^2 \frac{\sigma_j^2}{n_j}$$

where n_j and N_j are the sample size and population size of stratum j and σ_j is the variance in stratum j. (Note: Finite population correction factors have been ignored.)

Proof

Because of the independence of the results from the various strata and the squaring of the weights, we have

$$\begin{aligned}\text{Var}\left(\sum \frac{N_j}{N} \bar{Y}_j\right) &= \sum \text{Var}\left(\frac{N_j}{N} \bar{Y}_j\right)\\ &= \sum \left(\frac{N_j}{N}\right)^2 \text{Var}(\bar{Y}_j)\\ &= \sum \left(\frac{N_j}{N}\right)^2 \frac{\sigma_j^2}{n_j}\end{aligned}$$

Up to now, we have been considering properties of the sample mean. Now we turn to the sample variance s^2. Recall that

$$s^2 = \frac{\sum (y_i - \bar{y})^2}{n - 1}$$

The following identity is the key to proving results about s^2.

Result 7.6

$$\sum (y_i - \bar{y})^2 = \sum (y_i - \mu_{\text{pop}})^2 - n(\bar{y} - \mu_{\text{pop}})^2$$

Proof

The proof is another expand-the-square job, this time around μ_{pop}.

$$\begin{aligned}\sum (y_i - \bar{y})^2 &= \sum (y_i - \mu_{\text{pop}} + \mu_{\text{pop}} - \bar{y})^2\\ &= \sum (y_i - \mu_{\text{pop}})^2 + 2(\mu_{\text{pop}} - \bar{y}) \sum (y_i - \mu_{\text{pop}})\\ &\quad + \sum (\mu_{\text{pop}} - \bar{y})^2\end{aligned}$$

$$= \sum (y_i - \mu_{\text{pop}})^2 + 2(\mu_{\text{pop}} - \bar{y})n(\bar{y} - \mu_{\text{pop}})$$

$$+ \sum (\mu_{\text{pop}} - \bar{y})^2$$

[because $\Sigma (y_i - \mu_{\text{pop}}) = \Sigma y_i - n\mu_{\text{pop}} = n(\bar{y} - \mu_{\text{pop}})$].

$$= \sum (y_i - \mu_{\text{pop}})^2 - 2n(\bar{y} - \mu_{\text{pop}})^2 + n(\bar{y} - \mu_{\text{pop}})^2$$

$$= \sum (y_i - \mu_{\text{pop}})^2 - n(\bar{y} - \mu_{\text{pop}})^2$$

From this result and some basic properties of expected values, we can derive the expected value of the sample variance. Note that $E(Y_i - \mu)^2 = \sigma^2_{\text{pop}}$ and $E(\bar{Y} - \mu)^2 = \text{Var}(\bar{Y})$.

Result 7.7

If the sample is taken with replacement,

$$E(S^2) = \sigma^2_{\text{pop}}$$

If the sample is taken without replacement,

$$E(S^2) = \frac{N}{N-1}\sigma^2_{\text{pop}}$$

Proof

From Result 7.6, it follows that

$$E(S^2) = \frac{1}{n-1}E\left[\sum (Y_i - \mu_{\text{pop}})^2 - n(\bar{Y} - \mu_{\text{pop}})^2\right]$$

$$= \frac{1}{n-1}[n\sigma_{\text{pop}} - n\text{Var}(\bar{Y})]$$

When sampling is done with replacement, Result 7.2 indicates that $\text{Var}(\bar{Y}) = \sigma^2_{\text{pop}}/n$, so

$$E(S^2) = \frac{1}{n-1}\left(n\sigma^2_{\text{pop}} - \frac{n\sigma^2_{\text{pop}}}{n}\right)$$

$$= \frac{1}{n-1}[(n-1)\sigma^2_{\text{pop}}]$$

$$= \sigma^2_{\text{pop}}$$

When sampling is done without replacement, Result 7.4 applies, and

$$E(S^2) = \frac{1}{n-1}\left[n\sigma^2_{\text{pop}} - n\left(\frac{\sigma^2_{\text{pop}}}{n}\right)\frac{N-n}{N-1}\right]$$

$$= \frac{1}{n-1}\sigma^2_{\text{pop}}\left(n - \frac{N-n}{N-1}\right)$$
$$= \frac{1}{n-1}\sigma^2_{\text{pop}}\left(\frac{nN - n - N + n}{N-1}\right)$$
$$= \frac{n-1}{n-1}\sigma^2_{\text{pop}}\left(\frac{N}{N-1}\right)$$

Chapter Exercises

7.42. It was proposed to gather data on student opinion of the ability of high school teachers by leaving questionnaires in the halls of the school, to be filled out by anyone who wished to do so. Explain why this is a bad idea.

7.43. A research project was underway to determine the effect, if any, of attendance at rock music concerts on hearing acuity. The proposed study plan was to send a research team to local concerts with equipment, and to obtain volunteers for testing from among the concert-goers. What's wrong with this plan?

7.44. A computer was programmed to compute means for random samples taken from a population with the following distribution:

Value	1	2	3	4	5
Frequency	1000	2000	4000	2000	1000

$\mu_{\text{pop}} = 3.00$ and $\sigma^2_{\text{pop}} = 1.20$. The output indicated:

Sample Size	4	10	30	60
Average	2.9820	3.0076	2.9986	2.9942
Standard Deviation	0.5513	0.3563	0.2006	0.1364

a. Do the average values approximate the desired value?

b. What values are being estimated by the standard deviations? How close do the estimates come?

7.45. A normal probability plot of the means obtained in Exercise 7.44 ($n = 30$) is shown in Figure 7.10.

a. What curvature, if any, is shown in the plot? What does this indicate about the quality of a normal approximation to the theoretical distribution of means?

b. How is the discrete nature of the population reflected in the normal plot?

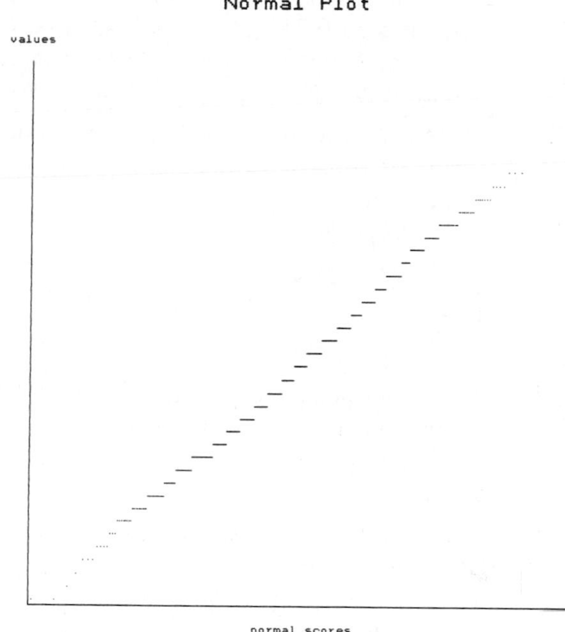

Figure 7.10 Normal plot for Exercise 7.45

7.46. Suppose that a researcher computed the smallest value (MIN) in a random sample of size 100 from a population with $\mu_{pop} = 100$ and $\sigma_{pop} = 15$, and said that the theoretical distribution of MIN should be normal with expected value 100 and standard error $15/\sqrt{100} = 1.5$. The claim is absurd on several grounds; explain why.

7.47. An experiment involved subjects' making decisions in one of three possible experimental conditions: no risk, low risk, or high risk.

a. The experimenter could obtain subjects from students who had completed a class in human decision-making. Explain why this would be a poor idea.

b. Alternatively, the experimenter could obtain subjects by randomly sampling the college population (and the experimenter was willing to regard college students' scores as the relevant population). The experimenter planned to allow each potential subject to choose the desired risk category. This, too, is a bad idea; explain why.

7.48. Chapter 2 indicated that the Groeneveld-Meeden "top-half skew" measure of tail-heaviness had expected value 0.27 when sampling was from a normal population. Suppose that under certain sampling conditions, the standard error is 0.042.

a. Explain the meaning of the numbers 0.27 and 0.042.

b. Find the probability that a sample value of "top-half skew" will exceed 0.312. Make any necessary assumptions.

c. What did you assume in answering part b? What do you need to know to decide if the assumptions are reasonable?

7.49. A sample was to be gathered by taking equal-sized random samples from each of three groups. Assume the information in Table 7.3 about the three (sub)populations.

	Group		
	A	B	C
Mean	50.0	50.0	50.0
Variance	100.0	80.0	120.0

Table 7.3

Assume that in the population, each group size is 1,000,000. It follows that $\mu_{pop} = 50.0$ and $\sigma_{pop} = 100.0$.

a. Find the expected value of $\bar{Y}_{str}$.

b. Find the standard error of $\bar{Y}_{str}$.

c. What advantage, if any, is there to stratifying in this situation? Explain why.

7.50. An experimenter proposed to obtain a sample of 40 scores by randomly choosing 8 subjects and obtaining 5 scores from each. Assuming that the subjects are in fact chosen randomly, that $\mu_{pop} = 100$, and that $\sigma_{pop} = 20$, the experimenter claimed that the expected value of the sample mean should be 100 and the standard error should be $20/\sqrt{40}$. Given the assumptions, which of these claims, if either, is suspect?

7.51. A computer was programmed to draw samples from a Laplace "two-tailed exponential" population, with $\mu_{pop} = 0$ and $\sigma^2_{pop} = 2$. The computer used sample sizes 30 and 60. The following results were obtained:

Sample Size	30	60
Average	0.0027	−0.0025
Standard Deviation	0.2639	0.1845

a. What should the average mean equal? Do the Monte Carlo results tend to confirm your belief?

b. What should the standard deviations equal? Are the Monte Carlo results close to your answers?

7.52. Refer to Exercise 7.51. The population from which the samples were drawn is symmetric, continuous, and somewhat outlier-prone. Normal probability plots for the distribution of sample means are shown in Figure 7.11.

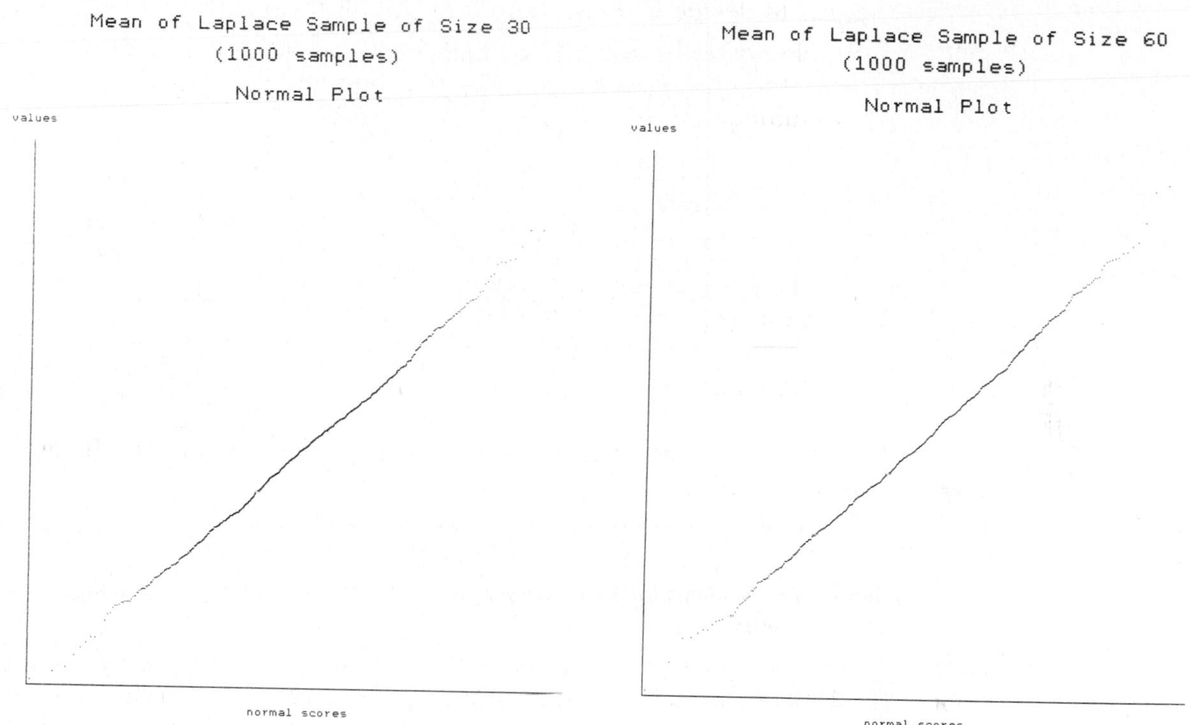

Figure 7.11 Normal plots for Exercise 7.52

a. How good should a normal approximation be in this case?

b. How critical is the heavy-tailness of the population in determining the quality of the normal approximation?

7.53. In the study of Exercise 7.51, sample medians were also computed. A normal plot of the resulting medians is shown in Figure 7.12.

a. Does it appear that the theoretical distribution of the median is also normal?

b. The standard deviations of the medians were 0.2070 for $n = 30$ and 0.1408 for $n = 60$. What quantity is being approximated by these values? How do these standard deviations compare to the standard deviations in Exercise 7.51?

7.54. A normal probability plot of a sample of data is shown in Figure 7.13, along with summary statistics for the sample.

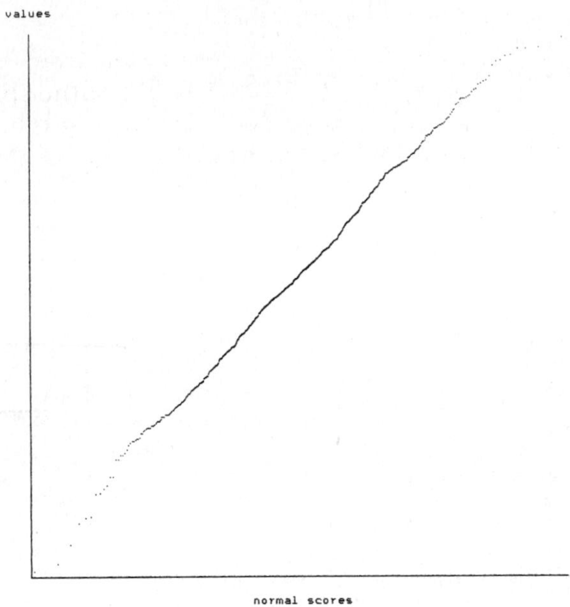

Figure 7.12 Normal plot for Exercise 7.53

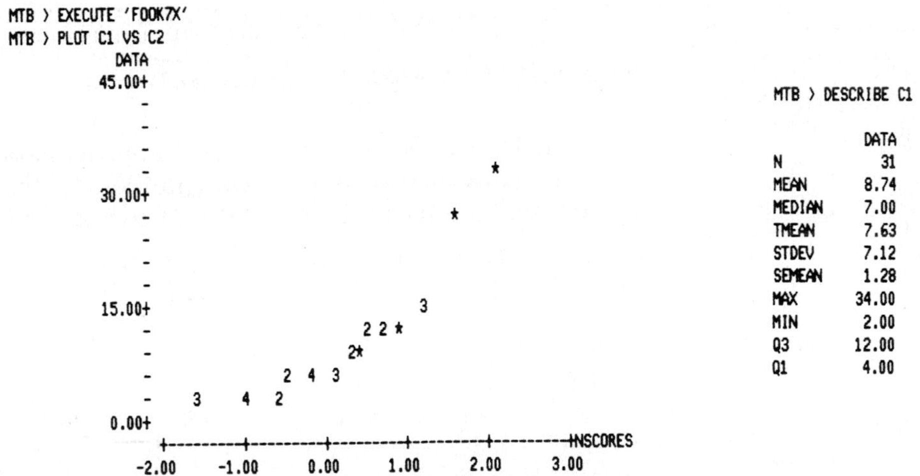

Figure 7.13 Output for Exercise 7.53

a. Does the plot indicate that the data are approximately normal? If not, what appears to be the shape of the data?

b. Use the summary statistics shown in the output to confirm your judgment made in part a.

7.55. Samples of data were taken from each of three populations: normal, skewed, and symmetric but outlier-prone. Normal probability plots and histograms, obtained from Minitab, are shown in Figure 7.14. Which plot should correspond to each of the three samples?

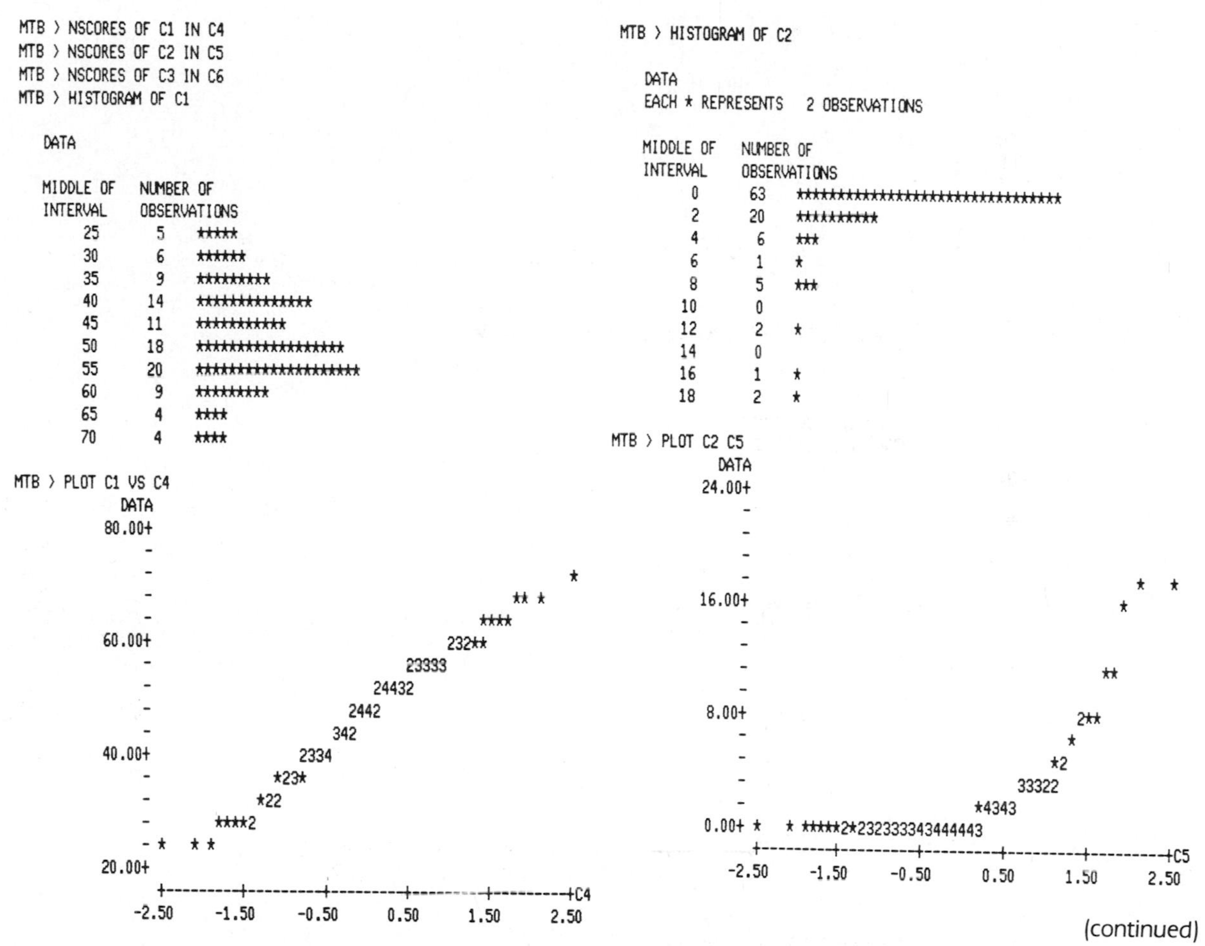

(continued)

Figure 7.14 Output for Exercise 7.55

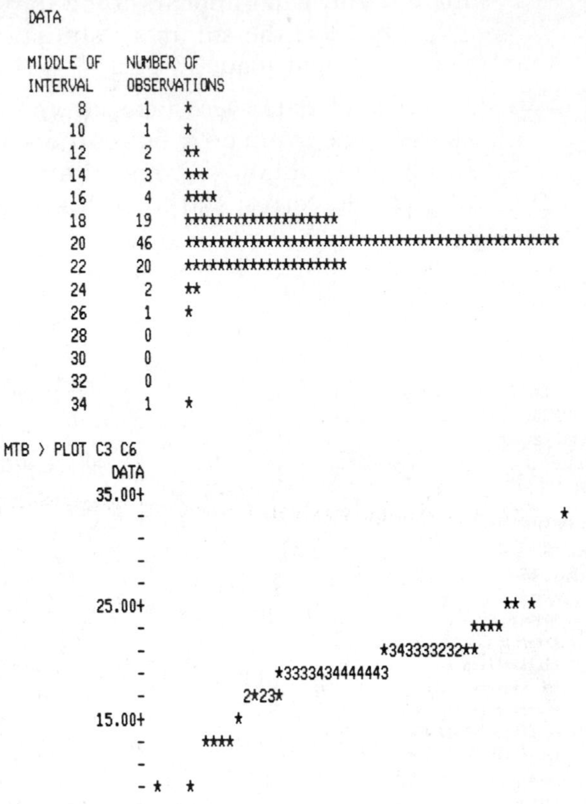

Figure 7.14 (continued)

Statistical Inference: Estimation

8

Now we can begin to discuss statistical inference. The most fundamental question is What summary figure, what statistic, should be calculated from the data? Section 8.1 sets out some general criteria for what constitutes a good statistic and discusses in particular when a sample mean is a good choice, as opposed to a sample median or some other statistic. Then Section 8.2 presents maximum likelihood estimation, a general method for selecting a reasonably good statistic, given some rather specific assumptions about the underlying population. Once a statistic has been chosen, the next step is to indicate the probable degree of random error reflected by the statistic. Section 8.3 introduces one of the two central forms of statistical inference, the confidence interval. (The other basic form of inference, the hypothesis test, shows up in Chapter 9.) Finally, in Section 8.4, we will take a first look at a basic problem in designing a statistical study: How big a sample is needed?

8.1 What Is a Good Estimation Method?

Suppose that you plan to take a sample from a certain population, and that you are willing to assume that the population is symmetric. You wish to estimate the population mean. One reasonable estimate is the sample mean. But if the population is symmetric, the population mean equals the population median, so the sample median is another reasonable estimate. The sample data will not be exactly symmetric, if only because of random variation, so the sample mean and sample median will not be equal. It matters, at least to some extent, which one is chosen. Also, to reduce the impact of outliers, you might want to use a 20% *trimmed mean,* in which you would discard the top and bottom 10% of the data and average the rest. (A common practice in judging sports such as competitive diving is to discard the highest and lowest judges' scores and to average the rest, precisely to avoid outlier problems.) Which is the best procedure? What does "best procedure" mean? In this section we will consider some standard criteria for best estimation, and try to answer the question of whether the mean, median, trimmed mean, or something else is "best" in a particular case.

The first criterion of estimation is that we want the sample statistic to estimate the population parameter correctly, on the average. We wouldn't use a sample 90th percentile to estimate a population median. Although in an occasional sample the 90th percentile might happen to be closer to the true median than was the sample median, generally the 90th percentile would tend to overestimate the population median. One desirable property of an estimating statistic is that it not systematically overestimate or underestimate the population parameter. If the long-run average (expected value) of a statistic equals the population parameter, the statistic is said to be an **unbiased estimator** of the parameter. We will use the generic symbol θ to indicate any particular population parameter, and the symbol $\hat{\theta}$ to indicate any particular **estimator** (sample statistic used in estimation).

unbiased estimator

estimator

Unbiased Estimator

A statistic $\hat{\theta}$ is an unbiased estimator of a parameter $\hat{\theta}$ if

$$E(\hat{\theta}) = \theta$$

See Figure 8.1 for examples of unbiased and biased estimators.

Unbiased estimators do exist. As stated in Chapter 7, the expected value of s^2 is σ^2_{pop}, so s^2 is an unbiased estimator of σ^2_{pop}. For any population, symmetric or not, $E(\bar{Y}) = \mu_{pop}$, so $\bar{Y}$ is an unbiased estimator of μ_{pop}. If the population can be assumed to be symmetric, the sample median and sample trimmed mean are also unbiased estimators of μ_{pop}. Sometimes it is appropriate to use estimators that are "just a little biased." The expected value of the sample standard deviation is not quite equal to σ_{pop}, but the discrepancy is so small that we use s anyway.

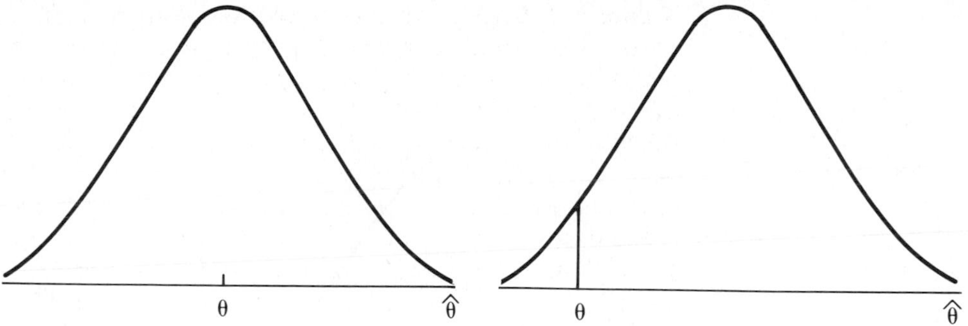

Figure 8.1 Unbiased and biased estimators

Example 8.1

In the Poisson distribution defined in Chapter 6, the true mean and variance are equal. Therefore, we might reasonably use either the sample mean or the sample variance to estimate the Poisson mean μ_{pop}. In any one sample, it's unlikely that the sample mean and sample variance will be equal, because of random variation. To study which method of estimation is preferable, a Monte Carlo study was done, using 1000 samples of size 50 from a Poisson population with mean and variance both equal to 4.0. The results were:

Statistic	$\bar{Y}$	s^2
Average Value	3.981	4.012

Do the statistics appear to be essentially unbiased?

Solution

Both averages are very close to the population value, 4.0. It is known theoretically that the expected values of both statistics are 4.0. Presumably, if an enormous number of samples had been taken, both averages would have been 4.000. □

Very often, several unbiased estimators of a parameter exist, so an additional criterion is needed to select among them. Figure 8.2 illustrates the theoretical distribution of two different estimators of a parameter θ. The expected values of both estimators equal θ, so both are unbiased. We would much prefer to use the $\hat{\theta}_2$ estimator. Its standard error is much lower, so it would be much more likely to come out close to the true value. Technically, $\hat{\theta}_2$ is more **efficient** than $\hat{\theta}_1$.

efficient

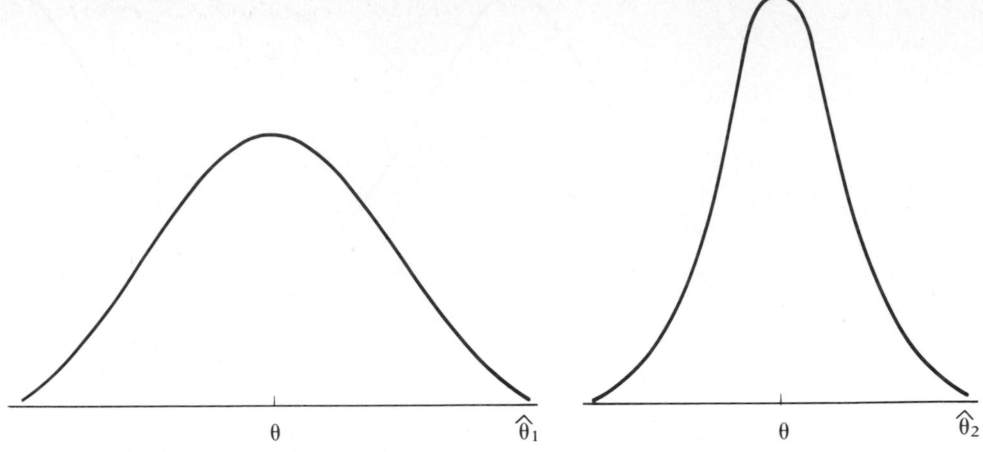

Figure 8.2 Less efficient ($\hat{\theta}_1$) and more efficient ($\hat{\theta}_2$) estimators

(Most) Efficient Estimator

An unbiased estimator $\hat{\theta}$ of a parameter θ is called most efficient if its theoretical variance (and therefore its standard error) is smaller than that of all other unbiased estimators of the parameter.

Finding most efficient estimators in a given situation is a problem in mathematical statistics. See, for example, Mood, Graybill, and Boes (1974). In particular, it has been proven that if *the population is exactly normally distributed,* then the sample mean is most efficient. In later chapters of this book we will often make the formal mathematical assumption that a population has a normal distribution. In general, the estimator indicated will be most efficient, given the assumption.

Example 8.2

(a) In a simulation study, 1000 samples of size 60 were drawn from a normal population with mean 50 and standard error 10. For each sample, the mean, trimmed mean, and median were calculated. Boxplots of the values of the three statistics are shown in Figure 8.3. (The trimmed mean is labeled as 80%, indicating that 80% of the data are included and 20% are excluded.) Which of the three estimators appears to be the most efficient?

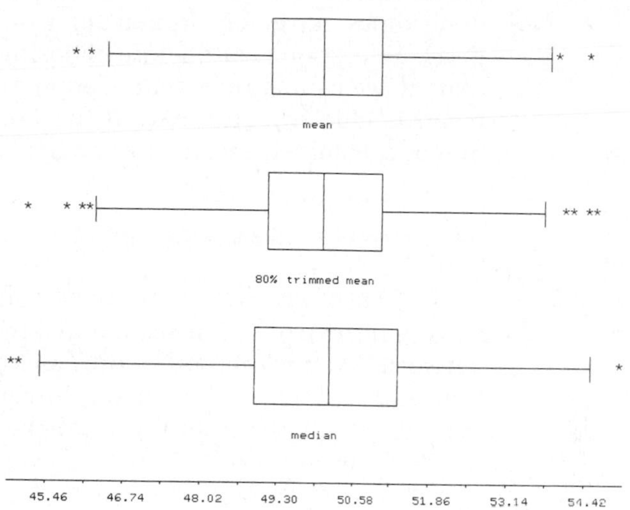

Figure 8.3 Boxplots for three estimators

(b) The averages and standard deviations (over the 1000 samples) of the three statistics were also calculated (Table 8.1).

Statistic	Average	Standard Deviation
Mean	50.1193	1.2959
Trimmed Mean	50.1200	1.3490
Median	50.1509	1.6417

Table 8.1

Do these numbers confirm your judgment in part a?

Solution

(a) The boxplot for the median is clearly the widest of the three, indicating that the median is more variable, and thus less efficient, in this situation. The boxplot for the mean appears slightly narrower than the one for the trimmed mean, indicating that the mean is the most efficient of the three.

(b) The average of all three statistics is just about 50, indicating that all three are essentially unbiased. The standard deviations, which are approximations of the respective standard errors, indicate that the mean is slightly more efficient than the trimmed mean, and much more efficient than the median, in the case of a normal population. □

There is a problem with the concept of efficiency. The mathematics of finding a most efficient estimator often depend very heavily on very specific assumptions. If the assumptions are wrong in practice, the estimator may well do a much poorer job than expected. Therefore, it becomes extremely important to check the validity of the assumptions, as best one can, before choosing which statistic to use. In particular, if we can assume that a population is nearly normal, the sample mean is efficient. However, if the population is symmetric but heavy-tailed, a trimmed mean or a median would be more efficient.

Example 8.3 Experience with a group of college students measured on an introversion/extroversion scale indicated that the bulk of students scored between 90 and 110, with scores as low as 20 in the extroversion direction and as high as 180 in the introversion direction. What does this indicate about the desirability of the sample mean as an estimator of the population mean?

Solution The description suggests that the distribution of introversion/extroversion scores is fairly symmetric, but heavy-tailed. In a sample, there are likely to be outliers, or long tails. In such a case, the sample mean will be more variable (less efficient) than a trimmed mean or a median. □

In practice, we can't verify the underlying assumptions exactly, because we don't know the exact population probabilities. (If we did, why would we take a sample?) All we will have is the sample data, plus perhaps a general sense of what the population should look like. The important thing is to *look at the data before doing anything else.* The histogram and stem and leaf diagram of Chapter 2 are very useful. These will give an indication of any gross violations of assumptions.

Example 8.4 Plomin, Foch, and Rowe (1981) report on a study of children's aggressive behavior. After instructions, each child had one minute to hit Bobo the Clown (a toy that bobs back up after being hit). The number of hits by each child was counted. Suppose that in a sample of 64 children, the following data were obtained:

13	16	11	15	7	14	24	16	15	26	14	13	12	11	23	14
13	17	10	19	24	19	11	8	8	2	12	15	6	14	6	12
25	12	18	15	12	14	10	11	13	15	15	13	23	13	19	24
13	12	7	19	17	24	19	10	13	11	12	8	12	11	14	16

Draw a histogram or stem and leaf display of the data. Is there reason to think that the sample mean will be an inefficient estimator of the population mean?

Solution The stem and leaf display in Figure 8.4 was constructed by hand.

```
0 | 2
0 |
0 | 6677
0 | 888
1 | 000111111
1 | 2222222233333333
1 | 444444555555
1 | 66677
1 | 899999
2 |
2 | 33
2 | 44445
2 | 6
```

Figure 8.4 Stem and leaf display for Example 8.4

There's no screaming nonnormality. The data are modestly right-skewed (the mean is 14.2 and the median is 13), and there are perhaps some slight outliers. The departure from normality is not so flagrant that a mean should be seriously inefficient. □

normal probability plot

For checking the specific assumption of normality, a so-called **normal probability plot** is very useful. This is a plot of the actual sample data, arranged from lowest to highest, against what would be expected under the assumption of normality. Many standard statistical computer packages will do this easily. Specially scaled graph paper called normal probability paper can also be used, but it's not as widely available as it used to be. If the population is normal, the normal probability plot will be a straight line except for random variation. If the data are badly skewed (indicating that the population is likewise skewed), the plot will be curved. Perhaps the greatest use of the normal probability plot is in checking for heavy-tailed distributions. In this case, the normal probability plot will show a pronounced S-shape. In Figure 8.5, histograms and normal probability plots (produced by the Minitab system) are shown for data drawn from, respectively, normal, skewed, and heavy-tailed populations.

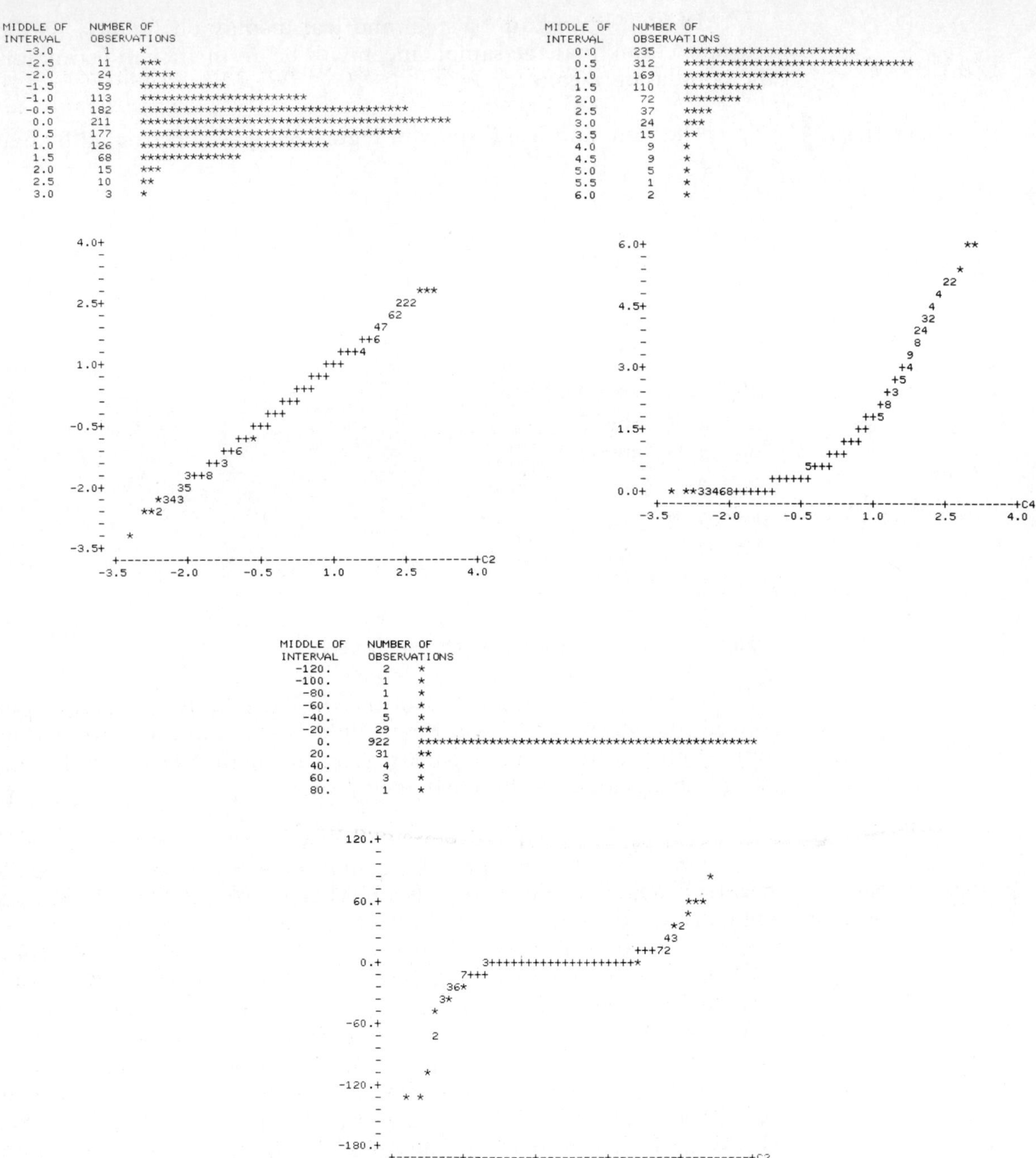

Figure 8.5 Histograms and normal probability plots

Example 8.5

Refer to the data of Example 8.4. A Minitab normal probability plot is shown in Figure 8.6. Do the sample data appear reasonably near normal?

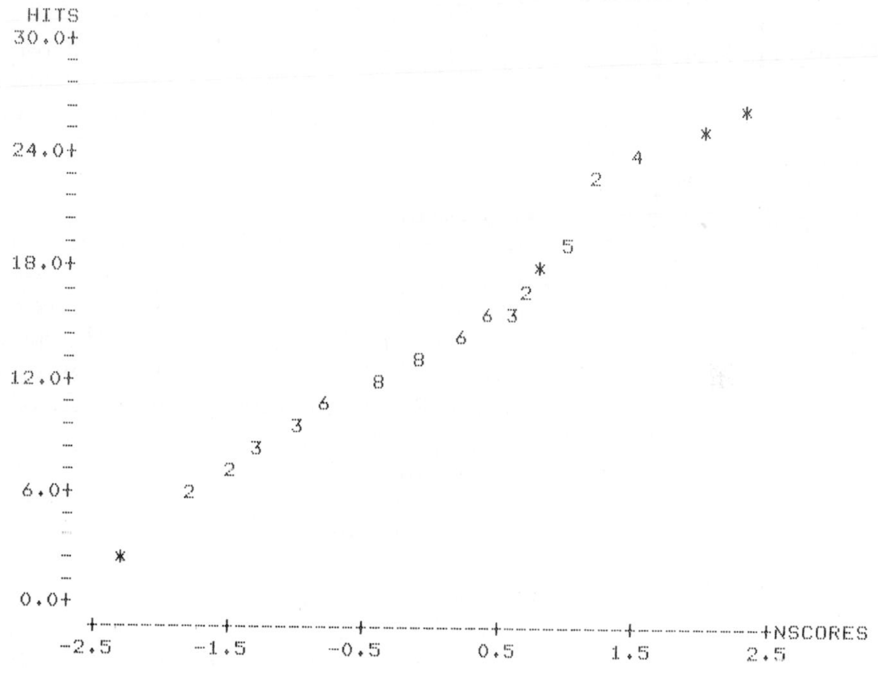

Figure 8.6 Normal probability plot for Example 8.5

Solution

There are certainly "wiggles and jiggles and bumps" in the plot, but no clear curvature or S-shape. Thus the data appear to have come from something close to a normally distributed population. □

If the picture of the data reveals severe nonnormality, what should a researcher do? There are many methods coming into use, such as calculating trimmed means or weighted means that effectively give less weight to extreme high and low values. A closely related group of estimators is the so-called M estimators, discussed in Devore (1982). These methods will be more widely used in the future; they are not yet generally available on standard packages, though some of them will be soon. Perhaps the simplest alternative to using a sample mean is to use a sample median. In several places in this book, we will present median-based alternatives to mean-type estimators. There is no law that says that only one statistic can be calculated from a given data set. Ideally, one tries several methods and they all give essentially the same result. If there is a conflict in the results of the several methods, then a plot of the data will usually indicate which result can be believed.

Example 8.6 Calculate the median and the 20% trimmed mean for the data of Example 8.4. Do these estimates differ substantially from the sample mean? On the basis of the shape of the data, which estimator would you select?

Solution For 64 scores, the median is the average of scores number 32 and 33, namely (13 + 13)/2 = 13. We might want to regard the scores as rounded to the nearest integer, as was done in Chapter 2. In that case, the interpolation idea leads to an approximate median of 12.5 + (1/8)(32 − 25) = 13.375. In either case, the median is below the mean, 14.219. To obtain the trimmed mean, we eliminate the top and bottom 10% of scores. Rather than try to drop 6.4 scores, we'll drop the top and bottom 6. The trimmed mean is the average of the remaining 52 scores, which comes out to 13.981. The estimators differ somewhat, reflecting a slight right skew in the data. Given that we assumed a near-normal population in the solutions to Examples 8.4 and 8.5, I'd use the mean. □

robust methods

nominal robustness

efficiency robustness

Alternatives such as use of trimmed means or weighted means are often referred to as **robust methods**. Generally speaking, a method is robust against a violation of an assumption if its performance is not affected much by moderate-to-substantial violation of that assumption. There are two very distinct concepts of robustness. First, there is what might be called **nominal robustness**—an estimator is robust in this sense if the theoretical probability distribution of the statistic is not much affected by violations of assumptions. For example, if the population distribution is normal, the theoretical probability distribution of the sample mean is also normal; according to the Central Limit Theorem, even if the population is nonnormal, the theoretical probabilities are still approximately normal, so long as the sample size is reasonably large. Therefore the sample mean is robust in the first, nominal sense. The second concept of robustness may be called **efficiency robustness**—an estimator is robust in this sense if it is nearly most efficient over a wide range of possible population distribution assumptions. We have said that the sample mean is most efficient if the population is normal. If, however, the population is symmetric but heavy-tailed, it has been established [see, for instance, Tukey (1977)] that the mean is much less efficient than, say, a trimmed mean. Therefore the mean is not robust in the second, efficiency sense. The distinction between the two ideas of robustness is important.

Example 8.7 Refer to Example 8.1. The property that the population mean equals the population variance holds for the Poisson distribution, but does not hold in general. What does this indicate about the robustness of the sample variance as an estimator of the population mean, if the Poisson assumption is dubious?

Solution If the population is not Poisson, the population variance may differ from the population mean. The sample variance would be estimating a parameter different from what it's supposed to be estimating. Therefore, it is not nominally robust with respect to violations of the Poisson assumption. □

In the social science literature, only the first, nominal type of robustness is mentioned much. The second, efficiency type of robustness is a relatively recent concept and hasn't yet become widely known among psychology and education researchers. Many times, researchers say, "The mean's so robust that with a sample of 200 subjects, I don't worry about normality." That's a mildly dangerous statement. With 200 subjects, it's quite likely that the probability statements made by the researcher will be *correct*, thanks to the Central Limit Theorem. But the researcher may not be getting the best use out of that expensive, hard-to-get data; that is, the data may not be being used as efficiently as they should be. It is *always* wise to plot data and check for violations of assumptions. Computer programs draw pictures very quickly. In fact, the computer's ability to plot data easily is just as important to good statistical analysis as the computer's ability to grind out numbers quickly.

Example 8.8 Refer to Example 8.7. A sample of 50 observations was taken from a population. To check on the Poisson assumption, the actual data and expected data assuming the Poisson distribution (with the same mean) were calculated:

Value	0	1	2	3	4	5	6	7	8	9	10
Actual	7	8	9	7	6	6	4	1	1	0	1
Expected	2.49	7.47	11.2	11.2	8.40	5.04	2.52	1.08	0.41	0.14	0.04

Is there an indication that the Poisson assumption might be wrong (and therefore that the sample variance is a nonrobust estimator of the population mean)?

Solution The actual frequencies aren't very similar to the expected frequencies. In particular, the actual frequencies are higher than expected at both ends and lower in the middle. Thus the Poisson assumption might well be wrong. □

Exercises for Section 8.1

8.1. The exponential density defined in Chapter 6 can be written as

$$f_Y(y) = 1/\mu e^{-y/\mu}, \quad \text{for } 0 < y < \infty$$

where μ is indeed the population mean. The population median can be shown to be $\mu/\log_e 2 = 1.4427\mu$. Two natural estimators of μ are $\hat{\mu}_1 = \bar{Y}$ and $\hat{\mu}_2 = 1.4427$ times the sample median. Suppose that a Monte Carlo study drew 1000 samples, each of size 40, and obtained the following:

Value of μ	0.5	1.0	1.5	2.0	2.5
Average $\hat{\mu}_1$	0.493	0.997	1.505	2.002	2.498
Average $\hat{\mu}_2$	0.504	1.004	1.503	1.988	2.492

What do these results suggest about the degree of bias of the two estimators?

8.2. The Monte Carlo study of Exercise 8.1 also involved the calculation of variances of the two estimators. The results:

Value of μ	0.5	1.0	1.5	2.0	2.5
Variance $\hat{\mu}_1$	0.00647	0.0241	0.0573	0.0986	0.1530
Variance $\hat{\mu}_2$	0.01340	0.0538	0.1159	0.2063	0.3164

What do these results indicate about the efficiency of the two estimators?

8.3. Refer to Exercise 8.1. The exponential distribution is right-skewed. The Groeneveld-Meeden skew measure is about 0.42.

a. Will the theoretical distribution of $\hat{\mu}_1$ be equally right-skewed? Why?

b. How could the Monte Carlo study discussed in Exercise 8.1 be used to determine whether normal-distribution probabilities were reasonably accurate?

8.4. Croft (1982) reported on various alternatives to the traditional spelling test form. In particular, a proofreading-and-correcting form was used. Croft reported the mean and standard deviation of the test scores. Assume that the actual data were as follows (mean 22.96, standard deviation 7.67):

3	7	8	12	16	17	17	18	19	19	19	20	20
20	20	20	21	21	21	21	22	22	22	22	23	23
23	23	23	23	23	24	24	24	24	25	25	26	26
26	26	28	29	29	30	33	35	36	40	50		

Draw a histogram or stem and leaf display of the data. Does the picture indicate that individual spelling scores are approximately normally distributed?

8.5. Refer to the data of Exercise 8.4. If a computer program is available to you, construct a normal probability plot of the data. What does this plot indicate about the normality of the spelling score distribution?

8.6. Calculate skewness and heavy-tailness statistics (from Chapter 2) for the data in Exercise 8.4. Do these statistics confirm the impression of the shape of the data given by the pictures drawn in Exercises 8.4 and 8.5?

8.7. For the data in Exercise 8.4, does it appear that the sample mean is the best available estimator of the population mean? (Note that the original article didn't present the actual data, so no criticism is intended.)

8.8. Refer to Exercise 8.1. If the underlying population is not exponential, is $\hat{\mu}_2$ still a reasonable estimator? What does your answer indicate about the robustness of that method?

8.9. An experiment involved measuring the times needed by 31 subjects to master a certain task. It seemed likely that the times would be severely right-skewed, because a few subjects would take an excessively long time to master the task. Assuming that right-skewness did hold, how robust would normal probabilities about $\bar{Y}$ be? Which sense of robustness is at issue here?

8.10. Refer to Exercise 8.9. The sample times and a normal probability plot of the data are shown in Figure 8.7. Did the times in fact indicate right-skewness in the population?

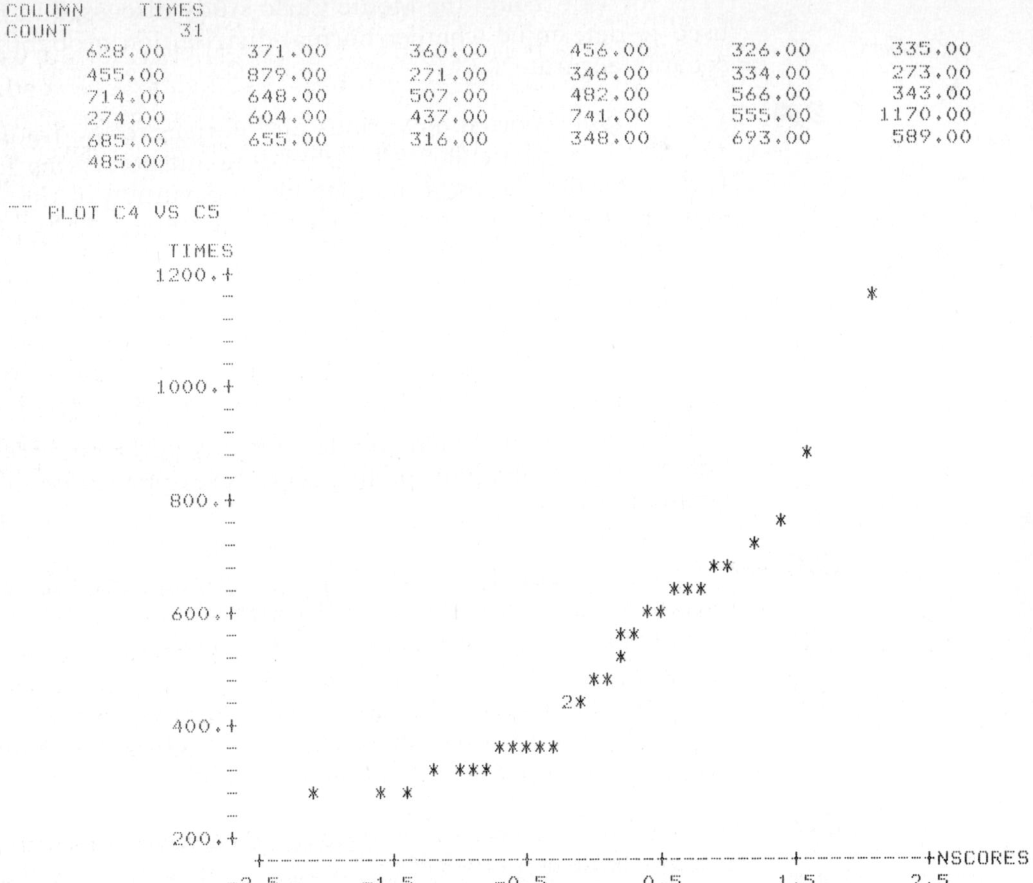

COLUMN TIMES
COUNT 31

628.00	371.00	360.00	456.00	326.00	335.00
455.00	879.00	271.00	346.00	334.00	273.00
714.00	648.00	507.00	482.00	566.00	343.00
274.00	604.00	437.00	741.00	555.00	1170.00
685.00	655.00	316.00	348.00	693.00	589.00
485.00					

Figure 8.7 Normal probability plot for Exercise 8.10

8.2 Maximum Likelihood Estimation

Section 8.1 set out principles for good estimation: unbiasedness and efficiency. It did not indicate how to achieve those goals. This section discusses a general principle that almost always yields a good estimator, given specific assumptions about a population distribution.

This general principle, called maximum likelihood, is best introduced by an example. Suppose that in an experiment, 20 lab rats may each exhibit one of two behaviors. Consider two questions. First, if the probability that any one rat exhibits behavior A is .40, what is the most probable number of rats exhibiting behavior A? Second, if 8 of the 20 rats exhibit behavior A, what is the most plausible estimate of the probability of behavior A? Both questions may be answered by looking at Table 1 of binomial probabilities in the appendix. All the binomial assumptions should be valid in this case. To answer the first question, we look down the $\pi = .40$ column for $n = 20$. The probabilities increase until $y = 8$, then decrease. Therefore, 8 is the most probable number of behavior A rats. To answer the second question, we look *across* the $y = 8$ row for $n = 20$. The probability that our actually observed $y = 8$ would occur is very low (.0000, to 4 places) when $\pi = .05$. Therefore .05 is a very implausible estimate for π. The probability that the actually observed $y = 8$ would occur increases as π increases, up to a maximum of .1797 at $\pi = .40$. Then it begins to decrease. Therefore the most plausible estimate of π is .40.

Example 8.9 Assume that a sample of three independent observations have been drawn from a process that follows Poisson probabilities. The observations are $y_1 = 7$, $y_2 = 1$, and $y_3 = 10$. Find the probability of this combination of values for $\mu = 5.0, 5.5, 6.0, 6.5,$ and 7.0. Which of these μ values is the most plausible estimate?

Solution Assuming independence, the probability of observing these values is $f_Y(7)f_Y(1)f_Y(10)$, where $f_Y(y)$ is the Poisson probability distribution found in Table 2 of the appendix. Recall that Table 2 contains cumulative probabilities, so the individual probabilities are found by subtraction. The probabilities are as shown in Table 8.2.

μ	5.0	5.5	6.0	6.5	7.0
$f_Y(7)$	.1044	.1235	.1377	.1463	.1490
$f_Y(1)$	.0337	.0225	.0149	.0098	.0064
$f_Y(10)$	.0181	.0285	.0413	.0558	.0710
Product	.000064	.000079	.000085	.000080	.000067

Table 8.2

The largest probability corresponds to $\mu = 6.0$, so 6.0 is the most plausible estimate of μ. The product probabilities are all very small, simply because the probability of such a very specific result as $y_1 = 7$,

$y_2 = 1$, or $y_3 = 10$ must be small. The issue is which probability is relatively largest. ☐

likelihood function

The **likelihood function** is defined to be the probability of obtaining the actually obtained data, as a function of the unknown population parameter(s). For the behavior A rats, the likelihood function is the binomial probability that $y = 8$ when $n = 20$, as a function of the unknown probability of behavior A. To find the **maximum likelihood estimate (MLE),** find the value(s) of the parameter(s) that maximize(s) the likelihood function.

maximum likelihood estimate (MLE)

We have already indicated one way to find a maximum likelihood estimate—namely, by computing the likelihood function for many values of the parameter(s) and picking the value with the largest likelihood. This is a valuable technique, particularly when the likelihood function is mathematically messy. It is not hard to write computer programs to do the calculations. There is, however, a technical objection that one can't be sure that one has the exact maximum without doing an infinite number of calculations; in the binomial example, we should have calculated the likelihood for $\pi = .39$ and .41 and .397 and so on.

Another way to find a maximum likelihood estimate (MLE) is to use calculus. (Those who don't know calculus should skip to the paragraph after Example 8.10.) Recall that to find the maximum of a function, you take its derivative, set the derivative equal to 0, and solve for the maximizing value. Technically, we should also check that the second derivative is negative at this value. It is usually more convenient in finding the MLE to work with the natural logarithm of the likelihood; the maximizing value will be the same for likelihood and log-likelihood functions. (As a function increases to its maximum, the logarithm of that function also increases to its maximum.) For the binomial example with $y = 8$ and $n = 20$, the log-likelihood is

$$\log \frac{20!}{8!12!} + 8 \log \pi + 12 \log (1 - \pi)$$

and its derivative with respect to π is

$$\frac{8}{\pi} - \frac{12}{1 - \pi}$$

Then, setting the derivative to 0 and solving, we find

$$8(1 - \pi) = 12\pi$$
$$8 = 20\pi$$

Therefore the MLE of π is $8/20 = .40$, as we found numerically. Note that the numerical calculations indicate that we have indeed found a maximum, not a minimum.

Example 8.10

Find the MLE of Example 8.9 mathematically.

Solution

Recall from Chapter 6 that the Poisson probability distribution is

$$f_Y(y) = \frac{e^{-\mu}(\mu)^y}{y!}$$

The likelihood for $y_1 = 7$, $y_2 = 1$, $y_3 = 10$ is

$$\frac{e^{-\mu}(\mu)^7}{7!}\frac{e^{-\mu}(\mu)^1}{1!}\frac{e^{-\mu}(\mu)^{10}}{10!} = \frac{e^{-3\mu}(\mu)^{18}}{7!1!10!}$$

and the log-likelihood is

$$-3\mu + 18 \log \mu - \log(7!1!10!)$$

The derivative of the log-likelihood with respect to μ is $-3 + 18/\mu$. Setting this to 0 and solving, we find that the MLE is $\hat{\mu} = 18/3 = 6.0$, which was the estimate we found most plausible in Example 8.9. Either by calculus or by our numerical calculations, it is clear that 6.0 represents a maximum, not a minimum. □

The maximum likelihood principle is reasonable in itself. It also yields estimators that do reasonably well in terms of unbiasedness and efficiency, given specific assumptions about the population. Under some technical conditions on the likelihood function, a maximum likelihood estimator is almost unbiased and almost most efficient. It is almost unbiased in that any bias goes to 0 as n goes to infinity, and almost most efficient in that the ratio of its standard error to that of any other almost unbiased estimator will never be larger than 1 as n goes to infinity. These results are proved in mathematical statistics textbooks; see Mood, Graybill, and Boes (1974). The consequence for a researcher is that maximum likelihood is a good general method for estimating parameters, *if* one can safely make assumptions about the underlying population.

Example 8.11

The general MLE for a sample from a Poisson distribution can be shown to be $\bar{Y}$, the sample mean, by the same calculus technique used in Example 8.10. Does Example 8.2 indicate that $\bar{Y}$ is relatively efficient?

Solution

In Example 8.2 we found that the sample mean had a lower standard error (was more efficient) than the sample variance, another plausible estimator. That finding agrees with the theoretical result that MLE's are (almost) most efficient. □

The catch to using maximum likelihood is that it requires very specific assumptions about the nature of the underlying population. If the assumptions are badly wrong, the MLE may be a poor method of estimation. Once again, if one is making assumptions, it is always best to plot the data to check those assumptions.

Example 8.12 Refer to the data of Example 8.8. Can we safely assume that the underlying population is Poisson, and therefore safely use $\bar{Y}$?

Solution The histogram of Example 8.8 deviates quite a bit from Poisson. Thus $\bar{Y}$ may not be the most efficient method of estimating μ. □

When the underlying population can be assumed to have a normal distribution, the maximum likelihood estimator of the population mean is the sample mean, which is unbiased and most efficient. The MLE of the population variance is $s^2[(n - 1)/n]$, not s^2. This MLE is slightly biased, so we use the unbiased estimator s^2. Later in the book we will encounter other, more complicated estimation problems involving normal-population assumptions; the estimators will in general be MLE's or, as in the case of s^2, almost MLE's.

Exercises for Section 8.2

8.11. Assume that a simple random sample of four observations is drawn from the exponential population described in Exercise 8.1, and that the observed values are $y_1 = 0.82$, $y_2 = 0.53$, $y_3 = 1.87$, and $y_4 = 0.38$. Calculate the likelihood function for $\mu = 0.80, 0.85, 0.90$, and 0.95. Which value seems to be closest to the maximum likelihood estimate?

8.12. Use calculus methods to find the maximum likelihood estimate in Exercise 8.11.

8.13. Define Y = the proportion of allocated time for an aptitude test. Assume that

$$f_Y(y) = (\theta + 1)y^{\theta}, \quad \text{for } 0 < y < 1$$

A simple random sample of six observations yielded scores of .98, .93, .99, .79, .93, and .89. Calculate the likelihood for $\hat{\theta}$ values ranging from 5 to 12. Can you estimate very roughly what the maximum likelihood estimate is? What calculations would be needed to give a more accurate idea of the MLE?

8.14. Use calculus methods to find the MLE in Exercise 8.13.

8.15. One of many possible population distributions that reflect right-skewness is a gamma distribution. A special case of this population distribution is

$$f_Y(y) = (.5\theta^{-3})\, y^2 e^{-y/\theta}, \quad \text{for } y > 0$$

Suppose that a random sample of size 2 is taken from this population distribution, yielding $y_1 = 2.40$ and $y_2 = 1.62$. Calculate the likelihood function for $\theta = .50, .60, .70$, and $.80$. Roughly where is the MLE?

8.16. Refer to Exercise 8.15. Use calculus to find the MLE.

8.3 Confidence Intervals

point estimation

interval estimation

Sections 8.1 and 8.2 discussed **point estimation.** The goal was to give the best single guess for the value of a population parameter. That is a limited goal. A statement such as "The best estimate of mean reaction time under these conditions is 1.024 seconds" is of limited scientific value, because it gives no indication of the potential degree of estimation error. It has no "plus or minus" attached to it. An estimate 1.024 ± .010 seconds indicates that the estimate is quite reliable, whereas an estimate such as 1.024 ± .410 seconds indicates a very imprecise estimate. This section introduces the fundamental statistical procedure of **interval estimation,** which does involve a "plus or minus" allowance for estimation error.

Suppose that we are trying to estimate the population mean reaction time of subjects under a certain experimental condition, based on a simple random sample of 64 subjects. Also suppose that we're willing to assume that the population of reaction times is nearly normally distributed, so the sample mean will be a reasonably efficient estimator. That might be a risky assumption. In many cases, reaction time data will be right-skewed, because of the catatonic few. Furthermore, suppose that we are willing to assume that the population standard deviation is 0.20 second. That, too, might well be a risky assumption. (Chapter 10 will introduce methods that don't require an assumption about the value of the population standard deviation.) We aim to specify not only the sample mean, but also the probable amount of error—random variation—inherent in that mean.

The theoretical distribution of the sample mean is used to answer the question of amount of error. Recall from Chapter 7 that the theoretical distribution is normal—exactly so if the population is exactly normal, or a close approximation because of the Central Limit Theorem. The expected value is the unknown μ_{pop}, and the standard error is

$$\sigma_{\bar{Y}} = \frac{\sigma_{pop}}{\sqrt{n}} = \frac{0.20}{\sqrt{64}} = 0.025$$

Therefore, there is a .95 probability that the sample mean will be within 1.96 (.025) = .049 of the true mean.

$$P(\mu_{pop} - 1.96(.025) \leq \bar{Y} \leq \mu_{pop} + 1.96(0.025)) = .95$$

The value 1.96 comes from normal tables; see Figure 8.8.

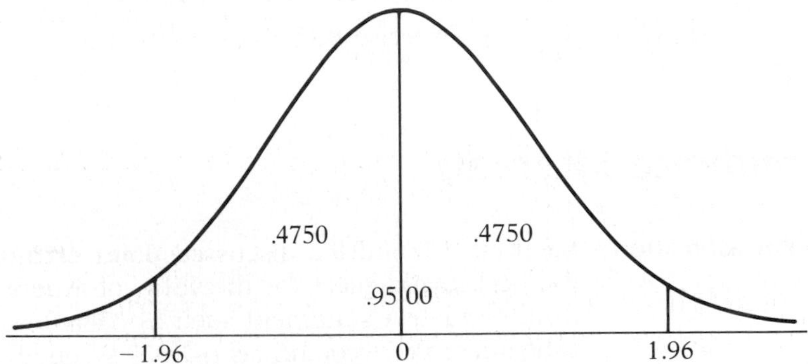

Figure 8.8 Normal distribution of sample mean

If there is a .95 probability that the sample mean will be within .049 second of the true mean, there is equally a .95 probability that the true population mean will be within .049 second of the sample mean. That's just two different ways of stating the same event.

$$P(\bar{Y} - 1.96(.025) \leq \mu_{pop} \leq \bar{Y} + 1.96(0.025)) = .95$$

Now suppose that the sample is taken and the sample mean comes out 1.024. The calculation of probable error of the sample mean indicates that the 95% probable error is ±.049. We can be 95% confident that the population mean lies within the range 1.024 ± .049, or 0.975 to 1.073 seconds.

Let's examine how we got that range. The value 1.024 was the sample mean $\bar{y}$. The standard error, .025, was calculated, assuming simple random sampling, as $\sigma_{pop}/\sqrt{n} = 0.20/\sqrt{64}$. The value 1.96 was obtained from the normal distribution table, and corresponds to 95% confidence. To develop a general formula for a confidence interval, we need some notation. Standard notation for the confidence level (e.g., 95%) is $100(1 - \alpha)\%$; for 95%, we would take $\alpha = .05$, because $100(1 - .05)\% = 95\%$. For 90%, we would take $\alpha = .10$; for 99%, $\alpha = .01$; etc. Conventionally, the normal-table number that cuts off a *right tail* area equal to a is denoted by z_a. See Figure 8.9. With this notation, we can generalize the example to a confidence interval formula.

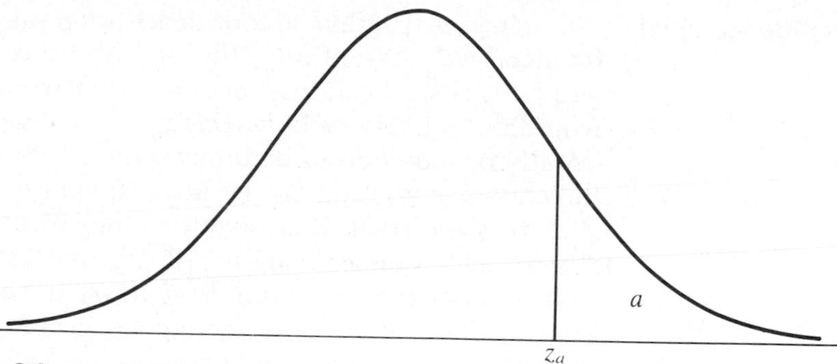

Figure 8.9 z_a notation

> **Confidence Interval for a Population Mean (σ_{pop} assumed known)**
>
> $$\bar{y} - z_{\alpha/2}\frac{\sigma_{pop}}{\sqrt{n}} \leq \mu_{pop} \leq \bar{y} + z_{\alpha/2}\frac{\sigma_{pop}}{\sqrt{n}}$$
>
> *NOTE:* The Central Limit Theorem must be applicable for the confidence level to be approximately correct. A heavy-tailed population may make this interval an inefficient method.

The reason for the $\alpha/2$ subscript in $z_{\alpha/2}$ is that the confidence probability is two-tailed. For 95% confidence ($\alpha = .05$), we cut off a .025 right tail above 1.96, and a .025 left tail below -1.96. The z table number for 90% confidence is $z_{.05} = 1.645$; for 99% confidence, it's $z_{.005} = 2.576$.

Example 8.13

Isbitsky and White (1981) measured obese children on their percentage recall of a list of food-related items. Assume that the population standard deviation is 8.0, which is essentially what Isbitsky and White found in their sample. If the sample mean is 37.0, with $n = 14$, what is a 95% confidence interval for the true, population mean? Assume that the sample can be regarded as a simple random sample.

Solution

The interval is

$$37.0 - 1.96\frac{8.0}{\sqrt{14}} \leq \mu_{pop} \leq 37.0 + 1.96\frac{8.0}{\sqrt{14}}$$

or

$$32.81 \leq \mu_{pop} \leq 41.19$$

□

confidence level

In constructing a confidence interval, one must choose the **confidence level**—essentially, the probability that the confidence interval will be correct. Certainly, one wouldn't want a 50% interval, because it would be as likely to be wrong as to be right. A 99.99% interval would usually be too wide to be informative. A 99.99% confidence interval for the class mean grade on the next 100-point exam would be from 4.4 to 99.2, or some such. Reasonable values would seem to be from perhaps 80% to 99%. Conventionally, 90, 95, and 99% are used. Some journals have a policy requiring that 95% be used, which is raising arbitrariness to an art form. If possible, it's a good idea to use lower confidence levels like 90% with small sample sizes, such as 20, and 99% with very large sample sizes, such as several hundred. Using a smaller confidence level with small sample sizes helps to compensate for the large standard error. The relatively low confidence level also may subconsciously remind readers that the conclusion is based on very limited data.

Example 8.14

For the data in Example 8.13, calculate confidence intervals corresponding to 90% and 99% confidence.

Solution

Replace 1.96 by 1.645 (for 90%) or by 2.576 (for 99%) in the calculations of Example 8.13. The resulting confidence intervals are

$$33.48 \leq \mu_{\text{pop}} \leq 40.52$$

and

$$31.49 \leq \mu_{\text{pop}} \leq 42.51$$

for 90% and 99%, respectively. Note that the higher the confidence level is, the wider the interval is. □

There is a small technical problem in interpreting the confidence level. In the reaction-time example, it is tempting to say that

$$P(0.975 \leq \mu_{\text{pop}} \leq 1.073) = .95$$

The problem is that nothing in that "probability" statement is a random variable. μ_{pop} is an unknown constant, and 0.975 and 1.073 are just numbers. Therefore a slightly different interpretation is needed.

A confidence interval is a random interval. Under repeated sampling, the confidence interval will jump around. Sometimes it will include μ_{pop}, sometimes not. Suppose that 10,000 samples, each of size 64, are taken from the reaction-time population we've been considering. The situation might be as shown in Figure 8.10.

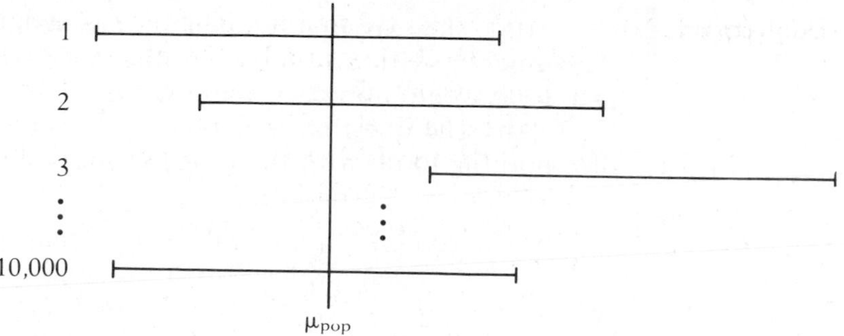

Figure 8.10 Repeated confidence intervals

In Figure 8.10, the confidence intervals from samples 1 and 2 include the true population mean, whereas the interval for sample 3 does not. Although one can't know in practice whether one specific interval has caught the true mean, one can know that, in the long run, 95% of the 95% confidence intervals will include the true, population value of the parameter.

Some care is needed to avoid misinterpreting what a confidence interval means. A famous trick question is "What is the probability that a sample mean is included in a 95% confidence interval for a mean?" The *sample* mean is always included in the confidence interval, because the interval has the form "sample mean plus or minus allowance for error." The confidence level, such as 95%, refers to the long-run proportion of intervals that include the *population* mean. Another trick question is "What proportion of the observations in a sample are included in a 95% confidence interval for a mean?" This is a nonsensical question, because the confidence level refers to the population *mean*, not to individual values. If the sample size is small, the interval will be wide and may well happen to include many individual values. Conversely, if n is large, the interval will be narrow and possibly may include relatively few individual values. The confidence level has nothing to do with individual values.

Example 8.15

Refer to Examples 8.13 and 8.14. Suppose that 1000 samples, each of size 14, were taken from the same population, and that 90%, 95%, and 99% confidence intervals for μ_{pop} were calculated for each sample.

(a) Would it be surprising to discover that 889 of the 90% intervals, 958 of the 95% intervals, and 993 of the 99% intervals actually included the population mean?

(b) Would it be surprising to discover that only 38 of the 90% confidence intervals included the entire range of data in the sample?

Solution (a) The expected frequencies are 900, 950, and 990, respectively. Presumably, with an infinitely long series of samples, the frequencies per 1000 would approach these values.

(b) The question is irrelevant. A confidence interval for a mean has nothing to do with the range of individual values in a sample. □

The confidence interval for μ_{pop} that we are using depends critically on two assumptions. We are assuming that σ_{pop} is a known number, not estimated from the data. In practice, σ_{pop} is not likely to be known. The t distribution methods introduced in Chapter 10 should be used when σ_{pop} isn't known and is estimated from the data. Further, we assume that the theoretical (sampling) distribution of the sample mean is normal. Recall from Chapter 7 that the Central Limit Theorem guarantees that the theoretical distribution will be approximately normal when n is large. Roughly, a sample size of 30 is large enough to make the normal an adequate approximation unless the population is skewed or lumpy (consisting of very few possible values). A sample size of 30 may not, however, be enough to make the width of the interval adequately small.

The mathematical form of the confidence interval we've been considering is typical of many others which we'll see later. It is centered on an estimate (here $\bar{y}$, generically $\hat{\theta}$) of a parameter (here μ_{pop}, generically θ). The plus-or-minus term is of the form "table value times standard error." In our example, normal tables apply, as they will in many, but by no means all, other situations. The standard error formula here has been $\sigma_{pop}/\sqrt{n}$. Many, many other standard error formulas may apply, depending on the statistic and the sampling method.

Typical Form of Confidence Interval

$$\hat{\theta} - z_{\alpha/2}(\text{std. error}) \leq \theta \leq \hat{\theta} + z_{\alpha/2}(\text{std. error})$$

NOTE: Each different estimator $\hat{\theta}$ will have a different standard error formula, which will also depend on the sampling method used. The validity of this form requires that the theoretical distribution of $\hat{\theta}$ be approximately normal. The rule of thumb for use of a normal approximation will vary from one estimator (and sampling method) to another.

Example 8.16 A statistic called Fisher's Z is closely related to the correlation coefficient r_{YX} discussed in Chapter 14. Under some very specific assumptions, the theoretical distribution of Fisher's Z is approximately normal with expected value "true Z" and standard error $1/\sqrt{n-3}$. In a sample of 38 observations, Fisher's Z equaled 1.283. Find a 95% confidence interval for "true Z."

Solution

$$1.283 - 1.96 \frac{1}{\sqrt{38-3}} \leq \text{true } Z \leq 1.283 + 1.96 \frac{1}{\sqrt{38-3}}$$

or

$$0.9517 \leq \text{true } Z \leq 1.6143$$

Note that for this confidence interval to have an actual 95% confidence level, the assumptions underlying the theoretical normal distribution must hold. Note also that the standard error for Fisher's Z is different from the standard error for $\bar{Y}$. □

The confidence intervals that we have been considering so far have all been "two-sided." The risk of error has been the combined probability in both tails of the theoretical distribution of the statistic. This is by far the most commonly used form of confidence interval, but one can also calculate a "one-sided" confidence interval. Suppose that, in sampling reaction times (again with $n = 64$ and $\sigma_{pop} = 0.20$), a sample mean of 1.024 seconds is observed. A **95% upper-bound confidence interval** can be calculated as

95% upper-bound confidence interval

$$\mu_{pop} \leq 1.024 + 1.645 \frac{0.20}{\sqrt{64}}$$

95% lower-bound confidence interval

or $\mu_{pop} \leq 1.065$. Similarly, a **95% lower-bound confidence interval** is

$$\mu_{pop} \geq 1.024 - 1.645 \frac{0.20}{\sqrt{64}}$$

or $\mu_{pop} \geq 0.983$. The number 1.645 comes from the normal table; see Figure 8.11. The generalization to any confidence level follows easily.

One-Sided Confidence Interval for μ (σ_{pop} assumed known)

Upper-bound interval:

$$\mu_{pop} \leq \bar{y} + z_\alpha \frac{\sigma_{pop}}{\sqrt{n}}$$

Lower-bound interval:

$$\mu_{pop} \geq \bar{y} - z_\alpha \frac{\sigma_{pop}}{\sqrt{n}}$$

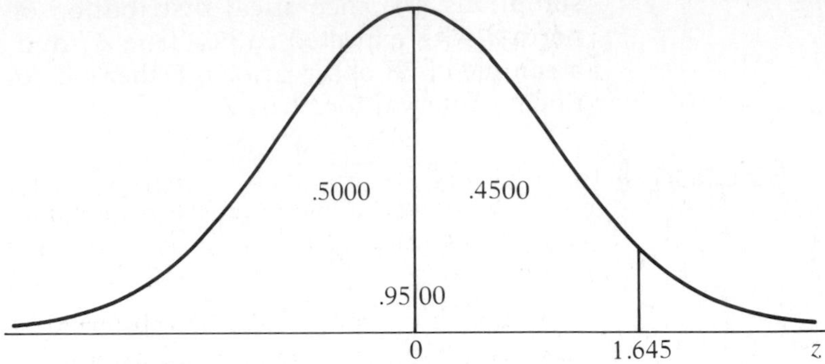

Figure 8.11 z table value for one-sided confidence interval

Example 8.17 Referring to Example 8.16, calculate a 95% lower-bound interval for "true Z."

Solution Because the relevant statistic is Fisher's Z, the relevant standard error is $1/\sqrt{n-3}$; the z table value cutting off a .05 area is 1.645.

$$\text{true } Z \geq 1.283 - 1.645 \frac{1}{\sqrt{38 - 3}}$$

or true $Z \geq 1.005$. □

The word "interval" may be a bit forced when applied to one-sided procedures. Is an "interval" allowed to stretch to infinity, as one-sided intervals do? In keeping with technical mathematical usage, we'll use the word even in the one-sided case. Those who think that intervals should have a finite beginning and end might want to say "lower-bound confidence set" rather than interval.

When the population is skewed, a one-sided confidence interval for a mean is somewhat more dubious than a two-sided one. Recall from Chapter 7 that one-tailed probabilities are not approximated as well as two-tailed ones in the case of skewness. Therefore the nominal confidence interval (95% or whatever) is likely to be a poorer approximation in the one-sided confidence interval.

Exercises for Section 8.3

8.17. Refer to Exercise 8.4. Make the somewhat heroic assumption that the population standard deviation exactly equals the sample standard deviation.

a. Find a 90% confidence interval for the population mean.
b. Give a careful interpretation of the meaning of this interval.

8.18. For the data of Exercise 8.4, calculate a 95% lower-bound confidence interval for the mean.

8.19. Refer to Exercises 8.4 and 8.5. What do the data plots indicate about the efficiency of using the sample mean as an estimator of the population mean? What effect would using a more efficient estimator have on the confidence interval?

8.20. Consider the Isbitsky and White study mentioned in Example 8. Assume that the population standard deviation is 8.0 and that the data (number of items recalled) are

23 26 26 30 32 35 37 37 38 41 44 46 49 54

a. Construct a histogram or stem and leaf display of the data. Is there evidence that the underlying population distribution is seriously nonnormal? If so, what is the most prominent departure from normality?

b. If an appropriate computer program (or normal probability paper) is available to you, construct a normal probability plot of the data. Does this plot convey the same impression that you had in part a?

8.21. Refer to the data of Exercise 8.11. In that exercise, the maximum likelihood estimator of μ_{pop} was shown to be the sample mean $\bar{y}$. It was argued that this justified using the methods of this section to calculate confidence intervals for μ_{pop}.

a. Calculate a (nominally) 95% confidence interval for the population mean.

b. Calculate a (nominally) 95% upper-bound confidence interval for the population mean.

c. Are the nominal 95% confidence levels likely to be reasonably close to the actual confidence level? If not, which nominal 95% is more suspicious?

8.22. Refer to Exercise 8.21. If the sample size had been 40, would either or both of the nominal confidence levels have been close approximations to the actual level? Why? Which confidence level should be a better approximation?

8.4 Determining the Sample Size

An important decision in planning any study or experiment is the specification of what quantity of data is to be collected. A study with a too-small sample size isn't likely to give much information. An experiment with a too-big sample size is likely to be very expensive. Thinking in terms of confidence intervals is a useful way of thinking about how big a sample size is needed.

Suppose that a study of spatial-relations ability is to be conducted on a group of eighth-grade girls. Their scores on a standard 100-point scale will be recorded. Assume that the population of scores is reasonably close to normally distributed, so the sample mean is a reasonably efficient estimator of the true, population means. Also assume that the test is known to have $\sigma_{\text{pop}} = 10.0$. (Again, in Chapter 10 there are methods that don't require such an assumption.) Also, suppose that a 99% confidence interval (two-sided) is to be calculated. Ideally, that interval would be very narrow, perhaps $\pm .1$ unit, so that the mean would be very accurately estimated. Unfortunately, to get such an accurate estimate would certainly require a hideously large sample. (In fact, it would require sampling 66,500 girls!) On the other hand, a very small sample would yield a confidence interval that was so wide as to be useless. Suppose that after some consideration, we decide to try for a 99% confidence interval with a width of 4 units, therefore a plus-or-minus of 2 units. Assuming simple random sampling, the actual plus-or-minus will be $2.58(10.0/\sqrt{n})$, which should equal the desired plus-or-minus, namely 2. We simply set the actual equal to the desired and solve.

$$2.58\frac{10.0}{\sqrt{n}} = 2$$

$$\sqrt{n} = 12.9$$

$$n = 166.4, \text{ rounded up to } 167$$

Although a formula can be given for this specific case, the basic process of setting the actual plus-or-minus equal to the desired one and solving for the required n applies much more generally.

Selecting a Sample Size via Confidence Intervals

Specify a confidence level and an acceptable plus-or-minus, E. Set the actual plus-or-minus of the confidence interval for the parameter equal to the specified E and solve for n. In the particular case of simple random sampling and estimation of μ_{pop} with σ_{pop} known, the result is

$$n = \frac{(z_{\alpha/2})^2 (\sigma_{pop})^2}{E^2}$$

where $z_{\alpha/2}$ is the appropriate entry of the normal table and E is the specified plus-or-minus desired for the confidence interval.

Example 8.18 Refer to Example 8.13. If a 95% confidence interval with plus-or-minus equal to 2 is desired, what sample size is required?

Solution σ_{pop} was assumed to be 8.0, and $z_{.025}$ (for a two-sided 95% confidence level) is 1.96. Set

$$1.96 \frac{8.0}{\sqrt{n}} = 2$$

and solve to get $\sqrt{n} = 7.84$, or $n = 61.5$, rounded up to 62. Of course, the formula in the box gives the same answer. □

From the formula given in the box above Example 8.18, we can note the consequences of changing the specification of the problem. Increasing the confidence level, say from 99% to 99.9%, will require a larger $z_{\alpha/2}$ value, and therefore a larger sample size. A larger assumed value of σ_{pop} will require use of a larger sample size to obtain a confidence level with the desired plus-or-minus, E. Decreasing the allowable value for E will also require use of a larger n.

Example 8.19 Refer to Example 8.18. Find the sample sizes required for

(a) $E = 2$, 99% confidence
(b) $E = 1.5$, 95% confidence

Solution **(a)** The confidence level has changed, so the required table z value is now 2.58.

$n = (2.58)^2(8.0)^2/(2)^2 = 106.2$, rounded up to 107

(b) The tolerable plus-or-minus, E, has decreased.

$n = (1.96)^2(8.0)^2/(1.5)^2 = 109.3$, rounded up to 110 □

A good way to specify an acceptable plus-or-minus, E, is to state it in units of σ_{pop}. An error of, say, $.6\sigma_{pop}$ would be quite large, whereas an error of $.02\sigma_{pop}$ would be very small. See Figure 8.12. The sample sizes (for 95% confidence intervals) corresponding to these errors are, respectively, 11 and 9604.

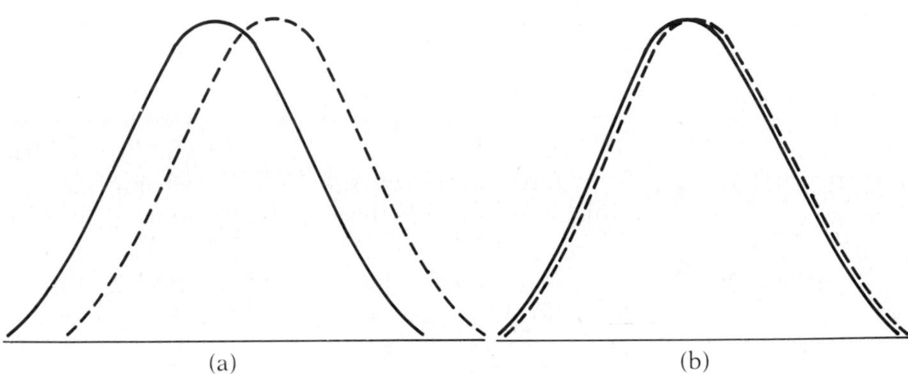

Figure 8.12 Populations estimated with errors: (a) $.6\sigma_{pop}$, (b) $.02\sigma_{pop}$

Example 8.20 Refer to Example 8.18. Suppose that we want a 95% confidence interval with a plus-or-minus of $.1\sigma_{pop}$. What sample size is required?

Solution

$$n = (1.96)^2 \frac{(\sigma_{pop})^2}{(.1\sigma_{pop})^2} = 384.2, \text{ or } 385$$

Note that this answer is correct regardless of the value of σ_{pop}. □

It should be noted that there are two issues in determining how large a sample size should be—a scientific one and a technical, statistical one. The scientific issue is "How big a sample must I take to get the scientific information I need?" That is the issue that has been discussed in this section, in terms of acceptable plus-or-minus of a confidence interval. There is also the technical issue of whether the sample size is large enough to justify the underlying statistical assumptions. In the case of a mean, the technical issue is whether the sample size is large enough to justify invoking the Central Limit Theorem. Rules of thumb for the sample size required for a good normal approximation to the theoretical distribution of the sample mean were given in Chapter 7. Usually, the scientific question is the dominant one. Barring severe skewness in the population, a sample size that is large enough to give scientifically interesting results will be amply large enough to justify using a normal theoretical distribution of the sample mean. However, there can be exceptions. Furthermore, a large sample size will not make an inefficient method suddenly more efficient. As always, it is a good idea to look at the data and consider whether the procedure is really appropriate.

Exercises for Section 8.4

8.23. Refer to Exercise 8.4. Assuming that the population standard deviation equals the sample standard deviation, find a 95% confidence interval for the mean. What sample size would be required to yield a 95% confidence interval accurate to ± 2 units?

8.24. **a.** Would the sample size found in Exercise 8.23 be adequate to ensure that the nominal 95% probability would be close to correct? Explain why.

b. Would the sample size found in Exercise 8.19 improve the relative efficiency of the sample mean as an estimator of the true, population mean (as compared to, say, a trimmed mean)?

8.25. Keller and Tetlow (1980) measured a sample of 18 backstage workers in a community theater on the Meyers-Briggs Type Indicator scale, and in particular on extroversion/introversion. The sample mean was 98.38 and the sample standard deviation 21.23. Once again, we boldly assume that the sample standard deviation equals the population standard deviation.

a. Calculate a 90% confidence interval for the population mean.

b. Calculate a 99% interval for the mean. How much wider is it?

8.26. In Exercise 8.25, how big a sample size would be needed to obtain a 90% confidence interval accurate to $\pm 1/8$ of a standard deviation? A 99% confidence interval with the same plus-or-minus?

8.27. Refer to Exercises 8.25 and 8.26.

a. If the actual sample data had been given, how would you decide if the 90% confidence level could be relied on? How could you decide if the procedure used was the most effective possible?

b. Would the sample size found in Exercise 8.26 be enough to allay any concerns you expressed in part a?

Chapter Exercises

8.28. A researcher obtained scores on a dominance scale from a sample of 42 subjects. The scores, ordered from lowest to highest, were

69	76	79	79	82	84	86	86	87	92	95
96	97	98	99	99	100	100	102	103	103	104
105	106	106	107	107	109	109	110	111	111	112
113	116	116	116	118	125	127	130	137		

The sample mean is 102.55. The researcher was willing to assume that the population standard deviation was 15.0. (The sample standard deviation is very close to 15 as well.)

a. Construct a stem and leaf display of the data. Is there evidence of considerable skewness or wild outliers?

b. For these data, the Groeneveld-Meeden skewness measures (as defined in Chapter 2) are −0.085 for the whole data set, −0.469 for the bottom half of the data, and 0.471 for the top half of the data. Do these values confirm your judgment made in part a?

c. Does the "fences" method defined in Chapter 2 turn up any outliers?

d. Is there compelling evidence that the sample mean would be a poor estimator of the population mean in this situation? What does "poor estimator" mean in this problem?

8.29. Using the sample mean found in Exercise 8.28 and the assumed population standard deviation, calculate a 95% confidence interval for the population mean. How wide is it?

8.30. Refer to Exercise 8.28. The researcher wanted to undertake a larger follow-up study. In order to obtain a 95% confidence interval for the population mean with a width of 4 (and therefore a plus-or-minus of 2), how large a sample would be needed?

8.31. The Groeneveld-Meeden skewness statistic defined in Chapter 2 is a sample statistic and therefore has a theoretical (sampling) distribution. Suppose that in a random sample of 204 individuals, the skewness measure equals 0.079, indicating slight right-skewness. Also, suppose that a computer program has calculated the standard error of the statistic to be 0.063. Calculate a 90% confidence interval for the population value of the skewness measure.

8.32. In Exercise 8.31, what assumptions did you make to obtain the interval?

8.33. In an experiment, subjects tried to move a cursor on a computer monitor into a target area while avoiding obstacles that randomly appeared on the screen. The subjects could only see the screen through the eye corresponding to the nondominant hand. For each subject, the researcher recorded Y = the number of trials until a success occurred. Under reasonable assumptions, the probability distribution of Y is

$$f_Y(y) = \pi(1 - \pi)^{y-1}, \quad \text{for } y = 1, 2, \ldots$$

where π is the probability of success in any one trial. Four subjects obtained scores $y_1 = 3$, $y_2 = 12$, $y_3 = 4$, $y_4 = 6$. Find the maximum likelihood estimator of π for these data.

8.34. Refer to Exercise 8.33. It can be proven that the long-run average value of the maximum likelihood estimator in this situation is slightly larger than π. What desirable property of an estimator does not hold for the maximum likelihood estimator in this situation?

8.35. A computer was programmed to draw 1000 samples of size 30 from a "Laplace" population. This population is symmetric but somewhat outlier-prone. In the computer program, the population mean was set to be 0. The program computed a mean, a trimmed mean, and a median for each sample. Boxplots of the results are shown in Figure 8.13. What do these plots suggest about the bias and efficiency of the three estimators when sampling from a Laplace population? (The 80% figure for the trimmed mean is the amount of the data included in the average; the top and bottom 10% of the data were trimmed, and the remainder averaged.)

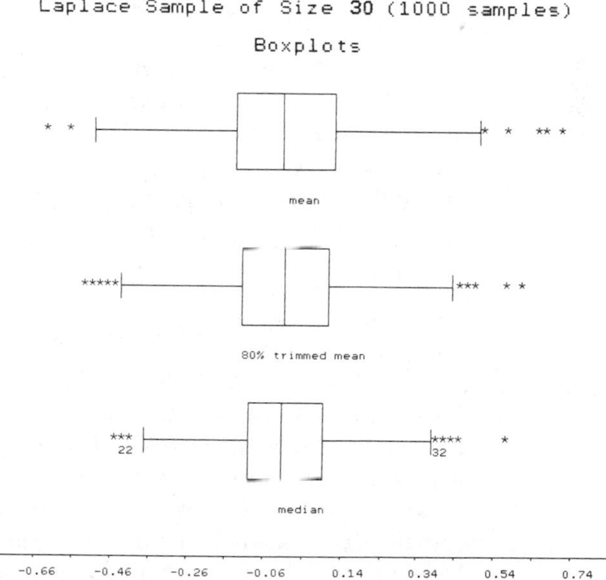

Figure 8.13 Boxplots for Exercise 8.35

8.36. The computer program in Exercise 8.35 also computed averages and standard deviations of the three estimators, over the 1000 samples. The results were as shown in Table 8.3.

	Average	Standard Deviation
Mean	0.0027	0.2639
Trimmed Mean	0.0032	0.2289
Median	0.0082	0.2070

Table 8.3

What do these results indicate about the bias and efficiency of the three estimators in this situation?

8.37. In a study, a sample of 142 women aged 60 to 70 were interviewed 6 months after the deaths of their husbands. A score on a standard depression scale was obtained for each woman. The sample mean was 16.24. The population standard deviation was assumed to be 8.0 (and the sample standard deviation was close to 8.0).

a. Construct a 99% confidence interval for the population mean.

b. Construct a one-sided, 99% upper-limit confidence region for the population mean.

8.38. A normal probability plot of the data for Exercise 8.37, not shown here, did not look like a straight line. Instead, it started upward then curved off to the right.

a. What would such a plot indicate about the shape of the data?

b. Does this finding indicate that the nominal 99% confidence of the intervals calculated in Exercise 8.37 might be seriously wrong? If so, is the two-sided confidence or the one-sided confidence more likely to be in error?

8.39. Suppose that another study, similar to that of Exercise 8.37, was planned. How large a sample would be needed to obtain a 99% confidence interval for the mean with a width of 2? Alternatively, how large a sample would be needed to obtain a 99% confidence interval with a width equal to one-fourth of a population standard deviation?

8.40. The γ statistic defined in Chapter 3 measures the relation between two ordinal variables. It is positive if there is an increasing relation, negative if there is a decreasing relation. It has been proved that, when γ is calculated from a random sample, it has an approximately normal theoretical (sampling) distribution for large n. Suppose that for a sample of 147 individuals, the SAS package indicates that γ equals 0.426 and that "ASE1" equals 0.126. ASE1, in SAS, means approximate standard error.

a. Calculate a 95% confidence interval for the population value of γ.

b. Based on this interval, can you reasonably conclude that there is an increasing relation in the population?

c. Because we are largely interested in the lower limit of the γ value in this case, a one-sided confidence interval might be more useful. Calculate a 95% lower-bound interval for the population value of γ.

8.41. Refer to Exercise 8.40. The standard error of the γ statistic can be shown to have the form

$$\frac{\text{long messy expression}}{\sqrt{n}}$$

where the long messy expression doesn't involve the sample size. Will increasing the size of the sample in Exercise 8.40 to 294 individuals cut the width of the confidence interval in half?

8.42. A computer program drew random samples of sizes 30 and 60 from a population that is symmetric around 0 but severely outlier-prone. For each sample, the mean, trimmed mean (20% trimmed, middle 80% averaged), and median were calculated. Boxplots of the results, based on 1000 samples, are shown in Figure 8.14.

a. What do the boxplots indicate about the relative efficiency of the three estimators in this case?

b. Does your conclusion in part a depend on whether the sample size is 30 or 60?

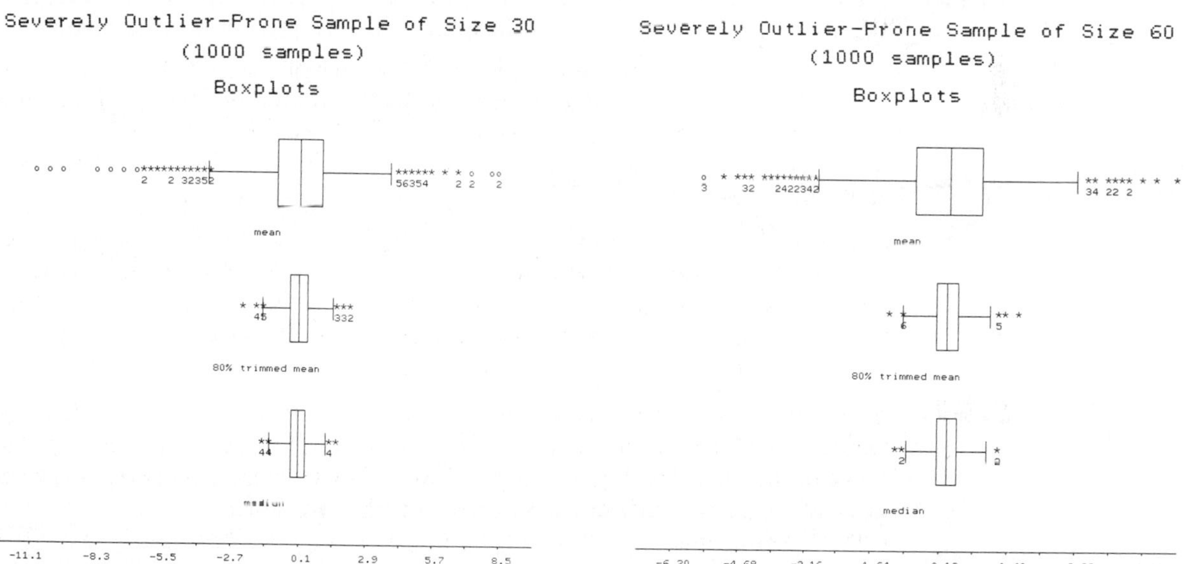

Figure 8.14 Boxplots for Exercise 8.42

8.43. The computer program of Exercise 8.42 also computed averages and standard deviations of the three estimators, over 1000 samples. The results were as shown in Table 8.4.

$n = 30$

	Average	Standard Deviation
Mean	0.0228	1.8757
Trimmed mean	0.0081	0.5667
Median	0.0148	0.4510

$n = 60$

	Average	Standard Deviation
Mean	−0.0285	1.3165
Trimmed mean	−0.0014	0.3699
Median	0.0003	0.3211

Table 8.4

Do these results confirm your opinion about the relative efficiency of the three estimators for this population?

8.44. A researcher planned to take a random sample from a population. Previous experience indicated that the population distribution was very close to normal.

a. What sample size is needed to obtain a 99% confidence interval with a width of two-tenths of a population standard deviation, and thus a plus-or-minus of one-tenth of σ_{pop}?

b. Could one get away with a smaller sample size by using a more efficient method than the sample mean?

8.45. Refer to Exercise 8.44. Suppose that we change the specifications as indicated below. What should happen to the required sample size?

a. Make the confidence level 95% rather than 99%.

b. Leave the confidence level at 99% but change the required width to three-tenths of σ_{pop}.

8.46. The Pareto probability density

$$f_Y(y) = \theta y^{-(\theta+1)}$$

is sometimes useful as a model for extremely skewed data. Suppose that a random sample yielded $y_1 = 9.9$, $y_2 = 6.7$, $y_3 = 47.6$, $y_4 = 5.2$, and $y_5 = 84.8$. Find the maximum likelihood estimator of the population parameter θ.

8.47. Suppose that a computer program was available that would draw repeated random samples from various Pareto distributions and calculate maximum likelihood estimators. How could you use this program to assess whether the estimator was essentially unbiased?

9 Hypothesis Testing

In Chapter 8 we considered two forms of statistical inference, point estimation and confidence intervals. In this chapter we turn to the most widely used, and grievously abused, type of statistical inference: hypothesis testing. The basic hypothesis-testing strategy is a form of proof by contradiction. This strategy is introduced with a simple binomial example in Section 9.1. This section defines most of the critical concepts of hypothesis testing. Section 9.2 contains a definition of the general binomial hypothesis test, along with a discussion of its properties. In Section 9.3 we turn to testing a single mean. The concepts of p-value and statistical significance are discussed in Section 9.4; understanding these concepts is crucial to thoughtful hypothesis testing. There is a close relation between hypothesis tests and confidence intervals; this relation is considered in Section 9.5. Just as Chapter 8 ended with a discussion of planning a study via confidence intervals, the main part of Chapter 9 closes with a discussion of planning via hypothesis tests. An appendix discusses an important mathematical result, the Neyman-Pearson Lemma, which plays a key role in the theory of choosing a good test statistic.

9.1 The Basic Strategy of Hypothesis Testing

Anyone who has read research reports in psychology, education, or other sciences is familiar with statements of the form "such-and-such was statistically significant (p = so-and-so)." Such statements are the results from statistical hypothesis testing, often called significance testing. This section presents the basic approach of hypothesis testing, as usual beginning with an example.

It has been claimed that minority school children make exceptional progress when using computerized materials, perhaps because they perceive the computer to be unprejudiced. Suppose that we try to establish this by developing a computerized English grammar and vocabulary unit to be tested on a sample of Hispanic fourth-graders. In this study, 40 children are randomly paired. Then one child from each of the 20 pairs is chosen, also at random, to study the materials using the computerized unit. The other child in each pair studies the same materials in a conventional workbook and class exercise unit. After the children complete the unit, an observer (who is "blind" to the assignment of the children to study type) rates each pair, specifying which child has a better command of grammar and vocabulary. No numerical rating is given, only a judgment of better or worse.

We would like to prove that Hispanic children learn English grammar and vocabulary better with the computer system than with the conventional system. Unfortunately, we can hardly ever *prove* a statistical phenomenon. Even if all 20 computer-taught children did better than all 20 conventionally taught children, that doesn't *prove* that all or even most Hispanic children would do better. Even if we tried the experiment with 200,000 pairs instead of 20, and even if all the computer-taught children did better, we wouldn't have absolute proof. What we *can* hope for is compelling evidence of the superiority of the computer system.

What would constitute compelling evidence of superiority? If the computer-taught child did better in 10 of the pairs and the conventionally taught child did better in the other 10, there would be no evidence at all of the superiority of a computer-based teaching method. If the computer-taught child did better in all 20 of the pairs, that would be the most compelling evidence one could get in a 20-pair experiment. A success rate of 19 out of 20 would be almost as compelling, 18 of 20 a bit less so. How can we decide what would or would not be reasonable evidence in favor of the computer-based system? We need to employ probability theory to answer this question.

Binomial probabilities apply in this situation. Each pair of children constitutes a trial, with success being defined as, say, the computer-taught child's doing better. The probability of success should be the same across all pairs, given the random assignment of children. Independence across pairs certainly should hold; there is no reason why the result in any one pair should change the probability of success in another pair. Finally, we are interested in the number of successes in a fixed number (20) of trials, without regard to order. What we don't know is the probability of success on any particular trial.

The most relevant possible probability of success is .5. Suppose we assume that the computer-based system is no better, and no worse, than the conventional system. Then the probability that the computer-taught child will do better than the conventionally taught child is .5 (assuming that one or the other is always judged as better). Suppose we assume that the probability is .5, but then find that in all 20 pairs, the computer-taught child does better. If $\pi = .5$, the probability of 20 successes in 20 trials is $(.5)^{20}$, or .000000953, just about one in a million! It would still be logically possible that the true probability was .5 and that the 20 successes in 20 trials result was a fluke. But almost no one would really believe that $\pi = .5$ in the face of that evidence. Even 19 successes in 20 pairs (probability .000019) or 18 successes (probability .000181) would be enough to convince all but the most hardened skeptic.

research hypothesis

null hypothesis

test statistic (T.S.)

This example illustrates three of the basic components of hypothesis testing. There is a **research hypothesis,** denoted H_a, that the researcher hopes to support; in the example, the research hypothesis is that the computer-taught child is more likely to do better in grammar and vocabulary than the conventionally taught child, that is, that $\pi > .5$. There is a **null hypothesis,** denoted H_0, which is essentially the opposite of the research hypothesis; for the time being, let's ignore the possibility that the conventional system is better and take as the null hypothesis that the two are equally good, $\pi = .5$. And there is a **test statistic,** abbreviated as **T.S.**, a summary figure calculated from the sample data and used to decide between the two hypotheses. In the example, the test statistic is simply the number of successes in 20 pairs.

Example 9.1

Results from a behavior-modification treatment for obesity have indicated that about 40% of all patients "relapse" within six months, returning to their previous, excessive weight. Suppose that a modified treatment is to be tested on a sample (presumed to be a simple random sample) of 50 new patients at the treatment center. These patients will be followed through the six months after treatment, and the number of relapsing patients counted. Formulate null and research hypotheses and indicate what the test statistic is.

Solution

Presumably, the intent of the modified treatment is to reduce the relapse rate below the current 40%. Thus, we should take as the research hypothesis H_a: $\pi < .40$, where π is the long-run proportion of relapsing patients (not just the proportion within the current sample) using the modified treatment. The opposite of the research hypothesis is H_0: $\pi \geq .40$; temporarily, we'll ignore the possibility that $\pi > .40$ and write H_0: $\pi = .40$. □

The null hypothesis H_0 is usually a negative one, as the name indicates. Typical null hypotheses are that there is no change associated with a treatment, that there is no correlation between two variables, or that there is no difference between two groups in mean response. There is no logical necessity that the null hypothesis have a "something = 0" quality, but in practice that is often the case.

The research hypothesis H_a can be *directional,* or *one-sided,* as is H_a: $\pi > .5$. In such a case, the intent of the experiment is to show that the experimental procedure has an effect, in a way that can be specified *a priori.* A more conservative strategy is to specify a *nondirectional* research hypothesis, also called a *two-sided* hypothesis, such as H_a: $\pi \neq .5$. In this conservative case, the researcher only claims that *something* should happen, without specifying in advance what that something should be. There has been a long dispute over the propriety of one-sided vs. two-sided hypotheses. The next section will have more to say about this issue.

Example 9.2

For the experiment in Example 9.1, why might a researcher prefer to formulate a two-sided research hypothesis?

Solution

There is no guarantee that the consequences of the new treatment will be beneficial. In fact, the new treatment might have unexpected adverse effects on the relapse rate. It would be important to detect such adverse effects, as well as beneficial ones. □

The basic strategy of hypothesis testing is to support the research hypothesis by "disproving" the null hypothesis. In the computerized-learning example, we assumed the null hypothesis value $\pi = .50$. If, for instance, we got 20 successes in 20 trials, for all practical purposes we would "disprove" the null hypothesis. The observed result would be so unlikely, given the null hypothesis, that we could either reject H_0 or believe that a major-league miracle had occurred. The null hypothesis H_0 is "disproven" if the test statistic's observed value is highly unlikely under H_0, and relatively more likely under the research hypothesis H_a. Of course, we must keep "disprove" in quotes, because we cannot absolutely, finally, logically rule out H_0. We can only discredit it beyond a reasonable doubt.

Example 9.3

Refer to Example 9.1 and the hypotheses formulated there. Assume that the experiment is carried out and that 2 of the 50 patients under the new treatment relapse. Does this result support the research hypothesis H_a: $\pi < .40$ by "disproving" H_0: $\pi = .40$?

Solution

Binomial probabilities should apply in this situation. Assume that H_0 is true, so the long-run relapse probability is .40. The binomial probability of obtaining 2 successes in 50 trials (with $\pi = .40$) is shown in Table 1 of the appendix as .0000. The same is true for long-run relapse rates $\pi = .45$, .50, etc. The observed result is extremely unlikely given H_0, whether that hypothesis is stated as $\pi = .40$ or as $\pi \geq .40$. Such a result supports H_a beyond a reasonable doubt. □

There is a serious problem in deciding the borderline between "disproving" and "not disproving" a null hypothesis H_0. Pretty clearly, 20 successes in 20 trials "disproves" H_0: $\pi = .50$, whereas 10 successes in 20 trials doesn't disprove it at all. Where is the boundary between "disproving" and "not disproving" H_0? Is it at 11 successes? 19? The problem of specifying a borderline is the essential problem of specifying a **rejection region (R.R.)**—the potential values of the test statistic that are taken to reject ("disprove") the null hypothesis, and therefore support the research hypothesis H_a. To specify a borderline intelligently, we must consider the errors that a statistical hypothesis test might make.

rejection region (R.R.)

The borderline could logically be set at any number of successes; the reasonable possibilities are 11, 12, ..., 20. If we decide to reject H_0 only if the test statistic (number of successes in 20 trials) equals 20, we will protect ourselves most amply against a statistical fluke. If, in fact, H_0 is true, there is only about one chance in a million that we will reject it. The economists' principle, TANSTAAFL (There Ain't No Such Thing As A Free Lunch), indicates that we must pay something for this ample protection. What we pay is that we make it very difficult to reject H_0 and therefore to support H_a, even if H_a is true. Suppose that in our sample of 20, we demand that all 20 computer-taught children do better before we reject H_0. Also suppose that in the population, 70% of Hispanic children will do better using a computer-based system. Then the true probability of success is $\pi = .7$ and the correct decision would be to reject H_0 and support H_a. The probability of rejecting H_0 in this situation turns out to be an unacceptably low value, .000798. Therefore, by protecting ourselves nicely against one kind of error—rejecting H_0 when it is correct—we have opened up a huge risk of another kind of error—accepting H_0 when it is false. We need a rejection region that balances the risks of these errors better.

In hypothesis testing, there are two potential errors. One error is to reject H_0 (and therefore support H_a) when H_0 is in fact true. This error is unmemorably known as **Type I**; it might also be called a "false positive," because we would be coming to a positive conclusion (that the computerized system works better) when in fact that conclusion was

Type I

Type II wrong. The other error is to accept (or perhaps we should say retain) H_0 when the research hypothesis H_a is true. This error is known as **Type II**; it might be called a "false negative," because we would be coming to a negative conclusion (no difference in effectiveness) that was in fact wrong. Table 9.1 shows the possibilities.

		Conclusion	
		Reject H_0	Retain H_0
Actually	H_0	Type I error	Correct
Correct Hypothesis	H_a	Correct	Type II error

Table 9.1

Example 9.4 What would be the Type I and Type II errors in the context of Example 9.1?

Solution A Type I error occurs when one supports the research hypothesis (and rejects H_0) when in fact H_0 is true. It would be a Type I error to claim that the new treatment reduced the long-run relapse rate if in fact the new-treatment relapse rate was the same as the previous rate. A Type II error occurs when one accepts the null hypothesis when in fact H_a is true. It would be a Type II error to claim that the new treatment was no better than the old if in fact the new treatment's relapse rate was better than the old. □

The specification of the rejection region, and particularly the borderline value, determines the probabilities of the two kinds of error. In the example of the computerized teaching of Hispanic children, if we specify the rejection region as "reject H_0 if and only if all 20 trials are successes," the risk of a false positive (Type I) error is .000000953, about one in a million.

α and β Probabilities

The probability of rejecting H_0, calculated assuming that H_0 is true, is denoted α (alpha):

$$\alpha = P(\text{reject } H_0 | H_0 \text{ true})$$

The probability of accepting H_0, calculated assuming that H_a is true, is denoted as β (beta):

$$\beta = P(\text{accept } H_0 | H_a \text{ true})$$

Thus, for the all-20 rejection region in our example of computerized teaching, α = .000000953, a very low risk. Previously, we calculated that if the true probability of success were .7, the probability of 20 consecutive successes would be .000798. Therefore, assuming that $\pi = .7$,

$$\beta = P(\text{we do } not \text{ get 20 successes})$$
$$= 1 - .000798 = .999202$$

For this rejection region, α, the risk of Type I error, is very low, but β, the risk of Type II error (assuming $\pi = .70$), is intolerably high.

Example 9.5 In the obesity relapse experiment of Example 9.1, suppose that the rejection region is to reject H_0 if Y, the number of relapses, is 0, 1, 2, or 3. Find the α probability. Find the β probability, assuming that the long-run probability of relapse under the new treatment is .25.

Solution To calculate the α probability, consult the binomial tables for $n = 50$ and $\pi = .40$ (the H_0 value). The probabilities for $y = 0, 1, 2,$ and 3 are all shown as .0000. Thus it appears that α is 0 to four decimal places. It might be, say, .0001 because of roundoff in the table, but in any case it's very small. To calculate β assuming $\pi = .25$, we find that the probabilities of $y = 0, 1, 2,$ and 3 in the .25 column are shown as .0000, .0000, .0001, and .0004, totaling .0005. The probability of rejecting H_0 given that $\pi = .25$ is .0005. So

$$\beta = P(\text{do not reject } H_0 | H_a \text{ is true})$$
$$= 1 - .0005 = .9995$$

□

The relative probabilities of the two types of error may be changed by changing the rejection region. In the example, suppose we specify the rejection region as "reject H_0 if and only if there are 15 or more successes." To find α, we add up the binomial probabilities ($n = 20$, $\pi = .50$) of 15, 16, ..., 20 successes and find $\alpha = .0207$. To find β, assuming that $\pi = .70$, add up all the probabilities of 0, 1, ..., 14 successes to find $\beta = .5837$.

Example 9.6 Refer to Example 9.5. What α probability results from choosing R.R.: reject H_0 if $Y \le 14$? What is the β probability, assuming $\pi = .25$?

Solution To find α, add the probabilities for $n = 50$ and $\pi = .40$ in Table 1, for $y = 0$ to 14; the result is $\alpha = .0539$. To find β probabilities, we could add the probabilities for $y = 0$ to 14 in the $\pi = .25$ column and subtract the sum from 1. Alternatively, we could add the probabilities of the nonrejection y values, $y = 15$ to 50. Either way, we get (to within roundoff error) $\beta = .2520$. □

Ideally, the relative probabilities of Type I and Type II error should reflect the relative costs of the two kinds of error. The relatively high-cost error should have the relatively low probability of occurring. Unfortunately, in most social science situations, the costs of the two kinds of error aren't measurable. Therefore, social scientists have established a conventional approach to the specification of error probabilities.

Conventionally, the rejection region is specified so that α, the risk of Type I error, is a small value such as .10, .05, or .01. In some journals, α is taken by dogma as .05. The standard operating procedure is to specify a value for α, then determine the rejection region to achieve that α value as closely as possible. Let's take $\alpha = .10$ in our example. From Table 1, the binomial probability of 14 or more successes is .0577, and the probability of 13 or more successes is .1316. Standard practice is to specify the rejection region so that the actual α is the closest possible value *below* the target. Therefore we would specify the rejection region as R.R.: reject H_0 if Y, the number of successes, is 14 or more.

Example 9.7 Refer to Example 9.6. How should the rejection region be chosen so that α will equal (at most) .10?

Solution In Example 9.6, the R.R. of $Y \leq 14$ had $\alpha = .0539$. An R.R. of $Y \leq 15$ has $\alpha = .0954$. ($Y \leq 16$ has α larger than .20.) The desired rejection region is R.R.: reject H_0 if the number of relapses in the sample is 15 or fewer. □

Everything that we have done so far can, in principle, be done before we ever gather the data. We can specify the null hypothesis (H_0), the research hypothesis (H_a), the test statistic (T.S.), and the rejection region (R.R.) in advance. Then we can gather the data and come to a conclusion. Assume that the 20 pairs of Hispanic children are rated, and that in 16 of the 20 pairs, the computer-taught child is judged to do better in grammar and vocabulary than the conventionally taught child. The value 16 falls in the rejection region. We reject H_0 and support the research hypothesis H_a that the computer-based system is more effective. The entire hypothesis-testing procedure is specified in the five steps outlined in the box below.

Five Steps of a Binomial Hypothesis Test

1. H_0: $\pi = .50$
2. H_a: $\pi > .50$
3. T.S.: Y = the number of successes in 20 trials
4. R.R.: for $\alpha = .10$, reject H_0 if $Y \geq 14$
5. Conclusion: $Y = 16$; reject H_0 and support H_a

Example 9.8 Refer to Example 9.7. Suppose that there are 17 relapses in the sample of 50. State the five steps of the hypothesis test.

Solution

1. H_0: $\pi = .40$
2. H_a: $\pi < .40$
3. T.S.: Y = the number of relapses in a sample of 50
4. R.R.: reject H_0 (at $\alpha = .10$) if $Y \leq 15$
5. Conclusion: $Y = 17$; do not reject H_0

The conclusion of any hypothesis test must be interpreted carefully. In the computerized-teaching example, when Y came out to be 16, we rejected H_0. That does *not* mean that we have absolutely proved that H_a is true; it means that the observed result was so highly unlikely to have occurred by chance, assuming that H_0 is true, that we have compelling evidence against H_0 and for H_a. Conversely, suppose that Y had come out 13, a value that's not in the rejection region. Formally, we would "accept" H_0. However, $y = 13$ would mean that in 65% (13 out of 20) of the trials, the computer-taught child did better than the conventionally taught one, which suggests that the computer-based system is better. The reason we "accepted" H_0 was that 13 of 20 could within reason have occurred by chance, assuming $\pi = .50$. A result such as $y = 13$ is ambiguous. It could reasonably be a chance variation from the expected 10, but it could also be an indicator of the superiority of the computer-based system. The phrase "accept H_0" does not convey this ambiguity. Hereafter, we'll say "retain H_0," to indicate that both the null and the research hypothesis remain as reasonable possibilities.

Example 9.9 State the conclusion of Example 9.8 carefully.

Solution It is possible that the new treatment reduces the relapse rate. Our best estimate is that the new rate is 17/50 = .34. But the degree of reduction is within reasonable random variation, so we must also retain the possibility that the new rate is still .40 (or even higher). □

Exercises for Section 9.1

9.1. Pharis and Manosevitz (1981) reported on the beliefs of first-pregnancy women about the sex of the fetus. The sample consisted of 24 women who had an opinion about the likely sex of the fetus. Define π to be the long-run, population proportion of women who believe that the fetus will be a girl.

a. Why would $\pi = .5$ be a natural null hypothesis?

b. What argument would lead to a one-sided research hypothesis? (It might be interesting to compare answers to this question, to see if everyone chose the same side.)

c. What argument would lead to a two-sided H_a?

9.2. In Exercise 9.1, what would be the meaning of a Type I error? Type II?

9.3. Binomial probabilities for $n = 24$ and $\pi = .5$ can be calculated, for example by the Minitab package.

MTB > BINOMIAL PROBS FOR N = 24 AND PI = .50

X	PROB(X)	PROB(≤X)
0	.0000	.0000
1	.0000	.0000
2	.0000	.0000
3	.0000	.0000
4	.0002	.0002
5	.0011	.0013
6	.0040	.0053
7	.0120	.0173
8	.0292	.0465

etc.

a. Suppose that a rejection region for testing the two-sided H_a of Exercise 9.1 is R.R.: reject H_0 if Y, the number of women in the sample who believe that the child will be a girl, is 6 or less, or 18 or more. What is the associated α value?

b. If the maximum allowable α value is .05, what R.R. should be used?

9.4. Refer to Exercises 9.1 and 9.3. The authors report that 5 of the 24 women believed that the child would be a girl. What conclusion can be reached, using $\alpha = .05$?

9.5. Dovidio (1982) had student volunteers ask other, randomly chosen students to lend them money to copy a paper that was due the next day. It was found that 50% of students would lend 10 cents. Then another 50 students were asked to lend 30 cents.

a. Formulate null and research hypotheses. Should H_a be one-sided or two-sided?

b. What would a Type I error be in this situation?

9.6. **a.** In the situation of Exercise 9.5, define Y to be the number of students, in the sample of 50, who do lend 30 cents. If H_0 is rejected whenever $Y \leq 20$, what is the α probability?

b. Define an appropriate R.R. to yield an α probability of at most .05.

9.7. Refer to part b of Exercise 9.6. If the actual Y value is 16, what conclusion should be drawn?

9.2 Properties of the Binomial Test

Section 9.1 introduced a binomial test of a statistical hypothesis. In this section, we'll discuss the general form of the test and its statistical properties. The focus will be on α and β, the probabilities of Type I and Type II (false positive and false negative) errors.

A binomial test is a test about a population proportion, denoted π. Characteristically, we want to support a research hypothesis about this parameter by "disproving" its opposite hypothesis, the null hypothesis. The first step in hypothesis testing is formulation of hypotheses.

The research hypothesis in the computer-learning example of Section 9.1 was directional, or one-sided. That is, the research hypothesis specified that the population proportion would be *greater* than .5. Logically, we might alternatively have specified a nondirectional, two-sided research hypothesis. We could have hypothesized that the true proportion would be either greater than or less than (but not equal to) .5; H_a: $\pi \neq .5$. In the latter case, we would be interested in more than random discrepancies in learning grammar and vocabulary *in either direction* from the value predicted by the null hypothesis. The argument for using a one-sided hypothesis is that the burden of proof falls on the proposed system. If fewer children learn under the proposed system than under the conventional system, then the proposed system simply fails the test, and no further analysis is needed. The argument for using a two-sided test is that all discrepancies in data are interesting. If a computer-based system did less well than a conventional system (to a more than random extent), there would be evidence against the prevailing wisdom, and such evidence is worth following up. There has been a recurring debate in the social science literature about the one-sided vs. two-sided hypothesis question. There seems to be general agreement at least about one point. *If an investigator wants to use a one-sided research hypothesis, there must be a compelling reason to choose the direction of the hypothesis before gathering the data.* Furthermore, any scientific report of the study should indicate that a one-sided research hypothesis has been employed. Unless there is a good reason to restrict scientific inference to a one-sided research hypothesis, a two-sided H_a should be used.

Example 9.10

A group of marriage counselors found that historically 30% of the couples visiting them separated within one year of entering counseling. A sample of 16 couples is to be taken randomly from all those entering counseling; couples in the sample will be treated by an experimental approach, to see if there is a marked change in the separation rate. Formulate null and research hypotheses. What are the arguments favoring one-sided and two-sided H_a?

Solution

The natural null hypothesis is that the experimental approach has no effect on the long-run separation rate; H_0: $\pi = .30$. One argument is that the intent of the new approach is to decrease the separation rate, so H_a should be one-sided; H_a: $\pi < .30$. The counterargument (which in this case seems rather compelling) is that a lower separation rate is not necessarily better if it merely results in more "prolonging the agony." Further, the intent and the actual result of a treatment may not be the same. This argument leads to a two-sided research hypothesis; H_a: $\pi \neq .30$. □

When a one-sided research hypothesis H_a is specified, a technical problem arises in specifying the opposite, null hypothesis H_0. If H_a is $\pi > .5$, should H_0 be $\pi = .5$ or $\pi \leq .5$? The only way in which the choice matters is in the calculation of α, the probability of Type I error (rejecting the null hypothesis when it is in fact true). As a very general principle, *only the boundary value of the null hypothesis need be considered in calculating the maximum probability α of Type I error.* In the computer-learning example of the previous section, we calculated that a rejection region (R.R.) consisting of $Y \geq 14$ had an α probability of .0577, assuming that the true value of π was the boundary value, .50. For any other value of π that is "on the other side of the boundary" from H_a, the α probability will be even smaller. For example, for $\pi = .45$, α is .0214, and for $\pi = .40$, $\alpha = .0065$. (These probabilities are obtained by adding up the probabilities for $y = 14, 15, \ldots, 20$ in the binomial probability table, Table 1 of the appendix.) Hereafter, whenever a one-sided research hypothesis is used, we will only have to consider the boundary null hypothesis value; the specified maximum probability of Type I error will always be an overstatement for any nonboundary null hypothesis value.

Example 9.11

Refer to Example 9.10. Taking the research hypothesis as one-sided and the rejection region as R.R.: reject H_0 if Y, the number of separating couples in the sample, is 2 or fewer, calculate α for $\pi = .30$, $\pi = .35$, and $\pi = .40$.

Solution

Consult the appropriate columns of Table 1 of binomial probabilities, and add the probabilities for $y = 0$, 1, and 2. The results:

π	.30	.35	.40
α	.0993	.0450	.0183

Note that α is largest at the H_0 boundary value, $\pi = .30$. ☐

Given a specification of H_0, H_a, and α, one may specify a rejection region as in Section 9.1. Conventionally, the R.R. is chosen to yield a maximum α risk (probability of a Type I error given that H_0 is true) equal to a given number. This convention unfortunately ignores the probability of committing a Type II (false negative) error, given that H_a is true. Another way of saying the same thing is to say that the convention ignores the power of a statistical test.

Power of a Test

The power of a statistical hypothesis test is the probability that the null hypothesis is rejected. Given that H_a is true,

$$\text{power} = 1 - \beta$$

Because the smallest possible value of β is desired, the largest possible power is desired.

Computing the power (or the β probability) of a hypothesis test requires that one specify a particular H_a value of the parameter. In fact, the power of a test is *not* a number; rather it is a function, depending on the value specified for the population parameter. In Section 9.1 we calculated a β probability (by adding up binomial probabilities from Table 1 of the appendix) assuming $\pi = .70$, and obtained $\beta = .3920$; therefore, assuming $\pi = .70$, power $= .6080$. But if instead we assume $\pi = .60$, power $= .2499$; for $\pi = .80$, power $= .9133$. Note that, for R.R.: $Y \geq 14$,

$$\text{power} = P(\text{reject } H_0) = P(Y \geq 14)$$

may be calculated by adding up the binomial probabilities for $y = 14, 15, \ldots, 20$. Power can be computed for all values of π. A curve that relates power to the assumed value of π is shown in Figure 9.1. This curve is called (reasonably enough) the **power curve** or **power function** of the hypothesis test.

power curve
power function

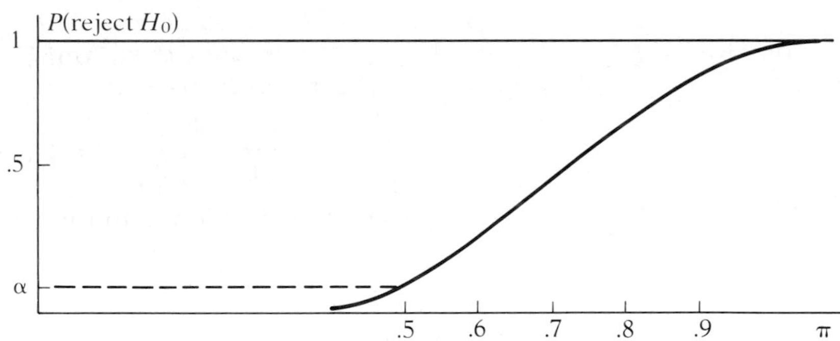

Figure 9.1 Power curve of a one-sided binomial test

Example 9.12

Refer to Example 9.10. Assume that a two-sided research hypothesis is used.

(a) Find a rejection region having maximum α probability .05.
(b) Find the power of the test for π = .20, .25, .35, and .40.

Solution

(a) The rejection region should include very small and very large y values, because H_a: $\pi \neq .30$. When H_0 is true, $\pi = .30$; adding binomial probabilities, we find that R.R.: reject H_0 when $Y \leq 1$ or when $Y \geq 9$ has $\alpha = .0261 + .0256 = .0517$, with the probability nearly equal in each tail of the R.R. This is so slightly above the (arbitrary) maximum α probability .05 that we will use it.

(b) The power is $P(\text{reject } H_0|\pi)$. Having just specified the R.R., we must add probabilities for the indicated π values, and y = 0, 1, 9, 10, 11, 12. The results:

π	.20	.25	.35	.40
Power	.1421	.0709	.0768	.1456

Note that all these power probabilities are quite small, presumably because the sample size is only 16. □

For any particular research hypothesis about a proportion, the power of a binomial test, and therefore β, the probability of Type II error, depends on three things: the probability of Type I error that is allowed, the sample size, and the assumed value of π under H_a. Once again, let's consider our computer-learning illustration. First suppose that we still have 20 pairs of children, but that we specify a maximum allowable α of .01, instead of .10. To achieve this we must change the rejection region to R.R.: $Y \geq 16$, because (assuming H_0: $\pi = .50$ is true) $P(Y \geq 15) = .0207$ and $P(Y \geq 16) = .0059$. The power of the test therefore is $P(Y \geq 16)$, evaluated for a specified value of π. For $\pi = .70$, the power of the $\alpha = .01$ test is .2374, compared to the power .6080 found for α = .10. *Decreasing the maximum allowable Type I error probability, α, will always increase β, the risk of Type II error, and decrease the power of the test,* if all else is equal. Again, TANSTAAFL (There Ain't No Such Thing As A Free Lunch); to gain something (reduced risk of a false positive), we must give up something (power to reject H_0 if it is false).

Example 9.13

Refer to Example 9.12. If the maximum allowable α is increased to .10, find a two-tailed R.R. and the revised power at π = .40.

Solution

From Table 1, $P(Y \leq 1 \text{ or } Y \geq 8) = .1004$, which surely is close enough to the maximum allowable .1000. For R.R.: reject H_0 if $Y \leq 1$ or $Y \geq 8$, the power can be found by adding the corresponding binomial probabilities in the $\pi = .40$ column. The power is .2873, almost double the power found in Example 9.12. □

For any specified value of α and of π, one can increase the power of the test by increasing the amount of information—that is, by increasing n, the sample size. In the computer-learning illustration, suppose that there had been 50 pairs of children instead of 20. For a maximum allowable α of .10, the desired rejection region is R.R.: $Y \geq 31$ (with an actual α of .0593); assuming $\pi = .70$, power $= P(Y \geq 31) = .8595$, as compared to .6080 when $n = 20$. *Increasing n will increase the power of the test,* all else being equal.

Example 9.14 Still assuming $\pi = .40$, compare the power found in Example 9.13 to the power for a test with $n = 50$, $\alpha = .10$.

Solution To find the power, we must first find an appropriate rejection region. Once again we turn to Table 1, and find that R.R.: reject H_0 if $Y \leq 9$ or $Y \geq 21$ has $\alpha = .0880$ and that no other reasonable R.R. has α closer to .10. Using this R.R., we add binomial probabilities for the $\pi = .40$ column and obtain power $= .4397$, which is substantially higher than the power found in Example 9.13. Of course, n was increased from 16 to 50 to obtain this increase in power. □

Finally, the power curve in Figure 9.1 indicates that *the power of the test increases as the difference between the H_0 and assumed H_a values increases,* all else being equal. Thus, in our illustration with $n = 20$ and maximum allowable $\alpha = .10$, the power was .2499 at $\pi = .60$, .6080 at $\pi = .70$, and .9133 at $\pi = .80$. It is easier to detect a large discrepancy between actual and null-hypothesis values of a parameter than a small discrepancy, so the power will be larger for parameter values far from the null hypothesis.

These considerations—choice of α, sample size, and difference between assumed and null-hypothesis values of the parameter—apply to any specified research hypothesis. The choice of one-sided vs. two-sided research hypothesis also affects power, indirectly. *For all practical purposes, the power of a test of a two-sided research hypothesis is equal to the power of a test of a one-sided test using a maximum allowable Type I error probability of $\alpha/2$.* In the computer-learning example, if we had taken H_a: $\pi \neq .50$ and used a (large) α value of .20, the power of the test would have been essentially the same as the power computed for the one-sided, $\alpha = .10$ test.

Example 9.15 Refer to Example 9.14. Find the power of a test of H_a: $\pi > .30$, assuming $\alpha = .05$ and $\pi = .40$.

Solution Here the rejection region will consist of only the right tail (large values) of Y. The desired rejection region is R.R.: $Y \geq 21$, with an actual α of .0477. The power is .4390, almost exactly the same as the .4397 reported in Example 9.14. □

The ideas discussed in this section (one-sided vs. two-sided H_a, power, and the relation among power, α, n, and the assumed H_a value of the parameter) apply to all the hypothesis-testing methods discussed in this book. They will be particularly important in discussing how to plan statistical studies and experiments.

Exercises for Section 9.2

9.8. Refer to Exercise 9.6, part b. What is the power of the test when π equals .45? .40? .35? Sketch the power function.

9.9. **a.** Recalculate the power function of Exercise 9.8, assuming that $\alpha = .01$. What happens to the power of the test for any particular value of π?

b. Recalculate the power function, assuming $\alpha = .05$ but $n = 100$. How does this power function compare to that of Exercise 9.8?

9.10. Refer to Exercise 9.3. What information do you need to calculate the power of this test for a true population proportion equal to .70? Explain how to use this information to calculate the power.

9.11. Refer to Exercise 9.3 again. Suppose that the number of women sampled is increased to 50.

a. Find the R.R. corresponding to $\alpha = .05$.

b. Find the power of the test for $\pi = .70$.

c. Which power should be higher, the one found in part b or the one of Exercise 9.10? Why?

9.3
Hypothesis Testing about a Mean

The most common hypothesis tests in psychology, education, and related fields involve mean values. The basic principles of hypothesis testing stated in the previous sections apply to means, as well as to proportions and to other statistical parameters. This section will introduce some of the basic ideas for testing means; Chapter 10 will carry the discussion further.

Suppose that a national survey of college students yielded a mean "authoritarianism" score of 52.8 and a standard deviation of 10.5. The sample size was large enough that for all practical purposes these values could be regarded as population values. Researchers at a state university decided to sample the students at that university to see if these students differed in authoritarianism, on the average. They planned to choose student names at random from the student directory, and to improve the response rate by offering a modest dollar reward for participation. In this way, they felt confident that they could obtain a good approximation to a random sample. The study was deliberately designed to yield a small but near-random sample rather than a large, probably biased, convenience sample. The problem was what to do with the random sample data, once the data were obtained.

To formulate the problem in hypothesis-testing terms, the researchers needed to specify null and research hypotheses. In this case, as in many, it's easy to specify a null hypothesis as a "no-difference" hypothesis—namely, that the mean authoritarianism score of the entire population (of students enrolled at the university) equaled the national mean, 52.8. To formulate the research hypothesis, they had to consider one-sided vs. two-sided hypotheses. They had no advance reason to suspect or to be interested in a particular direction of difference from the mean of 52.8, so they took the research hypothesis to be two-sided; H_a: $\mu \neq 52.8$.

Example 9.16 Feldman and Newcomb (1969) measured college sophomores on a standard conservatism scale. They obtained data for essentially the entire population of sophomores at a university. The mean was 61.9 and the standard deviation was 7.77. Suppose that a simple random sample of 40 sophomores at the same university is to be taken, for the purpose of testing whether the mean conservatism score has changed. Formulate null and research hypotheses.

Solution The natural null hypothesis is that the mean has not changed; H_0: $\mu = 61.9$. There is no obvious reason to expect that any change would be in a specified direction, and change in either direction would be of interest. It seems natural to take a two-sided research hypothesis; H_a: $\mu \neq 61.9$. □

The next issue in testing the hypothesis is the selection of a test statistic. Recall that the test statistic is a summary figure based on the sample data used to test the hypothesis. The most obvious choice to test a population mean is the sample mean. However, if the population can be assumed to be symmetric, a sample median or trimmed mean could be used instead. The subject of choice of an appropriate statistic arose also in Chapter 8, where the issue was efficiency. The more efficient a statistic, the smaller its standard error, hence the narrower its confidence interval. In hypothesis testing, the question of choice of a statistic can be phrased in terms of the power of the test. Recall that the power of a test is the probability that the null hypothesis is rejected when it should be (that is, when the research hypothesis is true). By definition, the ideal test statistic gives the largest possible power for a test with a specified value of α. In the appendix to this chapter, we discuss the Neyman-Pearson Lemma, which in certain common situations yields the best test statistic. Here, we need only note that *if* the population has a normal distribution, the sample mean is the best, most powerful, test statistic. If, however, the population is markedly nonnormal, some other statistic, such as the median, may be more powerful.

Sometimes, enough is known about the population that a choice of test statistic can be made before the data are taken. In the authoritarianism study, the scale ranged only from 0 to 100, with the mean fairly near the middle. The researchers were reasonably sure that the population wasn't horrendously skewed. Therefore, a sample size of 50 should be adequate to appeal to the Central Limit Theorem and to permit assertion that the theoretical distribution of the sample mean would be normal, to an excellent approximation. Further, the scale should not yield extreme outliers, so a sample mean should be a fairly efficient, powerful statistic.

Other times, however, not much is known about the shape of a population. In such cases, looking at a plot of the data can help one decide on a test statistic. Look for gross skewness in the data, which indicates a skewed population. Unless the sample size is rather large (say 100 or more), the Central Limit Theorem effect may not adequately "de-skew" the theoretical distribution of the sample mean. The problem is particularly acute for one-tailed tests. Look also for outliers, indicating a heavy-tailed population. For such populations, a median or trimmed mean is a more powerful statistic. There is one danger in using the data to guide the choice of test statistic. Don't compute several tests and report the one that comes closest to supporting your preconceived idea!

Example 9.17 The researchers in Example 9.16 obtained a sample of 40 sophomores. The conservatism scores are shown in the stem and leaf display in Figure 9.2.

```
4 | 7 9
5 | 0 0 0 1 3 4 4
5 | 5 6 6 8 9 9
6 | 0 0 1 2 2 3 3 4
6 | 5 5 7 8 8 9 9
7 | 0 0 3
7 | 5 5 5 7 9
8 | 0 4
```

Figure 9.2 Stem and leaf display for Example 9.17

Is it reasonable to base a test about the population mean μ on the sample mean?

Solution There is no flagrant skewness, so a sample size of 40 should be adequate to allow use of a theoretical normal distribution. There are no outliers or heavy tails. Indeed, if anything, the tails are rather light. The mean is an eminently reasonable choice as a test statistic. □

Once a test statistic has been chosen, the rejection region must be specified. Typically, this issue can be broken down into two questions. Should the rejection region consist of one tail of the theoretical distribution of the statistic, or both? What value of α should be selected? The research hypothesis dictates the choice of one tail or two tails. In the authoritarianism study, the research hypothesis was two-sided, so values of the sample mean far above *or* far below 52.8 (the H_0 value) support H_a. The choice of α is inevitably somewhat arbitrary. Ideally, the relative costs of the two kinds of errors (and the relative prior plausibility of the two hypotheses) should be carefully weighed and the α and β levels chosen accordingly. In practice, though, the costs are usually impossible to quantify, so some conventional standard must be employed. Typical α values are .10, .05, and .01. Choosing a small α value like .01 puts more of a burden of proof on H_a; this should be done if H_a is a highly novel idea or if mistaken support for H_a could be intellectually or economically very expensive. In the authoritarianism study, no such concern seemed necessary, so α was set at .05.

critical value The desired value of α is obtained by setting the **critical value,** or cutoff point, for the test statistic, using what is known about its theoretical probability distribution. In the case of a sample mean, assuming that the Central Limit Theorem applies adequately, the theoretical distribution is normal, with expected value equal to the population mean and standard error equal to $\sigma_{pop}/\sqrt{n}$. Using the strategy of trying to "contradict" H_0, take μ_{pop} to be μ_0, the null hypothesis mean. In the authoritarianism study, the researchers would take μ_{pop} to be 52.8. To calculate the exact standard error, one must assume a value of σ_{pop}. In the authoritarianism study, the researchers assumed that the national standard deviation, 10.5, applied to their single-university population as well. For $n = 50$, the standard error then would be $10.5/\sqrt{50} = 1.485$. (In practice, the t methods discussed in Chapter 10 can be used to avoid making unsupported assumptions about σ_{pop}.) All this information is conveniently combined in a ***z* statistic.** ***z* statistic**

***z* Statistic**

$$\text{T.S.: } z = \frac{\bar{Y} - \mu_0}{\sigma_{pop}/\sqrt{n}}$$

where μ_0 is the null hypothesis value of the population mean and $\sigma_{pop}/\sqrt{n}$ is the standard error of $\bar{Y}$.

The z statistic is convenient because its values can be referred directly to Table 3 of normal distribution probabilities. In the authoritarianism study, the test was to be two-tailed and the desired α was to be .05. A picture of the situation is shown in Figure 9.3. The desired α of .05 is split into two equal tails, each with area .025. The normal table yields a z-score of 1.960, cutting off a right tail of .025, so -1.96 cuts off a left tail of .025. For this study, 1.96 and -1.96 are the critical values for the test statistic.

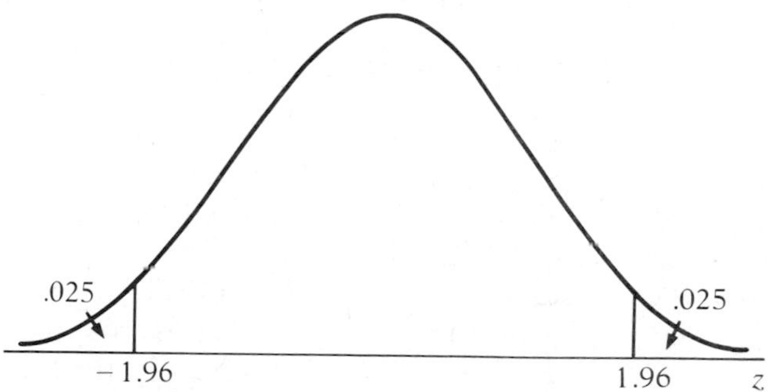

Figure 9.3 Critical values for the authoritarianism study

Example 9.18 Define the z statistic in Example 9.17. What are the critical values of this statistic corresponding to $\alpha = .10$?

Solution To use a z statistic, we must assume a value for σ_{pop}. If we are willing to assume that the variability in conservatism scores hasn't changed, we can use the previously reported $\sigma_{pop} = 7.77$. Therefore,

$$z = \frac{\bar{Y} - 61.9}{7.77/\sqrt{40}}$$

For a two-tailed probability $\alpha = .10$, each tail should have area .05. Table 3 shows that $z = 1.64$ yields a tail area of .0505, and $z = 1.65$ yields a tail area of .0495. We might as well split the difference and call the critical values ± 1.645. In fact, more extensive normal tables yield the value 1.645. □

The first four steps of a formal hypothesis test—specification of H_0 and H_a, choice of test statistic, and determination of rejection region—can be done before the data are collected. (However, the data should be plotted to avoid choosing a very unwise T.S.) Once the data have been collected, the actual value of the test statistic may be computed and compared to the critical value, and a conclusion reached. Assume that in the authoritarianism study the sample of 50 students yielded a sample mean of 50.4. The actual z statistic then equals $(50.4 - 52.8)/1.485 = -1.61$. This value does *not* lie in the R.R. shown in Figure 9.3. The conclusion of the study must be to retain H_0. It is reasonably possible that the university mean equals the national mean, 52.8. It is, of course, also possible that it differs somewhat from 52.8. Retaining H_0 characteristically is an inconclusive, "not proved" result. The entire five-step hypothesis-testing procedure is summarized in the following box.

***z* Test for a Population Mean**

General Expression	Authoritarianism Study
1. H_0: $\mu = \mu_0$	H_0: $\mu = 52.8$
2. H_a: (i) $\mu > \mu_0$ (ii) $\mu < \mu_0$ (iii) $\mu \neq \mu_0$	H_a: $\mu \neq 52.8$
3. T.S.: $z = \dfrac{\bar{Y} - \mu_0}{\sigma_{pop}/\sqrt{n}}$	T.S.: $z = \dfrac{\bar{Y} - 52.8}{10.5/\sqrt{50}}$

4. R.R.:	R.R.: $z > 1.96$ or
(i) $z > z_\alpha$	$z < -1.96$
(ii) $z < -z_\alpha$	
(iii) $z > z_{\alpha/2}$ or	
$z < -z_{\alpha/2}$	
5. Conclusion: use R.R.	Conclusion: $z = -1.61$; retain H_0

NOTE: It is assumed that n is large enough or the population near enough to normally distributed that normal probabilities are a good approximation to the theoretical probability distribution of $\bar{Y}$. σ_{pop} is assumed to be known. z_a denotes the z-table value that cuts off a right-tail area equal to a.

Example 9.19

The sample data of Example 9.17 have a mean of 63.125. Carry out the five steps of a hypothesis test.

Solution

1. H_0: $\mu = 61.9$
2. H_a: $\mu \neq 61.9$
3. T.S.: $z = \dfrac{63.125 - 61.9}{7.77/\sqrt{40}}$
4. R.R.: reject H_0 if $|z| > 1.645$
5. Conclusion: $z = 1.00$; retain H_0

□

Power considerations apply to hypothesis tests about a mean, just as they do to tests of a proportion described in the preceding sections. Power for mean tests will be discussed in Section 9.6, in the context of planning a study.

Discussions of hypothesis testing necessarily involve some subtle technical definitions and distinctions. These should not be allowed to obscure the rather simple basic idea of hypothesis testing. We try to support a research hypothesis by discrediting its opposite, the null hypothesis; H_0 is discredited if the sample statistic comes out far enough away from the H_0 values that the difference can't reasonably be attributed to random variation.

Exercises for Section 9.3

9.12. Otis (1979) interviewed moviegoers waiting to see the "space aliens" film *Close Encounters of the Third Kind.* Each moviegoer was asked to state his or her degree of agreement with the statement "Life on earth is being observed by intelligent aliens," on a scale from 1 = strongly disagree to 7 = strongly agree, with 4 = not sure at all. Assume that the population standard deviation is 1.00.

a. Why would $\mu = 4.0$ be a natural null hypothesis?

b. The purpose of the study was to test Otis's assertion that individuals selected movies that they were predisposed to believe. Formulate Otis's assertion as a research hypothesis.

9.13. Suppose that most (80%+) of the data from Otis's study had been concentrated in the 4.5 to 5.5 range, but that the rest of the scores had straggled out to 1 and to 7. Why might one want to base the test on the sample median, rather than the sample mean?

9.14. In the study described in Exercise 9.12, a sample of 127 moviegoers yielded a mean of 5.11. Perform all parts of a statistical hypothesis test.

9.15. Tellevik (1981) studied the time required for deaf and normal-hearing individuals to perform certain tasks. For one task, the mean for normal-hearing subjects was 427 seconds, and the standard deviation was 223.2 seconds. Treat these quantities as population parameters. (In fact they came from a sample, so the data were treated using the two-sample methods discussed in Chapter 11.) A random sample of 22 deaf subjects was obtained. The parameter of interest was the mean time required to complete the task in the population of (potential) deaf subjects.

a. What is a natural null hypothesis?

b. Should the research hypothesis be one-sided or two-sided? Defend your choice.

9.16. In Exercise 9.15, define a rejection region corresponding to $\alpha = .10$.

9.17. Refer to Exercise 9.15 again. The author reported that the mean for the 22 deaf subjects was 472 seconds. (The standard deviation was very close to the assumed population standard deviation.) Carry out the five hypothesis-testing steps, assuming an α of .10.

9.18. The mean and standard deviation reported in Exercise 9.15 suggest that the underlying population of times will be substantially skewed. If so, would the conclusion reached in Exercise 9.17 be cast into doubt?

9.4

p-Values and Statistical Significance

A major problem with formal hypothesis testing is the specification of α, the maximum allowable risk of Type I error. As has been indicated, the choice of an α value is an uncomfortably arbitrary, conventional process. This section will introduce the concept of p-value, which can sometimes be used to avoid the arbitrariness of a preselected α value. This concept leads to the related concept of statistical significance, the most widely used and misused concept in statistical research in the behavioral sciences.

The concept of a p-value can be illustrated in the context of the authoritarianism study of Section 9.3. In that section a z statistic of -1.61 was obtained, forcing retention of the null hypothesis (with $\alpha = .05$). Other sample mean and z statistic values might be obtained in such a study. In particular, consider two possibilities: $\bar{y} = 49.8$ or $\bar{y} = 48.6$. If $\bar{y}$ comes out 49.8, $z = (49.8 - 52.8)/1.485 = -2.02$, whereas if $\bar{y} = 48.6$, $z = -2.83$. In either case, the value of z falls beyond the two-tail $\alpha = .05$ critical value, -1.96. According to formal hypothesis-testing methods, we would come to exactly the same conclusion in the two cases: reject H_0 at $\alpha = .05$. The formal procedure fails to recognize one big difference between the two possible cases. If $\bar{y} = 49.8$ and $z = -2.02$, z is just barely in the rejection region and therefore the rejection of H_0 is not really conclusive; but if $\bar{y} = 48.6$, $z = -2.83$, which is well out in the tail of the rejection region, and the rejection of H_0 is more emphatic. To say the same thing another way, for a z of -2.02, the precise choice of α is important; if α had been set at a slightly smaller value, H_0 would have been retained. On the other hand, for a z of -2.83, H_0 would be rejected for any α value within the range of conventional ones.

p-value

The ***p*-value,** also called the prob-value or the attained significance level, for a test statistic is a *conclusiveness index*. It allows researchers to assess how conclusively they may reject H_0 and therefore support H_a. Use of p-values also reduces, but doesn't entirely eliminate, the reliance on arbitrary choices of the α value.

p-Value of a z Statistic

The p-value corresponding to an observed z statistic value, z_{obs}, is the probability that Z will be as extreme or more extreme than z_{obs}. If the rejection region is R.R.: $z > z_\alpha$, then

$$p\text{-value} = P(Z > z_{obs})$$

If the rejection region is R.R.: $z < -z_{\alpha}$,

$$p\text{-value} = P(Z < z_{\text{obs}})$$

If the R.R. is two-tailed,

$$p\text{-value} = 2P(Z > |z_{\text{obs}}|)$$

In our illustration, the rejection region is two-tailed. To be "more extreme than" an observed z of -2.02 is to be less than (more negative than) -2.02 *or* greater than 2.02. As shown in Figure 9.4, the two-tailed area more extreme than -2.02 is $.0217 + .0217 = 2(.0217) = .0434$. (Note that we needn't worry about exactly equaling -2.02 or 2.02; for continuous random variables such as z, the probability of exactly equaling a specified number can be taken to be 0.) For an observed z of -2.83, the two-tailed p-value is $2(.0023) = .0046$; to verify this statement, draw a picture like Figure 9.4.

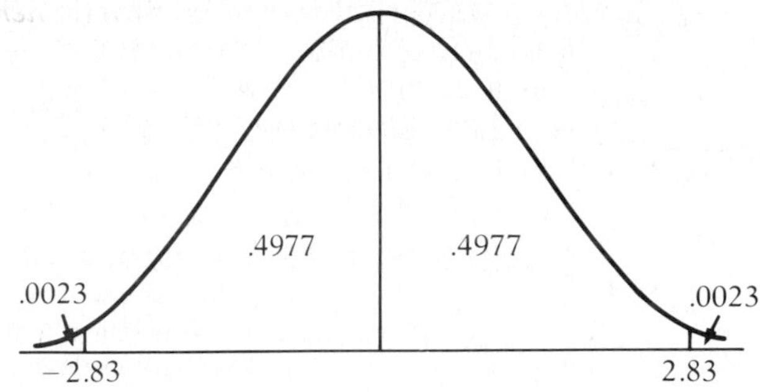

Figure 9.4 p-value for a z statistic

Example 9.20 Find the p-value for Example 9.19.

Solution The observed z statistic in that example was $z_{\text{obs}} = 1.00$. From Table 3, $P(Z > 1.00) = .1587$. For a two-tail test as was performed in Example 9.19, the p-value is $2(.1587) = .3174$, or about .32. □

conclusiveness index

The p-value is a **conclusiveness index** for the rejection of H_0 and support of H_a. Small p-values indicate conclusive rejection of H_0; large values indicate inconclusiveness. The p-value for a given test statistic value is the probability that, by sheer random variation, we would obtain as extreme or more extreme a test statistic value. If the p-value is small, random variation (assuming the null hypothesis) is not a plausible explanation of the value, and H_0 should be rejected. If the p-value is large, random variation is plausible and H_0 must be retained. In the authoritarianism illustration, a p-value of .0434 indicates moderately conclusive support of H_a, and a p-value of .0046 indicates very conclusive support of H_a.

Example 9.21 What does the p-value found in Example 9.20 indicate?

Solution The difference between the sample mean 63.125 and the hypothetical mean 61.9 is well within reasonable random variation. There is nothing approaching conclusive evidence of a change in means. ☐

The p-value should not be confused with α, the maximum allowable risk of Type I error. Both can be interpreted as probabilities, and both are computed under the assumption that H_0 is true. But *α is a long-run property of the test procedure, whereas the p-value is a property of the data obtained in one particular test.* There is, however, a relation between the two. If in fact α has been prespecified, a very simple rejection region can be defined. The idea is so general that we can call the region the **Universal Rejection Region.**

Universal Rejection Region

Universal Rejection Region
If α, the maximum allowable risk of Type I error, has been specified, reject H_0 if (and only if) p-value $< \alpha$.

Thus for $z = -2.02$ and p-value $= .0434$, H_0 can be rejected at $\alpha = .05$, but not at $\alpha = .04$, for instance. For $z = -2.83$, p-value $= .0046$, H_0 could be rejected not only at $\alpha = .05$ but also at $\alpha = .01$ or even at $\alpha = .005$.

Example 9.22 Refer to Example 9.20. Can the null hypothesis be rejected at $\alpha = .10$?

Solution Not at all. The p-value of .32 is substantially larger than $\alpha = .10$. ☐

The idea of p-value applies to any test statistic, not just z. It applies to the binomial test discussed in Sections 9.1 and 9.2. To find the

p-value in a binomial test, assume that the H_0 probability of success is correct and add up the binomial probabilities for the observed number of successes and anything more extreme. For two-tailed tests, include occurrences in the opposite direction.

Example 9.23 In a binomial test of H_0: $\pi = .50$ against a two-sided H_a, there were 19 successes in 24 trials. What p-value should be reported?

Solution Binomial probabilities for $n = 24$ and π are shown below, to four decimal places.

Value	3	4	5	6	...	19	20	21	22
Probability	.0001	.0006	.0025	.0206		.0025	.0006	.0001	.0000

The probability of 19 or more successes is thus $.0025 + .0006 + .0001 = .0032$. The test should be two-tailed, so we should also include probabilities in the opposite direction in the p-value. We note that the actual 19 successes is 7 more than the number expected under the null hypothesis. The corresponding result in the opposite direction would be 7 lower than expected, namely $12 - 7 = 5$. The probability of 5 or fewer successes is also $.0025 + .0006 + .0001 = .0032$. The reported two-tailed p-value should be $.0032 + .0032 = .0064$. ☐

Most statistical computer programs report at least approximate p-values for all statistical tests performed. The user of such programs can then simply compare the p-value to any desired α value. Conventionally, the programs report two-tailed p-values, unless instructed to do otherwise. To obtain a one-tailed p-value, divide the reported value by 2. This assumes that the sample mean will at least be in the direction specified by the research hypothesis. If the sample mean goes in the opposite direction, there is obviously no basis for supporting the research hypothesis.

Example 9.24 In a test of the null hypothesis H_0: $\mu = 50$ against the research hypothesis H_a: $\mu > 50$, the data yielded a sample mean of 54.2 and a z statistic of 2.60. Computer output indicated that the two-tailed p-value was .0094. What p-value should be reported?

Solution For a one-sided research hypothesis, we want a one-tailed p-value. First, we note that the sample mean does go in the direction specified by H_a; it is larger than 50. Thus, the reported p-value should be $.0094/2 = .0047$.

□

statistical significance

The idea of p-value is closely related to the idea of the **statistical significance** of a test statistic. *To say that a result is statistically significant is only to say that a null hypothesis is rejected.* Thus, in the authoritarianism illustration, if the sample mean had been 48.6 for a z value of -2.83, the researchers would have reported that "the sample mean was statistically significantly different from the national mean of 52.8 ($p = .0046$)." This merely means that the research hypothesis that the university mean differs from the national mean is supported by the data, at the indicated p-value. If the research hypothesis is one-sided, the result is reported by saying that the mean is "significantly less than" or "significantly greater than" the H_0 value.

Example 9.25

How would the findings of Examples 9.19 and 9.20 be reported?

Solution

The difference between the current sample mean and the previous mean (treated as a population mean) was not statistically significant ($p = .32$).

□

The phrase "statistically significant" has unfortunate connotations. In ordinary language "significant" is synonymous with "important." Not so in statistics. Rejection of a null hypothesis indicates that a test statistic deviates from its expected value by an amount that is too large to be attributed to mere random variation. Thus a statistically significant difference is one that is too large to be reasonably attributed to randomness. A better phrase would be "statistically detectable." A "statistically detectable" difference would be one that could be detected as more than reasonable random variation.

In particular, *a statistically significant result is not necessarily scientifically significant, nor is a nonsignificant result necessarily scientifically uninteresting.* Statistical significance measures the "beyond-randomness" of an observed statistic. In very large samples, even trivial and uninteresting discrepancies might be more than random, hence statistically significant. For example, suppose that the authoritarianism researchers had managed to get a random sample of 5000 students from the enormous population of the university, and had found a sample mean of 52.3. The corresponding z statistic would be

$$z = \frac{52.3 - 52.8}{10.5/\sqrt{5000}} = -3.37$$

which would be highly statistically significant, with a p-value of less than .001. Nevertheless, a discrepancy of half of one point on a hundred-point scale, with a standard deviation of 10.5 points, would most likely be considered a very small and uninteresting difference. Granted this small difference is beyond reasonable random variation; that doesn't make it much more interesting scientifically.

The reverse case, a possibly interesting result that is not statistically significant, requires careful interpretation. Suppose that a simple random sample of only 5 students had yielded $\bar{y} = 44.8$ and

$$z = \frac{44.8 - 52.8}{10.5/\sqrt{5}} = -1.70$$

The p-value is $2(.0446) = .0892$, indicating that the result is not statistically significant at $\alpha = .05$ (but is significant at $\alpha = .10$). The discrepancy between the sample mean, 44.8, and the hypothetical population mean, 52.8, is rather large, indicating that possibly this university has markedly fewer authoritarian people than the population as a whole. However, the sample is so small that no firm conclusion can be reached. Such a result usually indicates a possibility that is worth pursuing further.

Example 9.26 In Example 9.19, would you say that there was a numerically large change in mean conservatism scores (even though the result was not statistically significant)?

Solution The change was 1.225 points on a hundred-point scale, which certainly seems modest. The original standard deviation was only 7.77 points, so the shift was about 1/6 of a standard deviation, still not a startling change. □

In this section p-values have been calculated assuming that the theoretical distribution of the test statistic is exactly a standard normal (z) distribution. In practice, the theoretical (sampling) distribution will only be approximately normal, by way of the Central Limit Theorem. As discussed in Chapter 7, an appeal to the Central Limit Theorem is most risky in dealing with one-tailed probabilities in the face of skewness. It is also known that probabilities far out in the tail of the distribution are poorly approximated. Thus one-tailed p-values, especially very small ones, must be regarded as rough approximations.

Example 9.27 In Example 9.20, a *p*-value was calculated assuming that the theoretical distribution of the *z* statistic was indeed normal. Is there any reason to believe that the *p*-value is a poor approximation?

Solution No. In Example 9.17 we saw that the distribution of scores in the sample was not too far from normal. Therefore we may reasonably assume that we are sampling from a near-normal population. Furthermore, a sample size of 40 allows the Central Limit Theorem effect to make the theoretical distribution of the sample mean very close to normal. □

Exercises for Section 9.4

9.19. Refer to the Tellevik study described in Exercise 9.15.

a. Calculate an appropriate *p*-value.
b. State the conclusion in "statistical significance" language.
c. State carefully exactly what this result says about the difference in means between hearing and deaf subjects.

9.20. The *p*-value calculation in Exercise 9.19 assumes that the theoretical distribution of the sample mean is normal, to a good approximation.

a. How reliable is the approximation?
b. Is two-decimal-place accuracy of the approximation crucial to the answers to Exercise 9.19?

9.21. Refer to the Otis study in Exercise 9.12.

a. Should the *p*-value for this hypothesis test be one-tailed or two-tailed?
b. Calculate the appropriate *p*-value.
c. Is the result "statistically significant"?

9.22. In Example 9.21, is there any reason to doubt the accuracy of the computed *p*-value?

9.23. Suppose that you obtain data on 1264 subjects, indicating changes on an introversion-extroversion scale after a certain treatment. From past experience, the population standard deviation can be assumed to be about 6.25 points. The sample mean change is 0.44 point. Nummsculle, your research assistant, calculated a *z* statistic of 2.51 and a one-tailed *p*-value of .006. He interpreted this result as showing that the treatment had had a large effect on introversion-extroversion scores.

a. Should the p-value be one-tailed or two-tailed?

b. Calculate the appropriate p-value.

c. Nummsculle has a gift for making inappropriate interpretations of statistical results. Has he done so here?

9.24. A data set indicated changes in Scholastic Aptitude Test (SAT) scores of a sample of five subjects who went through an intensive coaching course. The sample mean was 52.21; make the (heroic and undesirable) assumption that the population standard deviation is 80.0, which is the sample standard deviation. Nummsculle calculated $z = 1.46$, and reported a p-value of .1442. He said that the result was not statistically significant and that the study had proved that there was no value to the coaching course.

a. Is the z statistic correct?

b. Is the p-value correct?

c. Is the result not statistically significant, at the usual α levels?

d. Where did Nummsculle go wrong in the interpretation?

9.5 Confidence Intervals and Hypothesis Testing

Now we have two basic methods for statistical inference—hypothesis tests and confidence intervals. Both are methods that allow for a degree of random variation within the data. It shouldn't be surprising that there is a connection between them.

For example, suppose that a sample of 68 developmentally normal children had been tested to determine the age at which they understood a certain Piagetian "conservation" concept, and that the sample mean had been 8.62 years. (Note that there would be serious problems in defining "developmentally normal" and in assessing just when a child understood the concept.) For the time being, assume that the population standard deviation was known to be 1.10 years. A 95% confidence interval for the true, population mean age would be

$$8.62 - 1.96\frac{1.10}{\sqrt{68}} < \mu_{pop} < 8.62 + 1.96\frac{1.10}{\sqrt{68}}$$

or

$$8.36 < \mu_{pop} < 8.88$$

Suppose that it had been claimed that the true, population mean was 9.1 years. This value may be taken as a null hypothesis to be tested. The confidence interval indicates that this null hypothesis should be rejected. The value 9.1 is not included within the "reasonable range" given by the confidence interval. Indeed, the z statistic would be

$$z = \frac{8.62 - 9.10}{1.10/\sqrt{68}} = -3.60$$

which is far more negative than the borderline rejection value of -1.96. Therefore the claimed mean age of 9.1 would be rejected (and the difference declared statistically significant at a small p-value). On the other hand, suppose that the claimed mean had been 8.8 years. The value 8.8 falls within the indicated 95% confidence interval, so it is a plausible value for μ_{pop}. The z statistic would be

$$z = \frac{8.62 - 8.80}{1.10/\sqrt{68}} = -1.35$$

which falls far short of the borderline -1.96 value. This example illustrates a very general principle.

A 95% confidence interval contains all H_0 mean values that would be retained by a hypothesis test using $\alpha = .05$. A claimed mean of 9.1 years would be rejected at $\alpha = .05$, and a claimed mean of 8.8 years would be retained at $\alpha = .05$. Of course, there's nothing magic about .05; for example, a 99% confidence interval corresponds to $\alpha = .01$.

Example 9.28

Lopez (1981) reported on a study of the performance rating of telephone operators when a reward structure was introduced into their work. Assume that the mean performance rating before the reward structure was 22.32 and the standard deviation 4.90 (the values reported by Lopez). Further, treat these values as population values (though in the study the data were more appropriately analyzed by methods to be discussed in Chapter 11). The sample mean, after introduction of the reward structure, was 27.93; assume that the standard deviation didn't change. The sample size was 186.

(a) Calculate a 99% confidence interval for the mean rating under the new structure.

(b) At $\alpha = .01$, can one claim a statistically significant change in the mean?

Solution

(a) $\bar{y} = 27.93$. The required z-table value corresponding to 99% confidence is 2.58. The interval is

$$27.93 - 2.58\frac{4.90}{\sqrt{186}} < \mu_{pop} < 27.93 + 2.58\frac{4.90}{\sqrt{186}}$$

or

$$27.00 < \mu_{pop} < 28.86$$

(b) The confidence interval doesn't include the null hypothesis (no change) mean, 22.32. Thus at $\alpha = .01$, the change is statistically significant. (In fact, the z statistic comes out 15.6!) ☐

We have emphasized the connection between the confidence level (e.g., 95%) and the maximum tolerable probability α of a Type I error (e.g., $\alpha = .05$). There is also a general relation between the width of the confidence interval and β, the probability of Type II error. A wide confidence interval indicates that the statistical inference isn't very precise (because of a small sample size or large variability). In such a case, β, the risk of Type II error, must be poor (which is to say large). Conversely, a relatively narrow confidence interval indicates a relatively low value of β. Useful guidelines are as follows: If the difference between μ_0, the null hypothesis mean, and μ_a, the mean under H_a, is equal to E, the plus- or -minus term of the confidence interval, the power equals roughly .5. If the difference equals $2E$, the power of the test equals $1 - \alpha/2$ (two-tailed test). These guidelines follow from the power formula to be presented in the next section, and hold to a good approximation for the t methods to be discussed in Chapters 10 and 11, as well.

Example 9.29 For the Lopez study of Example 9.28, $E = 2.58(4.90/\sqrt{186}) = 0.927$. Find the power of the test if $\mu_a = 23.3$ and if $\mu_a = 24.3$.

Solution If $\mu_a = 23.3$,

$$\mu_a - \mu_0 = 23.3 - 22.32 = 0.98$$

which is just about equal to E, so the power is just about .5. If $\mu_a = 24.3$,

$$\mu_a - \mu_0 = 1.98$$

which is about $2E$, so the power is just about $1 - .01/2 = .995$. (Recall that the confidence level was 99% in Example 9.28.) ☐

Confidence-interval inference is easier to understand, for most people, than is hypothesis testing. None of the heavy machinery of hypothesis testing (types of errors, power calculations, etc.) is required for confidence intervals. Because there is a close relation between confidence intervals and tests, it's a bit surprising that hypothesis tests are so much more widely used than confidence intervals. One legitimate reason is that confidence intervals aren't useful in determining p-values, except by a tedious trial-and-error method. Another reason is that some statistical procedures (such as the rank tests to be discussed in the next couple of chapters) are more easily understood in a hypothesis-testing framework. A slightly cynical view holds that the reason for the preference for hypothesis testing is a preference for impressive-looking results. It is possible to set up a mildly absurd null hypothesis, and to reject it with an impressively low (conclusive) p-value, while still having a very wide, unimpressive confidence interval for the parameter of interest. The small p-value looks good, whereas the wide confidence interval looks bad. Therefore, a cynical position asserts, report the p-value!

Example 9.30 What does the confidence interval of Example 9.28 indicate about the magnitude of the (two-tailed) p-value for testing $H_0 : \mu = 22.32$?

Solution In Example 9.28, we rejected H_0 at $\alpha = .01$, because 22.32 was not in the 99% confidence interval. Therefore the p-value must be less than .01, according to the Universal Rejection Region. In fact, the p-value is 0 to many decimal places. □

Implicitly, the discussion of the relation between hypothesis tests and confidence intervals has assumed a two-sided test. The usual confidence intervals, which involve a plus-or-minus term, are naturally two-sided. However, the one-sided confidence intervals discussed in Section 8.3 do correspond to one-tailed tests. A lower-limit confidence interval can be used to test H_a: $\mu > \mu_0$, and an upper-limit interval tests H_a: $\mu < \mu_0$. In the example involving Piagetian conservation, suppose that a researcher believes that the claimed mean age of 9.1 for acquiring the concept is too high. In hypothesis-testing terms, a one-sided formulation could be stated as H_0: $\mu = 9.1$ vs. H_a: $\mu < 9.1$. The 95% upper-limit confidence interval, based on a sample mean of 8.62, a sample size of 68, and an assumed σ_{pop} of 1.10, would be

$$\mu < 8.62 + 1.645 \frac{1.10}{\sqrt{68}}$$

or

$$\mu < 8.84$$

The upper-limit 95% interval doesn't include 9.1, so H_0 would be rejected at $\alpha = .05$. Although we won't use one-sided intervals much more in this book, you should be aware that it is perfectly possible to do so.

Example 9.31 Calculate a 99% lower-bound confidence interval for the population mean, using the data of Example 9.28. Does it support the one-sided research hypothesis H_a: $\mu > 22.32$?

Solution The required $z_{.01}$ is 2.33, from Table 3. The interval is

$$\mu > 27.93 - 2.33 \frac{4.90}{\sqrt{168}}$$

or

$$\mu > 27.05$$

The null hypothesis value 22.32 isn't close to this interval, so H_0 can be rejected and H_a supported.

Exercises for Section 9.5

9.25. Refer again to the Tellevik data of Exercise 9.15.

a. Calculate a 95% confidence interval for the mean.
b. Does the interval include the null hypothesis mean?
c. What does this finding indicate about the statistical significance of the deviation of the sample mean from the null hypothesis mean?

9.26. Refer to the Otis data of Exercise 9.12.

a. Calculate a 99% lower-bound confidence interval for the mean.
b. Does the interval include the H_0 value?
c. Is the sample mean significantly larger than the H_0 mean, in the statistical sense?

9.27. Refer to Exercise 9.26. What does the result of that exercise, by itself, say about the size of the p-value?

9.28. Refer to Exercise 9.23.

a. Calculate a 95% confidence interval for the mean.
b. Show that H_0 may be rejected at $\alpha = .05$.

c. The original introversion-extroversion scale had a mean of 100 and a standard deviation of 15. In absolute terms, how large is the mean change?

9.29. Refer to Exercise 9.24.

a. Calculate a 90% confidence interval for the mean change.
b. In the context of SAT scores, how wide is this interval?
c. What is an appropriate conclusion about the value of the coaching course, based on this interval?

9.6 Power of a z Test and Planning a Study

In Section 9.2 the concept of the power of a statistical hypothesis test was introduced in the context of a binomial test. The power of a test is the probability that H_0 will be rejected, given that H_a is true; power should be made as large as possible. For a binomial test, power may be calculated by adding up the appropriate binomial table probabilities, assuming a particular value (in H_a) of π, the hypothesized population proportion.

The calculation of power for a z test isn't much harder. For example, let's reconsider the authoritarianism study of Section 9.3. Recall that n was 50, σ_{pop} was 10.5, α was .05, the H_0 mean, μ_0, was 52.8, and H_a was two-sided. To calculate the power of the test, we must specify a particular value for μ_{pop} within H_a; call this number μ_a. For illustration, we'll compute the power of the test when μ_a is 50.7. Power is, by definition, the probability that the test statistic, which in this case is

$$z = \frac{\bar{Y} - 52.8}{10.5/\sqrt{50}}$$

falls in the rejection region, which is $z > 1.96$ or $z < -1.96$. A little algebra shows that $z > 1.96$ is the same event as

$$\bar{Y} > 52.8 + 1.96\frac{10.5}{\sqrt{50}}$$

or

$$\bar{Y} > 55.71$$

Similarly, $z < -1.96$ is the same as $\bar{Y} < 49.89$. Thus, for $\mu_a = 50.7$,

$$\text{power} = P(\bar{Y} - < 49.89 \text{ or } \bar{Y} > 55.71 | \mu_a = 50.7)$$

Figure 9.5 shows the situation.

We assume that the theoretical (sampling) distribution of $\bar{Y}$ is close to normal; barring gross population skewness, a sample size of 50 should be adequate to invoke the Central Limit Theorem. As always in computing normal-distribution probabilities, we convert to z scores. In this power computation, we are assuming that $\mu_{pop} = 50.7$.

$$\text{power} = P\left(z < \frac{49.89 - 50.7}{10.5/\sqrt{50}} \text{ or } z > \frac{55.11 - 50.7}{10.5/\sqrt{50}}\right)$$

$$= P(z < -0.55 \text{ or } z > 3.37)$$

From Table 3, $P(z < -0.55) = .2912$ and $P(z > 3.37)$ is negligibly small. Therefore, for this particular test, with this particular μ_a value, power is only about .29.

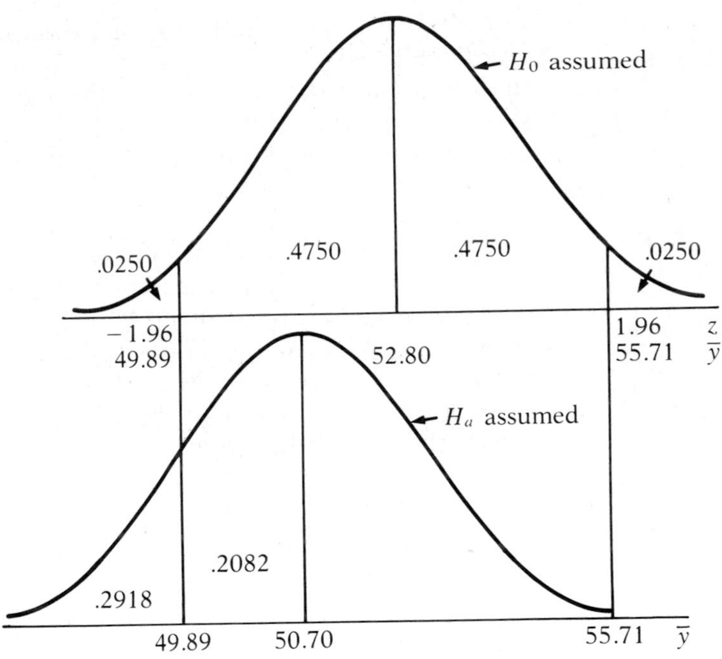

Figure 9.5 α and power for a z test

Example 9.32

Refer to Example 9.16. Suppose another follow-up study is planned of the mean conservatism level of sophomores. Another random sample of size 40 is to be taken. The researchers believe that the mean conservatism level has increased since the original survey was conducted, and plan a one-tailed test. If α is to be .05, n is 40, σ_{pop} is 7.77, μ_0 is 61.9, and in fact the mean conservatism score is 63.4, what is the probability that the research hypothesis will be supported?

Solution

The rejection region is R.R.: reject H_0 if

$$z = \frac{\bar{Y} - 61.9}{7.77/\sqrt{40}} > 1.645$$

or, equivalently, if

$$\bar{Y} > 61.9 + 1.645\frac{7.77}{\sqrt{40}} = 63.92$$

The probability that the research hypothesis will be supported, given that the true population mean is 63.4, is by definition the power of the test for $\mu_a = 63.4$.

$$\begin{aligned} P(\text{reject } H_0|\mu_a = 63.4) &= P(\bar{Y} > 63.92|\mu_a = 63.4) \\ &= P\left(z > \frac{63.92 - 63.4}{7.77/\sqrt{40}}\right) \\ &= P(z > 0.42) = .5000 - .1628 = .3372 \end{aligned}$$

It isn't necessary to go through all these steps every time a power is calculated. The steps can be condensed into a formula.

Power of a z Test

One-tailed test:

$$\text{power} = P\left(z > z_\alpha - \frac{|\mu_a - \mu_0|}{\sigma_{\text{pop}}/\sqrt{n}}\right)$$

Two-tailed test:

$$\text{power} = P\left(z > z_{\alpha/2} - \frac{|\mu_a - \mu_0|}{\sigma_{\text{pop}}/\sqrt{n}}\right)$$

NOTE: z_a is the Table 3 z value cutting off a right-tail area equal to a. It is assumed that the theoretical distribution of the sample mean is approximately normal.

Thus in the authoritarianism study,

$$\text{power} = P\left(z > 1.96 - \frac{|50.7 - 52.8|}{10.5/\sqrt{50}}\right)$$
$$= P(z > 0.55) = .2912$$

as before.

Example 9.33 Recalculate the power of the test in Example 9.32, using the power formula for a z test.

Solution

$$\text{power} = P\left(z > z_{.05} - \frac{|\mu_a - \mu_0|}{\sigma_{\text{pop}}/\sqrt{n}}\right)$$
$$= P\left(z > 1.645 - \frac{|63.4 - 61.9|}{7.77/\sqrt{40}}\right)$$
$$= P(z > 0.42) = .3372$$

once again. ☐

The power formula reflects all the factors that affect the power of a test. First, as α increases, z_α and $z_{\alpha/2}$ decrease, so power increases. Second, as n gets larger, the standard error $\sigma_{\text{pop}}/\sqrt{n}$ decreases, and the power increases. Third, as the distance between the hypothesized μ_a and μ_0 increases, it becomes easier to detect the difference between them, and power increases. Fourth, if σ_{pop} decreases, the standard error decreases, and power increases.

Example 9.34 Refer to Example 9.32. What will happen to the power of the test if α is taken as .10, n is 60, the true mean is 64.0, and σ_{pop} is 7.00?

Solution Compared to Example 9.32, the changes in α, n, μ_a, and σ_{pop} all point toward an increase in power.

$$\text{power} = P\left(z > 1.28 - \frac{|64.0 - 61.9|}{7.00/\sqrt{60}}\right)$$
$$= P(z > -1.04) = .85$$

☐

A scientist may be able to control some of these factors. The obvious controllable factor is the sample size n. It's always desirable to have a large sample size (for technical, normal-approximation reasons as well as for scientific, power, or accuracy reasons). Yet gathering data can be expensive and time-consuming. For example, a study of the effect on laboratory rats of an electrical stimulus applied to a certain part of the brain might involve half a working day per rat. A sample size of 800 rats would require 400 working days, which is the best part of two years of five-day weeks. Perhaps in some cases such an effort is justified, but it shouldn't be undertaken unnecessarily.

The level of α also may, in principle, be varied. To the extent that α can be made large, the test's power will be increased. In practice, there are limits to the manipulation of α. Some scientific journals insist on $\alpha = .05$. In any case, an α value of, say, .50 is virtually never used. In fact, α values above .10 are rare in the scientific literature.

Occasionally, the other factors, σ_{pop} and $|\mu_a - \mu_0|$, can also be controlled. The population standard deviation of measurements, σ_{pop}, reflects not only actual individual differences but also such things as random measurement error. Careful and accurate measurement often reduces σ_{pop} somewhat. Also, careful choice of subjects may reduce σ_{pop}. A good reason for using rats from the same strain in a study is to minimize genetic variation and therefore to reduce σ_{pop}. Furthermore, paired experiments of the before-after type often eliminate much of the effect of individual differences, thus reducing σ_{pop}; such experiments will be discussed further in Chapter 11.

size of experimental effect

The $|\mu_a - \mu_0|$ factor is often called the **size of experimental effect**. In this interpretation, μ_0 is a baseline standard (such as a national mean on an authoritarianism study) and μ_a is the mean under a certain experimental treatment (such as a course in evaluating propaganda messages). To the extent that the size of experimental effect can be increased, the power of the statistical test will be improved. For example, a study of the effect of an electrical stimulus on a certain part of a rat's brain will be more effective if the stimulus is substantial rather than barely noticeable. Obviously, there are limits to how large an effect can be generated, but some consideration of increasing the experimental effect is often valuable.

Example 9.35

In a study of campus conservatism, where will a hypothesis test have greater power to detect a change in mean of a specified number of points—at a university with a very heterogeneous student body or at one with a homogeneous student body?

Solution With a heterogeneous student body, σ_{pop} will be large, thus decreasing the power of the test (the ability to detect a change). Because there will be less variability obscuring the change, it will be easier to detect in the homogeneous campus. □

The concept of power is especially useful in planning a scientific study. If the power of a statistical test (under reasonable assumptions) is very poor in a particular experiment, the experiment should be redesigned. To calculate power, it is necessary to specify trial values of α and n. It is also necessary to specify $|\mu_a - \mu_0|$ and σ_{pop}, or at least their ratio,

$$\frac{|\mu_a - \mu_0|}{\sigma_{pop}}$$

This ratio is the size of experimental effect expressed as a fraction of a population standard deviation. Making this specification is the hardest part of a power calculation. One method is to specify the ratio using estimated values obtained in other studies. If such studies are available, it's a good idea to reduce the estimated value of $|\mu_a - \mu_0|/\sigma_{pop}$ slightly, to be on the conservative side.

Example 9.36 Refer to Example 9.33. Suppose that a study of a similar university showed an increase of .6 standard deviation in the mean conservatism score. The researchers hypothesized that the mean had increased only .5 standard deviation to get a conservative power estimate (which after all was only appropriate for this study). What power does the test have?

Solution Recall that $z_{.05} = 1.645$ and that $n = 40$ was planned. The hypothesis is that

$$\frac{|\mu_a - \mu_0|}{\sigma_{pop}} = .50$$

$$\text{power} = P(z > 1.645 - .5\sqrt{40}) = P(z > -1.52) = .94$$ □

If there are no trustworthy previous studies available, one can sometimes hypothesize a value for $|\mu_a - \mu_0|/\sigma_{pop}$ by considering what potential size of effect would be scientifically important. This is a scientific, not a statistical, issue, and judgments must be made. Very broadly speaking, an effect that is, say, only 2% of a population standard deviation is not usually of great interest, whereas an effect that is 2 full standard deviations would be of great interest (in the unlikely event that such an effect could be found). As a "when all else fails" rule of thumb, Cohen (1977) has suggested a value of 0.80 as a large value for the effect $|\mu_a - \mu_0|/\sigma_{pop}$, 0.50 as a moderate effect, and 0.20 as a small effect.

Example 9.37 If the sample size is 40, α is .05 (one-tailed), and, in Cohen's rule of thumb, the effect size is moderate, what is the power of the test?

Solution This problem is identical to that of Example 9.36, where an effect size of .5 standard deviation was assumed. Once again, power will be .94.

□

One way to plan an experiment is to specify a sample size (and an α value and size of effect) in advance. Then one can find out if the power is acceptably high. Another "when all else fails" rule of thumb is that a power of .9 (with $\alpha = .05$) is adequate. Alternatively, it's possible to determine what sample size is needed to obtain a specified power, by formula.

Required n to Obtain Power $1 - \beta$, z Test

One-tailed test:

$$n \geq \frac{(z_\alpha + z_\beta)^2(\sigma_{pop})^2}{(\mu_a - \mu_0)^2}$$

For a two-tailed test, replace z_α by $z_{\alpha/2}$.

Example 9.38 Refer to Example 9.32. What sample size is needed to obtain power .90?

Solution

$$n \geq \frac{(1.645 + 1.28)^2(7.77)^2}{(63.4 - 61.9)^2} = 230$$

□

It's usually a good idea to consider power issues in planning an experiment or study. A rough power calculation may indicate that far too few observations have been planned, so the experiment will probably be indecisive. Conversely, it may indicate that the planned study is far too ambitious, involving an unnecessarily large sample size. In either case, adjustments can often be made to improve the study.

Exercises for Section 9.6

9.30. Suppose that you are planning a followup to the Otis study of Exercise 9.12, using patrons of a similar space-age movie. A sample size of 40 is planned.

a. What is the power of a one-sided z test against $\mu_a = 4.25$? Assume $\alpha = .05$.

b. What sample size is needed to make this power equal .99?

9.31. Is a sample of size 30 enough to have a probability of detecting a "large" effect (in Cohen's sense) equal to .75, when α is .01 (two-tailed)? What sample size is needed to make the power be .90?

9.32. A study of human reaction times indicated that the population mean under specified conditions was about 420 milliseconds and that the standard deviation was about 280 milliseconds. The distribution of scores was very right-skewed. It was planned to test a new sample of subjects under similar conditions, but with a distraction added. A "guesstimate" of the effect of the distraction was that it would increase the mean time by 40 milliseconds, but not change the standard deviation. A two-tailed test was planned, using $\alpha = .05$.

a. If the "guesstimate" is correct, what is the power of the test when the sample size is 40?

b. What sample size is needed to make the power equal .90?

9.33. In part b of Exercise 9.32, is the sample size large enough that one needn't worry about the effect of population skewness?

9.34. Refer to Exercise 9.32.

a. Solve the problem assuming $\alpha = .01$, not .05.

b. Solve the problem assuming $\alpha = .05$, but the mean change is only 30 milliseconds.

c. Solve the problem assuming $\alpha = .05$ and the mean change is 40 milliseconds, but the population standard deviation is 350 milliseconds.

d. What is the effect of each of these alternative assumptions on the power of the test when the sample size is 40?

Appendix: Most Powerful Tests

Chapter 9 indicated that the ideal test statistic was one that yielded the maximum possible power (for a specified α, n, and size of effect). Given some very specific assumptions about the population distribution, an important theorem of mathematical statistics, the Neyman-Pearson Lemma, can be used to find the best, most powerful test statistic. This lemma, which is proved in most mathematical statistics books, also helps to clarify some basic hypothesis-testing ideas. See Larsen and Marx (1981), Mood, Graybill, and Boes (1974), or any other mathematical statistics book for details.

At the beginning of Chapter 9, we said that H_0 should be rejected if the sample data are much more likely given that H_a is true than given that H_0 is true. The Neyman-Pearson Lemma is based on this idea. Recall from Chapter 8 that the likelihood function is defined as the probability (or probability density) of the actually observed data, looked at as a function of the unknown population parameter. For example, assume that a random sample of size $n = 2$ is taken from a population with a Poisson distribution. If the observed values are 4 and 2, for example, then the likelihood is

$$L(\mu) = e^{-\mu}\frac{\mu^4}{4!}e^{-\mu}\frac{\mu^2}{2!}$$

In general, denote the observed values by y_1 and y_2. The likelihood function in this situation is

$$L(\mu) = e^{-\mu}\frac{\mu^{y_1}}{y_1!}e^{-\mu}\frac{\mu^{y_2}}{y_2!}$$

The Neyman-Pearson Lemma is most easily stated in the case where both H_0 and H_a consist of only one possible value for a single population parameter. Recall that the generic notation for a parameter is θ, where θ may stand for a population mean, median, standard deviation, or any other parameter. The simplest case tests H_0: $\theta = \theta_0$ vs. H_a: $\theta = \theta_a$.

likelihood ratio

Neyman-Pearson Lemma for a Most Powerful Test Statistic

Given an assumption about the form of the population distribution, the best (most powerful) test of H_0: $\theta = \theta_0$ vs. H_a: $\theta = \theta_a$ is to reject H_0 if and only if the **likelihood ratio**

$$\frac{L(\theta_a)}{L(\theta_0)} \geq c$$

where c is a constant chosen to yield the desired α value.

In the Poisson example, assume that H_0 is $\mu = 5.0$ and that H_a is $\mu = 3.5$. We can list a few possible outcomes of the two observations, together with their probabilities (likelihoods) under each of the hypotheses; from these likelihoods, we can calculate likelihood ratios. (See Table 9.2.)

	(y_1, y_2)										
	(0, 0)	(0, 1)	(1, 0)	(0, 2)	(1, 1)	(2, 0)	(0, 3)	(1, 2)	(2, 1)	(3, 0)	...
$L(3.5)$	.000912	.003182	.003182	.005585	.011171	.005585	.006516	.019548	.019548	.006516	...
$L(5.0)$	.000045	.000227	.000227	.000567	.001135	.000567	.000946	.002837	.002837	.000946	...
Ratio	20.3	14.1	14.1	9.85	9.85	9.85	6.89	6.89	6.89	6.89	...

Table 9.2

Notice that in this case the likelihood ratio is the same for all outcomes having a particular sum, and that the ratio decreases as the sum of the observations increases. R.R.: reject H_0 if the likelihood ratio ≥ 9.85 is therefore equivalent to R.R.: reject H_0 if $y_1 + y_2 \geq 2$. According to the Neyman-Pearson Lemma, this test is the best (most powerful) test of this H_0 and H_a for

$$\alpha = .000045 + 2(.000227) + 2(.000567) + .001135 = .002768$$

Note that we do *not* reject H_0 for all outcomes that have a likelihood ratio larger than 1. Instead the likelihood ratio must exceed some large number in order for H_0 to be rejected. The choice of the cutoff number is based on the desired value of α.

Example 9.39

Six Bernoulli trials are to be performed and the resulting number of successes recorded. If H_0 is $\pi = .20$ and H_a is $\pi = .30$, use Table 1 to calculate likelihood ratios for all possible y values; find the best (Neyman-Pearson) hypothesis test with $\alpha = .0170$.

Solution

	y						
	0	1	2	3	4	5	6
$L(.30)$	.1176	.3025	.3241	.1852	.0595	.0102	.0007
$L(.20)$	.2621	.3932	.2458	.0819	.0154	.0015	.0001
Ratio	0.449	0.769	1.32	2.26	3.86	6.64	11.4

Table 9.3

(Note: The ratios for $y = 5$ and 6 were calculated using more decimal places than given in Table 1.)

The best test is to reject H_0 if the ratio is ≥ 3.86, or equivalently if $y \geq 4$. Note that

$$P(Y \geq 4 | \pi = .20) = .0154 + .0015 + .0001 = .0170$$

□

In many (but not all) cases, the Neyman-Pearson Lemma applies to more general hypotheses. Suppose that, in the Poisson example, we have H_0: $\mu = 5.0$ vs. H_a: $\mu < 5.0$. If we take *any* particular value of μ in H_a, not necessarily 3.5, the Neyman-Pearson rejection region for $\alpha = .002768$ is always to reject H_0 if $y_1 + y_2 \leq 2$. This can be shown by trying various values for μ_a, or by algebra. Therefore, this test (rejection region) is most powerful, uniformly for all values of $a < 5.0$. The test is called **uniformly most powerful (UMP)** for this situation. Note however that this "best test" applies only if certain assumptions (in the example, the Poisson assumptions) are met; if the assumptions fail, there is no guarantee that the Neyman-Pearson test will remain optimal.

uniformly most powerful (UMP)

Example 9.40 Refer to Example 9.39. The likelihood ratio can be calculated for any value of $\pi > .20$, not just $\pi = .30$. In every case, the likelihood ratio is smallest for $y = 0$ and increases as the y value increases. What does this fact imply about the test in Example 9.39?

Solution Because of this property, the best test will always select $y = 6$, then $y = 5$, then $y = 4$ for inclusion in the R.R. Thus R.R.: reject H_0 if $Y \geq 4$ is UMP for testing H_0: $\pi = .20$ vs. H_a: $\pi > .20$ at $\alpha = .0170$. □

Typically, UMP tests apply only to one-sided tests with a single unknown parameter value. There are some extensions to the theory of UMP statistical tests; see Lehmann (1959) for a highly mathematical presentation. Alternatively, it is possible to generalize the idea of likelihood ratio tests to provide very good, if not necessarily optimal, test statistics. We'll take up the latter idea in an appendix to the next chapter.

Chapter Exercises

9.35. Student folklore holds that answer b is very often the correct one in the conventional multiple-choice exam, where each question has five possible answers. To test whether this belief holds for a certain instructor, the students plan to examine the next 20 multiple-choice questions asked by the instructor.

a. Would the binomial assumptions (with success being that b is the correct answer, failure being anything else) be reasonable in this situation?

b. Assuming that only one answer to any question is regarded as correct, what is the natural null hypothesis to be tested?

c. Should the research hypothesis be one-sided or two-sided? (You might get an argument from others on this one.)

9.36. Refer to Exercise 9.35. The students decided to use a one-sided research hypothesis and to assume that the binomial assumptions were valid. The rejection region was chosen as R.R: reject the null hypothesis if there are 8 or more successes in the next 20 questions.

a. Find α, assuming that the long-run probability that answer b is correct is 1/5.

b. Without any calculations, what can be said about the magnitude of α if the long-run probability that b is correct is .15?

c. Find the probability that the null hypothesis will be rejected if in fact the long-run probability that b is correct is .35. What is this probability called?

9.37. Refer to Exercise 9.36. The study was performed, and it was found that 7 of the 20 questions had b as the correct answer.

a. Carry out the five steps of a statistical test of hypotheses.

b. One student said that, because the null hypothesis wasn't rejected, the test proved that this teacher had no bias toward answer b. Is this statement correct?

9.38. An experiment was designed to see whether laboratory rats would, on the average, eat more when confined to a crowded cage or an uncrowded one. To minimize the effect of genetic differences, two rats were chosen from each of 30 litters. One of each pair of rats was randomly assigned to a crowded cage, the other to an uncrowded one. The data to be collected were the 30 differences in amount of food consumed between the "crowded" and the "uncrowded" rat from each litter.

a. What is a natural null hypothesis for this experiment?

b. Should the research hypothesis be directional? If so, which direction should be chosen?

9.39. Suppose that a nondirectional research hypothesis has been chosen for the study in Exercises 9.38. Also, suppose that the population standard deviation of the differences may be assumed (on the basis of previous studies) to be 12.50. Find a rejection region so that the probability of rejecting the null hypothesis, when it's true, is .01.

9.40. Refer to the rejection region found in Exercise 9.39. Find the probability that the null hypothesis will not be rejected if the population mean difference is 5.0. What is the name of this probability?

9.41. Refer again to Exercise 9.39. In the actual sample data, the mean difference was 6.41 (and the sample standard deviation was close to the assumed population standard deviation).

a. Carry out the five steps of statistical hypothesis testing. What is the conclusion?

b. Find the *p*-value. Should it be one-tailed or two-tailed?

9.42. For the data of Exercise 9.41, calculate a 99% confidence interval for the population mean difference. Does this interval indicate that H_0 should be rejected?

9.43. Examination of the data of Exercise 9.41 showed that most of the differences were between 5.0 and 8.0; however, three of the differences were large negative numbers and two of the differences were large positive values. What does this finding indicate about the desirability of basing the test on the sample mean?

9.44. Assume that the historic success rate for conventional therapy for a particular psychiatric disorder is 30%. A modified behavioral therapy has been proposed. To test the therapy, a sample of 16 patients have been given the behavioral therapy. Define π to be the long-run success rate of the behavioral therapy.

a. What randomization and/or "blinding" should be done in gathering the data? Why?

b. What is the natural null hypothesis about π?

c. Define the appropriate (one-sided or two-sided) research hypothesis. Again, you may get an argument about which form of H_a is appropriate.

9.45. Suppose that, for better or worse, the researchers doing the study of Exercise 9.44 chose H_a: $\pi > .30$. What would be the meaning and the consequences of Type I and Type II errors in this study?

9.46. Refer to Exercises 9.44 and 9.45.

a. Find an R.R. that has a maximum α of .10.

b. Calculate the probability that the null hypothesis will be rejected if $\pi = .25$. How does this probability compare to the maximum allowable α?

9.47. For the rejection region specified in Exercise 9.46, what is the probability that the research hypothesis will be supported, assuming that $\pi = .50$. What is this probability called?

9.48. Rework Exercises 9.46 and 9.47, assuming a maximum allowable α of .05. How is the probability of rejecting H_0 affected by reducing α?

9.49. Refer to Exercises 9.44–9.48. The actual sample yielded 12 successes.

a. Carry out the five parts of a statistical test, using $\alpha = .10$.

b. Would the same conclusion be reached using $\alpha = .05$?

c. What do the answers to parts a and b indicate about the *p*-value for the experiment?

9.50. In a certain school system, the mean performance of the population of fourth-grade students on a test of arithmetic skills had been 3.62 and the standard deviation had been 0.65. A sample (carefully randomized) of 94 fourth-grade students were assigned to a new arithmetic curriculum emphasizing concrete representations of arithmetic operations, in the hope that average performance on the test would be higher.

a. What is the population parameter to be tested by a formal statistical test?

b. Formulate null and (directional) research hypotheses.

c. Is there an argument in favor of using a nondirectional research hypothesis?

9.51. In Exercise 9.50, assume that a directional hypothesis was chosen. Also assume that the population standard deviation will not change.

a. Formulate a rejection region so that $\alpha = .05$.

b. Recalculate the R.R. so that $\alpha = .01$.

9.52. Refer to Exercise 9.51.

a. Calculate the probability that H_0 will be retained, assuming a population mean score of 3.90 under the new curriculum and assuming $\alpha = .05$.

b. Recalculate this probability assuming $\alpha = .01$. Which of the probabilities is larger? Why should this be so?

c. Are the probabilities you calculated in parts a and b *p*-values?

9.53. Refer to Exercise 9.51. The actual sample mean ($n = 94$) came out 3.79. Carry out the five steps of a formal hypothesis test. Assume $\alpha = .05$.

9.54. Refer to Exercise 9.53. Calculate a 95% lower-bound confidence interval for the mean under the new curriculum. Can the null hypothesis be rejected by the interval?

9.55. For the data of Exercise 9.53, find the *p*-value. Does it matter to the conclusion of the test whether α was set to be .05 or .01?

9.56. The researchers in the study of Exercises 9.50–9.55 reported that the change in scores was statistically significant and that therefore the study had proved that the new curriculum caused a large improvement in arithmetic performance.

a. Can the result be called statistically significant?

b. Can the researchers legitimately claim that the new curriculum caused a large improvement?

9.57. A new test instrument was designed to measure a certain trait. After preliminary administration of the instrument, a scoring system was developed that supposedly would yield a population mean of 500 and a population standard deviation of 100. A random sample of 628 adults were administered the instrument, with primary interest being in the mean score.

a. Formulate H_0 and a two-sided H_a.
b. Find the rejection region corresponding to $\alpha = .10$.

9.58. In Exercise 9.57, what is the probability that the null hypothesis will be retained, assuming the population mean is 490?

9.59. Refer to Exercise 9.57. The actual sample mean came out 493.1. (It was assumed that the population standard deviation, 100, was correct.) Show that the research hypothesis is supported.

9.60. Calculate a 90% confidence interval for the population mean, based on the data of Exercise 9.59. Show that this confidence interval leads to support of the research hypothesis defined in Exercise 9.57.

9.61. The conclusion of the study was reported as "the discrepancy from the hypothesized mean, 500, was statistically significant ($p < .05$). Therefore, we know that the actual mean differs greatly from the hypothesized mean."

a. Is the claim of statistical significance correct?
b. Is the "differs greatly" claim warranted?

t Distribution Methods

In the last two chapters, we have repeatedly noted that it is unrealistic to assume that the population standard deviation is known when the population mean isn't known. In this chapter, we seek to restore some realism by assuming that both the population mean and the population standard deviation are unknown quantities.

When σ_{pop} isn't known, it must be estimated from the data. Assuming simple random sampling and a somewhere-near-normal population, the natural estimate of σ_{pop} is s, the sample standard deviation defined back in Chapter 2.

$$s = \sqrt{\frac{\Sigma (y_i - \bar{y})^2}{n - 1}}$$

Therefore, the natural estimate of the (simple random sampling) standard error of the sample mean,

$$\sigma_{\bar{Y}} = \frac{\sigma_{pop}}{\sqrt{n}}$$

estimated standard error

is the **estimated standard error** of the sample mean, $s/\sqrt{n}$. Whenever the true standard error is replaced by the estimated standard error, an additional source of random error is introduced.

In this chapter the consequences of this additional random error are investigated. The basic consequence is that the z probabilities of Chapters 8 and 9 must be replaced by what are universally called t probabilities. In Section 10.1 the t statistic, t distribution, and t tables are defined and discussed. Confidence interval methods based on t are discussed in Section 10.2, and the parallel hypothesis-testing methods appear in Section 10.3. In Section 10.4, an important kind of statistical experiment, the paired-sample experiment, is discussed. An alternative approach to t methods, the median test, is defined in Section 10.5. This procedure is useful when data are highly skewed or outlier-prone. Finally, in Section 10.6, the power and sample-size questions raised in the last two chapters are raised again in the context of t methods. There are two appendixes—the first involving the mathematical definition of the t distribution, the second giving the mathematical basis for why one uses t methods at all.

10.1 The *t* Distribution

The basic confidence interval and hypothesis-testing methods about a population mean, discussed in the last two chapters, have been based on a z statistic.

$$z = \frac{\bar{Y} - \mu_{\text{pop}}}{\sigma_{\text{pop}}/\sqrt{n}}$$

When the population standard deviation isn't known (which is to say, almost always), the true standard error (of the sample mean, under simple random sampling) must be replaced by the estimated standard error $s/\sqrt{n}$. The resulting statistic is universally known as a t statistic:

$$t = \frac{\bar{Y} - \mu_{\text{pop}}}{s/\sqrt{n}}$$

More generally, *any time a true standard error is replaced by an estimated standard error, the statistic is denoted by t rather than z.* The use of an estimated standard error introduces another source of random variation. The additional randomness is most serious when the sample size is small, because the estimated standard error is most uncertain for small sample sizes. The distinction between the t methods of this chapter and the z methods of the previous chapter is most critical for small sample sizes.

The t distribution (of the t statistic) looks much like the z distribution of Table 3. It is symmetric around 0; the sign of either z or t is determined by the difference $Y - \mu_{\text{pop}}$, not by the true or estimated standard error. The variability of t is larger than the variability of z, because t involves an additional source of random variation. The variability is much larger when n, the sample size, is small, because in that case the random error in the estimated standard error is large. Figure 10.1 shows t distribution curves (relative to z curves) for two different sample sizes.

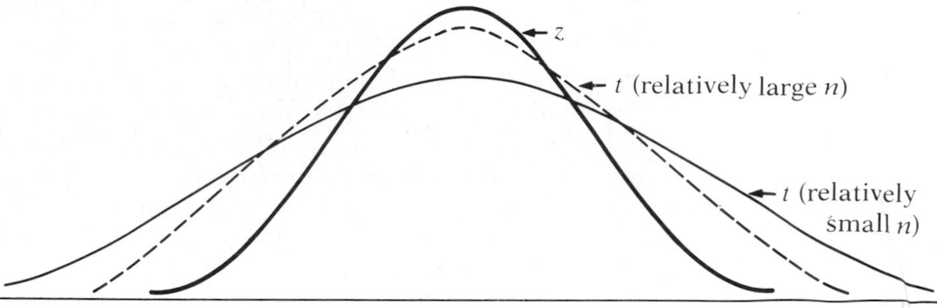

Figure 10.1 *t distribution vs. z distribution*

The distribution of the t statistic must depend on the sample size, n. It turns out to be more convenient to index the t distribution by its degrees of freedom. For a single, simple random sample, the number of degrees of freedom is $n - 1$. For more complicated sampling, the degrees of freedom formula is also more complicated. The name **degrees of freedom** refers to the number of pieces of information available to estimate the sample variance, s^2. The sample variance is based on the sample deviations from the mean, $y_i - \bar{y}$. The sum of these deviations, $\Sigma (y_i - \bar{y})$, is always 0. Thus, only the first $n - 1$ deviations are free to vary, and therefore carry any information. If $n = 4$ and the first three deviations are -3, 2, and 6, the remaining deviation must be -5 to make the deviations add to 0; there are only $4 - 1 = 3$ degrees of freedom. Note that in defining the sample variance, we divided by the degrees of freedom, $n - 1$, not by the sample size n. Degrees of freedom are denoted by ν (the Greek letter nu) or by df.

degrees of freedom

Example 10.1

How many degrees of freedom are available in the following sample to estimate the sample variance?

y_i: 25 20 30 40 35

Solution

The sample size is 5, so there must be $5 - 1 = 4$ df. The mean is 30, so the first four deviations are -5, -10, 0, and 10. The remaining deviation is automatically 5, so only the first four deviations are informative in estimating the variability of the sample data. □

Tables of the t distribution are always based on degrees of freedom. Conventional t tables such as Table 4 of the appendix are, in a sense, the reverse of z tables like Table 3. In the z table, Table 3, the cutoff number is given and the area (probability) is found. In t tables like Table 4, the area is given, and the cutoff number is found. Specifically, Table 4 gives the t number corresponding to a given *right-tail* area. For example, suppose that there are 19 degrees of freedom (corresponding to a simple random sample of size 20). Table 4 indicates that the t number 1.729 cuts off a right-tail area of .05. Symmetrically, $t = -1.729$ cuts off a left-tail area of .05. See Figure 10.2.

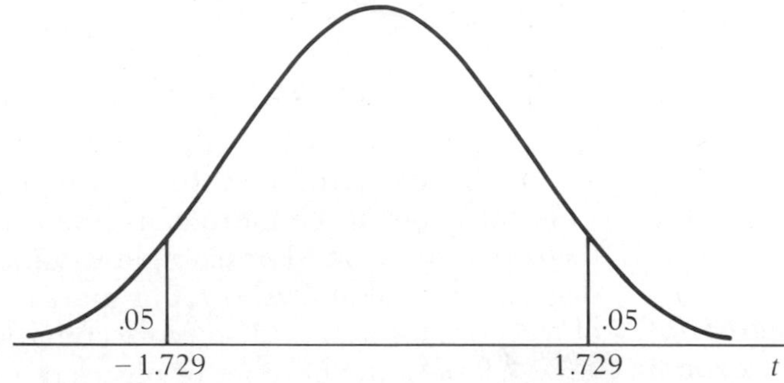

Figure 10.2 *t table value for 19 df*

Example 10.2

For 28 df, find t-table values that include the middle 98% of probability.

Solution

For probability .98 in the middle, there must be probability .01 in each tail. The .01 entry in the t table (Table 4) for 28 df is 2.467.

$$P(-2.467 < t < 2.467) = .98$$

□

In Table 4, the t-distribution cutoff points for a given area a decrease as the degrees of freedom increase. In the limit, as the degrees of freedom become infinite, the t numbers approach the corresponding numbers for the z table. This is certainly a reasonable result. The difference between t and z statistics is in the use of an estimated standard error rather than a true standard error. When the degrees of freedom are very large, the estimated standard error and true standard error will be virtually identical, so t and z probabilities will be virtually identical. Note that Table 4 only goes up to 120 degrees of freedom. For any larger df, the z (normal) values are adequate approximations. Some degrees of freedom between 30 and 120 are not shown in Table 4. Linear interpolation is adequate for most purposes to obtain an appropriate t value; alternatively, many computer programs are able to calculate t probabilities for any df.

Example 10.3 Find approximate t values cutting off a right-tail area of .025 for 50 df, 55 df, and 150 df.

Solution In Table 4 we find that 2.021 cuts off .025 for 40 df and 2.000 cuts off .025 for 60 df. Because 50 df is halfway between 40 and 60 df, take the t value to be the (unweighted) average of the table entries:

$$t_{.025,50\,\mathrm{df}} = \frac{2.021 + 2.000}{2} = 2.0105$$

perhaps rounded up to 2.011. (There's no reason to worry about exactly which digit should be in the third decimal place.) For 55 df, the following sketch might be useful:

40	55	60	df
2.021	2.005	2.000	$t_{.025}$

Note that 55 df is 15/20 of the way from 40 to 60 df. Thus we want to find a t number that is 15/20 of the way from 2.021 to 2.000. One way to do this is to take a weighted average.

$$t_{.025,50\mathrm{df}} = \frac{5}{20}2.021 + \frac{15}{20}2.000 = 2.005$$

For 150 df, we might as well use the z number, 1.960. Any error will be negligible in most situations. ☐

The theoretical basis of the t distribution is sketched in the first appendix to this chapter. The t distribution is fundamental to many of the applicable statistical methods described hereafter. In Section 10.2, we turn to the most basic of these methods.

Exercises for Section 10.1

10.1. Find the t-table values corresponding to a right-tail area of .025 for 10, 20, and 30 df. What is the effect on the table values of increasing df?

10.2. How do t-table values for 120 df compare to the corresponding z-table values? For what tail areas are the values farthest apart?

10.3. Suppose that, for a sample of 245 subjects, you obtained a sample mean of 55.2 and a sample standard deviation of 10.1. To calculate a 95% confidence interval for the population mean, you would, in principle, need a t-table value for 244 df. How much of an error would you be making if you used a z-table value instead?

10.4. Interpolate in the t table to find the value cutting off a right-tail area of .01 for 72 df.

10.2 Confidence Intervals for a Mean

Chapter 8 introduced methods for calculating a confidence interval for a mean, based on a z statistic. At that time, we had to make the artificial assumption that σ_{pop} was known, so that we could use the true standard error of the sample mean,

$$\sigma_{\bar{y}} = \frac{\sigma_{\text{pop}}}{\sqrt{n}}$$

In practice, it is almost always the case that σ_{pop} is unknown, so we must use the estimated standard error $s/\sqrt{n}$ instead. Thus our confidence interval is usually based on a t statistic and we use t tables rather than z tables.

100(1 − α)% Confidence Interval for a Mean (σ_{pop} Unknown)

$$\bar{y} - t_{\alpha/2}\frac{s}{\sqrt{n}} < \mu_{\text{pop}} < \bar{y} + t_{\alpha/2}\frac{s}{\sqrt{n}}$$

NOTE: $t_{\alpha/2}$ is the value found in Table 4 that cuts off a right-tail area $a = \alpha/2$.

Note that this confidence interval has the form "statistic plus or minus table value times standard error." The difference between this interval and the one found in Chapter 8 is that this interval uses the estimated standard error, and therefore uses the t table.

For example, suppose that a random sample of 20 college students, all female, are asked to give a rating to the statement "The fact that a candidate for student body president is an avowed homosexual is irrelevant to the candidate's ability to do the job." Responses range from 0 = "Homosexuality absolutely disqualifies the candidate" to

100 = "Homosexuality is a major positive reason to elect the candidate." The responses (sorted in increasing order) are

12 21 30 41 43 43 44 44 46 47
47 48 50 50 50 51 53 65 76 90
($\bar{y} = 47.55, s = 16.80$)

To find a 95% confidence interval for μ_{pop}, we need a t-table number for $20 - 1 = 19$ df, cutting off a right-tail area of $.05/2 = .025$. From Table 4, that number is 2.093. The estimated standard error of the sample mean is $16.80/\sqrt{20} = 3.76$. The 95% confidence interval is

$$47.55 - 2.093(3.76) < \mu_{pop} < 47.55 + 2.093(3.76)$$

or

$$39.7 < \mu_{pop} < 55.4$$

Example 10.4 An experiment with a sample of 18 nursery-school children involved the elapsed time required to put together a small jigsaw puzzle. The times were

3.1 3.2 3.4 3.6 3.7 4.2 4.3 4.5 4.7
5.2 5.6 6.0 6.1 6.6 7.3 8.2 10.8 13.6
(mean 5.783, standard deviation 2.784)

Calculate a 99% confidence interval for the population mean.

Solution We need a t-table number for a right-tail area of .005, corresponding to a 99% confidence level. There are $18 - 1 = 17$ df; the t table indicates that the required number is 2.898. The 99% confidence interval is therefore

$$5.783 - 2.898\frac{2.784}{\sqrt{18}} < \mu_{pop} < 5.783 + 2.898\frac{2.784}{\sqrt{18}}$$

or

$$3.88 < \mu_{pop} < 7.68$$

□

The interpretation of a t confidence interval is the same as that of any other confidence interval. A 95% confidence interval is constructed so that, in the long run, 95% of the intervals will include the true value of the population parameter.

Like any other statistical procedure, the t confidence interval is based on certain assumptions and should not be used when those assumptions are wrong. The formal mathematical assumption is that the data arise from a simple random sample from a population that has exactly a normal distribution of values. As always, it is important to exclude biases from the data gathering, as far as can possibly be done. The assumption of a normal population is also critical. If the underlying population is badly skewed, the actual theoretical distribution of the t statistic will not in fact be t. The Central Limit Theorem works for t statistics just as for z statistics. Thus, if the population distribution is symmetric (or nearly so), t probabilities are quite accurate approximations, for sample sizes as small as 15 or so. If there is clear but not overwhelming skewness in the population, a sample size of at least 40 makes t probabilities adequate. If the population is violently skewed, a sample size of 100 or more may be needed before t probabilities become accurate approximations. Like z probabilities, two-tailed probabilities (as are used in the plus-or-minus confidence interval) are better approximations than are one-tailed probabilities.

A reasonable way to check the normal-population assumption is to plot the sample data. The shape of the sample data will reflect the shape of the population, at least roughly. Severe skewness with a small sample size is a signal that the nominal 95% (or whatever) confidence level may be a poor approximation. The data for the campus survey can be pictured using a stem and leaf display, as shown in Figure 10.3.

```
1 | 2
2 | 1
3 | 0
4 | 1  3  3  4  4  6  7  7  8
5 | 0  0  0  1  3
6 | 5
7 | 6
8 |
9 | 0
```

Figure 10.3 Stem and leaf display of campus survey data

The data appear reasonably symmetric, and the sample size is 20. Thus the confidence level of the interval should be just about 95%.

Example 10.5 Plot the data of Example 10.4. Is there any indication that the supposed 99% confidence interval might not be a good approximation?

Solution

A histogram or stem and leaf display of the data (or, indeed, a careful look at the ordered data) indicates that the data are right-skewed. The sample size is only 18. Thus, the use of t probabilities is quite dubious. □

inefficient

Another possible violation of the normal-population assumption is heavy-tailness (outlier proneness). When the population is heavy-tailed, the sample mean may be **inefficient** as an estimator of the population mean. In that case, the plus-or-minus of the t confidence interval may be unnecessarily wide. This problem has nothing to do with sample size—an estimator that is inefficient for $n = 20$ will be just about as inefficient for $n = 200$. In the language of Chapter 8, the *nominal robustness* of the t interval improves as n increases, but the *efficiency robustness* of the t interval does not. The stem and leaf display of the survey data, shown in Figure 10.3, indicates that most of the data are concentrated in the 40's and 50's, with a few "stragglers" off each end. The 1.5 IQR rule of thumb described in Chapter 2 indicates that the values 12, 21, 30, 65, 76, and 90 are all candidate outliers. Looking at the normal probability plot described in Chapter 8 is another good way to detect heavy-tailness. The normal probability plot of the survey data shown in Figure 10.4 has a pronounced S-shape to it; this S-shape is characteristic of heavy-tailed data. Thus, the t interval does not seem to be an efficient use of the survey data.

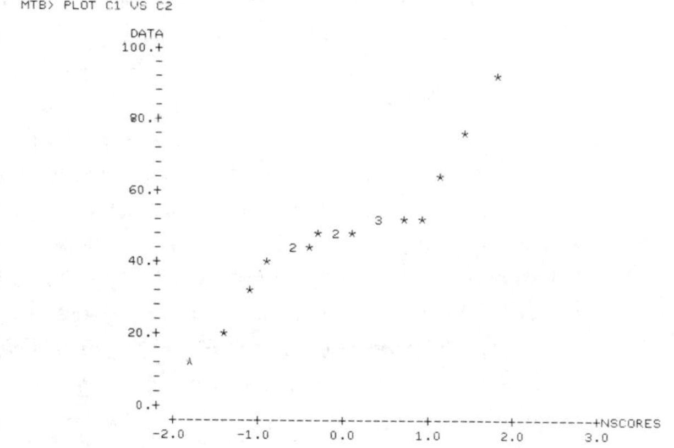

Figure 10.4 Normal plot for campus survey data

Example 10.6

Obtain a normal probability plot of the data in Example 10.4. Do the data appear to have arisen from a normal population distribution?

Solution In Example 10.5, we noted that the data were right-skewed. A normal probability plot would show a curve that clearly differed from the straight-line pattern of normally distributed data. □

There are several alternatives to the t interval for use with heavy-tailed, outlier-prone data. Some of these are based on trimmed means or other estimators, and require special-purpose computer programs. A simple alternative is to calculate a confidence interval for the median, as discussed in Section 10.5.

The t interval can be used to determine the required sample size for a study, with slightly more difficulty than the z interval of Chapter 8. A desired width (plus-or-minus) and confidence level can be specified, and the required n calculated. The minor difficulties are that the width of the interval depends on the sample standard deviation s (which, of course, won't be known until the sample is taken) and that the sample size affects not only the (estimated) standard error, but also the t-table value, because the required df $= n - 1$. The sample standard deviation can be guessed at in advance, perhaps based on prior studies, or the width of the interval can be expressed as a fraction of s. The df problem can be handled by making the tentative assumption that the resulting n will be large enough to justify use of the appropriate z-table value (e.g., 1.960 for 95%) as an approximate t-table value. The required n may be calculated by setting the actual plus-or-minus equal to the desired plus-or-minus, E, and solving for n:

$$z_{\alpha/2}\frac{s}{\sqrt{n}} = E$$

$$n = \frac{(z_{\alpha/2})^2 s^2}{E^2}$$

If the resulting n is not larger than 120, increase it by 1 or 2 and calculate the resulting plus-or-minus. Keep going until the required n is found.

For example, suppose that we planned to take a new survey of women regarding homosexuality as an issue in campus elections. Suppose that we wanted a 90% confidence interval accurate to 1/4 of a (sample) standard deviation. The value $z_{.10/2} = 1.645$ from Table 3 or the "inf." line of Table 4. The first approximation to the required n is

$$n = \frac{(1.645)^2 s^2}{(.25s)^2} = 43.3$$

rounded up to 44. For $n = 44$, t-table values should still be used, so we increase n to, say, 45. The t-table value for 44 df, right-tail area .05, can be interpolated roughly as

$$\frac{16}{20}1.684 + \frac{4}{20}1.671 = 1.681$$

For $n = 45$, the plus-or-minus would be

$$\frac{1.681s}{\sqrt{45}} = 0.2506s$$

This value is very close to the desired $0.2500s$. If one felt absolutely compelled to get the plus-or-minus under $0.25s$, one could use $n = 46$. Note that the t interval is likely to be inefficient in this situation, because the population seems outlier-prone. An alternative interval might well allow use of a smaller sample to obtain the same accuracy.

Example 10.7 For the situation of Example 10.4, what sample size is required to yield a 95% confidence interval with a plus-or-minus of .5 minutes? Assume that the sample standard deviation will not change.

Solution The sample standard deviation was 2.784. We initially assume that n will be so large that the appropriate normal table point (t with infinite df) will be a good approximation. For 95% confidence the required table value is 1.96. Set the actual plus-or-minus equal to the desired value and solve for n:

$$1.96\frac{2.784}{\sqrt{n}} = .05$$

$$\sqrt{n} = 1.96\frac{2.784}{0.5} = 10.913$$

or $n = (10.913)^2 = 119$. For this n, we could use the t-table value with 120 df (close to the actual 118 df), namely 1.98. The resulting plus-or-minus is .05053, a bit too large. With $n = 121$, the plus-or-minus is

$$1.98\frac{2.784}{\sqrt{121}} = 0.501$$

Although one could keep on trying increased n values, $n =$ about 120 would seem to be adequate. □

One-sided (lower-bound or upper-bound) t intervals can also be constructed. Merely take the definitions in Section 8.3, replace the true standard error $\sigma_{pop}/\sqrt{n}$ by the estimated standard error $s/\sqrt{n}$, and use t tables instead of z tables. Note that these intervals use one-tailed probabilities, which are poorer approximations than two-tailed probabilities when the population is seriously skewed.

Example 10.8

Refer to the data of Example 10.4. Calculate a 95% lower-bound confidence interval for the population mean.

Solution

The table value for left-tail area .05 and 18 − 1 = 17 df is 1.740. The interval is

$$\mu_{\text{pop}} > 5.783 - 1.740\frac{2.784}{\sqrt{18}}$$

or

$$\mu_{\text{pop}} > 4.64$$

□

Example 10.9

Refer to Examples 10.4 and 10.8. How believable is the supposed 95% confidence level?

Solution

Not very. The sample data are right-skewed, suggesting a right-skewed population. The sample size is small, and the probability is one-tailed. The confidence level could be in error. □

Exercises for Section 10.2

10.5. Plomin, Foch, and Rowe (1981) report on a study of children's aggressive behavior. Each child was instructed that it was okay to hit a Bobo the Clown doll. The elapsed time before each child hit the doll was recorded. For $n = 216$, the mean was 3.1 and the standard deviation was 2.4. Calculate a 90% confidence interval for the true, population mean.

10.6. Latency (waiting) times tend to be severely right-skewed. In fact, the mean and standard deviation reported in Exercise 10.5 suggest right-skewness; it is impossible to have negative waiting times, yet the standard deviation is almost as big as the mean. Does this fact indicate that the confidence level of the interval calculated in Exercise 10.5 will in fact be greatly different from 90%?

10.7. In a study of the self-concept of regular practitioners of Transcendental Meditation, a random sample of 11 regular meditators was obtained. The self-concept scores were

21.1 36.4 40.3 41.5 43.1 45.3 46.9 48.1 49.7 50.0 51.6
($\bar{y} = 43.09$, $s = 8.64$)

Calculate a 95% confidence interval for the population mean.

10.8. Calculate a 95% upper-bound confidence interval for the true (population) mean, for the data of Exercise 10.7.

10.9. For the data of Exercise 10.7, is there any indication of a violation of assumptions? If so, which confidence level is a poorer approximation, the one in Exercise 10.7 or the one in Exercise 10.8?

10.3 Testing Hypotheses about a Mean Using *t*

In Chapter 9, hypothesis tests about a mean involved the unrealistic assumption that σ_{pop} was known when μ_{pop} was unknown. In this section we remove that assumption. Not surprisingly, the true standard error $\sigma_{pop}/\sqrt{n}$ will be replaced by the estimated standard error $s/\sqrt{n}$, and t tables will be used instead of z tables.

Suppose that an experimental program to improve sixth-graders' language skills has been developed, and it is to be tested on a random sample of 30 sixth-graders in a large district. The children take a pre-test measuring correct understanding of sentence structure, go through the experimental program, then take a post-test. The data are the changes (post-test − pre-test) in test scores. The design can be criticized. Even if the average change is clearly positive, the change may not be due to the program. It may be due to familiarity with the test, or merely to increased maturity over the two-month span of the project. Further, some other program might have resulted in an even better improvement. Suppose that the changes are

−16	−7	−3	−3	−2	−2	−1	0	1	3
3	3	4	6	6	6	7	8	8	9
10	12	14	14	16	16	19	20	23	29

(mean 6.767, standard deviation 9.555)

Is the sample mean too far above 0 to be attributed to sheer random variation?

The problem may be formulated in the five steps of a hypothesis test. The natural null hypothesis is that the population mean change is

0; H_0: $\mu_{pop} = 0$. In this case, the intent is to improve the average score, so one might well take the one-sided research hypothesis H_a: $\mu_{pop} > 0$. We no longer can use z as a test statistic, because we no longer assume that σ_{pop} is known. Using the estimated standard error of the sample mean

$$s_{\bar{y}} = \frac{s}{\sqrt{n}}$$

we obtain a t statistic as our T.S. The rejection region therefore will be based on a t-table value. If the study uses $\alpha = .01$, we will need the t-table (Table 4) value cutting off a one-tail area of .01 for $n - 1 = 29$ df, namely 2.462. The five parts of the test can be shown, as in the box below.

Hypothesis Test about a Population Mean (σ_{pop} Unknown)

	General	*Example*
1.	H_0: $\mu_{pop} = \mu_0$	H_0: $\mu_{pop} = 0$
2.	H_a: (i) $\mu_{pop} > \mu_0$ (ii) $\mu_{pop} < \mu_0$ (iii) $\mu_{pop} \neq \mu_0$	H_a: $\mu_{pop} > 0$
3.	T.S.: $t = \dfrac{\bar{Y} - \mu_0}{s/\sqrt{n}}$	T.S.: $t = \dfrac{\bar{Y} - 0}{s/\sqrt{30}}$
4.	R.R.: (i) $t > t_\alpha$ (ii) $t < -t_\alpha$ (iii) $\lvert t\rvert > t_{\alpha/2}$	R.R. ($\alpha = .01$): $t > 2.462$
5.	Conclusion	Conclusion: $t = \dfrac{6.767 - 0}{9.555/\sqrt{30}} = 3.88$; reject H_0

Note that the only difference between the t test of a mean and a z test is that the estimated standard error replaces the true standard error, necessitating the use of t tables.

Example 10.10 Touliatos and Bedeian (1982) reported a study of achievement motivation in children. A sample of 67 children did a task and then could choose another version of the same task (scored from easiest, 0, to hardest, 4). Each child's score was the sum of the scores on four tasks. The sample mean was 2.04, and the sample standard deviation was 1.26. A simple-minded null hypothesis is that each child would choose a moderate (score 2) version of the same task, achieving on average a score of 8. Can this null hypothesis be rejected?

Solution

The test statistic is

$$t = \frac{2.04 - 8}{1.26/\sqrt{67}} = -38.71$$

We don't have t tables for $67 - 1 = 66$ df, but by comparison with, say, 60 df, it's obvious that the value of the statistic is far, far beyond the table value. The null hypothesis is rejected at any reasonable α value.

□

Like the t confidence interval, the t test is based on the assumption of a normal population distribution. The change-in-understanding data, when plotted, appear reasonably near symmetric, with no outliers. Thus the normal assumption appears reasonable in this case. If the sample data are skewed, suggesting that the underlying population is also skewed, the nominal probability may be a poor approximation, particularly for one-tailed tests with small n. If the sample data show outliers, suggesting a heavy-tailed population, the t test may not be efficient. Inefficiency in a hypothesis test indicates an unnecessarily high β probability for a specified α. In such a case, one might perhaps test the median instead of the mean. The median test is described in Section 10.5. One might also base the test on a trimmed mean, if one had the special-purpose computer program required to compute a standard error.

Example 10.11

To assess the effects of nonnormal populations on the t test, two computer simulations were performed. In the first simulation, 1000 samples of size 30 were drawn from a substantially right-skewed population having a mean of 50. In the second simulation, 1000 samples of size 30 were drawn from a moderately outlier-prone population also with mean 50. In both simulations, the number of t tests that rejected H_0: $\mu = 50$ at various α values was obtained. The results in Table 10.1 were obtained.

What do these results indicate about the sensitivity of the nominal α value in a t test to nonnormality of the underlying population?

Population shape is right skewed.

	number of times H0: "mean is 50" is rejected in favor of		
alpha	"mean > 50"	"mean < 50"	total (alpha doubled)
0.100	40	237	277
0.050	6	189	195
0.025	0	156	156
0.010	0	120	120
0.005	0	99	99

average t is -0.5794 with variance of 2.780990

Population shape is moderately outlier prone.

	number of times H0: "mean is 50" is rejected in favor of		
alpha	"mean > 50"	"mean < 50"	total (alpha doubled)
0.100	104	95	199
0.050	51	51	102
0.025	28	24	52
0.010	7	6	13
0.005	4	3	7

average t is 0.0077 with variance of 1.086943

Table 10.1

Solution

First, note that the simulations were done with $\mu = 50$, that is, with H_0 being true. Thus, every time H_0 was rejected, a Type I (false positive) error was committed. The expected number of such errors is 1000 α; when the nominal $\alpha = .100$, we expect $1000(.100) = 100$ errors in the simulation. For the outlier-prone population, the actual numbers are very close to expected. For example, in the .025 row, the actual frequencies are 28 and 24 (total 52) compared to the expected 25 and 25 (total 50). Thus, for this outlier-prone population, the nominal error rate appears to be a very close approximation. But for the skewed population, the observed error rates are very far from the nominal rates. Thus, for a sample of size 30 from this population, the nominal α is a very poor approximation. □

Chapter 9 introduced the p-value, which acts as a conclusiveness index, indicating how conclusively we reject the null hypothesis and therefore support the research hypothesis. In Chapter 9 we could compute the p-value merely by looking up the observed z statistic value in the z table. Our t table is nowhere near as extensive as the z table. For any given df, we have only five numbers, instead of a full-page table. Therefore, using the table, we can only put bounds on the p-value. Most computer programs that calculate a t statistic also calculate a p-value. The actually observed value of the t statistic, t_{obs}, should (if possible) be bracketed between two t-table numbers; then the one-tailed p-value is between the corresponding areas. In our first example $t = 3.88$ with 29 df. The largest t-table value for 29 df is 2.756, corresponding to a tail area of .005. Thus the tail area beyond 3.88 (that is, the one-tailed p-value) is something less than .005. Had the t statistic come out equal to 2.640, it would have been bracketed between t-table numbers 2.462 and 2.756 (areas .01 and .005). Thus the p-value would have been between .005 and .01. See Figure 10.5. For a two-tailed test, double the indicated areas.

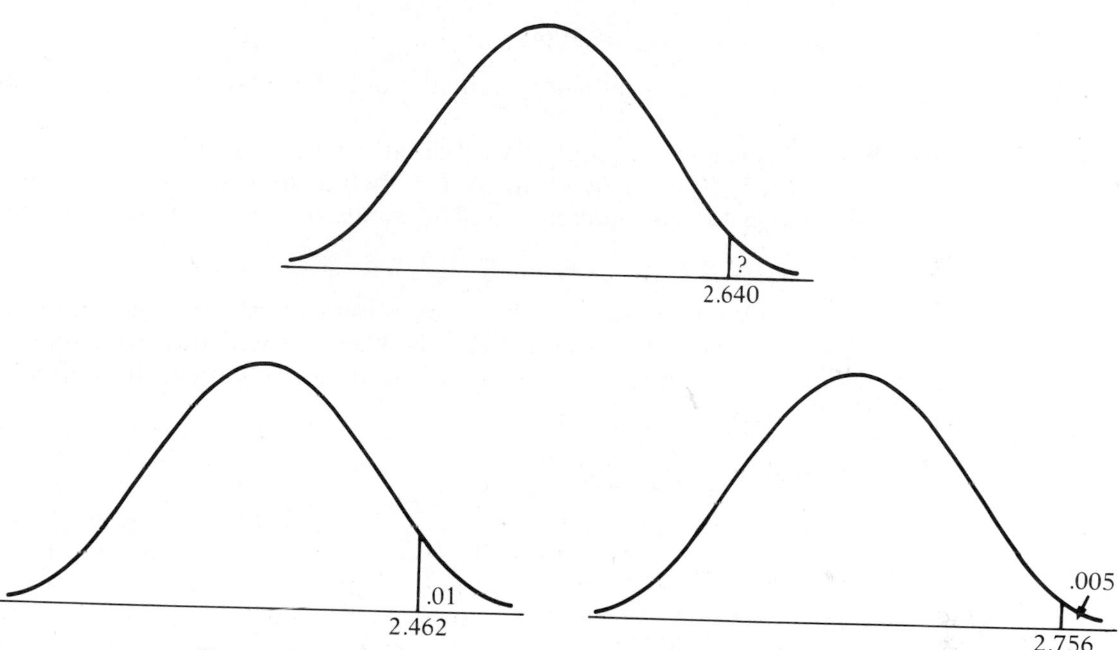

Figure 10.5 Bracketing p-values for a t test

Example 10.12 Refer to Example 10.10. Place bounds on the p-value, assuming a two-tailed test.

Solution The test statistic came out $t = -38.7$, which is far beyond the largest value in the table for 66 df (read as the closest value, 60 df, or interpolated between 60 and 120 df), corresponding to the smallest one-tailed area, .005. Thus the two-tailed p-value is (much) less than $2(.005) = .01$. □

As discussed in Chapter 9, there is a relation between hypothesis tests and confidence intervals. A null hypothesis value that falls outside a $100(1 - \alpha)\%$ confidence interval may be rejected at the indicated α level (.05 for 95% confidence, etc.). One-sided confidence intervals may be used in conjunction with one-tailed tests.

Example 10.13

Suppose that a 95% confidence interval for a population mean, based on a sample of size 25, is

$$88.2 < \mu_{pop} < 106.8$$

What does this indicate about the null hypothesis H_0: $\mu_{pop} = 100$?

Solution

The null hypothesis value falls within the confidence interval. At $\alpha = .05$, H_0 must be retained. For such a small sample size, it would be unwise to really *accept* H_0. The confidence interval is wide for small n. □

We now have the basic principles of inference based on a single random sample, using t methods. Next we will turn to a scientifically important application of the ideas in this chapter, the paired-sample experiment.

Exercises for Section 10.3

10.10. Refer to Exercise 10.5. Test the null hypothesis that the population mean is 3.5, against a two-sided alternative. Use $\alpha = .10$.

10.11. What is the p-value for the test of Exercise 10.10?

10.12. Does the conclusion of the test in Exercise 10.10 agree with the confidence interval found in Exercise 10.5? Explain why it should agree.

10.13. Refer to Exercise 10.7. Test H_0: $\mu_{pop} = 48$ vs. H_a: $\mu_{pop} < 48$, using $\alpha = .05$.

10.14. Put bounds on the p-value for the test of Exercise 10.13. Can you be confident that the p-value is a good approximation?

10.4 Paired-Sample Experiments

The most basic social science experiment involves the comparison of a treatment group to a control group. There are many reasons for using a control group as a comparison standard. For example, consider the experiment of the previous section, where we measured the change in verbal understanding of sixth-grade students after they went through an experimental program. We discovered that the mean change was significantly positive (in both the statistical and the practical sense). But we really can't attribute that change entirely to the effect of the program. The effects of that program are **confounded** with (mixed up with) the effects of other learning that took place over the two-month span of the program, and with the fact that the students had some experience with the verbal understanding exam. One major reason for using a control group as a comparison standard is to eliminate confounding as much as possible.

confounded

In many experiments, subjects are paired or matched on the basis of some characteristic, then randomly assigned to treatment or control. *An experiment is a* **paired-sample experiment** *if each score in the control group is related to one specific score in the experimental group, by the design of the experiment.* For instance, suppose that researchers wanted to see if a change in writing style in psychology journal articles improved readability. The treatment group read journal articles that had been rewritten to minimize jargon and reduce the length of sentences and paragraphs. The control group read the original articles. The data were the times required by all subjects (senior psychology majors) to read the articles. One way to do the experiment would be to assign 15 subjects to treatment and the other 15 to control, completely at random. Methods for analyzing such experiments are discussed in the next chapter. Another way would be to match subjects on the basis of a pre-test of reading speed. The two slowest readers would form one pair, the next two slowest would form another pair, on up to the two fastest readers. Then one subject in each pair would be randomly assigned to treatment, the other to control. This is a paired-sample experiment, because each control-group score is related to a score in the treatment group, namely the score of the subject in the same initial reading-speed pair.

paired-sample experiment

Example 10.14

Croft (1982) reports on a study of alternative ways to test spelling ability. Suppose that 24 children are tested using both the traditional dictation test and also a proofreading task (where the problem is to find all misspellings). Suppose that the data are as shown in Table 10.2. What is the basis of pairing in this example?

	Student											
	1	2	3	4	5	6	7	8	9	10	11	12
Traditional	10	8	13	15	19	20	19	17	20	22	18	15
Proofreading	8	12	18	16	15	22	24	11	30	29	24	22

	Student											
	13	14	15	16	17	18	19	20	21	22	23	24
Traditional	18	20	25	24	26	31	36	40	43	37	41	46
Proofreading	19	24	22	32	30	39	32	42	40	46	43	50

Table 10.2

Solution The scores are paired by student. Each student has both a traditional score and a proofreading score. ☐

The analysis of paired-sample experiments is based on the *differences* of pair scores. In the readability illustration, the analysis would begin with computation of the difference between the time required by a control-group subject and the time required by the matching treatment-group subject. Suppose that the times (in minutes) had been as shown in Table 10.3.

	Pairs														
	1	2	3	4	5	6	7	8	9	10	11	12	13	14	15
Control	309	298	264	281	265	220	247	251	223	208	201	187	199	203	176
Treatment	287	306	280	239	245	229	226	233	210	197	199	183	190	186	183
C − T	22	−8	−16	42	20	−9	21	18	13	11	12	4	9	17	−7

Table 10.3

By working with the differences, we reduce the problem to a single-sample problem. If we're interested in making inferences about the mean difference, as we often are, the t methods of the previous sections may be used on the difference data.

To obtain a 95% confidence interval, we need the mean $\bar{d}$ and standard deviation s_{diff} of the difference scores. The symbol $\bar{d}$ is used instead of $\bar{y}$ merely as a reminder that the data are in fact differences. In the readability differences, $\bar{d} = 9.9333$ and $s_{\text{diff}} = 15.126$. There are 15 data points (differences), so there are 14 df. The required t-table value (14 df, right-tail area $.05/2 = .025$) is 2.145. The estimated standard error is

$$\frac{s_{\text{diff}}}{\sqrt{n}} = \frac{15.126}{\sqrt{15}} = 3.906$$

The 95% confidence interval is

$$9.333 - 2.145(3.906) < \mu_{\text{diff}} < 9.333 + 2.145(3.906)$$

or

$$0.955 < \mu_{\text{diff}} < 17.711$$

The method is shown in the box below.

100(1 − α)% Confidence Interval for a Mean Difference

$$\bar{d} - t_{\alpha/2}\frac{s_{\text{diff}}}{\sqrt{n}} < \mu_{\text{diff}} < \bar{d} + t_{\alpha/2}\frac{s_{\text{diff}}}{\sqrt{n}}$$

NOTE: n is the number of pairs (differences). $t_{\alpha/2}$ is the Table 4 value cutting off a right-tail area equal to $\alpha/2$.

Example 10.15 Calculate a 99% confidence interval for the mean difference, based on the data of Example 10.14.

Solution The (proofreading − traditional) differences are

Student	1	2	3	4	5	6	7	8	9	10	11	12
Difference	−2	4	5	1	−4	2	5	−6	10	7	6	7
Student	13	14	15	16	17	18	19	20	21	22	23	24
Difference	1	4	−3	8	4	8	−4	2	−3	9	2	4

(mean 2.79167, standard deviation 4.53948)

For df = 24 − 1 = 23, the right-tail .005 cutoff point is found from the t table to be 2.807. Thus the 99% confidence interval is

$$2.79167 - 2.807\frac{4.53948}{\sqrt{24}} < \mu_{\text{diff}} < 2.79167 + 2.807\frac{4.53948}{\sqrt{24}}$$

or

$$0.19 < \mu_{\text{diff}} < 5.39$$

□

The t test of Section 10.3 can be applied to paired-sample experiments. The typical null hypothesis is H_0: $\mu_{\text{diff}} = 0$. Any value D_0 may be used instead of 0. The research hypothesis H_a can be one-sided or two-sided. The test statistic is written as

$$t = \frac{\bar{d} - D_0}{s_{\text{diff}}/\sqrt{n}}$$

but the test is exactly that of Section 10.3. It is recapitulated in the box below.

Testing a Mean Difference

1. H_0: $\mu_{\text{diff}} = D_0$ (Usually, $D_0 = 0$.)
2. H_a:
   (i) $\mu_{\text{diff}} > D_0$
   (ii) $\mu_{\text{diff}} < D_0$
   (iii) $\mu_{\text{diff}} \neq D_0$
3. T.S.: $t = \dfrac{\bar{d} - D_0}{s_{\text{diff}}/\sqrt{n}}$
4. R.R.:
   (i) $t > t_\alpha$
   (ii) $t < -t_\alpha$
   (iii) $|t| > t_{\alpha/2}$
5. Conclusion

In the readability illustration, we take H_0: $\mu_{\text{diff}} = 0$, the hypothesis that rewriting the articles makes no difference in average reading speed. The test statistic is

$$t = \frac{9.333 - 0}{15.126/\sqrt{15}} = 2.390$$

For a two-tailed test with $\alpha = .05$, the critical t-table value is $t_{.025} = 2.145$. Because $2.390 > 2.145$, H_0 is rejected and we conclude that the mean difference is not 0. In practice, we conclude that the mean difference is positive; the probability (given that the actual population mean difference is negative) of obtaining such a t statistic is very small. Note that we could also have rejected H_0 by noting that $\mu_{\text{diff}} = 0$ is not included in the previously calculated 95% confidence interval.

Example 10.16 For the difference data of Example 10.15, is the research hypothesis that the true (population) mean of the differences is positive conclusively supported? Use $\alpha = .005$.

Solution We found in Example 10.15 that $\bar{d} = 2.79167$ and $s_{\text{diff}} = 4.53948$. Thus

$$t = \frac{2.79167 - 0}{4.53948/\sqrt{24}} = 3.013$$

a value larger than the .005 value for 23 df, 2.807. Thus H_a is supported. □

The p-value for a t test in a paired-sample experiment can be obtained by the bracketing process explained in Section 10.3. In the readability illustration, the observed $t = 2.390$ is bracketed between $t_{.025} = 2.145$ and $t_{.01} = 2.624$. Thus the one-tailed p-value is between .01 and .025, and the two-tailed p-value is between .02 and .05. See Figure 10.6.

Example 10.17 What can be said about the p-value for Example 10.16?

Solution In Example 10.16 we found that H_0 could be rejected at $\alpha = .005$ (one-tailed), so the one-tailed p-value is less than .005. Because the t table contains no smaller tail areas, we can say no more. □

As always, it's important to check assumptions. A key assumption is that the population of *difference* scores is approximately normal. Skewness and heavy-tailness of the population of difference scores cause the same problems in paired-sample experiments as in other t tests. Skewness (particularly with small n) makes the t probabilities suspect; heavy-tailness makes the t test inefficient. A plot of the readability differences indicates that the data aren't terribly skewed, but might possibly have some outlier-proneness.

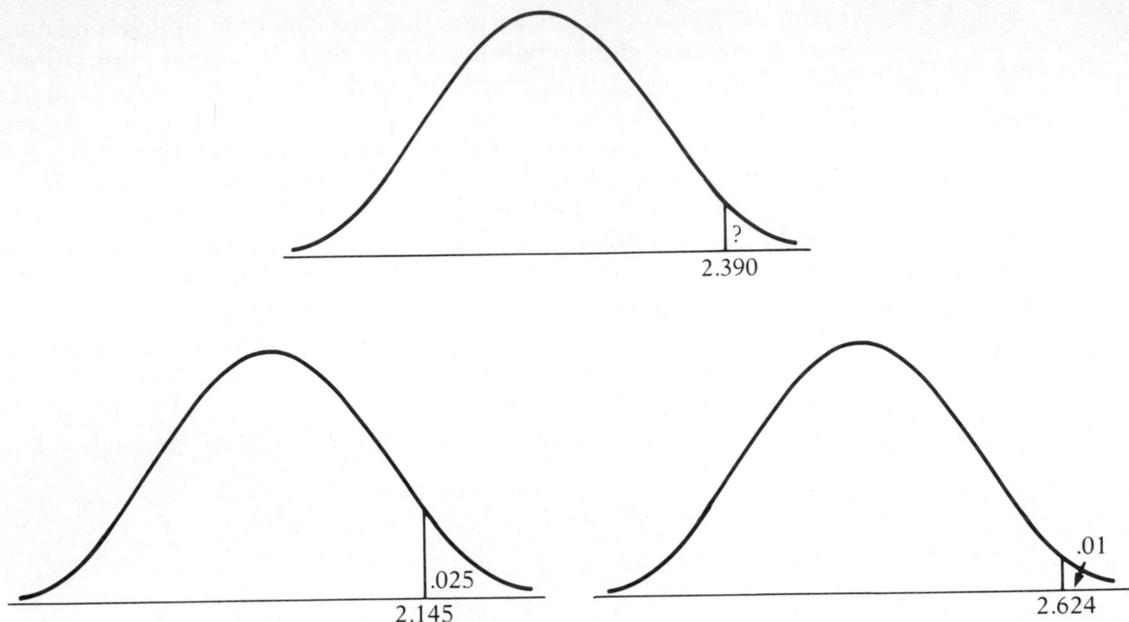

Figure 10.6 *p*-value for a *t* test

Example 10.18 For the differences of Example 10.15, are there any gross violations of the normality assumption?

Solution No. A plot of the differences indicates no excessive skewness and no great outliers. It appears that the data come from a tolerably near normal population. □

signed-rank test

Previous sections of this chapter have suggested testing the median if one is faced with skewness or outliers. That can be done with difference scores also. Alternatively, a test called the Wilcoxon **signed-rank test** may be carried out. As the name indicates, this test is based on ranking the difference scores. The scores are ranked from lowest absolute value (1) to highest absolute value (n). Each rank is given the sign of the corresponding difference. Any 0 differences are discarded before ranking, and the sample size n reduced. The differences from the readability data are shown below, along with the first few signed ranks.

Pair	1	2	3	4	5	6	7	8	9	10	11	12	13	14	15
Difference	22	−8	−16	42	20	−9	21	18	13	11	12	4	9	17	−7
Signed Rank		−3										1			−2

Thus 4 is the smallest difference (in magnitude). It is assigned rank 1 and a plus sign. The next smallest magnitude difference is −7, which gets signed rank −2; −8 gets signed rank −3. There is a tie between −9 and 9 for ranks 4 and 5. In case of ties, assign all tied scores the **midrank,** the average of the ranks in question. Thus our −9 and 9 differences are ranked (4 + 5)/2 = 4.5, with appropriate sign. Had the −8 been a −9, creating a three-way tie for ranks 3, 4, and 5, all three scores would have been assigned a rank of (3 + 4 + 5)/3 = 4, with appropriate sign. The full set of signed ranks is shown below.

midrank

Pair	1	2	3	4	5	6	7	8	9	10	11	12	13	14	15
Difference	22	−8	−16	42	20	−9	21	18	13	11	12	4	9	17	−7
Signed Rank	14	−3	−9	15	12	−4.5	13	11	8	6	7	1	4.5	10	−2

Example 10.19 Find the signed ranks for Example 10.15.

Solution There are no 0 differences to discard.

Student	1	2	3	4	5	6	7	8	9	10	11	12
Difference	−2	4	5	1	−4	2	5	−6	10	7	6	7
Signed Rank	−4.5	11.5	15.5	1.5	−11.5	4.5	15.5	−17.5	24	19.5	17.5	19.5

Student	13	14	15	16	17	18	19	20	21	22	23	24
Difference	1	4	−3	8	4	8	−4	2	−3	9	2	4
Signed Rank	1.5	11.5	−7.5	21.5	11.5	21.5	−11.5	4.5	−7.5	23	4.5	11.5

The Wilcoxon signed-rank test is based on the sums of the negative ranks T_- and of the positive ranks T_+. In the readability example,

$$T_- = -3 + -9 + -4.5 + -2 = -18.5$$

and

$$T_+ = 14 + 15 + 12 + 13 + 11 + 8 + 6 + 7 + 1 + 4.5 + 10 = 101.5$$

If there are no negative ranks, take $T_- = 0$, and similarly for T_+. The remainder of the test procedure is outlined in the box.

Wilcoxon Signed-Rank Test

1. H_0: $\mu_{\text{diff}} = 0$
2. H_a:
   (i) $\mu_{\text{diff}} > 0$
   (ii) $\mu_{\text{diff}} < 0$
   (iii) $\mu_{\text{diff}} \neq 0$
3. T.S.:
   (i) $T = |T_-|$
   (ii) $T = T_+$
   (iii) T = smaller of T_+, $|T_-|$
4. R.R.: if $n \leq 50$, reject H_0 at the given α level if $T \leq$ the value found in Table 5; if $n > 50$, compute

$$z = \frac{T - n(n + 1)/4}{\sqrt{n(n + 1)(2n + 1)/24}}$$

and refer the result to z tables (Table 3).

NOTE: If H_0: $\mu_{\text{diff}} = D_0$, subtract D_0 from all differences before ranking.

In the readability illustration, assume that we want to perform a two-tailed test using $\alpha = .05$. Then

$$T = \text{smaller of } |-18.5| \text{ and } 101.5 = 18.5$$

The Table 5 value (n = column 15, two-tailed α = row .05) is 25. Because $18.5 \leq 25$, we reject H_0 at $\alpha = .05$. In fact, the $\alpha = .02$ value is 20, so H_0 is also rejected at $\alpha = .02$; the $\alpha = .01$ value is 16, so H_0 is not rejected at $\alpha = .01$. By the Universal Rejection Region (reject H_0 when and only when the p-value is less than the specified α), the two-tailed p-value is between .02 and .01.

Example 10.20 Perform the signed-rank test, with H_a being that the differences tend to be positive, for the data of Example 10.19. As in that example, use $\alpha = .005$.

Solution The sum of the negative ranks is $T_- = -60$, so the test statistic is $T = |T_-| = 60$. In the table of critical values for the signed-rank test (with $n = 24$), the critical value for $\alpha = .005$, one-tailed, is 61. Thus $T \leq$ the table value, and H_0 is rejected at $\alpha = .005$. □

The signed-rank test is more efficient than the t test of differences when the population of differences is heavy-tailed. The readability data have a potential outlier in the 42 difference. Whether this difference had been 42 or 4200, it would merely be ranked largest in the signed-rank test. Thus, that test is relatively unaffected by outliers. The theory behind Table 5 assumes that the population of differences is symmetric (though not necessarily normal). See Hollander and Wolfe (1974) or Gibbons (1983). Thus the signed-rank test should be used with some caution if the data are badly skewed. Note that the result of the signed-rank test for the readability data is somewhat more conclusive than the result of the t test, reflecting the greater efficiency of the signed-rank test in the presence of outliers.

Example 10.21

To compare the power of the t and signed-rank tests in the presence of outliers, a simulation was done by taking 1000 samples of size 30 from a population of differences. The population was somewhat outlier-prone. (The Groeneveld-Meeden skewness of the top half of the data was .41; for a normal population the value is .27.) The results were as shown in Table 10.4.

```
Simulation of One Sample t-test (1000 samples)

Population shape is somewhat outlier prone.

        number of times H0:  "mean is    0" is rejected in favor of
alpha     "mean >    0"    "mean <    0"            total (alpha doubled)
0.100        913                 0                  913
0.050        831                 0                  831
0.025        745                 0                  745
0.010        629                 0                  629
0.005        537                 0                  537

average t is 2.8985 with variance of    1.412205

Simulation of Wilcoxon Signed Rank Test (1000 samples)

Population shape is somewhat outlier prone.

        number of times H0: "symmetric around   0" rejected in favor of
alpha     "tend to be >   0"   "tend to be <   0"     total (alpha doubled)
0.100            954                  0                  954
0.050            907                  0                  907
0.025            839                  0                  839
0.010            723                  0                  723
0.005            630                  0                  630
```

Table 10.4

What do these results indicate about the behavior of the two tests for this outlier-prone population?

Solution First, note that the mean is 5.00, so H_a is true. Thus, whenever H_0 is rejected, a correct decision is made. (Of course, if H_0 had been rejected in favor of "mean < 0," we would have had a "rejection in the wrong direction," but this never happened.) The output gives us estimates of the power of the test

$$\text{power} = P(\text{reject } H_0 | H_a \text{ is true})$$

Note that for any particular α, the signed-rank test rejects H_0 more frequently than does the t test. Thus, for this particular population, the signed-rank test is more powerful than the t test. □

Exercises for Section 10.4

10.15. In a maze learning experiment, rats were tested under either a standard reinforcement method or an experimental one. Two rats were picked from each of 12 litters; one of the two was randomly assigned to the standard method, the other to the experimental method. A learning score was obtained for each rat (Table 10.5).

	Litter											
	1	2	3	4	5	6	7	8	9	10	11	12
Standard	56	47	39	41	55	61	54	60	36	49	60	60
Experimental	59	54	62	43	61	43	59	60	46	51	64	63
Difference (E − S)	3	7	23	2	6	−18	5	0	10	2	4	3

Table 10.5

a. What is the reason for regarding this study as a paired-sample experiment?

b. Calculate a 90% confidence interval for the population mean difference.

c. Based on this interval, can one say that there is a statistically significant (α = .10) difference in means?

10.16. Find a two-sided p-value for the data of Exercise 10.15.

10.17. Is there any indication in the data of Exercise 10.15 that the assumptions underlying a t test have been violated?

10.18. Perform a signed-rank test using the data of Exercise 10.15. Again, use α = .10. Does the conclusion of this test differ from that of the t test? If so, why do you suppose that the difference occurred?

10.19. A group of supervisors were retrained, in an effort to achieve better productivity among supervised workers. Productivity gains (productivity after training − productivity before training) were calculated for 25 supervisors. The gains were

−3.2	−1.9	−1.0	−0.8	0.6	1.2	1.3	1.9	2.2
2.3	2.5	2.6	2.8	2.9	3.0	3.3	3.4	3.9
4.6	5.0	5.6	5.8	6.7	7.4	8.2		

a. Why should this study be treated as a paired-sample experiment?

b. Calculate a 99% lower-bound confidence interval for the mean gain.

c. Does this interval indicate that the mean gain is statistically significant at $\alpha = .01$?

10.20. Calculate the value of the t statistic for the data of Example 10.19. Find the p-value. Does the p-value agree with the conclusion in part c of Exercise 10.19?

10.21. Is there evidence of a violation of assumptions in Exercise 10.19?

10.22. Perform a signed-rank test for the data of Exercise 10.19. Use $\alpha = .01$. Does the conclusion agree with that of the t test?

10.5 Inferences about a Median

Sections 10.2 and 10.3 suggested that making inferences about the population median instead of the population mean would be more efficient if the population data were badly skewed or outlier-prone. In this section we discuss how to make inferences about the median.

Testing a hypothesis about a population median can be reduced to the binomial test discussed in Sections 9.1 and 9.2. For example, let us reconsider the data of Section 10.2 on student attitudes toward an avowed homosexual's running for student body president. The data were

12	21	30	41	43	43	44	44	46	47
47	48	50	50	50	51	53	65	76	90

Note that most of the data are concentrated in the 40's and 50's, with some outliers. Again, 50 is a neutral attitude score and a natural null hypothesis. Now, we shall test H_0: $\text{median}_{\text{pop}} = 50$, against the two-sided H_a: $\text{median}_{\text{pop}} \neq 50$. The test statistic is the number of scores falling above the H_0 median. If our H_0 is true, the probability that any particular score falls above 50 is .50, by definition of the median. The binomial assumptions hold, assuming that the scores are obtained independently. There's one small problem: What do we do with scores that fall exactly on the H_0 median? The simplest solution is to discard them and reduce the sample size accordingly. In our example, we discard the three 50 scores and take the sample size to be 17. The test statistic is Y = the number of scores above the hypothesized median. In the example, $y = 5$. The expected value of Y is $.5n = 8.5$. To see if the actual y value is significantly (more than randomly) different from expected, we need binomial probabilities. The probability of 5 or fewer successes in 17 trials can be calculated using binomial probabilities as .0717. Thus the two-tailed p-value is $2(.0717) = .1434$. The null hypothesis that the population median is 50 can't be rejected, even at $\alpha = .10$; the sample result is not statistically significant—it is within reasonable random variation.

Testing a Hypothesis about a Median

1. H_0: $\text{median}_{\text{pop}} = m_0$
2. H_a:
   (i) $\text{median}_{\text{pop}} > m_0$
   (ii) $\text{median}_{\text{pop}} < m_0$
   (iii) $\text{median}_{\text{pop}} \neq m_0$
3. T.S.: Y = the number of scores larger than m_0 out of n = the number of scores not exactly equal to m_0
4. R.R.:
   (i) Compute the binomial probability that there will be y or more successes in n trials, with $\pi = .50$. This probability is the p-value. Reject H_0 if p-value $< \alpha$.
   (ii) Same as (i), but for y or fewer successes.
   (iii) If $y < .5n$, compute the probability of y or fewer successes; if $y > .5n$, compute the probability of y or more successes. The p-value is two times the resulting probability. Reject H_0 if p-value $< \alpha$.

NOTE: n is the number of observations different from m_0.

Example 10.22

In a human-factors study, a control panel in a factory was rearranged in an effort to make it more easily understood. A sample of 20 potential users of the panel were tested before and after the rearrangement, and the change in comprehension scores was measured. The changes were

$$\begin{array}{cccccccccc} -21.1 & -2.2 & -1.1 & 0.2 & 0.6 & 0.7 & 0.9 & 1.3 & 1.8 & 1.9 \\ 2.2 & 2.4 & 2.6 & 2.8 & 2.9 & 3.5 & 4.0 & 4.7 & 6.3 & 18.1 \end{array}$$

(mean 1.625, standard deviation 6.728)

Test the research hypothesis that the median difference is positive.

Solution The null hypothesis is H_0: median $= 0$. Out of 20 trials (users), there are 17 scores above 0 and none exactly equal to 0. From the binomial table, the probability of 17 or more successes in 20 trials, assuming $\pi = .5$, is .0013. This p-value is very small, so H_a is conclusively supported. ☐

Alternatively, it's possible to compute a confidence interval for the true (population) median. The basis for the inference is once again binomial probabilities.

100(1 − α)% Confidence Interval for a Median

Using binomial probabilities (with n = the sample size and $\pi = .50$), find the value k such that

$$P(Y \leq k) \leq \frac{\alpha}{2} \quad \text{and} \quad P(Y \leq k + 1) > \frac{\alpha}{2}$$

Arrange the data in order from lowest to highest. The confidence interval is

$$(k + 1)\text{st value} \leq \text{median}_{\text{pop}} \leq (n - k)\text{th value}$$

NOTE: If $n > 20$, take $k = .5n - z_{\alpha/2}\sqrt{.25n}$, rounded down.

For the sample data on attitudes toward a homosexual's running for campus office, $n = 20$. From Table 1 of binomial probabilities, $P(Y \leq 5) = .0207$ and $P(Y \leq 6) = .0577$. Thus to obtain a 95% confidence interval, take $k = 5$. (The actual α is $2(.0207) = .0414$, making the actual confidence level 95.86%.) The 6th value is 43, and the $20 - 5 = 15$th value is 50. The interval is $43 \leq \text{median}_{\text{pop}} \leq 50$.

Example 10.23 For the data of Example 10.22, calculate a 95% confidence interval for the population median.

Solution The probability of 5 or fewer successes in $n = 20$ trials is .0207, and the probability of 6 or fewer successes is .0577. Thus we take $k = 5$ and the 95% confidence interval to range between the 6th and $20 - 5 = 15$th values, namely

$$0.7 \leq \text{median} \leq 2.9$$

□

If the population is symmetric, the population mean and median are identical, so we can make inferences about either. When the population is symmetric and heavy-tailed, the methods of this section are more efficient. Note, for instance, that the confidence interval we obtained for the median attitude is considerably narrower than the interval for the mean found in Section 10.2. When the population is skewed, the t and median tests are making inferences about different population parameters. It's important to state clearly which parameter is being considered.

Example 10.24 The t test for the data of Example 10.22 comes out $t = 1.08$, not significant at any reasonable α level, yet the median test for a 0 difference comes out significant at very small α values. Explain why.

Solution The sample data of Example 10.22 contain a couple of major outliers, suggesting that the population of differences is heavy-tailed. The median test (and to some extent the signed-rank test) are more efficient for such populations than the t test on differences. □

The median test (and interval) can be used with paired-sample experiments. The typical null hypothesis is that the population median difference, which is *not* the same as the difference of population medians, is 0. The test statistic is the number of difference scores that are above 0. This number is looked up in the binomial tables as before. Because the test is based on the number of positive differences, it is often called the **sign test**.

sign test

Example 10.25 Perform a sign test for the data of Example 10.22.

Solution We already did. The data are differences, and the test in Example 10.22 is based on the 17 positive differences out of 20 trials. □

Exercises for Section 10.5

10.23. Refer to the data of Exercise 10.19. Perform a sign test using $\alpha = .05$. What is the two-tailed p-value?

10.24. For the same data, calculate a 95% confidence interval for the median change. Does it include 0? Should it include 0, according to the test performed in Exercise 10.23?

10.25. Refer to Exercise 10.5. Assume that 65 of the 216 children had times larger than 3.5 (and that none had times exactly equal to 3.5). Test H_0: median = 3.5 vs. a two-sided research hypothesis, using $\alpha = .10$.

10.26. Compare the results of the tests in Exercises 10.10 and 10.25. Are they testing the same null hypothesis?

10.6 Power of *t* Tests

The power of a test—that is, the probability that the research hypothesis H_a will be supported when it is in fact true—is a crucial concept in planning experiments and studies. The power of a z test was studied in Section 9.6. The power of a t test is very similar. However, the extra variability introduced by the use of an estimated standard error in a t test makes the power of a t test somewhat lower, and also somewhat harder to calculate. This section will show how to calculate power, at least approximately, for a single-sample t test.

When H_a is true, the t statistic

$$t = \frac{\bar{Y} - \mu_0}{s/\sqrt{n}}$$

is no longer symmetric around 0. The wrong value for the population mean has been used. Instead of subtracting μ_0, we should have subtracted μ_a. In this case, the t distribution is centered around some number not equal to 0. Therefore, when H_a is true, the distribution of t is called **noncentral *t*.** The noncentral t distribution is the basis for the calculation of the power of a t test.

noncentral *t*

The noncentral t distribution depends on the degrees of freedom, just as the t distribution does. For a single random sample, df = $n - 1$. It also depends on the difference between μ_a and μ_0. If μ_a is close to μ_0, then t is "only slightly noncentral," but if the difference is large, the

noncentrality parameter

noncentrality is large. Conventionally, the difference between μ_a and μ_0 is expressed as a fraction of σ_{pop} in the **noncentrality parameter,**

$$\delta = \frac{\mu_a - \mu_0}{\sigma_{pop}}$$

Note: Some texts define the noncentrality parameter as $\delta\sqrt{n}$.

The power of a one-sample t test is, by definition, the probability that the t statistic falls in the rejection region.

$$\text{power} = P(t \text{ falls in R.R.} | \mu = \mu_a)$$

where, by hypothesis, the t statistic has a noncentral t distribution with $n - 1$ df and noncentrality parameter δ. These power probabilities have been extensively tabulated; see, for example, Cohen (1977).

For example, suppose that a survey is being planned of student attitudes toward an avowed homosexual's running for student body president, as in Section 10.2. Attitudes are to be measured on a 100-point scale, where 50 is a neutral point. A natural null hypothesis is that $\mu_{pop} = 50$. If the sample size is 30, and if in fact the campus-wide population mean attitude score is 1/2 of a standard deviation below 50, then df $= 30 - 1 = 29$, and the noncentrality parameter is $-.50$ (μ_a is 1/2 of a standard deviation below the H_0 value). The power of a two-tailed, $\alpha = .05$ test can be calculated by entering the power tables in Cohen (1977) using 29 df, $\alpha = .05$, and noncentrality parameter $\delta = -0.5$. In fact, these tables indicate that the power is between .7 and .8.

Example 10.26

Suppose that a paired-sample experiment is being planned, based on a sample of size 20. If α is taken to be .01, how can one find the probability that the research hypothesis H_a: $\mu_{diff} > 0$ will be supported, assuming that the population mean change is 2 and the population standard deviation of changes is 10?

Solution

The probability that the research hypothesis will be supported is by definition the power of the test. By assumption, the noncentrality parameter is

$$\delta = \frac{2 - 0}{10} = 0.2$$

because H_0 is that $\mu_{diff} = 0$. Thus we must enter the power tables at df $= 19$, $\alpha = .01$ (one-tailed), and noncentrality $= 0.2$. From those tables, the power is very low (less than .1). □

Power charts are somewhat inconvenient to use. A rough approximation to the power can be had by using an idea in Scheffé (1959). This approximation will be adequate for most purposes. Extreme accuracy is not usually important in planning an experiment, which is the primary use of the power concept. The power can be approximated using the formulas in the following box.

Approximate Power of a *t* Test

One-tailed test:

$$\text{power} = P\left(z > \frac{t_\alpha - \sqrt{n}(\delta)}{1 + t_\alpha^2/2(\text{df})}\right)$$

Two-tailed test:

$$\text{power} = P\left(z > \frac{t_{\alpha/2} - \sqrt{n}(\delta)}{1 + t_{\alpha/2}^2/2(\text{df})}\right)$$

In the survey example, we need $t_{.025, 29\,\text{df}} = 2.045$. The noncentrality parameter is assumed to be $\delta = 0.50$ and n is 30. For a two-tailed test, power is approximately

$$P\left(z > \frac{2.045 - \sqrt{30}(.5)}{1 + (2.045)^2/2(29)}\right) = P(z > 0.67) = .75$$

Thus, if the population mean is in fact 1/2 of a standard deviation below the neutral point 50, the test has about a 75% chance of asserting that the mean differs from 50.

This approximation might be rather rough. However, in many cases a precise power calculation is not necessary. If the approximate power is .21 for an alternative of interest, then whether the true power is .19 or .23 doesn't really matter. The power is in any case too low for the test to be useful.

Example 10.27 Find the approximate power in Example 10.26.

Solution We use $\alpha = .01$ (one-tailed), $\delta = 0.2$, and $n = 20$. The desired t-table value is 2.539.

$$\text{power} = P\left(z > \frac{2.539 - \sqrt{20}(0.2)}{1 + (2.539)^2/2(19)}\right) = P(z > 1.41) = .0793$$

from z tables. The power is very low. □

The power of a t test depends on exactly the same factors as does the power of a z test. Each of the following is a way to increase the power of a t test:

1. Increase α, the probability of Type I error.
2. Increase n, the sample size, and therefore increase df $= n - 1$.
3. Increase $\mu_a - \mu_0$, the effect size.
4. Decrease σ_{pop}.

As with z tests, the power of a t test is not a number, but rather is a function. In particular, it depends on the value of the noncentrality parameter δ. As δ increases (either because of an increase in $\mu_a - \mu_0$ or because of a decrease in σ_{pop}), the power increases, for any specified sample size.

As an alternative to power calculations, the width of confidence intervals may be taken as a rough guide. If the plus-or-minus of a confidence interval equals the anticipated size of effect, $\mu_a - \mu_0$, then the power is roughly equal to .5.

Example 10.28 In Example 10.27, the power was found to be very low for the effect size of interest. What does this fact imply about the width of the confidence interval?

Solution The interval will be very wide relative to the effect size of interest. □

Exercises for Section 10.6

10.27. A follow-up study to the study in Exercises 10.7 and 10.13 was planned. Assume that the true, population mean and standard deviation are equal to the sample mean and standard deviation found in Exercise 10.7. The planned sample size is 46, and the planned $\alpha = .05$ (one-tailed test).

a. Interpolate to find the needed table value.

b. What is the noncentrality parameter? H_0 is still that the population mean is 48.

10.28. Calculate the power of the test in Exercise 10.27 using the z approximation.

10.29. In planning a paired-sample experiment, it was conjectured that the true mean difference would be of the order of 15 and that the standard deviation would be about 50. A two-tailed test was to be run at $\alpha = .01$. A power of at least .8 is desired. Is a sample size of 25 big enough?

10.30. Refer to Exercise 10.29. What is the power if $n = 61$?

Appendix A: Chi-Squared and *t* Distributions

This appendix presents a brief sketch of the mathematics underlying the t distribution and the related chi-squared distribution. This math is the basis for the methods of the preceding chapter and of the next five chapters. It also should help in clarifying the degrees-of-freedom concept.

The formal mathematical assumption is that there are n random variables $Y_1, Y_2, \ldots, Y_n$, each having a normal distribution with mean μ and standard deviation σ; in particular, the assumptions hold for a random sample from a normal population. The corresponding z scores are

$$Z_1 = \frac{Y_1 - \mu}{\sigma}, \ldots, Z_n = \frac{Y_n - \mu}{\sigma}$$

First we define the chi-squared (χ^2) distribution with n degrees of freedom.

The Chi-Squared Distribution with *n* df

The distribution of

$$W = Z_1^2 + \cdots + Z_n^2 = \frac{\sum_i (Y_i - \mu)^2}{\sigma^2}$$

is the chi-squared distribution with n df.

Two properties of the chi-squared distribution follow directly from the definition:

1. If $\bar{Y} = \Sigma_i Y_i/n$, then $n(\bar{Y} - \mu)^2/\sigma^2$ has a chi-squared distribution with 1 df.

2. If W_1 and W_2 are independent random variables having chi-squared distributions with n_1 and n_2 df, respectively, then $W_1 + W_2$ has a chi-squared distribution with $n_1 + n_2$ df.

The first property follows because $(\bar{Y} - \mu)/(\sigma/\sqrt{n})$ is a z statistic, as shown in Chapter 7. Thus $[(\bar{Y} - \mu)/(\sigma/\sqrt{n})]^2 = n(\bar{Y} - \mu)^2/\sigma^2$ is a Z^2 statistic and can be regarded as a sum with only one term. The definition of a chi-squared distribution implies that $n(\bar{Y} - \mu)^2/\sigma^2$ has a chi-squared distribution with 1 df.

The second property follows because both W_1 and W_2 are sums of Z^2 terms, so $W_1 + W_2$, as a sum of sums, is itself a sum of $(n_1 + n_2)$ Z^2 terms. The assumption of independence of W_1 and W_2 can be shown to allow the assumption that all the Z^2 terms are independent, so $W_1 + W_2$ has the defining properties of a chi-squared distribution with $n_1 + n_2$ df.

These results allow us to relate the chi-squared distribution to the sampling distribution of s^2, the sample variance.

Sampling Distribution of s^2

For a random sample of size n from a normal position, the statistic

$$(n - 1)\frac{s^2}{\sigma^2}$$

has a chi-squared sampling distribution with $n - 1$ df.

Recall that $s^2 = \Sigma(Y_i - \bar{Y})^2/(n - 1)$, so $(n - 1)(s^2/\sigma^2) = \Sigma(Y_i - \bar{Y})^2/\sigma^2$. The proof of the sampling distribution is based on the ideas already developed in this appendix. Although we cannot provide a rigorous proof within the scope of this text, we can indicate the basic idea.

The key to the proof is the identity

$$\frac{\sum(Y_i - \mu)^2}{\sigma^2} = \frac{\sum(Y_i - \bar{Y})^2}{\sigma^2} + \frac{n(\bar{Y} - \mu)^2}{\sigma^2}$$

which can be proved by writing $\Sigma(Y_i - \mu)^2$ as $\Sigma[(Y_i - \bar{Y}) + (\bar{Y} - \mu)]^2$, expanding the square, and noting that

$$\sum_{i=1}(\bar{Y} - \mu)(Y_i - \bar{Y}) = (\bar{Y} - \mu)\sum_{i=1}(Y_i - \bar{Y}) = 0$$

The quantity $\Sigma(Y - \mu)^2/\sigma^2$ has, by definition, a chi-squared distribution with n df. As we have noted, the $n(\bar{Y} - \mu)^2/\sigma^2$ term has a chi-squared distribution with 1 df. Because independent chi-squared statistics add to form chi-squared statistics with more degrees of freedom, it is plausible that the other term $\Sigma(Y - \bar{Y})^2/\sigma^2$ also has a chi-squared distribution. Further, since degrees of freedom are additive, this term should have $n - 1$ df. The missing step of the argument is the proof that the two terms on the right side of the equation are independent; this fact is proved only in advanced texts.

These preliminary results on chi-squared distributions are used in developing the t distribution. We first define the t distribution with ν df.

The t Distribution with ν df

The statistic

$$t = \frac{Z}{\sqrt{W/\nu}}$$

has a t distribution with ν df if Z has a standard normal distribution ($\mu = 0, \sigma = 1$) and W a chi-squared distribution with ν df. Z and W are required to be independent.

We take Z to be $(\bar{Y} - \mu)/(\sigma/\sqrt{n})$ and W to be $(n - 1)(s^2/\sigma^2)$, which was just shown to have a chi-squared distribution with $\nu = n - 1$ df. (It can be shown that Z and W are independent.) By definition,

$$t = \frac{\dfrac{(\bar{Y} - \mu)}{\sigma/\sqrt{n}}}{\sqrt{(n - 1)\dfrac{s^2}{\sigma^2}\bigg/(n - 1)}} = \frac{\bar{Y} - \mu}{s/\sqrt{n}}$$

which is the t statistic used in this chapter.

Appendix B: Generalized Likelihood Ratio Tests and the t Test

In the appendix to Chapter 9 we discussed the Neyman-Pearson Lemma, which uses the idea of likelihood ratio to obtain best, uniformly most powerful tests in certain situations. Unfortunately, the Neyman-Pearson Lemma only applies to a limited number of situations, typically those involving only one unknown parameter. In situations where more than one parameter is unknown—as in the t test situation, where both μ_{pop}

Generalized Likelihood Ratio (GLR) principle

and σ_{pop} are unknown—a more general principle is needed. The **Generalized Likelihood Ratio (GLR) principle** applies widely, and in common situations often yields a very good, if not absolutely the best, test. In this appendix, the idea of a GLR test is sketched and then it is shown that the *t* test of Chapter 10 is a GLR test, given some formal mathematical assumptions.

likelihood

In Chapter 8 we defined the **likelihood** as the probability or probability density of the actually observed data, regarded as a function of the unknown population parameter(s). In Section 8.2 we defined the **maximum likelihood estimator (MLE)** of a parameter as the value of the parameter that maximizes the likelihood function. These ideas are combined with the likelihood ratio principle of the Neyman-Pearson Lemma to yield the GLR principle.

maximum likelihood estimator (MLE)

Neyman-Pearson tests are based on the ratio

$$\frac{\text{likelihood assuming } H_a}{\text{likelihood assuming } H_0}$$

In the case of a random sample from a normal population with mean μ and standard deviation σ_{pop}, the likelihood is

$$\frac{1}{(2\pi)^{n/2}\,\sigma_{pop}^2}\, e^{\{-.5\Sigma(y_i-\mu)^2/\sigma_{pop}^2\}}$$

If σ_{pop} is known, and if both H_0 and H_a define specific values for μ, the likelihood can be computed for any particular sample $y_1, \ldots, y_n$. Therefore, the Neyman-Pearson test can be computed.

However, it is much more likely that σ_{pop} will be unknown. Also, if H_a is two-sided, H_a: $\mu \neq \mu_0$, there will be no specific value for μ to plug into the likelihood assuming H_a. Situations like this, where there are several population parameters and/or there is a two-sided H_a, are appropriate for the GLR principle.

Typically, H_0 will specify the value of one or more parameters. First, we compute the maximum likelihood estimators of any remaining parameters, assuming that the null hypothesis is true. In the normal distribution case, with H_0: $\mu = \mu_0$, the maximum likelihood estimator of σ_{pop} turns out to be $\Sigma(y_i - \mu_0)^2/n$. When we replace any unknown parameters in the H_0 likelihood by their estimators, we have the maximum possible likelihood, assuming that H_0 is true.

Next, we compute the maximum likelihood estimators of all the population parameters, regardless of what hypothesis is true. In the normal distribution case, the maximum likelihood estimator of μ is $\bar{y}$ and the MLE of σ_{pop} is $\Sigma(y_i - \bar{y})^2/n$. When we replace all parameters in the likelihood by their MLE's, we obtain the maximum possible likelihood over all hypotheses.

The Generalized Likelihood Ratio test statistic is

$$\lambda = \frac{\text{maximum possible likelihood, assuming } H_0}{\text{maximum possible likelihood for any hypothesis}}$$

(Note that this use of the symbol λ has nothing to do with the measure defined in Chapter 3.) As in the Neyman-Pearson Lemma, the statistic is a ratio of likelihoods. Conventionally, λ is defined with the H_0 likelihood in the numerator rather than in the denominator.

Recall that the basic strategy of hypothesis testing is to reject the null hypothesis H_0 if the data are very unlikely assuming H_0. For such data, the likelihood assuming H_0, even the maximum possible likelihood, will be small compared to the maximum possible likelihood over all possible hypotheses. Thus the GLR test rejects the null hypothesis when λ is a small number.

In the normal-population case, with H_0: $\mu = \mu_0$, the GLR statistic λ is equivalent to the t statistic. Assuming that H_0 is true, we replace μ by μ_0 and σ_{pop} by its MLE, $\sqrt{\Sigma(y_i - \mu_0)^2/n}$, in the likelihood function. After some algebra, it follows that

$$\text{maximum possible likelihood, assuming } H_0$$

$$= \frac{n^n e^{-n/2}}{(2\pi)^{n/2}\left[\sum(y_i - \mu_0)^2\right]^{n/2}}$$

Similarly, if we maximize the likelihood without regard to which hypothesis might be true, we find that the MLE of μ is $\bar{y}$ and the MLE of Σ. Upon putting these estimates into the likelihood and doing some more algebra, we find that

$$\text{maximum possible likelihood for any hypothesis}$$

$$= \frac{n^n e^{-n/2}}{(2\pi)^{n/2}\left[\sum(y_i - \bar{y})^2\right]^{n/2}}$$

It follows that

$$\lambda = \frac{\dfrac{n^n e^{-n/2}}{(2\pi)^{n/2}\left[\sum(y_i - \mu_0)^2\right]^{n/2}}}{\dfrac{n^n e^{-n/2}}{(2\pi)^{n/2}\left[\sum(y_i - \bar{y})^2\right]^{n/2}}}$$

or, after cancellation,

$$\lambda = \left(\frac{\sum(y_i - \bar{y})^2}{\sum(y_i - \mu_0)^2} \right)^{n/2}$$

Upon expanding

$$\sum(y_i - \mu_0)^2 = \sum(y_i - \bar{y})^2 + n(\bar{y} - \mu_0)^2$$

and doing more algebra, we have that

$$\lambda = \frac{1}{\{1 + [n/(n-1)]t^2\}^{n/2}}$$

As t^2 gets larger, λ gets smaller. Therefore the GLR test strategy of rejecting H_0 when λ is small is equivalent to the strategy of rejecting H_0 when t^2 is large, or, equivalently, when $|t|$ is large. Thus, for a two-sided research hypothesis about μ in a normal population, the Generalized Likelihood Ratio test is equivalent to the t test discussed in this chapter.

In the following chapters of this book, many hypothesis tests will be introduced that are based on a normal-population assumption. In every case, such tests will be GLR tests. Provided that the population really is normal, these tests will be nearly most powerful. However, the GLR principle relies very heavily on assumptions about the population shape. If the population shape is not normal, as perhaps suggested by nonnormality of the sample data, then the GLR tests may not be nominally robust (the claimed α and β probabilities may be in error) or they may not be efficiency robust (another test may be more effective).

Evidently, the mathematics of GLR tests can be rather heavy going. We have only sketched the most basic ideas. For a more complete discussion, see mathematical statistics texts such as Mood, Graybill, and Boes (1974).

Chapter Exercises

10.31. Wold (1982) did a study involving, among other things, the number of adjectives correctly associated with nouns, 12 seconds after presentation. Under one set of conditions, the mean was 13.70. Even though the mean resulted from a sample, let us hypothesize it as a population mean. Suppose that we attempt to repeat Wold's experiment with a sample of size 20, and that we obtain a sample mean of 13.06 and a sample standard deviation of 1.52.

a. Calculate a 95% confidence interval for the population mean.

b. Does this interval include the mean found by Wold? If that mean is taken as the null hypothesis value, what conclusion would be reached by a hypothesis test?

10.32. Refer to Exercise 10.31. Perform a formal five-step test of hypotheses. Does the conclusion agree with the answer to part b of Exercise 10.31?

10.33. Put bounds on the (two-tailed) p-value in Exercise 10.32.

10.34. The data underlying Exercise 10.31 were roughly symmetric with no serious outliers. However, the data were only the integer numbers 10, 11, ..., 16. Does this present any problem with the answers to Exercises 10.31–10.33?

10.35. Data were collected from mothers of preschool children concerning the age at which each child would be able to perform a certain task. In fact, the mean age was known to be 7.0 years. In a sample of size 61, the mean was 6.20 and the standard deviation was 2.43.

a. Calculate a 95% upper-limit confidence interval for the population mean expectation.

b. Does this interval indicate that a population mean expectation of 7.0 years is a plausible value?

10.36. Refer to the data of Exercise 10.35. Perform a formal hypothesis test of the research hypothesis that the mean expectation is less than 7 years, using $\alpha = .05$. State bounds on the p-value.

10.37. The data underlying Exercises 10.35 and 10.36 had most values in the range 5.5 to 6.0, with no values below that range, but some as high as 12.0. Does this fact indicate that the nominal α and p-value may be in error?

10.38. A computer program drew 1000 samples of size 10 from a population having a normal distribution with a population mean of 50. The results in Table 10.6 were obtained for the t test.

a. In this simulation, is H_0 true or false? What probabilities are being approximated in the simulation?

b. Is there any indication that the nominal probabilities are grossly wrong?

	number of times H0: "mean is 50" is rejected in favor of		
alpha	"mean > 50"	"mean < 50"	total (alpha doubled)
0.100	109	93	202
0.050	53	36	89
0.025	22	20	42
0.010	10	7	17
0.005	6	2	8

Table 10.6

10.39. Another computer run involved drawing 1000 samples of size 10 from a normal population having a mean of 55 and a standard deviation of 10, and performing t tests on each sample. The results in Table 10.7 were obtained.

a. In this simulation, we are approximating the power of a t test. Explain why this is true.

b. Calculate the approximate power of the t test for $\alpha = .05$, one-tailed, where the null hypothesis mean is 50, the population mean is 55, the population standard deviation is 10, and the sample size is 10. How well does the approximation agree with the simulation result?

	number of times H0: "mean is 50" is rejected in favor of		
alpha	"mean > 50"	"mean < 50"	total (alpha doubled)
0.100	601	4	605
0.050	437	0	437
0.025	307	0	307
0.010	177	0	177
0.005	110	0	110

Table 10.7

10.40. In a study, children were first given an IQ test. The two lowest-scoring children were randomly assigned, one to a "noun-first" task, the other to a "noun-last" task. The two next-lowest IQ children were similarly assigned, one to "noun-first" task, the other to a "noun-last" task, and so on until all children were assigned. The data (scores on a word-recall task) are shown here, listed in order from lowest to highest IQ score:

Noun-first	13	21	12	16	20	39	26	29	30	35	38	34
Noun-last	10	12	23	14	16	8	16	22	32	13	32	35

a. The experiment involves paired samples. What is the basis of the pairing?

b. Calculate the mean and standard deviation of the differences.

c. Test the null hypothesis that the population mean difference is 0, using a two-sided research hypothesis. Assume $\alpha = .10$.

d. State bounds on the p-value.

10.41. Refer to the data of Exercise 10.40. Perform a signed-rank test of the hypothesis that the population of differences is symmetric around 0. Again, use $\alpha = .10$.

10.42. For the data of Exercise 10.40, is a t test or a signed-rank test more appropriate?

10.43. For the data of Exercise 10.40, test the null hypothesis that the median difference is 0 against a two-sided research hypothesis. Again, use $\alpha = .10$.

10.44. In a simulation study, a computer program drew 1000 samples of size 30 from a population with a Laplace distribution, a moderately outlier-prone distribution. The population mean was 0 and the population standard deviation was 10. The program performed both *t* tests and signed-rank tests for each sample. The results in Table 10.8 were obtained.

a. What probabilities are being estimated by this simulation?

b. Do the simulation probabilities agree reasonably closely with the nominal probabilities?

Simulation of One Sample t-test (1000 samples)

	number of times H0: "mean is 0" is rejected in favor of		
alpha	"mean > 0"	"mean < 0"	total (alpha doubled)
0.100	104	95	199
0.050	51	51	102
0.025	28	24	52
0.010	7	6	13
0.005	4	3	7

Simulation of Wilcoxon Signed Rank Test (1000 samples)

	number of times H0: "symmetric around 0" rejected in favor of		
alpha	"tend to be > 0"	"tend to be < 0"	total (alpha doubled)
0.100	100	95	195
0.050	53	48	101
0.025	24	16	40
0.010	10	5	15
0.005	6	2	8

Table 10.8

10.45. Another simulation involved *t* and signed-rank tests performed on 1000 samples of size 30. This time the population mean was 5 and the population standard deviation was 10. The results in Table 10.9 were obtained.

Simulation of One Sample t-test (1000 samples)

	number of times H0: "mean is 0" is rejected in favor of		
alpha	"mean > 0"	"mean < 0"	total (alpha doubled)
0.100	913	0	913
0.050	831	0	831
0.025	745	0	745
0.010	629	0	629
0.005	537	0	537

Simulation of Wilcoxon Signed Rank Test (1000 samples)

	number of times H0: "symmetric around 0" rejected in favor of		
alpha	"tend to be > 0"	"tend to be < 0"	total (alpha doubled)
0.100	954	0	954
0.050	907	0	907
0.025	839	0	839
0.010	723	0	723
0.005	630	0	630

Table 10.9

a. Show that this simulation is approximating the power of the two tests.

b. Which test appears to have better power for this situation?

10.46. In Exercise 9.38, 30 pairs of rats were compared for food consumption in crowded and in uncrowded cages. In that exercise, we used an assumed population standard deviation. The sample mean (of differences) was 6.41. Suppose that the sample standard deviation of the differences was 13.24. Calculate a 95% confidence interval for the population mean based on an estimated standard error.

10.47. Refer to Exercise 10.46.

a. Carry out a formal test of the null hypothesis that the mean of the differences is 0, against a two-sided research hypothesis. Use $\alpha = .05$.

b. Show that the confidence interval of Exercise 10.46 leads to the same conclusion as the formal test.

c. State bounds on the *p*-value of the test statistic.

10.48. Suppose that a sample of 19 children were tested on arithmetic skills. The data were

1.3	3.6	3.8	3.9	4.0	4.3	4.3	4.5	4.5	4.5
4.5	4.6	4.6	4.7	4.7	4.7	4.8	6.9	7.7	

The null hypothesis was that the population mean was 4.0. The research hypothesis was that the mean was larger.

a. Perform a *t* test based on the data. Use $\alpha = .05$.

b. Subtract the null hypothesis value from all scores to obtain differences.

c. Assign signed ranks to all differences. What should be done about the 0 difference?

d. Perform a signed-rank test, with $\alpha = .05$.

10.49. Refer to Exercise 10.48.

a. Perform the signed-rank test, varying α from .05 down to .005.

b. Based on the results of these tests, assign bounds to the *p*-value from the signed-rank test.

10.50. For the data of Exercise 10.48, does the *t* test or the signed-rank test seem more appropriate? Explain why.

10.51. Calculate a 95% confidence interval for the median, based on the data of Exercise 10.48. Does this interval indicate that a median score of 4.0 is a reasonable possibility?

10.52. A computer program drew 1000 samples of size 30 from a highly skewed population with mean 0. For each sample, a signed-rank test of the null hypothesis that the distribution is symmetric around 0 was performed.

For a nominal $\alpha = .100$, H_0 was rejected 748 times, and for a nominal $\alpha = .050$, H_0 was rejected 631 times. What do these results indicate about the effect of skewness on the signed-rank test?

10.53. In an experiment, a sample of subjects performed a proofreading task under each of two conditions. Condition A was an "environmental control" situation, in which the subject could control, within limits, the intensity of background noise. Condition B was a no-control situation. A score was obtained for each subject under each condition.

a. What is the basis for treating this experiment as a paired-sample experiment?

b. A portion of the SAS output is shown in Figure 10.7. Locate the value of the t statistic.

c. Locate the p-value for the t test. Is it one-tailed or two-tailed?

d. What conclusion can be reached, based on the t test?

```
                                   PLOTS OF SAMPLE DIFFERENCES

VARIABLE=DIFF          DIFFERENCE (A - B) OF SCORES BY PAIRS

T:MEAN=0     3.85941   PROB>|T| .000277142        RANGE        30
SGN RANK       477.5   PROB>|S|     0.0001        Q3-Q1         5
NUM ^= 0          55                              MODE         -1
     STEM LEAF                     #          BOXPLOT                     NORMAL PROBABILITY PLOT
       28 0                        1             *              29+                                   *
       26                                                         |
       24                                                         |
       22                                                         |
       20 0                        1             *                |                                 *
       18                                                         |
       16                                                         |
       14                                                         |                                   +
       12 0                        1             0              13+                              * +
       10 0                        1             |                |                             *+
        8 00                       2             |                |                          +**
        6 00000                    5             |                |                        ++***
        4 000000                   6          +-----+             |                     ++****
        2 0000000000000           13          | + |               |                  ++*****
        0 000000                   6          *-----*             |                ******
       -0 0000000000000000000     19          +-----+             | ****************+
       -2 0000000                  7             |             -3+ +++++++++++++
          ----+----+----+----+                                     +----+----+----+----+----+----+----+----+----+----+
                                                                       -2        -1       +0        +1        +2
                            PAIRED SAMPLE T TEST OF DISTRA - DISTRB

        VARIABLE   LABEL                                     MEAN        STD ERROR        T      PR>|T|
                                                                          OF MEAN
```

Figure 10.7 Output for Exercise 10.53

10.54.

a. Locate the value of the signed-rank statistic in Figure 10.7.

b. Locate the p-value of the signed-rank test.

c. Which of the two tests appears to be more conclusive?

10.55. What is the apparent shape of the data in Exercise 10.53? What does this shape indicate about the reliability of the tests in Exercises 10.53 and 10.54?

10.56. In an experiment, the number of times that a subject spoke during a ten-minute period was counted, once in a college seminar setting and once in a party setting. Not astonishingly, most subjects spoke a great deal more frequently in the party setting. One theory, based on similar previous experiments, held that the average difference should be 14 words. Minitab output for the data is shown in Figure 10.8. In Minitab, WTEST refers to the signed-rank test.

a. What conclusion can be reached from the t test on the differences?

b. What conclusion can be reached from the signed-rank test?

```
MTB > SUBTRACT C1 FROM C2 PUT RESULTS IN C3
MTB > NAME C1 'SEMINAR' C2 'PARTY' C3 'DIFF'
MTB > HISTOGRAM OF C3

 DIFF

  MIDDLE OF   NUMBER OF
  INTERVAL    OBSERVATIONS
        -5        1   *
         0        2   **
         5        1   *
        10        7   *******
        15       14   **************
        20       14   **************
        25        2   **
        30        0
        35        1   *

MTB > TTEST OF MU=14 DATA IN C3

TEST OF MU = 14.0 VS MU N.E. 14.0

               N      MEAN    STDEV   SE MEAN       T    P VALUE
DIFF          42     15.38     7.19       1.1    1.24       0.22

MTB > WTEST OF MU = 14 DATA IN C3

TEST OF CENTER = 14.00 VERSUS CENTER N.E. 14.00

                   N FOR   WILCOXON              ESTIMATED
               N    TEST   STATISTIC  P-VALUE      CENTER
DIFF          42      40       548.5    0.064       16.00
```

Figure 10.8 Output for Exercise 10.56

10.57. **a.** What is the reason for plotting the difference data in Exercise 10.56, rather than the seminar data and the party data?

b. Do there appear to be outliers in the difference data? What does your answer indicate about the relative desirability of the two tests employed in Exercise 10.56?

11 Two-Sample Methods for Means

Section 10.4 pointed out the importance of having a control group as a comparison standard in an experiment. The analysis in Section 10.4 was appropriate for *paired* sampling, in which each measurement in the treatment group was associated with a specific measurement in the control group. It is not always possible or desirable to use paired samples. In this chapter we discuss methods for analyzing two separate, independent samples; the two samples may constitute treatment and control groups, or may reflect some other grouping.

An assumption critical to the use of these methods is that the results from the two samples are statistically independent of each other, both within and between samples. These methods will give seriously incorrect answers when results from the samples are dependent. In particular, use of these methods when the data arise from paired sampling is wrong.

In Section 11.1, we look at the theoretical distribution of the difference of sample means. This theoretical distribution is the basis for most of the methods in the chapter. Then in Section 11.2 we discuss the pooled-variance t statistic, which requires an assumption about the underlying population variances. Section 11.3 presents an approximate t method that doesn't require the variance assumption. In Section 11.4 the Mann-Whitney-Wilcoxon rank sum test is presented. As a rank test, it requires no assumption about the shape (normal or otherwise) of the underlying population distributions. Section 11.5 discusses the power of the various tests. Finally, in Section 11.6 we compare independent-sample and paired-sample methods, in order to find the relative advantages of each.

11.1
Theoretical Distribution of the Difference of Means

To make inferences in the two-independent-sample situation, we need to have the appropriate theoretical (sampling) distributions. In this section we focus on the difference of the two sample means. Recall from Chapter 8 that, when the underlying population is (close to) normal, the sample mean is (close to) the most efficient estimator of the population mean. It follows that when two separate populations are each (nearly) normal, the differences in sample means will be (nearly) most efficient for estimating the difference in population means. In order, we'll consider the expected value, standard error, and shape of the theoretical distribution of the difference of sample means.

First, in Chapter 7 we showed that the expected value (long-run average) of the sample mean equals the population mean:

$$E(\bar{Y}) = \mu_{\text{pop}}$$

In the language of Chapter 8, $\bar{Y}$ is an *unbiased estimator* of the population mean. In Chapter 5 we proved that the expected value of a difference of any two random variables is the difference of their expected values. Combining these mathematical facts, we find

$$E(\bar{Y}_1 - \bar{Y}_2) = \mu_1 - \mu_2$$

where $\bar{Y}_1$ and $\bar{Y}_2$ are sample means for two separate samples and μ_1 and μ_2 are the respective population means. Thus, $\bar{Y}_1 - \bar{Y}_2$ is an unbiased estimator of $\mu_1 - \mu_2$. Note that this result is valid whether or not the two samples are independent.

The independence assumption becomes crucial in the calculation of the standard error of the difference of means. Recall that the standard error of any statistic is the standard deviation of its theoretical probability distribution. The properties of variances discussed in Chapter 5 indicate that *if two random variables are independent,* the variance of their difference is the sum of the variances. Notice that it's the variances that are added, not the standard deviations. Thus, assuming independent random samples,

$$\text{Var}(\bar{Y}_1 - \bar{Y}_2) = \text{Var}(\bar{Y}_1) + \text{Var}(\bar{Y}_2) = \frac{\sigma_1^2}{n_1} + \frac{\sigma_2^2}{n_2}$$

where σ_1^2 and σ_2^2 are the respective true, population variances and n_1 and n_2 are the respective sample sizes. The standard error of $\bar{Y}_1 - \bar{Y}_2$ is the square root of this variance.

$$\sigma_{\text{diff of means}} = \sqrt{\frac{\sigma_1^2}{n_1} + \frac{\sigma_2^2}{n_2}}$$

For example, suppose that we take samples from two populations, as in Table 11.1.

Population	Mean	Variance	Standard Deviation	Sample Size
1	40.0	16.0	4.0	8
2	32.0	25.0	5.0	10

Table 11.1

The expected value is

$$E(\bar{Y}_1 - \bar{Y}_2) = 40.0 - 32.0 = 8.0$$

the variance is

$$\text{Var}(\bar{Y}_1 - \bar{Y}_2) = \frac{16.0}{8} + \frac{25.0}{10} = 4.50$$

and the standard error is

$$\sigma_{\text{diff of means}} = \sqrt{4.50} = 2.121$$

Example 11.1

Suppose that two populations are sampled randomly, and that both populations have variance 30.0. A total of 30 observations are to be allocated to the combined sample. Compute the standard error of the difference of sample means in each of the following cases:

(a) $n_1 = 15, n_2 = 15$
(b) $n_1 = 20, n_2 = 10$
(c) $n_1 = 25, n_2 = 5$

In which case is the standard error smallest?

Solution

(a) $\sigma_{\text{diff of means}} = \sqrt{\frac{30.0}{15} + \frac{30.0}{15}} = 2.00$

(b) $\sigma_{\text{diff of means}} = \sqrt{\frac{30.0}{20} + \frac{30.0}{10}} = 2.12$

(c) $\sigma_{\text{diff of means}} = \sqrt{\frac{30.0}{25} + \frac{30.0}{5}} = 2.68$

The standard error is smallest when the sample sizes are equal, and increases as the sample sizes become more and more different. □

The shape of the theoretical distribution of $\bar{Y}_1 - \bar{Y}_2$ depends on the respective sample sizes. The Central Limit Theorem discussed in Chapter 7 says that if the sample size is large, the theoretical distribution of any one sample mean will be approximately normal. A theorem of mathematical statistics [see Mood, Graybill, and Boes (1974)] asserts that the difference of two independent, normally distributed random variables is itself normally distributed. *Therefore, if both sample sizes are large, the theoretical distribution of $\bar{Y}_1 - \bar{Y}_2$ will be approximately normal.*

The sample sizes needed to yield a good normal approximation depend on the skewness of the underlying populations. If the sample sizes are equal, the *total* sample size, $n_1 + n_2$, determines the quality of the normal approximation. The rules of thumb for using the Central Limit Theorem that were presented in Chapter 7 apply equally to this situation, where n is interpreted as the total sample size. If the populations are skewed in the same direction (say, both right-skewed), taking the difference in sample means tends to "cancel the skewness," so the normal approximation in this case will even be somewhat better than the Chapter 7 rules of thumb indicate. If the sample sizes are seriously unequal, the smaller sample size becomes the critical factor in determining the quality of the normal approximation. Thus, if one sample size is only, say, half the other, the Chapter 7 rules of thumb should be applied, with n interpreted as the smaller of the two sample sizes. Again, if both populations are similarly skewed, there will be some cancellation of skewness, so the normal approximation will be somewhat better than the Chapter 7 guidelines indicate.

Example 11.2 Refer to Example 11.1. Suppose that both populations are substantially right-skewed. In which of the three cases will the normal approximation be best? Poorest?

Solution In the first case, both sample sizes are equal, so the total sample size, 30, should give a fairly good approximation. The Chapter 7 rules of thumb for skewed populations would demand a somewhat larger total sample size, but the effect of taking a difference should help to improve the normal approximation. In the third case, one sample size is only 5, so the normal approximation might well be poor in that case. □

The assumption of independence of the two samples is critical to the correctness of these results. In particular, the formula for the standard error of the difference of sample means is seriously wrong when the samples are dependent. In addition, the quality of the normal approximation may be worsened by dependence in the samples.

Exercises for Section 11.1

11.1. Independent samples were to be taken from two populations. The assumed population means and standard deviations were as shown in Table 11.2.

Population	Mean	Standard Deviation
1	52.4	8.0
2	56.1	12.0

Table 11.2

Calculate the expected value and standard error of the difference of the sample means for each of the following sample sizes:

a. $n_1 = 15, n_2 = 15$
b. $n_1 = 12, n_2 = 18$
c. $n_1 = 9, n_2 = 21$

11.2. It is known that both populations are left-skewed. Which set of sample sizes yields the best standard error? Which set gives the poorest normal approximation to the theoretical distribution of sample means?

11.3. Separate samples were to be taken from two populations having different means but similar standard deviations. What are the reasons for taking equal-sized samples from both populations?

11.2 Pooled-Variance *t* Methods

The results of Section 11.1 are important for making inferences in the two-sample case. A problem is that the standard error of the difference of means depends on the underlying population variances. Typically the population variances will be unknown in practice. In the one-sample case, we solved the same problem by replacing the true standard error by the estimated standard error, and replacing z-table values by t-table values. In the two-sample situation, we will follow the same procedure, but in a somewhat unexpected way.

One would expect to replace σ_1^2 by s_1^2 and σ_2^2 by s_2^2 in the standard error formula. It turns out that this procedure does *not* lead to a t statistic. The mathematics involved is shown in the appendix. In effect, the mathematics says that, to obtain a true t statistic, one may only replace *one* unknown population variance by a sample variance. Trying to make two such replacements fails to yield a t statistic.

An additional assumption is necessary to obtain a true t statistic in the two-sample case. *We assume that the two unknown population variances are equal.* The two populations are assumed to have a common variance:

$$\sigma_1^2 = \sigma_2^2 = \sigma_{\text{common}}^2$$

homoscedasticity

The motivation for this assumption, called the assumption of **homoscedasticity**—homo for equal, and scedasticity from a Greek root meaning variability—is purely mathematical. We shall make the assumption, discuss its implications, and then consider the effect of violation of this assumption.

The first task is to combine the information of the two independent samples to estimate σ_{common}^2. For example, Tellevik (1981) reported the information in Table 11.3 on the time in seconds required to perform a certain task by samples of 28 hearing and 22 deaf students:

Sample	Mean	Standard Deviation
Hearing	427	223.2
Deaf	472	218.8

Table 11.3

The sample variances $(223.2)^2$ and $(218.8)^2$ are not equal. Even if the population variances were equal, the sample variances would differ, by sheer random variation. The sample variances aren't so far apart that the homoscedastity assumption is outrageous. To estimate σ^2_{common}, it seems very plausible to average the two sample variances, weighting the hearing variance more heavily because it is based on the larger sample size. The estimate of σ^2_{common} is a weighted average of the sample variances; the weights are the respective degrees of freedom.

Pooled Variance Estimate of σ^2_{common}

$$s^2_{\text{pooled}} = \frac{(n_1 - 1)s_1^2 + (n_2 - 1)s_2^2}{n_1 + n_2 - 2}$$

$$s_{\text{pooled}} = \sqrt{s^2_{\text{pooled}}}$$

For the Tellevik data,

$$s^2_{\text{pooled}} = \frac{(28 - 1)(223.2)^2 + (22 - 1)(218.8)^2}{28 + 22 - 2}$$

$$= 48967.39$$

$$s_{\text{pooled}} = \sqrt{48967.39} = 221.3$$

Example 11.3 Beez (1970) reports on a study in which children at a summer camp were assigned to supposed ability (low-expectation or high-expectation) groups, completely at random without regard to any actual ability. Teachers were told of the supposed level of ability of each child. One variable of interest was the number of complex symbols that teachers attempted to teach to each child. Suppose that the data had been

Low Expectation:	3	3	3	4	4	4	5	5	5
	6	6	6	6	7	7	8	9	11
High Expectation:	7	8	8	8	9	9	9	9	10
	10	11	11	11	12	12	13	14	17

Using the figures in Table 11.4, calculate the pooled variance estimate.

Expectation	Mean	Standard Deviation	Variance	n
Low	5.67	2.169	4.706	18
High	10.44	2.502	6.261	18

Table 11.4

Solution

$$s^2_{\text{pooled}} = \frac{(18 - 1)4.706 + (18 - 1)6.261}{18 + 18 - 2} = 5.483$$

□

The pooled variance is used in obtaining an estimated standard error. Because we are assuming that $\sigma_1 = \sigma_2$, we replace both σ_1^2 and σ_2^2 by s^2_{pooled} in the true standard error formula of Section 11.1, to obtain the estimated standard error:

$$\text{est. st. error}(\bar{Y}_1 - \bar{Y}_2) = \sqrt{\frac{s^2_{\text{pooled}}}{n_1} + \frac{s^2_{\text{pooled}}}{n_2}}$$

$$= s_{\text{pooled}}\sqrt{\frac{1}{n_1} + \frac{1}{n_2}}$$

For the Tellevik data,

$$\text{est. st. error }(\bar{Y}_1 - \bar{Y}_2) = 221.3\sqrt{\frac{1}{28} + \frac{1}{22}} = 63.05$$

Example 11.4

Find the estimated standard error of the difference of sample means for Example 11.3.

Solution

In Example 11.3 we obtained a pooled variance of 5.483; both sample sizes were 18. Thus

$$\text{est. st. error} = \sqrt{5.483}\sqrt{\frac{1}{18} + \frac{1}{18}} = 0.781$$

□

Once we've obtained the estimated standard error, inference procedures very much like those of Chapter 10 may be employed. In particular, confidence intervals follow the "statistic plus or minus table value times estimated standard error" format.

100(1 − α)% Confidence Interval for Difference of Means, Two Independent Samples

$$(\bar{y}_1 - \bar{y}_2) - t_{\alpha/2}(s_{\text{pooled}})\sqrt{\frac{1}{n_1} + \frac{1}{n_2}} \le \mu_1 - \mu_2$$

$$\le (\bar{y}_1 - \bar{y}_2) + t_{\alpha/2}(s_{\text{pooled}})\sqrt{\frac{1}{n_1} + \frac{1}{n_2}}$$

NOTE: $t_{\alpha/2}$ is the Table 4 value cutting off a right-tail area of $\alpha/2$ in a t distribution with $n_1 + n_2 - 2$ df. The df is the sum of the df for samples 1 and 2.

For the Tellevik data, a 95% confidence interval requires that we find the $t_{.025}$ value for 28 + 22 − 2 = 48 df. The table values for 40 and 60 df are 2.021 and 2.000, respectively. Interpolation yields

$$t_{.025} = \frac{12}{20}2.021 + \frac{8}{20}2.000 = 2.013$$

The confidence interval is

$$(427 - 472) - 2.013\left(221.3\sqrt{\frac{1}{28} + \frac{1}{22}}\right) \le \mu_1 - \mu_2$$

$$\le (427 - 472) + 2.013\left(221.3\sqrt{\frac{1}{28} + \frac{1}{22}}\right)$$

or

$$-171.9 \le \mu_1 - \mu_2 \le 81.9$$

Example 11.5 Refer to Examples 11.3 and 11.4. Calculate a 99% confidence interval for the difference of population means (say, as low expectations minus high expectations).

Solution There are 18 + 18 − 2 = 34 df. By rough interpolation, the a = .005 table value needed for a two-sided 99% interval is 2.73. The estimated standard error was found in Example 11.4 to be 0.781. The difference in means in Example 11.3 is 5.67 − 10.44 = −4.77. Thus the 99% confidence interval is

$$-4.77 - 2.73(0.781) < \mu_L - \mu_H < -4.77 + 2.73(0.781)$$

or

$$-6.90 < \mu_L - \mu_H < -2.64$$ □

It is also possible to define lower-bound or upper-bound confidence intervals, similar to those defined in Chapter 8. The lower-bound interval is

$$\mu_1 - \mu_2 > (\bar{y}_1 - \bar{y}_2) - t_\alpha(s_{\text{pooled}})\sqrt{\frac{1}{n_1} + \frac{1}{n_2}}$$

and the upper-bound interval is

$$\mu_1 - \mu_2 < (\bar{y}_1 - \bar{y}_2) + t_\alpha(s_{\text{pooled}})\sqrt{\frac{1}{n_1} + \frac{1}{n_2}}$$

Note that the form for the lower-bound interval is "estimate minus table value times estimated standard error" and the form for the upper-bound interval is "estimate plus table value times estimated standard error."

Example 11.6 Refer again to Examples 11.3 and 11.4. Calculate a 99% upper-bound interval for the difference (low minus high) of means.

Solution By rough interpolation, the $a = .01$ value for $18 + 18 - 2 = 34$ df is about 2.44. The interval is

$$\mu_L - \mu_H < -4.77 + 2.44(0.781)$$

or

$$\mu_L - \mu_H < -2.86$$ □

Hypothesis tests using pooled-variance t also follow a generic format. The test statistic is

$$\frac{(\text{estimate} - H_0 \text{ value})}{\text{est. std. error}}$$

For the Tellevik data, the estimate of the difference in means is $427 - 472 = -45$, the natural null hypothesis expected value is 0 (corresponding to a null hypothesis that the mean times are equal), and the estimated standard error was found previously to be 63.05. Thus a t statistic is

$$t = \frac{-45 - 0}{63.05} = -0.714$$

The t-statistic value should be compared to a t-table value with $28 + 22 - 2 = 48$ df. A t statistic equal to -0.714 doesn't even come close to t-table values for any reasonable α, so the null hypothesis would be retained.

Hypothesis Test for a Difference of Sample Means, Two Independent Samples

1. H_0: $\mu_1 - \mu_2 = D_0$ (D_0 will often be 0.)
2. H_a:
   (i) $\mu_1 - \mu_2 > D_0$
   (ii) $\mu_1 - \mu_2 < D_0$
   (iii) $\mu_1 - \mu_2 \neq D_0$
3. T.S.: $t = \dfrac{(\bar{y}_1 - \bar{y}_2) - D_0}{s_{\text{pooled}}\sqrt{\dfrac{1}{n_1} + \dfrac{1}{n_2}}}$
4. R.R.:
   (i) reject H_0 if $t > t_\alpha$
   (ii) reject H_0 if $t < -t_\alpha$
   (iii) reject H_0 if $|t| > t_{\alpha/2}$
5. Conclusion

Example 11.7 For the data of Example 11.3, test the research hypothesis that the mean number of symbols taught to the low-expectation children is less than the mean number for the high-expectation children. Use $\alpha = .01$.

Solution Taking group 1 as the low-expectation group, we have H_a: $\mu_L < \mu_H$. The t statistic equals

$$t = \frac{5.67 - 10.44}{0.781} = -6.121$$

Note that the estimated standard error 0.781 was calculated in Example 11.4. The t statistic is less than (more negative than) the t-table value for a left-tail area of .01 and 34 df, which is roughly -2.44. Thus the null hypothesis is rejected and the research hypothesis supported.

Alternatively, note that the upper-bound 99% interval calculated in Example 11.6 doesn't include 0. Thus, again, H_0 is rejected at $\alpha = .01$. □

The assumptions underlying the two-sample t procedure may be summarized as follows:

1. The populations are not seriously skewed, especially if the sample sizes are small. Each set of sample data should be plotted. If one or both of the samples reveal substantial skewness, t probabilities may be poor approximations for small sample sizes. In such a case the rank sum test to be defined in Section 11.4 is an alternative procedure that is not sensitive to skewness.

2. The populations are not heavy-tailed (outlier-prone). If plots of the samples indicate outliers or a long tail of straggler points, the sample means may not be efficient as estimators or test statistics. Again, the rank sum test of Section 11.4 should be considered, this time on grounds of efficiency rather than correctness.

3. The population variances are equal. Violation of this assumption turns out to be damaging if the sample sizes are substantially different. In particular, if one population variance is substantially larger, and if the corresponding sample size is smaller, the nominal t probability may be too small and the inferences too optimistic. If the sample sizes are not nearly equal, the t' test to be defined in Section 11.3 is less sensitive to the homoscedasticity assumption.

Example 11.8 Refer to the data of Example 11.3. Do there appear to be any serious violations of assumptions in the use of the pooled-variance t test?

Solution Clearly, the data are right-skewed. For equal-sized samples of 18 each, giving a total n of 36, the nominal t probabilities shouldn't be too terrible. Perhaps an alternative procedure might be even more effective than a pooled-variance t test (which rejected H_0 at $\alpha = .01$). The sample variances aren't terribly far apart, and the sample sizes are equal, so there should be no problem with the constant-variance assumption. □

balanced design

The results of the last two sections have one important implication for the planning of an experiment. There are major advantages to a **balanced design**—equal sample sizes in the two samples. If the variances are in fact equal, a balanced design yields the smallest standard error for a given total number of observations. If the variances are in fact unequal, a balanced design protects the pooled-variance t against serious error. Finally, a balanced design seems to give the Central Limit Theorem its maximum effect in protecting against skewness in the underlying population. It isn't always possible to achieve a balanced design in practice; individuals may drop out of the study or be lost for other reasons. Data should not be thrown away to achieve balanced design. It's hard enough to obtain good data; data shouldn't be wasted. But a balanced design is a good research objective in most cases. Only if the experimenter somehow knows that one population has much more variability than another should the design be unbalanced. In such a case, the standard error of the difference of sample means can be made smaller by taking unequal sample sizes in proportion to the unequal standard deviations.

Example 11.9 Refer to Example 11.3 again. Suppose that another res[illegible]her is planning a repeat study with new samples of children. Is th[illegible]re[illegible] son not to balance the design?

Solution The only reason to unbalance a design deliberately is if o[illegible] ulation is known to be much more variable than the other. T[illegible] ability in the two original samples was quite similar, so a balanced de- sign seems best. □

Given the assumptions of independence, normal populations, and equal variances, the pooled-variance t test may be derived from the Generalized Likelihood Ratio principle stated in Appendix B of Chapter 10. See Mood, Graybill, and Boes (1974).

11.3 An Approximate t Test for Unequal Variances

The pooled-variance t inferences discussed in Section 11.2 were based on the assumption (among others) that the underlying population variances were equal. That assumption was made out of mathematical necessity, to guarantee that the resulting statistic would follow a t distribution. In some situations, particularly when the sample sizes are unbalanced, it is better to avoid making the gratuitous homoscedasticity assumption, at the price of having only approximate t probabilities. The methods discussed in this section are approximately t-distribution methods, but do not depend on the equal-variance assumption.

The problem goes back to the true standard error of the difference in sample means, as derived in Section 11.1.

$$\text{true std. error} = \sqrt{\frac{\sigma_1^2}{n_1} + \frac{\sigma_2^2}{n_2}}$$

heteroscedasticity

Suppose that we are not willing to make the equal-variance, homoscedasticity assumption, but rather assume unequal population variances. (The unequal-variance assumption is called **heteroscedasticity**, or, as a student once suggested, "antihomoscedasticity.") Then the reasonable way to estimate the standard error is to replace each σ^2 by the corresponding s^2.

$$\text{est. std. error} = \sqrt{\frac{s_1^2}{n_1} + \frac{s_2^2}{n_2}}$$

As indicated in Section 11.2, this substitution does not lead to a t statistic. However, Welch (1938) showed that a reasonable approximation was to treat the resulting statistic as t distributed with a certain approximate df. The resulting hypothesis-testing method is shown in the box below.

Hypothesis Test for a Difference of Means (Variances Not Assumed to Be Equal)

1. H_0: $\mu_1 - \mu_2 = D_0$ (D_0 often is 0.)
2. H_a:
   (i) $\mu_1 - \mu_2 > D_0$
   (ii) $\mu_1 - \mu_2 < D_0$
   (iii) $\mu_1 - \mu_2 \neq D_0$
3. T.S.: $t' = \dfrac{\bar{y}_1 - \bar{y}_2}{\sqrt{(s_1^2/n_1) + (s_2^2/n_2)}}$
4. R.R.:
   (i) reject H_0 if $t' > t_\alpha$
   (ii) reject H_0 if $t' < -t_\alpha$
   (iii) reject H_0 if $|t'| > t_{\alpha/2}$

   where t is based on

   $$\text{df} = \frac{(n_1 - 1)(n_2 - 1)}{(n_2 - 1)c^2 + (n_1 - 1)(1 - c)^2}$$

   where

   $$c = \frac{s_1^2/n_1}{(s_1^2/n_1) + (s_2^2/n_2)}$$
5. Conclusion

For the Tellevik data of Section 11.2 (with H_0 being that the difference of population means is 0),

$$t' = \frac{(427 - 472) - 0}{\sqrt{(223.2)^2/28 + (218.8)^2/22}} = -0.716$$

which is virtually identical to $t_{\text{pooled}} = -0.714$ found in Section 11.2. The df approximation is

$$c = \frac{(223.2)^2/28}{(223.2)^2/28 + (218.8)^2/22} = .4498$$

$$\text{df} = \frac{(28 - 1)(22 - 1)}{(22 - 1)(.4498)^2 + (28 - 1)(1 - .4498)^2} = 45.6$$

instead of the 48 found in Section 11.2. In principle, we should obtain an interpolated t value for 45.6 df, but $t' = -0.716$ will be nowhere near statistically significant for any reasonable value.

Example 11.10 Loucks et al. (1979) reported on Jackson Personality Research Form scores for female and male medical students. Assume that the results in Table 11.5 were obtained (the sample sizes and means are those given in the article). Perform a t' test based on these data.

Sex	Mean	Standard Deviation	n
F	55.6	11.62	68
M	50.6	8.47	178

Table 11.5

Solution

$$t' = \frac{55.6 - 50.6}{\sqrt{(11.62)^2/68 + (8.47)^2/178}} = 3.235$$

To calculate approximate df, note that

$$c = \frac{(11.62)^2/68}{(11.62)^2/68 + (8.47)^2/178} = 0.8313$$

$$\text{df} = \frac{(68 - 1)(178 - 1)}{(178 - 1)(.8313)^2 + (68 - 1)(1 - .8313)^2} = 95.5$$

For 60 df or 120 df, shown in the t table, the value 3.235 is "off the table." The two-tailed p-value is less than $2(.005) = .01$. □

Confidence intervals can also be constructed using the t' method. The estimated standard error is based on replacing each population variance by the corresponding sample variance, and the t-table value is based on the approximate df.

100(1 − α)% Confidence Interval for a Difference of Means (Variances Not Assumed to Be Equal)

$$(\bar{y}_1 - \bar{y}_2) - t_{\alpha/2}\sqrt{\frac{s_1^2}{n_1} + \frac{s_2^2}{n_2}} \leq \mu_1 - \mu_2$$

$$\leq (\bar{y}_1 - \bar{y}_2) - t_{\alpha/2}\sqrt{\frac{s_1^2}{n_1} + \frac{s_2^2}{n_2}}$$

NOTE: The df for $t_{\alpha/2}$ are based on the approximate df given in the hypothesis test (above).

Example 11.11 Refer to Example 11.10. Calculate a 90% confidence interval for the difference of female and male means, using the unequal-variances method.

Solution In Example 11.10 we found that there were approximately 95.5 df; for this df, the t-table value (one-tail area .05, corresponding to 90% confidence) is about 1.664. The interval is

$$(55.6 - 50.6) - 1.664\sqrt{\frac{(11.62)^2}{68} + \frac{(8.47)^2}{178}} < \mu_F - \mu_M$$

$$< (55.6 - 50.6) + 1.664\sqrt{\frac{(11.62)^2}{68} + \frac{(8.47)^2}{178}}$$

or

$$2.43 < \mu_F - \mu_M < 7.57$$

□

The major difference between the approximate t' methods and the pooled-variance t methods of Section 11.2 is that the t' method does not require the assumption of equal population variances. Neither should be used when the n's are small and the populations badly skewed, because then all t probabilities are suspect. Nor should they be used when the data contain outliers, because then the sample mean is not an efficient method of estimating the population mean. In either of these cases, the rank sum method of Section 11.4 is a more desirable approach.

When the samples are reasonably near symmetric and not plagued by outliers, the choice between t and t' depends on the (sample) variances and the sample sizes. If the sample sizes are equal, then algebraically $t = t'$; the approximate df typically are slightly less than the df $= n_1 + n_2 - 2$ used in Section 11.2. As a result, confidence intervals are slightly wider using t', in the balanced design case, and tests have slightly less power. When the sample sizes are not equal, the t_{pooled} inferences are not nominal-robust to the assumption of equal variances; if the sample sizes and variances are unequal, the nominal confidence level and probability of the t_{pooled} method can be seriously wrong. Therefore, the t_{pooled} method is slightly preferred for balanced designs, but the t' method is substantially preferred for unbalanced designs. Often, as for the Tellevik data, essentially identical conclusions are reached by the two methods.

Example 11.12

A computer simulation study compared the performance of pooled-variance t and t' tests. In the study, the respective sample sizes were 10 and 30, the two population means were both equal to 50, and the respective population standard deviations were 10.0 and 5.0. Both populations were somewhat outlier-prone. The results in Table 11.6 were obtained for 1000 t and t' tests.

Of the two tests, which one seems preferable? Why?

using the pooled variance

alpha	number of times H0: "mu1-mu2 > 0"	"mu1-mu2 is 0" is rejected in favor of "mu1-mu2 < 0"	total (alpha doubled)
0.100	166	156	322
0.050	116	96	212
0.025	74	62	136
0.010	42	33	75
0.005	24	16	40

using separate variances (t′)

alpha	number of times H0: "mu1-mu2 > 0"	"mu1-mu2 is 0" is rejected in favor of "mu1-mu2 < 0"	total (alpha doubled)
0.100	94	94	188
0.050	47	43	90
0.025	19	17	36
0.010	6	5	11
0.005	3	2	5

Table 11.6

Solution In this simulation, both population means are equal, so H_0 is true. Every rejection of the null hypothesis is a Type I (false positive) error. Thus, the question becomes Which of the two tests yields a more accurate approximation to the nominal α? Consistently, there are too many Type I errors when the pooled-variance t test is used. For example, where nominally $\alpha = .10$ so we would expect 100, 100, and a total of 200 rejections, we actually got 166, 156, and a total of 322. The number of rejections for the t' test is consistently slightly smaller (more conservative) than the nominal α would indicate. The t' test is clearly preferable here. Notice that the sample sizes and the standard deviations are both unequal. □

Exercises for Sections 11.2 and 11.3

11.4. Data were obtained on the length of stay in a state mental hospital for patients suffering from a certain disorder and treated by one of two methods. The times (to the nearest month) were

Method A:	0	0	1	1	1	2	2	3	4
	4	6	8	8	8	9	10	10	11
	12	13	16	18	19	25	29	36	41

(mean 11.00, standard deviation 10.94)

Method B:	0	1	1	5	17	35	58

(mean 16.71, standard deviation 22.17)

a. Calculate the pooled variance estimate.

b. Calculate the estimated standard error.

c. Calculate the pooled-variance t statistic for the null hypothesis of equal population means. Is the result statistically significant at $\alpha = .10$ (two-tailed test)?

d. Find the two-tailed p-value.

11.5. For the data of Exercise 11.4, calculate a pooled-variance, 90% confidence interval for the difference of true population means. Does the interval include 0? Should it, according to the test in part c of Exercise 11.4?

11.6. Calculate the t' statistic for the data of Exercise 11.4. Find the approximate df. Is the t' result statistically significant at $\alpha = .10$?

11.7. In the data of Exercise 11.4, is there evidence of a violation of assumptions for a t test? Which test, t_{pooled} or t', is preferred? Is either one appropriate for this situation?

11.8. It has been claimed that teenage boys are much more interested in microcomputers than are teenage girls. Separate random samples of 12 boys and 12 girls, aged 16 years, were obtained, and scores on a scale of eagerness/resistance found.

Boys:	94	80	102	140	88	96
	128	110	116	125	105	112
Girls:	110	64	80	92	96	89
	120	90	93	101	86	71

a. Calculate a 95% confidence interval for the difference of means, using the pooled variance.

b. According to this interval, is the difference statistically significant at $\alpha = .05$?

11.9. For the data of Exercise 11.8, calculate t_{pooled}. Find bounds on the p-value.

11.10. For the data of Exercise 11.8, calculate t'. Find bounds on the p-value. Does it matter much whether t_{pooled} or t' is used?

11.11. Do the data of Exercise 11.8 suggest that the assumptions underlying the pooled-variance t test are seriously violated?

11.12. Calculate a 95% confidence interval for the difference of sample means based on the data of Exercise 11.8, using the t' method. How does the width of this interval compare to the width of the interval in Exercise 11.8?

11.4 Two-Sample Rank Methods

The t and t' methods of Sections 11.2 and 11.3 are applicable when the underlying population distributions are not terribly skewed or outlier-prone. But when the sample data are badly skewed or heavy-tailed, a reasonable inference is that the populations are skewed and/or heavy-tailed, so t methods are incorrect (the stated probabilities are poor approximations) or inefficient (better use can be made of the data). This section introduces rank methods that are not at all sensitive to skewness, and are more efficient than t methods in the face of heavy-tailed data.

For example, suppose that an experiment involves random assignment of subjects to treatment or control groups in a memory task. The treatment group attempts to learn a series of words with the aid of some theoretically useful mnemonics. The control group tries to learn the same words without mnemonics. The data are the number of words recalled the next day. The data, for samples of size 10 each, are

Treatment: 17 27 33 38 43 47 50 52 54 55
($\bar{y} = 41.600$, $s = 12.686$)

Control: 7 13 25 29 30 36 37 41 45 46
($\bar{y} = 30.700$, $s = 13.008$)

The question is whether the treatment group has a higher mean number of recalled words than does the control group. Within each group, the data are somewhat left-skewed. The total sample size is only 20, so a normal or t approximation is very suspect. The (supposedly) t statistic equals 1.90, but the assumptions underlying a t test do not appear to be met. A rank test, based on the relative values of the data but not on their magnitudes, is in order.

The null hypothesis H_0 of this rank test is that the population distributions from which the samples are drawn are identical. The research hypothesis H_a is that one population is shifted over from the other. H_a may be stated that the difference of means is not 0 or that the difference of medians is not 0. As Figure 11.1 indicates, if in fact the two populations differ only by being shifted over, one from the other, then the difference of means equals the difference of medians. Thus it really doesn't matter much how the hypotheses are stated. The point is that the rank sum test is primarily sensitive to difference of location (mean or median), not difference of variability.

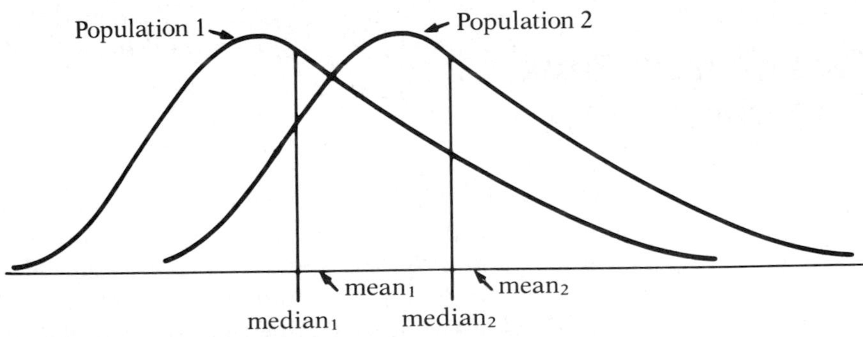

Figure 11.1 Shifted populations

The Wilcoxon rank sum test is based on the ranks of the individual data values, ordered from lowest to highest. The lowest value in the combined data is assigned rank 1, the next lowest rank 2, etc. For the word-recall data:

Treatment	17	27	33	38	43	47	50	52	54	55
Rank	3	5	8	11	13	16	17	18	19	20

Control	7	13	25	29	30	36	37	41	45	46
Rank	1	2	4	6	7	9	10	12	14	15

As the name suggests, the rank sum test is based on the sum of the ranks in each sample. For the word-recall data, the rank sums are 130 for the treatment scores and 80 for the control scores. The total of the rank sums is simply the sum of the numbers

$$1 + 2 + \cdots + (n_1 + n_2) = (n_1 + n_2)\frac{(n_1 + n_2 + 1)}{2}$$

so knowledge of one rank sum determines the other. Thus the test can be based on T = the sum of the ranks in the first group. The mechanics of the rank sum test are defined in the box below.

Rank Sum Test for Means (or Medians), Two Independent Samples

1. H_0: $\mu_1 - \mu_2 = D_0$ (D_0 often is 0.)

2. H_a:

(i) $\mu_1 - \mu_2 > D_0$
(ii) $\mu_1 - \mu_2 < D_0$
(iii) $\mu_1 - \mu_2 \neq D_0$

3. T.S.: T = the total (sum) of ranks in sample 1

4. R.R.: if n_1 or $n_2 < 10$, consult tables in Hollander and Wolfe (1974); if both n_1 and $n_2 \geq 10$, calculate

$$z = \frac{T - (n_1)(n_1 + n_2 + 1)/2}{\sqrt{n_1 n_2 (n_1 + n_2 + 1)/12}}$$

Reject H_0 if

(i) $z > z_\alpha$
(ii) $z < -z_\alpha$
(iii) $|z| > z_{\alpha/2}$

NOTE: The formal null hypothesis is that the two populations have identical shapes. The rank sum test is sensitive largely to differences in means (or medians), as opposed to (say) differences in variances. If $D_0 \neq 0$, subtract D_0 from all sample 1 values before ranking.

For the example, $T = 130$ and both n's are 10. Thus

$$z = \frac{130 - 10(21/2)}{\sqrt{(10)(10)(21)/12}} = 1.89$$

The one-tailed p-value for $z = 1.89$ is .0294.

Example 11.13 A study of math aptitude among seventh-grade children involved assigning 15 children randomly to an experimental program, and another 15 children to a control, standard program. At the end of the school year, all the children were tested for math aptitude, yielding the following scores:

Experimental:	42	76	89	94	98	106	108	109
	111	112	114	119	126	151	175	
Control:	33	55	62	75	79	84	88	90
	92	93	96	99	107	120	146	

(a) Assign ranks to each score.
(b) Find the rank sums.
(c) Calculate the rank sum z statistic, and state the conclusion to be drawn from the statistic.

Solution **(a)** The data are arranged in order within each sample, which is the first step in ranking. After ordering the data, one simply assigns ranks from low to high:

Experimental:	2	6	10	14	16	18	20	21
	22	23	24	25	27	29	30	
Control:	1	3	4	5	7	8	9	11
	12	13	15	17	19	26	28	

(b) The rank sums are obtained by simple addition. For the first sample, the sum is 287, and for the second sample, the sum is 178.

(c)

$$z = \frac{287 - 15(31/2)}{\sqrt{(15)(15)(31)/12}} = 2.26$$

The one-tailed p-value corresponding to $z = 2.26$ is .0119, so the two-tailed p-value is .0238. There is fairly conclusive evidence that there is a more than random difference in mean scores between the two groups. □

An equivalent test to the rank sum test is the Mann-Whitney test, based on U = the number of times that a score from the first sample is smaller than the score for the second sample. In the word-recall illustration, the 17 score in the first sample is larger than two scores (7 and 13) in the second sample, the 27 score in the first is larger than three scores in the second sample, and so on. The total number of times is $U = 2 + 3 + \cdots = 75$. U is linearly related to T, the sum of the ranks in the first sample.

$$U = n_1 n_2 + \frac{n_1(n_1 + 1)}{2} - T$$

so the two statistics yield equivalent tests. Some tables and computer outputs use U (or U' = the number of times that a score in the first sample is *larger* than a score in the second sample). The easy thing to do is to convert U to T, or vice versa.

Example 11.14 Find the value of the Mann-Whitney U statistic for the data of Example 11.13.

Solution Both n_1 and n_2 equal 15.

$$U = 15(15) + \frac{15(16)}{2} - 287 = 58$$

Alternatively, the 42 score in the experimental group is smaller than 14 of the 15 control scores (all but the 33 score). The 76 score is less than 11 of the 15 control scores, and so on up to the 175 score, which is less than 0 control scores. The total number of experimental scores that are less than control scores is $14 + 11 + \cdots + 0 = 58$. □

Up to now, the values being ranked have all been different. There is a possibility that two or more values in the combined sample may be

ties equal, so there are **ties** in the ranking. In this situation, assign the average rank to all tied scores. For example, suppose that the word-recall data had been

Experimental:	17	27	33	38	43	47	50	52	52	52
Control:	7	17	27	27	30	36	38	41	45	46

The lowest value, 7, gets rank 1. The next two scores are both 17, tied for ranks 2 and 3; they are both assigned rank $(2 + 3)/2 = 2.5$. The next three scores are all 27, tied for ranks 4, 5, and 6; they are assigned rank $(4 + 5 + 6)/3 = 5$. The complete set of ranks is

Experimental	17	27	33	38	43	47	50	52	52	52
Rank	2.5	5	8	10.5	13	16	17	19	19	19
Control	7	17	27	27	30	36	38	41	45	46
Rank	1	2.5	5	5	7	9	10.5	12	14	15

Once the ranking has been done, the test proceeds as before. Some texts [e.g., Hollander and Wolfe (1974)] show a correction to the z statistic to account for ties. Unless there are an enormous number of ties, such as might occur if there were a large number of observations on a 1-to-5 scale, the effect of the correction is negligible.

Example 11.15 Perform a rank sum test on the data of Example 11.3.

Solution There are many ties.

LOW EXPECTATION

Value	3	3	3	4	4	4	5	5	5	6	6	6	6	7	7	8	9	11
Rank	2	2	2	5	5	5	8	8	8	11.5	11.5	11.5	11.5	15	15	18.5	23	29.5

HIGH EXPECTATION

Value	7	8	8	8	9	9	9	9	10	10	11	11	11	12	12	13	14	17
Rank	15	18.5	18.5	18.5	23	23	23	23	26.5	26.5	29.5	29.5	29.5	32.5	32.5	34	35	36

The sum of ranks in the low-expectation group is 192.0. The test statistic is

$$z = \frac{192.0 - 18(18 + 18 + 1)/2}{\sqrt{(18)(18)(18 + 18 + 1)/12}} = -4.46$$

Noting that both n's are larger than 10, we refer -4.46 to the normal table and find that the result is far below the lowest (most negative) table value. The discrepancy is statistically significant with a very small p-value. The ties correction increases the value of z slightly. There's no need to show the ties correction; we merely mention that in this example, the ties-corrected $z = -4.48$, as compared to the uncorrected $z = -4.46$. □

As indicated previously, use of the rank sum test is indicated when the underlying populations are substantially different from normal. When the underlying populations are skewed and one or both of the sample sizes are small, use of t probabilities will produce a poor approximation, particularly for a one-sided test. In this case, one can use the rank sum test, which requires no assumption about the shape of the populations. (In some situations, it is possible to assume a specific, nonnormal shape to the populations and base the test on a distribution different from t. We shall not pursue this possibility.) The word-recall data shows rather substantial skewness; one can have more faith in the z probability computed from the rank sum statistic than in the t probability.

When the populations are nearly symmetric but outlier-prone, the use of means in a t test is not very efficient. In hypothesis-testing language, the t test will not be as powerful as the rank sum test if the populations are outlier-prone. It is desirable to look at the shapes of the two samples (separately) to decide the most appropriate test.

Thus there is a basic question as to what is the most effective (efficient) test in a given situation. There is a very useful technical definition of efficiency that helps in deciding which test to use.

Efficiency of a Test Relative to Another Test

Hypothesis test A has $x\%$ efficiency relative to hypothesis test B if the power of test A for a certain (large) sample size n is the same as the power of test B for a sample of size xn. Thus, if test A is 95% efficient relative to test B, the power of test A for a sample of size 100 is the same as the power of test B for a sample of size $(.95)100 = 95$.

It can be shown [as in Hollander and Wolfe (1974)] that the efficiency of the rank sum test relative to the t test is about 95%, even if the normal-population assumptions are met. For whatever population shape, the efficiency of the rank sum test relative to the t test is never worse than 86.4%. Thus, regardless of the population shape, the rank sum test is never much less effective than the t test. For many nonnormal populations, particularly outlier-prone populations, the rank sum test may be much more efficient than the t test.

Example 11.16 Refer to Examples 11.3, 11.7, and 11.15. Which test (t_{pooled} or the rank sum test) is more believable? More effective?

Solution The data of Example 11.3 are highly skewed, but the sample size is 36. The rank sum test doesn't rely on population normality, so it is slightly more believable. The rank sum test isn't necessarily more effective, yielding a z of -4.46 as compared to a t of -6.121. Of course, there's a real question as to whether one can believe t probabilities, given the skewness and the modest sample size. □

The rank sum method also can be used to obtain a confidence interval for the difference in means. The basic idea is that any difference of (medians or) means $\mu_1 - \mu_2 = D_0$ that would be accepted by the rank sum test (say, at $\alpha = .05$) should be included in the (say, 95%) confidence interval, whereas any D_0 that would not be accepted would not be included. To test H_0: $\mu_1 - \mu_2 = D_0$, subtract D_0 from each observation in the first sample before ranking. Hodges and Lehmann (1963) developed methods for performing the computations; normally they are done by a computer program. When the data are skewed or heavy-tailed, the resulting interval is more reliable or shorter than the t interval.

Example 11.17 A Minitab computer package analysis of the data of Example 11.13 yielded the following results:

```
LOWEXP  N = 18  MEDIAN =  5.500
 HIEXP  N = 18  MEDIAN = 10.000

A POINT ESTIMATE FOR ETA1 - ETA2 IS -5.0.
A 95.2 PERCENT C.I. FOR ETA1 - ETA2 IS ( -6.0, -3.0)
```

In Minitab, ETA stands for the population median. Note that if one population is shifted over from the other, the difference of medians equals the difference of means. Can the null hypothesis of equal medians (or equal means) be rejected at $\alpha = .05$ (or, rather, at $\alpha = .048$)?

Solution The 95.2% confidence level corresponds to $\alpha = .048$. The 95.2% confidence interval doesn't include 0. Indeed, it doesn't come close. Thus we may conclusively reject the null hypothesis that the difference of location parameters (medians or means) is 0. This conclusion is the same as that reached in Example 11.13. □

Exercises for Section 11.4

11.13. Perform a rank sum test for the data of Exercise 11.8. Find the two-tailed p-value.

11.14. How does the conclusion of the rank sum test in Exercise 11.13 compare to the conclusions of the t_{pooled} and t' tests? For these data, does it appear that the rank sum test is more believable than the two t tests?

11.15. Perform a rank sum test for the data of Exercise 11.4. Assume that the z approximation is applicable. Can the null hypothesis of equal distribution of hospital stays be rejected using $\alpha = .10$?

11.16. Compare the p-values for the tests of Exercises 11.4, 11.6, and 11.15. Is there reason to believe that any of these p-values is not a good approximation?

11.17. Find the Mann-Whitney statistic for the data of Exercise 11.4.

11.5 Size of *t* Confidence Intervals and Power of a *t* Test

The preceding three sections have focused on the actual use of some standard statistical methods. Now we turn to some more theoretical questions about these methods. The issues discussed in this section are particularly important in planning studies and experiments involving two independent, random samples.

In Section 10.2 we considered the width of a confidence interval for a single mean, based on t-distribution methods. Now we consider the same issue for the two-sample problem. First, suppose that we are willing to make the constant-variance (homoscedasticity) assumption, so that pooled-variance t methods may be employed. The plus-or-minus of the confidence interval is

$$t_{\alpha/2}\left(s_{\text{pooled}}\sqrt{\frac{1}{n_1}+\frac{1}{n_2}}\right)$$

where $t_{\alpha/2}$ is the appropriate t-table value for $n_1 + n_2 - 2$ df. The width of the interval increases as the α value decreases, because the t value increases. Evidently, the width also increases if s_{pooled} gets larger. The width decreases if either n_1 or n_2 gets larger. It is not so obvious what happens if one n increases but the other decreases by an equal amount. It can be shown that, for any fixed value of $n_{\text{total}} = n_1 + n_2$, the width of the confidence interval is smallest when $n_1 = n_2$ (in other words, when the design is balanced).

Example 11.18

Suppose a followup to the Tellevik study discussed in this chapter is being planned. For planning purposes, it's reasonable to assume that the pooled standard deviation will be the same value found in the original study, $s_{\text{pooled}} = 221.3$. A total sample size of 30 is planned. It is possible to have 15 hearing and 15 deaf subjects or to have 13 hearing and 17 deaf subjects. Calculate the width of a 90% confidence interval for each possibility. Which interval is shorter?

Solution

In either case the df will be $30 - 2 = 28$. The appropriate t-table value for 90% confidence is 1.701. For $n_1 = n_2 = 15$, the plus-or-minus will be

$$1.701(221.3)\sqrt{\frac{1}{15} + \frac{1}{15}} = 137.45$$

For $n_1 = 13$ and $n_2 = 17$, the plus-or-minus will be

$$1.701(221.3)\sqrt{\frac{1}{13} + \frac{1}{17}} = 138.69$$

In either case, the width is twice the plus-or-minus value. The interval with $n_1 = n_2 = 15$ is slightly narrower than the interval with slightly unbalanced sample sizes. □

The situation is slightly more complicated for the t'-based confidence interval. In particular, when one is not willing to assume that the variances are equal, it is not true that the optimal (shortest interval) way to choose the sample sizes is to make them equal. Instead, it can be shown that the sample sizes should be chosen in proportion to the standard deviations. Thus, if the first population's standard deviation is twice the second one's, then the first sample size should be twice the second one, to minimize the width of the confidence interval for a specified n_{total}. In practice, one would not be likely to know the relative sizes of the standard deviations, so a certain amount of "guesstimating" would be needed. If one is convinced that one of the standard deviations is clearly larger than the other, it is desirable to allocate a larger sample size to the larger standard deviation population. In the usual case in which one doesn't know in advance which standard deviation is larger, the safest procedure is to balance the design as closely as possible.

Example 11.19

Refer to Exercise 11.18. For planning purposes, assume that the sample standard deviations for hearing and deaf children will be 200 and 240, respectively. Calculate the widths of the t' confidence intervals for $n_1 = n_2 = 15$ and for $n_1 = 13$, $n_2 = 17$.

Solution If $n_1 = n_2 = 15$, the approximate df for t' works out to 27.1, which may be rounded down to 27. The corresponding table value is 1.703. The plus-or-minus will be

$$1.703 \sqrt{\frac{240^2}{15} + \frac{200^2}{15}} = 137.37$$

If $n_1 = 17$ and $n_2 = 13$, the approximate df is 27.7. It doesn't really matter whether we use the table value for 27 df, 1.703, or for 28 df, 1.701. To be conservative, we round down and use 1.703. The plus-or-minus will be

$$1.703 \sqrt{\frac{240^2}{17} + \frac{200^2}{13}} = 136.93$$

Of course, the width of the interval will be twice the plus-or-minus value. Note that the slightly unbalanced design yields a very slightly smaller width for the confidence interval. □

If a balanced design is planned, it's not too hard to determine the overall sample size required to yield a confidence interval with a width equal to a specified fraction of s_{pooled}. The procedure is very similar to the method in Chapter 10 for the one-sample problem. Specify a confidence level and a desired plus-or-minus, as a multiple of s_{pooled}. Initially, assume that the resulting df will be large enough that z-table values apply. Set the actual plus-or-minus equal to the desired plus-or-minus, and solve for the n of each sample. If the resulting df are small enough that z probabilities are not an adequate approximation to t probabilities, increase each n equally, and proceed by trial and error. For example, suppose that a 95% confidence interval with a plus-or-minus of $.3s_{\text{pooled}}$ is desired. Initially, assume that the z-table value for 95% confidence, 1.96, is an adequate approximation. Set

$$1.96 s_{\text{pooled}} \sqrt{\frac{1}{n} + \frac{1}{n}} = .3 s_{\text{pooled}}$$

where n represents the size of each sample in a balanced design. Solve to obtain

$$\frac{1.96\sqrt{2}}{.3} = \sqrt{n}$$

or

$$n = 85.4$$

rounded up to $n = 86$. The resulting df $= 86 + 86 - 2 = 170$ is large enough that 1.96 is very close to the exact t-table value, so no further work need be done. Had the required n's been smaller, we could have increased them by 1 or 2 and calculated the resulting plus-or-minus, until the desired value had been reached.

Example 11.20 Refer to Example 11.18. Are sample sizes of 15 each large enough to yield a 90% confidence interval with a plus-or-minus equal to $.4s_{\text{pooled}}$? If not, what sample sizes are needed?

Solution In Example 11.18 we found the required t-table value to be 1.701. The plus-or-minus will be

$$1.701s_{\text{pooled}}\sqrt{\frac{1}{15}+\frac{1}{15}} = .621s_{\text{pooled}}$$

Therefore samples of size 15 each aren't sufficient. To find the required n's, we assume that the sample sizes will be large enough for the normal table value, 1.645, to apply. We solve

$$\frac{1.645s_{\text{pooled}}\sqrt{2}}{.4s_{\text{pooled}}} = \sqrt{n}$$

and obtain $n = 34$ in each sample. For $n_1 = n_2 = 34$, there are 66 df. By rough interpolation, the t-table value is about 1.670. The plus-or-minus of the interval will be

$$1.670s_{\text{pooled}}\sqrt{\frac{1}{34}+\frac{1}{34}} = .405s_{\text{pooled}}$$

If the sample sizes are increased to 35, the t-table value will still be about 1.67, and the plus-or-minus will be about $.399s_{\text{pooled}}$. Thus samples of size 35 each are needed. □

Just as the methods for establishing the width of a confidence interval are extensions of Chapter 10 methods, so are methods for obtaining the power of t tests. When the research hypothesis H_a is true, the power of the pooled-variance t test is determined by the **noncentral t distribution.** For balanced designs, the noncentrality parameter is given by

noncentral t distribution

$$\delta = \frac{\mu_1 - \mu_2}{\sqrt{2\sigma^2_{\text{common}}}}$$

and the size of each sample by n. The 2 factor occurs because, given a common population variance and a balanced design, the true standard error is

$$\sqrt{\frac{\sigma^2_{\text{common}}}{n} + \frac{\sigma^2_{\text{common}}}{n}} = \sqrt{\frac{2}{n}\,\sigma^2_{\text{common}}}$$

The tables in Cohen (1977) may be used, or the Scheffé (1959) approximation may be used to obtain noncentral t probabilities.

Approximate Power of a Two-Sample t Test

One-tailed test:

$$P\left(z > \frac{z_\alpha - \sqrt{n}\,(\delta)}{1 + t^2_\alpha/2(\text{df})}\right)$$

Two-tailed test:

$$P\left(z > \frac{z_{\alpha/2} - \sqrt{n}\,(\delta)}{1 + t^2_{\alpha/2}/2(\text{df})}\right)$$

NOTE: n is the common sample size of both samples.

For example, suppose that a two-tailed pooled-variance t test is to be run (using $\alpha = .01$) with both sample sizes equal to 21. Further suppose that the true, population mean for the first group is 1/2 of a standard deviation larger than the population mean for the second group. What is the probability that the research hypothesis will be supported?

The question is exactly the question of the power of the t test, assuming $\alpha = .01$ (two-tailed) and

$$\frac{\mu_1 - \mu_2}{\sigma_{\text{common}}} = 0.5$$

with $n = 21$. Thus $\delta = .5/\sqrt{2} = 0.35355$. For $n_1 = n_2 = 21$, there are 40 df, so the table value $t_{\alpha/2} = t_{.005} = 2.704$. The z-table value is $z_{.005} = 2.58$. The power is approximately

$$P\left(z > \frac{2.58 - \sqrt{21}\,(0.35355)}{1 + (2.704)^2/2(40)}\right) = P(z > 0.88) = .1894$$

For this case the power is very low.

Example 11.21 A pooled-variance t test is to be done on a study involving independent experimental and control samples, each of size 121. If the true, population mean of the experimental group is 3/10 of a standard deviation above the population mean of the control group, and if α is taken as .10 (two-tailed test), what is the power of the test?

Solution The noncentrality parameter is assumed to be $\delta = .3/\sqrt{2} = 0.212132$. There are $121 + 121 - 2 = 240$ df. The appropriate t- and z-table values are 1.651 and 1.645. The power is approximately

$$P\left(z > \frac{1.645 - \sqrt{121}\,(0.212132)}{1 + (1.651)^2/2(240)}\right) = P(z > -0.68) = .75$$

□

The factors influencing the power of the test are not surprising. As α, the Type I error probability, decreases, so does the power of the test. As σ_{common} increases, the amount of "noise" in the data increases, and the power decreases. As the hypothesized difference between population means, $\mu_1 - \mu_2$, decreases, it becomes harder to detect the smaller difference, and again the power decreases. Finally, as the sample size (balanced over the two samples) decreases, the power decreases.

Example 11.22 Refer to Example 11.21. Recalculate the power assuming that

(a) the difference of the means is 2/10 of a standard deviation, $\alpha = .05$ (two-tailed), and both sample sizes are 121

(b) the difference of the means is 3/10 of a standard deviation, $\alpha = .10$ (two-tailed), and both sample sizes are 61.

Solution **(a)**

$$P\left(z > \frac{1.645 - \sqrt{121}\,(.2)/\sqrt{2}}{1 + (1.651)^2/\,2(240)}\right) = P(z > 0.09) = .46$$

The difference of means is smaller in this example than in Example 11.21, and α is also smaller. Both factors reduce the power of the test.

(b)

$$P\left(z > \frac{1.645 - \sqrt{61}\,(.3)/\sqrt{2}}{1 + (1.658)^2/2(120)}\right) = P(z > 0.00) = .50$$

□

For the t' test, the theory behind a power calculation becomes difficult. The statistic doesn't have a noncentral t distribution when H_a is true, any more than it has a t distribution when H_0 is true. A rough approximation to power, based on the Scheffé approximation, is

$$P\left(z > \frac{z_a - \dfrac{|\mu_1 - \mu_2|}{\sqrt{\sigma_1^2/n_1 + \sigma_2^2/n_2}}}{1 + t_a^2/2(\text{df})}\right).$$

where the df are those used in the t' test and a is α for a one-tailed test and $\alpha/2$ for a two-tailed test.

Example 11.23 Refer to Example 11.10. Another study is planned, using 88 female and 216 male subjects. For planning purposes, assume that the difference of population means is 5, that the population standard deviations are equal to the sample standard deviations found in Example 11.10, namely 11.62 for females and 8.47 for males, and that a two-tailed test with $\alpha = .05$ will be used. Is there a high probability of finding a significant difference?

Solution The probability of finding a significant difference is the power of the test, by definition. Assuming that the sample standard deviations will also stay constant, the degrees of freedom can be found.

$$c = \frac{(11.62)^2/\,88}{(11.62)^2/88 + (8.47)^2/216} = 0.822$$

$$\text{df} = \frac{(87)(215)}{215(.822)^2 + 87(.178)^2} = 126.3$$

The t-table value will be close to the 120 df value, 1.980. Thus, as a rough approximation,

$$\text{power} = P\left(z > \frac{1.960 - \dfrac{5}{\sqrt{(11.62)^2/88 + (8.47)^2/216}}}{1 + (1.980)^2/2(126)}\right)$$

$$= P(z > -1.65)$$

which is about .95. Thus in this case the power of the test is quite good. □

Exactly the same factors affect power in the t' test as in the t test. A decrease in α, a decrease in $|\mu_1 - \mu_2|$, a decrease in either n, or an increase in either standard deviation results in a decrease in power. If there is reason to believe that one variance will be larger than another, the best power for the t' test occurs when the sample sizes are proportional to the standard deviations.

Exercises for Section 11.5

11.18. A follow-up study to the survey of Exercise 11.8 was being planned. A 95% confidence interval with a half-width (plus-or-minus) of 4 points was desired. Assuming a balanced design, what sample sizes are needed? Assume that the pooled variance in the new study will equal that found in Exercise 11.8.

11.19. Refer to Exercise 11.18. Assume that the population means are equal to the sample means found in Exercise 11.8 and that the common population variance equals the pooled sample variance found there. Will the sample sizes found in Exercise 11.18 be large enough to give good power to an $\alpha = .05$ test?

11.20. Independent samples of size 15 apiece are to be taken from two populations. Assume that the populations have equal variability. Find the power ($\alpha = .01$) for $\delta = .1$, .3, and .5. Use the z approximation.

11.21. Refer to Exercise 11.20. How will the power change (say, for $\delta = .3$) if

a. $\alpha = .05$
b. both n's are 21
c. $n_1 = 15$, $n_2 = 25$

11.22. Assuming a balanced design, what sample sizes are needed to yield a pooled-variance t confidence interval with a plus-or-minus equal to $.3s_{\text{pooled}}$?

11.6 Independent-Sample vs. Paired-Sample Methods

One choice often faced in the design of a study is whether to use paired-sample or independent-sample methods. The choice must be made before the data are gathered. It is totally inappropriate to analyze paired-sample data as if they were independent-sample data, or vice versa. In this section, we'll discuss the basic considerations in making this choice.

Pairing (matching) samples is effective when the basis for pairing is a source of a great deal of variability in the data. For example, IQ is thought to be an important predictor of scores on a memory task. One way to pair subjects for assignment to treatment or control groups, where the response of interest is memory score, would be to give a preliminary IQ test. The two lowest scorers would be paired, with one assigned randomly to treatment, the other to control. Then the two next-lowest IQ subjects would form a second pair, with one assigned randomly to treatment, the other to control, and so on up to the two highest-IQ subjects. Suppose that 20 subjects are matched in this way, and that the resulting memory scores are as shown in Table 11.7.

	Pair											
	1	2	3	4	5	6	7	8	9	10	$\bar{y}$	s
Treatment	16	21	25	23	28	32	30	35	40	41	29.10	8.14
Control	12	22	20	26	26	30	31	32	33	38	27.00	7.51
T − C	4	−1	5	−3	2	2	−1	3	7	3	2.10	3.03

Table 11.7

There is a marked tendency for the memory scores in either group to increase as the pair number (hence the IQ) increases. Thus IQ does seem to be an important factor in determining score on a memory task. Pairing on the basis of IQ seems to have been worthwhile.

To see the gain to pairing, compare the width of a 95% confidence interval for $\mu_T - \mu_C$ obtained using the paired-data method of Chapter 10 with the width of a confidence interval obtained by (inappropriately) using the pooled-variance t method of Section 11.2. The paired-sample interval is

$$-0.07 < \mu_{\text{diff}} < 4.27$$

The pooled-variance t interval is much wider:

$$-5.26 < \mu_T - \mu_C < 9.46$$

Thus the paired-sample approach has given us a much better inference (narrower confidence interval) than would be obtained using independent samples. Because the power of a test and the width of a confidence interval are closely related (the narrower the interval, the better the power), the power of a paired-sample t test would be much better in this situation than would the power obtained by an independent sample study.

Another way to see the potential usefulness of pairing is to think of the score of an individual in a treatment group as

$$\text{score} = \text{effect of individual} + \text{effect of group} + \text{error}$$

Then the difference of scores in a pair is

score(indiv. i_1, group T) − score(indiv. i_2, group C)

= effect of i_1 + effect of T + error − (effect of i_2 + effect of C + more error)

When pairing is effective, the effects of i_1 and i_2 will be very similar, so taking the difference of their scores will cancel out the individual effects, leaving the effect of T minus the effect of C plus relatively small random error.

Conversely, pairing subjects on the basis of an irrelevant characteristic does no good, and indeed can be mildly harmful to the precision of the statistical inference. Suppose that subjects for a memory experiment are paired on the basis of height, and that height is uncorrelated with memory skill. Suppose that the data are as shown in Table 11.8.

	Pair											
	1	2	3	4	5	6	7	8	9	10	$\bar{y}$	s
Treatment	27	43	26	29	30	32	25	33	21	9	29.50	5.89
Control	23	34	25	42	22	22	37	17	25	24	27.10	7.87
T − C	4	9	1	−13	8	10	−12	16	−4	5	2.40	9.51

Table 11.8

Note that in these data there is no particular trend of scores increasing (or decreasing) with pair number (the pairs are numbered from shortest height to tallest height). Thus height doesn't seem to be useful in predicting memory score, for these data—hardly a surprising conclusion! The paired-t confidence interval is $-4.41 < \mu_{\text{diff}} < 9.21$, and the pooled-variance t interval (still inappropriate—the data were in fact paired, however unsuccessfully) is $-4.13 < \mu_T - \mu_C < 8.93$.

If the pairing is totally useless, then s^2_{diff} will be approximately equal to $s_1^2 + s_2^2$; this fact follows because, in the population, the variance of a difference of independent quantities is the sum of the variances of the quantities. The plus-or-minus for the paired-t confidence interval will be

$$\frac{t_{\alpha/2}s_{\text{diff}}}{\sqrt{n}}$$

whereas the plus-or-minus for the pooled-variance t would have been

$$t_{\alpha/2}s_{\text{pooled}}\sqrt{\frac{1}{n_T}+\frac{1}{n_C}}$$

The only difference is that the t-table value for the paired-t confidence interval is based on $n_{diff} - 1 = 9$ df, whereas the df for the pooled-variance interval is based on $n_1 + n_2 - 2 = 18$ df. As the df increase in a t table (for a specified right-tail area) the magnitude of the table value decreases. Thus the confidence interval for the independent-sample t procedure will tend to be slightly shorter than that of the paired-t procedure, when the pairing is done on the basis of an irrelevant characteristic.

Example 11.24

A study compared schools using a traditional workbook-based method for teaching grammar to schools using a writing-oriented approach. The schools were paired as closely as possible based on family incomes of their students. Specifically, the schools were paired based on the percentage of students participating in free school lunch programs. For each school, the median score on a grammar test was obtained. The data were as shown in Table 11.9.

Pair	Writing-Based Method	Traditional Method	W − T
1	67.0	43.9	23.1
2	60.9	45.5	15.4
3	60.0	54.9	5.1
4	75.3	60.3	15.0
5	59.2	40.7	18.5
6	39.6	45.3	−5.7
7	58.0	62.4	−4.4
8	43.7	42.5	1.2
9	41.6	51.1	−9.5
10	77.8	66.7	11.1
11	62.8	62.0	0.8
12	41.8	54.2	−12.4
13	39.0	56.3	−17.3
14	46.9	43.5	3.4
15	65.1	58.4	6.7
16	62.2	48.6	13.6
17	75.7	53.3	20.4
18	46.3	33.9	12.4
Mean	56.828	51.417	5.411
Standard Deviation	12.974	8.923	11.804

Table 11.9

Perform t tests based on the differences (appropriately) and on a pooled-variance approach (inappropriately). Is there an indication that the pairing of schools made the test sharper?

Solution The t statistic based on differences comes out to equal 1.945, with a two-tailed p-value of .0685, whereas the pooled-variance t equals 1.458, with a two-tailed p-value of .1540. (The p-values were computed using a computer program.) The test based on differences is more conclusive, though still not overwhelmingly clear. Thus pairing has been of some value, but the result is still not conclusive. Note that

$$s_1^2 + s_2^2 = (8.923)^2 + (12.974)^2 = 247.94$$

whereas

$$s_{\text{diff}}^2 = (11.804)^2 = 139.33$$

Again, the pairing seems to be useful, because s_{diff}^2 is smaller than what we would estimate under independence. □

Exercises for Section 11.6

11.23. Refer to Exercise 10.15 in Chapter 10. Test the hypothesis of equal means assuming (incorrectly) that the samples had been taken independently.

11.24. Does the conclusion of Exercise 11.23 differ from that of Exercise 10.15? What does this indicate about the effectiveness of pairing?

11.25. Female subjects in an experiment were matched by height, then assigned randomly to one of two groups. Results are in Table 11.10.

Sample	$\bar{y}$	s	n
Treatment	106.21	10.64	31
Control	102.94	9.98	31
T − C	3.27	5.14	31

Table 11.10

a. Calculate a 95% confidence interval for the difference of means based on the sample differences.

b. Calculate a 95% confidence interval assuming that the samples were independent.

11.26. How do the widths of the confidence intervals in Exercise 11.25 compare? What does this indicate about the effectiveness of pairing by height?

11.27. For the data of Exercise 11.25, calculate t_{diff} and t_{pooled}. Which statistic yields a more conclusive result? Does your answer agree with your finding in Exercise 11.26?

Appendix: Homoscedasticity and the *t* Distribution

In developing the pooled-variance t statistic, we made the seemingly arbitrary assumption that the population variances were equal. The chi-squared and t-distribution results of the appendix to Chapter 10 can be used to indicate why this assumption was made.

The definition of a t statistic is

$$t = \frac{z}{\sqrt{W/\nu}}$$

where W has a chi-squared distribution with ν df. As the true standard error of $\bar{Y}_1 - \bar{Y}_2$ is $\sqrt{\sigma_1^2/n_1 + \sigma_2^2/n_2}$, the desired Z statistic is

$$Z = \frac{(\bar{Y}_1 - \bar{Y}_2) - (\mu_1 - \mu_2)}{\sqrt{\dfrac{\sigma_1^2}{n_1} + \dfrac{\sigma_2^2}{n_2}}}$$

Because $(n_1 - 1)s_1^2/\sigma_1^2$ and $(n_2 - 1)s_2^2/\sigma_2^2$ are (independently) chi-squared-distributed, as argued in the appendix to Chapter 10, the appropriate chi-squared statistic is the sum

$$W = (n_1 - 1)\frac{s_1^2}{\sigma_1^2} + (n_2 - 1)\frac{s_2^2}{\sigma_2^2}$$

with $(n_1 - 1) + (n_2 - 1) = n_1 + n_2 - 2$ df.

Substituting these expressions for Z and W into the definition of a t statistic yields the following result:

$$t = \frac{\dfrac{(\bar{Y}_1 - \bar{Y}_2) - (\mu_1 - \mu_2)}{\sqrt{\dfrac{\sigma_1^2}{n_1} + \dfrac{\sigma_2^2}{n_2}}}}{\sqrt{\left((n_1 - 1)\dfrac{s_1^2}{\sigma_1^2} + (n_2 - 1)\dfrac{s_2^2}{\sigma_2^2}\right) \Big/ (n_1 + n_2 - 2)}}$$

Without further assumptions, this can't be simplified, and the t statistic depends on the unknown population variances. To solve this problem we assume $\sigma_1^2 = \sigma_2^2$. Then, the common variance can be factored out of the square roots. With a little algebra, the t definition becomes

$$t = \frac{(\bar{Y}_1 - \bar{Y}_2) - (\mu_1 - \mu_2)}{\sqrt{\dfrac{(n_1 - 1)s_1^2 + (n_2 - 1)s_2^2}{n_1 + n_2 - 2}}\sqrt{\dfrac{1}{n_1} + \dfrac{1}{n_2}}}$$

$$= \frac{(\bar{Y}_1 - \bar{Y}_2) - (\mu_1 - \mu_2)}{s_{\text{pooled}}\sqrt{\dfrac{1}{n_1} + \dfrac{1}{n_2}}}$$

which is the form used in Chapter 11. In summary, the assumption that $\sigma_1^2 = \sigma_2^2$ is made so that the defining condition for t yields a t statistic that does not depend on unknown population variances.

Chapter Exercises

11.28. Suppose that we perform an experiment similar to that of Porac and Coren (1981). Female subjects indicate a preference for hearing sounds through the right ear or the left. A series of words is heard by each subject in the right ear. The variable of interest is the number of words correctly heard. Suppose that the data are as shown in Table 11.11.

Group	Mean	Standard Deviation	n
Right ear preferred	38.21	6.68	31
Left ear preferred	33.22	8.83	11

Table 11.11

Assume that both samples yielded roughly normal histograms.

a. Calculate the pooled estimate of variance. Show that it is closer to the variance of the right-ear-preferred group than to that of the left-ear-preferred group, and explain why.

b. Use the pooled estimate of variance to calculate the estimated standard error of the difference of sample means.

c. Use this standard error to calculate a 99% confidence interval for the difference of means.

11.29. Calculate a 99% confidence interval for the difference of means, using the separate-variance method.

11.30. **a.** Perform a pooled-variance t test of the null hypothesis of equal means, using the data of Exercise 11.28. Put bounds on the p-value.

b. Repeat part a, using the separate-variance (t') test.

c. Does the conclusion depend on which test you use? If so, which test is more believable?

11.31. Meyer and Thompson (1956) studied a sixth-grade classroom for a period of time. They recorded the number of "disapproval contacts" the teacher had with each child in the room. Suppose that we repeat the study and obtain the following data:

Boys:	1	3	4	5	7	11	15	19	21	27	33	41		
Girls:	0	1	2	2	2	3	4	4	4	5	5	8	10	15

a. Should the research hypothesis be one-sided or two-sided? (If this question is to be discussed in class, you might well wish to wear protective clothing.)

b. Perform the five steps of the pooled-variance t test for the null hypothesis of equal means. Use $\alpha = .05$. Assume a two-sided research hypothesis.

c. Is there any indication that one or more assumptions have been violated?

11.32. Does the t' test for the data of Exercise 11.31 reach the same conclusion as the t_{pooled} test?

11.33. **a.** Perform a rank sum test for the data of Exercise 11.31.

b. Of the three tests performed in Exercises 11.31–11.33, which one is the most appropriate in this situation?

11.34. A computer simulation was performed by drawing independent samples of sizes 10 and 30 from normal populations. The summary output in Table 11.12 was obtained for 1000 tests.

```
Popn        Mu        Sigma       n
  1       50.000     10.0000     10
  2       50.000      5.0000     30

using the pooled variance

         number of times H0:   "mu1-mu2 is    0" is rejected in favor of
alpha     "mu1-mu2 >   0"     "mu1-mu2 <    0"    total (alpha doubled)
0.100          189                  178                  367
0.050          131                  115                  246
0.025           82                   73                  155
0.010           40                   45                   85
0.005           23                   27                   50

using separate variances (t')

         number of times H0:   "mu1-mu2 is    0" is rejected in favor of
alpha     "mu1-mu2 >   0"     "mu1-mu2 <    0"    total (alpha doubled)
0.100          112                   94                  206
0.050           53                   38                   91
0.025           20                   19                   39
0.010            9                    9                   18
0.005            5                    2                    7
```

Table 11.12

a. What hypothesis about the means is true in this simulation? What probability is being estimated?

b. What assumption about the pooled-variance *t* test is violated in the simulation?

c. Which test should have closer approximations to the nominal probabilities? Does it?

11.35. Suppose that a study is planned involving two random samples. Assume that both sample sizes will be equal to 30, α will be .01 (two-tailed), both populations will have the same standard deviation, and the difference of population means will be equal to 1 standard deviation.

a. What is the probability that the test will find a statistically detectable (significant) difference of the samples means?

b. Is the probability computed in part a a *p*-value?

c. What sample size is required for each sample so that the width of a 99% confidence interval, based on the pooled variance, will be 1/2 of a standard deviation?

11.36. A computer program involved taking independent samples of size 30 each from two normally distributed populations. Summary output for 1000 tests included the results in Table 11.13.

Popn	Mu	Sigma	n
1	60.000	10.0000	30
2	50.000	10.0000	30

using the pooled variance

alpha	number of times H0: "mu1-mu2 > 0"	"mu1-mu2 is 0" is rejected in favor of "mu1-mu2 < 0"	total (alpha doubled)
0.100	992	0	992
0.050	983	0	983
0.025	964	0	964
0.010	919	0	919
0.005	884	0	884

using separate variances (t′)

alpha	number of times H0: "mu1-mu2 > 0"	"mu1-mu2 is 0" is rejected in favor of "mu1-mu2 < 0"	total (alpha doubled)
0.100	992	0	992
0.050	983	0	983
0.025	964	0	964
0.010	919	0	919
0.005	883	0	883

Table 11.13

a. What hypothesis about the means is true in this simulation? What probability is being estimated?

b. How important is it whether the pooled-variance t or the t' test is used?

c. Does the simulation result roughly agree with the probability that you calculated in part a of Exercise 11.35?

11.37. A sample of 12 subjects were tested on ability in a spatial rotation task, both before and after a training period. The resulting data, listed from subject 1 to subject 12, were

Before: 25 11 39 26 21 28 20 24 47 25 26 30
(mean 26.75, standard deviation 9.14)

After: 28 14 46 33 25 37 18 32 49 30 31 34
(mean 31.42, standard deviation 10.00)

The researchers reported that the pooled-variance t statistic equaled 1.54, not statistically significant at $\alpha = .05$. Why was the choice of test statistic very poor?

11.38. Show that a more appropriate hypothesis test for the data of Exercise 11.37 yields a very different conclusion. Explain why the conclusion is so different.

11.39. Suppose that a study measured the time required by patients in two distinct groups to solve a particular puzzle. There was no pairing of patients. Suppose that the data were

Group S: 25 28 31 33 35 39 40 43
48 55 64 71 83 88 106
(mean 52.60, standard deviation 24.62, $n = 15$)

Group B: 37 44 46 49 61 70 85 97 116 152
(mean 75.70, standard deviation 37.02, $n = 10$)

a. Perform a pooled-variance t test of H_a: $\mu_B > \mu_S$. State bounds on the p-value.

b. Using the pooled-variance t procedure for these data is unwise. Explain why.

11.40. Perform a more appropriate hypothesis test for the data of Exercise 11.39. Does the conclusion differ from that of Exercise 11.39?

11.41. A computer simulation involved drawing two independent samples of size 20 each from highly outlier-prone populations. Pooled-variance t and rank sum tests were performed for each pair of samples. The results, based on 1000 of each type of test) were as shown in Table 11.14.

```
Popn        Mu         Sigma        n
  1       60.000      10.0000      20
  2       50.000      10.0000      20

using the pooled variance

        number of times H0:  "mu1-mu2 is   0" is rejected in favor of
alpha     "mu1-mu2 >   0"    "mu1-mu2 <   0"    total (alpha doubled)
0.100          944                  0                  944
0.050          910                  0                  910
0.025          872                  0                  872
0.010          822                  0                  822
0.005          766                  0                  766

Results of Wilcoxon Rank Sum Test using Z as test statistic

       number of times H0:  "two populations are identical" rejected in favor of
alpha    Popn1 rt of Popn2    Popn1 left of Popn2    total (alpha doubled)
0.100        1000                    0                    1000
0.050        1000                    0                    1000
0.025        1000                    0                    1000
0.010        1000                    0                    1000
0.005         998                    0                     998
```

Table 11.14

a. Is the null hypothesis true or false in this case?

b. What assumption underlying the t test is violated in this simulation?

c. Which test appears to be better in this case? What does "better" mean here?

11.42. Another computer simulation comparing the pooled-variance t test to the rank sum test involved taking two independent samples from an exponential population, which is a sharply right-skewed population. The summary figures in Table 11.15 were obtained, based on 1000 of each type of test.

Popn	Mu	Sigma	n
1	1.000	1.0000	25
2	1.000	1.0000	5

using the pooled variance

	number of times H0: "mu1-mu2 is 0" is rejected in favor of		
alpha	"mu1-mu2 > 0"	"mu1-mu2 < 0"	total (alpha doubled)
0.100	82	108	190
0.050	28	62	90
0.025	8	32	40
0.010	1	18	19
0.005	1	7	8

Results of Wilcoxon Rank Sum Test using Z as test statistic

	number of times H0: "two populations are identical" rejected in favor of		
alpha	Popn1 rt of Popn2	Popn1 left of Popn2	total (alpha doubled)
0.100	114	90	204
0.050	53	36	89
0.025	27	18	45
0.010	6	4	10
0.005	3	3	6

Table 11.15

a. Is the null hypothesis true or false in this case?

b. What assumption underlying the t test is violated in this simulation?

c. Are the two-tailed results in the "total" column roughly consistent with the nominal α probabilities?

d. Are the one-tailed results roughly consistent with the nominal α probabilities?

11.43. Subjects learning to use a microcomputer were matched by age. One of the two youngest was randomly assigned to a system operated by choosing from a menu of commands, and the other was assigned to a system operated by choosing among "icons"—pictures symbolizing particular operations. The two next-youngest were randomly assigned, one to menu, the other to icon, and so on. Suppose that the data, time in minutes required to perform a particular task, were

Menu	31	22	55	40	46	38	43	37	40	48	62	33
Icon	41	54	40	36	47	74	52	45	58	63	49	60
I − M	10	32	−15	−4	1	36	9	8	18	15	−13	27

Results are as shown in Table 11.16.

	Mean	Standard Deviation	n
Menu	41.25	10.73	12
Icon	51.58	10.92	12
I − M	10.33	16.44	12

Table 11.16

a. Compute a 95% confidence interval for $\mu_I - \mu_M$ using a pooled-variance method.

b. Compute a 95% confidence interval for $\mu_I - \mu_M$ using a paired-sample method.

c. Which interval is appropriate for this study?

11.44. There is little difference in the widths of the confidence intervals calculated in parts a and b of Exercise 11.43. What does this fact indicate about the effectiveness of matching by age?

11.45. A study was planned of the performance of two independent groups of subjects on a tracking task. A pilot study indicated that the standard deviation of scores in group A would be about 20 points, and the B standard deviation would be about 12 points. One proposal for the study was to take samples of size 50 from each group; another proposal was to take a sample of 65 from group A and a sample of 35 from group B. Assuming that the standard deviations are the correct population standard deviations, which proposal will lead to a smaller standard error of the difference of sample means?

11.46. In addition to the assumptions made in Exercise 11.45, assume that the A population mean is 10 points larger than the B population mean.

a. If the sample sizes are each 50, what is the probability that a pooled-variance t test will find a statistically detectable (significant) difference, using $\alpha = .05$?

b. If the respective sample sizes are 65 and 35, what is the probability that a t' test will find a statistically detectable difference, again using $\alpha = .05$?

11.47. A computer simulation study of the effects of within-sample dependence on two-sample tests involved computer generation of dependent measurements. The first 10 measurements were assigned to sample 1, and the second 10 were assigned to sample 2. Both pooled-variance t and t' tests were performed on the results. Summary figures, based on 1000 of each test, were as shown in Table 11.17. The null hypothesis was in fact true in this study.

using the pooled variance

alpha	number of times H0: "mu1-mu2 > 0"	"mu1-mu2 is 0" "mu1-mu2 < 0"	is rejected in favor of total (alpha doubled)
0.100	197	225	422
0.050	139	165	304
0.025	100	117	217
0.010	67	82	149
0.005	36	61	97

using separate variances (t′)

alpha	number of times H0: "mu1-mu2 > 0"	"mu1-mu2 is 0" "mu1-mu2 < 0"	is rejected in favor of total (alpha doubled)
0.100	195	224	419
0.050	137	163	300
0.025	98	115	213
0.010	64	79	143
0.005	32	61	93

Table 11.17

a. Are the simulation results tolerably close to the nominal α values?

b. How much difference does it make whether the pooled-variance t or the separate-variance t' test is used, when there is within-sample dependence in the data?

11.48. Another simulation study like that of Exercise 11.47 was performed, but with sample sizes of 30 each instead of 10 each. The results for 1000 pooled-variance t tests were as shown in Table 11.18. Again, the null hypothesis was true in this study.

a. Are the results consistent with the nominal α values?

b. Compare the results of Exercises 11.47 and 11.48. Does it appear that increasing the sample size is helpful in dealing with the problem of within-sample dependence?

using the pooled variance

one-tail:

alpha	number of times H0: "mu1-mu2 is 0" is rejected in favor of "mu1-mu2 > 0"	"mu1-mu2 < 0"	total (alpha doubled)
0.100	248	242	490
0.050	182	182	364
0.025	132	129	261
0.010	90	87	177
0.005	65	67	132

Table 11.18

11.49. Samples of 27 liberal arts students and 27 preprofessional students at a particular university were obtained. A measurement of each student's anxiety over course performance was obtained. A portion of the output from SAS is shown in Figure 11.2.

```
                        2 SAMPLE T-TEST OF ANXIETY SCORES OF COLLEGE STUDENTS

                                          TTEST PROCEDURE

VARIABLE: SCORE

SCHOOL              N         MEAN     STD DEV   STD ERROR      MINIMUM       MAXIMUM  VARIANCES        T     DF  PROB > |T|

LIBERAL ARTS       27   9.37037037  3.93356511  0.75701496   1.00000000   20.00000000  UNEQUAL    -3.7072   49.6      0.0005
PRE-PROFESSIONAL   27  12.96296296  3.14375427  0.60501579   2.00000000   17.00000000  EQUAL      -3.7072   52.0      0.0005

FOR HO: VARIANCES ARE EQUAL, F'=    1.57 WITH 26 AND 26 DF      PROB > F'= 0.2596


                        RANK SUM TEST OF ANXIETY SCORES OF COLLEGE STUDENTS
                      ANALYSIS FOR VARIABLE SCORE CLASSIFIED BY VARIABLE SCHOOL

                                 AVERAGE SCORES WERE USED FOR TIES
                                 WILCOXON SCORES (RANK SUMS)
                                        SUM OF   EXPECTED    STD DEV       MEAN
                  LEVEL            N    SCORES   UNDER HO   UNDER HO      SCORE

                  PRE-PROFESSIONAL 27   970.00     742.50      57.54      35.93
                  LIBERAL ARTS     27   515.00     742.50      57.54      19.07
                        WILCOXON 2-SAMPLE TEST (NORMAL APPROXIMATION)
                        (WITH CONTINUITY CORRECTION OF .5)
                        S=  970.00      Z= 3.9453      PROB >|Z|=0.0001
                        T-TEST APPROX. SIGNIFICANCE=0.0002
                        KRUSKAL-WALLIS TEST (CHI-SQUARE APPROXIMATION)
                        CHISQ=  15.63     DF=  1   PROB > CHISQ=0.0001
```

Figure 11.2 Output for Exercise 11.49

a. Locate the values of the pooled-variance t statistic and the t' statistic. Why are they equal?

b. Can one conclude, using any of the usual α levels, that there is a statistically detectable (significant) difference between the means of the two groups?

11.50. **a.** Locate the value of the rank sum statistic in Figure 11.2.

b. Does the rank sum test lead to the same conclusion as either t test?

12 Inferences about Variances and Standard Deviations

The inference methods discussed in the past four chapters have largely been devoted to inferences about means. Although inferences about means are widely used in psychology and other social sciences, other parameters are of interest as well. In this chapter we'll discuss methods for making inferences about variability, as measured by standard deviations or variances.

The methods of this chapter involve two new theoretical distributions, the chi-squared (χ^2) and F distributions. Each distribution is introduced, then its use for inference is described. These distributions are used in a great many other inference procedures as well. Many of the methods described in the rest of the book involve these distributions. The chi-squared distribution is introduced in Section 12.1. Its use in one-sample inference about the population variance or standard deviation is described in Section 12.2. The F distribution is described in Section 12.3. Its use in two-sample inference is discussed in Section 12.4.

Inferences about variances and standard deviations differ from those about means in an important way. Violations of the assumption of normal populations have extremely serious effects on the standard χ^2 or F methods. Furthermore, these effects are as serious for large samples as for small ones. In this chapter we discuss a rank test alternative and a new, promising method called jackknifing. Both of these approaches are far less sensitive to nonnormality and deserve wider use.

12.1 The Chi-Squared Distribution

The chi-squared distribution is used in making inferences about variances and standard deviations. It is also the relevant theoretical distribution for many of the inference methods about proportions described in Chapter 13. This section is devoted to a discussion of the properties of chi-squared distributions.

Like the t distribution, chi-squared distributions are indexed by degrees of freedom. The chi-squared distribution with specified df, sometimes denoted ν (greek letter nu) df, is defined by one version of the χ^2 statistic

$$\chi^2 = z_1^2 + z_2^2 + \cdots + z_{\text{df}}^2$$

where $z_1, z_2, \ldots, z_{\text{df}}$ are *independent* random variables, each having a standard normal (mean 0, variance 1) distribution. From this definition, some basic properties of chi-squared statistics follow immediately. The properties are stated in the box below. Informal arguments will then be given as to why some of the properties are true.

Properties of a Chi-Squared Statistic

1. The expected value of χ^2 is its df; $E(\chi^2) = \text{df}$.

2. The variance of χ^2 is twice its df; $\text{Var}(\chi^2) = 2(\text{df})$.

3. If χ_1^2 and χ_2^2 are *independent* chi-squared statistics with respective degrees of freedom df_1 and df_2, their sum $\chi_1^2 + \chi_2^2$ is also a chi-squared statistic with $\text{df} = \text{df}_1 + \text{df}_2$.

The expected value property follows from the fact that $E(z^2) = \text{Var}(z) = 1$. This expected value is correct even if the assumption that z has a normal distribution is false. The variance property depends critically on the normal distribution assumption. The variance of z^2 involves squared deviations of squared z scores, hence fourth powers. If, in fact, the distribution of z is exactly normal, it can be shown that $\text{Var}(z^2) = 2$. With independent z statistics, variances may be added, so the variance property follows. If, however, the distribution of z is not normal, the variance of z^2 will not equal 2, and the variance of χ^2 will not equal 2(df). The third, additive property of chi-squared statistics follows from the definition. A sum of sums of independent z^2 terms can be written out as a long sum of independent z^2 terms—that is, as a χ^2 statistic. Note that the df for the individual χ^2 statistics can be added.

Example 12.1

If a random variable Y_1 has a chi-squared distribution with 8 df and if Y_2 has a chi-squared distribution with 12 df independent of Y_1, find the expected value and variance of Y_1 and of $Y_1 + Y_2$. Which of these results are sensitive to the underlying assumption of normality?

Solution

From the properties of the chi-squared distribution listed in the box above, we have that $E(Y_1) = 8$, $\text{Var}(Y_1) = 2(8) = 16$, $E(Y_2) = 12$, and $\text{Var}(Y_2) = 2(12) = 24$. The sum of independent chi-squared random variables again has the chi-squared distribution, so

$$E(Y_1 + Y_2) = 8 + 12 = 20$$

and

$$\text{Var}(Y_1 + Y_2) = 2(8 + 12) = 40$$

The expected value properties are valid regardless of the underlying normality of the population, but the variance properties fail to hold if the underlying population isn't normal. □

Chi-squared distributions are *not* symmetric. A chi-squared statistic can't be negative, but can be any positive value. Thus chi-squared tables must indicate both left- and right-tail areas. Table 6 in the appendix gives values of χ^2_a, values cutting off a *right*-tail area equal the specified a. We denote such a table value by χ^2_a. For 20 df and $a = .025$, the value $\chi^2_{.025}$ is 34.1696. Note that the table values are numerically much different from t- or z-table values. To find a value cutting off a left-tail area such as .025, enter the table for the complementary right-tail area, such as .975. For 20 df, the value cutting off a left-tail area .025 is found in the $a = .975$ column: $\chi^2_{.975} = 9.59083$. See Figure 12.1.

The chi-squared tables cannot include all possible df. For large df, it can be shown that a good approximation to the table values can be had by treating

$$\sqrt{2\chi^2} - \sqrt{2(\text{df}) - 1}$$

as if it were a z statistic. By straightforward algebra, this leads to approximating the tabled χ^2 value corresponding to a right-tail area of a as

$$\chi^2_a = \left(\frac{z_a}{\sqrt{2}} + \sqrt{(\text{df} - .5)}\right)^2$$

Example 12.2 If Y has a chi-squared distribution with 12 df, find values that cut off right- and left-tail areas equal to .05. Use the large-df approximation to find approximate values. How well does the approximation work in this small-df situation?

Solution From the chi-squared table, the right-tail value is $\chi^2_{.05} = 21.0261$. The left-tail value $\chi^2_{.95}$ is found by entering the table in the .95 column; it is 5.22603. For the large-df approximation, recall that $z_{.05} = 1.645$; by the symmetry of the z distribution, $z_{.95} = -1.645$. Approximately,

$$\chi^2_{.05} = \left(\frac{1.645}{\sqrt{2}} + \sqrt{12 - .5}\right)^2 = 20.74$$

and

$$\chi^2_{.95} = \left(\frac{-1.645}{\sqrt{2}} + \sqrt{12 - .5}\right)^2 = 4.964$$

Even for very small df, the large-df approximation is fairly close. □

The distribution of the sample variance

$$s^2 = \frac{\sum (y_i - \bar{y})^2}{n - 1}$$

is closely related to the chi-squared distribution.

Theoretical Distribution of the Sample Variance

If a random sample is taken from a normally distributed population, then

$$\chi^2 = \frac{(n - 1)s^2}{\sigma^2_{\text{pop}}}$$

has a chi-squared distribution with $n - 1$ df.

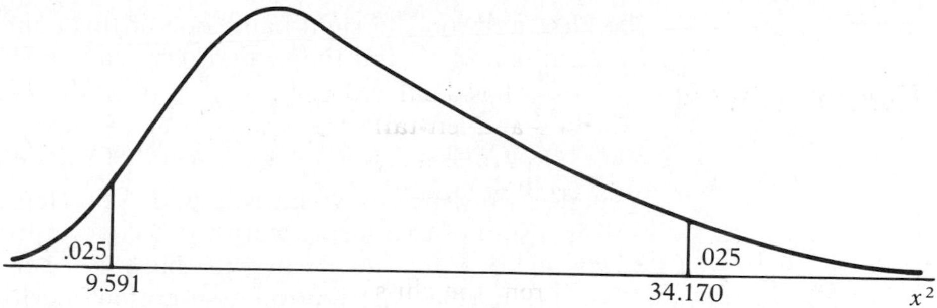

Figure 12.1 Chi-squared distribution with 20 df

Example 12.3 If a random sample of size 13 is taken from a normal population having mean 50 and standard deviation 10, and if s^2 is the resulting sample variance, find the probability that the sample variance will exceed 263.27625.

Solution

$$P(s^2 > 263.27625) = P\left(\frac{(13-1)s^2}{(10)^2} > \frac{12(263.27625)}{10^2}\right)$$
$$= P(\chi^2 > 21.0261) = .05$$

Obviously the value 263.27625 was chosen to make the probability come out .05. □

The reason there are $n - 1$ df, even though s^2 involves a sum of n terms, is that the terms are not independent. The fact that the deviations $y_i - \bar{y}$ must always sum to 0 means that the nth deviation is completely dependent on the values of the preceding $n - 1$ deviations. Thus only $n - 1$ of the deviations are free to vary. Alternatively, we can expand the square around y in

$$\chi^2 = \sum z_i^2 = \sum \frac{(y_i - \mu)^2}{\sigma^2_{\text{pop}}}$$

and obtain

$$\chi^2 = \sum \frac{(y_i - \bar{y})^2}{\sigma^2_{\text{pop}}} + \sum \frac{(\bar{y} - \mu)^2}{\sigma^2_{\text{pop}}}$$

The first term on the right-hand side of this equation is $(n - 1)s^2/\sigma^2_{pop}$ by definition of s^2, and the second term can be shown to be

$$\frac{(\bar{y} - \mu)^2}{(\sigma^2_{pop}/\sqrt{n})} = z^2$$

A single z^2 statistic is a χ^2 statistic with 1 df. Hence the χ^2 statistic with n df is a sum of two terms, with the second term having 1 df. Because df are additive, the first term must have $n - 1$ df, as it does.

The normal distribution assumption is critical to the theoretical distribution of the sample variance. There is no theorem comparable to the Central Limit Theorem for sums and means, claiming that for non-normal populations the distribution is "approximately χ^2" for large n. The reason is that the theoretical variance of s^2 depends on fourth powers. If the population is skewed or outlier-prone, the fourth-power properties will be very different from those of a normal population. Thus if the population isn't normal, the variability of s^2 will not be the variability implied by the chi-squared distribution.

Example 12.4 In a Monte Carlo study, 1000 samples of size 13 were taken from a population having a highly right-skewed distribution. The population standard deviation was 10. It was found that 146 of the sample variances exceeded 263.27625. Is this Monte Carlo result compatible with values in the χ^2 table?

Solution No. In Example 12.3 we found that, if the population standard deviation is 10 and the sample size is 13, the probability that the sample variance will exceed 263.27625 is .05. Thus, if chi-squared probabilities applied, we should have 1000(.05) = 50 such cases. In fact, we have almost three times that many. □

Exercises for Section 12.1

12.1. Find the values that cut off left- and right-tail areas equal to .025 for the chi-squared distribution with 16 df.

12.2. If $\sigma^2_{pop} = 100$ and $n = 26$, find the value exceeded by the sample variance s^2 with probability .10.

12.3. Refer to Exercise 12.2. Would it matter much if the population from which the sample is drawn were substantially right-skewed?

12.2 Inferences about Variance and Standard Deviation in One Sample

Statistical inference methods for a variance (and standard deviation) in a single random sample are applications of the chi-squared distribution defined in the previous section. In this section, we describe χ^2-based methods for hypothesis testing and confidence intervals. In Section 12.1 we noted that the theoretical χ^2-related distribution of s^2 was unpleasantly sensitive to nonnormality of the population. This section describes an alternative approach that is far less sensitive to population nonnormality.

For example, it has been claimed that the increase in the proportion of high school students going on to college has lead to a decrease in average Scholastic Aptitude Test (SAT) scores. It is also reasonable to suppose that the inclusion of low-aptitude students will increase the standard deviation of scores. Assume that a college counselor obtains the scores of 21 students on the SAT verbal test, and that these 21 can be regarded as a random sample from the current population of college-going students. Also assume that, in the past, the standard deviation of SAT verbal scores has been 85.0. The data:

290 370 370 390 400 420 420 440 450 450 470
480 500 520 560 560 580 600 620 630 770
(mean 490.00, standard deviation 111.53)

Is the sample standard deviation significantly (at $\alpha = .05$, say) higher than the past standard deviation, 85.0?

Any significance (hypothesis) test requires specification of null and research hypotheses. The null hypothesis is that the current population standard deviation equals the old one, and the research hypothesis is that it is larger. The test statistic is based on the idea that $(n - 1)s^2/\sigma^2_{\text{pop}}$ is a χ^2 statistic. The procedure is shown in the box. It's conventionally stated in terms of the variance rather than the standard deviation.

Hypothesis Test for a Variance

	General	*Example*
1.	H_0: $\sigma^2 = \sigma_0^2$	H_0: $\sigma = (85.0)^2$
2.	H_a: (i) $\sigma^2 > \sigma_0^2$ (ii) $\sigma^2 < \sigma_0^2$ (iii) $\sigma^2 \neq \sigma_0^2$	H_a: $\sigma > (85.0)^2$
3.	T.S.: $\chi^2 = \dfrac{(n-1)s^2}{\sigma_0^2}$	T.S.: $\chi^2 = \dfrac{(21-1)s^2}{(85.0)^2}$
4.	R.R.: reject H_0 if (i) $\chi^2 > \chi_\alpha^2$ (ii) $\chi^2 < \chi_{1-\alpha}^2$ (iii) $\chi^2 < \chi_{\alpha/2}^2$ or $\chi^2 < \chi_{1-\alpha/2}^2$	RR: reject H_0 at $\alpha = .05$ if $\chi^2 > 31.4104$
5.	Conclusion	Conclusion: $\chi^2 = \dfrac{20(111.53)^2}{(85.0)^2}$ $= 34.436$ > 31.4104; reject H_0

Example 12.5

Rats raised in normal laboratory conditions show a standard deviation of 5.23 in their ability to learn a certain task, according to an established measurement scale. A sample of 30 rats were raised in relatively more crowded conditions. The following data were obtained on the same learning scale:

10.9	11.2	12.0	12.6	12.7	12.9	13.0	13.6	13.9	13.9
14.0	14.0	14.0	14.1	14.2	14.3	14.6	14.9	15.0	15.3
15.8	16.0	16.9	18.0	19.9	22.3	26.1	29.8	35.0	39.0

(mean 16.663, standard deviation 7.3252, $n = 30$)

Do the data indicate a statistically significant ($\alpha = .05$) change in the variability of scores, as measured by the standard deviation?

Solution

H_0 is that the standard deviation is still 5.23, and H_a should be taken as two-sided, because we are looking for any change in variability. The test statistic is

$$\chi^2 = \frac{(30 - 1)(7.3252)^2}{(5.23)^2} = 56.8897$$

with 30 − 1 = 29 df. For a two-tailed $\alpha = .05$ test, the relevant cutoff values are the right-tail and left-tail .025 points, found from the table to be 45.7222 and 16.0471, respectively. Because the value of the T.S. is $\chi^2 = 56.8897 > 45.7222$, we reject H_0 and conclude that the change in variability is statistically significant at $\alpha = .05$. □

The procedure for confidence intervals is also derived from the chi-squared distribution. It is shown in the next box.

100(1 − α)% Confidence Interval for a Variance

Two-sided interval:

$$\frac{(n - 1)s^2}{\chi^2_{\alpha/2}} < \sigma^2 < \frac{(n - 1)s^2}{\chi^2_{1-\alpha/2}}$$

Lower-bound interval:

$$\sigma^2 > \frac{(n - 1)s^2}{\chi^2_{\alpha}}$$

Upper-bound interval:

$$\sigma^2 < \frac{(n - 1)s^2}{\chi^2_{1-\alpha}}$$

NOTE: To obtain a confidence interval for a standard deviation, take square roots throughout.

For the SAT data, a 95% lower-bound interval for the new value of the population variance is

$$\sigma^2 > \frac{(21 - 1)(111.53)^2}{31.4104}$$

or

$$\sigma^2 > 7290.94$$

Taking square roots, we obtain

$$\sigma > 89.00$$

Note that this interval is entirely above the old standard deviation of 85.0. The corresponding hypothesis test found that the new standard deviation was significantly larger than 85.0.

Example 12.6 Refer to Example 12.5. Calculate a 95% confidence interval for the standard deviation of learning scores under crowded conditions. What does this interval say about the null hypothesis that the standard deviation is 5.23?

Solution The confidence interval for the variance is

$$\frac{(30-1)(7.3252)^2}{45.7222} < \sigma^2 < \frac{(30-1)(7.3252)^2}{16.0471}$$

or

$$34.0338 < \sigma^2 < 96.9707$$

Taking square roots, the interval is

$$5.834 < \sigma < 9.847$$

which excludes the null hypothesis value 5.23. Thus H_0 is rejected at $\alpha = .05$, corresponding to the 95% interval. □

As we noted in Section 12.1, the theoretical distribution of the sample variance (and standard deviation) is extremely sensitive to population nonnormality. Therefore, the inference procedures of this section are very sensitive to nonnormality. In the language of Chapter 8, these methods are *not robust* if the population is nonnormal. The problem is one of nominal robustness; the nominal probability may be seriously wrong if the population is nonnormal. It is, at a minimum, imperative to plot the data. Clear evidence of nonnormality makes these procedures very suspect. In the SAT verbal score data, there is a suggestion of heavy-tailness. The highest score, 770, is right on the boundary of outlierdom, according to the 1.5 IQR rule of thumb. Although the data aren't flagrantly nonnormal, the sensitivity of χ^2-based procedures to nonnormality makes them of dubious value in this problem.

Example 12.7 Refer to the data of Example 12.5. Is the conclusion of that example (and Example 12.6) reliable?

Solution No. The procedure is based on a method that has nominal probabilities that are highly sensitive to nonnormality of the underlying population. The data of Example 12.5 are badly skewed to the right, strongly suggesting that the underlying population is also right-skewed. Thus the conclusion is suspect. □

jackknife

An alternative approach to this and many other robustness problems is the **jackknife** method. It has been shown [see Miller (1974)] to be less sensitive to population nonnormality than are the χ^2-based methods. The jackknife method is not a cureall; when sampling is from a highly skewed or outlier-prone population, jackknife probabilities may also be in error for small sample sizes. However, the nominal robustness of jackknife inferences tends to improve as n increases, whereas the nominal robustness of χ^2-based methods does not. The essential problem with χ^2-based methods for a variance is that there is no data-based estimate of the variability of the statistic. The χ^2 test statistic involves only s^2, not any measure of the variability of s^2. The jackknife procedure involves deleting one observation at a time from the sample and recomputing the sample variance. This procedure gives an indication of the variability of s^2; if the recomputed s^2 values vary widely, the original s^2 value must have a great deal of uncertainty in it. We shall not go into further detail about the jackknife computations, because the procedure will always be done by computer. See Miller (1974). For the SAT verbal score data, a 95% lower-bound confidence interval for the standard deviation, based on jackknifing, is

$$\sigma > 82.33$$

Note that this interval does include the old value, 85.0. Therefore, the somewhat more believable jackknife procedure comes to a different conclusion than does the dubious chi-squared interval.

Example 12.8 A computer drew 1000 samples, each of size 51, from a mildly outlier-prone population. The population variance was 64.8, and the null hypothesis was that the population variance was 64.8. Both chi-squared and jackknife tests for the variance were run, and the results in Table 12.1 were obtained.

(a) What numbers would be expected in Table 12.1 if the nominal α values were good approximations?

(b) How good are the nominal robustness properties of the two tests?

using jackknife test

alpha	number of times H0: "variance > 64.8"	"variance is 64.8" is rejected in favor of "variance < 64.8"	total (alpha doubled)
0.100	90	216	306
0.050	39	166	205
0.025	17	134	151
0.010	6	108	114
0.005	2	88	90

using chi-square variance test

alpha	number of times H0: "variance > 64.8"	"variance is 64.8" is rejected in favor of "variance < 64.8"	total (alpha doubled)
0.100	205	289	494
0.050	162	221	383
0.025	127	171	298
0.010	106	111	217
0.005	87	86	173

Table 12.1

Solution

(a) The computer was programmed so that H_0 was in fact true in the simulation. Thus, in the $\alpha = .100$ row, the expected frequencies should be $1000(.100) = 100$, another 100, and a combined 200. The $\alpha = .050$ row should have 50, 50, and 100, to within the random error of 1000 simulations, and so on.

(b) The nominal robustness is bad in both cases, but really dreadful in the χ^2 case. Especially in the χ^2 case, the actual frequencies are greatly different from the desired frequencies. The situation is most absurd for small α values; for the χ^2 test with a nominal $\alpha = .005$, so the desired frequencies were 5, 5, and 10, we obtain 87, 86, and 173 erroneous rejections of H_0! □

Exercises for Section 12.2

12.4. Refer to Exercises 10.5 and 10.6, where a random sample yielded $n = 216$, $\bar{y} = 3.1$, and $s = 2.4$. Calculate a 90% confidence interval for σ_{pop}.

You will need to use the large-df approximation for the appropriate χ^2-table values.

12.5. In Exercise 10.6, it was indicated that the underlying population distribution of waiting times was skewed to the right. It was argued that the sample size (216) was so large that the confidence level should be almost exactly correct. Is this argument valid?

12.6. Refer to Exercise 10.7, where a sample of $n = 11$ subjects gave a sample standard deviation $s = 8.64$. Test the research hypothesis that $\sigma_{\text{pop}} < 12$, using $\alpha = .05$.

12.7. In Exercise 12.6, the sample data showed a rather large negative value of the Groeneveld-Meeden skew measure. Is that a reason to doubt that the nominal $\alpha = .05$ is correct?

12.8. Refer to Exercise 10.15. Calculate a 95% confidence interval for the population standard deviation of the differences.

12.9. In Exercise 12.8, can one be reasonably certain that the confidence level is approximately correct?

12.3 The F Distribution

In this section we discuss the last of the widely used theoretical probability distributions, the *F* distribution. The *F* distribution will be used extensively in Chapters 14 and 15 and is probably the distribution most often used in the psychological literature. It is particularly important in the study of designed experiments.

The *F* distribution is defined in terms of χ^2 statistics. The *F* distribution is the theoretical distribution of the defining *F* statistic:

$$F = \frac{\chi_1^2/\text{df}_1}{\chi_2^2/\text{df}_2}$$

where χ_1^2 and χ_2^2 are *independent* χ^2 statistics and df_1 and df_2 are their respective degrees of freedom. From this definition follow some properties, as shown in the box.

Properties of the F Distribution

1. An F statistic is never negative, but can attain any positive value.

2. The F distribution is indexed by two df numbers, df_1 and df_2, corresponding to numerator and denominator, respectively.

3. The expected value of an F statistic is slightly greater than 1, and approaches 1 as df_2 gets large:

$$E(F) = \frac{df_2}{df_2 - 2}$$

4. The variance of an F statistic is a complicated function of the degrees of freedom.

Tables of the F distribution can get bulky, because there are two df indices to handle. In the appendix of this book, Table 7 contains values cutting off a right-tail area a, for the indicated df, and $a = .25$, .10, .05, .025, .01, .005, and .01. The numerator df is shown along the top of the table, indexing columns, and the denominator df is shown along the side, indexing rows. We denote the table value by F_{a,df_1,df_2} to emphasize the role of the two df numbers. For example, the value cutting off a .05 right-tail area with $df_1 = 4$ and $df_2 = 20$ is $F_{.05,4,20} = 2.86$. Note that this is not the same as the $a = .05$ cutoff point for $df_1 = 20$ and $df_2 = 4$, namely, $F_{.05,20,4} = 5.80$. It's necessary to keep track of which df is which. See Figure 12.2.

Example 12.9

Assume that a statistic F is based on 8 (numerator) and 12 (denominator) df. Find the expected value, upper .025 point, and upper .01 point of F.

Solution

The expected value is

$$\frac{df_2}{df_2 - 2} = \frac{12}{12 - 2} = 1.2$$

From the F tables with areas .025 and .01, the desired cutoff points are 3.51 and 4.50. Note that these values would not apply for 12 numerator and 8 denominator df. □

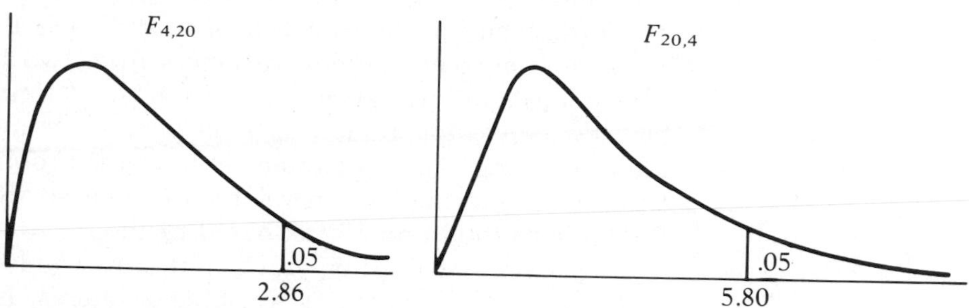

Figure 12.2 *F distributions*

The F distribution is based on a ratio of χ^2 statistics, and it is known that the actual distribution of χ^2 statistics is very sensitive to population nonnormality. Thus it is logical, and true, that the actual distribution of an F statistic is also sensitive to nonnormality. The reason is identical to the reasoning for the χ^2 statistic, namely that there is no data-based estimate of variability in the F statistic. This sensitivity to nonnormality holds, in principle, for any sample sizes.

Example 12.10 Refer to Example 12.9. Assume that 1000 sample F statistics have been computed, but that the underlying populations are not normal, but rather outlier-prone. Of these statistics, 68 exceed 3.51. Is this result close to what would be expected?

Solution No. In Example 12.9 we found that the probability that F exceeds 3.51, given normal populations, should be .025. Thus we'd expect 1000(.025)= 25 F statistics to exceed 3.51, not 68. The result indicates that the distribution of F can be sensitive to nonnormal populations. □

12.4 Comparing Sample Variances

The F distribution described in Section 12.3 is the basis for inference methods comparing sample variances from two independent samples. The methods follow very directly from the definition of the F ratio. In fact, the only even slightly difficult part of the methods is reading the tables properly. These methods do have the problem that they are sensitive to nonnormality. Therefore, we suggest alternative analyses—a jackknife method and a rank-based method.

Consider a situation in which rats are randomly assigned to treatment or control groups. The variable of interest is latency time to next feeding. It is suspected that rats in the treatment group will show greater variability, as well as a higher mean. The data:

Treatment:	12.1	13.0	13.6	13.9	14.0
	14.7	16.3	18.9	21.4	26.1
	($\bar{y} = 16.4$, $s = 4.449$)				
Control:	11.8	12.0	12.3	12.7	13.2
	13.9	14.7	15.0	17.1	20.3
	($\bar{y} = 14.3$, $s = 2.666$)				

To see whether the variability in the treatment group is significantly larger, we need a hypothesis-testing procedure for comparing variances (or standard deviations) from independent samples. The procedure is based on the definition of an F statistic as the ratio of independent χ^2 statistics, each divided by its df, and on the statistic

$$\chi^2 = \frac{(n-1)s^2}{\sigma^2_{\text{pop}}}$$

The definition yields the statistic

$$F = \frac{\chi_1^2/\text{df}_1}{\chi_2^2/\text{df}_2} = \frac{\dfrac{(n_1-1)s_1^2/\sigma_1^2}{n_1-1}}{\dfrac{(n_2-1)s_2^2/\sigma_2^2}{n_2-1}} = \frac{s_1^2/\sigma_1^2}{s_2^2/\sigma_2^2}$$

Under the null hypothesis H_0: $\sigma_1^2 = \sigma_2^2$, the F statistic reduces (by cancellation of the σ^2 factors) to $F = s_1^2/s_2^2$, the ratio of the sample variances. The formal procedure is shown in the box. The only difficulty is the proper use of F tables.

Testing the Equality of Variances (Two Independent Samples)

	General	Example
1.	H_0: $\sigma_1^2 = \sigma_2^2$	H_0: $\sigma_T^2 = \sigma_C^2$
2.	H_a: (i) $\sigma_1^2 > \sigma_2^2$ (ii) $\sigma_1^2 < \sigma_2^2$ (iii) $\sigma_1^2 \neq \sigma_2^2$	H_a: $\sigma_T^2 \neq \sigma_C^2$
3.	T.S.: $F = s_1^2/s_2^2$	T.S.: $F = s_T^2/s_C^2$
4.	R.R.: reject H_0 if (i) $F > F_{\alpha,n_1-1,n_2-1}$ (ii) $F < 1/F_{\alpha,n_2-1,n_1-1}$ (iii) $F > F_{\alpha/2,n_1-1,n_2-1}$ or $F < 1/F_{\alpha/2,n_2-1,n_1-1}$	R.R. ($\alpha = .05$): reject H_0 if $F > 4.03$ or $F < 1/4.03 = 0.248$
5.	Conclusion	Conclusion: $F = (4.449)^2/(2.666)^2$ $= 2.79$; retain H_0

The only problem with this test procedure is the table reading. Note that the df sometimes are reversed. For $n_1 = 6$ and $n_2 = 11$,

$$F_{.05,6-1,11-1} = 3.33 \quad \text{but} \quad F_{.05,11-1,6-1} = 4.74$$

Example 12.11

Porac and Coren (1981) present data on subjects' ability to correctly recall words presented to left and right ears. They obtained the results in Table 12.2.

Preferred Ear	Sample Size	Standard Deviation
Left	106	12.76
Right	121	8.80

Table 12.2

Is there a statistically significant difference ($\alpha = .05$) in the standard deviations?

Solution The null hypothesis is that the true, population standard deviations are equal. We shall use a two-sided research hypothesis. The test statistic is

$$F = \frac{(12.76)^2}{(8.80)^2} = 2.1025$$

Our F tables do not include 105, 120 df. For 60, 120 df, the $\alpha/2 = .025$ cutoff point is 1.53, whereas for 120, 120 df the .025 point is 1.43. Presumably, the desired 105, 120 df point is somewhere between these two numbers. The F statistic exceeds either table value. Thus, the null hypothesis is rejected and the difference is statistically detectable (significant). □

Confidence-interval inference is also a mere problem of table-reading. Confidence-interval procedures are also based on the F distribution and on ratios of variances.

100(1 − α)% Confidence Interval for Ratio of Variances

Two-sided interval:

$$\frac{s_1^2}{s_2^2}\frac{1}{F_{\alpha/2,n_1-1,n_2-1}} < \frac{\sigma_1^2}{\sigma_2^2} < \frac{s_1^2}{s_2^2}F_{\alpha/2,n_2-1,n_1-1}$$

Lower-bound interval:

$$\frac{s_1^2}{s_2^2}\frac{1}{F_{\alpha,n_1-1,n_2-1}} < \frac{\sigma_1^2}{\sigma_2^2}$$

Upper-bound interval:

$$\frac{\sigma_1^2}{\sigma_2^2} < \frac{s_1^2}{s_2^2}F_{\sigma,n_2-1,n_1-1}$$

For the latency-time data, n_1 and n_2 are both 10. To calculate a two-sided 95% confidence interval for the ratio of variances, we need $F_{.025,9,9} = 4.03$. The interval is

$$\frac{(4.449)^2}{(2.666)^2}\frac{1}{4.03} < \frac{\sigma_T^2}{\sigma_C^2} < \frac{(4.449)^2}{(2.666)^2}4.03$$

or

$$0.69 < \frac{\sigma_T^2}{\sigma_C^2} < 11.22$$

Taking square roots throughout, we get a confidence interval for the ratio of standard deviations,

$$0.83 < \frac{\sigma_T}{\sigma_C} < 3.35$$

Because the sample sizes are small, the interval is wide. Note that the 95% confidence interval for the ratio of variances (or the ratio of standard deviations) contains the value 1.00. Thus at $\alpha = .05$ we would retain the null hypothesis that the variances were equal. Of course, equal variances mean that the ratio is 1.00.

Example 12.12 Calculate a 95% confidence interval for the ratio of population standard deviations in Example 12.11. Does the interval indicate that the null hypothesis of equal standard deviations should be rejected?

Solution By rough interpolation, the .025 F-table value for 105, 120 df is about 1.46, and the .025 F-table value for 120, 105 df is about 1.57. Thus the confidence interval for the ratio of variances is

$$\frac{(12.76)^2}{(8.80)^2}\frac{1}{1.46} < \frac{\sigma_1^2}{\sigma_2^2} < \frac{(12.76)^2}{(8.80)^2}1.57$$

or

$$1.440 < \frac{\sigma_1^2}{\sigma_2^2} < 3.301$$

Taking square roots, we obtain the 95% interval for the ratio of standard deviations,

$$1.20 < \frac{\sigma_1}{\sigma_2} < 1.82$$

Note that a ratio of 1.00 is not included within this interval. As in Example 12.11, we conclude that the true standard deviations are not equal. □

The F-based procedures for comparing variances (or standard deviations) from two independent samples lean rather heavily on the assumption of normal populations. When the assumption is wrong, the inferences can also be wrong, regardless of the sample sizes. The F test for variances is known to be far more sensitive to nonnormality than is the t test for means discussed in Chapter 11. Therefore, attention must be paid to the shape of the data. The latency-time data that we have been studying in this section are right-skewed, and it is reasonable to assume that latency-time population data will have the same shape. The conclusions of the F test and interval are suspect.

Example 12.13 Refer to Example 12.11. The sample sizes are quite large. Does that mean that the conclusion of the test statistic is quite reliable?

Solution No. Large sample sizes help make procedures for means robust in the face of nonnormal populations. That is not true for inferences about variances. In Example 12.11, the value of the F statistic came out quite a bit larger than the critical value, so we may be reasonably sure that there is a significant difference. However, the confidence interval in Example 12.12 may not be very accurate. □

Ansari-Bradley test

There are alternative procedures that are much less susceptible to violations of the normal-population assumption. One is a rank test called the **Ansari-Bradley test**. To perform the Ansari-Bradley test, first find the deviations of each value from its own sample median. Then rank the positive and negative deviations separately, in order of absolute value, *from largest to smallest*. Thus the largest positive deviation and the most negative deviation both are assigned rank 1; the second largest and second most negative, rank 2; and so on. In case of ties, follow the usual procedure of assigning the average rank to all the tied deviations. For the latency data, the treatment median is 14.35 and the control median is 13.55. The deviations from medians and their Ansari-Bradley ranks are

Treatment	−2.25	−1.35	−0.75	−0.45	−0.35	0.35	1.95	4.55	6.65	11.65
Rank	1	4	7	8	9.5	9.5	6	4	3	1
Control	−1.75	−1.55	−1.25	−0.85	−0.35	0.35	1.15	1.45	3.55	6.75
Rank	2	3	5	6	9.5	9.5	8	7	5	2

The test is based on the sum of the ranks in the first sample. If the first population is more variable than the second, then the first sample will tend to have more extreme values and thus lower Ansari-Bradley ranks. Thus a low value for the sum of the ranks in the first sample indicates that the first population is more variable than the second. For small sample sizes, the sum of ranks is compared to values found in special tables in Hollander and Wolfe (1974). If both n's are 10 or larger, the test statistic can be regarded as a z statistic. The expected value and the standard error of this statistic depend on whether the total sample size is odd or even. The reason is that if the total sample size is odd, the last value ranked will be the only one with that rank, whereas if the total sample size is even, the last two values ranked will have the same rank. For an odd total sample size, the expected value and standard error of the rank sum, assuming that H_0 is true, are

$$\frac{n_1(n_1 + n_2 + 1)^2}{4(n_1 + n_2)} \quad \text{and} \quad \sqrt{\frac{n_1 n_2(n_1 + n_2 + 1)[3 + (n_1 + n_2)^2]}{48(n_1 + n_2)^2}}$$

For an even total sample size, the H_0 expected value and standard error are

$$\frac{n_1(n_1 + n_2 + 2)}{4} \quad \text{and} \quad \sqrt{\frac{n_1 n_2(n_1 + n_2 + 2)(n_1 + n_2 - 2)}{48(n_1 + n_2 - 1)}}$$

For the latency-time data, the total sample size $10 + 10 = 20$ is even, so the expected value is $10(22)/4 = 55$ and the standard error is

$$\sqrt{\frac{(10)(10)(22)(18)}{48\ (19)}} = \sqrt{43.421}$$

Thus

$$z = \frac{53 - 55}{\sqrt{43.421}} = -0.30$$

The value of z is nowhere close to statistically significant. We retain the null hypothesis of equal variability, as we did with the F test. However, the Ansari-Bradley test is more believable in the case of highly skewed latency data than is the F test.

Another alternative procedure is based on the jackknife idea discussed in Section 12.2. The jackknife method involves leaving out one sample value at a time and computing "pseudo-values." The details are technical and can be incorporated into a computer program. For the latency-time data, the 95% jackknife confidence interval is

$$0.495 < \frac{\sigma_T}{\sigma_C} < 5.345$$

This very wide interval for the ratio of the standard deviations does include 1.00, so once again we would retain the null hypothesis of equal standard deviations.

Both the Ansari-Bradley method and the jackknife method are much more nominally robust than the F method when the populations are not normal. Unfortunately, neither method is, as yet, widely available in standard statistical computer packages.

Example 12.14

A computer simulation study compared the nominal robustness of F and jackknife tests for comparing standard deviations. Independent samples of size 20 each were drawn from mildly outlier-prone popula-

```
using variance ratio test

one-tail:

           number of times H0:  "variances equal" is rejected in favor of
alpha      "var1 > var2"    "var1 < var2"       total (alpha doubled)
0.100          194               196                  390
0.050          139               134                  273
0.025           97                93                  190
0.010           63                65                  128
0.005           41                43                   84


using the jackknife test

one-tail:

           number of times H0:  "mu1-mu2 is    0" is rejected in favor of
alpha      "mu1-mu2 >     0"  "mu1-mu2 <    0"     total (alpha doubled)
0.100             112               117                  229
0.050              64                75                  139
0.025              37                45                   82
0.010              14                16                   30
0.005               9                 8                   17
```

Table 12.3

tions. In both populations, the variance was 2.0, so the null hypothesis was true. Both the variance-ratio (F) and jackknife tests were performed 1000 times. The results were as shown in Table 12.3.

Which of the two tests has better nominal robustness in this situation?

Solution The nominal α probabilities for the variance-ratio test are too low. In the α = .005 case, where we should have had 5 and 5 (total 10), we actually get 41 and 43 (total 84) Type I errors out of 1000 trials. □

Exercises for Sections 12.3 and 12.4

12.10. For each of the following pairs of df (numerator df first), find the value that cuts off a right-tail area of .05 in the F distribution.

a. 6, 12
b. 12, 6
c. 6, 6
d. 12, 12

12.11. If two independent samples of size 16 each are taken from two populations with equal variability, find the probability that the ratio $s_1^2/s_2^2 > 2.86$.

12.12. Refer to Exercise 11.8. Calculate a 95% confidence interval for the ratio of the population standard deviations. Does this interval indicate that there is a statistically significant ($\alpha = .05$) difference in standard deviations?

12.13. Find the F ratio for the sample variances of Exercise 11.8. Place bounds on the two-tailed p-value.

12.14. Is there reason to believe that the confidence level and p-value found in Exercises 12.12 and 12.13 might be poor approximations?

12.15. Refer to Exercise 11.4. Test the hypothesis of equal population standard deviations against a two-sided alternative. Use $\alpha = .10$.

12.16. In Exercise 12.15, place bounds on the p-value.

12.17. Do the data of Exercise 11.4 indicate that there is a potentially serious violation of the assumptions underlying the test of equal population standard deviations?

Chapter Exercises

12.18. Suppose that a study of a sample of 24 entering freshmen at a very large state university found a standard deviation of 14.63 points on a scale measuring preparation for a college English composition course.

a. Calculate a 95% confidence interval for the population standard deviation.

b. Calculate a 95% lower-bound confidence interval for the population standard deviation.

12.19. **a.** Using the data of Exercise 12.18, test the research hypothesis that the population standard deviation is less than 20. Use $\alpha = .05$.

b. Which, if either, of the intervals in Exercise 12.18 could also be used to test this hypothesis?

12.20. The scale used in Exercises 12.18 and 12.19 yields a moderate number of scores in the 90's, a large number of scores in the 80's, then decreasing numbers in the 70's, 60's, and 50's. A few scores are down in the 20's.

What does this fact indicate about the reliability of the procedure you used in Exercises 12.18 and 12.19?

12.21. In Exercise 11.39, independent samples yielded the results in Table 12.4.

Group	Mean	Standard Deviation	n
S	52.60	24.62	15
B	75.70	37.02	10

Table 12.4

a. Calculate a 99% confidence interval for the ratio of population standard deviations.

b. Does the interval indicate that there is a statistically detectable difference of population standard deviations, at $\alpha = .01$?

12.22. The data underlying Exercise 12.21 are quite right-skewed in both samples. Does this pose any problem for the inferences made in Exercise 12.21?

12.23. A sample of 420 subjects yielded a sample variance of 193.727.

a. Find approximate values for the table entries $\chi^2_{.05}$ and $\chi^2_{.95}$.

b. Use these values to calculate a 90% confidence interval for the population variance.

12.24. A "homebrew" computer program performed a jackknife analysis of the data of Exercise 12.23 and obtained as a 90% confidence interval for the population variance

$$99.090 < \sigma^2_{\text{pop}} < 214.93$$

Is this interval similar to the interval calculated in Exercise 12.23?

12.25. The data underlying Exercises 12.23 and 12.24 contained several values far above and far below the main part of the data. It was argued that the large sample size made the χ^2 confidence interval of Exercise 12.23 reliable. Is the argument valid?

12.26. A computer simulation study of the behavior of the χ^2 test and the jackknife procedure for variances involved 1000 samples of size 11 taken from an extremely outlier-prone population. The results were as shown in Table 12.5. Note that H_0 is true in this simulation.

a. How accurate are the nominal α probabilities in this situation?

b. Which method gives a better (or perhaps "less worse") approximation to the nominal α probabilities?

using jackknife test

alpha	number of times H0: "variance is 99.01" is rejected in favor of "variance > 99.01"	"variance < 99.01"	total (alpha doubled)
0.100	39	488	527
0.050	18	431	449
0.025	8	379	387
0.010	5	328	333
0.005	4	290	294

using chi-square variance test

alpha	number of times H0: "variance is 99.01" is rejected in favor of "variance > 99.01"	"variance < 99.01"	total (alpha doubled)
0.100	130	712	842
0.050	122	670	792
0.025	111	626	737
0.010	100	572	672
0.005	90	517	607

Table 12.5

12.27. Another computer simulation of the χ^2 and jackknife procedures involved 1000 samples of size 31 from the same extremely outlier-prone population as in Exercise 12.26. The summary results in Table 12.6 were obtained. Note that H_0 is true in this simulation, as well.

using jackknife test

alpha	number of times H0: "variance is 99.01" is rejected in favor of "variance > 99.01"	"variance < 99.01"	total (alpha doubled)
0.100	56	395	451
0.050	26	348	374
0.025	20	306	326
0.010	8	262	270
0.005	6	239	245

using chi-square variance test

alpha	number of times H0: "variance is 99.01" is rejected in favor of "variance > 99.01"	"variance < 99.01"	total (alpha doubled)
0.100	200	627	827
0.050	186	605	791
0.025	175	584	759
0.010	163	553	716
0.005	151	529	680

Table 12.6

a. Compared to the results shown in Exercise 12.26 for the jackknife procedure, are the nominal α probabilities better approximations to the simulation probabilities?

b. Same question, for the χ^2 procedure.

c. Does increasing the sample size always decrease the effect of a nonnormal population?

12.28. A study involving two independent random samples yielded the summary statistics in Table 12.7.

Sample	Variance	Sample Size
1	1.176	41
2	0.778	31

Table 12.7

a. Test the null hypothesis that the population variances are equal against the research hypothesis that the variance in population 1 is larger. State bounds on the p-value.

b. The data in both samples had a pronounced right skew. Does this fact indicate that the p-value found in part a might be inaccurate?

12.29. A "homebrew" computer program for performing jackknife tests yielded the following results:

$$t' = 4.756 \text{ with } 45.1 \text{ df}$$

Assuming that t' is a t statistic, place bounds on the p-value. Are these bounds similar to those found in Exercise 12.28?

12.30. The data underlying Exercises 12.28 and 12.29 were sharply right-skewed, in both samples. What does this indicate about the reliability of the p-values found in Exercises 12.28 and 12.29?

12.31. A computer simulation was done to compare the performance of F and jackknife tests for comparing variances. Independent samples of sizes 10 and 30 were drawn from a Laplace population, a mildly outlier-prone population. Both the variance-ratio (F) and the jackknife test were performed. The results for 1000 of each test were as shown in Table 12.8. Note that H_0 is true in this simulation.

a. How close to the nominal α probabilities are the simulation results for the F test?

b. How close are they for the jackknife test?

```
Popn       Mu        Sigma       n
  1       0.000      1.4142     10
  2       0.000      1.4142     30

using variance ratio test

one-tail:

        number of times H0:  "variances equal" is rejected in favor of
alpha    "var1 > var2"    "var1 < var2"    total (alpha doubled)
0.100        175              200               375
0.050        121              135               256
0.025         99               93               192
0.010         59               54               113
0.005         49               31                80

using the jackknife test

one-tail:

        number of times H0:  "mu1-mu2 is  0" is rejected in favor of
alpha    "mu1-mu2 >  0"   "mu1-mu2 <  0"   total (alpha doubled)
0.100        132              128               260
0.050         75               74               149
0.025         40               47                87
0.010         14               26                40
0.005          8               14                22
```

Table 12.8

12.32. In Exercise 11.49, we compared samples, each of size 27, of liberal arts and preprofessional students. A portion of the SAS output from that exercise is reproduced in Figure 12.3. Is there a statistically detectable difference in the variances, assuming $\alpha = .05$?

```
                    2 SAMPLE T-TEST OF ANXIETY SCORES OF COLLEGE STUDENTS

                                      TTEST PROCEDURE

VARIABLE: SCORE

SCHOOL              N      MEAN       STD DEV     STD ERROR    MINIMUM     MAXIMUM     VARIANCES      T       DF   PROB > |T|

LIBERAL ARTS       27   9.37037037  3.93356511  0.75701496  1.00000000  20.00000000   UNEQUAL    -3.7072   49.6      0.0005
PRE-PROFESSIONAL   27  12.96296296  3.14375427  0.60501579  2.00000000  17.00000000   EQUAL      -3.7072   52.0      0.0005

FOR H0: VARIANCES ARE EQUAL, F'=   1.57 WITH 26 AND 26 DF    PROB > F'= 0.2596
```

Figure 12.3 Output for Exercise 12.32

Methods for Qualitative Data

13

Up to now, almost all the discussion of statistical inference has been oriented to *quantitative* data—values measured along a legitimate numerical scale. In this chapter we consider some basic methods for *qualitative* data—variables having values that are merely categories, possibly in order from low to high. Such variables arise very frequently in psychology and social sciences generally, so the methods in this chapter are widely—and often poorly—used.

Section 13.1 is devoted to methods for a yes/no (binomial) variable. The key idea is that a normal approximation may be used in performing hypothesis tests and confidence-interval inference. Then, in Section 13.2 we turn to the chi-squared goodness-of-fit test, which is designed to test a certain hypothesized set of probabilities. In Section 13.3 another chi-squared test, this one to establish the statistical significance of an apparent statistical association, is discussed. Finally, Section 13.4 very briefly discusses a newer approach, log-linear models.

13.1 Normal Distribution Methods for Inference about a Proportion

Chapter 9 introduced the basic ideas of hypothesis testing with binomial examples. We were concerned about a variable that could take on only two possible values—"success" or "failure." Such a variable is the simplest qualitative variable, in that the result merely falls into a category, rather than being an actual measurement. Of course, one can arbitrarily assign a number 1 to success and 0 to failure, but the very arbitrariness of the choice suggests that other methods might be better.

The typical hypothesis-testing situation for such a variable involves a null hypothesis that the population proportion of successes, π, equals a certain value. In Chapter 9 we referred the sample number of successes (denoted Y) to a binomial table. Alternatively, binomial probabilities can be approximated by normal probabilities, as we saw in Chapter 7, so a nearly equivalent test may be based on the z table.

Recall that if n, the number of binomial trials, is large, then Y, the number of successes in n trials, has approximately a normal distribution with expected value $n\pi$ and standard error $\sqrt{n\pi(1 - \pi)}$.

$$z = \frac{Y - n\pi}{\sqrt{n\pi(1 - \pi)}}$$

Equivalently, if n is large, then $\pi = Y/n$, the sample *proportion* of successes is approximately normal with expected value π and standard error $\sqrt{\pi(1 - \pi)/n}$.

$$z = \frac{\hat{\pi} - \pi}{\sqrt{\pi(1 - \pi)/n}}$$

The two values of z will always be equal. The approximation is reasonably close provided that $n\pi$, the expected number of successes, and $n(1 - \pi)$, the expected number of failures, both are at least 5.

For example, a just noticeable difference (jnd) between two different weights is often defined to be a difference such that a subject can choose the heavier one with probability .75. If a subject is given 50 supposed jnd's, what is the probability of getting fewer than 30 right (or, equivalently, that the proportion right will be less than $30/50 = .60$)? The expected number of correct answers is $50(.75) = 37.5$, and the standard error is

$$\sqrt{50(.75)(1 - .75)} = 3.026$$

Thus, using a normal approximation,

$$P(Y < 30) = P\left(z < \frac{30 - 37.5}{3.062}\right) = P(z < -2.45) = .0071$$

According to Table 1 in the appendix, the exact binomial probability is .0061. Note that the expected number of successes is 37.5 and the expected number of failures is 12.5. Both are well above the rule-of-thumb value, 5.

Example 13.1

A random sample is to be taken of 100 people teaching computer usage in elementary schools. One question to be answered is whether each teacher uses "turtle graphics" in computer work. Assume that "turtle graphics" are used by 20% of the teachers in all elementary schools.

(a) Should binomial probabilities apply in this situation?

(b) Using binomial probabilities, find the probability that 25 or more teachers will respond "yes."

(c) Find the same probability using a normal approximation.

Solution

(a) We have a series of 100 trials (teachers), each of which is either a success (using "turtle graphics") or a failure (not using them). The probability of success is constant at $\pi = .20$. Neglecting the very minor issue of sampling without replacement from the very large population, we may assume independence in the sample. Finally, the number of trials is fixed and Y = the number of successes obtained, without regard to order. Thus all binomial assumptions are met.

(b) In Table 1, find binomial probabilities for $n = 100$ and $\pi = .20$. The probability of 25 or more successes is

$$P(Y = 25) + P(Y = 26) + \cdots = .0439 + .0316 + \cdots = .1313$$

(c) The expected value of Y is $100(.20) = 20$, and the standard error is

$$\sqrt{100(.20)(.80)} = 4.00$$

Thus the probability that Y is 25 or more is

$$P\left(z \geq \frac{25 - 20}{4.00}\right) = P(z \geq 1.25) = .1056$$

from Table 3. Equivalently, we want the probability that the proportion of successes in the sample is at least .25. The expected value of the sample proportion is $\pi = .20$, and the standard error is

$$\sqrt{\frac{(.20)(.80)}{100}} = .0400$$

Thus the desired probability is

$$P\left(z \geq \frac{.25 - .20}{.0400}\right) = P(z \geq 1.25) = .1056$$

once again. □

The normal approximation to binomial probabilities leads immediately to an approximate z test for a hypothesized proportion. We state it in terms of a sample proportion, but it could equally be stated in terms of Y, the sample number of successes.

z Test of a Proportion

1. H_0: $\pi = \pi_0$
2. (i) H_a: $\pi > \pi_0$
   (ii) H_a: $\pi < \pi_0$
   (iii) H_a: $\pi \neq \pi_0$
3. T.S.: $z = \dfrac{\hat{\pi} - \pi_0}{\sqrt{\pi_0(1 - \pi_0)/n}}$
4. R.R.:
   (i) $z > z_\alpha$
   (ii) $z < -z_\alpha$
   (iii) $|z| > z_{\alpha/2}$

For instance, suppose that a sample of 50 subjects tried to determine which of two weights was heavier, where the difference in weights was supposedly a just noticeable difference, so 75% of subjects should be able to pick the heavier weight correctly. A two-sided research hypothesis seems appropriate, assuming that it would be of interest to see whether the difference was too noticeable or not noticeable enough. Suppose that 31 of the 50 subjects (proportion .62) made the correct choice. We have $\pi_0 = .75$, and

$$z = \frac{.62 - .75}{\sqrt{(.75)(1 - .75)/50}} = -2.12$$

with a p-value of about .0340. (The p-value is obtained in the usual way; see Figure 13.1.) There is fairly conclusive, though not overwhelming, evidence that too few people notice the difference; that is, that the difference in weights is less than a jnd.

Example 13.2

In Example 13.1 we assumed that the population proportion of successes was .20. Suppose that 31 of the teachers sampled reported using "turtle graphics." Is this evidence that the assumed .20 value is wrong?

Solution We take .20 as our null hypothesis value; H_0: $\pi = .20$. There is no reason to believe that the study was done with a directional, one-sided hypothesis in mind, so we must take the research hypothesis to be two-sided; H_a: $\pi \neq .20$. We have

$$z = \frac{(31/100) - .20}{\sqrt{(.20)(.80)/100}} = 2.75$$

From the normal table, we have an approximate p-value of $2(.0030) = .0060$. This small p-value indicates that we can reject the hypothesis that $\pi = .20$ rather conclusively. □

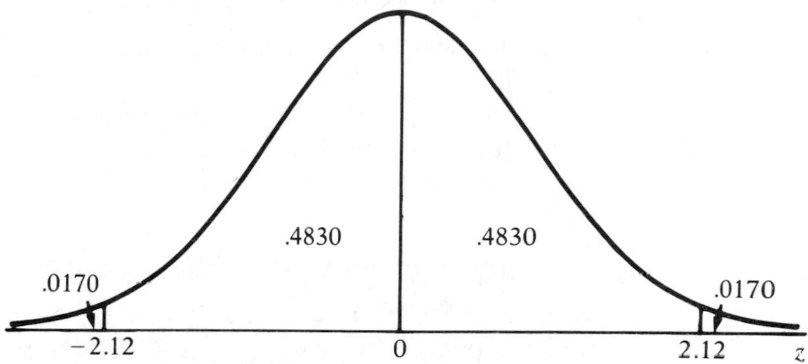

Figure 13.1 p-value for $z = -2.12$

Rather than testing a specific null hypothesis, one can calculate a confidence interval for the population proportion. Recall that a $100(1 - \alpha)\%$ confidence interval consists of all values of the population parameter that couldn't be rejected as null hypotheses (using the particular α value). A computationally elaborate confidence interval can be obtained by solving the inequality

$$\frac{|\hat{\pi} - \pi_0|}{\sqrt{\pi_0(1 - \pi_0)/n}} < z_{\alpha/2}$$

for π_0. The mathematics is to square each side, then use the quadratic formula to find the boundaries of the interval. The result (omitting the 0 subscript) is

$$\frac{\hat{\pi} + k/2 - \sqrt{k\hat{\pi}(1 - \hat{\pi}) + (k/2)^2}}{1 + k} < \pi < \frac{\hat{\pi} + k/2 - \sqrt{k\hat{\pi}(1 - \hat{\pi}) + (k/2)^2}}{1 + k}$$

where $k = z^2_{\alpha/2}/n$.

In the jnd example, we have $n = 50$ and $\hat{\pi} = .62$. For a 95% confidence interval, the desired table value is $z_{.025} = 1.96$. The value of k is $(1.96)^2/50 = .076832$. Thus the confidence interval is

$$\frac{.62 + .076832/2 - \sqrt{.076832(.62)(.38) + (.076832/2)^2}}{1.076832}$$

$$< \pi < \frac{.62 + .076832/2 + \sqrt{.076832(.62)(.38) + (.076832/2)^2}}{1.076832}$$

which (eventually) works out to

$$.4815 < \pi < .7414$$

Note that this interval does not contain the rejected null hypothesis value $\pi = .75$.

A computationally simpler, approximate confidence interval is often used instead. It fits in the familiar "statistic plus-or-minus table value times standard error" format.

Approximate 100(1 − α)% Confidence Interval for a Proportion

$$\hat{\pi} - z_{\alpha/2}\sqrt{\frac{\hat{\pi}(1-\hat{\pi})}{n}} < \pi < \hat{\pi} + z_{\alpha/2}\sqrt{\frac{\hat{\pi}(1-\hat{\pi})}{n}}$$

NOTE: The nominal confidence level may be a poor approximation if n is small, particularly if the estimated proportion is near 0 or near 1.

In the jnd example, with $n = 50$, $\hat{\pi} = .62$, and $z_{.025} = 1.96$, the approximate 95% confidence interval is

$$.62 - 1.96\sqrt{\frac{(.62)(.38)}{50}} < \pi < .62 + 1.96\sqrt{\frac{(.62)(.38)}{50}}$$

or

$$.4855 < \pi < .7545$$

Note that this interval isn't much different from the interval derived from the test statistic. Unfortunately, the approximate confidence interval and the test statistic don't always agree; it is possible for the test statistic to reject a certain π value while the confidence interval includes that value. This happens with the jnd data. According to the hypothesis test, we (barely) rejected $\pi = .75$ at $\alpha = .05$, but $\pi = .75$ is (barely) included in the approximate 95% confidence interval.

Example 13.3 Refer to Example 13.2. Calculate a 99% confidence interval for π using both methods. Do the resulting intervals give the same conclusion about H_0: $\pi = .20$, the hypothesis that was rejected at $\alpha = .01$ (p-value .0060) in Example 13.2?

Solution In Example 13.2, $\hat{\pi}$ was .31 and n was 100. The required normal table value for 99% confidence is $z_{.005} = 2.58$, so $k = (2.58)^2/100 = .066564$. The interval based on the test statistic is

$$\frac{.31 + .066564/2 - \sqrt{.066564(.31)(.69) + (.066564/2)^2}}{1 + .066564}$$

$$< \pi < \frac{.31 + .066564/2 + \sqrt{.066564(.31)(.69) + (.066564/2)^2}}{1 + .066564}$$

or

$$.206 < \pi < .438$$

This interval excludes, and therefore rejects, the value $\pi = .20$.

The approximate 99% confidence interval is

$$.31 - 2.58\sqrt{\frac{(.31)(.69)}{100}} < \pi < .31 + 2.58\sqrt{\frac{(.31)(.69)}{100}}$$

or

$$.191 < \pi < .429$$

which includes, and therefore retains, the value $\pi = .20$ □

For large n, the difference in the two confidence interval methods is negligible. For small n, both methods (as well as any other conceivable methods) are unsatisfactory in that they yield very wide confidence intervals. For example, if n is 20 and there are 12 successes (sample proportion .60), the 95% confidence interval derived by the approximate method in the box is

$$.385 < \pi < .815$$

and the equivalent-to-hypothesis testing interval is

$$.387 < \pi < .781$$

Remember that the true proportion can only range from 0 to 1. Whichever way one calculates the interval, it is so wide as to be useless.

Typically, an inference based on a certain sample size obtained using qualitative data "feels" less satisfactory than does an inference based on the same sample size obtained using quantitative data. For example, a confidence interval for a proportion based on $n = 100$ often seems excessively wide, whereas a confidence interval for a mean based on $n = 100$ often seems adequately narrow. The reason is that qualitative data give very little information per observation. At the extreme, a binomial variable only gives a "yes" or "no" per observation; it is far less informative than a specific numerical value obtained by a measurement. Typically, much larger sample sizes are needed with less informative, qualitative variables than with measured, quantitative variables.

One approach to determining the sample size needed to achieve a certain accuracy is to use the approximate confidence interval for a proportion. For instance, how large a sample is required to estimate a proportion with 95% confidence and a plus-or-minus term equal to .04 (4 percentage points)? Using the approximate interval, the plus-or-minus term for 95% confidence is found to be

$$1.96\sqrt{\frac{\hat{\pi}(1-\hat{\pi})}{n}}$$

The problem, obviously, is that $\hat{\pi}$ isn't known until the sample is actually taken. A conservative approach is to assume that $\hat{\pi}$ will equal .50, because the plus-or-minus term is largest at that $\hat{\pi}$ value. Given that assumption, the plus-or-minus will equal

$$1.96\sqrt{\frac{.5(1-.5)}{n}} \quad \text{or} \quad \frac{0.98}{\sqrt{n}}$$

Setting that plus-or-minus to the desired .04 yields

$$\sqrt{n} = \frac{0.98}{.04} = 24.5 \quad \text{or} \quad n = 600.25$$

rounded up to 601. If researchers have some idea of the likely size of $\hat{\pi}$, they can make alternative assumptions. For instance, if it is thought that $\hat{\pi}$ will be somewhere between .10 and .20, assume that it will be .20, the value in that range yielding the largest plus-or-minus. (In general, take the value in the likely range that is closest to .50.) For 95% confidence and a desired plus-or-minus of .04, set

$$1.96\sqrt{\frac{.2(1-.2)}{n}} = .04$$

and solve to get $n = 385$. Note that bounding the reasonable values away from .50 allows for a substantial reduction in the required sample size.

Example 13.4 It has been observed that when a person gets on an elevator occupied by one other person, the entering person will often choose to stand on the opposite side of the elevator. It was desired to estimate the population proportion of "opposite-siders" with 90% confidence, to an accuracy of $\pm .04$. If it is assumed that $\hat{\pi}$ will be somewhere between .70 and .90, find the required sample size. If no further assumptions are made, find the required sample size.

Solution Given the assumption that $\hat{\pi}$ is between .70 and .90, we will take $\hat{\pi} = .70$, the closest value to $\hat{\pi} = .50$. The required table value is $z_{.05} = 1.645$. We solve

$$1.645\sqrt{\frac{.70(1 - .70)}{n}} = .04$$

and obtain $n = 355$. If we weren't willing to assume anything about $\hat{\pi}$, we would take the worst case, $\hat{\pi} = .50$.

$$1.645\sqrt{\frac{.50(1 - .50)}{n}} = .04$$

which yields $n = 423$, a larger sample size. □

Alternatively, the selection of sample size can be based on the desired power of a significance (hypothesis) test.

Power of a z Test for a Proportion

One-tailed test:

$$\text{power} = P\left(z > z_\alpha \frac{\sqrt{\pi_0(1 - \pi_0)}}{\sqrt{\pi_a(1 - \pi_a)}} - \frac{|\pi_a - \pi_0|}{\sqrt{\pi_a(1 - \pi_a)/n}}\right)$$

Two-tailed test: Replace z_α by $z_{\alpha/2}$.

NOTE: π_0 and π_a refer to the hypothesized population proportions under H_0 and H_a, respectively.

As usual with power calculations, the main problem is to find a reasonable specification of an H_a value. If the study is a repetition of a previous study, the value of π found in the previous study is a reasonable choice for π_a. Alternatively, a rough, subjective guess at the value of π_a may be made. Or, one can consider "how much of a difference makes a difference." If, say, H_0 is that $\pi = .75$, it is not likely to be of much scientific interest if in fact π_a is .74 or .76, whereas a major difference from H_0, such as a π_a of .60 or .90, would most likely be important to detect.

Suppose that $n = 100$, H_0 is $\pi = .75$, $\alpha = .01$, and we hypothesize that $\pi_a = .60$. For a two-tailed test,

$$\text{power} = P\left(z > 2.58\frac{\sqrt{.75(1-.75)}}{\sqrt{.60(1-.60)}} - \frac{|.60-.75|}{\sqrt{.60(1-.60)/100}}\right)$$

$$= P(z > -0.78) = .72$$

Example 13.5 Students taught a skill by a conventional method have a 70% mastery rate. A new method is to be tested on a sample of students, to see if a higher proportion master the skill. A hypothesis test was to be run, using $\alpha = .05$. The researcher wanted a high probability of finding a statistically significant difference, assuming that π was .80 under the new method. Is a sample size of 400 enough for this purpose?

Solution Because the researcher is interested in increasing the mastery rate, a one-tailed test should be used. The table value is $z_{.05} = 1.645$, $\pi_0 = .70$, and $\pi_a = .80$. The probability of finding a significant difference is, by definition, the power of the test.

$$\text{power} = P\left(z > 1.645\frac{\sqrt{.70(.30)}}{\sqrt{.80(.20)}} - \frac{|.70-.80|}{\sqrt{.80(.20)/400}}\right)$$

$$= P(z > -3.12)$$

The value $z = 3.12$ isn't even contained in Table 3, indicating that $P(z > -3.12)$ is very high. Thus $n = 400$ does give a high probability of finding a significant effect. □

The *sample size required to yield a desired power* $1 - \beta$ can be found by trial and error, or by solving

$$\sqrt{n} = \frac{z_\alpha\sqrt{\pi_0(1-\pi_0)} + z_\beta\sqrt{\pi_a(1-\pi_a)}}{|\pi_a - \pi_0|}$$

To obtain power $= .90$ ($\beta = .10$), with $\alpha = .05$ (one-tailed), $\pi_0 = .75$, and $\pi_a = .60$, we use the z-table values $z_{.05} = 1.645$ and $z_{.10} = 1.28$. We must solve

$$\sqrt{n} = \frac{1.645\sqrt{.60(1-.60)} + 1.28\sqrt{.75(1-.75)}}{|.60-.75|}$$
$$= 9.0676$$

or $n = (9.0676)^2 = 82.2$, rounded up to 83.

Example 13.6 Refer to Example 13.5. Find the sample size required to give a power of .80 under the assumed conditions.

Solution In Example 13.5 we specified $\pi_0 = .70$, $\pi_a = .80$, and $\alpha = .05$ (one-tailed), so $z_{.05} = 1.645$. To obtain power $= .80$ or, equivalently, $\beta = .20$, we use the table value $z_{.20} = 0.84$. We must solve

$$\sqrt{n} = \frac{1.645\sqrt{.70(1-.70)} + 0.84\sqrt{.80(1-.80)}}{|.70-.80|}$$
$$= 10.898$$

or $n = 119$. Recall that the sample size $n = 400$ used in Example 13.5 gave a much higher power than .80. ☐

Typically, the sample sizes required to give reasonable scientific accuracy (in either the confidence interval or the power sense) will be more than large enough to assure that the technical issue of using a z approximation to binomial probabilities will not be a difficulty. The rule of thumb for using the z tables (expected numbers of successes and failures each at least 5) typically requires a smaller n than does reasonable scientific accuracy.

Exercises for Section 13.1

13.1. Teaching materials for a computerized learning system were to be tested on fourth- and fifth-grade children. The desire was to have more than 80% of the children able to use the system within one hour. A random sample of 50 children were selected; the actual number of children who could use the system within an hour was to be recorded.

a. Formulate null and research hypotheses. Should the research hypothesis be one-sided or two-sided? (Warning: There might be an argument on this question.)

b. Formulate the rejection region for a one-sided test, using $\alpha = .05$ and a z approximation.

13.2. Refer to Exercise 13.1. Use the binomial probability tables to find an exact rejection region. Does it make a big difference whether one uses the exact or approximate tables?

13.3. Refer to Exercise 13.1. Suppose that 45 of the 50 children are able to use the system within one hour. Can the null hypothesis that the true proportion is .80 be rejected at $\alpha = .05$? Assume a one-sided research hypothesis.

13.4. Refer to Exercise 13.3.

a. Calculate a 90% confidence interval for the population proportion of children who could use the system within an hour. (Note that the 90% confidence of a two-sided interval corresponds to a one-sided α of .05.) Use both the hypothesis-test-equivalent interval and the approximate interval.

b. How much difference does it make which interval is used?

13.5. Refer to Exercise 13.3. From basic principles, what is a 99% lower-bound confidence interval for the population proportion?

13.6. A researcher on the sense of taste developed a scale of spiciness. According to this scale, the median rating of a particular food should be 60. A sample of 88 subjects rated the food as to its spiciness.

a. Formulate null and research hypotheses in terms of the proportion π in the population rating the food as "above 60" in spiciness. Should the research hypothesis be one-sided or two-sided?

b. Specify a rejection region (R.R.) for a two-tailed test. Use a normal approximation.

c. Is there any reason to assume that the approximation in part b is a poor one? Explain.

13.7. Refer to Exercise 13.6. If, in fact, the proportion of subjects who rate the food above 60 is either .40 or .60, it is desired that the probability of rejecting H_0 should be at least .90.

a. Is a sample size of 90 adequate for this purpose?

b. If not, what sample size is needed?

13.8. Refer to Exercise 13.6. In fact, 31 of the 88 subjects rated the food above 60 in spiciness. Can the null hypothesis be rejected at typical α values? State an approximate p-value.

13.9. Refer to Exercises 13.6 and 13.8. Calculate 95% confidence intervals for π using both the test-equivalent and the approximate method. Does it make much difference which method is used?

13.10. A sample of agoraphobics (those who are afraid to leave familiar surroundings and panic in public places) were treated by a desensitization therapy. The proportion of agoraphobics who were judged to have become functional (e.g., able to hold a full-time job) was the parameter of interest. In the sample, 19 of 33 subjects were judged to have become functional.

a. Calculate a 95% confidence interval for π. Use the hypothesis-test-equivalent method.

b. Do the same using the "plus-or-minus table value times estimated standard error" method.

13.11. Refer to Exercise 13.10. In your opinion, are the confidence intervals narrow enough to be scientifically useful?

13.12. Refer to Exercise 13.10. Assuming that the estimated proportion having become functional is somewhere in the neighborhood of .50, what sample size is needed to obtain a 95% confidence interval with a width of .10?

13.13. Refer to Exercise 13.10. Based on the confidence interval, can one reject the null hypothesis that $\pi = .40$, using a two-sided $\alpha = .05$ test?

13.14. Refer to Exercises 13.10, 13.11, and 13.13.

a. What, generally, does the width of the interval indicate about the power of the test?

b. Calculate the power of the test assuming $\pi = .60$.

13.15. In Exercise 13.14, what sample size is needed to have the power of the test equal to .80?

13.2 Goodness-of-Fit Tests

In Section 13.1 we considered the binomial situation in which each observation fell in one of only two categories. Now we consider situations in which each observation may fall in one of several categories. For example, Inhelder and Piaget (1964) reported that 65% of all 5-year-old children had no anticipation of how a new object would be classified, 25% had some anticipation, and 10% had complete anticipation. In fact these figures are from a sample, but for illustration assume that they are population figures. Suppose that a random sample of 80 5-year-olds has been obtained. (Note that considerable care would be needed to minimize bias.) The sample frequencies are as follows: no anticipation, 38; some anticipation, 26; and complete anticipation, 16. Are these frequencies significantly (in the sense of more than randomly) different from those reported by Inhelder and Piaget?

To perform the significance test, we need a test statistic. We do *not* want to perform separate z tests as in Section 13.1 for each category. The reason is that one cannot know the overall α value if each z test is run at, say, $\alpha = .05$. The overall α is not $3(.05) = 15$; by that reasoning, if there had been 21 tests, the overall α would have been $21(.05) = 1.05$! With three tests, one can only know that the probability of rejecting some part of the null hypothesis is somewhere between .05 and .15. Instead, we need a single test statistic that summarizes all the discrepancies between the actual and theoretical proportions. The most widely used statistic is the **chi-squared goodness-of-fit** statistic.

chi-squared goodness-of-fit

This statistic is based on the difference between the actually observed frequencies and the *expected frequencies*, assuming that the null hypothesis probabilities hold. In the Inhelder and Piaget illustration, the assumed probabilities for no, some, and complete anticipation are .65, .25, and .10, respectively. In 80 observations, we would expect $80(.65) = 52$ children with no anticipation, $80(.25) = 20$ with some anticipation, and $80(.10) = 8$ with complete anticipation. The mechanics of the test follow.

Chi-Squared Goodness-of-Fit Test (Completely Specified Probabilities)

1. H_0: probabilities for each category are correctly specified

2. H_a: probability of at least one category is incorrectly specified

3. T.S.: $\chi^2 = \Sigma(\text{Obs} - \text{Exp})^2/\text{Exp}$, where Obs is the actual count (frequency) in the category and Exp is obtained by multiplying n, the number of observations, by the specified probability for that category

4. R.R.: reject H_0 if $\chi^2 > \chi^2_\alpha$, where χ^2_α is the right-tail cutoff point for the chi-squared distribution with $K - 1$ df, where K is the number of categories

NOTE: The test is based on the observed counts, not on proportions or percentages. The use of chi-squared tables is an approximation. The approximation may be poor if any Exp is less than 1 or if more than 20% of the Exp values are less than 5. The df depends on the number of categories, not on the sample size.

In the Inhelder and Piaget illustration, the calculations are as shown in Table 13.1.

	No	Some	Complete	Total
Observed frequency (Obs)	38	26	16	80
Expected frequency (Exp)	52	20	8	80
$(\text{Obs} - \text{Exp})^2/\text{Exp}$	3.769	1.800	8.000	13.569

Table 13.1

Thus the χ^2 statistic equals 13.569. There are three categories, so there are $3 - 1 = 2$ df. The right-most entry in the chi-squared table is about 10.6, corresponding to a right-tail area of .005; therefore, even for α as small as .005, we would reject H_0. (Thus the p-value is less than .005.) The observed frequencies are significantly different from those shown in the Inhelder and Piaget study. Note that all the Exp values are above 5, so the use of the chi-squared tables should be a good approximation.

Example 13.7 Suppose that, before the deinstitutionalizing movement in mental health care, it was known that 62% of all patients of a certain type took prescribed medication regularly, 30% usually, and 8% sporadically or never. After deinstitutionalization, a sample of 70 patients were followed regularly. Of the 70, 36 were rated as regular, 26 as usual, and 8 as sporadic or never. Are these sample frequencies within random variation (using $\alpha = .10$) of the expected frequencies?

Solution The expected frequencies are $70(.62) = 43.4$, $70(.30) = 21.0$, and $70(.08) = 5.6$. All Exp are greater than 5.

$$\chi^2 = \frac{(36 - 43.4)^2}{43.4} + \frac{(26 - 21.0)^2}{21.0} + \frac{(8 - 5.6)^2}{5.6} = 3.481$$

There are $3 - 1 = 2$ df. The χ^2-table value corresponding to a right-tail area of .10 is 4.605. The test statistic does not fall in the rejection region, so the sample frequencies are within reasonable random variation of the expected frequencies. □

multinomial assumptions

The assumptions underlying this procedure are sometimes called **multinomial assumptions.** The experiment consists of a series of trials, with each trial yielding one of K mutually exclusive and exhaustive categories. The probabilities of falling in the various categories are assumed to be constant over trials, and the trials are assumed to be independent. In the multinomial, the number of trials is assumed to be fixed.

Example 13.8 Refer to Example 13.7. Do the multinomial assumptions seem reasonable in that situation?

Solution Yes. Each patient constitutes a trial with three possible outcomes. Because the patients are randomly chosen, the probabilities of regular, usual, and sporadic should be constant. There is no obvious dependence from one patient to another. The number of trials was fixed at 70. ☐

The chi-squared goodness-of-fit test can also be used to see if a certain probability distribution is a good model. To illustrate, suppose that rats in a T-maze are rewarded with food whether they turn left or right. After a series of "warm-up" trials where presumably each rat learns that it doesn't matter which way it turns, each of 48 rats is allowed four trials, and the number of left turns is recorded. One plausible assumption is that each rat has a 50-50 chance of turning left or right on each run, independently of past runs. Then the number of left turns should be a binomial random variable, with probability of success (left turn) .50 per trial. From the binomial tables, the probabilities of 0, 1, 2, 3, and 4 successes are .0625, .2500, .3750, .2500, and .0625, respectively. Suppose that the following observed frequencies are found:

Number of left turns	0	1	2	3	4
Observed number of rats	4	5	12	17	10
Expected number of rats	3.0	12.0	18.0	12.0	3.0

The chi-squared statistic may be computed as usual, and comes out 24.8333; there are five categories, so there are 4 df. The actual chi-squared statistic is far beyond even the .005 table value of 14.8602. Thus the binomial, $\pi = .50$ model is not "a good fit" to the observed frequencies. It might be objected that two of the expected values are less than 5, so the use of the chi-squared tables involves a potentially bad approximation. Had the actual chi-squared statistic come out somewhat close to a table value, we would have had to be cautious in reporting a p-value. As it is, the actual value is so far beyond the table that the conclusion doesn't depend on a good approximation.

Example 13.9 A study attempted to evaluate whether high school students were "test-wise"—that is, could guess answers to questions merely from the way questions and answers were posed, rather than from actual knowledge. A sample of 200 students were given a four-question multiple-choice quiz on a completely unfamiliar topic. Each question had five possible answers. The observed frequencies for the number of correct answers were

Number correct	0	1	2	3	4
Observed frequency	73	78	41	7	1

(a) Under the null hypothesis of no "test wisdom," what are the probabilities of 0 correct, 1 correct, etc.?

(b) Test this null hypothesis against a general research hypothesis, and state the *p*-value.

Solution **(a)** For any one student, the quiz is a four-trial binomial experiment. If there is no "test wisdom," the probability of a correct answer is $1/5 = .20$. Assuming that the four questions are unrelated, it is fair to assume that the correctness of the answers is independent from one question to the next. From Table 1 of binomial probabilities ($n = 4$, $\pi = .20$), the probabilities are

Number correct	0	1	2	3	4
Probability	.4096	.4096	.1536	.0256	.0016

(b) To obtain expected frequencies, we multiply the probabilities by the sample size, $n = 200$.

Number correct	0	1	2	3	4
Expected frequency	81.92	81.92	30.72	5.12	0.32

We note that the expected frequency of four correct answers is well below 1.0, the minimum standard for using the χ^2 table. To avoid a poor approximation, we combine the 3 and 4 categories:

Number correct	0	1	2	3–4
Observed frequency	73	78	41	8
Expected frequency	81.92	81.92	30.72	5.44

Now

$$\chi^2 = \frac{(73 - 81.92)^2}{81.92} + \frac{(78 - 81.92)^2}{81.92} + \frac{(41 - 30/72)^2}{30.72} + \frac{(8 - 5.44)^2}{5.44}$$
$$= 5.804$$

For $4 - 1 = 3$ df, the tabulated χ^2_α value is 6.251 for $\alpha = .10$. Thus we can't reject the null hypothesis at $\alpha = .10$, and the p-value is greater than .10. □

A slight modification of the goodness-of-fit test can be used to test *probabilities specified except for unknown parameters.* The computation of expected values is longer, and the df changes. Otherwise the mechanics of the test are the same.

Chi-Squared Goodness-of-Fit Test (Probabilities Specified Except for Some Unknown Parameters)

1. H_0: The probability distribution is a particular one (e.g., binomial, Poisson, etc.) specified except for unknown population parameters (e.g., the probability of success, the average rate of occurrence, etc.).

2. H_a: The probability distribution is something other than the specified one.

3. T.S.: First, use the frequencies to estimate the unknown parameters. Then, calculate the probabilities and expected values, using the estimated parameter values. Then find

$$\chi^2 = \sum \frac{(\text{Obs} - \text{Exp})^2}{\text{Exp}}$$

The df is $K - 1 - P$, where K is the number of categories and P is the number of separately estimated parameters.

NOTE: Again, the chi-squared tables will be a poor approximation if any Exp is less than 1, and may be poor if more than 20% of the Exp are less than 5.

For the T-maze frequencies, suppose that we assume that the turns are independent from trial to trial, and that the probability of a left turn is constant, but not necessarily .50. (It's been reported, for instance, that rats are more likely to turn toward a light source than away from one.) Thus the probability of success is an unknown parameter that must be estimated from the data before a goodness-of-fit test is done. We note that with 4 runs for each of 48 rats, there are $4(48) = 192$ runs. There are a total of $0(4) + 1(5) + 2(12) + 3(17) + 4(10) = 120$ left turns. Thus an estimate of π, the probability of a left turn, is $\hat{\pi} = 120/192 = .625$. Our binomial tables don't include $\pi = .625$, but binomial probabilities can be calculated either by computer or by hand calculator to be .0198, .1319, .3296, .3662, and .1526. Now the chi-squared statistic can be computed as before.

Number of left turns	0	1	2	3	4
Observed frequency	4	5	12	17	10
Expected frequency	0.9492	6.3281	15.8203	17.5781	7.3242

The chi-squared statistic comes out 12.0034. There are $K = 5$ categories and $P = 1$ parameter estimated (namely, π, the probability of a left turn). Thus there are $5 - 1 - 1 = 3$ df. The observed statistic falls between the .01 and .005 table values, 11.3449 and 12.8381. Thus the nominal p-value is between .01 and .005, indicating that the binomial model is highly dubious.

Note that the expected frequency for 0 left turns is less than 1. Thus the chi-squared approximation may well be poor. One might combine the 0 and 1 categories to get an observed frequency of $4 + 5 = 9$ and an expected frequency of $0.9492 + 6.3281 = 7.2773$. In this case, the chi-squared statistic comes out 2.3270 with 2 df, well within random variation, and the binomial model could be declared a "good fit." We have a problem; without collapsing the 0 and 1 categories together, we have a "bad fit," but when we collapse, we have a "good fit"! Combining categories causes some information to be lost and should only be done when the expected frequencies are very low. In our illustration, the expected number of 0 left turns is close to 1. The p-value of the uncollapsed chi-squared statistic may not be a great approximation, but even so indicates that the discrepancies from binomial frequencies are suspiciously large.

Example 13.10 Refer to Example 13.9. Test the null hypothesis that the probabilities are binomial, not necessarily with $\pi = .20$.

Solution We must first estimate the probability of a correct answer. Recall that we collapsed the 3-correct and 4-correct categories to get

Number correct	0	1	2	3+
Observed frequency	73	78	41	8

If we assume that all the entries in the 3+ category represent students scoring exactly 3 right, then there were $4(200) = 800$ questions and $0(73) + 1(78) + 2(41) + 3(8) = 184$ correct answers. The natural estimate of π, the probability of a correct answer, is $184/800 = .23$. Binomial probabilities assuming $\pi = .23$ may be calculated by hand or using a computer. The results are shown below.

Number correct	0	1	2	3+
Probability	.3515	.4200	.1882	.0403
Expected frequency	70.30	84.00	37.64	8.06
Observed frequency	73	78	41	8

The χ^2 statistic comes out 0.8327, which is not nearly significant at any reasonable α value, with $4 - 1 - 1 = 2$ df. Thus the binomial hypothesis is compatible with the data. □

The chi-squared goodness-of-fit test has a major problem in applications. Often a researcher will develop a probabilistic theory, obtain some data, obtain a nonsignificant chi-squared statistic, and triumphantly declare a "good fit" to the theory. The problem is that the actual research hypothesis (the probabilistic theory) is formulated as the *null* hypothesis of the chi-squared test. Thus the usual strategy of hypothesis testing—trying to contradict a null hypothesis—is turned upside down. The researcher hopes not to reject H_0, but to retain it. One good way to improve the chances of retaining a null hypothesis is to use a test having very low power. One good way to have a low-power test is to take a very small sample size. Thus, use of a goodness-of-fit test often encourages a researcher to take small samples, rather than large ones! In assessing goodness of fit, it's important to look not only at the test statistic, but also at the actual discrepancies between observed and expected frequencies. If the discrepancies seem substantial but the statistic comes out nonsignificant, the only conclusion that can be justifiably drawn is that there is a suggestion that the null hypothesis is substantially wrong, but that suggestion could conceivably be merely the result of randomness.

Example 13.11 Refer to Example 13.9, where we retained the hypothesis that students had no "test wisdom." Have we proved that hypothesis?

Solution No way! The sample size is only 200. Within that sample size, the data (observed frequencies) *might* be compatible with the "no test wisdom" hypothesis. They might also be compatible with other hypotheses. The conclusion is ambiguous, at best. □

The chi-squared goodness-of-fit test is sometimes used to test data for normality. There are several major problems with this procedure. First, the data must be grouped into categories, and the choice of groups is highly arbitrary. Second, the mean and standard deviation usually must be estimated from the grouped frequencies rather than the raw data. (Using the raw-data mean and standard deviation turns out to destroy the df formula.) Third, the power of the test against the most crucial alternatives—skewness and outlier-proneness—is not as good as the power of alternative tests. There are a number of other goodness-of-fit tests; see Hollander and Wolfe (1974). One simple test for normality that seems much more effective than the chi-squared test is to draw a normal probability plot, then have a computer program calculate the correlation (as defined in Chapter 14) between the actual data values and the "normal scores." See Ryan and Joiner (1974). If the population is normal, the correlation should be extremely high. The hypothesis of normality is rejected if the correlation falls below the value given in Table 9.

Example 13.12

A random sample of 200 scores was entered into Minitab and a normal probability plot obtained. The normal plot had an S-shape, indicating that the data were outlier-prone. The correlation between the actual data values and the normal scores was computed by Minitab to be .962. Can we conclude that the data came from a nonnormal population?

Solution

Table 9 does not go up to $n = 200$. We note that the table values increase as n increases, and that a correlation of .962 is smaller than the table values for $n = 75$. Thus we may reject the hypothesis of a normal population and conclude that the data came from a nonnormal population. □

The power of a chi-squared goodness-of-fit test is based on the *noncentral chi-squared distribution,* just as the power of a t test was determined by the noncentral t distribution. Although the mathematical form of the noncentral chi-squared distribution is cumbersome, an approximation due to Patnaik (1949) makes it possible to compute a good approximation to the power. The basic ideas are shown in the following box.

Power of the Chi-Squared Goodness-of-Fit Test

The power of the test is determined by a noncentral χ^2 distribution with $K - 1$ df, in the case of no estimated parameters, and noncentrality parameter

$$\delta^2 = \sum \frac{(n\pi_i - n\pi_{i,0})^2}{n\pi_{i,0}} = n\sum \frac{(\pi_i - \pi_{i,0})^2}{\pi_{i,0}}$$

where π_i represents the actual probability of an observation's falling in category i and $\pi_{i,0}$ represents the null hypothesis probability of that category.

The Patnaik approximation to the power is

$$\text{power} = P\left(z > \sqrt{\frac{2\chi^2_\alpha(\text{df} + \delta^2)}{\text{df} + 2\delta^2}} - \sqrt{\frac{2(\text{df} + \delta^2)^2}{\text{df} + 2\delta^2} - 1}\right)$$

For example, suppose that our null hypothesis is that each of four categories has probability .25, but that in fact categories 1 and 2 have probabilities .3 each, and categories 3 and 4 have probabilities .2 each. The sample size is 100. The noncentrality parameter is

$$\delta^2 = 100\left(\frac{(.3 - .25)^2}{.25} + \frac{(.3 - .25)^2}{.25} + \frac{(.2 - .25)^2}{.25} + \frac{(.2 - .25)^2}{.25}\right)$$
$$= 4.00$$

For $\alpha = .05$ and 3 df, the χ-table value is 7.8143. The probability that the null hypothesis will be rejected is approximately

$$P\left(z > \sqrt{\frac{2(7.8143)(3 + 4.00)}{3 + 2(4.00)}} - \sqrt{\frac{2(3 + 4.00)^2}{3 + 2(4.00)} - 1}\right)$$
$$= P(z > 0.34) = .5 - .1331 = .3669$$

Some general properties of the power of the chi-squared test have been proved. As usual, the power of the test increases as either the sample size or the α value increases. Also, it's known that the power of the test decreases as the number of categories increases. All else being equal, power is best for the minimum number of categories, namely 2. These results are formal, mathematical ones, and often point in a direction opposite to what one wants to do practically. In particular, the power results push one in the direction of combining categories as much as possible, but in practice too much combining may obscure exactly the features of practical interest in the data. The power results do indicate that a chi-squared goodness-of-fit test will have little value in trying to detect small differences of small probabilities among many categories, unless the sample size is enormous.

Example 13.13

A Monte Carlo study was done to investigate the effect of the number of categories on a goodness-of-fit test. Two runs were made. In the first, there were four categories. The null hypothesis stated that all four had equal probabilities, but in the actual run the probabilities were .3, .3, .2, and .2. In the second run, there were eight categories, also having equal probabilities under the null hypothesis, but with actual probabilities .15, .15, .15, .15, .10, .10, .10, and .10. In each run, 1000 samples were drawn, each sample of size 100. The output indicated the following:

RUN 1

Table value	6.25139	7.8143	11.3449
Number of samples with χ^2 larger than table value	480	354	165

RUN 2

Table value	12.0170	14.0671	18.4753
Number of samples with χ^2 larger than table value	350	229	89

What do the Monte Carlo results indicate about the power of the chi-squared goodness-of-fit test?

Solution

The table values correspond to right-tail areas of .10, .05, and .01 in the χ^2 tables with the appropriate df (3 for run 1, 7 for run 2). In both runs, the null hypothesis is false, so every sample that leads to rejection of H_0 is a correct decision. To the accuracy of the simulation results for 1000 samples, the power of an $\alpha = .10$ χ^2 test in the run 1 situation is $480/1000 = .48$. Note that the run 2 situation is equivalent to splitting each run 1 category in half, thus increasing the number of categories. In run 2, the indicated power is lower for every α.

Note also that we computed the Patnaik approximation to the power for run 1, $\alpha = .05$ to be .3669. The output estimates the power to be $354/1000 = .354$, very close to the approximation. □

Exercises for Section 13.2

13.16. In an experiment, laboratory rats were presented with a choice of eight paths fanning out from a central point. Only one of the paths led to food. It was hypothesized that with as many as eight paths, the rat would not be able to remember failure paths, so the choice of paths would be random. Thus the number of trials Y until the rat found the correct one should be geometrically distributed, with probabilities

y	0	1	2	3	4	5	6	7	8	9	10	11+
$p_Y(y)$	.125	.109	.096	.084	.073	.064	.056	.049	.043	.038	.033	.230

A sample of 100 rats were tested in this situation and the value of Y recorded for each rat. The data were

y	0	1	2	3	4	5	6	7	8	9	10	11+
Frequency	10	10	13	12	10	11	9	6	7	5	3	5

Are the sample frequencies compatible with the hypothesis of random choice of paths? Use, say, $\alpha = .01$.

13.17. In Exercise 13.16, are the expected frequencies large enough for one to be confident that the chi-squared tables are a good approximation? How critical is a precise approximation to your conclusion?

13.18. In a study of borderline-personality patients at a clinic, two symptoms (A and B, say) were observed. The frequencies observed were

Symptoms	Neither	A only	B only	Both	Total
Frequency	5	12	15	8	40

Test the hypothesis that the four categories are equally likely, using $\alpha = .05$.

13.19. In Exercise 13.18, would it be reasonable to report that the data gave strong support to the "equally likely" theory? Does the fact that the expected frequencies all are 10 matter in answering this question?

13.20. Assume that the true, population proportions in the four categories of Exercise 13.18 are .15, .35, .35, and .15.

a. Calculate the power of the test performed in Exercise 13.18, using Patnaik's approximation.

b. What sample size is needed to make the power at least .80?

13.21. Refer to Exercise 13.10. Suppose that after treatment the subjects were rated as nonfunctional, partially functional, or fully functional by an observer who was "blind" to the therapy used. Historically, 50% of agoraphobics treated by other methods have wound up nonfunctional, 30% partially functional, and 20% fully functional. The frequencies for the new treatment were

Nonfunctional: 14
Partially functional: 8
Fully functional: 11
Total: 33

a. Show that the χ^2 goodness-of-fit statistic equals 3.677 for these data.

b. Is it reasonable to conclude that the data clearly indicate that the treatment does no better than historical treatments?

13.22. Refer to Exercise 13.21. Assume that, under the treatment used, the three categories are equally likely. Calculate the power of the test, using the Patnaik approximation. Assuming that it is scientifically worthwhile to detect a change in proportions of this magnitude, is the power adequate? If not, why not?

13.23. A diagnostic method for classifying sixth-grade school children was tested in a nationwide study. It was intended that the proportions of children classified as gifted, normal, and slow be .10, .80, and .10. In fact, 1273 children were classified as gifted, 9442 as normal, and 1285 as slow.

a. Verify that the chi-squared goodness-of-fit statistic is 13.062.

b. Is this value of χ^2 statistically significant at usual α levels? Place a bound on the p-value.

c. Is there any reason to be concerned that nominal chi-squared probabilities are inaccurate in this situation?

13.24. Refer to Exercise 13.23. Would the researcher conducting the study be justified in reporting that "because $\chi^2 = 13.062$, a highly significant value, the data indicate that the actual proportions in the three categories are far from the intended proportions"?

13.25. A computer was programmed to draw 1000 samples, each of size 50. Each observation could be 1, 2, 3, 4, or 5, with programmed probabilities .02, .10, .10, .10, and .68, respectively. For each sample, a χ^2 goodness-of-fit statistic was calculated, using the same probabilities for H_0.

a. How many of the 1000 χ^2 statistics would be expected to exceed $\chi^2_{.10}$?

b. What are the expected frequencies in any one sample of the five values? Do these expected frequencies indicate that the chi-squared probabilities will be good approximations?

13.26. Refer to Exercise 13.25. The results for the 1000 samples were

α	.10	.05	.01
Number of times the statistic exceeded χ^2_α	114	64	13

Do these results indicate that the nominal chi-squared probabilities are in fact good approximations in this situation?

13.27. A sample of 100 subjects was obtained. Each subject was measured on a newly developed scale. The developers intended that scores on the scale be normally distributed with a mean of 50 and a standard deviation of 10. The 100 scores obtained were grouped into six classes: up to 40.3, 40.3 up to 45.7, 45.7 up to 50.0, 50.0 up to 54.3, 54.3 up to 59.7, 59.7 and higher. (Theoretically, any subject with a scale score right on one of the dividing lines would be put in the higher category, but in fact no subject fell on any of these values.)

a. Assuming that the scores on this scale are in fact normally distributed with mean 50 and standard deviation 10, what are the probabilities of falling in each of the six classes?

b. The actual frequencies in the six classes were 17, 11, 21, 20, 15, and 16, respectively. Can H_0: normal distribution (mean 50, standard deviation 10) be rejected at $\alpha = .05$?

13.28. Refer to Exercise 13.27. A normal plot of the data was obtained (using Minitab).

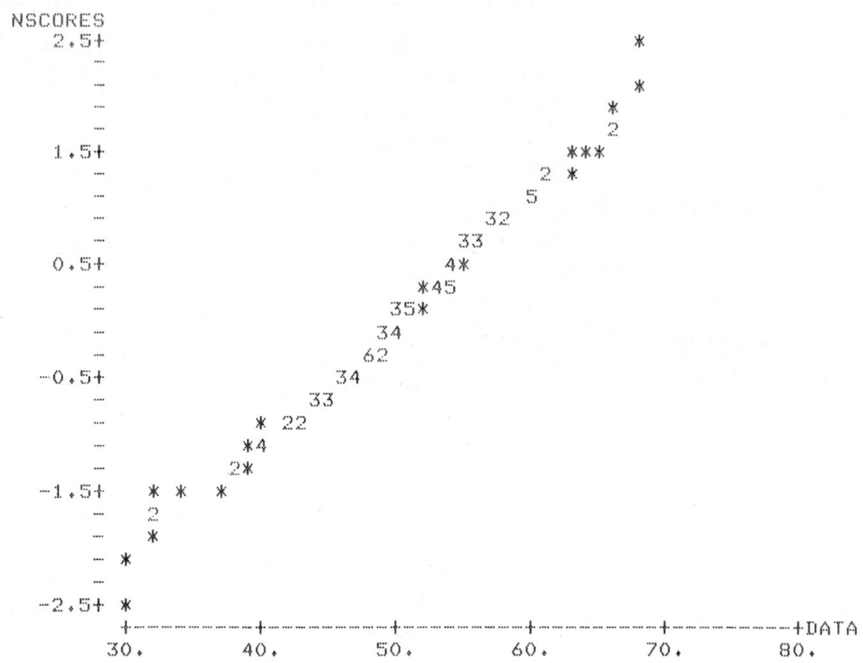

Figure 13.2 Output for Exercise 13.28

a. Does the plot appear to be essentially a straight line, indicating normality?

b. The correlation between the actual scores and the "normal scores" (scores expected given the hypothesis of a normal distribution) was 0.990. Can the null hypothesis of a normal distribution be rejected on the basis of this fact?

13.3 Testing Independence

Another chi-squared test is available for testing the statistical significance of apparent dependence or association between two qualitative variables. It has a form similar to that of the goodness-of-fit test, but the practical issues in applying it are somewhat different.

Consider the frequencies reported by Lindgren (1969) in a study of 40 successful and 40 unsuccessful college students (Table 13.2).

	Successful	Unsuccessful	Total
Almost Always Attends Class	34	19	53
Sometimes Absent from Class	3	3	6
Often Absent from Class	3	18	21
Total	40	40	80

Table 13.2

contingency table
cross-tabulation
cell

These frequency data are arranged in what is called a **contingency table** or **cross-tabulation** (cross-tab). The combination of each row and column of the contingency table defines a **cell.** Conventionally, the number of rows is denoted R and the number of columns C. Here $R = 3$ and $C = 2$.

There appears to be a relation between success and class attendance. Most of the successful students almost always attend class, and many unsuccessful students are often absent. The sample sizes are fairly small; is it possible that the apparent relation is merely the result of random variation?

The null hypothesis is that there is no real relation. It can be stated in at least two ways: that the row and column categories are statistically independent, or that the probability distribution in one column is the same as the probability distribution in any other column. In the first interpretation the test is called a test of independence, and in the second it's called a test of homogeneity of proportions. The difference isn't terribly important. The research hypothesis is quite general—namely, that there is some sort of relation (or dependence or association) between the two variables.

The test statistic once again compares observed frequencies (not proportions or percentages) to expected frequencies. The expected frequencies must be estimated from the data, because the null hypothesis only specifies statistical independence, not the exact numerical values of the probabilities. In Chapter 4 we defined statistical independence of events A and B by $P(A \text{ and } B) = P(A)P(B)$. Assuming independence, the expected frequency in a particular cell (particular row and column) is

$$nP(\text{cell}) = nP(\text{row})P(\text{column})$$

$P(\text{row})$ is estimated by (row total)/n; thus we estimate the row probabilities in the class attendance illustration as 53/80, 6/80, and 21/80. Similarly, $P(\text{column})$ is estimated by (column total)/n. The estimated expected frequency in any cell is then

$$\begin{aligned} \text{Exp}_{i,j} &= n\left(\frac{\text{row total}}{n}\right)\left(\frac{\text{column total}}{n}\right) \\ &= \frac{(\text{row total})(\text{column total})}{n} \end{aligned}$$

In fact, the calculation of estimated expected frequencies is exactly the same one that was done in calculating ∇ in Section 3.2. For the class attendance illustration, we have the estimated expected frequencies shown in Table 13.3.

	Successful	Unsuccessful
Almost Always Attends Class	26.5	26.5
Sometimes Absent from Class	3.0	3.0
Often Absent from Class	10.5	10.5

Table 13.3

Now the chi-squared test proceeds much the same way as the goodness-of-fit test.

Chi-Squared Test of Independence

1. H_0: row and column variables are independent

2. H_a: row and column variables are associated (dependent)

3. T.S.: $\chi^2 = \Sigma\,(\text{Obs} - \text{Exp})^2/\text{Exp}$, where the sum is taken over all cells of the table

4. R.R.: reject H_0 if $\chi^2 > \chi^2_\alpha$, where χ^2_α cuts off area α in the right tail of a chi-squared distribution with $(R - 1)\cdot(C - 1)$ df. Recall that R and C are the respective numbers of rows and columns.

NOTE: The use of chi-squared tables is an approximation. The approximation will be poor if any Exp is less than 1 or if more than 20% of all Exp are less than 5.

For the Lindgren class-attendance data,

$$\chi^2 = \frac{(34 - 26.5)^2}{26.5} + \frac{(19 - 26.5)^2}{26.5} + \frac{(3 - 3)^2}{3} + \cdots + \frac{(18 - 10.5)^2}{18.5}$$
$$= 14.960$$

with $(3 - 1)(2 - 1) = 2$ df. The $\alpha = .005$ table value for 2 df is 10.5966, so the null hypothesis of independence may be rejected conclusively. The apparent relation is not plausibly the result of mere randomness.

Example 13.14

Researchers did a study in which subjects were measured initially on their attitude toward a particular question. The subjects were divided into those committed on the question and those uncommitted. All subjects heard a presentation favoring one side of the question and were then measured again on their attitude toward the question. Each subject was recorded as moving toward the advocated position, being unchanged, or moving away from the advocated position. The frequencies in Table 13.4 were observed.

(a) Calculate a table of estimated expected frequencies.

(b) Find the value of the chi-squared statistic. Can we say conclusively that there is a relation between prior commitment and movability?

	Toward	Unchanged	Away	Total
Committed	19	34	16	69
Uncommitted	48	22	22	92
Total	67	56	38	161

Table 13.4

Solution (a) The row totals and column totals are given. For the (committed, toward) cell, the estimated expected frequency is (69)(67)/161 = 28.714. Similar calculations for all cells yield the Exp table (Table 13.5).

	Toward	Unchanged	Away	Total
Committed	28.714	24.000	16.286	69.000
Uncommitted	38.286	32.000	21.714	92.000
Total	67.000	56.000	38.000	161.000

Table 13.5

(b)

$$\chi^2 = \frac{(19 - 28.714)^2}{28.714} + \cdots + \frac{(22 - 21.714)^2}{21.714} = 13.051$$

The contingency table has $R = 2$ rows and $C = 3$ columns. The table value for $(2 - 1)(3 - 1) = 2$ df is 10.5966, so the null hypothesis of independence may be emphatically rejected. There is good evidence that there is a relation between prior commitment and mobility. Comparing observed and expected frequencies, we note that the committed subjects show a higher than expected number of unchanged opinions, and the uncommitted subjects show higher than expected frequencies of both movement toward the advocated position and movement away from it. □

The assumptions underlying the chi-squared test of independence are very similar to those of the goodness-of-fit test. We assume that each observation yields an outcome in exactly one cell, that the probability of landing in any cell remains constant over trials, and that the observations are independent. Note that the assumption that there is independence between one observation and another is different from the null hypothesis that there is independence between rows and columns. Independence of observations means that the row and column results for one observation have no predictive value for other observations. Independence of rows and columns means that the row results for one observation have no predictive value for the column result *of that same observation.*

There are at least two possible ways that the data can be gathered. The total number of observations may be prespecified, in which case both the row and column totals—often called row and column **marginal totals**—are random. In this case, the test is often called a test of independence. Or the researcher may specify how many observations will be in each column (say), in which case the column marginal totals are fixed or predetermined. In this case, the chi-squared test is sometimes called a test of homogeneous distributions.

marginal totals

Example 13.15

Refer to Example 13.14.

(a) Is it plausible that the row totals should be regarded as predetermined?

(b) Is there any reason to think that any assumptions may have been violated?

Solution

(a) No. Which row a subject belonged in was determined by a test administered during the experiment, not when the subject was selected for the sample. It is much more reasonable to treat the overall sample size as the only predetermined number.

(b) There are six possible outcomes (cells) for each subject. Given random sampling, there is no reason to think that the probabilities would change. Unless subjects announced their choices aloud or some such, there's no reason to think that the results for one subject should depend on the results for another. The assumptions seem quite reasonable. □

The use of chi-squared tables is, once again, approximate, with the approximation improving as the expected frequencies increase. In the Lindgren example, two of the six expected frequencies (33.3%) are less than 5, so the approximation may not be wonderful. However, the test statistic fell so far beyond even the nominal .005 value that the result is clearly conclusive. If expected frequencies are extremely small, rows and/or columns may be collapsed (combined) together. However, information on the specific row or column distribution is lost in the collapsing. Alternatively, there is a test called **Fisher's Exact Test,** which can be used (and gives exactly correct p-values) when the expected frequencies are small. Typically, a computer program is used to perform Fisher's Exact Test. It should be noted that another difficulty with small expected frequencies is that the power of the test will be low, whether the chi-squared test or Fisher's Exact Test is used.

Fisher's Exact Test

Example 13.16 Refer to Example 13.14. Is there any reason to think that Fisher's Exact Test is needed?

Solution No. All the estimated expected frequencies are well above 5. In addition, even if the table values were somewhat inaccurate, the result of the test would still be highly conclusive. □

In practice, researchers using the chi-squared test of independence usually are trying to establish that there's dependence. Thus the usual hypothesis-testing strategy applies. Occasionally, researchers will have, as a real research hypothesis, the formal null hypothesis of independence. In this case, the same problem arises as we found with the goodness-of-fit test—namely, that the researcher, wanting to "prove" that the null hypothesis is true (which, of course, can't ever be done completely), is impelled to take the smallest possible sample! More often, the researcher wants to reject H_0, so a large sample size is desired.

Sometimes, users of the chi-squared test confuse a statistically significant (more than random) result with evidence of a strong relation. *The chi-squared test is only useful in testing whether an apparent relation is within random variation; it does not measure how strong the relation is.* There are a number of ways to measure how strong the relation between two variables appears to be. In fact, many of the measures defined in Chapter 3 apply to contingency table (cross-tab) data. It is perfectly possible to have a very weak and uninteresting relation be statistically detectable (significant) according to the chi-squared test. With very large samples, virtually any relation at all will be detected as more than a chance relation. Conversely, with very small samples, even an apparently strong relation may be nonsignificant. That only means that the apparent relation *might* have arisen by chance.

There are many measures of the strength of a relation in contingency table data. The ones discussed in Chapter 3 have the advantage that they can be easily interpreted as error-reduction measures. There are several other measures like the contingency coefficient and Cramer's measure that are derived from the chi-squared statistic. These measures have the overwhelming drawback that the actual numerical value has no particular meaning. Thus, a value such as .43 for such a measure doesn't really mean anything in particular and doesn't really indicate whether a relation is strong, moderate, or weak.

Example 13.17

Refer to Example 13.14. Find the value of ∇ for the prediction that committed subjects will be unchanged and uncommitted subjects will move toward or away from the advocated position. What does the resulting number indicate about the relation?

Solution

There are 19 + 16 + 22 = 57 observed errors. Adding the corresponding entries in the expected frequency table, we have 28.714 + 16.286 + 32.000 = 77.000 expected errors. Thus

$$\nabla = \frac{(77 - 57)}{77} = .260$$

Although the chi-squared test indicates that there is a relation, the ∇ statistic indicates that it is not overwhelmingly strong. □

The power of the chi-squared independence test is also determined by the noncentral chi-squared distribution. The power can be approximated using Patnaik's approximation given in Section 13.2. The noncentrality parameter δ^2 becomes

$$\delta^2 = n \sum \frac{(\pi_{ij} - \pi_{i.}\, \pi_{.j})^2}{\pi_{i.}\, \pi_{.j}}$$

where π_{ij} is the actual probability in cell (i, j), $\pi_{i.}$ is the true row probability, and $\pi_{.j}$ is the true column probability. The df becomes $(R - 1)(C - 1)$.

Qualitatively, it is known that the power tends to decrease as the number of cells in the contingency table increases, unless the sample size is increased to compensate. Thus, for large contingency tables, very large sample sizes are needed to achieve adequate power of the test. Note that this is a very different issue from needing most of the expected frequencies to exceed 5; that's a problem of having an adequate probability approximation when using the table. The power problem is one of having enough information to support a research hypothesis, assuming that it is in fact true.

Example 13.18

A Monte Carlo study of the chi-squared independence test involved drawing random samples from two different probability distributions. The probabilities were as shown in Table 13.6.

DISTRIBUTION A

		Column 1	Column 2	Total
Row	1	.30	.20	.50
	2	.20	.30	.50
	Total	.50	.50	1.00

DISTRIBUTION B

		Column 1	Column 2	Column 3	Column 4	Total
Row	1	.15	.15	.10	.10	.50
	2	.10	.10	.15	.15	.50
	Total	.25	.25	.25	.25	1.00

Table 13.6

From each population, 1000 samples of size 80 were drawn. The results in Table 13.7 were obtained.

	Number of times table value exceeded	
α	Population A	Population B
.10	529	409
.05	427	284
.025	298	206
.01	191	114
.005	140	77

Table 13.7

What is approximated by probabilities such as 529/1000 = .529? What do these results indicate about the test?

Solution

In both populations, the rows and columns are dependent, so the research hypothesis is true. Whenever the test statistic exceeds the table value, H_0 is rejected. Thus, to the approximation involved in taking 1000 samples instead of infinitely many,

$$.529 = P(\text{reject } H_0 | H_a \text{ true}) = \text{power of the test}$$

when $\alpha = .10$ and population A holds. In every case, the power of the test is higher in population A, which involves fewer df than population B.

We can check the quality of the Patnaik approximation to power using these results. For example, in population B, four of the $\pi_{ij} = .15$, and the other four $\pi_{ij} = .10$. Both of the $\pi_{i.} = .50$ and all the $\pi_{.j} = .25$, so all the $\pi_{i.}\ \pi_{.j}$ values $= (.5)(.25) = .125$. The noncentrality is

$$\delta^2 = 80\left(\frac{4(.15 - .125)^2}{.125} + \frac{4(.10 - .125)^2}{.125}\right) = 3.2$$

The table value for $\alpha = .05$ and $(2 - 1)(4 - 1) = 3$ df is 7.8143. The approximate probability is

$$P\left(z > \sqrt{\frac{2(7.8143)(3 + 3.2)}{3 + 2(3.2)}} - \sqrt{\frac{2(3 + 3.2)^2}{3 + 2(3.2)} - 1}\right)$$

$$= P(z > 0.53) = .5 - .2019 = .2981$$

which is quite close to $284/1000 = .284$, shown in the output for population B, $\alpha = .05$. □

Other hypotheses can also be tested within a contingency table. In particular, suppose that subjects are tested on their ability to carry out a logical task, both before and after being taught some computer programming concepts, and are rated poor, fair, or good. The data could be arranged in a contingency table like Table 13.8.

		Before			
		Poor	Fair	Good	Total
	Poor				
After	Fair				
	Good				
	Total				

Table 13.8

The research hypothesis of dependence isn't very interesting here. It's quite obvious that those with good ability before will tend to have good ability after, and so on. A more relevant research hypothesis is that there will be a change in the marginal (total) distribution, ideally in the direction of higher scores. A general method for testing many hypotheses in contingency tables has been developed by Grizzle, Starmer, and Koch (1969); the method is discussed in Kleinbaum and Kupper (1978).

Exercises for Section 13.3

13.29. Buss, Iscoe, and Buss (1979) asked the parents of children whether the children would show signs of embarrassment. The frequency of each answer was tabulated according to the child's age; see Table 13.9.

		Age in Years								
		3&4	5	6	7	8	9	10	11&12	Total
Embarrassment	Yes	9	19	32	38	37	36	26	28	225
	No	26	13	12	13	20	18	10	16	128
	Total	35	32	44	51	57	54	36	44	353

Table 13.9

a. Calculate a table of (estimated) expected frequencies under the null hypothesis of independence.

b. Is there conclusive evidence that there is a relation between age and showing embarrassment?

13.30. Refer to Exercise 13.29. Suppose we combine all children aged 6 and above into one age group called "school age." (Note, however, that this combining might be considered cheating, because these are also the ages with a high frequency of "yes" responses.)

a. Recalculate the frequencies in a 2-row, 3-column table.

b. Recalculate the statistic for testing the null hypothesis of no relation between age and embarrassment.

c. Is the result of this test more conclusive or less conclusive than the result in Exercise 13.29? What does the answer indicate about the power of the test in large and in small tables?

13.31. Recall the ∇ statistic defined in Section 3.2.

a. Calculate ∇ for the frequencies in Exercise 13.29, using the prediction "3- and 4-year-olds will not show embarrassment, 5-year-olds may or may not, those 6 years old and older will." Does the result indicate a fairly strong relation between age and embarrassment?

b. Recalculate ∇ for the frequencies in Exercise 13.30. How much difference does it make in ∇ whether or not the older ages are combined?

13.32. Goodwin and Michel (1981) performed a study of newborn infants, recording birth position (left or right) and head position preference (whether the child consistently looked right, had a bias toward looking right, had a bias toward looking left, or consistently looked left). The observed frequencies were as shown in Table 13.10.

		Head Position Preference				
		Consistent Right	Biased Right	Biased Left	Consistent Left	Total
Birth Position	Left	29	8	8	4	49
	Right	18	7	12	13	50
	Total	47	15	20	17	99

Table 13.10

Is there a statistically significant relation between head position preference and birth position? Put bounds on the p-value.

13.33. Refer to Exercise 13.32. The categories for head position preference and birth position were regarded as ordinal, with "left" arbitrarily being taken to be higher, "right" lower (no political meaning implied). The gamma statistic (see Section 3.3) equals −0.429. Is there a strong relation between the two variables, whether or not the relation is statistically detectable (significant)?

13.34. Goodwin and Michel (1981) also studied a different sample of infants aged 19 weeks, observing head position preference and handedness. They observed the frequencies shown in Table 13.11.

		Head Position Preference				
		Consistent Right	Biased Right	Biased Left	Consistent Left	Total
Handedness	Consistent Right	11	1	6	1	19
	Biased Right	16	8	2	1	27
	Biased Left	3	4	4	1	14
	Consistent Left	3	3	2	3	11
	Total	33	16	14	8	71

Table 13.11

a. Calculate the χ^2 statistic for testing the hypothesis of independence of head position preference and handedness.

b. Place a bound on the p-value associated with the statistic.

c. Are the expected frequencies so small that there is reason to be concerned about the accuracy of the χ^2-table approximation? Is a good approximation critically important with these data?

13.35. Refer to Exercise 13.34. Calculate the λ statistic (Section 3.1) for predicting head position preference from handedness. Does the value of λ indicate that there is a strong relation?

13.36. Pharis and Manosevitz (1981) interviewed expectant parents of first-born children. Among other questions, each parent was asked to state a preference for a boy or a girl. The frequencies in Table 13.12 were obtained.

	Preference			
	Boy	Girl	Either	Total
Father	9	5	11	25
Mother	8	5	12	25
Total	17	10	23	50

Table 13.12

a. Calculate the expected frequencies.

b. Is it necessary to calculate a χ^2 statistic to test whether the data indicate dependence of preference on gender of parent? Why?

13.37. Pharis and Manosevitz (1981) also obtained frequencies for the belief of each parent about the sex of the child. (See Table 13.13.)

	Belief			
	Boy	Girl	Don't Know	Total
Father	10	5	10	25
Mother	19	5	1	25
Total	29	10	11	50

Table 13.13

a. Calculate the expected frequencies.

b. Without further calculation, does there appear to be a relation of belief and gender of parent?

c. Perform a test of independence. Does the result confirm your answer to part b?

13.38. A computer was programmed to draw 1000 samples of 100 observations each and perform a χ^2 test of independence for each sample. There were two rows and three columns. The programmed probabilities for the first row were .02, .08, and .30, and the programmed probabilities for the second row were .03, .12, and .45. (With these probabilities, rows and columns are in fact independent.) In 12 of the 1000 samples, the χ^2 statistic was not calculated because certain Expected entries were 0. Of the 988 calculated χ^2 statistics, 71 were statistically significant at $\alpha = .10$, 32 were significant at $\alpha = .05$, and 4 were significant at $\alpha = .01$.

a. Do the results of this study indicate that the claimed α probabilities are lower than the actual α probabilities, in this situation?

b. The average value of the χ^2 statistics was reported to be 1.943. How close is this to the theoretical expected value of the statistic?

13.39. As a follow-up to the study in Exercise 13.38, 1000 samples of 200 observations each were drawn. There were two rows and five columns in the table. The probabilities in the first row were .01, .01, .04, .04, and .30, and the probabilities in the second row were .015, .015, .06, .06, and .45. Independence holds for these probabilities. Of the 990 χ^2 statistics that could be calculated, 82 had significant values at $\alpha = .10$, 40 were significant at $\alpha = .05$, and 3 were significant at $\alpha = .01$. Do these results indicate that the nominal α probabilities understate the true probabilities?

13.4 Log-Linear Models for Qualitative Data

In this chapter we have discussed chi-squared methods for testing goodness of fit and independence. More complicated issues can be dealt with using chi-squared tests. In particular, the idea of conditional independence, developed in Chapter 4, can be tested with frequency data, using so-called log-linear models. Log-linear models have become popular in recent years, particularly in sociological journals. In this section we briefly discuss some of the ideas and pitfalls of these models. A general discussion is found in Bishop, Fienberg, and Holland (1975).

The language of log-linear models is an analog to the language of experimental design, which will be discussed in the next chapter. Therefore, we can't be very precise or complete in this section. Some of the simpler log-linear models can be considered, though. The simplest case is a contingency table, with rows indicated by the i subscript and columns by the j subscript. An arbitrary cell (row i, column j) probability can be denoted π_{ij}. It is always possible to write

$$\log \pi_{ij} = u_0 + u_{1(i)} + u_{2(j)} + u_{12(ij)}$$

where u_0 is the average of all the logarithms (all logarithms are natural logarithms, taken to the base e) of cell probabilities. u_0 has little meaning or interest. The $u_{1(i)}$ and $u_{2(j)}$ terms refer to row and column averages and are of relatively little concern. The primary interest is in what are called the **interactions** $u_{12(ij)}$. It can be proved that if all these interactions are 0, then the row and column variables are independent. A test statistic for the null hypothesis of independence within the log-linear context is

interactions

$$X^2 = 2 \sum \text{Obs} \log (\text{Obs}/\text{Exp})$$

If H_0: 0 interactions (independence) is true, the X^2 statistic has approximately a chi-squared distribution with $(R - 1)(C - 1)$ df. Thus it is an alternative test statistic to the χ^2 statistic discussed in Section 13.3. In fact, the X^2 statistic is approximately equal to the chi-squared statistic in large samples. In the class attendance data of Section 13.3, the X^2 statistic is equal to 16.1912, somewhat larger than the chi-squared statistic. In either test, however, there is conclusive evidence of dependence. For a simple test of independence, there is some evidence [Odoroff (1970)] that the χ^2 statistic gives a more accurate approximation to the chi-squared tables than does the X^2 statistic. Thus, there is little reason to use the log-linear model for a simple independence test.

The real benefit to log-linear models occurs when there are more than two qualitative variables under consideration. In particular, conditional independence of the i variable and the j variable given a third variable indexed by k can be formulated as

$$\log \pi_{ijk} = u_0 + u_{1(i)} + u_{2(j)} + u_{3(k)} + u_{13(ik)} + u_{23(jk)}$$

Note that there is not a $u_{12(ij)}$ term in the model, nor a u_{123} term, so there is no "direct dependence" between the i and j (1 and 2) variables. However, variables 1 and 2 both interact with variable 3, so there is an "indirect dependence." The model is pictured in Figure 13.3; this figure is precisely the diagram of conditional independence in Chapter 3.

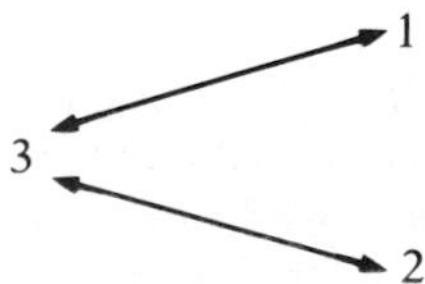

Figure 13.3 Diagram of conditional independence, log-linear model

Consider a study of problem solving such as one performed by Kaufmann and Bengtson (1980). There were three variables: result (solved or not), sex of subject (F or M), and instruction condition (abstract or concrete). One relevant question is For a given instruction condition, is there a sex-based difference in the probability of a solution? To say "no" to that question is to take as a model conditional independence of solution and sex, given instruction condition. Assume that the observed frequencies were as shown in Table 13.14.

ABSTRACT

	F	M
Solved	21	11
Not Solved	39	29

CONCRETE

	F	M
Solved	17	35
Not Solved	23	25

Table 13.14

There are two ways to attack the problem. One is to run two separate χ^2 tests of independence, one for the abstract condition and the other for the concrete condition. Our model says that conditional independence should hold, given instruction condition, so we should have independence in each of the two subtables. The other procedure is to use the log-linear model approach. Two serious problems with the separate-χ^2 approach are that the overall α value will be larger than the α value for either test alone, and that the sample size in either sample will be small, yielding a relatively low power test. The log-linear model approach gives a single test statistic to test the null hypothesis of conditional independence of sex and result, given instruction. A computer output yielded $X^2 = 3.0446$, 2 df. (It's not worth worrying about the df formula; it can be programmed into the computer.) The result is well below right-tail chi-squared-table values with 2 df, such as $\chi^2_{.05} = 5.991$, indicating that the data are a "good fit" to the conditional independence model.

Example 13.19

Suppose that a study related three qualitative variables—whether or not a child exhibited prejudiced behavior, the ethnic category of the child, and the degree of racial integration of the child's school. Assume that the frequencies were as shown in Table 13.15.

A computer program calculated a test statistic for the null hypothesis of no interaction between ethnicity and prejudice, given integration. The result was $X^2 = 12.953$, with 6 df. Can the null hypothesis be rejected at $\alpha = .01$?

LITTLE INTEGRATION

		Ethnicity		
		B	H	W
Prejudice	Y	34	14	20
	N	7	15	7

SOME INTEGRATION

		Ethnicity		
		B	H	W
Prejudice	Y	7	9	15
	N	12	6	12

HIGH INTEGRATION

		Ethnicity		
		B	H	W
Prejudice	Y	10	4	15
	N	15	6	35

Table 13.15

Solution The chi-squared-table value for 6 df and $\alpha = .01$ is 16.81. The test statistic does not exceed this value, so H_0 must be retained. □

Some very elaborate models of behavior can be stated in the log-linear language and tested using X^2 statistics. For example, we might have six different qualitative variables measured on a number of different subjects. A model might include the interactions suggested by Figure 13.4. Note that there are at least indirect relations among all six variables. However, the relations between variables 1 and 2, on the one hand, and 5 and 6, on the other, are indirect, by way of 3 and 4. Thus Figure 13.4 is indicating a complicated conditional independence model. Given both variables 3 and 4, variables 1 and 2 are conditionally independent of variables 5 and 6. This model can be obtained by assuming that there is a 1234 interaction and a 3456 interaction, but no 1256 interaction.

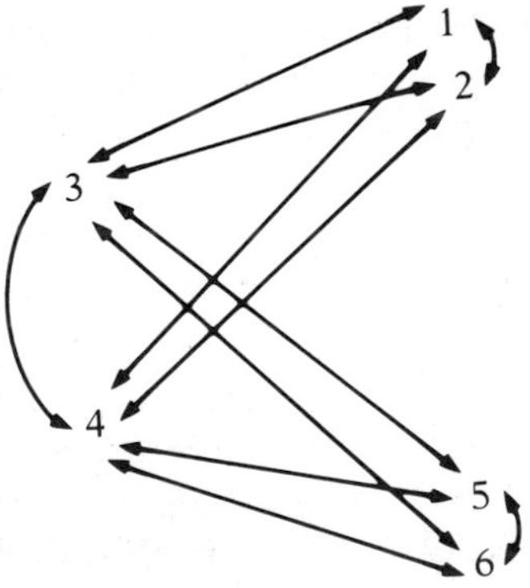

Figure 13.4 A complex relation among six qualitative variables

It's possible to proceed step by step in selecting a log-linear model that appears to be consistent with the data. See Bishop, Fienberg, and Holland (1975). There are some dangers in these procedures, though. One is that the X^2 tests are used as goodness-of-fit tests, with all the dangers of such tests. A "good fit" conclusion may merely indicate that there aren't enough data to contradict the assumed log-linear model. Good log-linear model programs will print out the observed and expected (assuming the model) frequencies; it's a good idea to look for substantial differences in these models before accepting a "good fit" conclusion. If there are substantial differences but the X^2 test statistic is not significant, one should retain the model as one possibility, but not accept it as revealed truth. Oler (1978) has shown that the power of the X^2 tests can be quite low, even for rather large sample sizes. Once again, the low-power problem is most acute in the case of many categories and large tables.

Example 13.20 Refer to Example 13.19. Is it fair to say that the data prove that there is no relation between ethnicity and prejudice, given integration?

Solution No, indeed! One can never prove a null hypothesis. Further, the sample size in Example 13.19 was fairly small. All that can be said is that the data do not disprove H_0. □

Recently, there's been some enthusiastic use of log-linear methods. However, there hasn't been enough recognition that these statistical methods, like all others, have limitations. We've already mentioned the difficulty that any goodness-of-fit test incurs. Further, it isn't always easy to understand exactly what a given log-linear model is assuming. The language of log-linear models is borrowed from the language of experimental design, which is perhaps unfortunate. The analogy is largely a formal, mathematical one; the scientific content of a log-linear model may differ from the content of an experimental design model. In addition, it should be recognized that the definition of categories can have major effects on the statistical analysis of log-linear models. For example, it is perfectly possible that a model may be rejected when the categories "Don't know" and "Don't care" are kept separate, yet be a good fit when the two categories are combined. As with any other statistical method, understanding and critical analysis are important in thoughtful use of log-linear methods.

Exercises for Section 13.4

13.40. Two different observers rated third-grade children on, respectively, their degree of autonomy and their ability to read new materials. The ratings frequencies were as shown in Table 13.16.

		Autonomy					
		VH	H	M	L	VL	Total
	H	18	12	16	9	5	60
Reading Ability	M	13	23	31	26	9	102
	L	7	6	13	12	16	54
	Total	38	41	60	47	30	216

Table 13.16

a. Does there appear to be relation, just looking at the frequencies?

b. Calculate the likelihood-ratio (X^2) statistic. Is there a statistically significant relation, using $\alpha = .05$?

c. Put bounds on the p-value for these data.

13.41. Refer to Exercise 13.40. Calculate the ordinary (Pearson) χ^2 statistic. Do you get the same conclusion?

13.42. Refer to Exercise 13.41. Is there any reason to be concerned that the use of χ^2 tables may be a poor approximation? Why?

13.43. A study of newborn white rats focused on their competitive ability to obtain food. The interest was in developmental qualities—do rats that begin as effective competitors stay that way? Rats were labeled and observed for three consecutive weeks. Each rat was rated as a poor, fair, or good competitor, three times. The frequencies were as shown in Table 13.17.

a. Sum over the week 2 frequencies to obtain a cross-tabulation of week 1 vs. week 3. Does it appear that there is a relation?

b. A null hypothesis states that the week 1 and week 3 probabilities are conditionally independent, given the result in week 2. What does this hypothesis assert about the relation between week 1 and week 3 competitiveness?

c. A computer program yielded a chi-squared statistic of 3.636, with 12 df, for testing the null hypothesis stated in part b. Should the null hypothesis be rejected?

		Week 2								
		P			F			G		
		Week 3								
		P	F	G	P	F	G	P	F	G
	P	10	6	2	8	10	6	1	5	6
Week 1	F	8	10	5	10	15	11	3	9	6
	G	6	8	5	6	7	7	2	10	18

Table 13.17

13.44. Refer to Exercise 13.43. The estimated expected frequencies were computed to be the values shown in Table 13.18.

		Week 2								
		P			F			G		
		Week 3								
		P	F	G	P	F	G	P	F	G
	P	7.2	7.2	3.6	7.2	9.6	7.2	1.2	4.8	6.0
Week 1	F	9.2	9.2	4.6	10.8	14.4	10.8	1.8	7.2	9.0
	G	7.6	7.6	3.8	6.0	8.0	6.0	3.0	12.0	15.0

Table 13.18

Is there reason to think that the conclusion reached in part c of Exercise 13.43 is seriously wrong?

13.45. An anthropologically oriented social psychologist observed the behavior of young adult males in a singles bar. Each subject either did or did not exhibit three behaviors, discreetly labeled A, B, and C. A total of 300 subjects were observed, and the frequencies in Table 13.19 were obtained.

		Behavior B			
		Y		N	
		Behavior C			
		Y	N	Y	N
Behavior A	Y	47	37	27	18
	N	27	9	45	90

Table 13.19

For the hypothesis that behaviors A and C were conditionally independent given B, the estimated expected frequencies were as shown in Table 13.20.

		Behavior B			
		Y		N	
		Behavior C			
		Y	N	Y	N
Behavior A	Y	51.8	32.2	18.0	27.0
	N	22.2	13.8	54.0	81.0

Table 13.20

a. Calculate the likelihood-ratio (X^2) statistic.

b. It can be shown that the chi-squared statistic has 2 df. Can the hypothesis of conditional independence be rejected at $\alpha = .10$?

13.46. Refer to Exercise 13.45. Is there any problem with using the χ^2 approximation for these data?

Chapter Exercises

13.47. Pharis and Manosevitz (1981) collected data on a sample of 25 expectant mothers and 25 expectant fathers of first-born children. Data were reported on sex of parent, belief as to the sex of the child, and preference as to the sex of the child. The data were analyzed by the SPSS-X system to test the null hypothesis that parents' sex and preference were independent, given the belief. Output is shown. Can the null hypothesis be rejected at the usual α levels?

```
Goodness-of-Fit test statistics

 Likelihood Ratio Chi Square =    3.04812    DF = 6  P =  .803
         Pearson Chi Square =    3.77305    DF = 6  P =  .707
```

13.48. Most expected frequencies in Exercise 13.47 are less than 5. How accurate an approximation can one expect for the χ^2 tables? How critical is accuracy of approximation in this example?

13.49. A survey was planned of a sample of elementary-school teachers in a large urban school district. The topic of most interest was the intellectual expectation of minority children. One question asked, "Do you think that teachers expect too little of minority children?" Assume that the responses were 1 = not a problem, 2 = minor problem, 3 = major problem. For planning purposes, assume that 50% of the responding teachers would be white and the other 50% black (neglecting the relatively few teachers of other races in the district). Also assume that in the population of teachers, the white teachers would divide 40%, 30%, 30% in response to the question, and black teachers would divide 30%, 30%, 40% in response to the same question.

a. What are the joint probabilities for race of teacher and response to the question?

b. Formulate a null hypothesis to be tested. Is that hypothesis true in the assumptions?

c. Calculate a noncentrality parameter assuming a sample size of 200.

d. Calculate the power of a χ^2 test of independence, assuming a sample size of 200 and $\alpha = .05$. Use the Patnaik approximation.

13.50. Refer to Exercise 13.49. What sample size is needed to have power = .90, under the given conditions? Use the Patnaik approximation.

13.51. Refer to Exercise 13.49. Suppose that response 3 to the question had been divided up into 3a = major problem and 3b = extreme problem. Suppose further that of the white teachers, 20% responded 3a and 10% 3b, and of the black teachers, 20% responded 3a and 20% 3b.

a. Recompute the noncentrality parameter and power for the hypothesis test.

b. Recalculate the power of the test for $n = 200$ and $\alpha = .05$.

13.52. Refer to Exercises 13.49 and 13.51. Which power is larger? Is this result compatible with the general power principles discussed in Chapter 13 for χ^2 tests?

13.53. A sample of psychotherapists were polled regarding the preferred treatment for chronic depression. A very large survey taken two years earlier had shown that 48.2% preferred treatment A, 36.7% preferred B, and 15.1% preferred C. Assume that this survey accurately reflected the opinion of the population of psychotherapists two years earlier. In the current sample of size 200, 84 preferred A, 75 preferred B, and 41 preferred C.

a. Formulate the analysis problem as a χ^2 goodness-of-fit problem. What does the null hypothesis mean?

b. Verify that the χ^2 statistic equals 5.492.

c. Can the null hypothesis be rejected at the conventional $\alpha = .05$?

13.54. Refer to Exercise 13.53. The researcher in charge of the project stated that "we have proved that there has been no change over the past two years in professional opinion as to preferred treatment." Is this conclusion justified?

13.55. Refer to Exercise 13.53.

a. Calculate a 95% confidence interval for the population proportion of therapists preferring treatment C. Use the large-sample approximate confidence interval.

b. Recalculate the confidence interval in part a using the method that's equivalent to hypothesis testing.

c. Do the confidence intervals calculated in parts a and b allow you to reject the null hypothesis that the population proportion is still .151? Does it matter which interval is used?

13.56. Refer to Exercises 13.53 and 13.55. The procedure in Exercise 13.55 can be regarded as "lumping" procedures A and B together in comparison to procedure C. Are the results of these exercises consistent with the arguments in this chapter about the effects of combining categories?

13.57. A study of the expressed fears of retarded adults resulted in a cross-tabulation of type of fear by sex. A portion of the output from SPSS-X is shown in Figure 13.5.

a. Is there conclusive evidence that there is a relation between sex and type of fear? Locate the appropriate p-value.

b. Does there appear to be a severe problem of small expected values?

c. Does there appear to be a strong relation, whether or not it's statistically significant?

```
                EXPRESSED FEARS OF RETARDED ADULTS

 - - - - - - - - - - - - - - - - - - - - -  C R O S S T A B U L A T I O N   O F  - - - - - - - - - - - - - - - - - -
     FEAR     STATED TYPE OF FEAR                                   BY  SEX
 - - - - - - - - - - - - - - - - - - - - - - - - - - - - - - - - - - - - - - - - - - - - - - - - - - -  PAGE  1 OF  1

                          SEX
                  COUNT  I
                 EXP VAL IMALE      FEMALE     ROW
                 COL PCT I                     TOTAL
                         I       1I       2I
 FEAR            --------+--------+--------+
                      1  I    14  I    18  I    32
    SUPERNATURAL         I  14.9  I  17.1  I  21.9%
                         I  20.6% I  23.1% I
                         +--------+--------+
                      2  I    17  I    33  I    50
    ANIMALS              I  23.3  I  26.7  I  34.2%
                         I  25.0% I  42.3% I
                         +--------+--------+
                      3  I    20  I    11  I    31
    PHYSICALINJURY       I  14.4  I  16.6  I  21.2%
                         I  29.4% I  14.1% I
                         +--------+--------+
                      4  I     7  I     9  I    16
    PSYCH   STRESS       I   7.5  I   8.5  I  11.0%
                         I  10.3% I  11.5% I
                         +--------+--------+
                      5  I     3  I     1  I     4
    EGOCENTRIC           I   1.9  I   2.1  I   2.7%
                         I   4.4% I   1.3% I
                         +--------+--------+
                      6  I     7  I     6  I    13
    NOTHING              I   6.1  I   6.9  I   8.9%
                         I  10.3% I   7.7% I
                         +--------+--------+
                 COLUMN      68       78       146
                  TOTAL    46.6%    53.4%    100.0%

 CHI-SQUARE    D.F.       SIGNIFICANCE        MIN E.F.       CELLS WITH E.F.< 5
 ----------    ----       ------------        --------       ------------------

   8.91671       5           0.1124              1.863       2 OF      12 ( 16.7%)
                                                  WITH FEAR               WITH SEX
      STATISTIC                     SYMMETRIC    DEPENDENT               DEPENDENT
      ---------                     ---------    -------------           -------------

 LAMBDA                              0.09146        0.03125                 0.17647
 SOMERS' D                          -0.11719       -0.14913                -0.09652
```

Figure 13.5 Output for Exercise 13.57

13.58. A study obtained qualitative measures of students' skill in a science class, liking for it, and preparation for it. One theory was that skill and liking would both be related to preparation, but that there was no direct relation between skill and liking. This model was tested using the SPSS-X package. A portion of the output is shown in Figure 13.6.

LIKING SCIENCE CLASS AS A FUNCTION OF PREPARATION AND LAB SKILL

************************ L O G   L I N E A R   A N A L Y S I S ************************

FACTOR Information

Factor	Level	Label
PREP	4	RATED QUALITY OF PREVIOUS WORK
LIKING	3	ATTITUDE TOWARD COURSE
SKILL	2	PREVIOUSLY ACQUIRED LAB SKILL

Observed, Expected Frequencies and Residuals

Factor	Code	OBS. count & PCT.	EXP. count & PCT.	Residual	Std. Resid.	Adj. Resid.
PREP	POOR					
LIKING	LIKE					
SKILL	YES	1.00 (.37)	1.17 (.43)	-.1692	-.1565	-.1920
SKILL	NO	3.00 (1.10)	2.83 (1.04)	.1692	.1006	.1920
LIKING	NEUTRAL					
SKILL	YES	2.00 (.74)	1.75 (.64)	.2462	.1859	.2319
SKILL	NO	4.00 (1.47)	4.25 (1.56)	-.2462	-.1195	-.2319
LIKING	DISLIKE					
SKILL	YES	16.00 (5.88)	16.08 (5.91)	-.0769	-.0192	-.0581
SKILL	NO	39.00 (14.34)	38.92 (14.31)	.0769	.0123	.0581
PREP	FAIR					
LIKING	LIKE					
SKILL	YES	4.00 (1.47)	4.07 (1.50)	-.0698	-.0346	-.0478
SKILL	NO	6.00 (2.21)	5.93 (2.18)	.0698	.0286	.0478
LIKING	NEUTRAL					
SKILL	YES	20.00 (7.35)	18.72 (6.88)	1.2791	.2956	.5629
SKILL	NO	26.00 (9.56)	27.28 (10.03)	-1.2791	-.2449	-.5629
LIKING	DISLIKE					
SKILL	YES	11.00 (4.04)	12.21 (4.49)	-1.2093	-.3461	-.5569
SKILL	NO	19.00 (6.99)	17.79 (6.54)	1.2093	.2867	.5569
PREP	GOOD					
LIKING	LIKE					
SKILL	YES	12.00 (4.41)	8.18 (3.01)	3.8182	1.3348	2.2523
SKILL	NO	3.00 (1.10)	6.82 (2.51)	-3.8182	-1.4623	-2.2523
LIKING	NEUTRAL					
SKILL	YES	16.00 (5.88)	15.27 (5.61)	.7273	.1861	.3638
SKILL	NO	12.00 (4.41)	12.73 (4.68)	-.7273	-.2039	-.3638
LIKING	DISLIKE					
SKILL	YES	8.00 (2.94)	12.55 (4.61)	-4.5455	-1.2833	-2.3582
SKILL	NO	15.00 (5.51)	10.45 (3.84)	4.5455	1.4058	2.3582
PREP	EXC					
LIKING	LIKE					
SKILL	YES	19.00 (6.99)	12.13 (4.46)	6.8727	1.9735	3.7631
SKILL	NO	4.00 (1.47)	10.87 (4.00)	-6.8727	-2.0843	-3.7631
LIKING	NEUTRAL					
SKILL	YES	8.00 (2.94)	8.96 (3.30)	-.9636	-.3219	-.5632
SKILL	NO	9.00 (3.31)	8.04 (2.95)	.9636	.3399	.5632
LIKING	DISLIKE					
SKILL	YES	2.00 (.74)	7.91 (2.91)	-5.9091	-2.1012	-3.5835
SKILL	NO	13.00 (4.78)	7.09 (2.61)	5.9091	2.2191	3.5835

- -

Goodness-of-Fit test statistics

Likelihood Ratio Chi Square = 27.95108 DF = 8 P = .000

Pearson Chi Square = 25.85210 DF = 8 P = .001

Figure 13.6 Output for Exercise 13.58

a. Locate the relevant value of the χ^2 statistic.

b. Are the data consistent with the model, within reasonable random variation?

c. Is there a severe problem of small expected values?

14 Design and Analysis of Experiments

The analysis of controlled experiments accounts for an enormous part of the use of statistical reasoning in psychology and education. In this chapter we discuss some of the most important ideas about designing and analyzing experiments. The discussion proceeds from the simplest experiments to more complicated ones. Section 14.1 considers the case in which only one experimental factor is involved and indicates how to test hypotheses within such experiments. Then in Section 14.2 we consider methods for comparing specific mean scores arising within a single-factor experiment. The topic in Section 14.3 is descriptive: how one can explain the variation among means in the experiment. The logical extension to two experimental factors occurs in Section 14.4, where the key concept of interaction is introduced. The methods of Sections 14.2 and 14.3 are extended to two-factor experiments in Section 14.5. The design and analysis of many-factor experiments is briefly discussed in Section 14.6. Then in Section 14.7 we consider a different type of experiment—"repeated measures" experiments in which subjects are measured many times under varying conditions. In planning experiments, it's very useful to be able to calculate the power of statistical tests used in experimental design; that's the topic of Section 14.8. Finally, an appendix contains some elementary proofs of mathematical results used in the chapter.

14.1
Completely Randomized Experimental Design

The methods of this chapter are most widely used for designed experiments. For example, suppose that a psychologist decides to test the effectiveness of using computer packages to teach statistical methods. The plan is to take a class of 30 students. Of the 30, 10 will work all problems with a hand calculator, 10 will work half the problems with a hand calculator and half with a computer, and the remaining 10 will work all the problems with a computer. All students will then be graded on a test that stresses critical understanding, not computation. The question is whether a statistically detectable (significant) difference in mean scores will result.

completely randomized design

This experimental design is called a **completely randomized design**. As the name indicates, students should be assigned randomly to the three groups. Randomization is an important principle of experimental design. Suppose that, rather than being randomly assigned to groups, students selected their own method. All sorts of potential biases would arise. For example, students who found the material easy might well opt for the computer group, to avoid the drudgery of hand computation. High scores for the computer-only group might well reflect only the high ability of the students selecting that group, not the effectiveness of the computer method. Random assignment to groups avoids such systematic biases. Difference in ability among students becomes one part of random error, rather than a bias.

factor

levels

The experiment is called "completely" randomized because no restrictions, other than that there be 10 students per method, are placed on the randomization. For instance, no requirement is made that there be equal numbers of men and women in each group. The design is also called a one-factor design. A **factor** is an independent variable, often qualitative, that is systematically varied in an experiment; the possible values of the factor are called **levels.** The only variable that is systematically varied in our example is the problem-solving method; hence it is a one-factor design. The levels of the method factor are 1 = calculator only, 2 = half calculator, half computer, and 3 = computer only. In this chapter, one-factor and two-factor designs are considered in detail; some more general experiments are briefly considered. Through most of the chapter, we'll be implicitly assuming that we're dealing with "fixed factors." See Section 14.6 for the definition of such factors.

Example 14.1

Keller and Tetlow (1981) did a study on subjects who participated in an amateur theater group. The subjects were divided into three categories: backstage-only participants, onstage-only participants, and combination backstage-onstage participants. A Myers-Briggs type indicator extroversion-introversion score was obtained for each subject.

(a) Identify the factor(s) in this study.

(b) What form of experimental randomization would you expect in this study?

Solution

(a) The dependent variable is the Myers-Briggs score. The only independent variable (factor) is participation category. The factor has three levels—backstage, onstage, and combination.

(b) In this study, we would *not* expect subjects to be randomly assigned to levels. Instead, the subjects would be observed in the categories they chose for themselves. Systematic natural differences in extroversion-introversion are what we'd like to investigate. Because of the natural lack of randomization, this study is observational, rather than experimental. □

mathematical model

response variable

In discussing the analysis of data from an experiment, it is convenient to begin with a **mathematical model** incorporating the notation and assumptions of the analysis. In the experiment, there is a **response variable,** or dependent variable, such as test score. It is denoted by Y. To distinguish the scores, we let i be a subscript referring to individuals and j be a subscript referring to factor levels. Y_{ij} means the score of the ith individual within the jth level of the factor; thus Y_{92} refers to the 9th score within level 2 of the factor. Using this notation, we write the model as

$$Y_{ij} = \mu + \alpha_j + \epsilon_{ij}$$

grand mean

The letter μ indicates a **grand mean** or "superpopulation mean." Imagine the entire population of all potential subjects of the experiment, divided among the levels of the factor in the same proportions as in the sample. The mean of this conceptual population is μ

The α_j term is the *true effect* of the jth level of the experimental factor. It is the difference between the population mean for the jth level of the factor and the grand mean:

$$\alpha_j = \mu_j - \mu$$

Note that the word "effect" is now being used to represent the difference between a specific mean and a grand mean, rather than the difference between two specific means. Because the sum of deviations from a mean is always 0, it follows that

$$\sum n_j\alpha_j = 0$$

For *balanced designs,* having equal sample sizes for each level of the factor, the simpler equation

$$\sum \alpha_j = 0$$

holds. Note that α-with-subscript, the effect of a level, means something quite different from α-without-subscript, the probability of a Type I error.

random error

The ϵ_{ij} term is a **random error** term. It includes all the other sources of variation among the scores. Sources of random variation include individual differences, measurement error, and the effect of all the factors not explicitly controlled in the experiment.

Example 14.2 Interpret the terms of the mathematical model for the experiment in Example 14.1.

Solution In principle, the population would be all individuals currently participating in amateur theater groups, not just the ones in the study. We assume that the division into categories in the population would be in proportion to the size of the categories in the sample. The μ would be the mean extroversion-introversion score for the entire population. If backstage workers were denoted by $j = 1$, α_1 would be the difference between the mean of the backstage population and the overall population mean. The ϵ_{ij} error term would capture all the reasons why individual scores would vary around subpopulation means. □

ANalysis Of VAriance (ANOVA)

The standard method for analyzing data from designed experiments is called the **ANalysis Of VAriance (ANOVA),** despite the fact that it is an analysis of means, not variances. The analysis of variance is an extension of the pooled-variance t test described in Chapter 11, so it's not surprising that the assumptions are the same as those of the pooled-variance t test. The assumptions are stated in the box.

Assumptions for Analysis of Variance

1. The observations (and therefore the errors) are independent.

2. The population variance of Y_{ij} scores is constant for all levels j of the factor.

3. The population distribution of values within each level of the factor is normal.

Assumptions 2 and 3 should be checked, as usual, by plotting the data. Plot the data for each level separately, using any method you prefer—stem and leaf displays, box plots (good when there are many levels), histograms, or normal probability plots. Look for gross skewness or heavy tails. Also look for wild differences in variability for different levels of the factor.

Example 14.3 Refer to Example 14.1. Scores on the introversion-extroversion scale typically have a mean near 100 and a standard deviation of 20 or so. Keller and Tetlow report that the scores ranged from 11 to 161. What does this suggest about the data?

Solution A score of 11 would be more than 4 standard deviations below the mean, and 161 would be 3 standard deviations above the mean. The data are outlier-prone. □

We will consider methods for dealing with violations of the assumptions after looking at standard analysis of variance methods.

The null hypothesis to be tested is that the population means for all levels of the factor are equal. This null hypothesis may be stated in terms of the mathematical model by saying that the α_j effects are all 0.

$$H_0\text{: all } \alpha_j = 0$$

H_0 asserts that all the population means equal the grand mean, and so are all the same, and that differences among sample means are attributable purely to random variation. The research hypothesis is always two-sided: that the means are not all equal. It follows that some of the true effects aren't 0.

H_a: some (or all) $\alpha_j \neq 0$

The appropriate hypothesis test, given the assumptions, can be derived using the likelihood ratio technique discussed in the appendix to Chapter 10. The same test can be derived using the ideas of variation between levels and variation within levels. **Variation between levels** refers to how far apart the dependent variable means are. It is measured by summing the squared difference between the dependent variable mean (for each level j) and the grand mean, weighted by the number of observations at that mean:

variation between levels

$$\text{SS(Between)} = \sum n_j(\bar{y}_{.j} - \bar{y}_{..})^2$$

where

$$\bar{y}_{.j} = \sum_i \frac{y_{ij}}{n_j}$$

is the sample mean for the jth level of the factor and

$$\bar{y}_{..} = \frac{\sum y_{ij}}{\sum n_j} = \frac{\sum n_j \bar{y}_{.j}}{\sum n_j}$$

is the grand mean of the data. Note that a dot (.) replacing a subscript indicates that we have averaged over the subscript. In calculating the mean for the jth level $\bar{y}_{.j}$, we have averaged y_{ij} values over individuals i.

variation within levels

Variation within levels is measured by adding up all the squared deviations of individual values from group (level) means:

$$\text{SS(Within)} = \sum (y_{ij} - \bar{y}_{.j})^2 = \sum (n_j - 1)s_j^2$$

where s_j^2 is the sample variance within the jth group (level).

As an illustration, assume that we have the following data as test scores from three different experimental groups:

Group 1:	25	33	37	41	43	45	49	50	58	67
Group 2:	36	41	46	46	53	57	61	68	73	79
Group 3:	28	30	36	44	46	48	52	56	61	66

The means and variances are shown in Table 14.1.

Group	Sample Mean	Sample Variance
1	44.8	146.8444
2	56.0	202.4444
3	46.7	160.4556

Table 14.1

The grand mean is

$$\frac{10(44.8) + 10(56.0) + 10(46.7)}{30} = 49.1667$$

Thus

$$\begin{aligned} SS(\text{Between}) &= 10(44.8 - 49.1667)^2 + 10(56.0 - 49.1667)^2 \\ &\quad + 10(46.7 - 49.1667)^2 \\ &= 718.4667 \end{aligned}$$

and

$$\begin{aligned} SS(\text{Within}) &= 9(146.8444) + 9(202.4444) + 9(160.4556) \\ &= 4587.7000 \end{aligned}$$

Example 14.4

Refer to Example 14.1. The article reported the data in Table 14.2.

Group	Mean	Standard Deviation	n
Backstage	109.00	26.56	18
Onstage	89.71	17.93	17
Combination	96.93	19.32	30

Table 14.2

Find SS(Between) and SS(Within).

Solution

First, we need the grand sample mean $\bar{y}$.

$$\bar{y} = \frac{18(109.00) + 17(89.71) + 30(96.93)}{18 + 17 + 30} = 98.38415$$

Then

$$\begin{aligned}\text{SS(Between)} &= 18(109.00 - 96.38415)^2 + 17(89.71 - 96.38415)^2 \\ &\quad + 30(96.93 - 96.38415)^2 \\ &= 3371.06 \\ \text{SS(Within)} &= 17(26.56)^2 + 16(17.93)^2 + 29(19.32)^2 = 27960.74\end{aligned}$$

□

As proved in the appendix to this chapter, these two SS add up to SS(Total), where

$$\text{SS(Total)} = \sum (y_{ij} - \bar{y}_{..})^2$$

These sums of squares, for our example, are shown in Table 14.3, a so-called analysis of variance (ANOVA) table.

Source	SS
Between	718.4667
Within	4587.7000
Total	5306.1667

Table 14.3

We can use these SS to assess the predictive value of a factor for predicting a dependent variable. If the means for all levels of the factor are very similar, the factor has little predictive value and also SS(Between) will be small. Conversely, if the means differ greatly for different levels of the factor, the factor is very useful in predicting Y and also SS(Between) will be large. For any fixed sample sizes, the better the predictive value of the factor, the less likely it is that the apparent predictive power is attributable to random variation. Thus, if SS(Between) is large relative to SS(Total)—or, equivalently, relative to SS(Within)—we should reject the null hypothesis and conclude that the difference in means is statistically significant (more than merely random).

To see exactly what test statistic and table to use, we need a result known as Cochran's Theorem [see Scheffé (1959)]. Essentially, this theorem says that if SS(Total) can be expanded into K sums of squares:

$$\text{SS(Total)} = \text{SS}(1) + \text{SS}(2) + \cdots + \text{SS}(K)$$

then each of the component SS, divided by the population error variance σ_ϵ^2, has a chi-squared distribution, and the various SS are statistically independent of one another. The definition of an F statistic is

$$F = \frac{\chi_1^2/\text{df}_1}{\chi_2^2/\text{df}_2}$$

where the chi-squared statistics are independent of each other.

The chi-squared statistics will be of the form SS/σ_ϵ^2. Cochran's Theorem indicates that the chi-squared statistics are independent. All we need are the appropriate degrees of freedom. SS(Between) is based on a sum of J terms, one for each level of the factor. Because

$$\sum n_j(\bar{y}_{.j} - \bar{y}_{..}) = 0$$

only $J - 1$ of the deviations from the grand mean are free to vary, and SS(Between) has $J - 1$ df. SS(Within) is based on a sum of sample variances, s_j^2, with respective df $n_j - 1$. Because degrees of freedom add, SS(Within) must have

$$\sum (n_j - 1) = n_{\text{overall}} - J \text{ df}$$

where n_{overall} is the total sample size (30 in our example). The df for SS(Total) is the familiar $n_{\text{overall}} - 1$. Note that for our example $J - 1 = 2$ and $n_{\text{overall}} - J = 30 - 3 = 27$; the total df is

$$n_{\text{overall}} - 1 = 29 = 2 + 27$$

mean square (MS)

Conventionally, the F statistic is reported as a ratio of **mean square (MS)** values, where

$$\text{MS(Anything)} = \frac{\text{SS(Anything)}}{\text{df}_{\text{Anything}}}$$

From Cochran's Theorem,

$$F = \frac{[\text{SS(Between)}/\sigma_\epsilon^2]/(J - 1)}{[\text{SS(Within)}/\sigma_\epsilon^2]/(n_{\text{overall}} - J)} = \frac{\text{MS(Between)}}{\text{MS(Within)}}$$

because the σ_ϵ^2 values cancel out.

The calculations of the F statistic are almost always displayed in an analysis of variance table, as shown in Table 14.4.

Source	SS	df	MS	F
Between	718.4667	2	359.3333	2.114
Within	4587.7000	27	169.9148	
Total	5306.1667	29		

Table 14.4

The rejection region for the F test is R.R.: reject H_0 if the calculated F statistic is larger than the F-table value with the indicated degrees of freedom. In the example the F statistic equals 2.114 with 2 numerator and 27 denominator df. In many journals this is reported in shorthand as $F(2, 27) = 2.114$. The F-table value, for 2 and 27 df with $\alpha = .05$, is 3.35. Thus the result is not close to statistically significant, and there is not enough evidence to claim that the true, population means are different.

Example 14.5

Refer to Example 14.4. Compute the ANOVA table. What conclusion may be reached by the F test?

Solution

In Example 14.4 we computed SS(Between) and SS(Within). There are $J = 3$ levels of the factor, and $n_{\text{overall}} = 18 + 17 + 30 = 65$. Table 14.5 is the ANOVA table.

Source	SS	df	MS	F
Between	3371.06	2	1685.53	3.737
Within	27960.74	62	450.98	
Total	31331.80	64		

Table 14.5

Our F table does not include 62 denominator df, but the values shown for 60 denominator df will be very close. The $\alpha = .05$ table value is 3.15, and the $\alpha = .025$ value is 3.93. These two values bracket the observed F statistic. Thus, by the Universal Rejection Region, the p-value is between .025 and .05. We have fairly conclusive evidence of a real, more than random difference among means. □

The computations for the F test are often done by a standard computer package. If computations for the F table are to be done by hand, one should compute means and variances separately for each level of the factor and then use the definitions of SS(Between) and SS(Within). The means and standard deviations will be useful later on. A summary of the F test is shown below.

Hypothesis Test for One-Factor ANOVA

1. H_0: all $\alpha_j = 0$
2. H_a: some (or all) $\alpha_j \neq 0$
3. T.S.: Compute

$$\text{SS(Between)} = \sum n_j(\bar{y}_{.j} - \bar{y}_{..})^2$$

where $\bar{y}_{.j}$ is the mean for the jth level of the factor and $\bar{y}_{..}$ is the grand mean.

$$\text{SS(Within)} = \sum (n_j - 1)s_j^2$$

where s_j^2 is the jth level sample variance,

$$F = \frac{\text{SS(Between)}/(J - 1)}{\text{SS(Within)}/(n_{\text{overall}} - J)} = \frac{\text{MS(Between)}}{\text{MS(Within)}}$$

4. R.R.: reject H_0 if $F > F_\alpha$ with $J - 1$ and $n_{\text{overall}} - J$ df

The F statistic can be related to a measure of predictability similar to those defined in Chapter 3. If we knew the level of a factor, we could predict that the dependent variable Y would equal $\bar{y}_{.j}$, the mean for that level. The squared prediction error for all such predictions would be

$$\sum (y_{ij} - \bar{y}_{.j})^2 = \text{SS(Within)}$$

If we did not know the level of the factor, we could only predict the grand mean $\bar{y}_{..}$. The squared error of all such predictions would be

$$\sum (y_{ij} - \bar{y}_{..})^2 = \text{SS(Total)}$$

How much better we do knowing the factor level than not knowing it is measured by η^2 (eta-squared), the proportionate reduction in error:

$$\eta^2 = \frac{\text{SS(Total)} - \text{SS(Within)}}{\text{SS(Total)}} = \frac{\text{SS(Between)}}{\text{(Total)}}$$

It follows by algebra that

$$\eta^2 = \frac{(\text{df}_1)F}{(\text{df}_1)F + \text{df}_2}$$

where $\text{df}_1 = J - 1$ is the numerator df and $\text{df}_2 = n_{\text{overall}} - 1$ is the denominator df. Because η^2 is not often reported and F and both df usually are, this is a convenient way to calculate η^2 without the raw data. Solving for F, we have

$$F = \frac{(df_2)\eta^2}{(df_1)(1 - \eta^2)}$$

Example 14.6 In Example 14.5 we reported $F = 3.737$ with 2 and 62 df. Calculate η^2.

Solution

$$\eta^2 = \frac{2(3.737)}{2(3.737) + 62} = 0.108$$

From Example 14.4 we have SS(Between) = 3371.06 and SS(Total) = 31331.80, so

$$\eta^2 = \frac{3371.06}{31331.80} = 0.108$$

again. This value of η^2 indicates that the squared error in predicting Y is reduced by only 10.8% by knowing the factor level. The factor is not extremely useful in predicting Y. □

The F statistic has exactly an F distribution if the assumptions are exactly satisfied. Of course, the assumptions never will be exactly satisfied, in practice. The most critical assumption is the assumption of independence. In the appendix to Chapter 5 we noted that dependence among the observations can have a large effect on the variability (standard error) of sample means, and thus on the inferences made from those means. If the data are collected in some order over time, there is always the possibility of a sequential dependence from one observation to the next; this problem is called **autocorrelation.** The presence of autocorrelation to any serious degree invalidates the use of F tables. A visual way to check for autocorrelation is to plot the data from each sample in the order taken. If the independence assumption holds, there should be no pattern at all. The existence of a trend or some other pattern in the plot suggests that autocorrelation is a problem. There are several formal tests for autocorrelation available. The most common one is the Durbin-Watson statistic discussed in Chapter 15. If there is a suspicion that there may be carry-over dependence from one observation to the next, this statistic should be computed. Most computer packages will do this computation. If the data are collected from a cross-section at a given time as in our example, dependence should not be a problem.

autocorrelation

Example 14.7 Should autocorrelation be a problem in the study in Example 14.1?

Solution No. There is no reason why any one individual's score should have any predictive value for other individuals' scores. □

heteroscedasticity

If constant variability in response over levels doesn't hold, there is a problem of **heteroscedasticity.** A major advantage of balanced-design experiments (equal n's for all levels) is that heteroscedasticity has next to no effect on the correctness of F inferences. Even when the largest population variance is 5 times the smallest one, a nominal $\alpha = .05$ test will have an actual α of perhaps .052 or .049, as long as the design is balanced. F methods may not be as efficient as some kind of weighted method, but the loss of efficiency is probably not too serious unless the heteroscedasticity is wild. When the design is badly unbalanced, and particularly when the larger variances are associated with the smaller n's, F inferences may be seriously incorrect.

Example 14.8 Refer to Example 14.4. Is there reason to believe that nonconstant variance is a problem?

Solution Perhaps there is a modest problem. Of course, we know only the sample variances, not the population variances. The sample standard deviation of the relatively small backstage group is quite a bit larger than the standard deviations of the other groups. Thus the inferences made must be viewed with some caution. □

If the populations of responses at various levels of the factor are not normal, there are potential problems of correctness and of efficiency. Skewness and/or heavy tails can indicate trouble. However, *nonnormality of populations is far less serious in making inferences about means, as in ANOVA, than in making inferences about variances.* The analysis of variance is in fact based on sample means; therefore the Central Limit Theorem effect is at work. As with the two-sample t test, it seems to be the total sample size, n_{overall}, that determines the rate of convergence to F-ness. Barring gross skewness in the populations, an overall sample size of 30 or so will probably be enough to make F inferences reasonably close to correct. Thus F tests are quite robust in the correctness sense. They are not so robust in the efficiency sense. When there is substantial skewness and/or heavy tails in the population, methods other than the normal-theory F methods may be more effective. In the example, plots of the three samples show no serious nonnormality.

Example 14.9 In Example 14.3 we noted that the data showed some rather severe outliers. Does this fact invalidate the use of an F test?

Solution No, the F test still is a legitimate one. However, an alternative procedure may be more efficient. □

The same issues of robustness arose in the context of the two-sample t test. There we introduced the rank sum test as an efficient competitor. There is an extension of the rank sum test, called the Kruskal-Wallis test, which applies to the one-factor ANOVA situation. To perform this test, rank the data from lowest to highest in the overall sample, just as in the rank sum test. Calculate the sums T_j of ranks at each level of the factor. The test statistic is

$$H = \frac{12}{n(n+1)}\left(\sum \frac{T_j^2}{n_j}\right) - 3(n+1)$$

where n is n_{overall}, the total sample size. The statistic is to be compared to the chi-squared table with $J - 1$ df. The data, ranks, and rank sums for our three-group illustration are shown here.

Group 1	25	33	37	41	43	45	49	50	58	67	
Rank	1	4	7	8.5	10	12	17	18	23	27	$T_1 = 127.5$
Group 2	36	41	46	46	53	57	61	68	73	79	
Rank	5.5	8.5	14	14	20	22	24.5	28	29	30	$T_2 = 195.5$
Group 3	28	30	36	44	46	48	52	56	61	66	
Rank	2	3	5.5	11	14	16	19	21	24.5	26	$T_3 = 142.0$

Thus

$$H = \frac{12}{(30)(31)}\left(\frac{(127.5)^2}{10} + \frac{(195.5)^2}{10} + \frac{(142.0)^2}{10}\right) - 3(31) = 3.3013$$

The right-tail .10 value in the chi-squared tables with $3 - 1 = 2$ df is 4.605. The value of H is not significant even at $\alpha = .10$. Note that the conclusion of the Kruskal-Wallis test agrees with that of the F test for this illustration.

It certainly isn't obvious, but it is true that $H = 0$ when the average rank is the same for all levels of the factor, and H increases as the discrepancies increase among average ranks for the various levels of the factor. There is a correction to the H statistic that can be used when there are many ties [see Ott (1984)]; unless there are overwhelmingly many ties, the effect of the correction is minor. Ignoring the correction is always conservative.

Example 14.10 Suppose that the rank sums in Table 14.6 had been calculated for the data underlying Example 14.5.

Group	T_j	n_j
Backstage	765.5	18
Onstage	462.5	17
Combination	917.0	30

Table 14.6

What is the conclusion of the Kruskal-Wallis test and how does it differ from the conclusion of the F test (p-value between .025 and .05)? What might explain the difference?

Solution

$$H = \frac{12}{65(66)}\left(\frac{(765.5)^2}{18} + \frac{(462.5)^2}{17} + \frac{(917.0)^2}{30}\right) - 3(65)$$
$$= 9.664$$

For $3 - 1 = 2$ df, this value is between table values 9.21034 and 10.5966, corresponding to $\alpha = .01$ and .005, respectively. Thus the Kruskal-Wallis test is more conclusive, assuming these rank sums. A reason for this result is that there were a number of outliers reported in the data. □

The Kruskal-Wallis test is a strong competitor of the F test. Even when all the assumptions of the F test are met, the Kruskal-Wallis test is only a little worse; its relative efficiency is about 95.5%. When the normal-population assumption fails, the Kruskal-Wallis test is often more efficient. It should be noted that the Kruskal-Wallis assumptions differ from the F assumptions only in not requiring normality; violation of the other assumptions is still a potential problem. There is some reason to believe that heteroscedasticity (nonconstant variance) has less effect for the Kruskal-Wallis test than for the F test. Given balanced design, the effect should be minimal for either statistic. Dependence over time of the observations remains the most serious problem for both statistics.

transform

Rather than use the Kruskal-Wallis approach to grossly nonnormal data, one might **transform** the data to a new, "nicer" scale. For right-skewed data, transforming to square roots will often produce more symmetric data plots; logarithms (either natural or base 10) are used in highly skewed situations. See Tukey (1977) for a discussion of a whole series (called a "ladder") of possible transformations.

The problem with the transformation strategy is not statistical, but scientific. If data are analyzed after some exotic transformation, what do the results mean, scientifically? Suppose that latency times, transformed as the inverse of the square root of the logarithm of the measured time, are longer on average for each of the two treatment groups than for the control group. What does that say in terms of the original latency? It can be argued that a complicated transformation that yields a "nice" (symmetric, near-normal) data set is the reasonable scale for that data set; but if statistics is to serve science and not vice versa, I'd prefer to state conclusions in meaningful units. Some transformations are meaningful directly. The inverse of a time-to-next-behavior (latency) measure is a behaviors-per-unit-time (speed) measure. Other transformations, such as the logarithmic one, have some useful mathematical properties and have appeared enough times in the psychological literature to be well understood. More exotic transformations (such as a cube root) should be used with caution.

Exercises for Section 14.1

14.1. Jacobs (1981) reported a study in which 27 subjects were measured on the number of "empathic responses" made in eight situations. There were three groups of subjects with 9 subjects per group. Besides a control group, there was a group given "ordinary" training in empathy and a group given "special" training. The data were

Control:	0	0	0	0	0	0	0	0	0
Ordinary:	1	2	3	3	3	3	4	4	7
Special:	5	5	6	7	7	8	8	8	8

a. Compute means and standard deviations for each group. The computations for the control group may not be overly difficult.

b. Compute sums of squares within and between groups.

c. Find the value of the F statistic. Is it statistically significant at $\alpha = .01$?

d. State bounds on the p-value for the F statistic found in part c.

14.2. Refer to Exercise 14.1. Are there any evident violations of assumptions? In this case, how critical are those violations to the conclusion of the F test?

14.3. Perform a Kruskal-Wallis test for the data of Exercise 14.1. Does this test give essentially the same conclusion as does the F test?

14.4. Suppose that we attempt to replicate a study done by Davis and Unruh (1980) in which the dependent (response) variable of interest was the number of words freely recalled from a list. The subjects were divided among those with no history of depression (group 1), those with only short-term depression (group 2), and those with long-term depression (group 3). Assume that the data are

Group 1: 0 0 1 1 1 1 2 2 2 2 2 2 2 2 3
3 3 3 3 3 3 3 3 4 4 4 7 8 9 15
(mean 3.167, standard deviation 2.937, $n = 30$)

Group 2: 1 2 2 3 3 3 4 4 5
5 5 6 7 7 9 10 13 17
(mean 6.444, standard deviation 4.592, $n = 18$)

Group 3: 0 0 0 1 2 4 4 5 5 7 10 18
(mean 4.667, standard deviation 5.211, $n = 12$)

a. Calculate the grand mean.

b. Construct an ANOVA table.

c. Perform the F test for the null hypothesis of equal means. Place bounds on the p-value.

14.5. Refer to Exercise 14.4.

a. Is there evidence of violations of assumptions?

b. Are any assumption violations you found in part a potentially serious? Are they likely to cast doubt on the results of the F test found in part c of Exercise 14.4?

14.6. Refer to Exercise 14.4.

a. Perform a Kruskal-Wallis test on the data. State bounds on the p-value.

b. Which test, the Kruskal-Wallis or the F, is more believable in this situation? How critical is the choice?

14.7. An experiment was designed to measure how long subjects could track a moving target on a computer screen. The target moved erratically according to a predetermined program. As the experiment was designed, the first 6 subjects (volunteers recruited by signs along the main campus walkway) were to be tested under a control condition with no distractions, the next 12 were to be tested under a flashing-lights distraction, and the next 18 were to be tested under a loud-noise distraction.

a. The experimental design is dreadful. Explain why.

b. Suggest improvements to the design.

14.8. Refer to Exercise 14.7. The experiment was carried out as designed, despite our objections. The data (times in seconds) were as follows (listed in the sequence from first to last subject):

No distraction:	56 53 54 49 40 42
	(mean 49.00, variance 44.00)
Light distraction:	44 42 43 40 39 35
	37 38 32 50 30 32
	(mean 38.50, variance 33.55)
Sound distraction:	34 40 30 31 29 27 28 26 26
	27 25 24 20 22 19 17 10 31
	(mean 25.89, variance 46.10)

a. Compute an ANOVA table.
b. Find the value of the *F* statistic.
c. What is the nominal *p*-value for the *F* statistic?

14.9. Refer to the data of Exercise 14.8. Is there evidence of violations of assumptions? If so, are the apparent violations serious?

14.10. Perform a Kruskal-Wallis test on the data of Exercise 14.8. Does the use of this test ameliorate any assumption violations that you claimed in Exercise 14.9?

14.2 Comparing Means

The *F* test in Section 14.1 is designed to test an "omnibus" null hypothesis, namely that all levels of the factor have the same population mean response. It does not indicate which specific means (if any) differ, nor does it yield confidence intervals. Methods for testing specific means and for getting confidence intervals are the subject of this section. There are many, many methods in the literature, of varying degrees of justification. In this section only some of the most popular ones having reasonable justification are discussed. For a discussion of other ones, see Howell (1982).

comparisons
pairwise

These methods are based on the idea of **comparisons,** also known as contrasts. A comparison may be **pairwise**—a difference of two specific means—or more general. For example, in the experiment involving different methods of doing statistics problems discussed in Section 14.1, a pairwise comparison would be the difference in average score between those doing all problems with a hand calculator and those doing all problems with a computer. A nonpairwise comparison would be the average of the means of those doing some or all problems with a computer minus the mean of those doing all problems with a hand calculator. By definition, any comparison involving means for more than two levels of the factor is not pairwise.

Formally, any true (population) comparison can be written as a weighted sum of true means:

$$L = \sum c_j\mu_j$$

where $\Sigma\, c_j = 0$, and any sample comparison as

$$\hat{L} = \sum c_j\bar{y}_{.j}$$

where $\Sigma\, c_j = 0$.

For example, a pairwise comparison between sample means of groups 1 and 3 would be

$$\begin{aligned}\hat{L} &= \bar{y}_{.1} - \bar{y}_{.3}\\ &= (1)\bar{y}_{.1} + (0)\bar{y}_{.2} + (-1)\bar{y}_{.3}\end{aligned}$$

where the weights (c_j's) are 1, 0, and -1, which add to 0. The difference between the average of means 2 and 3 and mean 1 is

$$\begin{aligned}\hat{L} &= \frac{\bar{y}_{.2} + \bar{y}_{.3}}{2} - \bar{y}_{.1}\\ &= (-1)\bar{y}_{.1} + \frac{1}{2}\bar{y}_{.2} + \frac{1}{2}\bar{y}_{.3}\end{aligned}$$

where the weights are -1, 1/2, and 1/2, which add to 0.

Example 14.11

Suppose that subjects were assigned randomly to one of four experimental methods for learning to take a certain test. The test was supposedly a general aptitude test, and therefore not coachable. Each subject took the test, went through one of the learning methods, then retook the test. The response (Y) value of interest was the change in score, post-test minus pre-test, of each subject. Treatments A and C were two different "control" methods. Treatments B and D were experimental, with the intent of improving subjects' scores. What is being measured by each of the following comparisons?

$$\hat{L}_1 = \bar{y}_B - \bar{y}_D$$

$$L_2 = .5(\mu_B + \mu_D) - .5(\mu_A + \mu_C)$$

Solution

First note that both *are* comparisons. The weights c_A, c_B, c_C, and c_D are 0, 1, 0, and -1 for the first comparison and $-.5$, .5, $-.5$, and .5 for the second. In both cases, the sum of the c_j's is 0. The first comparison compares the two sample means of the experimental groups. The second comparison compares the average population mean in the experimental groups to the average population mean in the control groups. □

The condition that $\Sigma c_j = 0$ guarantees two things: that the comparison is "free of the grand mean," and that a comparison of means is also a comparison of the level effects α_j. Positive and negative weights attached to various means can just as well be thought of as being attached to the level effects

$$\begin{aligned}\sum c_j\mu_j &= \sum c_j(\mu + \alpha_j)\\ &= \mu \sum c_j + \sum \alpha_j\\ &= \sum c_j\alpha_j\end{aligned}$$

because $\Sigma c_j = 0$.

A simple, but dubious, way to test for statistically detectable differences among pairs of means would be to run two-sample t tests, as in Chapter 11, on all possible pairs of means. However, there's a problem; if each specific comparison is made at $\alpha = .05$, say, what is the overall α? There's no easy answer. We can't add all the .05's because the events in question are not mutually exclusive; nor are they independent. In the language of comparisons, the **per-contrast error rate** (the probability of making an error on one specific comparison) is .05, but the **experimentwise error rate** (the probability of making at least one error in testing all the comparisons) is unknown.

per-contrast error rate

experimentwise error rate

per-experiment error rate

There is also a concept of **per-experiment error rate**—the expected number of errors in all the comparisons. It differs from the experimentwise error rate because the per-experiment rate is an expected value, not a probability, and it is possible to make more than one error within a given set of comparisons. If 10 t tests are run, each at $\alpha = .05$, the per-experiment error rate is 10(.05) = .50, but the experimentwise error rate is unknown.

A multiple-comparison approach that only concerns itself with per-comparison error rates rewards experimenters who make massive numbers of comparisons, relevant or not, in hopes of finding something or other that passes for statistical significance. Control of the experimentwise error rate has the opposite effect; it rewards experimenters who keep designs small and make only those comparisons which are most scientifically relevant. Therefore, it is better to work with procedures that control the experimentwise error rate.

Example 14.12

In Example 14.11, how many pairwise comparisons can be made? If t-test comparisons are made at $\alpha = .10$, what are the per-comparison, experimentwise, and per-experiment error rates?

Solution

Among four means, there are six comparisons to be made, namely AB, AC, AD, BC, BD, and CD. The per-comparison error rate is .10 by choice. The per-experiment error rate is $6(.10) = .60$. The experimentwise error rate is unknown, which is precisely the problem with multiple t tests. □

One procedure for multiple comparisons originated with Tukey (no date). It was designed for, and works best with, pairwise comparisons of means in balanced-design experiments. It is based on the theoretical (sampling) distribution of the **studentized range** statistic, which is the difference between the largest mean and the smallest one, divided by a standard error. In the case of a balanced design, with all $n_j = n_{\text{per mean}}$, calculate

studentized range

$$W = q(J, \text{df}_{\text{within}}) \sqrt{\frac{\text{MS(Within)}}{n_{\text{per mean}}}}$$

where $q(J, \text{df}_{\text{within}})$ is the entry in the studentized range table (Table 8) corresponding to the desired *experimentwise* error rate α, J is the number of levels of the factor, $n_{\text{per mean}}$ is the sample size corresponding to each mean, and $\text{df}_{\text{within}} = n_{\text{overall}} - J$ is the df for MS(Within). For three treatments with 27 denominator df, as in the illustration in Section 14.1, the q numbers must be interpolated between 24 and 30 df. They are approximately 3.51 (for $\alpha = .05$) and 4.50 (for $\alpha = .01$).

For confidence intervals with specified experimentwise confidence, simply take each pairwise difference in means plus or minus W. To test for statistical detectability (significance) of differences, declare a difference significant at the specified value if it exceeds (in absolute value) W. For the illustration of Section 14.1, there were three groups with respective means 44.8, 56.0, and 46.7; MS(Within) was 169.9148, with 27 df, and each n_j was 10. For an experimentwise 95% confidence, $W = 3.51\sqrt{169.9148/10} = 14.47$. The pairwise, Tukey confidence intervals are

$$(44.8 - 56.0) - 14.47 < \mu_1 - \mu_2 < (44.8 - 56.0) + 14.47$$
$$(44.8 - 46.7) - 14.47 < \mu_1 - \mu_3 < (44.8 - 46.7) + 14.47$$
$$(56.0 - 46.7) - 14.47 < \mu_2 - \mu_3 < (56.0 - 46.7) + 14.47$$

Note that none of the differences in means exceeds 14.47 in absolute value. Thus none of the pairwise differences is significant at an experimentwise error rate $\alpha = .05$.

Example 14.13 Refer to Example 14.11. The results in Table 14.7 were obtained from the sample data.

Treatment	Mean	Standard Deviation
A	6.40	7.214
B	16.20	7.525
C	−3.90	9.848
D	7.30	5.559

Table 14.7

Each sample size was 10. SS(Between) = 2029.0 and SS(Within) = 2129.0. Which means, if any, are significantly different, using an experimentwise error rate $\alpha = .01$?

Solution We have $n_{\text{per mean}} = 10$ and $J = 4$, so $n_{\text{overall}} = 40$. There is no entry in Table 8 for $40 - 4 = 36$ denominator df. We have

$$q(.01, 4, 30) = 4.80 \qquad \text{and} \qquad q(.01, 4, 40) = 4.70$$

Interpolating, we take $q(.01, 4, 36) = 4.76$.

$$\text{MS(Within)} = \frac{2129.0}{36} = 59.139$$

so

$$W = 4.76\sqrt{\frac{59.139}{10}} = 11.58$$

The only difference of means that is larger than 11.58, and therefore is significant, is $B - C = 16.20 - (-3.90) = 20.10$. □

The Tukey procedure is exactly correct (has exactly the nominal experimentwise error rate) if all the ANOVA assumptions are exactly correct and the design is balanced. If the assumptions appear reasonable but the design is somewhat unbalanced (say the largest n is no more than twice the smallest), a modification due to Kramer can be used. To compare means $\bar{y}_{.j}$ and $\bar{y}_{.j'}$, calculate

$$W_{j,j'} = q(J, \text{df}_{\text{within}}) \sqrt{(.5)\text{MS(Within)}\left(\frac{1}{n_j} + \frac{1}{n_{j'}}\right)}$$

Confidence intervals with the desired experimentwise error rate α are obtained as

$$\bar{y}_{.j} - \bar{y}_{.j'} \pm W_{j,j'}$$

and significance tests with the desired experimentwise error are obtained by declaring significance if

$$|\bar{y}_{.j} - \bar{y}_{.j'}| > W_{j,j'}$$

Example 14.14

In Example 14.5 we had three groups, 62 denominator df, MS(Within) = 450.98, and the means and sample sizes shown in Table 14.8.

Group	$\bar{y}_{.j}$	n_j
B	109.00	18
O	89.71	17
C	96.93	30

Table 14.8

Calculate experimentwise 95% confidence intervals for all pairwise differences of means, using the Tukey-Kramer method.

Solution

There is no entry in the studentized range table for 62 denominator df. We will use $q(.05, 3, 60) = 3.40$. Those who feel compelled to interpolate may note that $q(.05, 3, 120) = 3.36$ and may use 3.3987. For comparing B to O, calculate

$$W_{B,O} = 3.40 \sqrt{450.98(.5)\left(\frac{1}{18} + \frac{1}{17}\right)} = 17.27$$

The confidence interval is

$$(109.0 - 89.71) + 17.27 < \mu_B - \mu_O < (109.00 - 89.71) + 17.27$$

or

$$2.02 < \mu_B - \mu_O < 36.56$$

Note that this interval doesn't include 0, so we may declare the B mean significantly different (and in practice larger) than the O mean. Similar calculations yield two other intervals:

$$-3.15 < \mu_B - \mu_C < 27.29$$

$$-22.72 < \mu_O - \mu_C < 8.28$$

Both these intervals include 0, so the hypothesis of equal means must be retained in each case. □

Tukey procedures, like the *F* test, are quite sensitive to any dependence that can arise from carryover effects from one measurement to the next. Careful design of the experiment is the best way to avoid that problem. Unlike the *F* test, the Tukey procedure is somewhat sensitive to nonconstant variance (heteroscedasticity). If the population variance is not constant, MS(Within) will tend to underestimate the large variances and overestimate the small ones. In particular, confidence intervals and tests will be too generous (anti-conservative) for means corresponding to large variances. The problem is most acute when large variances are associated with small sample sizes. See Stoline (1981) for some possible methods of handling this problem.

The Tukey procedure is not equivalent to the *F* test. It is possible to have an *F* statistic significant at a specified α, yet no pairs of means differ significantly at the same α by the Tukey test. Typically, this occurs when there are several means bunched at the high end of the scale and others bunched at the low end. The reverse phenomenon (Tukey significant, *F* not) can also occur, typically with several means bunched in the middle and stragglers on each end.

Example 14.15 Refer to Example 14.14. Is there reason to be concerned about problems of nonconstant variance? Do the Tukey conclusions differ from the conclusions of the *F* test?

Solution There is reason for concern about the nonconstant variance. We expressed that concern in the examples of Section 14.1, and the Tukey procedure is more sensitive to this problem than is the *F* test. In this case, the *F* test calculated in Section 14.1 indicated a significant difference at $\alpha = .05$, and the Tukey method also found one significant difference at that α. □

The Tukey method can be sensitive to nonnormal populations. If the data exhibit severe skewness and/or heavy tails, nominal probabilities may be in error or the whole inference method may be inefficient. The Central Limit Theorem effect, mentioned in Section 14.1, has limited effect in pairwise comparisons, because only the combined size of the two samples applies, rather than the combined size of the entire sample. Section 14.1 suggested the Kruskal-Wallis rank test for cases of skewness and/or heavy tails. Hollander and Wolfe (1974) describe a method for pairwise comparisons using average ranks, using special tables to achieve the desired experimentwise error rate. As Hollander and Wolfe point out, some funny things can happen with this method. If one ranks only the two particular samples of interest, the average rank for level 1 (say) may be less than the average rank for level 2; but if one ranks all the data, the average rank for level 1 may be greater than that for level 2 because of "interference" from data for levels 3, 4, ...

Up to now in this section, the discussion has focused entirely on pairwise comparisons. Often, more complicated comparisons, such as the difference between the average of three "experimental" treatments and the average of two "control" treatments, are scientifically relevant. The standard approach to analyzing general comparisons is the Scheffé method. Recall that a sample estimate of a comparison value is

$$\hat{L} = \sum c_j \bar{y}_{.j}$$

where $\Sigma c_j = 0$. The c_j weights may be chosen any way the experimenter wishes, so long as the weights sum to 0. The most interesting weights would seem to yield comparisons of an average (possibly weighted) of one set of means to an average of another set. Curiously, there is a tradition in comparison analysis to compare sums rather than means. Although that's mathematically valid, it seems like a nuisance to interpret. The weights may be chosen in advance, while the experiment is being planned, or later, upon examination of the data.

Example 14.16 In an experiment having four levels of the factor, levels 1 and 3 were control groups and levels 2 and 4 were experimental groups. Define

$$\hat{L} = .5(\bar{y}_{.1} + \bar{y}_{.3}) - .5(\bar{y}_{.2} - \bar{y}_{.4})$$

Should the comparison be regarded as planned?

Solution Because the weights were specified to yield a comparison of the control groups with the experimental groups, it seems reasonable to regard the comparison as planned prior to data gathering. □

planned comparison
post-hoc comparison

The Scheffé method used depends on whether the comparison has been specified in advance—a **planned comparison**—or after a judicious look at the data—a **post-hoc comparison.** The reason is that post-hoc comparisons run the risk of "capitalizing on chance." The data are examined twice, once to select the feature of interest (the comparison) and again to estimate its value. There's a natural tendency to select comparisons that have large values in the particular data at hand, whether or not they actually have large values in the population. For example, in the three-group study of Section 14.1, one might make a post-hoc comparison of the largest mean (56.0) to the average of the two smallest means (44.8 and 46.7). The plus-or-minus attached to a comparison value selected post hoc must reflect this tendency. The Tukey method and its relatives, discussed previously in this section, were designed to give experimentwise error control over *all* pairwise comparisons, so the planned/post-hoc distinction wasn't needed. The Scheffé method provides an easy way to treat more general comparisons, both planned and post hoc.

For both types of comparisons, the basic principle is familiar: attach a plus-or-minus term to the sample estimate of the particular comparison. The plus-or-minus is calculated, once again, as a table value times a standard error. For either planned or post-hoc comparisons, the estimated standard error is

$$\sqrt{\text{MS(Within)} \sum \frac{c_j^2}{n_j}}$$

For a post-hoc comparison, the desired table value is

$$\sqrt{(J-1)F_\alpha}$$

where J is the number of levels of the factor (whether or not the comparison actually involves the means for all levels) and F_α is the F-table value for the desired α with $J - 1$ and $n_{\text{overall}} - J$ df (the same df used in the F test). Confidence intervals for the true value of the comparison, which have the property that the specified confidence holds experimentwise for all possible comparisons, are formed as

$$\hat{L} \pm S_{\hat{L}}$$

where

$$S_{\hat{L}} = \sqrt{(J-1)F_\alpha \text{MS(Within)} \sum \frac{c_j^2}{n_j}}$$

A comparison is declared detectably (significantly) different from 0 if

$$|\hat{L}| > S_{\hat{L}}$$

For the three-group study of Section 14.1, let's compare the mean for group 2 to the average of the means of the other two groups. Thus $c_2 = 1$, and $c_1 = c_3 = -.5$. The required F-table value for 95% confidence, with 2 and 27 df, is 3.35. All n's equal 10 and MS(Within) = 169.9148. A post-hoc 95% confidence interval is

$$[56.7 - .5(44.8 + 46.7)]$$

$$\pm \sqrt{(3-1)3.35\left[169.9148\left(\frac{(1)^2}{10} + \frac{(-.5)^2}{10} + \frac{(-.5)^2}{10}\right)\right]}$$

or 10.95 ± 13.07. Because $|10.95| < 13.07$, the value of the comparison is not significantly different from 0 at $\alpha = .05$.

Example 14.17

Compare the average of the A and C means to the average of the B and D means in Example 14.13. Temporarily regard the comparison as *post hoc*. Calculate a Scheffé 95% confidence interval for the population L value.

Solution

$$\begin{aligned}\hat{L} &= .5(\bar{y}_{.A} + \bar{y}_{.C}) - .5(\bar{y}_{.B} + \bar{y}_{.D}) \\ &= .5[6.4 + (-3.9)] - .5(16.2 + 7.3) = -10.5\end{aligned}$$

In Example 14.13, we calculated MS(Within) = 59.139. The df are 3 and 36, so the $F_{.05}$ value is about 2.87. Thus

$$S_{\hat{L}} = \sqrt{(4-1)(2.87)(59.139)\frac{2(.5)^2 + 2(-.5)^2}{10}} = 7.14$$

The confidence interval is -10.5 ± 7.14, or

$$-17.64 < L < -3.36$$

Note that the interval doesn't include 0, so the comparison of the average control means (A and C) to the average experimental means (B and D) is statistically significant at $\alpha = .05$. □

For planned comparisons, the table value $\sqrt{(J-1)F_\alpha}$ is replaced by $t_{\alpha/2}$, where $t_{\alpha/2}$ is the t-table value for the desired α and df = $n_{\text{overall}} - J$. Confidence intervals and significance tests then proceed as usual. Confidence level and Type I error probability are per-comparison for this method of handling planned comparisons.

Example 14.18 Refer to Example 14.17. Calculate the confidence interval regarding the comparison as planned.

Solution With 36 df, $t_{.025}$ is about 2.03. Thus

$$S_{\hat{L}} = 2.03\sqrt{59.139\frac{[2(.5)^2 + 2(-.5)^2]}{10}} = 4.94$$

The confidence interval becomes -10.5 ± 4.94, or

$$-15.44 < L < -5.56$$

The advantage to the researcher in using planned comparisons is that the confidence interval is narrower, and therefore the power of the significance test is greater. For large experiments, with many levels, the advantage can be substantial. The disadvantage is that there is a temptation to cheat. It's frighteningly easy to look at the data, find a particular comparison that looks good, and say, "Oh, I knew it all along; I'll treat it as planned." The consequence can be the publication of a supposedly significant result that can't be replicated. It can be very embarrassing to publish a paper containing some "significant" experimental results and then have thirteen other people publish papers saying that your results didn't happen when they tried to repeat them.

The Scheffé approach is equivalent to F tests, in a certain sense. If an F test for a set of data comes out significant, it is always possible to find weights c_j that lead to a significant comparison. [Take $c_j = n_j(\bar{y}_{.j} - \bar{y}_{..})$.] Conversely, if F is not significant, no comparison can possibly be significant (post hoc). Thus the Scheffé approach is better related to the F test than is the Tukey approach. However, the Scheffé method should not be used for pairwise comparisons with (nearly) balanced designs. The Scheffé confidence intervals are very wide and the power very low in this case.

Generally researchers feel that the Scheffé method is very conservative. If comparisons are selected post hoc from a very large set of possibilities, that feeling is undoubtedly justified. However, when more complex comparisons are desired, the Scheffé method is just about the only procedure available. The conservatism of the Scheffé post-hoc approach does give ample protection against anything but the most severe violations of assumptions. For the planned comparison case, the same principles apply to the Scheffé method as to the Tukey method. Be careful if the means entering the comparison are associated with relatively large variances or relatively small sample sizes. The Central Limit Theorem effect (which matters if the data are skewed or heavy-tailed) depends on how many observations are included in the means that are actually represented in the comparison.

Example 14.19 Refer to Examples 14.17 and 14.18. In which example is there more reason to worry about violations of assumptions? Is there any major reason to be skeptical of the confidence levels?

Solution The post-hoc comparison of Example 14.17 is conservative, thus giving considerable protection against violation of assumptions. The planned comparison of Example 14.18 is less conservative, hence more sensitive. However, with a balanced design and roughly equal sample variances, the homoscedasticity assumption seems reasonable. Unless there is some improbable dependence among the measurements, or severe skewness in the data, both confidence levels should be valid. □

14.3 Allocating SS(Between) among Comparisons

In the first two sections of this chapter the emphasis has been on inference, either by an omnibus F test or by more specific pairwise or general comparisons of means. In this section we first turn back to descriptive issues. We can account for a certain portion of the variability in a variable as attributable to differences in means. SS(Between) measures the variability attributable to the levels of a factor. Now we refine the description, asking how much of the explained variability, SS(Between), is attributable to specific comparisons among level means.

In particular, suppose that we are analyzing a one (fixed) factor experiment with $n_{\text{per mean}} = 8$ and suppose that the means are

Level j	1	2	3	4	5
$\bar{y}_{.j}$	36.50	50.25	49.75	51.00	62.50

The grand mean is $\bar{y}_{..} = 50.00$. SS(Within) can be calculated to be 2717. The question is Which differences among means account for the variability captured in SS(Between)? Looking at the means, we note that $\bar{y}_1$ is well below the grand mean and $\bar{y}_5$ is well above it; all the other means are close to the grand mean. Thus the difference between means 5 and

1 should account for most of the variability among the sample means, SS(Between).

For any comparison $\hat{L} = \Sigma c_j \bar{y}_{.j}$, we may define a sum of squares (SS) attributable to $\hat{L}$.

SS for a Comparison

$$SS(\hat{L}) = \frac{(\hat{L})^2}{\sum (c_j)^2/n_j}$$

In our example, if $\hat{L} = \bar{y}_5 - \bar{y}_1$,

$$SS(\hat{L}) = \frac{(62.5 - 36.5)^2}{1^2/8 + (-1)^2/8} = 2704$$

Thus in this example, virtually all the variability among means (2704/2717 = .9952 proportion of it) is attributable specifically to the difference between the largest and smallest means.

Example 14.20 In Example 14.13, we had SS(Between) = 2029.0 with $n_{\text{per mean}} = 10$. In Example 14.17, referring to the same data and to the comparison of the average of two control means to the average of two experimental means, we found $\hat{L} = -10.5$. Does this comparison account for most of SS(Between)?

Solution

$$\hat{L} = .5(\bar{y}_{.A} + \bar{y}_{.C}) - .5(\bar{y}_{.B} + \bar{y}_{.D}) = -10.5$$

$$SS(\hat{L}) = \frac{(-10.5)^2}{2(.5)^2/10 + 2(-.5)^2/10} = 1102.5$$

Thus this $\hat{L}$ accounts for a bit over half (1102.5/2029.0 = .5434) of the variability among means. □

There are infinitely many $\hat{L}$ comparisons that might be made among means. How can one select a comparison that will account for most of the variability among means? Generally, order the means from low to high. Take the average of the highest means minus the average of the lowest means. This $\hat{L}$ will account for a reasonably large share of SS(Within). In our five-means example, there was one very large mean and one very small one; all the rest of the means fell near the grand mean. Thus the comparison of the largest vs. the smallest mean accounted for virtually all the variability among means.

Example 14.21

Explain why the comparison defined in Example 14.20 did not account for more of SS(Within) than it did.

Solution

Looking back at Example 14.13, we note that there was one large mean (16.2), one small mean (−3.9), and two means close to the grand mean (6.4 and 7.3). The comparison defined in Example 14.20 did not compare only the large and small means. Thus it didn't capture most of SS(Within). □

Further, given several comparisons among means, how can one assess how much of SS(Between) is explained by the comparisons together? In general, that problem is almost insoluble. The variability explained by several comparisons is *not* the sum of the explained variations of each one. Thus it is not possible to say that so much of the variability in Y is attributable to this comparison, so much to that one, etc. There is one exception. We may add $SS(\hat{L})$ values for a set of **orthogonal comparisons.**

orthogonal comparisons

Orthogonal Comparisons

Comparisons $\hat{L}_1 = \Sigma c_{1j}\bar{y}_{.j}$ and $\hat{L}_2 = \Sigma c_{2j}\bar{y}_{.j}$ are orthogonal if

$$\sum \frac{c_{1j}c_{2j}}{n_j} = 0$$

Several comparisons are orthogonal if any pair is orthogonal.

Note that if all sample sizes are equal, the condition for orthogonality reduces to

$$\sum c_{1j}c_{2j} = 0$$

Example 14.22

In Example 14.21, we defined the comparison

$$\hat{L}_1 = .5(\bar{y}_{.A} + \bar{y}_{.C}) - .5(\bar{y}_{.B} + \bar{y}_{.D})$$

Consider two other comparisons:

$$\hat{L}_2 = \bar{y}_{.A} - \bar{y}_{.C}$$

and

$$\hat{L}_3 = \bar{y}_{.B} - \bar{y}_{.D}$$

The sample sizes all equaled 10. Are the comparisons orthogonal?

Solution

The c weights are shown in Table 14.9 .

Comparison	A	B	C	D
$\hat{L}_1$	.5	$-.5$	.5	$-.5$
$\hat{L}_2$	1.0	0	-1.0	0
$\hat{L}_3$	0	1.0	0	-1.0

Table 14.9

For comparisons 1 and 2, we have

$$\begin{aligned}\sum c_{1j}c_{2j} &= (.5)(1.0) + (-.5)(0) + (.5)(-1.0) + (-.5)(0)\\ &= .5 + 0 - .5 + 0 = 0\end{aligned}$$

Thus comparisons 1 and 2 are orthogonal. For comparisons 1 and 3, we have

$$\begin{aligned}\sum c_{1j}c_{3j} &= (.5)(0) + (-.5)(1.0) + (.5)(0) + (-.5)(-1.0)\\ &= 0 - .5 + 0 + .5 = 0\end{aligned}$$

and orthogonality holds again. Similarly, comparisons 2 and 3 are orthogonal. □

In ANOVA theory, SS(Between) can be thought of as the squared length of a line in a certain space. The dimension of the space is the df, $J - 1$. A comparison can be thought of as an axis in that space. If comparisons are orthogonal, they can be thought of as perpendicular axes. There are many ways to draw perpendicular axes. Even in two-dimensional space, there are infinitely many sets of perpendicular (orthogonal) comparisons. But in $J - 1$ dimensions, there are always exactly $J - 1$ perpendicular axes; thus, with J means, one can define only $J - 1$ orthogonal comparisons. There are many sets of $J - 1$ orthogonal comparisons, but one always can get $J - 1$ and only $J - 1$ orthogonal comparisons.

When the comparisons are orthogonal, the SS($\hat{L}$) can be added. For any set of $J - 1$ orthogonal comparisons, their SS adds to SS(Between); for orthogonal comparisons,

$$\text{SS(Between)} = \text{SS}(\hat{L}_1) + \text{SS}(\hat{L}_2) + \cdots + \text{SS}(\hat{L}_{J-1})$$

Example 14.23

In Example 14.22 we defined three orthogonal comparisons among four means. Show that their SS add to SS(Between).

Solution In Example 14.20 we found $SS(\hat{L}_1) = 1102.5$. Similar calculations show that $SS(\hat{L}_2) = 530.45$ and $SS(\hat{L}_3) = 396.05$. The sum of these SS is $1102.5 + 530.45 + 396.05 = 2029.0 = SS(\text{Between})$. □

The discussion in this section has been in a descriptive spirit, emphasizing how much of the variability among means may be captured by particular comparisons. The ideas of this section can also be used in statistical inference. In particular, the Scheffé tests defined in Section 14.2 are equivalent to tests based on $SS(\hat{L})$. A post-hoc comparison is declared significant at a specified α level if

$$F = \frac{SS(\hat{L})/(J-1)}{MS(\text{Within})} > F_{\alpha,J-1,n-J}$$

and a planned comparison is significant if

$$F = \frac{SS(\hat{L})/1}{MS(\text{Within})} > F_{\alpha,1,n-J}$$

Note that the only difference between planned and post-hoc comparisons is the numerator df.

Exercises for Sections 14.2 and 14.3

14.11. Refer to Exercise 14.1.

a. Calculate confidence intervals for all pairs of means. Use an experimentwise 99% confidence level.

b. Based on these intervals, can any null hypotheses of equal means be rejected at an experimentwise $\alpha = .01$?

14.12. In Exercise 14.2 you were asked to identify any potential violations of assumptions. Do any violations that you found cast doubt on your answers to Exercise 14.11?

14.13. Refer to Exercise 14.4.

a. A natural comparison is of the difference between the mean score for all depressed (short-term or long-term) subjects and the mean for all nondepressed subjects. Show that

$$\hat{L} = \frac{18\bar{Y}_2 + 12\bar{Y}_3}{30} - \bar{Y}_1$$

is a comparison between depressed and non-depressed subjects.

b. Is the value of this comparison statistically significant at α = .10?

14.14. Refer again to Exercise 14.4.

a. Perform a Tukey-Kramer test for the significance of the difference between the short-term and long-term means. Use α = .05.

b. Compare the same means at the same experimentwise α as in part a, but using the Scheffé method.

c. Do you get the same answer in parts a and b? If not, which answer should be believed?

14.15. An experiment involved samples of size 10 each in each of four experimental conditions. The data were reasonably near normal, and there were no time issues of concern. Summary figures are shown in Table 14.10.

Condition	Mean	Standard Deviation
C1	46.2	15.4
C2	46.9	7.3
E1	55.3	8.1
E2	56.8	16.4

Table 14.10

a. Perform an F test of the hypothesis of equal means. Can this hypothesis be rejected at α = .05?

b. Perform a test of pairwise equality of means, using α = .05 (experimentwise).

c. Do the tests in parts a and b give compatible answers? Why?

d. Define a contrast which is the difference between the average E mean and the average C mean. Is this comparison statistically significant at an experimentwise α = .05?

14.16. Refer to Exercise 14.15.

a. Is there any evidence of violations of assumptions?

b. If so, which parts of Exercise 14.15 are most suspect?

14.17. Refer to the comparison defined in Exercise 14.13. Find the associated sum of squares. Is this SS a large fraction of the sum of squares for groups found in Exercise 14.4?

14.18. Perform a post-hoc test of significance for the comparison defined in Exercise 14.13. Use α = .05.

14.19. Refer to Exercise 14.1.

a. Define two comparisons, the first being the pairwise comparison of the "ordinary" mean to the "special" mean, and the second being the comparison of the control mean to the average of the other two means.
b. Show that these comparisons are orthogonal.
c. Find the SS for each of the two comparisons.
d. Do these SS add up to SS(Between)? Should they?

14.20. Perform post-hoc ($\alpha = .05$) Scheffé tests of significance for the comparisons in Exercise 14.19.

14.21. Refer to Exercise 14.15. Define the comparison between the average E mean and the average C mean.

a. Find the sum of squares for this comparison.
b. Test the statistical significance ($\alpha = .01$) of the result, assuming that the comparison was planned.
c. Test the statistical significance ($\alpha = .01$) of the result, assuming that the comparison was post hoc.

14.22. Refer to Exercise 14.21. Define $\hat{L}_2 = \bar{Y}_{E2} - \bar{Y}_{E1}$ and $\hat{L}_3 = \hat{Y}_{C2} - \hat{Y}_{C1}$.

a. Show that the comparison defined in Exercise 14.21 is orthogonal to each of the comparisons $\hat{L}_2$ and $\hat{L}_3$ and that these two comparisons are orthogonal to each other.
b. Find the SS associated with each of the comparisons and verify that the SS add up to SS(Between).

14.4 Two-Factor Experiments

Up until now, we have been considering completely randomized, single-factor experiments. Now we turn to a slightly more complex type of experiment, involving two factors. Such experiments involve a new concept, the idea of interaction.

A fairly typical two-factor experiment might involve individuals from each of two groups; 10 individuals from each group are (randomly) assigned to each of three treatments. A single response (dependent variable value) is measured for each individual. Treatment is one controlled factor, and group the other.

(a) NO INTERACTION

		Treatment A	B	C	Mean
Group	1	73	79	70	74
	2	67	73	64	68
	Mean	70	76	64	71

(b) INTERACTION

		Treatment A	B	C	Mean
Group	1	63	86	73	74
	2	77	66	61	68
	Mean	70	76	67	71

Table 14.11

Two possible sets of response means are shown in Table 14.11. The means shown in Table 14.11(a) have a consistent pattern. The mean score for B is 6 units higher than the mean for A regardless of group (79 vs. 73 in group 1 and 73 vs. 67 in group 2); similarly the mean score for B is 9 units higher than the mean for C regardless of group. The overall averages for the three treatments (the 70, 76, and 67 shown at the bottom of the three columns) reflect these consistent differences. The consistency works in the other direction, too. Group 1 averages 6 units higher than group 2, both in overall average (74 vs. 68, at the right of the two rows) and within each treatment.

There is no such consistency in Table 14.11(b). In group 1, the B mean is 23 units higher than the A mean (86 vs. 63); but in group 2, the B mean is 11 units *lower* than the A mean. The B mean is 13 units higher than the C mean in group 1 but only 5 units higher in group 2. Table 14.11(b) is an example of means that exhibit **interaction** whereas Table 14.11(a) has no interaction.

interaction

Interaction

Two factors interact if the differences in dependent variable means for levels of one factor change over levels of the other factor.

Example 14.24

In an experiment, children from three different cultural groups (urban middle class, rural middle class, and mountain) were exposed to one of three methods for teaching computer literacy. (Children were assigned to methods completely at random.) After training, each child was measured on a 0–100 scale which measured the child's ability to perform moderately complex tasks using a computer. Suppose that the mean scores in Table 14.12 had been obtained.

		Method		
		1	2	3
Group	U	62.2	56.7	52.5
	R	58.7	50.1	45.1
	M	49.3	55.9	57.9

Table 14.12

Do these means suggest the presence of interaction?

Solution There is a strong indication of interaction. The responses of the M group clearly differ from those of the U and R groups. Note that the U and R scores decrease as one goes from method 1 to 2 to 3, whereas the M scores increase as one goes from method 1 to 2 to 3. Numerically, the difference between means under methods 1 and 2 is 5.5 for U children, but −6.6 for M children. Alternatively, U children are the highest scorers under methods 1 and 2, but M children are highest under method 3. □

profile plot

A good way to see interaction is to draw a **profile plot.** The levels of one factor—for instance, treatments—are plotted along the horizontal axis, and the observed means are plotted on the vertical axis. The means for a given level of the second factor—for instance, groups—are connected by lines. Figure 14.1 contains profile plots for the means of Table 14.11.

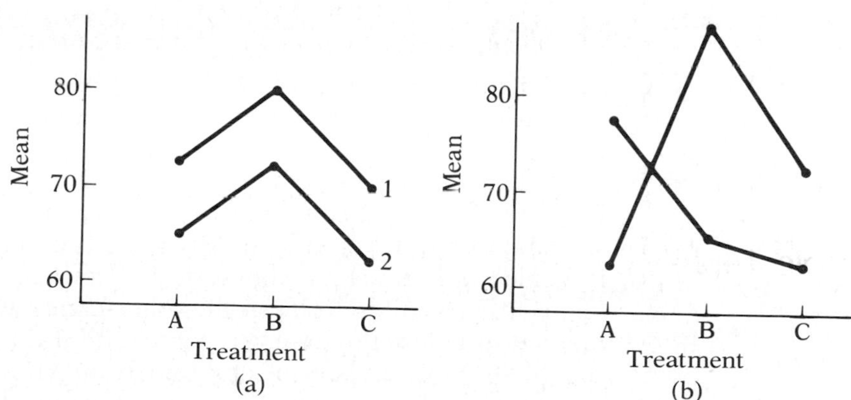

Figure 14.1 Profile plots for Table 14.11: (a) no interaction, (b) interaction

When there is no interaction, the differences between group means are by definition the same for various treatments; therefore the group profiles (the connected line segments) will be parallel, as in Figure 14.1(a). Strong interaction is indicated by extremely nonparallel profiles, as in Figure 14.1(b). Actual experimental data almost never yield perfectly parallel profiles, if only because of the random variation of sample means. If the profiles show crossing patterns and opposite slopes, as in Figure 14.1(b), there is an indication of substantial interaction. Later in this section, an F test for the statistical significance of interaction will be presented.

When the experimental means exhibit substantial interaction, the main effects of the factors may not be scientifically very interesting. For example, in Table 14.11(b), treatment B has a mean 6 units higher than A (76 vs. 70); if higher scores are better, we might conclude that B is a better treatment than A. However, the 6-unit difference is the average of a 23-unit difference for group 1 and an 11-unit difference in the *other* direction for group 2. Certainly treatment B is not *consistently* better than A. Some texts on experimental design state that tests on main effects should never be done if interaction is significant. The F tests presented later in this chapter remain mathematically valid if interaction is present; the question is whether they are scientifically relevant. In the experimental situation, is it relevant to compare overall treatment means averaged (with equal weights) across groups, even when severe interaction is present? There may be some situations in which the comparison is relevant, so it wouldn't be advisable to issue a flat prohibition against such tests. At minimum, though, you should be very careful about interpreting main effects tests when interaction is present.

Example 14.25

Refer to Example 14.24. Does it make sense to you to compare the average of the scores under the three methods (averaged across groups)?

Solution

In Example 14.24 we saw that there was interaction present. If we were choosing a computer literacy teaching method for a mountain school district, it would be mildly stupid to pick method 1 just because it yielded the best results on average; method 3 yielded the best results for mountain children. If, on the other hand, we had to choose a single method to be used in an entire state, *and* if the state were composed of roughly equal numbers of children from urban, rural, and mountain groups, then it would be sensible to consider the average results. □

With one exception,* the assumptions for analysis of variance for two-factor (and for many-factor) experiments are the same as those for one-factor experiments, although the notation becomes more cumbersome. The critical extra assumption is that of *balanced design.* The tests to be presented in this chapter are only valid when equal numbers of observations are made under each combination of factors. It was indicated in Section 14.1 that balanced designs were desirable because they protected against violation of the equal-variance assumption. Now balanced designs are even more desirable, because they allow use of the straightforward methods of this chapter. There are methods for analyzing multi-factor unbalanced data, but they are complicated and somewhat delicate. If you must deal with unbalanced data, either your own or someone else's, get expert help.

As in the single-factor experiment, we begin with a mathematical model. The notation for the two-factor model requires three subscripts; i will refer to an individual observation, j to the level of one factor (such as group), and k to the level of the second factor (such as treatment). For example if $i = 4$, $j = 2$, and $k = 1$, Y_{421} would mean the response (dependent variable) value for observation number 4 in group 2 under treatment 1. The subscript i runs from 1 to $n_{\text{per cell}}$, the number of observations for each combination of factor levels; j runs from 1 to J, and k from 1 to K. If there are 2 groups, 3 treatments, and 10 observations per group-treatment combination, $J = 2$, $K = 3$, and $n_{\text{per cell}} = 10$. In this notation the model is

$$Y_{ijk} = \mu + \alpha_j + \beta_k + \alpha\beta_{jk} + \epsilon_{ijk}$$

Here μ is the theoretical grand mean, the long-run average of large (and equal) numbers of observations under all experimental conditions. The main effect for level j of the first factor is

$$\alpha_j = \mu_j - \mu$$

where μ_j is the long-run (population) mean for level j, averaged with equal weights over all levels of k. Similarly,

$$\beta_k = \mu_k - \mu$$

The notation $\alpha\beta_{jk}$ is fairly standard notation for the interaction term. It does not necessarily imply multiplication in any literal sense. Because the sum of all deviations from a mean is 0 it follows that

*Implicitly, we are assuming that both factors are "fixed factors." See Section 14.6.

$$\sum \alpha_j = 0$$

$$\sum \beta_k = 0$$

$$\sum_j \alpha\beta_{jk} = 0$$

$$\sum_k \alpha\beta_{jk} = 0$$

The probabilistic assumptions are made on the random error terms ϵ_{ijk}. They are the same assumptions as were made in Section 14.1. The expected error is assumed to be 0, the error variances are assumed to be constant (over j and k), the distribution of errors is assumed to be normal, and the errors are assumed to be statistically independent. As in the previous chapter, violation of the independence assumption is most dangerous. The analysis of variance is a form of inference about means; the Central Limit Theorem and two-tailed tests make population normality a less critical assumption. With balanced designs, nonconstant variance is nearly irrelevant unless the discrepancy is huge. Dependent errors, however, can invalidate the ANOVA F tests quite seriously. Dependence is most likely to occur when measurements are taken successively over time. Any carryover from one observation to the next will cause dependence.* Good experimental practice involves great care in avoiding carryover dependence.

Example 14.26

An experiment was conducted to test the susceptibility of high school juniors to peer pressure. Twenty boys and twenty girls were chosen at random. Ten juniors of each sex were assigned to an experimental group, and the other ten of each sex to a control group. Each student was asked to estimate the length of a line projected on a screen, relative to a reference line, which was projected a foot below. In the experimental group, four "stooges" (the experimenter's assistants) announced estimates of the line that were 40 percent too large, then asked the student to make an estimate; in the control group, the estimates of the "stooges" varied slightly around the correct value. Each of the 40 students was tested separately.

*The methods of this section should not be used for repeated-measures designs in which many measurements are made on the same individual. In such designs the individual becomes a level of a controlled but "random effect" factor. See Section 14.7.

(a) Identify the controlled factors in this experiment.

(b) What violations of assumptions might be suspected in this experiment?

Solution

(a) The factors are sex and treatment.

(b) All the assumptions are suspect to some degree. It is possible that the responses in the experimental group might be nonnormal, and specifically bimodal, with susceptible subjects giving answers near the answers of the "stooges" and others giving answers near the correct value. Such a phenomenon would also imply that the variance of responses in the experimental group would be larger than that in the control group. There is even some possibility of dependence, if the "stooges" became systematically more (or less) effective as time went by. Fortunately, the most serious violation, dependence, seems least likely. □

The basic inference procedures for two-factor (and many-factor) analysis of variance are direct extensions of the procedures described in the previous chapter. Once again we will define sums of squares, degrees of freedom, and mean squares. F statistics will be ratios of mean squares and will be compared to appropriate F-table values. Naturally, as the experimental setup grows more complicated, the formulas also grow more complicated. (Very often the actual arithmetic for such analysis is done via a computer program.) The basic inference concepts remain the same.

In a two-factor experiment, SS(Total), the sum of squared deviations from the overall grand mean, can be split into four parts: explained by the main effect of factor 1, explained by the main effect of factor 2, explained by interaction, and unexplained (random error). To define these sums of squares some notation is needed. Let

y_{ijk} = actual observed value

$\bar{y}_{.jk}$ = cell mean for level i of factor 1, level j of factor 2

$$= \sum_i \frac{y_{ijk}}{n_{cell}}$$

$\bar{y}_{.j.}$ = mean for level j of factor 1

$$= \sum_k \frac{\bar{y}_{.jk}}{K}$$

$\bar{y}_{..k}$ = mean for level k of factor 2

$$= \sum_j \frac{\bar{y}_{.jk}}{J}$$

$$\bar{y}_{...} = \text{grand sample mean}$$

$$= \sum_j \frac{\bar{y}_{.j.}}{J} = \sum_k \frac{\bar{y}_{..k}}{K}$$

$$s_{jk}^2 = \text{variance in cell } (j,k)$$

$$= \sum_i \frac{(y_{ijk} - \bar{y}_{.jk})^2}{n_{\text{cell}} - 1}$$

Example 14.27 Suppose that the data of the experiment in Example 14.26 had been as follows:

Experimental Group, Females:	27	24	23	19	20
	27	20	21	25	21
Experimental Group, Males:	21	27	29	28	20
	26	27	20	26	19
Control Group, Females:	18	22	20	22	17
	21	19	19	20	20
Control Group, Males:	17	19	20	21	20
	18	16	18	22	19

The sample means are shown in Table 14.13, with sample standard deviations in parentheses.

		Sex		
		F	M	Average
Treatment	E	22.7 (2.9458)	24.3 (3.8312)	23.5
	C	19.8 (1.6193)	19.0 (1.8257)	19.4
	Average	21.25	21.65	21.45

Table 14.13

(a) Let factor 1 be treatment, and factor 2 be sex. Identify y_{812}.
(b) Identify $\bar{y}_{.1.}$, $\bar{y}_{..2}$, and $\bar{y}_{...}$.

Solution **(a)** y_{812} indicates treatment 1 (E), sex 2 (M), observation 8; this value is 20.

(b) $\bar{y}_{.1.}$ is the mean for the first treatment (E), namely 23.5. $\bar{y}_{..2}$ is the mean for sex 2 (M), namely 21.65. $\bar{y}_{...}$ is the grand mean, 21.45. □

With this notation, we can define the appropriate sums of squares.

Sums of Squares for a Two-Factor Experiment

$$\text{SS(Factor 1)} = n_{\text{per mean}} \sum_{j} (\bar{y}_{.j.} - \bar{y}_{...})^2$$

$$\text{SS(Factor 2)} = n_{\text{per mean}} \sum_{k} (\bar{y}_{..k} - \bar{y}_{...})^2$$

$$\text{SS(Interaction)} = n_{\text{per cell}} \sum_{j,k} (\bar{y}_{.jk} - \bar{y}_{.j.} - \bar{y}_{..k} + \bar{y}_{...})^2$$

$$\text{SS(Error)} = \sum_{i,j,k} (y_{ijk} - \bar{y}_{.jk})^2$$

$$= (n_{\text{per cell}} - 1) \sum_{j,k} s_{jk}^2$$

$$\text{SS(Total)} = \sum_{i,j,k} (y_{ijk} - \bar{y}_{...})^2$$

NOTE: For factor 1, $n_{\text{per mean}}$, the number of observations for each factor 1 mean, is $Kn_{\text{per cell}}$; for factor 2, $n_{\text{per mean}}$ is $Jn_{\text{per cell}}$.

The first three of these sums of squares have a similar pattern. Each is of the form "number of observations per effect times sum of squared effects." Each $(\bar{y}_{.j.} - \bar{y}_{...})$ term is the estimated effect of level j of factor 1; it is calculated from $n_{\text{per mean}} = Kn_{\text{cell}}$ observations. A similar argument holds for SS(Factor 2). For SS(Interaction), the squared term is an estimated interaction effect in cell (j, k), based on n_{cell} observations. The error sum of squares does not follow the same pattern, but rather is based on the sum of within-cell variances, multiplied by $n_{\text{cell}} - 1$.

Example 14.28 Calculate the sums of squares for the data of Example 14.27.

Solution First, note that J and K are both 2, and n_{cell} is 10. So

$$\text{SS(Factor 1)} = 2(10)[(23.5 - 21.45)^2 + (19.4 - 21.45)^2]$$
$$= 168.1$$

$$\text{SS(Factor 2)} = 2(10)[(21.25 - 21.45)^2 + (21.65 - 21.45)^2]$$
$$= 1.6$$

$$\begin{aligned}
\text{SS(Interaction)} &= 10[(22.7 - 23.5 - 21.25 + 21.45)^2 \\
&\quad + (24.3 - 23.5 - 21.65 + 21.45)^2 \\
&\quad + (19.8 - 19.4 - 21.25 + 21.45)^2 \\
&\quad + (19.0 - 19.4 - 21.65 + 21.45)^2] \\
&= 14.40 \\
\text{SS(Error)} &= (10 - 1)[(2.9458)^2 + (3.8312)^2 \\
&\quad + (1.6193)^2 + (1.8257)^2] \\
&= 263.8 \\
\text{SS(Total)} &= (27)^2 + (24)^2 + \cdots + (20)^2 + (19)^2 - (21.45)^2 \\
&= 447.9
\end{aligned}$$

Note that

$$\text{SS(Factor 1)} + \text{SS(Factor 2)} + \text{SS(Interaction)} + \text{SS(Error)} = \text{SS(Total)}$$

Therefore only four of the SS need be computed from the definition and the other one follows. Because the computation of SS(Interaction) is tedious, SS(Interaction) is often found by subtraction. □

Associated with each SS are the appropriate degrees of freedom (df) and mean square (MS). Because SS(Factor 1) is based on J estimated effects, and because these estimated effects automatically add to 0, SS(Factor 1) has $J - 1$ df; similarly, SS(Factor 2) has $K - 1$ df. SS(Interaction) is based on $(J - 1)(K - 1)$ df, by an argument very similar to that of Chapter 13 for the chi-squared test. Finally, the within-cell squared error, SS(Error), has $JK(n_{\text{cell}} - 1)$ df. In each case, the mean square (MS) is, as usual, the SS divided by the associated df.

These mean squares may be used to test hypotheses about the factor main effects and interactions. In each case, the appropriate F statistic is the mean square for the effect, divided by MS(Error).* The usual procedure is first to test the hypothesis of no interaction,

$$H_0\colon (\alpha\beta)_{jk} = 0, \quad \text{for all } j \text{ and } k$$

If this test indicates nonsignificance, and if a profile plot shows approximate parallelism, the differences in means for levels of one factor can be assumed to be reasonably consistent across levels of the other factor. A natural null hypothesis is that the row or column means are all equal, so main effects are 0. For factor 1,

*This statement is only true for fixed-factor analysis. Methods for dealing with random factors are discussed in Section 14.7.

H_0: $\alpha_j = 0$ for all j

and for factor 2,

H_0: $\beta_k = 0$ for all k

If the F test for interaction is significant at small α values or if the profile plot shows severe nonparallelism, some caution is needed in testing main effects, as indicated above. Before one tests main-effects hypotheses in the presence of interaction, one should ask if the null and research hypotheses are scientifically at all interesting. If so, the F tests remain mathematically valid. If not, so-called tests of simple effects may be done instead. See Winer (1971) for details.

The various summary and test statistics are conveniently laid out in an analysis of variance table. As usual the ANOVA table contains SS, df, MS, and F values. The df for the various F tests are the df for the numerator and denominator MS.

Analysis of Variance Table for a Two-Factor Experiment

Source	SS	df	MS	F
Factor 1				
Factor 2				
Interaction				
Error				
Total				

Example 14.29

(a) Complete the ANOVA table for the data of Example 14.27, using the SS found in Example 14.28.

(b) Is there a serious interaction present in the data?

(c) Is there a statistically significant treatment (factor 1) effect?

Solution

(a) Using the SS from Example 14.28, we have Table 14.14.

Source	SS	df	MS	F
Factor 1	168.1	1	168.1	22.94
Factor 2	1.6	1	1.6	0.22
Interaction	14.4	1	14.4	1.97
Error	263.8	36	7.327778	

Table 14.14

(b) The F for interaction is 1.97, which is not significant at α = .10 (but is significant at α = .25, using extensive F tables). The profiles can be plotted and shown to be fairly close to parallel. The interaction effect seems to be negligible. Even if it were not, we might test for main effects anyway, because an equally weighted average over the two sexes seems sensible.

(c) The F for treatments (factor 1) is 22.94. The F-table value for 1 and 36 df and α = .005 is about 9.0, so the computed F statistic has a p-value much below .005. The existence of real treatment effects is quite conclusively established. □

As always, it's a good idea to look at the data to see how reliable the F tests are. By this time, you may not be overwhelmingly astonished to learn that skewness and/or outliers in the data make the F tests suspect, if the overall sample is small. Separate plots of the data in each cell can be examined. If there are many cells, box plots are useful. Alternatively, the **residuals** (differences of actual values less cell means) can be combined and examined. Unfortunately no simple rank test exists to deal with skewness or outlier problems. Sometimes a transformation (such as square root or a logarithm) can be used to deal with extreme skewness. In theory, major differences in cell variances also make F tests suspect. However, for balanced designs, it's well established that such differences by themselves have little effect unless they are huge (such as the largest cell variance being five or more times the smallest). The most critical assumption, statistical independence of observations, is hard to check in the data, unless the residuals can be arranged in some kind of order, such as successively over time. If there is any pattern in these ordered residuals, the independence assumption is suspect.

residuals

Example 14.30

Examine the data of Example 14.27, and the ordered residuals shown below. Is there any indication of violation of assumptions? Is there reason to question the conclusions reached in Example 14.29?

0.0	−1.7	2.2	0.2	−4.3	1.0	2.7	2.2	−0.8	−2.7
1.2	−4.3	1.7	3.0	2.7	−5.3	−1.0	1.7	4.3	−1.7
0.0	1.0	2.0	−0.8	4.7	1.3	−2.7	3.7	0.2	−2.0
−2.8	−3.0	−3.7	−3.3	2.3	0.3	−1.8	4.3	−1.0	0.2

Solution There is some evidence of bimodality in the E data, as suggested in Example 14.26. You may wish to plot the data to see this. The variances in the E data are higher than those in the C data, also as suggested in Example 14.26. In the residuals, there is no evidence of long strings of negative or positive residuals, which would indicate dependence. With a balanced design, a total sample size of 40, and no evidence of dependence, there is no serious reason to doubt the conclusions of Example 14.29, particularly since the F value for the treatment effect was so conclusively large. □

Unfortunately, most journals do not publish the raw data underlying the reported statistical results. This practice makes it very hard for a reader to assess the appropriateness of the procedures that were used. It would be nice to know that the raw data had been provided to the referees of each article, so that *someone* besides the authors could check for appropriateness. Unfortunately, even this policy is not universal.

14.5 Multiple Comparisons in Two-Factor Experiments

The F tests of the previous section are tests of global null hypotheses, such as "the means for levels of factor 1, averaged over levels of factor 2, are all equal." If an F test of such a null hypothesis comes out significant, indicating that some of the means are different, a natural question is "which specific means can be declared different?" Also, if the F test does not yield significance, it's useful to know if confidence intervals for mean differences are wide (suggesting a "not proven" verdict) or narrow (suggesting a "no major effect" verdict). The Tukey and Scheffé multiple-comparison methods described in the previous chapter can be extended to two-factor experiments and used to answer these questions. The modifications required are minor.

Recall that the Tukey method was designed for pairwise comparison of means in balanced designs. Of course, balanced design is assumed in this entire section. The Tukey approach assigns the same plus-or-minus term to all pairwise differences. The formula for this term carries over to the two-factor case with the appropriate interpretation of symbols.

Pairwise Comparison of Factor 1 Means

$100(1-\alpha)\%$ overall confidence level intervals for true differences $\mu_j - \mu_{j'} = \alpha_j - \alpha_{j'}$ are given by

$$\bar{y}_{.j.} - \bar{y}_{.j'.} \pm q(J, \text{df}_{\text{error}}) \sqrt{\frac{\text{MS(Error)}}{n_{\text{per mean}}}}$$

where j and j' are indices for levels of factor 1, J is the number of levels of that factor, $\text{df}_{\text{error}} = JK(n_{\text{cell}} - 1)$ is the error df, and $n_{\text{per mean}} = Kn_{\text{cell}}$ is the number of observations for each factor 1 mean.

Interchanging j with k and J with K yields the formula for factor 2 means.

NOTE: The error rate α is experimentwise, for each set of effects separately. Thus the probability that all the confidence intervals for the factor 1 effect will be correct is $1 - \alpha$, as is the probability that all the factor 2 confidence intervals will be correct.

Example 14.31

Suppose that an experiment had been done to test the susceptibility of college students to political labels. Samples of 24 freshmen and 24 seniors from a university were given a list of statements supposedly made by a U.S. senator, and asked to indicate their degree of agreement (on a 0–4 scale) with each statement. For 8 students from each class, the "senator" was identified as a conservative; for another 8 of each, the label was moderate; and the remaining 8 of each class were told that the senator was a liberal. The statements were the same, regardless of label, and were not obviously right- or left-wing. An overall agreement score was obtained for each student by adding up the agreement scores on each question; possible scores ranged from 0 to 100. The data were not particularly skewed, nor were any outliers observed. The means were as shown in Table 14.15.

MEANS

	C	M	L
F	46.7500	51.2500	68.1250
S	48.2500	50.0000	50.0000

Table 14.15

The ANOVA table in Table 14.16 was calculated.

Source	SS	df	MS	F
Class	426.0208	1	426.0208	11.9018
Label	1144.7916	2	572.3958	15.9911
Interaction	903.2916	2	451.6458	12.6176
Error	1503.3749	42	35.7946	
Total	3977.4791	47		

Table 14.16

(a) Is there a major and statistically significant interaction effect?
(b) Which label means are statistically significant (at $\alpha = .05$) according to the Tukey criterion?

Solution

(a) The F value for interaction is "off the tables," indicating that an interaction has been conclusively established. It is clear from the table of means that the F response to the liberal label was much more approving than was any other response. There is some danger in looking at the average effect of labels, because the F and S responses to the liberal label differed markedly. We proceed with caution.

(b) The label means can be calculated easily by averaging the F and S means for each label. They are 47.5000, 50.6250, and 59.0625, for C, M, and L, respectively. The appropriate table value for 3 treatments and 42 error df is not given. The value for 3 treatments and 40 df, 3.44, will be very close, and slightly conservative. We have $J = 2$ and $n_{\text{per cell}} = 8$. Therefore, the desired value of W is

$$W = 3.44 \sqrt{\frac{35.7946}{16}} = 7.2765$$

The difference between C and M means is less than 7.2765, so this difference could possibly be attributed to random error. The difference between the L mean and the M mean is larger than 7.2765, as is the difference between the L and C means; these differences are statistically significant. Note, again, that such comparisons may be misleading scientifically, given the interaction effect. □

The Scheffé approach to more general comparisons also is a direct generalization of the approach in Section 14.2. This method applies both to main effects comparisons (differences of factor 1 means or of factor 2 means) or to interaction comparisons. Confidence intervals for main effects comparisons of the form $\hat{L} = \Sigma\, c_j \bar{y}_{.j.}$ are calculated just as in that section.

General Comparisons of Factor 1 Means

$100(1-\alpha)\%$ overall confidence level intervals for all possible comparisons of the form

$$\sum c_j \mu_j \text{ (with } \sum c_j = 0)$$

are given by

$$\sum c_j \bar{Y}_{.j.} \pm \sqrt{(J-1)F_\alpha \text{ MS(Error)} \sum \frac{c_j^2}{n_{\text{per mean}}}}$$

where F_α is the F-table value $[J-1$ and $JK(n_{\text{cell}}-1)$ df$]$ cutting of a right-tail area equal to α and $n_{\text{per mean}} = Kn_{\text{per cell}}$ is the number of observations for each factor 1 mean.

A similar formula holds for general comparisons of factor 2 means.

NOTE: The probability that the confidence intervals for all possible factor 1 comparisons are all correct is $1-\alpha$. The same is true for all factor 2 comparisons.

Example 14.32

Is the difference between the L mean and the average of the C and M means of Example 14.31 statistically significant at $\alpha = .01$?

Solution

Note that this is not a pairwise comparison. The Tukey approach does not apply. The F-table value (2 and 42 df) is less than the 2 and 40 df value, 5.18. Conservatively, then,

$$S = \sqrt{(3-1)5.18 \frac{1^2 + 2(-.5)^2}{16}}$$
$$= 5.896$$

The actual difference is

$$59.0625 - \frac{47.5000 + 50.6250}{2} = 10.0000$$

which exceeds the S value. Therefore, the difference is statistically significant at $\alpha = .01$. □

A comparison of the preceding formulas with those of Section 14.2 shows that they are effectively the same. The denominator mean square [MS(Within) or MS(Error)] is divided by the number of observations per mean. The q or F table is entered for the appropriate denominator df. The F-table value is the one used for the corresponding main effect test. The same principle applies to the more complicated experimental designs discussed in the next section.

interaction comparison

The concept of interaction, introduced in Section 14.4, is reflected in the concept of an **interaction comparison.** Recall that the means of Table 14.20b exhibited (a substantial amount of) interaction because the differences in factor 2 means were (substantially) different, depending on which level of factor 1 was chosen. For instance, the difference between treatment B and A means was $86 - 63 = 23$ in group 1 and $66 - 77 = -11$ in group 2. Thus interaction represents a **difference of differences** of cell means. The comparison

difference of differences

$$\hat{L} = (\bar{y}_{12} - \bar{y}_{11}) - (y_{22} - \bar{y}_{21}) = (86 - 63) - (66 - 77)$$
$$= (23) - (-11) = 34$$

indicates this difference of differences, and is an interaction comparison.

Interaction Comparison

A comparison among cell means

$$\hat{L} = \sum_{j,k} c_{jk}\bar{y}_{jk.}$$

is an interaction comparison if

$$\sum_{k} c_{jk} = 0 \quad \text{for all levels } j$$

and

$$\sum_{j} c_{jk} = 0 \quad \text{for all levels } k$$

The definition of interaction comparison is easily illustrated by showing the weights (c_{jk}'s) of a comparison in a two-way table like Table 14.17. The weights in Table 14.17(a) are those of the interaction comparison discussed in the preceding paragraph. The definition of an interaction comparison requires that the weights add to 0 across any row or down any column, as shown. The weights in Table 14.17(b) are those of a more complicated interaction comparison. The weights in Table 14.17(c) do not sum to 0 in each column, and therefore do not represent an interaction comparison. In fact, they represent a comparison of main effects for columns 1 and 2.

(a)

1	−1	0	0
−1	1	0	0
0	0	0	

(b)

.5	.5	−1	0
−.5	−.5	1	0
0	0	0	

(c)

.5	−.5	0	0
.5	−.5	0	0
1	−1	0	

Table 14.17

Example 14.33 In Example 14.31 we noted that the F response to the L label was quite different from the S response, and that this was the key to the apparent interaction. Define the comparison

$$\hat{L} = (\bar{y}_{F,L} - \bar{y}_{S,L}) - .5[(\bar{y}_{F,C} + \bar{y}_{F,M}) - (\bar{y}_{S,C} + \bar{y}_{S,M})]$$

Show that this comparison is an interaction comparison.

Solution The c_{jk} weights of this comparison are as shown in Table 14.18.

	C	M	L
F	−.5	−.5	1
S	.5	.5	−1

Table 14.18

The weights add to 0 across each row and down each column, and therefore define an interaction comparison. □

Interaction comparisons are never pairwise comparisons, so the Tukey approach is irrelevant to them. The Scheffé approach does apply.

Interaction Comparison Confidence Intervals and SS
$100(1-\alpha)\%$ overall confidence level intervals for all possible comparisons of the form

$$L = \sum c_{jk}\,\mu_{jk} \quad \left(\text{with} \sum_j c_{jk} = 0 \text{ and} \sum_k c_{jk} = 0\right)$$

are given by

$$\hat{L} = \sum c_{jk}\bar{y}_{.jk} \pm \sqrt{(J-1)(K-1)F_\alpha \,\text{MS(Error)} \sum \frac{c_{jk}^2}{n_{\text{per cell}}}}$$

where F_α is the F-table value [with $(J-1)(K-1)$ and $JK(n_{\text{cell}}-1)$ df] cutting off a right-tail area of α.

The SS for such a comparison is

$$\text{SS}(\hat{L}) = \frac{(\hat{L})^2}{\sum c_{jk}^2/n_{\text{per cell}}}$$

There is nothing really new in these formulas. In the confidence interval formula, the F-table value is once again the same one used in the F test. The denominator mean square, MS(Error), is divided by the number of observations per mean in the comparison. Since we are comparing cell means, the divisor is $n_{\text{per cell}}$.

Example 14.34

Refer to the comparison defined in Example 14.33. Calculate a 95% confidence interval for the true value of the comparison. Also, find $\text{SS}(\hat{L})$.

Solution

For the comparison,

$$\begin{aligned}\hat{L} &= (68.125 - 50.000) - .5[(46.750 + 51.250) \\ &\quad - (48.250 + 50.000)] \\ &= 18.375\end{aligned}$$

The interaction df is $(3-1)(2-1) = 2$. The required F-table value is about 3.23, MS(Error) $= 35.7946$, $n_{\text{per mean}} = 8$. There are two c_{jk} values equal to .5, two equal to $-.5$, one equal to 1, and one equal to -1. Thus

$$S = \sqrt{\frac{(3-1)(2-1)(3.23)(35.7946)[2(.5)^2 + 2(-.5)^2 + (1)^2 + (-1)^2]}{8}}$$
$$= 9.312$$

The desired 95% confidence interval is

$$18.375 - 9.312 < L < 18.375 + 9.312$$

or

$$9.053 < L < 27.687$$

The SS for the comparison may be calculated from the same information.

$$\text{SS}(\hat{L}) = \frac{(18.375)^2}{[2(.5)^2 + 2(-.5)^2 + (1)^2 + (-1)^2]/8} = 900.375$$

This comparison accounts for virtually all of SS(Interaction), which was 903.2916. □

14.6 Many-Factor Experiments

So far in this chapter we have considered only two-factor experiments. There is no logical or mathematical reason to restrict the number of experimentally manipulated factors to two. Several factors can be controlled, and their effects removed from the random error factor, at the cost of increasing complexity and computation.

Suppose an experiment consists of obtaining a recommended jail sentence for a "criminal." Each person involved in the study is shown the same videotape of a middle-aged, fairly well-dressed man confessing to a burglary, is given information about the past "criminal record" of the man, and is asked to specify a recommended jail sentence in months. The subjects are equally divided by race (identified as black or white), by sex (female or male) and by education (less than high school diploma, high school diploma but no more than one year of college, or two or more years of college). The aim of the experiment is to identify any differences in severity of sentences attributable to the three factors race, sex, and education. Actually, a more common experiment would be to have each subject recommend sentences for several "criminals" where the "criminals" had committed similar crimes but varied in race or perceived socioeconomic status. Such an experiment would be an example of a repeated-measures design, because each subject is measured (recommends a sentence) under several conditions. Repeated-measures designs are studied in the next section.

Example 14.35 An experiment was designed to see whether rats were "sociable" in their eating habits. Rats were deprived of food for either a short, medium, or long time, were placed in cages next to (and in sight of) other rats who were either hungry or satiated, and were rewarded with food after every 1 or 3 bar presses. The response variable of interest was the number of food pellets consumed in a 10-minute period. Each rat was measured under exactly one experimental condition. Identify the factors of this experiment.

Solution The primary factor of interest is the "neighbor" factor, measuring whether or not the neighboring rat was hungry. Other factors were deprivation time and effort (1 or 3 presses). □

The basic concepts of two-factor analysis extend directly to experiments with three or more factors. The same assumptions are made about the random error term (independence, constant variance, and normal distribution) with the same indication of relative importance. The model for a many-factor experiment is necessarily more complicated than for a two-factor experiment. For an experiment involving four factors, the model will include four main effects terms, interactions among every pair of factors, three-factor interactions, and even a four-way interaction. A two-factor interaction has been discussed before. The idea is that the difference in dependent variable means for a factor depends on the level of another factor. A three-factor interaction means that the degree of interaction of two factors changes with the levels of the third factor. It is more than a bit difficult to explain what a four-factor interaction means.

The basic analysis of variance and F tests for many-factor experiments proceed just as in the two-factor case. There are defining and short-cut formulas for sums of squares, degrees of freedom, and mean squares. In this computer era, it seems almost certain that analysis of such experiments will be done by standard computer programs, not by hand. Therefore, the formulas belong in the computer, not in the head. Or, see Winer (1971). The analysis depends critically on the assumption that all factors are "fixed factors" as opposed to the "random factors" discussed in the next section.

In the discussion of F tests for two-factor analysis of variance, the importance of exercising caution in interpreting tests of main effects in the presence of interaction was stressed. If the differential effect of some treatments varies greatly among groups, is it scientifically interesting to ask about the average (equally weighted over groups) treatment effects? The same, if not greater, caution is needed in a many-factor experiment. If there is an interaction between sex and education levels in determining the severity of sentencing recommendation, and if furthermore this pattern of interaction varies by race (a three-way race by sex by education interaction), is it relevant to establish that the average sentence for women (equally weighted over race and education) differs from that for men? At minimum, it would seem that a researcher should highlight such interactions when discussing main effects.

Example 14.36

In the hunger sociability experiment described in Example 14.35, data were collected on six rats for each combination of experimental conditions. No gross skewness or outliers were observed in the data, and the variability in food consumed was reasonably constant over conditions. The data in Table 14.19 were obtained; the ANOVA table in Table 14.20 was calculated. Is there any evidence of serious interaction?

Neighbor	Deprivation	Effort	Mean
H	S	1	35.000
		3	27.333
	M	1	39.167
		3	38.000
	L	1	54.333
		3	48.500
S	S	1	27.667
		3	19.000
	M	1	38.000
		3	35.167
	L	1	48.167
		3	42.833

Table 14.19

Source	SS	df	MS	F
N	561.1249	1	561.1249	21.5426
D	5150.3611	2	2575.1805	98.8658
E	435.1249	1	435.1249	16.7052
N * D	141.0833	2	70.5416	2.7082
N * E	8.6805	1	8.6805	0.3332
D * E	84.0833	2	42.0416	1.6140
N * D * E	9.3610	4	2.3402	0.0898
Error	1562.8333	60	26.0472	
Total	7952.6527	71		

Table 14.20

Solution The only interaction that even approaches statistical significance is the N * D interaction. The SS for this interaction is not overwhelmingly large, and a profile plot of the N * D means shows no severe crossing. Therefore there is no major interaction problem. □

The F tests and multiple-comparison methods discussed in this chapter extend to many-factor experiments. Tukey and Scheffé confidence intervals can be calculated in such experiments. The q or F table is entered for the appropriate number of degrees of freedom (for factor and for error). The standard error term for either the Tukey or the Scheffé involves MS(Error) divided by the number of observations per compared mean. The sum of squares attributable to a comparison can be found as usual, as

$$\frac{(\hat{L})^2}{\sum c^2/n_{\text{per mean}}}$$

The only thing to be careful of is to let the "n" term represent the number of observations per compared mean.

Example 14.37 **(a)** In Example 14.36, which D means differ by a statistically significant amount?

(b) Find SS($\hat{L}$) for $\hat{L} = \bar{y}_{.L.} - .5(\bar{y}_{.S.} + \bar{y}_{.M.})$.

Solution **(a)** First, we must compute the means. There were six observations in each experimental condition, so the deprivation means may be calculated as unweighted averages of the means shown in Example 14.36.

$$\bar{y}_{.S.} = \frac{35.000 + 27.333 + 27.667 + 19.000}{4}$$
$$= 27.25$$

Similarly, $\bar{y}_{.M.} = 37.58$ and $\bar{y}_{.L.} = 48.46$.

To perform all pairwise comparisons of these means, we note that each mean is based on 24 observations. (There were a total of 72 observations divided among the 3 deprivation levels.) In Example 14.36, MS(Error) was shown to be 26.0472, with 60 df. Suppose that we use $\alpha = .01$. The required value from the studentized range table (3 levels of treatment and 60 error df) is 4.28. Thus the plus-or-minus for the Tukey pairwise comparison method is

$$W = 4.28\sqrt{26.0472/24} = 4.46$$

All pairwise differences of the three means are larger than 4.46, so all differences are significant.

(b) First, compute

$$\hat{L} = 48.46 - .5(27.25 + 37.58) = 16.045$$

Again we note that $n_{\text{per mean}} = 24$. The c_j constants defining the comparison are 1, $-.5$, and $-.5$. Therefore,

$$SS(\hat{L}) = \frac{(16.045)^2}{[(1)^2 + (-.5)^2 + (-.5)^2]/24} = 4119.0724$$

This comparison accounts for roughly 80% of SS(Deprivation), found in Example 14.36 to be 5150.3611. □

Exercises for Sections 14.4, 14.5, and 14.6

14.23. Peters and Durding (1979) reported on a study of the speed of subjects' finger tapping. The factors were the subject's handedness and the finger being tapped. Suppose that there were 28 subjects per cell, that the data showed no severe skewness nor outliers, and that the means and standard deviations (of time per tap in thousandths of a second) were as shown in Table 14.21.

MEANS AND (STANDARD DEVIATIONS)

		Finger Tapped	
		Left	Right
Handedness	Left	441.4(54.2)	462.2(63.7)
	Right	471.0(71.4)	441.0(50.2)

Table 14.21

a. Calculate an ANOVA table of SS, df, and MS values.

b. Is there a significant interaction, at $\alpha = .01$?

c. Construct a profile plot. Does there appear to be a large interaction?

d. Is the SS(Interaction) large relative to the SS for main effects?

14.24. Refer to Exercise 14.23.

a. Is the comparison $\hat{L} = .5\bar{y}_{LL} + .5\bar{y}_{RR} - .5\bar{y}_{LR} - .5\bar{y}_{RL}$ an interaction comparison?

b. Find SS($\hat{L}$) for this comparison. How does it compare to SS(Interaction)? Why should this be true?

c. Test the null hypothesis that the population value of this comparison is 0.

14.25. Refer to Exercise 14.23. Instead of defining the finger-tapped factor as left or right, define it as dominant hand or opposite hand. Thus the means and standard deviations would be as shown in Table 14.22.

MEANS AND (STANDARD DEVIATIONS)

		Finger Tapped	
		Dominant	Opposite
Handedness	Left	441.4(54.2)	462.2(63.7)
	Right	441.0(50.2)	471.0(71.4)

Table 14.22

a. Recompute the ANOVA table. Why does SS(Error) not change from Exercise 14.23?

b. Is there a statistically significant ($\alpha = .05$) interaction?

c. Is there a statistically significant ($\alpha = .05$) effect of finger tapped?

14.26. Dey (1978) reported on an experiment in which the factors were anxiety level and number of categories. The response variable was the number of anagrams solved (maximum 12). Suppose that there were 12 subjects per cell. The cell means were as shown in Table 14.23.

		Number of Categories				
		1	2	3	4	Average
Anxiety	L	9.8	8.1	6.3	5.4	7.400
	H	8.3	8.2	7.7	6.9	7.775
	Average	9.05	8.15	7.00	6.65	7.5875

Table 14.23

Assume that SS(Error) = 109.73.

a. Compute an ANOVA table.

b. Draw a profile plot. Does it appear that there is a substantial interaction?

c. Does SS(Interaction) indicate that there is a substantial interaction?

d. In your opinion, is it meaningful to consider the overall average score for each number of categories?

14.27. Refer to Exercise 14.26. Define the comparison

$$\hat{L} = (\bar{y}_{L1} - \bar{y}_{L4}) - (\bar{y}_{H1} - \bar{y}_{H4})$$

a. Is the value of this comparison statistically significant by a Scheffé test? Use $\alpha = .10$.

b. Verify that $\hat{L}$ is an interaction comparison.

c. Find SS($\hat{L}$). Is it a substantial fraction of SS(Interaction)?

14.28. Refer to Exercise 14.26. Define the comparison

$$\hat{L} = .5\bar{y}_{.1} + .5\bar{y}_{.2} - .5\bar{y}_{.3} - .5\bar{y}_{.4}$$

a. Is this comparison an interaction comparison? If not, what is it?

b. Find SS($\hat{L}$).

c. Test the significance of this comparison, using $\alpha = .01$.

14.29. An experiment was undertaken involving methods for teaching calculus to first-year college students. Students were divided by prospective major (physical science, social science, or business). Also, two teaching methods (abstract first or concrete first) were used. Randomly, 10 students from each prospective major were assigned to each teaching method. The total examination score of each student was recorded. The data were as shown in Table 14.24.

		Teaching Method	
		Abstract First	Concrete First
Major	PS	261 298 301 310 321 335 342 356 367 383	205 247 255 267 291 300 310 316 326 334
	SS	198 242 255 261 264 279 291 299 311 325	243 266 287 299 316 322 334 345 357 368
	B	149 212 240 257 265 270 279 290 304 318	250 269 272 283 296 299 317 320 335 347

Table 14.24

a. Compute means and standard deviations in each cell.

b. Compute an ANOVA table. If a computer program is available for two-factor ANOVA, this is the place for it.

c. Does it appear that there are any violations of assumptions? How serious are the violations?

14.30. Refer to Exercise 14.29.

a. Does it appear that there is a substantial interaction?
b. Is the interaction effect significant?

14.31. Refer to Exercise 14.29.

a. Test the null hypothesis that there is no effect of teaching method, on average. Place bounds on the p-value.
b. In view of the results of Exercise 14.30, how meaningful is the test performed in part a?

14.32. Morelli, Andrews, and Morelli (1982) performed a two-factor experiment with 10 observations per cell. The factors were relevance and neuroticism, and the response was a score on an anxiety scale. The means were as shown in Table 14.25.

	Relevance		
	L	M	H
Neurotics	7.68	9.43	10.81
Stables	4.99	5.77	7.68

Table 14.25

a. Draw a profile plot.
b. Does it appear that there is a major interaction?

14.33. Refer to Exercise 14.32. Assume that SS(Error) = 226.8.

a. Calculate an ANOVA table.
b. Does SS(Interaction) indicate that there is a large interaction? Is the interaction statistically significant at the usual α levels?
c. Are there significant main effects of relevance and of neuroticism? Place bounds on the p-values.

14.34. Refer to Exercises 14.32 and 14.33.

a. Which pairs of relevance means are significantly different using $\alpha = .05$?
b. Define $\hat{L} = \bar{y}_H - \bar{y}_L$. Find the associated sum of squares for this comparison. What fraction of SS(Relevance) is accounted for by this comparison?

14.35. Refer to Exercises 14.32 and 14.33.

a. Define $\hat{L} = (\bar{y}_{NM} - \bar{y}_{SM}) - (.5\bar{y}_{NL} + .5\bar{y}_{NH} - .5\bar{y}_{SL} - .5\bar{y}_{SH})$. Show that this is a valid interaction comparison.

b. Find SS($\hat{L}$) for this comparison.

c. What is the result of a Scheffé post-hoc test of this comparison, using $\alpha = .05$?

14.36. Sarason (1981) reported an experiment involving subjects with three different levels (L, M, H) of test anxiety, with two levels (stressful or control) of instruction, and two levels (social support or control) of test-taking condition. Assume that 8 subjects were tested in each combination of anxiety, instruction, and condition. (All subjects were tested only once.) The mean test scores were as shown in Table 14.26.

Anxiety	Instruction	Condition	Mean
L	S	SS	9.3
		C	6.7
	C	SS	10.9
		C	9.9
M	S	SS	7.9
		C	5.7
	C	SS	9.9
		C	7.7
H	S	SS	6.2
		C	3.8
	C	SS	8.0
		C	6.4

Table 14.26

The ANOVA table in Table 14.27 was calculated.

Source	SS	df	MS
Anxiety	154.24	2	77.12
Instruction	116.16	1	116.16
Condition	96.00	1	96.00
A * I	0.64	2	0.32
A * C	0.64	2	0.32
I * C	3.84	1	3.84
A * I * C	2.56	2	1.28
Error	102.48	84	1.25

Table 14.27

a. Does the ANOVA table indicate that there are any serious interactions?

b. Are any interactions significant at the usual α levels?

c. Which main effects (anxiety, instruction, or condition) are significant? In each case, state a p-value.

14.37. Refer to Exercise 14.36.

a. Compute the mean values for all anxiety levels. Which pairs of means are significantly different (at $\alpha = .01$)?

b. For $\hat{L}_1 = \bar{y}_{L..} - \bar{y}_{H..}$, find $SS(\hat{L}_1)$.

14.38. Refer to Exercise 14.37. Define $\hat{L}_2 = \bar{y}_{M..} - .5(\bar{y}_{L..} + \bar{y}_{H..})$.

a. Show that $\hat{L}_1$ and $\hat{L}_2$ are orthogonal.

b. Show that the sums of squares for the two comparisons add to SS(Anxiety).

14.7 Random Factors, Nested Factors, and Other Experimental Designs

This chapter has introduced some of the critical ideas in the design and analysis of experiments. The whole field of experimental design is a huge one, and we can't cover all the modifications. In this section we'll discuss a few more of the essential ideas, more in the spirit of understanding concepts than technical analysis. The two fundamental ideas to be introduced are the concepts of random factor and nested factor.

repeated-measures experiment

A very common type of social science experiment is a **repeated-measures experiment.** A random sample of subjects is selcted, and a dependent variable value for each subject is measured (possibly several times) under each of several conditions. For example, a researcher investigating the effects of three kinds of distraction (A, B, or C) on learning of nonsense syllables might measure the number (out of 25) of syllables learned by a subject under each of the distractions. To get an idea of the random variation involved, each subject might be measured twice under each distraction. The data could be organized as in Table 14.28.

		Subject			
		1	2	⋯	10
	A	16	14		23
		20	12		17
Distraction	B	12	7	⋯	18
		9	8		15
	C	7	3		9
		10	5		13

Table 14.28

This is exactly the way the data are arranged in a two-factor experiment. Here the factors are distraction and subject. The point of a repeated-measures design is to control for variability from subject to subject, as in the paired-sample experiments of Chapter 10. In fact, such experiments are the simplest repeated-measures designs. However, there is a fundamental difference between the distraction factor and the subject factor. The three distraction conditions are the entire relevant population of distraction conditions, but the subjects in the experiment are only a random sample of the possible subjects. That is, inferences about learning are to be made only for the three distraction conditions used, not some other distractions D, E, F, ...; in contrast, inferences about learning are to be made for a whole population of (potential) subjects, not merely the subjects who happen to be involved in the particular experiment. Distraction is an example of a **fixed factor**; subjects are an example of a **random factor.**

fixed factor
random factor

Fixed and Random Factors

The levels of a fixed factor in an experiment are the only ones about which inferences are to be made. The levels of a random factor are a random sample from a larger population of possible levels.

A good test for whether a factor is fixed or random is to consider a replication (repetition) of the experiment. If a replication would use exactly the same levels or values for a factor, the factor is fixed. If a replication would use other values, particularly randomly chosen ones, the factor is random.

Example 14.38 Consider an experiment in which subjects were randomly assigned to one of three helplessness levels (low, medium, or high). Further, the subjects were randomly assigned to either a treatment or a control group. Each subject completed a battery of tests. There were 8 tests in the battery, so there were $8 \cdot 7 \cdots 1 = 40{,}320$ possible sequences of tests in the battery. Six sequences were selected arbitrarily, and each subject was given the test in one (and only one) of the sequences. Identify the fixed and random factors in the experiment.

Solution A factor is an independent variable that is systematically varied in an experiment. There are three factors: helplessness, group (treatment or control), and sequence. Subject is not a factor here, because each subject was measured only once, under specified conditions, rather than being measured under multiple conditions. Presumably, the group factor is fixed, as the treatment and control conditions are the only conditions of interest. The helplessness levels probably should also be treated as fixed. Someone trying to repeat the experiment should use the same helplessness conditions. Sequence should probably be treated as a random factor, although it would be better to select the sequences randomly, rather than arbitrarily. □

Experiments may involve only fixed factors (in what are called Model I experiments), only random factors (Model II), or some of each (mixed-model or Model III). Repeated-measures designs typically involve at least one random factor (such as subjects) and at least one fixed factor (such as treatment).

The inclusion of random factors in an experiment results in additional sources of random variation. In the experiments previously discussed in this chapter, only the error term was random. When random factors are involved, some main effects and interactions are also random. This additional randomness necessitates some changes in inference methods. In particular, the choice of a denominator (error) MS is critically affected by what factors are regarded as random. See Winer (1971) or Kirk (1984) for a discussion. You should be aware that statistical inference for repeated-measures designs is a bit different from the inference methods discussed in this chapter. In particular, the expected values (long-run averages) of these mean squares will reflect additional variation resulting from the additional randomness associated with random factors. The change in the probability distributions of various mean squares will, in turn, lead to changes in the appropriate F tests of various hypotheses. Once the appropriate denominator MS is chosen, the

F test proceeds in a natural way. Also, Tukey and Scheffé procedures are done as before, with the appropriate denominator MS replacing MS(Error) and with the appropriate change in denominator df.

Example 14.39 Refer to Example 14.38. Should the experiment be analyzed as a three-factor experiment as outlined in the previous section?

Solution No. The inclusion of the random sequence factor necessitates a different analysis. See texts on experimental design for a better analysis. □

nested factor

Another idea that is critical in the design and analysis of complex experiments is the idea of a **nested factor.** To illustrate, suppose that we are testing several different formats for third-grade readers, using—reasonably—third-grade children. We choose 4 boys and 4 girls. Each child reads material in five different formats (labeled A through E, say). A reading comprehension score is obtained after each reading. There are three experimental factors: format, sex, and subject. The data might be arranged as in Table 14.29.

		Sex							
		F				M			
		Subject							
		1	2	3	4	1	2	3	4
Format	A								
	B								
	C								
	D								
	E								

Table 14.29

The "subject" factor must be treated differently than we have treated factors so far in this chapter. It makes no sense to consider the average score for "subject 3," because girl 3 is a different person from boy 3. The subject factor is an example of a nested factor.

Nested and Crossed Factors

Experimental factor B is nested within factor A if any particular level of factor B occurs with only *one* level of factor A. Factor B is crossed with factor A if any level of factor B occurs with *all* levels of factor A.

Each subject is necessarily of one and only one sex (barring some experiments that will never get past the human subjects committee). Thus subject is nested within sex. Each subject is measured with each format, so subject and format are crossed factors. Finally, each sex is measured under each format, so sex and format are crossed factors. A good way to decide what factors are nested in other factors is to consider how one would record the data (as in Table 14.29). There is no reason why the children should be numbered 1, 2, 3, 4, 1, 2, 3, 4. They could equally well be numbered 1–8. Each child is associated with one sex, indicating that child is nested in sex. Note that in the table the subject (child) factor is naturally listed underneath the sex factor.

Example 14.40

In a study, four schools were randomly chosen from the schools in a large urban school system. Within each school, three classrooms were chosen, also at random. The children in each classroom were randomly divided into three groups of 10 students each. Group A learned geometry using a traditional textbook method, group B using a mix of traditional and computer-assisted methods, and group C using purely computer-assisted methods. The dependent variable of interest was each student's score on a standardized test of knowledge of geometry. Identify the fixed and random, crossed and nested factors.

Solution

Three factors are systematically varied in this study, and therefore are factors: school, classroom, and method. Both schools and classrooms are randomly chosen from a much larger potential set of schools and classrooms, so both should be regarded as random factors. The three methods are treatment factors, and should be regarded as fixed. Someone trying to repeat the experiment would use the same methods, but different schools and classrooms. Each school and each classroom used all three methods, so school and method, and classroom and method, should be regarded as crossed. However, each classroom occurs in only one school. Thus, classroom should be regarded as nested in school. □

Alternative language is often used in describing these more complicated experiments, particularly when repeated measurements are taken on specific subjects (so that "subject" becomes an experimental factor). Factors that are crossed with the subject factor are called **within-subjects factors,** because such factors are varied within the measurements taken on a particular subject. If the subject factor is nested within another factor (such as sex in the illustration), the other factor is called a **between-subjects factor.** A widely used set of computer programs (BMDP) calls within-subjects factors "trial factors," and between-subjects factors "grouping factors."

within-subjects factors

between-subjects factor

When there are nested factors, the SS arithmetic changes, as does the choice of error MS. Computer programs are available to handle the arithmetic. See Winer (1971) or Kirk (1984) for details about the choice of denominator MS. Once the appropriate ANOVA table (with SS, df, and MS) has been computed, and the denominator MS chosen, both F tests and Tukey and Scheffé comparisons proceed as usual.

Example 14.41

Refer to Example 14.40. Suppose that the data from the study had been analyzed by a computer package, and that the mean scores for the three methods were reported as A: 56.72, B: 66.40, and C: 58.04. Further, the "error" mean square appropriate to this design was reported as 94.66, with 16 df. Indicate how to perform a Tukey test at $\alpha = .05$ of the significance of the differences among means.

Solution

Each mean results from

(4 schools)(3 classrooms)(10 students per classroom)

or $n_{\text{per mean}} = 120$. According to the computer output, the denominator df is 16. (Exactly where 16 df comes from is not relevant here; we accept the computer's formula, whatever it is.) There are three treatments—that is, teaching methods. The required studentized range table value is $q_{.05}(3, 16) = 3.65$. Thus

$$W = 3.65\sqrt{\frac{94.66}{120}} = 3.24$$

The B − A and B − C differences are much larger than W, hence are statistically detectable according to the Tukey method. The A − C difference is smaller than W in magnitude; it is within reasonable random variation. □

Exercises for Section 14.7

14.39. A study of children's cooperation in problem-solving is to be undertaken. Eight elementary schools are chosen to participate (out of the several dozen elementary schools in a particular district). Eighteen fourth-grade children in each school are assigned to one of six teams. Each team tries to solve a geometrical puzzle, after hearing one of three sets of instructions. Set A emphasizes cooperation heavily, set B emphasizes individual effort, and set C merely states the task, with no comment about cooperation. Two teams from each school are given each set of instructions. The response variable of interest is the number of instances of cooperative behavior shown by each team in solving the problem.

a. What are the experimental factors in this study? Should "child" be regarded as a factor? Should "team" be a factor?

b. For each factor you specified in part a, should the factor be regarded as fixed or random?

c. In part a, which factors, if any, are nested within other factors?

14.40. Refer to Exercise 14.39. Would it be appropriate to analyze the data from this experiment as a factorial experiment, with the usual MS(Error) as the denominator of all F tests?

14.41. Berry (1985) reports on an experiment involving the time required for rats to escape from a maze while underwater. Four rats from each of 10 strains (four purebred strains and the six possible first-generation crosses of the purebreds) learned to escape from each of four mazes while on dry land (more accurately, dry lab table). Then each rat was placed in each maze and put under water; the time required to escape was recorded. The sequence of mazes was random and differed for various rats; there was no interest in sequence effects.

a. What are the experimental factors in this study?

b. For each factor you specified in part a, should the factor be regarded as fixed or random? (There is room for argument on this question.)

c. Which factors identified in part a are between-subjects factors, with the subjects being the rats? Which factors are within-subjects factors?

14.42. Assume that data from the experiment in Exercise 14.41 have been analyzed by a special set of computer programs for repeated-measures designs (OVAANOVA, perhaps), and that the results in Table 14.30 have been obtained.

Source	SS	df	MS
Strain	2086.381	9	231.82
Error1	306.012	30	10.20
Maze	216.719	3	72.24
Maze * Strain	112.500	27	4.17
Error2	417.032	90	4.63

Table 14.30

You may assume that the indicated df and error MS terms are the correct ones.

a. Is there a statistically detectable ($\alpha = .10$) interaction effect?

b. Are the main effects of strain statistically significant? What can be said about the p-value?

14.43. Refer to Exercises 14.41 and 14.42. The following means were obtained for each strain, based on $n_{\text{per mean}} = 16$.

Strain	P1	P2	P3	P4	C12	C13	C14	C23	C24	C34
Mean	24.50	23.25	26.75	25.50	18.88	19.00	17.12	16.25	18.75	18.00

Which pairs of means are significantly different? Use an experimentwise error rate of .05.

14.44. Refer to Exercises 14.42 and 14.43. A planned comparison for the experiment is the difference between the mean time of all the purebred rats and the mean time of the crossbred rats.

a. Define a comparison $\hat{L}$ that achieves that comparison.

b. Find the sum of squares for that comparison.

c. Test the significance of that comparison at $\alpha = .01$.

14.45. Burnham (1966) conducted a study in which some rats were operated on to produce a brain lesion. Other rats underwent a sham operation that left indistinguishable scars but did no damage to actual brain tissue. Suppose that 8 rats were sent to each of 5 experimenters, with the experimenters chosen more or less randomly. In each shipment, there were 2 lesioned rats labeled as lesioned, 2 nonlesioned rats labeled as lesioned, 2 lesioned rats labeled as nonlesioned, and 2 nonlesioned rats labeled as nonlesioned. The experimenters reported back the number of trials required by each rat to learn a T-maze discrimination problem.

a. Identify the factors in this study. Should rat be regarded as a factor?

b. Which factors should be regarded as fixed, which as random?

c. Are any factors nested in others?

14.46. Refer to Exercise 14.45. Suppose that the data were analyzed by a computer program for mixed-model ANOVA with the results shown in Table 14.31.

Source	SS	df	MS
Lesion	102.40	1	102.40
Error1	28.44	4	7.11
Label	209.71	1	209.71
Error2	50.08	4	12.52
Lesion * Label	6.22	1	6.22
Error3	20.64	4	5.16

Table 14.31

a. Is the interaction significant at the usual α levels?
b. Which main effects are statistically detectable at $\alpha = .05$?

14.47. Refer to Exercise 14.46. Which is the more important predictor of the dependent variable, the actual lesion or the label?

14.8 The Power of F Tests

Experimental design is one of the most important areas of statistical theory, and is most relevant to psychologists and other social scientists who want to plan scientific experiments. In particular, it is important for experimenters to understand how to plan an experiment so as to have a high probability of finding a real effect. For example, suppose that an educational researcher wants to test the effect of computerized programmed learning methods on the acquisition of spelling skills by third-graders. It's easy to imagine several experiments. One might have one experimental and one control group, each consisting of 10 children; or two experimental groups (differing slightly on the details of the experimental treatments) and two control groups (also differing slightly), each consisting of 5 children; or two experimental and two control groups, each consisting of 10 children. Obvious question: Which experiment is best?

Answering that question is a large order, and statistical principles are only part (but a major part) of the answer. From the statistical point of view, the best experiment is the one that has the highest probability of finding a significant effect, assuming that in fact there are real differences in mean response for various levels of the factor. That is, assume that the research hypothesis is true, so the true population means at various levels of the experimental factor are unequal; then we would want a high probability of rejecting the null hypothesis—we would want a high *power*. In this section we will consider those aspects of an experiment that can be varied to give high power, and therefore a precise experiment.

In previous sections we argued that the F statistic should have an F distribution when the null hypothesis is true. When we consider power, we want to make the opposite assumption—that the research hypothesis is true. It has been proved [as, for instance, in Scheffé (1959)] that when the research hypothesis is true, the F statistic has a *noncentral* F distribution. Consider the simplest case, the one-factor experiment. The idea of noncentrality is that the terms $\bar{y}_{.j} - \bar{y}_{..}$ which make up SS(Between) are estimates of the true effects α_j; when in fact the research hypothesis is true, and $\alpha_j \neq 0$, these terms will tend to be large (in either the positive or the negative direction), which in turn will tend to make the F statistic large and statistically significant. As usual, the *noncentrality parameter* measures the degree to which the research hypothesis is true—that is, the degree to which the true response means differ over levels of the factor. The power of an F test depends on the value of the noncentrality parameter.

The noncentrality parameter is closely related to the idea of expected mean squares, denoted E(MS). Recall that the expected value of any statistic is the long-run mean of the statistic, averaged over a very large number of samples taken under exactly the same conditions. It can be shown by algebra that

$$E(\text{MS(Between)}) = \sigma_\epsilon^2 + \sum n_j \frac{\alpha_j^2}{J-1}$$

and

$$E(\text{MS(Within)}) = \sigma_\epsilon^2$$

When the null hypothesis H_0: $\alpha_j = 0$, for all j, is true, $E(\text{MS(Between)}) = E(\text{MS(Within)})$. But when the research hypothesis is true, $E(\text{MS(Between)}) > E(\text{MS(Within)})$. The amount of the discrepancy between the two E(MS) values is measured by the noncentrality parameter. Specifically, the noncentrality parameter for a one-factor ANOVA is calculated as

$$\delta^2 = \frac{\sum n_j \alpha_j^2/(J-1)}{\sigma_\epsilon^2}$$

Example 14.42 Calculate the noncentrality parameter in the following experiment: The error standard deviation $\sigma_\epsilon = 10.0$, the sample sizes in each sample are 20, there are four populations being sampled, and the means are, respectively, 50, 55, 55, and 60.

Solution To calculate α_j, we first need the grand mean μ. For a balanced design, we can calculate the unweighted average of the (sub-)population means

$$\mu = \frac{50 + 55 + 55 + 60}{4} = 55$$

Thus the main effects α_j are, respectively, -5, 0, 0, and 5. All $n_j = 20$. Thus

$$\delta^2 = \frac{20\,[(-5)^2 + 0^2 + 0^2 + 5^2]/(4-1)}{(10.0)^2}$$
$$= 3.333$$

□

For a more general (but balanced) experimental design, the noncentrality parameter will be of the form

$$\delta^2 = n_{\text{per effect}} \frac{\sum (\text{true effect})^2/\text{df}_{\text{num}}}{\sigma^2_{\text{denom}}}$$

where $n_{\text{per effect}}$ is the sample size for each effect in the numerator MS [whether it is MS(row), MS(column), or MS(interaction)], df_{num} is the numerator df of the F test, and σ^2_{denom} is the denominator variance—which will be the error variance in fixed-factor studies but something more complicated when there are random factors.

Example 14.43 Suppose that a two-fixed-factor experiment is being planned. For planning purposes, it is assumed that the random error standard deviation will be about 8.0, that $n_{\text{per cell}}$ will be 5, that there will be four levels of factor 1 and three levels of factor 2. Further, hypothesize that the true, population effects of the three levels of factor 2 will be -6, -2, and 8. Calculate the noncentrality parameter for the factor 2 F test.

Solution First, σ_ϵ^2 is assumed to be $(8.0)^2 = 64.0$. Next, note that $n_{\text{per effect}}$ for the factor 2 test is

$$(4 \text{ levels of factor } 1)(5 \text{ observations per cell}) = 20$$

Thus

$$\delta^2 = 20\,\frac{[(-6)^2 + (-2)^2 + (+8)^2]/(3-1)}{64.0}$$
$$= 16.25$$

□

The power of an F test depends crucially on α, the numerator and denominator df, and the noncentrality parameter δ^2. Some basic observations about the power of an F test can be made, either from first principles or from more theoretical considerations outlined in Scheffé (1959).

1. The power of an F test, like the power of any test, increases as the significance level α increases.

2. The power of an F test increases as the noncentrality parameter δ increases. A larger noncentrality parameter means larger true effects α_j (or β_k or $\alpha\beta_{jk}$). If there are large true effects, they are relatively easily detected and the power is large.

3. The power of an F test increases as the denominator df increases. This is a fundamental principle of experimental design, which is conceptually easy. In a one-factor experiment, the denominator df is $n_{\text{overall}} - J$; increasing the denominator df is equivalent to increasing the overall sample size n_{overall}. Reasonably enough, increasing the sample size increases the precision (power) of the test. The principle applies even in less obvious situations. Statistical theory always indicates that increasing the denominator df is desirable.

4. The power of an F test increases as the error variance σ_ϵ^2 decreases. Naturally, as the random error component of an experiment gets smaller, the precision (power) of the test gets better. As far back as the discussion of the paired-sample t test, we have seen the desirability of decreasing the error variance.

5. The effect of increasing numerator df on the power of an F test is subtle. Numerator df are increased by adding new levels of the experimental factor (new groups or new treatments, for instance). If the new levels have no effect (e.g., $\alpha_j = 0$), then $\Sigma\, \alpha_j^2$ is not changed, while J increases. Therefore the noncentrality parameter δ^2 decreases and the power decreases. If the number of levels of the factor (number of groups or treatments) is increased but n_{overall} stays constant, the power decreases. If the number of levels of the factor increases and $n_{\text{per mean}}$ stays constant (which implies an increase in n_{overall}), the power increases somewhat (but not as much as it would if $n_{\text{per mean}}$ increased and the number of levels of the factor stayed constant).

Example 14.44 What will be the effect of each of the following changes in experimental design on the probability of finding a statistically detectable (significant) difference?

(a) Decreasing α from .05 to .01.

(b) Decreasing σ_ϵ from 10.0 to 8.0.

(c) Changing from $n_{\text{per mean}} = 8$ with 3 means, assumed equal to 50, 55, and 60, to $n_{\text{per mean}} = 6$ with 4 means, assumed equal to 50, 55, 55, and 60.

Solution **(a)** Decreasing α always decreases the power, or probability of finding a detectable difference.

(b) Decreasing the error variability always increases the power.

(c) n_{overall} stays constant at 24 while a new level of the factor is introduced with $\alpha_j = 0$. The noncentrality parameter decreases from

$$\delta^2 = 8\frac{[(-5)^2 + (0)^2 + (5)^2]/(3-1)}{\sigma_\epsilon^2} = \frac{200}{\sigma_\epsilon^2}$$

to

$$\delta^2 = 6\frac{[(-5)^2 + (0)^2 + (0)^2 + (5)^2]/(4-1)}{\sigma_\epsilon^2} = \frac{100}{\sigma_\epsilon^2}$$

Decreasing δ^2 decreases the power. □

Given assumptions about the sample sizes (and therefore the df), and the noncentrality parameter δ^2, the power of an F test can, in principle, be calculated. Charts of power can be found in Scheffé (1959) or Cohen (1977). A very rough approximation to power, which may be good enough for most purposes, can be had by extending the Patnaik (1949) z approximation to the noncentral F situation. The crude approximation is

$$\text{power} = P\left(z > \frac{k_\alpha - Q}{1 + k_\alpha^2/2(\text{df})}\right)$$

where

$$k_\alpha = \sqrt{2(\mathrm{df}_1)F_\alpha \frac{\mathrm{df}_1 + \delta^2}{\mathrm{df}_1 + 2\delta^2}}$$

and

$$Q = \sqrt{2\frac{(\mathrm{df}_1 + \delta^2)^2}{\mathrm{df}_1 + 2\delta^2} - 1}$$

Example 14.45 A Monte Carlo study was run, based on 1000 samples, each of size 8, from three normally distributed populations. All three populations had standard deviations equal to 10.0. The population means were, respectively, 50, 55, and 60. Of the 1000 computed F tests, 354 were significant at $\alpha = .05$ and 154 were significant at $\alpha = .01$.

(a) Compute the noncentrality parameter.

(b) Compute the extended Patnaik approximation to power in each case.

(c) How close is the Monte Carlo probability to the extended Patnaik approximation?

Solution **(a)** We have $n_{\text{per mean}} = 8$, and $\sigma_\epsilon^2 = (10.0)^2 = 100$. The grand mean is $(50 + 55 + 60)/3 = 55$, so the $\alpha_j = 50 - 55 = -5$, $55 - 55 = 0$, and $60 - 55 = 5$. Thus

$$\delta^2 = 8\frac{(-5)^2 + 0^2 + 5^2}{100} = 4.0$$

(b)

$$Q = \sqrt{2\frac{(2 + 4.0)^2}{2 + 2(4.0)} - 1} = 2.4900$$

For $\alpha = .05$, $F_{.05,2,21} = 3.47$, so

$$k_\alpha = \sqrt{2(2)3.47\frac{2 + 4.0}{2 + 2(4.0)}} = 2.8858$$

The approximate power is

$$P\left(z > \frac{2.8858 - 2.4900}{1 + 2.8858^2/2(21)}\right) = P(z > 0.33) = .5 - .1293 = .3707$$

For $\alpha = .01$, $F_{.01,2,21} = 5.78$, so

$$k_\alpha = \sqrt{2(2)\,5.78\,\frac{2 + 4.0}{2 + 2(4.0)}} = 3.7245$$

The approximate power is

$$P\left(z > \frac{3.7245 - 2.4900}{1 + 3.7245^2/2(21)}\right) = P(z > 0.93) = .5 - .3485 = .1515$$

(c) The power in the Monte Carlo study is $354/1000 = .354$ for $\alpha = .05$, and $154/1000 = .154$ for $\alpha = .01$. In both cases, the Monte Carlo probability is very close to the extended Patnaik approximation, probably closer than the approximation deserves. □

If, in the planning process, issues of power become critical, the best strategy is to compute an approximate power, recognizing that it is a rough and ready approximation. A given experiment can be evaluated based on the number of levels of the factor (groups or treatments), the denominator df, and the anticipated true effects (α_j's), relative to the error variance. If the calculated power is not adequate, increase the number of observations per level of the factor, find ways to decrease the error variance, or eliminate marginally interesting levels of the factor.

Example 14.46 Refer to Example 14.45. If the sample sizes for each population are increased to 20, will the power (say at $\alpha = .05$) be high?

Solution We may use the extended Patnaik approximation to power. We have

$$\delta^2 = \frac{20[(-5)^2 + 0^2 + 5^2]}{100} = 10.0$$

Thus

$$Q = \sqrt{2\,\frac{(2 + 10.0)^2}{2 + 2(10.0)} - 1} = 3.4772$$

The F-table value now is based on 2 numerator df and 57 denominator df; for $\alpha = .05$, the value is about 3.15. So

$$k_\alpha = \sqrt{2(2)(3.15)\,\frac{2 + 10.0}{2 + 2(10.0)}} = 2.6216$$

Therefore, approximately,

$$\text{power} = P\left(z > \frac{2.6216 - 3.4772}{1 + 2.6216^2/2(57)}\right) = P(z > -0.81)$$

which is only slightly less than .8. The power has increased; in Example 14.45, it was approximately .37. But the power still could not be called large. □

Exercises for Section 14.8

14.48. Suppose that a one-factor analysis of variance procedure is being planned. The design is to be balanced, and the factor is to have four levels. Suppose that samples of size 25 are taken at each level of the factor, and the population means for the four levels are 100, 92, 95, and 97. The error standard deviation is assumed to be 10.

a. Find the noncentrality parameter.
b. If the F test is to be done using $\alpha = .01$, what is the probability that the result will be statistically significant?

14.49. Refer to Exercise 14.48. What should happen to the probability computed in part b if

a. $\alpha = .05$, not .01
b. $\alpha = .01$, but the error standard deviation is 12 rather than 10
c. $\alpha = .01$, the error standard deviation is 10, but the means are 102, 90, 95, and 97

14.50. An experiment was planned in which samples of size 12 were to be taken from each of five groups. The true, population means for each group were assumed to be as follows:

A: At the grand mean
B: .5 standard deviation below the grand mean
C: .3 standard deviation below the grand mean
D: .3 standard deviation above the grand mean
E: .5 standard deviation above the grand mean

(The standard deviation referred to is the within-group, error standard deviation.)

a. Find the noncentrality parameter.

b. If the F test is run at $\alpha = .05$, find the probability that the null hypothesis of equal means will be rejected.

14.51. Refer to Exercise 14.50. How should the probability found in part b change if the population means for the groups are assumed to be the following?

A: At the grand mean
B: .8 standard deviation below the grand mean
C: .4 standard deviation below the grand mean
D: .4 standard deviation above the grand mean
E: .8 standard deviation above the grand mean

14.52. Refer to Exercise 14.50. An alternative experiment would involve splitting the five groups by sex, so there would be 10 groups—AF, AM, BF, etc. Then samples of size 6 would be taken from each of the 10 groups. Assuming that there are no real sex effects in this experiment, which procedure would be more likely to produce a statistically significant result, or would the two experiments be equivalent?

14.53. Refer to Exercise 14.47 in the previous section. An alternative design for the experiment would be to send 4 rats (instead of 8) to each of 10 investigators (instead of 5). It can be shown that all the error terms in Exercise 14.47 would have 9 df (instead of 4). Would the alternative design give a higher or a lower power, all else being equal?

14.54. An experiment involves two factors: treatment, with three levels, and group, with four levels. In each treatment-group combination, 8 subjects will be measured (once). Assume that the error standard deviation σ_ϵ will be 0.4, and that the effects of the treatments will be $-.2$, $-.1$, and .3, respectively.

a. Find the noncentrality parameter for the test of treatment effects. Note that there are 32 subjects per treatment.
b. Find the power of the F test, assuming $\alpha = .05$.

Appendix: The Mathematics of Sums of Squares

In this appendix we prove that the total sum of squared deviations is the sum of two other sums of squares. This result is proved by the same expand-the-square method used in Section 2.6. There are many results that one sum of squares equals a total of others; all such results may be proved the same way. In fact, all of them are variations on the Pythagorean theorem of geometry—the square of the hypotenuse of a right triangle equals the sum of the squares of the sides.

Result 14.1 SS(Total) = SS(Within) + SS(Between)

Proof Recall that

$$SS(Total) = \sum_{i,j} (y_{ij} - \bar{y})^2$$

$$SS(Within) = \sum_{i,j} (y_{ij} - \bar{y}_j)^2$$

$$SS(Between) = \sum_{j} n_j (\bar{y}_j - \bar{y})^2$$

Starting with the definition of SS(Total), we first get the group means $\bar{y}_j$ into the equation by adding and subtracting them. Then we expand the square around these means.

$$SS(Total) = \sum_{i,j} (y_{ij} - \bar{y}_j + \bar{y}_j - \bar{y})^2$$

$$= \sum_{i,j} (y_{ij} - \bar{y}_j)^2$$

$$+ 2 \sum_{i,j} (y_{ij} - \bar{y})(\bar{y}_j - \bar{y}) + \sum_{i,j} (\bar{y}_j - \bar{y})^2$$

Now the first term in the expansion is the definition of SS(Within). The last term is, in fact, SS(Between). Note that when the summation over the i subscript is done, we are summing the same $(\bar{y}_j - \bar{y})^2$ value n_j times; of course, that's the same as multiplying the value by n_j. Therefore, the last term becomes

$$\sum n_j (\bar{y}_j - \bar{y})^2 = SS(Between)$$

All that's left to do is to show that the middle term is 0. In this term, when the summation is done over the i subscript, $(\bar{y}_j - \bar{y})$ is a common factor.

$$\sum_{i,j} (y_{ij} - \bar{y})(\bar{y}_j - \bar{y}) = \sum_{j} (\bar{y}_j - \bar{y}) \sum_{i} (y_{ij} - \bar{y}_j)$$

But $\Sigma (y_{ij} - \bar{y}_j)$ is a sum of deviations of numbers from their mean. By Result 2.1, that sum is 0. Thus the middle term in the expansion of SS(Total) is 0, and the proof is complete.

Chapter Exercises

14.55. Sarason (1981) ranked student subjects on a test anxiety scale, and divided them into three groups. Group A consisted of those who scored in the lowest 25%; group B, those in the middle 50%; and group C, those in the highest 25%. Equal numbers of subjects were sampled from each of the three groups, even though there were twice as many potential subjects in group B. Why is it desirable to have equal numbers in the sample, even if group B becomes "underrepresented"?

14.56. Refer to Exercise 14.55. Suppose that we attempt to replicate Sarason's experiment. We draw 16 students randomly from each of the three groups and measure the number of correct anagram solutions in a specified period of time. The results:

Group A: 2 3 3 4 4 4 4 5 5 5 5 5 5 6 7 9
(mean 4.750, standard deviation 1.65)

Group B: 1 2 2 2 3 3 4 4 4 5 5 5 5 6 6 7
(mean 4.000, standard deviation 1.71)

Group C: 0 1 1 2 2 2 2 3 3 3 3 3 3 4 5 5
(mean 2.625, standard deviation 1.36)

a. Calculate an ANOVA table, including SS, df, MS, and F.

b. Is there a statistically detectable difference among the means, using $\alpha = .05$?

c. Place bounds on the p-value for the F statistic.

14.57. Refer to the data of Exercise 14.56. Is there any reason to think that any assumptions have been violated seriously? If so, which parts of your answer to Exercise 14.56 are most in question?

14.58. Refer to the data of Exercise 14.56.

a. Calculate confidence intervals for all differences of group means. Use an experimentwise 95% confidence level.

b. Which pairs of means, if any, are significantly different from each other, using an experimentwise α of .05?

14.59. Refer again to the data of Exercise 14.56. Define the comparison $\hat{L}_1 = .5\bar{y}_A + .5\bar{y}_B - \bar{y}_C$.

a. Find $SS(\hat{L}_1)$. Is it a large portion of the SS for groups?

b. Is the value of $\hat{L}_1$ significantly different from 0, assuming a post-hoc comparison using $\alpha = .05$?

c. Show that $\hat{L}_2 = \bar{y}_A - \bar{y}_B$ is orthogonal to $\hat{L}_1$.

14.60. Refer to Exercise 14.55. Sarason also divided students from each group randomly into two instruction conditions, a stressful condition and a control condition. There were 32 subjects in each cell. Sarason reported the results shown in Table 14.32.

	LOW TEST ANXIETY			MEDIUM TEST ANXIETY			HIGH TEST ANXIETY	
	Mean	Standard Deviation		Mean	Standard Deviation		Mean	Standard Deviation
S	4.46	2.55	S	5.21	1.94	S	4.96	2.02
C	5.31	2.10	C	3.78	1.82	C	4.06	1.92

Table 14.32

Does it appear visually that there is a substantial interaction between test anxiety and instruction condition?

14.61. Refer to Exercise 14.60. Verify that the ANOVA table for the given data is as shown in Table 14.33.

Source	SS	df	MS
Instruction	11.682	1	11.682
Anxiety	6.250	2	3.125
Interaction	45.556	2	22.778
Error	806.338	186	4.335

Table 14.33

14.62. Refer to Table 14.33.

a. What explains the most variation—the main effect of instruction, the main effect of anxiety, or the interaction? Does this result agree with your finding in Exercise 14.60?

b. Is the interaction statistically significant at $\alpha = .10$? Place bounds on the p-value.

14.63. A developmental psychologist administered a standard questionnaire to samples of students in grades 6, 8, 10, and 12 of an urban school district.

The resulting scale score was basically a "likeability" score measuring how likeable the student felt at the time. The data were

Grade 6: 50 75 79 81 84 87 89 90 90 92 94 96
($\bar{y} = 83.9167$, $s = 12.3764$)

Grade 8: 15 35 52 60 63 65 68 72 75 76 80 84
($\bar{y} = 62.0833$, $s = 19.9110$)

Grade 10: 12 21 46 50 56 59 61 62 66 71 74 79
($\bar{y} = 54.7500$, $s = 20.2849$)

Grade 12: 42 54 67 74 77 77 78 80 82 84 88 91
($\bar{y} = 74.5000$, $s = 14.1067$)

a. Calculate SS(Within), SS(Between), and SS(Total) for these data.

b. Is grade a very good predictor of scale score?

c. Test H_0: all μ are equal, using $\alpha = .01$. Place bounds on the p-value.

14.64. Refer to Exercise 14.63.

a. Use the unmodified Tukey method to test the statistical significance of all pairwise differences of means. Use an experimentwise error rate of .01.

b. Redo these tests using the Tukey-Kramer modification. Are any conclusions different in this particular data set?

14.65. Refer to Exercise 14.63.

a. Is there evidence in the data that the usual ANOVA methods are not appropriate?

b. Carry out the appropriate rank test for the data. Does the conclusion differ from that of Exercise 14.63?

14.66. Ghatala (1981) tested different subjects for recall of the objects of sentences. In the experiment, subjects were divided randomly among three encoding conditions and two test conditions. The means in Table 14.34 were obtained.

		Encoding		
		Generate	Judge	Read
Test Condition	Forward	74.6	73.3	59.8
	Backward	65.8	65.0	53.9

Table 14.34

Is there visual evidence of a serious interaction?

14.67. Refer to Exercise 14.66. Assume that the data yielded the analysis of variance table shown in Table 14.35.

Source	SS	df	MS
Encoding	2204.10	2	1102.05
Condition	881.67	1	881.67
Interaction	12.02	2	6.01
Error	403.92	54	7.48

Table 14.35

a. Compute the η^2 values for encoding, condition, and interaction.

b. Does the η^2 value for interaction indicate that there is a serious problem?

14.68. Refer to Exercise 14.67.

a. Is there a statistically significant (detectable) interaction at the usual α levels?

b. Are the main effects of encoding and condition significant?

14.69. Refer to Exercise 14.67. In your opinion, is it reasonable to regard the factors as fixed factors?

14.70. Platt and Cohen (1981) studied the reaction times of children performing a mental rotation task on objects rotated at various angles. Subjects were divided into untrained and trained groups. Each of the 10 subjects in each group was tested once on objects rotated at 30, 60, 120, and 150 degrees. (In fact, the authors also varied age between 5- and 8-year-olds; for simplicity, we consider only the 8-year-old subjects.)

a. Identify the factors in the experiment. Which factors, if any, should be considered random?

b. Identify the within-subjects and between-subjects factors.

c. Identify the crossing and nesting pattern of factors.

14.71. Refer to Exercise 14.70. Assume that the means in Table 14.36 have been calculated and the ANOVA table is as shown in Table 14.37.

MEANS

	Rotations (degrees)				
	30	60	120	150	Average
Untrained subjects	1.93	2.02	2.57	2.68	2.30
Trained subjects	1.71	1.84	2.51	2.74	2.20
Average	1.82	1.93	2.54	2.71	2.25

Table 14.36

Source	SS	df	MS
Training	0.200	1	0.200
Error1	0.992	18	0.055
Angle	11.660	3	3.887
Angle * Tr.	0.240	3	0.080
Error2	3.725	54	0.069

Table 14.37

a. Does it appear visually that there is a large interaction between training and angle?

b. Does the SS for interaction confirm your judgment in part a?

c. Is the interaction significant even at $\alpha = .25$?

14.72. Refer to Exercise 14.71. Which main effects are statistically detectable (significant) at $\alpha = .05$? In each case, place bounds on the p-value.

14.73. Refer to Exercise 14.71. Use the Tukey test to check the significance of pairwise differences among means. Assume an experimentwise α of .05.

14.74. Refer again to Exercise 14.71. The comparison $\hat{L} = 2\bar{y}_{150} + \bar{y}_{120} - \bar{y}_{60} - 2\bar{y}_{30}$ can be shown to capture the linear relation between the angle factor and the dependent variable.

a. Calculate the value for $\hat{L}$.

b. Find SS($\hat{L}$). What fraction of SS(Angle) is explained by this particular comparison?

c. Assuming that this comparison may be regarded as planned, is its value statistically significantly different from 0 at $\alpha = .01$?

14.75. Following Wold (1982), suppose that we tested adult subjects on a task involving associating adjectives with nouns in previously used sentences. The subjects were randomly assigned according to all combinations of three factors: sequence (noun first or noun last); interval between sentence presentations (4, 8, 12, or 16 seconds); and recall condition (incidental or intentional). For each combination of the three factors, 5 subjects were tested. The means (of the number of adjectives recalled in association with the correct nouns) in Table 14.38 and the ANOVA table in Table 14.39 were obtained.

a. Do any of the interactions involving the interval factor appear to be large?

b. Are any of the interactions involving the interval factor statistically detectable at $\alpha = .10$?

MEANS

	Interval								
	4		8		12		16		
	Recall								
	Int	Inc	Int	Inc	Int	Inc	Int	Inc	Average
Noun first	11.6	8.0	12.2	9.6	14.0	11.6	12.8	8.8	11.075
Noun last	5.6	4.4	5.6	5.4	7.4	7.2	6.2	5.6	5.925
Average	8.6	6.2	8.9	7.5	10.7	9.4	9.5	7.2	8.50

Table 14.38

Source	SS	df	MS
Sequence	530.45	1	530.45
Interval	74.50	3	24.83
Recall	68.45	1	68.45
S * I	1.85	3	0.62
S * R	33.80	1	33.80
I * R	5.05	3	1.68
S * I * R	2.10	3	0.70
Error	80.02	64	1.25

Table 14.39

14.76. Refer to Exercise 14.75.

a. Test the null hypothesis that there is no interval effect. Use $\alpha = .05$.

b. State bounds on the p-value for the test in part a.

14.77. Refer to Exercise 14.75.

a. Calculate the means for each of the four levels of interval.

b. Which pairs of these means are significantly different, using an experimentwise error rate $\alpha = .05$?

14.78. Refer again to Exercise 14.75. Define the comparison $\hat{L} = \bar{y}_{12} - \bar{y}_4$.

a. Calculate the value of $\hat{L}$.

b. Find $SS(\hat{L})$.

c. Treating the comparison as a post-hoc comparison, is its value significantly different from 0, using $\alpha = .05$?

14.79. Data were collected for 8 subjects at each of five levels of a factor. The data were analyzed using the Minitab package. Some output is shown in Figure 14.2.

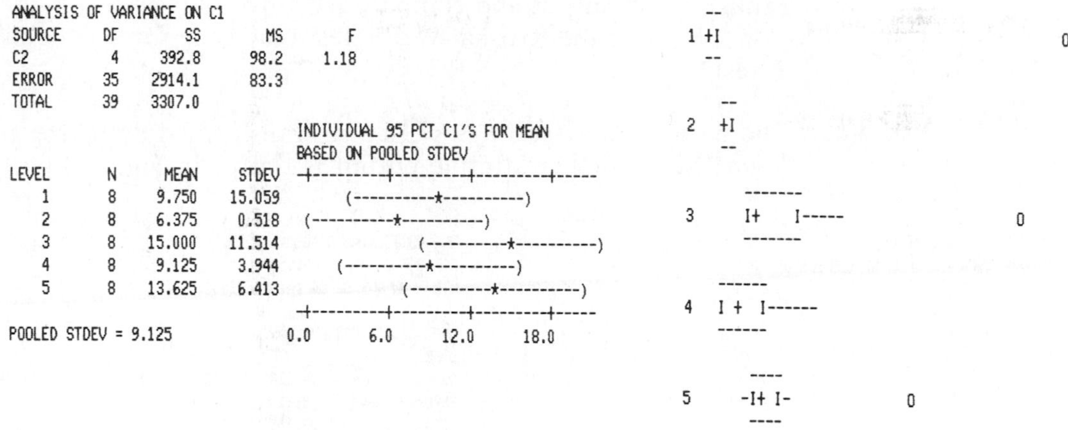

```
MTB > ONEWAY OF C1 DATA FOR LEVELS IN C2

ANALYSIS OF VARIANCE ON C1
SOURCE     DF        SS        MS       F
C2          4     392.8      98.2    1.18
ERROR      35    2914.1      83.3
TOTAL      39    3307.0
                                    INDIVIDUAL 95 PCT CI'S FOR MEAN
                                    BASED ON POOLED STDEV
LEVEL       N      MEAN     STDEV  -+---------+---------+---------+-----
    1       8     9.750    15.059       (----------*----------)
    2       8     6.375     0.518  (----------*----------)
    3       8    15.000    11.514               (----------*----------)
    4       8     9.125     3.944      (----------*----------)
    5       8    13.625     6.413             (----------*----------)
                                   -+---------+---------+---------+-----
POOLED STDEV = 9.125               0.0       6.0      12.0      18.0
```

```
MTB > BOXPLOTS OF C1 BY LEVELS IN C2

     --
  1 +I                                       0
     --

     --
  2  +I
     --

        -------
  3     I+    I-----                         0
        -------

       ------
  4    I + I------
       ------

        ----
  5     -I+ I-              0
        ----
```

Figure 14.2 Output for Exercise 14.79

a. Does an F test indicate that there is a statistically detectable difference among the five means, at usual α levels?

b. Test for the significance of all pairwise differences among means, using an experimentwise $\alpha = .05$.

14.80. In Figure 14.2, is there evidence of violation of assumptions? If so, would the violation have more effect on the F test or on the pairwise comparisons?

14.81. A rank procedure was also tried for the data of Exercise 14.79. Further Minitab output is shown in Figure 14.3.

```
MTB > NOTE DO ANOVA ON RANKS
MTB > RANK C1 PUT RANKS IN C3
MTB > ONEWAY OF RANKS IN C3 BY LEVELS IN C2

ANALYSIS OF VARIANCE ON C3
SOURCE     DF        SS        MS        F
C2          4    2958.2     739.5    11.23
ERROR      35    2305.8      65.9
TOTAL      39    5264.0
```

Figure 14.3 Output for Exercise 14.81

a. The Kruskal-Wallis statistic may be computed as

$$H = \frac{12}{n(n+1)} \text{SS(Between, ranks)}$$

where n is the total sample size and SS(Between, ranks) is SS(Between) computed for the ranks rather than the original data. Locate SS(Between, ranks) in the output and compute H.

b. Does the Kruskal-Wallis test reach the same conclusion as the F test?

14.82. The data of Exercise 14.79 were also transformed by taking the natural logarithms. Further Minitab output is shown in Figure 14.4.

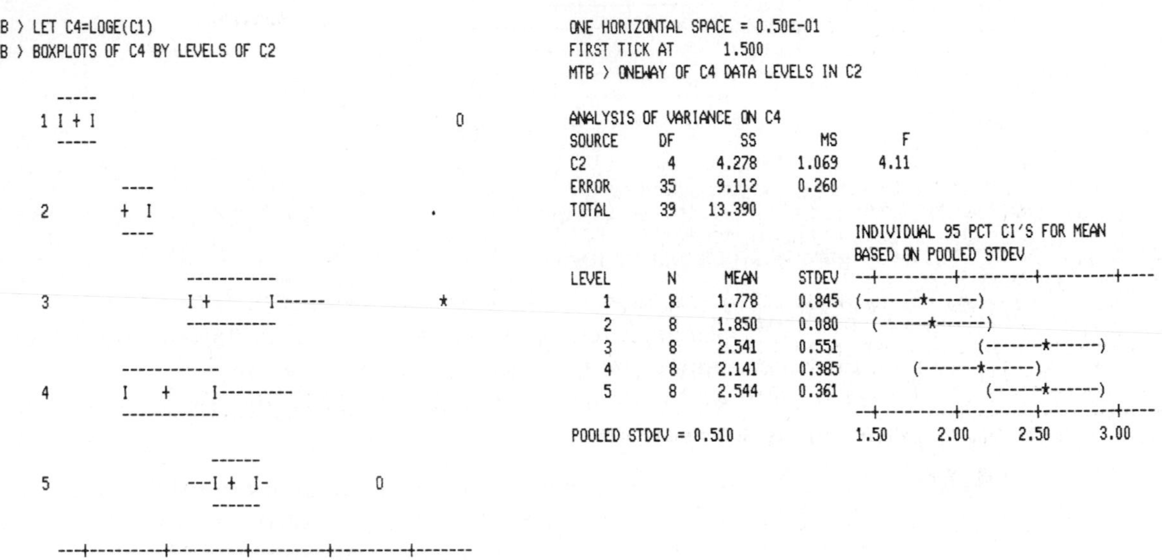

```
MTB > LET C4=LOGE(C1)
MTB > BOXPLOTS OF C4 BY LEVELS OF C2

       -----
  1    I + I                                      0
       -----

            ----
  2         + I                                   .
            ----

                 -----------
  3              I +       I------                *
                 -----------

           ------------
  4        I   +     I---------
           ------------

                    ------
  5              ---I + I-            0
                    ------

   ---+---------+---------+---------+---------+-------

ONE HORIZONTAL SPACE = 0.50E-01
FIRST TICK AT       1.500
MTB > ONEWAY OF C4 DATA LEVELS IN C2

ANALYSIS OF VARIANCE ON C4
SOURCE     DF       SS        MS        F
C2          4     4.278     1.069     4.11
ERROR      35     9.112     0.260
TOTAL      39    13.390
                                     INDIVIDUAL 95 PCT CI'S FOR MEAN
                                     BASED ON POOLED STDEV
LEVEL       N      MEAN     STDEV  --+---------+---------+---------+----
  1         8     1.778     0.845  (-------*------)
  2         8     1.850     0.080    (------*------)
  3         8     2.541     0.551               (-------*------)
  4         8     2.141     0.385        (-------*------)
  5         8     2.544     0.361                (------*------)
                                   --+---------+---------+---------+----
POOLED STDEV = 0.510                 1.50      2.00      2.50      3.00
```

Figure 14.4 Output for Exercise 14.82

a. Does the F test reach a different conclusion than in Exercise 14.79?

b. Has the transformation eliminated any violations of assumptions present in the original data?

14.83. An experiment had different subjects performing a task under different intensities of background noise. The data were analyzed using SAS. Some of the output is shown in Figure 14.5.

a. Is there a statistically detectable difference among the five means, using $\alpha = .01$?

b. Locate the p-value for the F test.

14.84. Which pairs of means are shown to be significantly different in Figure 14.5, using an experimentwise error rate of .05?

ANALYSIS OF VARIANCE OF SKILL AS A FUNCTION OF NOISE

ANALYSIS OF VARIANCE PROCEDURE

DEPENDENT VARIABLE: SKILL   SCORE ON A PROBLEM-SOLVING TASK

SOURCE	DF	SUM OF SQUARES	MEAN SQUARE	F VALUE	PR > F	R-SQUARE	C.V.
MODEL	3	3593.47500000	1197.82500000	12.38	0.0001	0.507741	17.3729
ERROR	36	3483.90000000	96.77500000		ROOT MSE		SKILL MEAN
CORRECTED TOTAL	39	7077.37500000			9.83742853		56.62500000

SOURCE	DF	ANOVA SS	F VALUE	PR > F
NOISE	3	3593.47500000	12.38	0.0001

TUKEY'S STUDENTIZED RANGE (HSD) TEST FOR VARIABLE: SKILL
NOTE: THIS TEST CONTROLS THE TYPE I EXPERIMENTWISE ERROR RATE
ALPHA=0.05 CONFIDENCE=0.95 DF=36 MSE=96.775
CRITICAL VALUE OF STUDENTIZED RANGE=3.809
MINIMUM SIGNIFICANT DIFFERENCE=11.8487
COMPARISONS SIGNIFICANT AT THE 0.05 LEVEL ARE INDICATED BY '***'

NOISE COMPARISON	SIMULTANEOUS LOWER CONFIDENCE LIMIT	DIFFERENCE BETWEEN MEANS	SIMULTANEOUS UPPER CONFIDENCE LIMIT	
50 - 40	-10.849	1.000	12.049	
50 - 60	-3.449	8.400	20.249	
50 - 70	11.851	23.700	35.549	***
40 - 50	-12.849	-1.000	10.849	
40 - 60	-4.449	7.400	19.249	
40 - 70	10.851	22.700	34.549	***
60 - 50	-20.249	-8.400	3.449	
60 - 40	-19.249	-7.400	4.449	
60 - 70	3.451	15.300	27.149	***
70 - 50	-35.549	-23.700	-11.851	***
70 - 40	-34.549	-22.700	-10.851	***
70 - 60	-27.149	-15.300	-3.451	***

Figure 14.5 Output for Exercise 14.83

14.85. The time required to master a certain task was measured for a number of laboratory rats under varied conditions. (Of course, each rat's time to mastery could only be measured once.) The data were analyzed using SPSS-X, yielding the output in Figure 14.6.

a. Does there appear to be a large interaction?
b. Is there a statistically detectable interaction?

14.86. In Figure 14.6, does there appear to be a relation between variances (or standard deviations) and means?

**********************ANALYSIS OF VARIANCE**********************

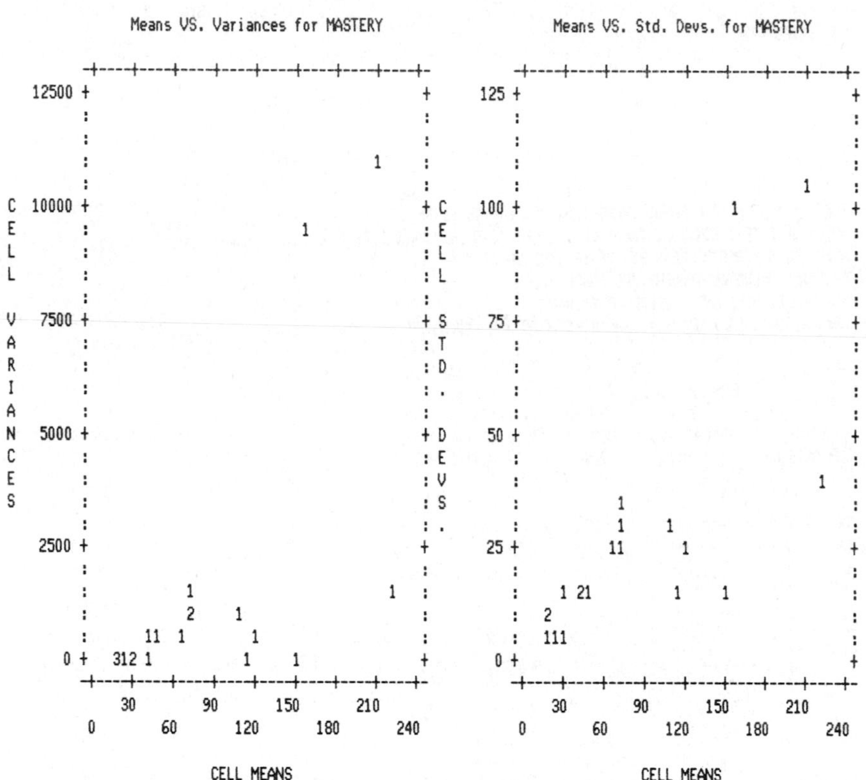

**********************ANALYSIS OF VARIANCE**********************

Tests of significance for MASTERY using SEQUENTIAL sums of squares

Source of Variation	Sum of Squares	DF	Mean Square	F	Sig. of F
WITHIN CELLS	86425.75000	60	1440.42917		
CONSTANT	529588.51250	1	529588.51250	367.66023	.000
SUBTASKS	218303.83750	3	72767.94583	50.51824	.000
DEADENDS	59941.42500	4	14985.35625	10.40340	.000
SUBTASKS BY DEADENDS	17465.47500	12	1455.45625	1.01043	.451

Figure 14.6 Output for Exercise 14.85

14.87. Natural logarithms of the data in Exercise 14.85 were also taken. Output from the analysis of the logarithmic data included that in Figure 14.7.

ANOVA OF LOGARITHMS OF MASTERY TIMES

********************ANALYSIS OF VARIANCE********************

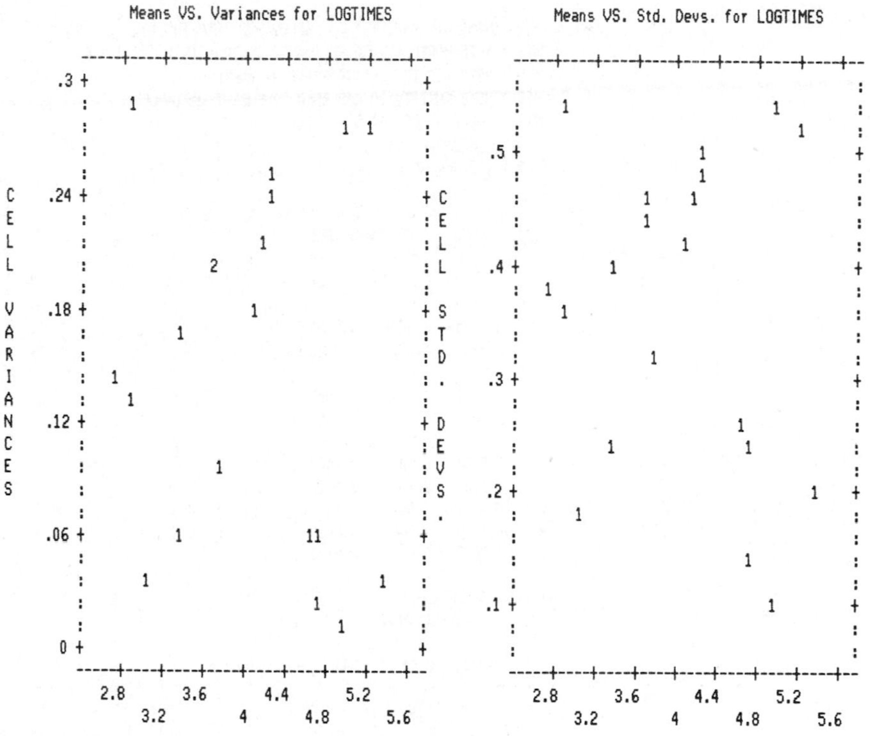

Tests of significance for LOGTIMES using SEQUENTIAL sums of squares

Source of Variation	Sum of Squares	DF	Mean Square	F	Sig. of F
WITHIN CELLS	8.87284	60	.14788		
CONSTANT	1309.26319	1	1309.26319	8853.51632	.000
SUBTASKS	40.10244	3	13.36748	90.39373	.000
DEADENDS	11.01709	4	2.75427	18.62498	.000
SUBTASKS BY DEADENDS	.65508	12	.05459	.36915	.969

Figure 14.7 Output for Exercise 14.87

a. In the transformed data, does interaction appear to be a serious problem?

b. Perform F tests for the main effects of number of subtasks and number of dead ends.

c. Does there appear to be a relation between means and variances (or standard deviations) in the transformed data?

14.88. Morelli, Andrews, and Morelli (1982) performed an experiment involving the effect of task relevance and subject's neuroticism on anxiety levels. Suppose that a repetition of the experiment yielded data that were analyzed by Minitab, and that the output in Figure 14.8 resulted.

```
MTB > NAME C1 'ANXIETY' C2 'RELEV' C3 'HEALTH'
MTB > NOTE MORELLI DATA RELEVANCE OF TASK IS 1=L 2=M 3=H
MTB > NOTE HEALTH IS 1=NEUROTIC 2=STABLE
MTB > NOTE ANXIETY IS A SCALED DEPENDENT VARIABLE
MTB > TABLE BY C2 AND C3;
MTB >   MEANS OF C1;
MTB >   STDEV OF C1.


ROWS: RELEV      COLUMNS: HEALTH

            1         2       ALL

  1    5.3000    2.8000    4.0500
       0.9487    1.3984    1.7313

  2    6.7000    4.0000    5.3500
       1.8886    2.2608    2.4554

  3    9.6000    5.7000    7.6500
       2.4585    2.7508    3.2326

ALL    7.2000    4.1667    5.6833
       2.5650    2.4507    2.9198

 CELL CONTENTS --
         ANXIETY:MEAN
                 STD DEV
MTB > TWOWAY OF C1 BY LEVELS OF C2 AND C3

ANALYSIS OF VARIANCE ON ANXIETY

SOURCE          DF        SS        MS
RELEV            2    132.93     66.47
HEALTH           1    138.02    138.02
INTERACTION      2      5.73      2.87
ERROR           54    226.30      4.19
TOTAL           59    502.98
```

Figure 14.8 Output for Exercise 14.88

a. Draw a profile plot. Does there appear to be a major interaction?

b. Is the interaction statistically significant at $\alpha = .10$?

c. Test for main effects of relevance and neuroticism. State bounds on the p-value.

14.89. Is there evidence in Figure 14.8 of a systematic difference in variability between neurotics and stables?

14.90. Which relevance means, if any, are significantly different in Exercise 14.88? Use an experimentwise α of .05.

14.91. An experiment yielded cooperativeness scores for elementary-school-aged boys and girls. SAS output of the data is shown in Figure 14.9.

ANOVA OF COOPERATIVENESS SCORES OF GRADE SCHOOL CHILDREN

ANALYSIS OF VARIANCE PROCEDURE

DEPENDENT VARIABLE: COOP SCORE ON A COOPERATIVENESS SCALE

SOURCE	DF	SUM OF SQUARES	MEAN SQUARE	F VALUE	PR > F	R-SQUARE	C.V.
MODEL	7	501.64375000	71.66339286	17.46	0.0001	0.445631	36.5497
ERROR	152	624.05000000	4.10559211		ROOT MSE		COOP MEAN
CORRECTED TOTAL	159	1125.69375000			2.02622607		5.54375000

SOURCE	DF	ANOVA SS	F VALUE	PR > F
GRADE	3	158.11875000	12.84	0.0001
SEX	1	310.80625000	75.70	0.0001
SEX*GRADE	3	32.71875000	2.66	0.0497

ANOVA OF COOPERATIVENESS SCORES OF GRADE SCHOOL CHILDREN

ANALYSIS OF VARIANCE PROCEDURE

MEANS

GRADE	N	COOP
2	40	4.37500000
3	40	4.75000000
4	40	6.65000000
5	40	6.40000000

SEX	N	COOP
FEMALE	80	6.93750000
MALE	80	4.15000000

SEX	GRADE	N	COOP
FEMALE	2	20	5.50000000
FEMALE	3	20	6.85000000
FEMALE	4	20	8.10000000
FEMALE	5	20	7.30000000
MALE	2	20	3.25000000
MALE	3	20	2.65000000
MALE	4	20	5.20000000
MALE	5	20	5.50000000

Figure 14.9 Output for Exercise 14.91

a. Does a profile plot indicate that interaction is large?

b. Is there statistically detectable interaction?

c. Are there statistically significant effects for grade or for sex?

14.92. **a.** For the situation in Exercise 14.91, define a comparison measuring the difference between the average fourth- or fifth-grader and the average second- or third-grader.

b. Is the value of this comparison significantly different from 0? Use $\alpha = .05$.

14.93. Subjects in an experiment were given a complex task to complete. Each subject was told how much time was allowed, and also the average time that other subjects had taken to complete the task. In fact, both the actual time and the announced average time were varied among subjects. (Literal truth and psychological experiments are not totally compatible.) The dependent variable was the error rate of the subject in the task. Minitab output of the data is shown in Figure 14.10.

a. Should the factors be regarded as fixed or random?

b. What fraction of SS(Total) is accounted for by interaction effects? Is the interaction statistically significant at $\alpha = .05$?

```
MTB > NAME C1 'ERRORPCT' C2 'ACTUAL' C3 'ANNOUNCD' C4 'TRIALNO'
MTB > TWOWAY ANOVA OF DATA IN C1 FACTORS IN C2 AND C3

ANALYSIS OF VARIANCE ON ERRORPCT

SOURCE        DF       SS       MS
ACTUAL         3       68       23
ANNOUNCD       2    31354    15677
INTERACTION    6     7660     1277
ERROR         84    13352      159
TOTAL         95    52434
```

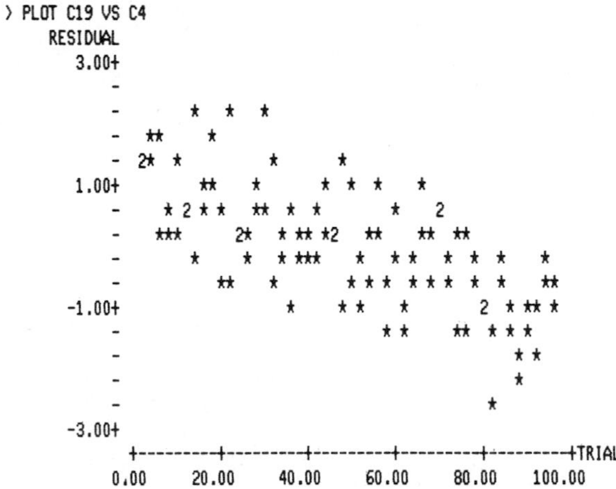

Figure 14.10 Output for Exercise 14.93

14.94. Figure 14.10 contains a plot of residuals (differences between actual and predicted) against the trial number. The trials were numbered in time order. What does the plot indicate about possible violations of assumptions?

15 Regression and Correlation Methods

In Chapter 3 we discussed ideas of prediction and relation for qualitative and ordinal variables. In this chapter we consider prediction and relation for quantitative variables. The methods used are called regression and correlation methods. Regression and correlation are among the most widely used statistical methods. Section 15.1 discusses the process of forming a regression (prediction) model involving one or several independent variables. Then the process of estimating the coefficients of the regression equation is discussed in Section 15.2. The ideas of correlation and "explained" sums of squares are discussed in Section 15.3. Then, in Section 15.4, we consider several forms of hypothesis tests used in regression and correlation analysis. Confidence and prediction interval inference is the topic of Section 15.5. Section 15.6 discusses possible violations of underlying assumptions and suggests some basic procedures to follow to protect against being misled. An appendix contains proofs of some of the basic mathematical results used in this chapter.

Regression and correlation studies are almost always performed using standard computer program packages. Thus computational issues are less important for us than are issues of interpretation and violation of assumptions.

15.1 Regression Models

regression analysis

A **regression analysis** is a method for predicting or explaining the variation of a *dependent variable* on the basis of variation in one or more *independent variables*. The first step in the analysis is to specify a mathematical form for the prediction equation. In this section we'll consider some of the issues involved in that specification.

simple linear regression The simplest situation is **simple linear regression,** in which there's only one independent variable and the prediction equation is a straight line. We write the equation as

$$\hat{y} = \hat{\beta}_0 + \hat{\beta}_1 x$$

intercept **slope** where $\hat{y}$ means the predicted value of Y, $\hat{\beta}_0$ is the **intercept** of the prediction equation, and $\hat{\beta}_1$ is the **slope** of the linear equation. For example, a study might relate Y = MLU (mean length of utterance, or number of words used in a single statement) of young children to X = age in months. $\hat{\beta}_1$ is the slope of the equation—that is, the predicted change in the dependent variable y for a one-unit change in the independent variable x. In the MLU illustration, suppose that

$$\hat{y} = -1.40 + 0.20x$$

The slope is 0.20, meaning that if one child is 1 month older than another (a 1-unit change in x), the predicted MLU is 0.20 words larger for the older child. In using a linear equation, one is assuming that the slope remains constant over the entire range of the independent variable. $\hat{\beta}_0$ **intercept** is the **intercept**—that is, the predicted y value when $x = 0$. In the MLU illustration, the intercept is -1.40, an impossible length of utterance. Very often, $x = 0$ is an impossible value or far outside the range of the data. Thus the intercept is often not very meaningful. In the illustration, it seems safe to say that few 0-month-old children talk a lot; $x = 0$ would be well outside the range of the data, and the intercept term wouldn't have a direct interpretation.

Example 15.1 Suppose that a regression study related Y = delinquency index of juveniles to X = IQ, as suggested in Kleinbaum and Kupper (1978). What would be the interpretation of the slope and intercept terms?

Solution The slope would be the predicted difference in delinquency indexes of two individuals differing by a single IQ point. The intercept would be the predicted delinquency index for a juvenile with an IQ of 0! □

The use of a straight-line equation is an assumption. A good way **scatter plot** to test the assumption is to construct a **scatter plot** of the data. Each (x, y) pair is represented by a point on the plot. Figure 15.1(a) shows a linear relation, and Figures 15.1(b) and 15.1(c) show nonlinear shapes.

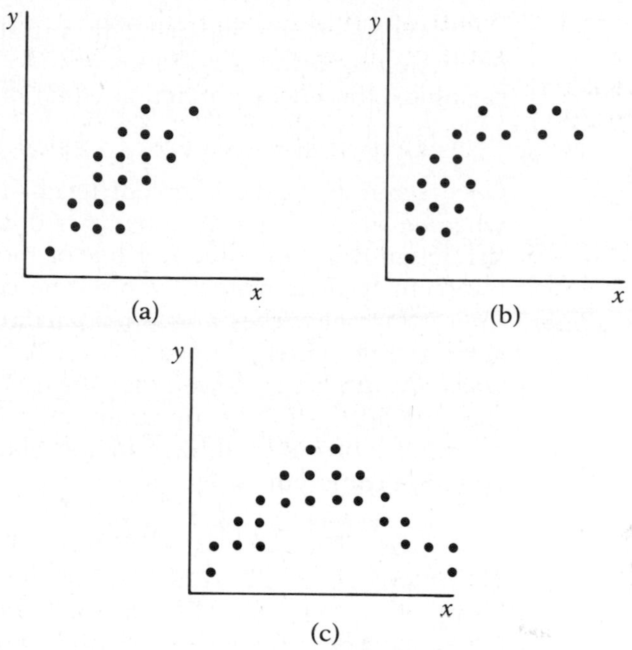

Figure 15.1 Linear (a) and nonlinear (b and c) scatter plots

Example 15.2 Suppose that a scatter plot of delinquency index and IQ is as shown in Figure 15.2. What relation seems appropriate?

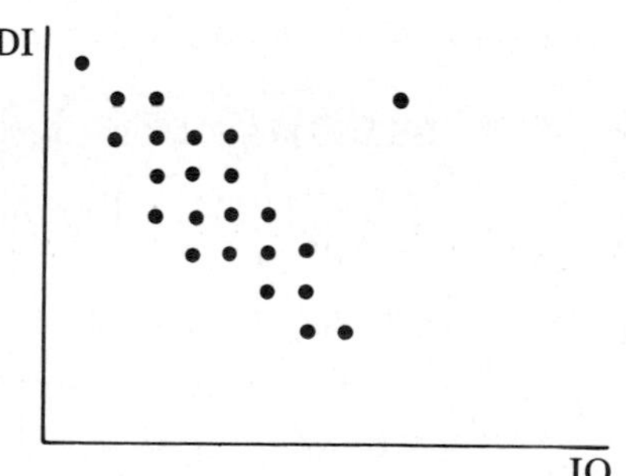

Figure 15.2 Scatter plot of delinquency and IQ scores

Solution Apart from the one extreme score in the upper right corner of the plot, the relation appears linear. □

When several independent variables are involved, the simplest prediction equation is again linear in form. When there are independent variables, the linear prediction equation is

$$\hat{y} = \hat{\beta}_0 + \hat{\beta}_1 x_1 + \hat{\beta}_2 x_2 + \cdots + \hat{\beta}_k x_k$$

Once again, $\hat{\beta}_0$ is the intercept term—that is, the predicted value of y when *all* the x's equal 0. Often, $x = 0$ is an impossible value or outside the actual range of values for one or more of the x's; in such a case, the intercept term is of little value. The coefficients $\hat{\beta}_1, \hat{\beta}_2, \ldots, \hat{\beta}_k$ are the interesting ones. They are called **partial slopes.** Each partial slope measures the predicted change in y for a 1-unit change in that x variable, *when all other x's in the equation are held constant.* Suppose, for example, that the MLU study involved not only X_1 = age in months, but also X_2 = number of older siblings in the house. Assume that the prediction equation came out

partial slopes

$$\hat{y} = -1.27 + 0.18x_1 + 0.13x_2$$

The intercept, -1.27, is of little interest; $x_2 = 0$ is a reasonable possibility, but $x_1 = 0$ is not. The partial slope for x_1, 0.18, can be interpreted by comparing two children with the same number of older siblings (x_2 constant) where the first child is one month older than the second. The first child is predicted to have an MLU 0.18 word larger than the second. Similarly, in a comparison of two children of the same age (x_1 constant), where the first child has one more older sibling than the second, the first child is predicted to have an MLU 0.13 word larger than the second.

Example 15.3

Bolton (1984) attempted to develop a regression model that would predict total SAT score (Y) from total PSAT score (X_1) and IQ (X_2). What would the coefficients of a multiple regression equation mean in this context?

Solution

The intercept term would mean nothing. Neither $X_1 = 0$ nor $X_2 = 0$ is meaningful in context. The coefficient $\hat{\beta}_1$ of X_1 is the predicted change in SAT score, given a 1-point difference in PSAT score and the same IQ score. The coefficient $\hat{\beta}_2$ of X_2 is the predicted difference in SAT score, given the same PSAT score, and a 1-point difference in IQ. □

The slope of a variable in multiple regression typically is not the same as the slope of that variable in simple regression. In the MLU illustration, the simple regression slope of the age variable is 0.20, but when x_2 = number of siblings is added to the equation, the coefficient of age becomes 0.18. The difference is the "all else constant" idea of multiple regression. It's one thing to compare two children who differ by one month in age, and a different thing to compare two children who *have the same number of siblings* and differ by one month in age. There is one case in which the simple and multiple regression slopes are equal. If there is no correlation at all between x's, a change in one x will leave the others constant, so that it doesn't matter whether we consider other x's as held constant or free to vary. We do not yet have a formal definition of correlation, but clearly two variables are uncorrelated if one has no value at all for predicting the other. If x's are uncorrelated, the simple and multiple regression slopes will be equal. In the MLU study, there shouldn't be a huge correlation between age and number of older siblings, but the correlation wouldn't be .00000, by random variation alone. Thus we'd expect the simple and multiple regression slopes to differ, if only slightly.

Example 15.4

Refer to Example 15.3. Would one expect that the coefficient of IQ in a multiple regression predicting SAT score from PSAT score and IQ would be the same as the coefficient of IQ in a regression predicting SAT score from only IQ?

Solution

Goodness, no. There almost surely will be a strong correlation between IQ and PSAT score. Thus there will be a large difference between the effect of a difference in IQ and the effect of a difference in IQ with PSAT held constant. □

It's harder to detect nonlinear patterns in multiple regression than in simple regression. Scatter plots of y against each x are somewhat useful, but variation in other x's may obscure a nonlinear relation. Later in this chapter we'll discuss the use of residual plots to detect some kinds of nonlinearity.

If nonlinearity appears to be present in a plot, it's a good idea to explore some possible *transformations* either of the dependent variable or of one or more independent variables. In the MLU illustration, one might expect a nonlinear relation. After all, the mean length of utterance can't increase forever as a person gets older (except possibly in the case of certain Senators and college professors). Also, it doesn't seem reasonable that having two older siblings would have twice the effect of having one older sibling. Thus a transformation of the X_2 variable might be useful. One can replace an independent variable x by something else such as $\sqrt{x}$, one can add an additional term such as x^2, or one can transform the dependent variable into, for example, log (y). If a linear prediction is used in any of the data sets underlying Figure 15.3, the prediction equation will be less effective than it should be. Often, though, a linear prediction rule will be an adequate approximation, unless the independent x variable takes values over a wide range.

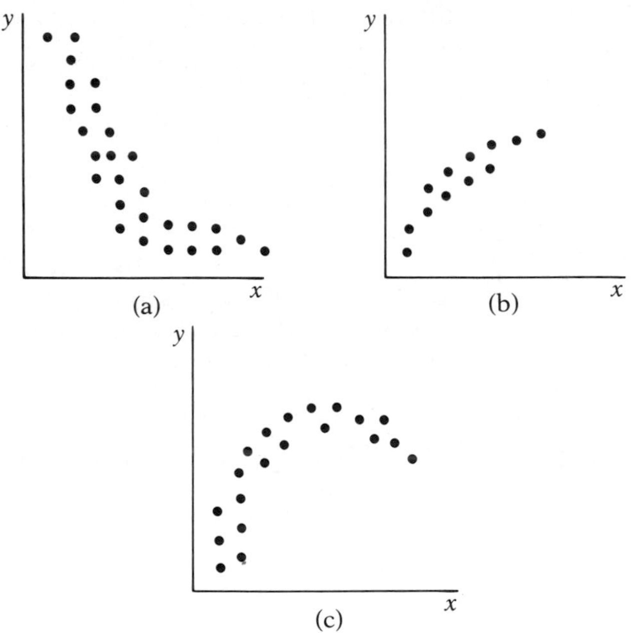

Figure 15.3 Scatter plots indicating need for transformations

Several possible transformations are discussed in Tukey (1977). A summary of reasonable transformation possibilities is shown in Figure 15.4.

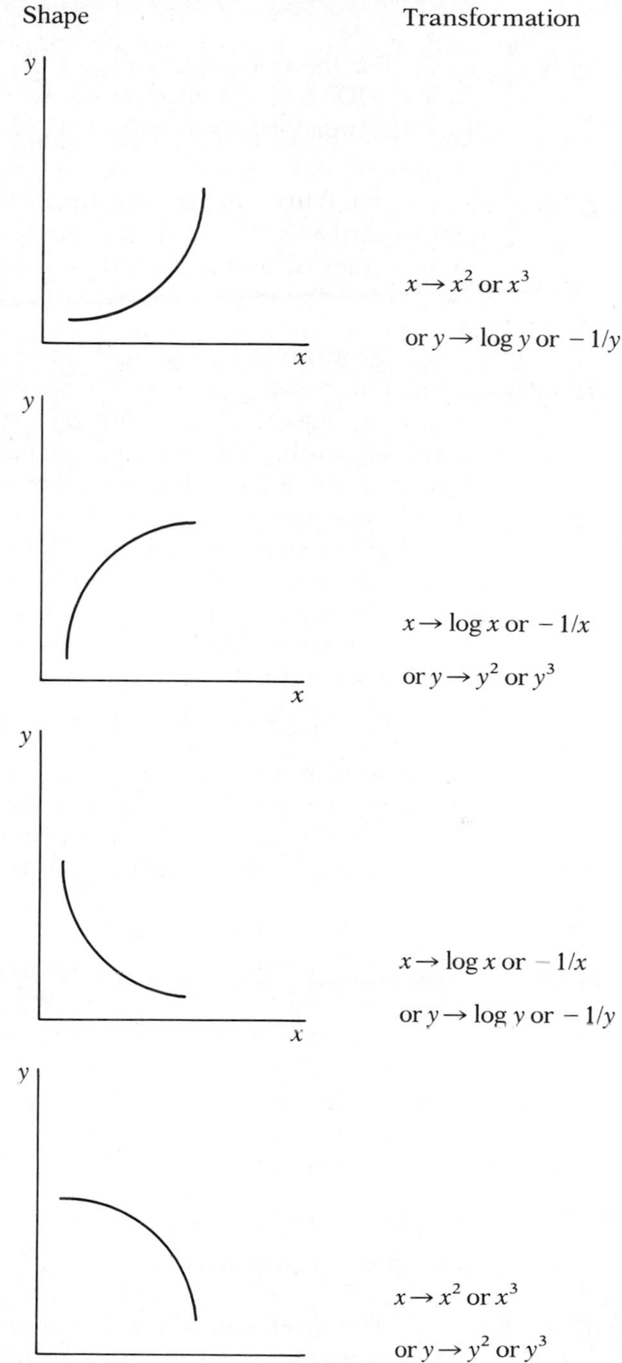

Figure 15.4 Reasonable transformations for various shapes of scatter plots

Example 15.5 For the data of Example 15.3, scatter plots of SAT vs. PSAT and SAT vs. IQ were obtained, as shown in Figure 15.5. Is there any reason to try a transformation on either X?

Solution No. Both plots show a linear relation. Although there might be a nonlinearity that is obscured in the plots, there is as yet no evidence that a transformation should be used. □

dummy variables

Qualitative (categorical) independent variables may be incorporated into a regression model by using **dummy variables,** also called indicator variables. The simplest example occurs when a variable has only two possible values (female/male, did not graduate from high school/did graduate, etc.). Define an indicator variable $x = 1$ if the observation falls in one category (e.g., female) and $x = 0$ if the observation falls in the other category (e.g., male). Then the coefficient of the dummy variable is the predicted change in y when x changes from the 0 category to the 1 category, with all the other x's held constant. For the MLU illustration, we might let $x_3 = 1$ if the child was 2 or more months premature, 0 otherwise. If the prediction equation is

$$\hat{y} = -1.29 + 0.19x_1 + 0.13x_2 - 0.37x_3$$

then the coefficient of x_3 indicates that a comparison of two children (of the same age, with the same number of siblings), one premature and the other not, the MLU of the premature child is predicted to be 0.37 word *less* (because of the negative sign) than that of the full-term child.

Example 15.6 Suppose that a study was done to compare Y = the calories consumed by subjects who were allowed to eat as much as they wished. The key independent variables were X_1 = amount overweight and X_2 = 1 if the subject ate from a normal-size plate, 0 if the subject ate from an undersized plate. Assume that the prediction equation was

$$\hat{y} = 562.5 + 22.7x_1 + 79.3x_2$$

Interpret the coefficients.

Solution The intercept, 562.5, is the predicted consumption of a person 0 units overweight, eating from an undersized plate ($x_2 = 0$). The coefficient of x_1, 22.7, is the predicted increase in caloric consumption per unit increase in overweight, for persons using a plate of any specified size.

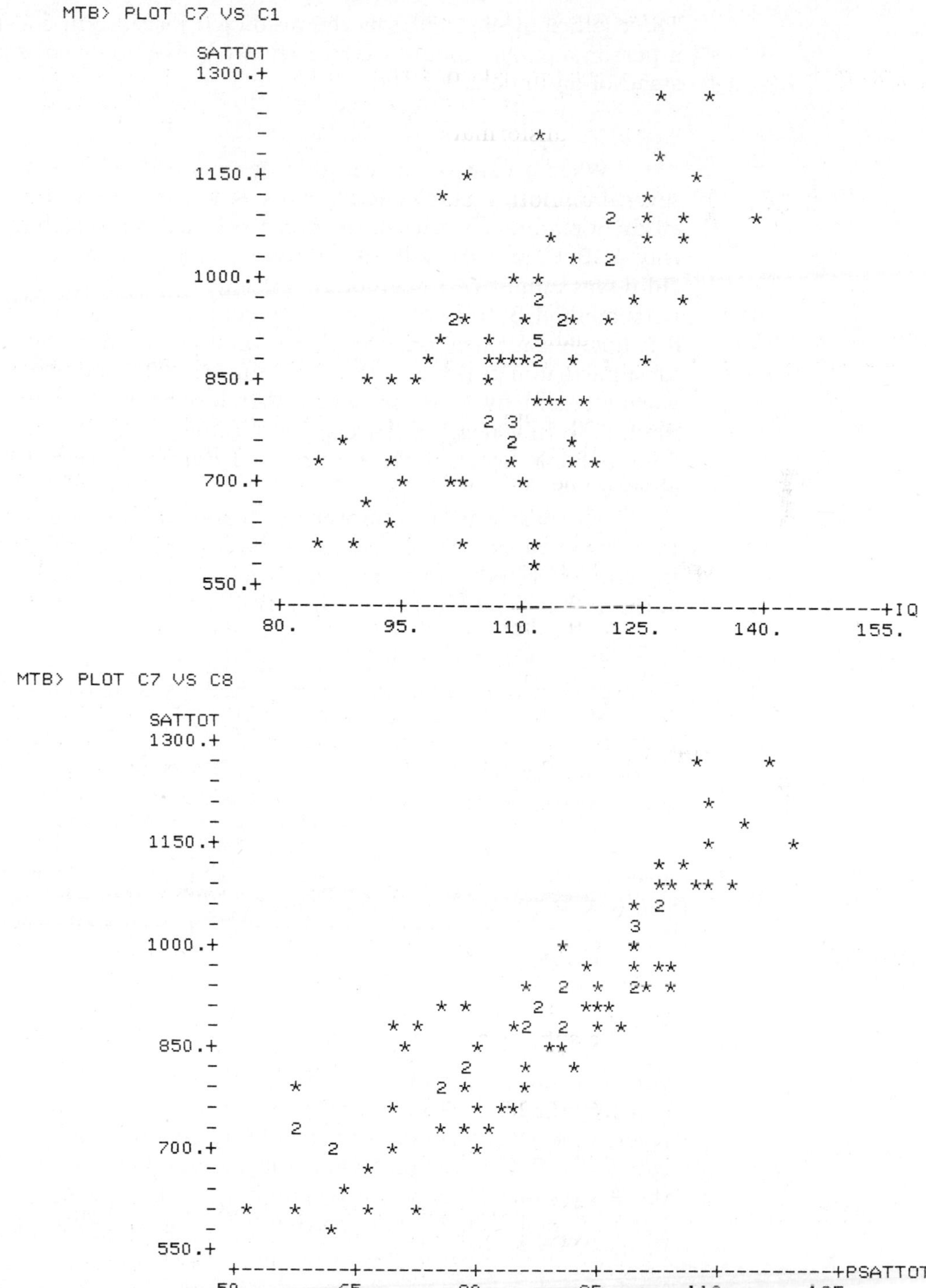

Figure 15.5 Scatter plots of Bolton data

The coefficient of x_2, 79.3, is the predicted increase in consumption for a person a given amount overweight who uses a normal-sized plate instead of an undersized plate. □

When a categorical variable has more than two possible values, several dummy variables must be used. For example, suppose that in a study of reading abilities of children, four different teaching approaches (say A, B, C, and D) were used. Define a dummy variable $x_1 = 1$ if the child was taught with method A, 0 if not; $x_2 = 1$ if the child was taught with method B, 0 if not; $x_3 = 1$ if the child was taught with method C, 0 if not. It isn't necessary to define another dummy for method D. If, for a particular child, $x_1 = 0$, $x_2 = 0$, and $x_3 = 0$, it follows that the child wasn't taught by A, B, or C, and therefore must have been taught by method D. In general, if there are c possible values of a qualitative variable, it's only necessary to define $c - 1$ dummy variables for that variable.

The best way to interpret the coefficients of dummy variables is to "plug in" appropriate 1's and 0's. For example, suppose that the reading study led to the following regression equation, where the dependent variable Y is a measure of reading skill:

$$\hat{y} = 43.6 + 2.4x_1 - 4.7x_2 + 5.3x_3$$

Then predicted Y values for each method are as shown in Table 15.1.

Method	x_1	x_2	x_3	$\hat{y}$
A	1	0	0	$43.6 + 2.4(1) - 4.7(0) + 5.3(0) = 46.0$
B	0	1	0	$43.6 + 2.4(0) - 4.7(1) + 5.3(0) = 38.9$
C	0	0	1	$43.6 + 2.4(0) - 4.7(0) + 5.3(1) = 48.9$
D	0	0	0	$43.6 + 2.4(0) - 4.7(0) + 5.3(0) = 43.6$

Table 15.1

Note that the intercept term is the predicted value for the category that has no dummy variable (method D in the example). Call this category the baseline category. The coefficient of a dummy variable represents the *difference* between the predicted value for the corresponding category and the predicted value for the baseline category. For example, the 2.4 coefficient of x_1, corresponding to method A, is the difference between the predicted value for method A, 46.0, and the predicted value for baseline method D, 43.6.

Example 15.7 Refer to Example 15.6. Suppose that the study was done using the same Y and X_1, but that there were three plate sizes: normal, moderate, and undersized. Define $X_2 = 1$ if normal-sized plates were used, 0 if not, and $X_3 = 1$ if moderate-sized plates were used, 0 if not. Assume that the regression equation came out

$$\hat{y} = 588.3 + 24.8x_1 + 62.1x_2 + 33.0x_3$$

Interpret the coefficients.

Solution We plug in 0 and 1 values as appropriate.

$$\begin{aligned}
\text{Normal } (x_2 = 1, x_3 = 0)\text{: } \hat{y} &= 588.3 + 24.8x_1 + 62.1(1) + 33.0(0) \\
&= 649.4 + 24.8x_1 \\
\text{Moderate } (x_2 = 0, x_3 = 1)\text{: } \hat{y} &= 588.3 + 24.8x_1 + 62.1(0) + 33.0(1) \\
&= 621.3 + 24.8x_1 \\
\text{Undersized } (x_2 = 0, x_3 = 0)\text{: } \hat{y} &= 588.3 + 24.8x_1 + 62.1(0) + 33.0(0) \\
&= 588.3 + 24.8x_1
\end{aligned}$$

In this regression model, it is assumed that the coefficient of x_1 stays constant, regardless of the plate size, at 24.8. The coefficients of the dummy variables x_2 and x_3 affect the intercept terms. $62.1 = 649.4 - 588.3$ is the difference of intercepts for the normal-sized case and the undersized case. $33.0 = 621.3 - 588.3$ is the difference of intercepts for the moderate case and the undersized case. Note that the "undummied case," the undersized plate, is the baseline against which the other cases are compared. □

More complicated looking models can also be simplified by plugging in values of 0 and 1 for the dummy variables. For example, suppose that a study of cocaine abusers, alcohol abusers, and a control group of nonabusers yielded the following regression model:

$$\hat{y} = 28.0 + 0.35x_1 + 3.6x_2 + 10.2x_3 - 0.40x_1x_2 - 0.80x_1x_3$$

where Y = score on a self-esteem scale; X_1 = age in years; $X_2 = 1$ if the subject is a cocaine abuser, 0 if not; $X_3 = 1$ if the subject is an alcohol abuser, 0 if not. (We assume that there are no subjects abusing both alcohol and cocaine.) Inserting 0 and 1 values, we find

Cocaine abusers:

$$\begin{aligned}
\hat{y} &= 28.0 + 0.35x_1 + 3.6(1) + 10.2(0) - 0.40x_1(1) - 0.80x_1(0) \\
&= 31.6 - 0.05x_1
\end{aligned}$$

Alcohol abusers:

$$\begin{aligned}\hat{y} &= 28.0 + 0.35x_1 + 3.6(0) + 10.2(1) - 0.40x_1(0) - 0.80x_1(1) \\ &= 38.2 - 0.45x_1\end{aligned}$$

Nonabusers:

$$\begin{aligned}\hat{y} &= 28.0 + 0.35x_1 + 3.6(0) + 10.2(0) - 0.40x_1(0) - 0.80x_1(0) \\ &= 28.0 + 0.35x_1\end{aligned}$$

The three prediction equations are shown in Figure 15.6.

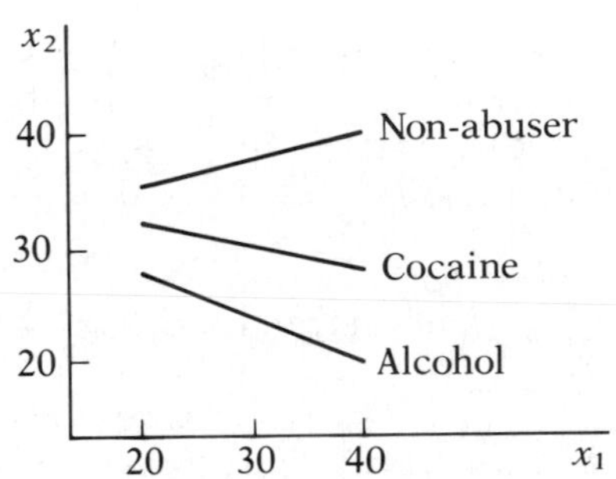

Figure 15.6 Prediction lines for substance-abuse model

Notice that the lines are not even close to parallel. For nonabusers, self-esteem is predicted to increase as age increases, whereas for cocaine abusers and especially for alcohol abusers, self-esteem is predicted to decrease with age. In the regression model, the nonparallelism is captured by the **product terms** involving x_1x_2 and x_1x_3. Figure 15.6 is very similar to the profile plots defined in Chapter 14 and used to look for interaction. In fact, nonparallelism in a regression model (nonzero coefficients of product terms) is a form of interaction.

product terms

Example 15.8

Suppose that the model in Example 15.7 was extended to include product terms and that the result comes out

$$\hat{y} = 584.7 + 24.7x_1 + 63.3x_2 + 34.1x_3 + 1.1x_1x_2 - 0.9x_1x_3$$

Find prediction equations for the three groups. Does there appear to be a large interaction effect?

Solution

For a normal-sized plate,

$$\hat{y} = 584.7 + 24.7x_1 + 63.3(1) + 34.1(0) + 1.1x_1(1) - 0.9x_1(0)$$
$$= 648.0 + 25.8x_1$$

For a moderate-sized plate,

$$\hat{y} = 584.7 + 24.7x_1 + 63.3(0) + 34.1(1) + 1.1x_1(0) - 0.9x_1(1)$$
$$= 618.8 + 23.8x_1$$

For an undersized plate,

$$\hat{y} = 584.7 + 24.7x_1 + 63.3(0) + 34.1(0) + 1.1x_1(0) - 0.9x_1(0)$$
$$= 584.7 + 24.7x_1$$

Although the slopes of x_1 are not identical, they are very similar. Thus there appears to be little interaction. Notice that the coefficients of the product terms are small relative to the coefficient of x_1, yielding a small interaction. □

The dummy-variable device is one way to incorporate the analysis of variance into a regression context. The main effects of any factors can be obtained by individual dummy variables, and interactions can be obtained by appropriate product terms. Thus everything said about regression applies as well to analysis of variance. For a more extended treatment of this idea, see Cohen and Cohen (1975).

general linear model

The **general linear model** incorporates not only the original independent variables, but also any additional terms that may arise from the use of transformations, dummy variables, or product terms. This model is indeed very general. It incorporates not only ordinary regression, but also the analysis-of-variance methods of Chapter 14, via dummy variables, and combinations of qualitative and quantitative independent variables. The phrase "linear model" is used in a slightly unusual way. The model

$$\hat{y} = \hat{\beta}_0 + \hat{\beta}_1 x + \hat{\beta}_2 x^2$$

is not a linear function of the independent variable x. But for any specified value of x (and therefore of x^2) it is a linear function of the $\hat{\beta}$'s. Any model that is linear in the $\hat{\beta}$'s for given values of the x's is a general linear model.

Example 15.9

Can the model

$$\hat{y} = \hat{\beta}_0 + \hat{\beta}_1 \log(\hat{\beta}_2 + x)$$

be incorporated in the general linear model?

Solution No. For a given value of x, the model is not a linear function of the coefficients. □

Once a model has been specified, it remains to estimate the coefficients of that model. That is the task to be taken up in Section 15.2.

Exercises for Section 15.1

15.1. A study was undertaken of the relation between X = score on a reading comprehension test and Y = score on a college aptitude test.

a. If a straight-line prediction equation is used, what assumption is being made?

b. In a straight-line model with aptitude test score as the dependent variable, what do the intercept and slope coefficients mean?

15.2. Suppose that the data for the study in Exercise 15.1 were

X	18	22	24	25	25	26	26	26	27	27	29	30	32	35	37	40	43
Y	28	30	36	34	37	33	39	35	38	36	39	42	38	41	49	54	52

Plot the data. Is there clear evidence of a nonlinear relation?

15.3. In Chapter 1 data were given from a study by Gallistel (1983) in which electric charges of various durations and intensities were administered to a certain portion of the brains of several laboratory rats and the running speeds of the rats were obtained. For rat S-61, the data in Table 15.2 were obtained.

$D = 0.4$		$D = 4.0$	
Intensity	Speed	Intensity	Speed
0.90	34	4.0	21
1.00	40	4.6	29
1.10	31	5.1	32
1.26	68	5.8	36
1.44	59	6.4	42
1.59	73	7.2	57
1.81	61	8.0	60
2.00	96	9.1	70
		10.1	87

Table 15.2

a. If a linear prediction rule is used for predicting speed from intensity, using only the $D = 4.0$ data, what is the interpretation of the slope?

b. Plot the $D = 4.0$ data. Is there evidence that a nonlinear prediction rule should be used? In judging linearity, you may find it helpful to draw a rather large scatter plot and to use a ruler to approximate the regression line.

15.4. The Gallistel data ($D = 4.0$) of Exercise 15.3 were transformed by taking logarithms (base 10) of both X and Y. A scatter plot of the transformed data is shown in Figure 15.7. Do the transformed data appear more nearly linear?

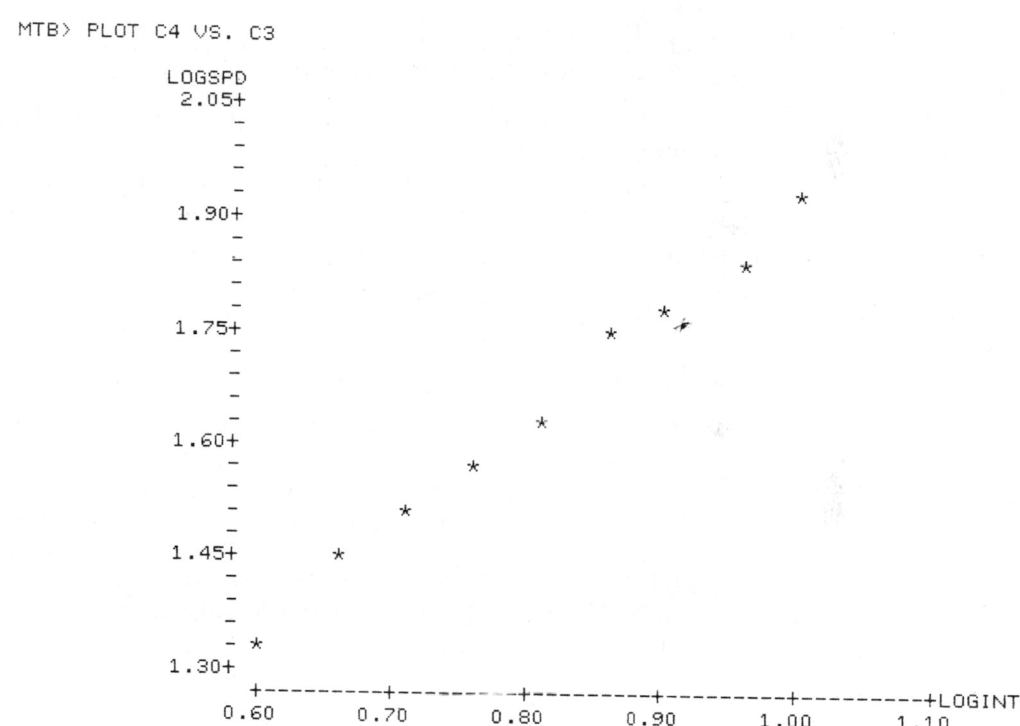

Figure 15.7 Plot of transformed data

15.5. The following data set is artificial, to illustrate a point:

X_1	−2	−2	−2	−1	−1	−1	0	0	0	1	1	1	2	2	2
X_2	−1	0	1	−1	0	1	−1	0	1	−1	0	1	−1	0	1
Y	5	4	9	2	9	13	6	10	14	10	13	13	12	14	16

a. Plot Y against X_1 and Y against X_2. Is there clear evidence of a curved relation?

b. The prediction equation comes out $\hat{y} = 10 + 2x_1 + 3x_2$. Interpret each of the coefficients of the model. Is the intercept meaningful given these data?

c. Plot X_1 against X_2. Does there seem to be a correlation between these variables?

d. The prediction equation using X_1 alone is $\hat{y} = 10 + 2x_1$. Why is the slope of x_1 the same in this model as in the multiple regression (x_1 and x_2) model?

15.6. The following is another artificial data set:

X_1	−2	−2	−2	−1	−1	−1	0	0	0	1	1	1	2	2	2
X_2	−1	−1	0	−1	0	0	−1	0	1	0	0	1	0	1	1
Y	1	6	5	4	8	9	7	12	11	9	13	17	14	21	13

a. Plot Y against X_1 and Y against X_2. Is there clear evidence of a nonlinear relation?

b. Note that when X_1 is negative, X_2 is negative or 0, and that when X_1 is positive, X_2 is positive or 0. What does this fact suggest about the correlation between X_1 and X_2?

c. The prediction equation using both X's turns out to be

$$\hat{y} = 10 + 2x_1 + 3x_2$$

and the equation using only X_1 is

$$\hat{y} = 10 + 3x_1$$

Does this fact confirm your opinion about the correlation of X_1 and X_2?

15.7. A study involved the time required for a subject to solve a small jigsaw puzzle. Each subject was told to use one of three strategies (A, B, or C). A picture of the solved puzzle was flashed on a screen for 1, 2, 4, or 8 seconds. Then Y = the number of seconds required to solve the puzzle was recorded. Suppose that a model

$$\hat{y} = 78.2 - 5.1x_1 - 4.7x_2 + 9.3x_3$$

was obtained, where X_1 = time of display of the solved puzzle; $X_2 = 1$ if strategy A was suggested, 0 if not; and $X_3 = 1$ if strategy B was suggested, 0 if not.

a. Interpret the coefficients. Is the intercept term at all meaningful?

b. A plot of the data indicated that points for $X_1 = 1$ or 8 were consistently below the predicted value, and points for $X_1 = 2$ or 4 were consistently above the predicted value. What does this indicate about the suitability of the model for the data?

15.8. Refer to Exercise 15.7. The model was extended through use of product terms. The result was

$$\hat{y} = 80.1 - 5.0x_1 - 4.4x_2 + 9.6x_3 + 0.2x_1x_2 - 0.3x_1x_3$$

a. Interpret the coefficients of this model.

b. Does it appear that there is a large interaction between display time and suggested strategy in determining solution time?

c. A plot of the data indicated that points for $X_1 = 1$ or 8 were once again consistently below the predicted value, and points for $X_1 = 2$ or 4 were again consistently above the predicted value. What does this indicate about the suitability of the model for the data?

15.9. A study of the ability to discriminate closely related tastes involved Y = score on a taste discrimination test, X_1 = age in years, X_2 = years of formal schooling, and $X_3 = 1$ if the subject received special training, 0 if not. A regression model

$$\hat{y} = \hat{\beta}_0 + \hat{\beta}_1x_1 + \hat{\beta}_2x_2 + \hat{\beta}_3x_3$$

was proposed.

Interpret the meaning of each coefficient. In particular, what does $\hat{\beta}_3$ mean?

15.10. It was proposed to add two terms, $\hat{\beta}_4x_1x_3$ and $\hat{\beta}_5x_2x_3$, to the model in Exercise 15.9. What is the purpose of these terms?

15.11. Refer to Exercises 15.9 and 15.10. Does either (or both) of these models fit the form of the General Linear Model?

15.2 Estimating Parameters of the General Linear Model

Section 15.1 defined the general linear model, which included ordinary linear regression, certain kinds of nonlinear (in the independent variables) regression models, and dummy-variable models. Whatever model is selected, it's necessary to use sample data to obtain estimates of the coefficients in the model, to estimate the variability of the actual data around the prediction equation, and to assess how useful the model is in predicting the dependent variable. In this section we'll consider the problems of estimating the coefficients and estimating variability around the equation.

least squares method **residuals** The traditional method for estimating the coefficients of a regression equation is the **least squares method.** This method is based on the **residuals**—prediction errors—of a regression equation. For each data point,

$$\text{residual} = \text{actual } y \text{ value} - \text{predicted } y \text{ value}$$

A scatter plot and a residual value are shown in Figure 15.8. Note that because we are concerned about errors in Y, the residual is measured parallel to the Y axis.

The coefficients of the regression equation are chosen to minimize the sum of squared prediction errors for the data. Thus, if the data points are $(y_1, x_1), \ldots, (y_n, x_n)$, the coefficients are chosen to minimize

$$\text{SS(Residual)} = \sum (y_i - \hat{y}_i)^2$$

A standard proof, provided in the appendix, shows that the desired coefficient estimates are as shown in the box.

Least-Squares Estimates, Simple Regression

$$\hat{\beta}_1 = \frac{S_{xy}}{S_{xx}}$$

where

$$S_{xy} = \sum (x_i y_i) - \frac{\left(\sum x_i\right)\left(\sum y_i\right)}{n}$$

and

$$S_{xx} = \sum (x_1^2) - \frac{\left(\sum x_i\right)^2}{n}$$

$$\hat{\beta}_0 = \frac{\sum y_i - \hat{\beta}_1 \sum x_i}{n}$$

NOTE: n is the number of (x, y) pairs in the sample data.

For example, Schumm, Bollman, and Jurich (1981) report on a study involving the relation between a marital conventionalization scale and certain other variables, among them years of education. Assume that

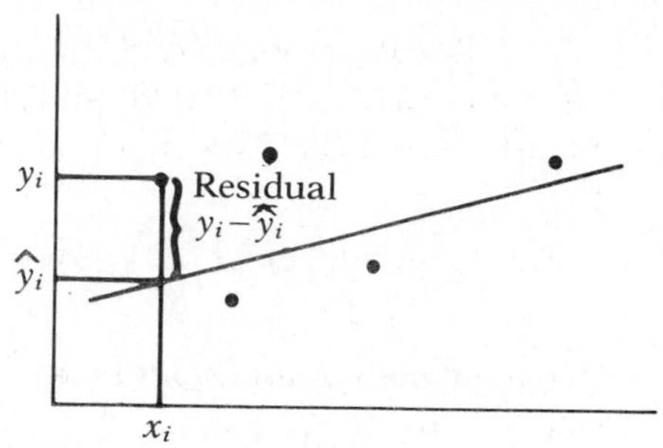

Figure 15.8 Scatter plot and a residual

a sample of 8 urban-dwelling, married men yielded the following scores (x = years of education, y = conventionalization score):

x	6	10	11	11	12	12	14	16
y	56	62	43	51	56	37	48	40

The scatter plot of the data in Figure 15.9 reveals a more or less linear, but rather weak, relation. There seems to be some tendency for y scores to decrease as x scores increase.

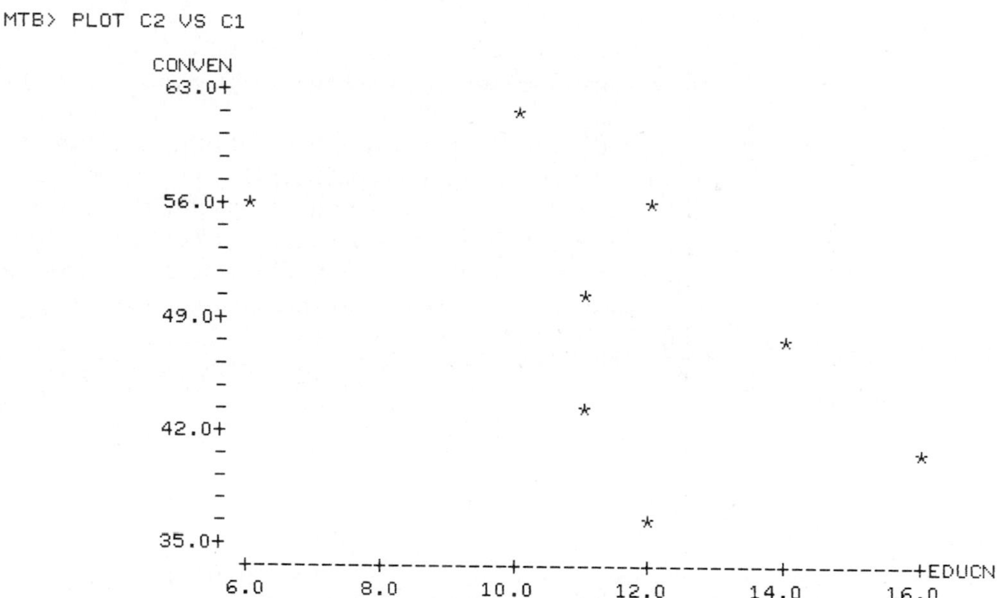

Figure 15.9 Scatter plot for marital conventionalization scores

For the marital conventionalization data, $n = 8$, $\Sigma\, x_i = 92$, $\Sigma\, x_i^2 = 1118$, $\Sigma\, y_i = 393$, and $\Sigma\, x_i y_i = 4418$. Thus

$$S_{xy} = 4418 - \frac{(92)(393)}{8} = -101.5$$

and

$$S_{xx} = 1118 - \frac{(92)^2}{8} = 60$$

The estimated coefficients are

$$\hat{\beta}_1 = \frac{-101.5}{60} = -1.691667$$

and

$$\hat{\beta}_0 = \frac{393 - (-1.691667)92}{8} = 68.579167$$

Note that there is no x value even close to $x = 0$, so the intercept term isn't directly meaningful.

Example 15.10 Lasky et al. (1981) report a study of developmental measures of infants at various ages. In particular, they measured boys on a motor-skills scale at ages 6 and 15 months. Assume that the following summary figures were obtained: $n = 221$, $\Sigma x = 6724$, $\Sigma y = 7519$, $\Sigma x^2 = 219{,}516$, $\Sigma y^2 = 269{,}645$, and $\Sigma xy = 236{,}336$, where X = score at 6 months and Y = score at 15 months. Find the least-squares linear prediction equation.

Solution

$$S_{xy} = 236{,}336 - \frac{(6724)(7519)}{221} = 7567.8733$$

and

$$S_{xx} = 219{,}516 - \frac{(6724)^2}{221} = 14936.0181$$

Therefore,

$$\hat{\beta}_1 = \frac{7567.8733}{14936.0181} = 0.506686$$

and

$$\hat{\beta}_0 = \frac{7519 - (0.506686)6724}{221} = 18.6065$$

The equation could be reported, with rounded-off coefficients, as

$$\hat{y} = 18.61 + 0.507x$$

□

correlation coefficient

The slope of a regression line is closely related to the **correlation coefficient.** The sample correlation coefficient is a measure of the strength and direction of a linear relation, given by

$$r_{xy} = \frac{S_{xy}}{\sqrt{S_{xx}}\sqrt{S_{yy}}}$$

where S_{xy} and S_{xx} are exactly the same as for the slope and

$$S_{yy} = \sum y_i^2 - \frac{\left(\sum y_i\right)^2}{n}$$

For the marital conventionalization data,

$$S_{yy} = 19{,}839 - \frac{(393)^2}{8} = 532.875$$

and so

$$r_{xy} = \frac{-101.5}{\sqrt{60}\sqrt{532.875}} = -0.568$$

The sign of the correlation always equals the sign of the slope. A negative correlation means that y tends to decrease as x increases—that is, that the slope is negative. In fact, it's easy to show algebraically that

$$\hat{\beta}_1 = r_{xy}\frac{s_y}{s_x}$$

where s_y and s_x are the standard deviations of Y and X, respectively. For the marital conventionalization data, $s_y = 8.725$ and $s_x = 2.928$. Thus

$$\hat{\beta}_1 = (-0.568)\frac{8.725}{2.928} = -1.69255$$

equaling the coefficient found previously, except for round-off error in the third decimal place.

Example 15.11 Refer to Exercise 15.10. Would you anticipate a high correlation between X and Y in this study? Find r_{XY}.

Solution There will be a high correlation if boys who are developmentally advanced at 6 months maintain the lead to age 15 months. The correlation will be low if there are large variations in developmental pace between those ages. Also, measurement error would reduce the correlation. To calculate the actual correlation, we need $S_{xx} = 14936.0181$ and $S_{xy} = 7567.8733$, as found in Example 15.10, as well as

$$S_{yy} = 269{,}645 - \frac{(7519)^2}{221} = 13828.8869$$

Thus

$$r_{XY} = \frac{7567.8733}{\sqrt{14936.0181}\,\sqrt{13828.8869}} = 0.5266$$

□

z-score form

Sometimes regression analyses are performed on data in **z-score form.** Recall that to convert a variable to z-scores, one subtracts the mean score and divides the difference by the standard deviation; z-scores always have mean 0 and standard deviation 1. If X and Y are in z-score form, so both standard deviations are 1, then the simple-regression slope equals the correlation. Further, because the mean in z-score form is 0, it follows that the intercept is also 0:

$$\hat{z}_y = r_{xy} z_x$$

Thus if the X variable is at its mean ($z_x = 0$), the predicted Y value will equal the Y mean ($\hat{z}_y = 0$). If the X variable is 1 standard deviation above the X mean ($z_x = 1$), the predicted Y value will be r_{xy} standard deviations from the Y mean ($\hat{z}_y = r_{xy}$). Note that when X is 1 standard deviation above the mean, we do *not* predict that Y will be 1 standard deviation above the mean, but rather something less than that. For example, if one tests a group of subjects twice (and if the correlation between the first and second tests is less than 1), subjects who score high on the first test are not predicted to score as high on the second test, but rather to "come back toward the average." This phenomenon is called **regression toward the mean**—whence the name regression.

regression toward the mean

Example 15.12 In Examples 15.10 and 15.11, is there evidence of regression toward the mean?

Solution Yes. The correlation found in Example 15.11 was only .5266. Thus a child who was 1 standard deviation above average at age 6 months would be predicted to be only .5266 standard deviation above average at age 15 months. □

Given certain assumptions that will be discussed in Section 15.4, the least-squares estimates of the slope and intercept are unbiased and most efficient. However, if the assumptions aren't correct, the least-squares estimates may not be good ones. In particular, these estimators are quite sensitive to **outliers**, especially those corresponding to extreme X values. Figure 15.10 shows a prediction line and one of the residuals. If an outlier observation had been at the point marked with a #, the residual from the indicated line would have been large, and the squared residual would have been huge. Once again, the effect of using squared error is to accept many small errors in order to avoid a few large ones; the least-squares line would be "twisted" toward the outlier. Outliers are particularly damaging if they are outlying not only in the Y direction, but also in the X direction. The point marked # is such a point. Because it is far from the middle of the data in the X direction, it is said to have large **influence** or **leverage.** Belsley, Kuh, and Welsch (1980) and Weisberg and Cook (1982) have developed ways to measure the influence of particular points. Those that have large influence should be checked to see if they may be distorting the regression equation. Once again, it's important to plot the data.

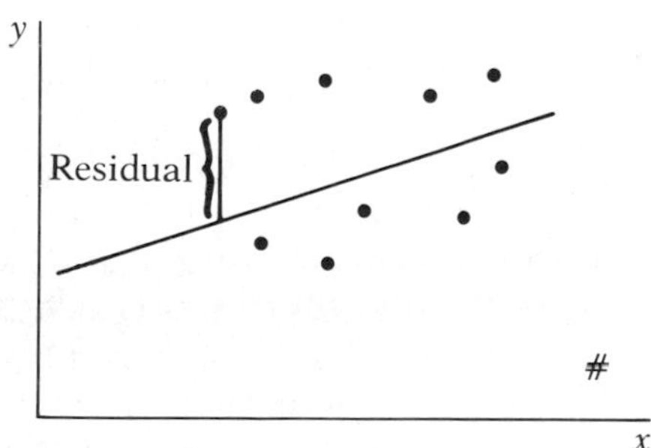

Figure 15.10 Residual, and an outlier point

Some computer packages (Minitab, for one) allow one to calculate **robust regression** (or resistant) lines. The robust estimates are much less affected by outlier values. As usual, it is a good idea to plot the data before deciding on the desired analysis. In Example 15.2, we noted that there was a very large outlier. The least-squares estimates of the slope and intercept would be severely affected by the outlier; the resistant line, much less so. If there's doubt about whether the assumptions are reasonable (e.g., whether certain points should be regarded as outliers), it's easy enough to do both forms of analysis and see if there's a serious difference in the results. Unless there's one or more outliers, the least-squares and robust regression lines will usually be about the same.

Example 15.13

Refer to Example 15.10. A resistant line calculated for the data yielded an intercept estimate of 17.8107 and a slope estimate of 0.5099. Are these estimates greatly different from the least-squares estimates? What does the result suggest about the desirability of using the least-squares estimates?

Solution

The least-squares slope was 0.5067, remarkably close to the resistant-line slope. The least-squares intercept was 18.607, not far from the resistant-line intercept. Thus, in this case, the least-squares line seems unperturbed by outliers. □

Up until now, we have been considering estimating the coefficients of a simple linear regression, with a single independent variable.

multiple regression

The same basic ideas apply to **multiple regression,** where there are several independent variables (possibly including dummy variables, product terms, powers or other transformations of other variables, etc.). When there are many coefficients to estimate, the arithmetic obviously becomes more complicated. Almost always, multiple regression is done by one of the literally thousands of available computer programs. For the mathematics of multiple regression, see Draper and Smith (1982), Hanushek and Jackson (1977), or any econometrics textbook. The more interesting question, for our purposes, is the relation between coefficients in simple and multiple regression.

collinearity

A key idea in understanding multiple regression is **collinearity**—correlation among independent variables. As we noted in Section 15.1, when there are two (or more) independent variables in a regression model, the slope coefficients are *partial slopes*—predicted change in Y corresponding to a 1-unit change in that x, with other x's held constant. When there's only one independent variable in the model, obviously there aren't any other variables to be held constant. If the correlations among independent variables are all 0, then there's no tendency for any one x to change as another changes. Thus the simple and multiple regression slopes are the same. The intercept may change, but that's usually a minor issue. But when there's collinearity among the independent variables, the simple and multiple regression slopes will be different.

The issue can be understood by considering z-score regression. Recall that when there is a single independent variable,

$$\hat{z}_Y = r_{YX} z_X$$

It is possible to prove that when there are two independent variables, the coefficient of z_{X_1} is

$$\frac{r_{YX_1} - r_{X_1X_2} r_{YX_2}}{1 - r^2_{X_1X_2}}$$

where the subscripts on r indicate which two variables are being correlated. If there is no collinearity at all, $r_{X_1X_2} = 0$, and the coefficient of z_{X_1} reduces to r_{YX_1}, as in the simple regression case. But if $r_{X_1X_2}$ is not 0, the coefficient of z_{X_1} will change in multiple regression.

The extreme case of collinearity occurs when one x is an exact *linear* function of other x's, as when $x_3 = x_1 - x_2$ or $x_4 = (x_1 + x_2 + x_3)/3$. Then as one x value changes, another necessarily changes in proportion in the data, and there is no information at all about the effect of changing one variable while holding others constant. In such a case, a computer program can't estimate the coefficients, and should produce some kind of error message. (Note that if $r_{X_1X_2}$ were equal to 1 or -1 in the z-score equation, we would be trying to divide by 0, an undefined operation.) Fortunately, it is legitimate to have one x as an exact, but nonlinear, function of another, as when $x_2 = x_1^2$. In such a case, coefficients can be estimated. The interpretation of the coefficients depends completely on the relation among x's.

Example 15.14 Refer to the Bolton study of Example 15.3, where we were trying to predict Y = total SAT score. Would it be wise to use PSAT verbal score, PSAT quantitative score, and PSAT total score as predictors in the same equation?

Solution No, indeed! By definition, the total score is the sum of the verbal and quantitative scores. Thus we would have perfect collinearity and the coefficients could not be estimated at all. □

So far in this section, the concern has been with estimating the coefficients ($\hat{\beta}$'s) of a specified model. There is another important parameter to be estimated, namely the standard deviation of individual values around the model's prediction. Ideally, the prediction is so good that the individual values all are very close to predicted values; in this ideal case, the standard deviation of individual values around the model equation will be nearly 0. Of course, in practice models will not give perfect predictions, and the standard deviation of prediction errors will not be 0.

The variance of prediction errors is based on the sum of squared residuals,

$$\text{SS(Residual)} = \sum (y_i - \hat{y}_i)^2$$

There are two ways (at least) to compute this SS. We could calculate all the predicted values $\hat{y}_i$ and all the residuals $y_i - \hat{y}_i$, then square the residuals and sum. Alternatively, it can be shown that, for simple regression,

$$\text{SS(Residual)} = \sum y_i^2 - \hat{\beta}_0\left(\sum y_i\right) - \hat{\beta}_1\left(\sum x_i y_i\right)$$

For the marital conventionalization data, SS(Residual) can be calculated either way. The residuals are −2.4092, 10.3375, −6.9708, 1.0292, 7.7288, −11.2792, 3.1042, and −1.5125, and the sum of the squared residuals is 361.17. By the alternative computational formula,

$$\begin{aligned}\text{SS(Residual)} &= 19839 - 68.579167(393) - (-1.691667)(4418)\\ &= 361.17\end{aligned}$$

Often, computer programs calculate SS(Residual) using both the shortcut, computational formula and the definition. If the reported values by both methods are nearly equal, then there should be little problem with round-off error, at least as far as the residuals are concerned.

To obtain MS(Residual), an estimate of the residual variance, we need the appropriate df. In simple regression (only one x), there are two model coefficients, the intercept and the slope, to be estimated, using up 2 df; the remaining $n - 2$ df are available for estimating variability. Thus, if $n = 8$ and $k = 1$ (simple regression), the residual df are $8 - (1 + 1) = 6$. In multiple regression with k independent variables (counting all dummy variables, product terms, and other transformations), there are $k + 1$ coefficients to be estimated (k partial slopes and the intercept), so in general there are $n - (k + 1)$ df for variability. If we had 25 observations, 2 original independent variables, 3 dummy variables, and 6 product terms, k would be $2 + 3 + 6 = 11$, and there would be $25 - (11 + 1) = 13$ df for SS(Residual). Notice that dummy variables and product terms can chew up a great many df; to estimate more complicated models, one naturally needs more data.

Residual Standard Deviation

Estimate of variance around the regression equation:

$$\text{MS(Residual)} = \frac{\text{SS(Residual)}}{n - (k + 1)}$$

Estimate of residual standard deviation:

$$s_{Y|1,2,\ldots,k} = \sqrt{\text{MS(Residual)}}$$

NOTE: The subscripts following the | indicate which independent variables have been used in the model.

For example, for the first artificial data set, a computer program would show that SS(Residual) = 49.33333. There are 6 residual df, so

$$\text{MS(Residual)} = \frac{49.33333}{6} = 8.22222$$

and the residual standard deviation would be

$$s_{Y|1,2} = \sqrt{8.22222} = 2.86744$$

Example 15.15 Refer to the Lasky et al. data of Example 15.10. Calculate SS(Residual) and the residual standard deviation.

Solution In Example 15.10, we had $\Sigma y^2 = 269{,}645$, $\Sigma y = 7519$, and $\Sigma xy = 236{,}336$. We calculated $\hat{\beta}_0 = 18.6065$ and $\hat{\beta}_1 = 0.506686$. Thus

$$\begin{aligned}\text{SS(Residual)} &= 269{,}645 - (18.6065)(7519) - (0.506686)(236{,}336)\\ &= 9994.58\end{aligned}$$

The sample size n was 221, so there are $221 - (1 + 1) = 219$ df.

$$\text{MS(Residual)} = \frac{9994.58}{219} = 45.6374$$

so

$$s_{Y|X} = \sqrt{45.6374} = 6.7555$$

□

standard error of estimate

In some computer outputs, the residual standard deviation is labeled the **standard error of estimate.** This name is a bit misleading, in that the residual standard deviation is obtained directly from data, and standard errors are properties of theoretical probability distributions. Therefore, we'll use the phrase "residual standard deviation," even though it's not in universal usage.

The residual standard deviation can be interpreted like any other standard deviation. By the Empirical Rule, about 68% of the residuals (prediction errors) will be within ±1 standard deviation of 0 (0 always being the mean of the residuals), and about 95% of the residuals will be within ±2 standard deviations of 0.

Example 15.16 In the Example 15.10 data, 150 of the residuals were within ±6.7555, and 211 of them were within ±13.5110. Are these values compatible with the Empirical Rule approximation?

Solution The plus-or-minus values given are 1 and 2 standard deviations, respectively. 150/221 is 67.9%, and 211/221 is 95.5%, both very close to the Empirical Rule approximation. □

The residual standard deviation is also like any other standard deviation in being highly sensitive to outliers. It's always a good idea to obtain a plot of the residuals to check for excessively large values. Section 15.6 will have more to say about residual plots.

Exercises for Section 15.2

15.12. The data for the study in Exercise 15.1 were

X	18	22	24	25	25	26	26	26	27	27	29	30	32	35	37	40	43
Y	28	30	36	34	37	33	39	35	38	36	39	42	38	41	49	54	52

a. Compute S_{xy}, S_{xx}, and S_{yy}.

b. Compute and interpret the coefficients of a straight-line regression model.

c. Compute SS(Residual).

d. Compute the residual standard deviation.

15.13. The data from Exercise 15.5 are reproduced here:

X_1	−2	−2	−2	−1	−1	−1	0	0	0	1	1	1	2	2	2
X_2	−1	0	1	−1	0	1	−1	0	1	−1	0	1	−1	0	1
Y	5	4	9	2	9	13	6	10	14	10	13	13	12	14	16

a. Find the slope and intercept for predicting Y from only X_1, using a straight-line model.

b. Calculate each of the residuals.

c. Find SS(Residual) and the residual standard deviation directly from the definitions.

d. Use the short-cut method to recalculate SS(Residual).

15.14. Refer to the data in Example 15.13.

a. Find the slope and intercept for predicting Y from only X_2, using a straight-line model.

b. Find the residual standard deviation for this model.

c. Which X variable appears to give more accurate prediction of Y?

15.15. Calculate correlations between all pairs of variables for the data set in Exercise 15.13.

a. Which X variable has a larger correlation with Y? How does your answer relate to the answer to Exercise 15.14?

b. What does the correlation between X_1 and X_2 indicate about the coefficients of X_1 and X_2 in a multiple regression model?

15.16. The data from Exercise 15.6 are reproduced here:

X_1	−2	−2	−2	−1	−1	−1	0	0	0	1	1	1	2	2	2
X_2	−1	−1	0	−1	0	0	−1	0	1	0	0	1	0	1	1
Y	1	6	5	4	8	9	7	12	11	9	13	17	14	21	13

a. Calculate the intercept and slope for predicting Y using only X_1.

b. Calculate the coefficients for predicting Y using only X_2.

15.17. Calculate the residual standard deviations for both models in Exercise 15.16, using either the shortcut or the definitions. Which variable seems to be the better predictor?

15.18. Calculate the correlations between Y and X_1 and between Y and X_2. Which correlation is larger? Does your answer confirm the answer to Exercise 15.17?

15.19. The data from Exercise 15.3 are shown again here in Table 15.3.

$D = 0.4$		$D = 4.0$	
Intensity	Speed	Intensity	Speed
0.90	34	4.0	21
1.00	40	4.6	29
1.10	31	5.1	32
1.26	68	5.8	36
1.44	59	6.4	42
1.59	73	7.2	57
1.81	61	8.0	60
2.00	96	9.1	70
		10.1	87

Table 15.3

a. Let X = intensity and Y = speed. Calculate the coefficients of a linear regression model for predicting Y given X, using the $D = 0.4$ data.

b. Do the same for the $D = 4.0$ model.

c. Calculate the residual standard deviations for both models.

15.20. The transformed data of Exercise 15.4 were put through the Minitab computer programs. The resulting equations (where Y' is the natural logarithm of Y and X' is the natural logarithm of X) were

$$D = 0.4: \hat{y}' = 3.64 + 1.21x', \; s_{Y'|X'} = 0.2129$$

$$D = 4.0: \hat{y}' = 1.06 + 1.46x', \; s_{Y'|X'} = 0.05731$$

a. Can the coefficients be compared directly to the coefficients of the linear models?

b. Can the residual standard deviations be compared directly?

15.21. The models resulting from Exercise 15.20 were transformed back to the original Y (speed) units. The sums of squared residuals for the resulting predictions were 848.12 (for the $D = 0.4$ data) and 50.921 (for the $D = 4.0$ data). How do these values compare to the SS(Residual) for the original, untransformed models in Exercise 15.19? Which model appears to yield a better prediction, the original model or the "retransformed" model?

15.3 Correlations and Predictability

In Section 15.2 we defined the correlation coefficient and showed that it was related to the slope of a regression line. We now turn to the issue of interpreting the numerical value of the correlation coefficient and assessing how well a model predicts a dependent variable.

Just as in Chapter 3, we will assess predictive value by comparing how well a dependent variable can be predicted when an independent variable is known to how well the same variable can be predicted when the independent variable is not known. In regression, we will use the squared error of prediction as our criterion. If an independent variable X is not known, the best we can do in predicting Y is to predict that it will be at the mean value $\bar{y}$. Recall that the sum of squared deviations of Y values around $\bar{y}$ is defined as

$$\text{SS(Total)} = \sum (y_i - \bar{y})^2$$

SS(Total) is the same thing as S_{yy}, in the notation of Section 15.1. This SS is the total squared error made in predicting Y values *without* any information about the independent variables (X's). The squared error made in predicting Y values *with* information about the independent variables is

$$\text{SS(Residual)} = \sum (y_i - \hat{y}_i)^2$$

squared correlation

as defined in Section 15.2. It can be proved that the **squared correlation** measures the proportionate reduction in (squared) prediction error obtainable by using the independent variable information. Thus, in simple regression with a single X variable, we have

$$r_{YX}^2 = \frac{\text{SS(Total)} - \text{SS(Residual)}}{\text{SS(Total)}}$$

Example 15.17

In Example 15.11, we specified the following summary values, where X = motor skills scores of boys aged 6 months and Y = motor skills of the same boys aged 15 months: $S_{xx} = 14936.0181$, $S_{yy} = 13828.8869$, and $S_{xy} = 7567.8733$. Calculate SS(Residual), SS(Total), and r_{YX}, and verify that

$$r_{YX}^2 = \frac{\text{SS(Total)} - \text{SS(Residual)}}{\text{SS(Total)}}$$

Solution In Examples 15.11 and 15.15, we computed SS(Residual) = 9994.58, SS(Total) = S_{yy} = 13828.8869, and $r_{YX} = .5266$. We have $r_{YX}^2 = (.5266)^2 = .2773$, and

$$\frac{\text{SS(Total)} - \text{SS(Residual)}}{\text{SS(Total)}} = \frac{3834.3069}{13828.8869} = .2773$$ □

When there are several X variables, possibly including dummy variables, transformed variables, or product terms, the computation of predictive value is virtually always done using a computer program. In

multiple correlation

multiple regression, the relevant correlation is the **multiple correlation,** defined as the correlation between actual and predicted values.

$$R_{Y|1,2,\ldots,k} = r_{Y\hat{Y}}$$

The interpretation of the multiple correlation coefficient is once again that the squared correlation measures the proportionate reduction in squared error:

$$R_{Y|1,2,\ldots,k}^2 = \frac{\text{SS(Total)} - \text{SS(Residual)}}{\text{SS(Total)}}$$

To calculate the error reduction, one can either calculate the correlation and square it, or calculate the appropriate SS. Virtually all computer programs report both the SS and the R^2 value. For example, the Minitab system yielded the output in Figure 15.11 for the two artificial data sets of Section 15.2.

The names for the sums of squares are not universal. SS(Total) is sometimes called SS(Corrected Total)—"corrected" only in the sense that the mean was subtracted before squaring. SS(Regression) = $\Sigma\,(\hat{y}_i - \bar{y})^2$ is also called SS(Model) or SS(Explained), among other names. SS(Residual) is also called SS(Error) or SS(Unexplained). There is, however, a universal convention; SS(Regression) is always listed first, SS(Residual) is always second, and SS(Total)—if it's listed at all—is always last.

```
MTB> REGRESS C3 ON 2 VARS IN C1 AND C2

THE REGRESSION EQUATION IS
Y =     10.0 +  2.00 X1 +  3.00 X2

                                            ST. DEV.    T-RATIO =
        COLUMN      COEFFICIENT             OF COEF.    COEF/S.D.
        --            10.0000               0.4216        23.72
X1      FIRSTX1        2.0000               0.2981         6.71
X2      FIRSTX2        3.0000               0.5164         5.81

THE ST. DEV. OF Y ABOUT REGRESSION LINE IS
S = 1.633
WITH (  15- 3) =  12 DEGREES OF FREEDOM

R-SQUARED = 86.8 PERCENT
R-SQUARED = 84.6 PERCENT, ADJUSTED FOR D.F.

ANALYSIS OF VARIANCE

 DUE TO      DF          SS        MS=SS/DF
REGRESSION    2      210.000        105.000
RESIDUAL     12       32.000          2.667
TOTAL        14      242.000

MTB> REGRESS C13 ON 2 VARS IN C11 AND C12

THE REGRESSION EQUATION IS
Y =     10.0 +  3.00 X1 +  2.00 X2

                                            ST. DEV.    T-RATIO =
        COLUMN      COEFFICIENT             OF COEF.    COEF/S.D.
        --            10.0000               0.6325        15.81
X1      SECONDX1       3.0000               0.4472         6.71
X2      SECONDX2       2.0000               0.7746         2.58

THE ST. DEV. OF Y ABOUT REGRESSION LINE IS
S = 2.449
WITH (  15- 3) =  12 DEGREES OF FREEDOM

R-SQUARED = 81.2 PERCENT
R-SQUARED = 78.0 PERCENT, ADJUSTED FOR D.F.

ANALYSIS OF VARIANCE

 DUE TO      DF          SS        MS=SS/DF
REGRESSION    2      310.000        155.000
RESIDUAL     12       72.000          6.000
TOTAL        14      382.000
```

Figure 15.11 Minitab output

Example 15.18 Computer output for the data of Example 15.10 is shown below. Find the r^2 value.

R-SQUARED = 27.7 PERCENT
R-SQUARED = 27.4 PERCENT, ADJUSTED FOR D.F.

ANALYSIS OF VARIANCE

DUE TO	DF	SS	MS = SS/DF
REGRESSION	1	3834.54	3834.54
RESIDUAL	219	9994.35	45.64
TOTAL	220	13828.89	

Solution There are two r^2, one plain and one "adjusted for df." Comparing the output with the calculation in Example 15.17, we note that the regular r^2 value, shown as 27.7 percent rather than as .2773, is the desired one. □

In multiple regression, it's natural to try to analyze the overall R^2 into pieces attributable to each X separately. For example, if $R^2_{Y|1,2} =$ .638, so X_1 and X_2 together account for 63.8% of the variation in Y, how much of that 63.8% is attributable to X_1 and how much to X_2? In general, there's no good way to answer that question because of the problem of collinearity (correlated X variables). At the extreme, suppose that we have two X variables: X_1 = subject's height in inches, and X_2 = subject's height in feet. If the two X's combine to explain 63.8% of the variation in Y = subject's weight, how much is attributable to height in inches and how much to height in feet?!? Obviously, the two X's will be perfectly correlated, and there's no way to separate out the predictive value of one from the other. When there is collinearity, as there usually is in observational studies, it is difficult to separate out the predictive value of separate X's in an exact way.

orthogonal predictors

There is, however, one case in which the predictive value of each variable can be defined unambiguously. That occurs in the so-called **orthogonal predictors** case, where all the X's are completely uncorrelated with one another (though correlated with Y). If all correlations among X's are 0, then

$$R^2_{Y|1,2,\ldots,k} = r^2_{Y,1} + r^2_{Y,2} + \cdots + r^2_{Y,k}$$

The first artificial data set of Section 15.2 had two X variables that were uncorrelated. For that data set, SS(Total) = 280 and SS(Residual) = 49.33333. Thus

$$R^2_{Y|1,2} = \frac{280 - 49.33333}{280} = .8238$$

Direct calculations show that $r^2_{Y1} = .5357$ and $r^2_{Y2} = .2881$; in this case, multiple R^2 is the sum of the simple r^2 values, .5357 + .2881. In the second artificial data set, multiple R^2 is .6900, which is not the sum of the simple r^2 values, .5357 + .6095.

Example 15.19 In the Bolton study discussed in Example 15.3, can one expect additivity of the R^2 values?

Solution No. The X variables (IQ and PSAT score) will almost certainly show positive correlations. □

partial correlation

An alternative analysis of a multiple-regression R^2 involves the idea of using independent variables in sequence, and the concept of **partial correlation.** The partial correlation between Y and X_k, adjusting for the effects of $X_1, \ldots, X_{k-1}$, measures the *additional* predictive value of X_k, after accounting for the predictive value of $X_1, \ldots, X_{k-1}$. In principle, though not in practice, it can be computed in three steps. First, find the residuals of Y in a regression using the first $k - 1$ X's (yielding the "unpredicted" part of each Y value). Second, find the residuals of X_k in a regression with X_k temporarily playing the role of dependent variable and the other $k - 1$ X's being independent variables. Third, find the correlation of the two sets of residuals. Thus the partial correlation relates the unexplained part of the Y values to the unexplained part of the X_k values. The key idea is not the computation of a partial correlation, but rather its interpretation. The error reduction interpretation of $r_{Yk|1,2,\ldots,k-1}$, the partial correlation of Y with X_k, adjusting for $X_1, \ldots, X_{k-1}$, is

$$r^2_{Yk|1,2,\ldots,k-1} = \frac{\text{SS(Residual after } X_1, \ldots, X_{k-1}) - \text{SS(Residual after } X_1, \ldots, X_k)}{\text{SS(Residual after } X_1, \ldots, X_{k-1})}$$

For example, if $r^2_{Y3|1,2} = .231$, variable X_3 accounts for 23.1% of the variation in Y that has *not* been explained by X_1 and X_2.

Example 15.20 What would be the interpretation of $r_{Y3|1,2} = .00$? Would such a result mean that X_3 was useless as a predictor of Y?

Solution In this case variable X_3 accounts for none of the *residual* variation in Y, after the effects of X_1 and X_2 are accounted for. X_3 may or may not be a useful predictor of Y in its own right; the partial correlation indicates that once X_1 and X_2 have been considered, X_3 loses any predictive value it may have had in the data. □

One way to compute a partial r^2 is to fit one regression model including the relevant X and another model excluding it. Then the reduction in residual SS can be computed by hand. Alternatively, some computer packages will compute the incremental amounts of Y variation explained by each independent variable, entered into the regression in a specified sequence. For example, suppose that a computer output indicates that

SS(Total) = 2000

and that the incremental explained (regression) SS for X_1, X_2, and X_3, entered in that order, are

X1 500
X2 700
X3 100

It follows that the regression and residual SS for various models are as shown in Table 15.4.

Model	SS(Regression)	SS(Residual)
X_1	500	2000 − 500 = 1500
X_1, X_2	500 + 700 = 1200	2000 − 1200 = 800
X_1, X_2, X_3	500 + 700 + 100 = 1300	2000 − 1300 = 700

Table 15.4

In this sequence,

$$r^2_{Y1} = \frac{2000 - 1500}{2000} = .25$$

$$r^2_{Y2|1} = \frac{1500 - 800}{1500} = .467$$

$$r^2_{Y3|1,2} = \frac{800 - 700}{800} = .125$$

Example 15.21 Computer output from a multiple regression analysis of the data for the Bolton study of Example 15.3 yielded the following SS for predicting SAT from PSAT, IQ, and QPA (high school grade average):

ANALYSIS OF VARIANCE

DUE TO	DF	SS	MS = SS/DF
REGRESSION	3	1669189	556396
RESIDUAL	78	400256	5131
TOTAL	81	2069445	

FURTHER ANALYSIS OF VARIANCE
SS EXPLAINED BY EACH VARIABLE WHEN ENTERED IN THE ORDER GIVEN

DUE TO	DF	SS
REGRESSION	3	1669189
PSAT	1	1621452
IQ	1	556
QPA	1	47182

Calculate $r^2_{\text{SAT,IQ}|\text{PSAT}}$ and $r^2_{\text{SAT,QPA}|\text{PSAT,IQ}}$.

Solution For $r^2_{\text{SAT,IQ}|\text{PSAT}}$, we need the incremental SS explained by adding IQ to PSAT as a predictor; this SS is shown in the output as 556. To find SS(Residual, PSAT), we note that SS(Total) = 2,069,445, and SS(Regression, PSAT) is shown as 1,621,452. So SS(Residual, PSAT) = 2,069,445 − 1,621,452 = 447,993. Thus

$$r^2_{\text{SAT,IQ}|\text{PSAT}} = \frac{556}{447{,}993} = .0012\ (!)$$

For $r^2_{\text{SAT,QPA}|\text{PSAT,IQ}}$, we note that

$$\begin{aligned}\text{SS(Residual, SAT and QPA)} &= 2{,}069{,}445 - (1{,}621{,}452 + 556)\\ &= 447{,}437\end{aligned}$$

and the additional SS explained by adding QPA is shown in the output as 47,182. Thus

$$r^2_{\text{SAT,QPA}|\text{PSAT,IQ}} = \frac{47{,}182}{447{,}437} = .1054$$

The interpretation is that adding IQ to the model after incorporating PSAT accounts for 0.12% of the unexplained variation, and adding QPA to the PSAT, IQ model accounts for 10.54% of the residual variation. Obviously, the IQ variable appears to be practically worthless. □

multiple R^2 adjusted for degrees of freedom

Some variations on these R^2 measures are often reported in the social science literature, or by computer packages. The most widely used is the **multiple R^2 adjusted for degrees of freedom.** Instead of the proportionate reduction in the total squared prediction error (which is the ordinary multiple R^2), the adjusted R^2 is the proportionate reduction in residual variance.

$$R^2_{Y|1,2,\ldots,k,\text{adj}} = \frac{\dfrac{\text{SS(Total)}}{n-1} - \dfrac{\text{SS(Residual)}}{n-(k+1)}}{\text{SS(Total)}/(n-1)}$$

If n, the number of observations, is large relative to k, the number of x terms, the unadjusted and adjusted R^2 values will be almost the same. If n is relatively small, the adjusted R^2 will be substantially less than unadjusted R^2 and can even be negative. The adjusted R^2 is closely related to the residual standard deviation. The adjusted R^2 is largest when the residual standard deviation is smallest. Although one could also adjust partial r^2 values for df, few computer programs or research reports do so.

Example 15.22

In the output of Example 15.18, the ordinary r^2 was .277, whereas the r^2 adjusted for df was .274. How critical is the choice of adjusted or unadjusted r^2?

Solution

The two figures are very similar. The X variable accounts for slightly more than 27% of the Y variability, whichever way r^2 is measured. □

part correlation

Occasionally, a **part correlation** is reported. The squared part correlation of Y with X_k, adjusting for $X_1, \ldots, X_{k-1}$, is the increment to the explained SS achieved by adding X_k to the model, but taken as a fraction of SS(Total). Partial correlation squared is the increment taken as a fraction of SS(Residual after $X_1, \ldots, X_{k-1}$). The meaning of "part r^2 of Y and X_3 controlling for X_1 and X_2 equals .216" is that adding X_3 to the model explains an additional 21.6% of the variation in Y. The problem is that, if the first two variables had explained 78.3%, the additional 21.6% would be almost the maximum possible, but if the first two variables had only explained 4.3%, the additional 21.6% would be only a modest part of the "still-explainable" variation. Therefore, we'll not use part correlation further.

The error reduction interpretation of correlations is a bit dangerous when the experimenter has some control over the values of the x variables. Typically, if the x variables take on values over an unnaturally wide range, correlations will be unnaturally large. Conversely, if the independent variables take on values over an unnaturally narrow range, correlations will be unnaturally small. Suppose that there is a single independent variable X, perhaps an aptitude score, correlated with a result Y. Further, suppose that the actual scores naturally ranged over the values shown in Figure 15.12, but that only the points with X scores to the right of the dotted line in Figure 15.12 were included in the study. This situation arises, for example, when X is an aptitude test score and Y is the grade point average of students admitted to a very selective university. Only those students with very high X (aptitude) scores are admitted in the first place, so only those students have Y scores to be analyzed in the study. Had the full range of X scores been included in the data, the correlation might have been rather high. But when there is a **truncated range** of X scores, the reported correlation will be substantially lower.

truncated range

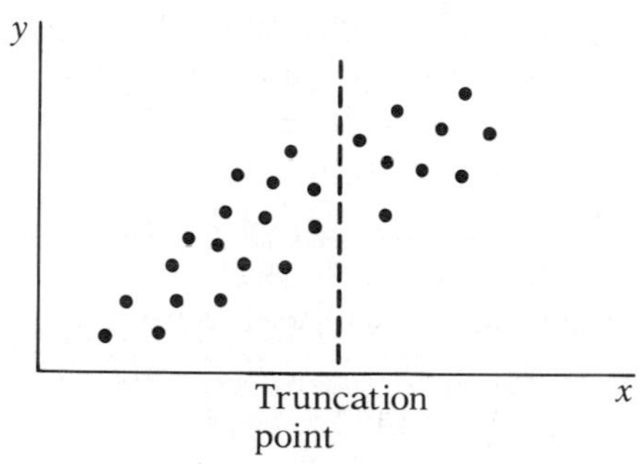

Figure 15.12 Effect of truncated range

regression experiments
correlation studies

Some textbooks make a distinction between **regression experiments,** where the values of the independent variables are fixed in advance by the researcher, and **correlation studies,** where the X values occur naturally. The distinction is a useful one, because R^2 values can be distorted by the choice of x values. The error-reduction interpretation of correlations is so useful that we won't abandon it, even when the X's are preset; however, it's a good idea to consider whether the values of the independent variables have been obtained in a natural way or in a distorted way. If the independent variables have been set in such a way as to exaggerate the natural range of variability, the reported correlations may be too optimistic.

Example 15.23

Refer to Example 15.3. Is there a danger that correlations computed from this experiment might be misleading?

Solution Presumably, the relevant population is all students of the school who will take the SAT tests. Assuming that the data are chosen randomly from the population of SAT-takers, there's no problem. A truncated-range problem might arise if the data had been taken only from "middling-quality" students, on grounds that top students and poor students have little need for an SAT forecast. □

Exercises for Section 15.3

15.22. Refer to the data of Exercise 15.12.

a. Calculate r_{YX}.

b. Calculate SS(Total) $= s_{yy}$ and SS(Residual).

c. Verify that the squared correlation is the proportional reduction in squared error for these data.

15.23. Refer to the data of Exercise 15.13. (The correlations for these data were calculated in Exercise 15.15.) Can $R^2_{Y|1,2}$ be easily calculated from these correlations alone? Why?

15.24. In an experiment, equal numbers of men and women were assigned to control and to experimental groups. The recorded variables were Y = number of trials required to solve a puzzle; $X_1 = 1$ if subject was in the experimental group, 0 if control; $X_2 = 1$ if subject was a woman, 0 if a man; $X_3 = X_1 X_2$. The data were

Y	10	15	13	15	12	5	5	13	11	12	12	12	5	10	12	10
X_1	1	1	1	1	0	0	0	0	1	1	1	1	0	0	0	0
X_2	1	1	0	0	1	1	0	0	1	1	0	0	1	1	0	0
X_3	1	1	0	0	0	0	0	0	1	1	0	0	0	0	0	0

(The sum of the y values is 172, and the sum of the squared y values is 2004.)

a. Find the correlation of Y and X_1. What does the number indicate about the predictive value of Y?

b. Find the correlation of Y with X_2 and the correlation of Y with X_3. Of the three variables, which one appears to be the best predictor of Y?

15.25. Refer to Exercise 15.24. A computer program was used to fit a regression model to the data. SS(Total) was found to be 155.00 and SS(Residual) was found to be 96.00.

a. Calculate $R^2_{Y|1,2,3}$.

b. Is this $R^2 = r_1^2 + r_2^2 + r_3^2$?

15.26. The data of Exercise 15.24 were also fit to a regression model with only X_1 and X_2 included. SS(Regression) was 58.00 and SS(Residual) was 97.00. Calculate $r^2_{Y3|1,2}$. What does the resulting number indicate about the predictive value of X_3, the product term?

15.27. Refer to Exercises 15.24 and 15.25. Compute the R^2 values, adjusted for degrees of freedom.

15.28. A study of boys playing soccer in a certain league yielded data on four variables: EXPECT = boy's pregame expectation of his performance, ABILITY = boy's self-rating of his own ability, ESTEEM = score on a self-esteem scale, and ANXIETY = score on a "competitive trait anxiety" scale. Selected parts of SAS output for the data are shown in Figure 15.13.

CORRELATIONS FOR CHILDREN'S SOCCER DATA

VARIABLE	N	MEAN	STD DEV	SUM	MINIMUM	MAXIMUM
EXPECT	187	4.86096257	1.97073348	909.00000000	0	9.00000000
ABILITY	187	5.27807487	2.26419074	987.00000000	0	10.00000000
ESTEEM	187	5.65240642	1.28348990	1057.00000000	2.00000000	8.00000000
ANXIETY	187	5.13903743	2.20992997	961.00000000	0	10.00000000

CORRELATION COEFFICIENTS / PROB > |R| UNDER H0:RHO=0 / N = 187

	EXPECT	ABILITY	ESTEEM	ANXIETY
EXPECT	1.00000	0.65453	0.59719	0.37357
PREGAME PERSONAL PERFORMANCE EXPECTATION	0.0000	0.0001	0.0001	0.0001
ABILITY	0.65453	1.00000	0.89001	0.34896
RATING OF OWN ABILITY	0.0001	0.0000	0.0001	0.0001
ESTEEM	0.59719	0.89001	1.00000	0.29576
SCORE ON SELF ESTEEM SCALE	0.0001	0.0001	0.0000	0.0001
ANXIETY	0.37357	0.34896	0.29576	1.00000
COMPETITIVE TRAIT ANXIETY	0.0001	0.0001	0.0001	0.0000

REGRESSION ANALYSIS OF CHILDREN'S SOCCER DATA

DEP VARIABLE: EXPECT PREGAME PERSONAL PERFORMANCE EXPECTATION

SOURCE	DF	SUM OF SQUARES	MEAN SQUARE	F VALUE	PROB>F
MODEL	3	327.829	109.276	50.684	0.0001
ERROR	183	394.556	2.156046		
C TOTAL	186	722.385			
ROOT MSE		1.468348	R-SQUARE	0.4538	
DEP MEAN		4.860963	ADJ R-SQ	0.4449	
C.V.		30.20694			

Figure 15.13 Output for Exercise 15.28

a. Locate the correlation coefficients of all variables. Which other variable predicts the dependent variable EXPECT best?

b. Locate R^2 for predicting EXPECT given ABILITY, ESTEEM, and ANXIETY. What does the number mean?

c. How much difference does it make whether or not the R^2 value is adjusted for degrees of freedom?

15.29. Refer to Exercise 15.28. Should the R^2 value be equal to

$$r^2_{\text{EXPECT,ABILITY}} + r^2_{\text{EXPECT,ESTEEM}} + r^2_{\text{EXPECT,ANXIETY}}$$

15.30. A regression study involved predicting perceived well-being as a function of age and education. Selected SPSS-X output is shown in Figure 15.14.

```
              PERCEIVED  WELL-BEING AS A FUNCTION OF AGE AND EDUCATION


              MEAN  STD DEV  LABEL

WELL       9.780    3.320  PERCEIVED WELL-BEING
EDUCN      8.150    3.316  YEARS OF FORMAL EDUCATION
AGE       52.430   17.792  AGE IN YEARS

N OF CASES =   100

CORRELATION:

               WELL       EDUCN        AGE

WELL          1.000       -.425        .202
EDUCN         -.425       1.000       -.476
AGE            .202       -.476       1.000


                                    **** M U L T I P L E   R E G R E S S I O N ****
VARIABLE LIST NUMBER 1   LISTWISE DELETION OF MISSING DATA
EQUATION NUMBER 1    DEPENDENT VARIABLE..  WELL     PERCEIVED WELL-BEING


VARIABLE(S) ENTERED ON STEP NUMBER  1..   AGE       AGE IN YEARS
                                    2..   EDUCN     YEARS OF FORMAL EDUCATION

-----------------------------------------------------------


MULTIPLE R             .42543        ANALYSIS OF VARIANCE
R SQUARE               .18099                          DF      SUM OF SQUARES      MEAN SQUARE
ADJUSTED R SQUARE      .16410        REGRESSION         2           197.49064         98.74532
STANDARD ERROR        3.03531        RESIDUAL          97           893.66936          9.21309

                                     F =      10.71794       SIGNIF F =  .0001
```

Figure 15.14 Output for Exercise 15.30

a. Locate the correlation coefficients. Which variable appears to be a better predictor of WELL?

b. Locate the multiple correlation and R^2 values.

c. Locate adjusted R^2. Is it similar in magnitude to unadjusted R^2?

15.31. Additional output for the study of Exercise 15.10 is shown in Figure 15.15.

```
------------------ VARIABLES IN THE EQUATION ------------------

VARIABLE              B        SE B        BETA          T  SIG T

AGE         -9.53299E-05     .01949 -5.109E-04      -.005  .9961
EDUCN            -.42614     .10459    -.42567     -4.075  .0001
(CONSTANT)      13.25808    1.64084                 8.080  .0000
```

Figure 15.15 Output for Exercise 15.31

a. Show that the coefficient of AGE is, for all practical purposes, 0. (The notation E-05 means to shift the decimal point 5 places to the left.)

b. What does this fact indicate about the partial correlation of AGE and WELL, controlling for EDUC?

15.4 Hypothesis Tests in Regression and Correlation Studies

So far in this chapter, the entire focus has been on descriptive measures in regression analysis. Thus, the data have been taken at face value, and no allowance has been made for random variability. Now we discuss basic inference ideas in regression and correlation studies. In this section we focus on hypothesis tests, and in the next section we consider confidence intervals.

Like all other inference procedures, regression-correlation inference methods are based on assumptions. The first assumption is simply that the mathematical form of the model is correct. Specifically, for any given values of the independent variables, the expected (mean) value of the dependent variable Y is assumed to be given by the regression model:

$$E(Y) = \beta_0 + \beta_1 X_1 + \cdots + \beta_k X_k$$

The model may include transformed or dummy variables in addition to linear terms. Very likely, the model will not be exactly, perfectly correct.

Rather, it should be considered as an approximation, to be made as close as possible. In simple regression, with a single independent variable, the model can be checked using a scatter plot. A pronounced curve in the plot signals a need for a transformation.

Additional assumptions have to do with the random variation around the modeled expected value. For observation i, the random error is denoted ϵ_i, so the observed value equals the modeled expected value plus random error:

$$Y_i = \beta_0 + \beta_1 X_{i,1} + \cdots + \beta_k X_{i,k} + \epsilon_i$$

The random error term contains the effect of variables that might have been included in the equation but weren't, plus truly random effects such as measurement error in measuring Y, a subject's not feeling up to par when being observed, and so on. The assumptions about the error term are very similar to those made in previous chapters, and are shown in the box below.

Regression Assumptions about Random Errors

1. The errors are statistically independent of each other.

2. The errors have expected value 0 and constant variance, regardless of the value of the independent variables.

3. The errors are normally distributed.

The independence, normality, and homoscedasticity (constant variance) assumptions all are formal assumptions in the analysis of variance procedures discussed in the last chapter, as well as the regression procedures discussed in this chapter. Since analysis of variance methods can be incorporated into the general linear model by using dummy variables, the similarity of assumptions shouldn't be surprising.

Example 15.24

Refer to Example 15.10, relating motor skills of boys at ages 6 months and 15 months. What do the regression assumptions mean in that context?

Solution

For any given value of X = motor skills at 6 months, there is a true, population mean value of Y = motor skills at 15 months. Individual boys vary around the mean because of individual developmental differences. It is assumed that the deviation of any one child from the mean is not related to the deviation of any other child (independence), that the variability of 15-month scores around the mean is the same regardless of 6-month scores (homoscedasticity), and that the deviations for any given X are bell-shaped (normality). □

bivariate normal distribution

These assumptions all hold in simple regression when X and Y have a **bivariate normal distribution.** This distribution occurs when both X and Y are sums or averages of many component pieces. In three dimensions (say with the x axis running east-west, the y axis running north-south, and frequencies going up and down) a bivariate normal distribution is a mound-shaped hill. The contour lines (lines along which the probability density equals some constant) are all ellipses with the same shape. A few contour lines for bivariate normal densities are shown in Figure 15.16. If the X and Y observations are drawn from a bivariate normal distribution, the expected value of Y, conditional on a given X, can be shown to be a linear function of X; furthermore, the errors around that regression line will be normally distributed, with a constant variance. As long as it can be assumed that pairs of observations are selected randomly and independently from a bivariate normal distribution, the basic regression assumptions will be met. However, it remains important to check these assumptions, and we will discuss ways of doing so later in this chapter.

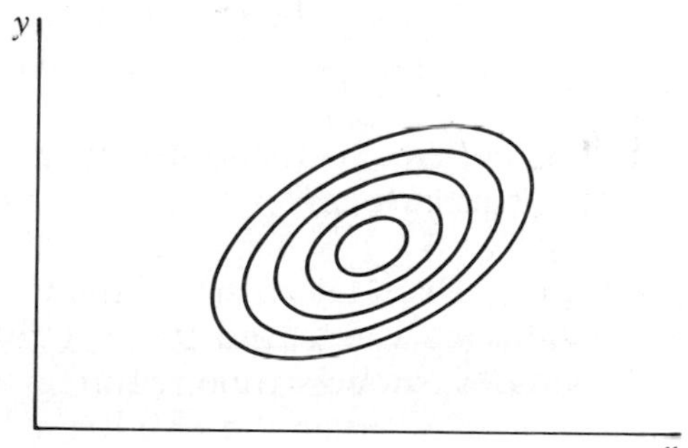

Figure 15.16 Contours of a bivariate normal density

multivariate normal distribution

In multiple regression, the same assumptions can be shown to hold if Y and all the X's jointly have a **multivariate normal distribution.** The mathematics of this density are discussed in multivariate analysis books such as Anderson (1958) and Morrison (1976).

One additional, and rather subtle, assumption must be made: that the values of the independent variables are measured reliably, without measurement error. If there is error in measuring an X value, both the slope of a regression line and the correlation coefficient can be biased toward a 0 value. This phenomenon is shown in Figure 15.17. Suppose that the correct values of X (and Y) are shown as points. The slope will be sharply positive and the correlation high. But now suppose that because of erroneous measurement of the X values, the points labeled with a * are recorded. Reasonably enough, additional random error tends to decrease the correlation between variables, and the correlation calculated using the *-marked points will be lower. Because the slope and correlation are closely related, the magnitude of the slope will also tend to be lower when there is unreliable measurement of X values. If the error variance of X measurement is large relative to the variance of Y around the regression equation, the amount of bias can be severe.

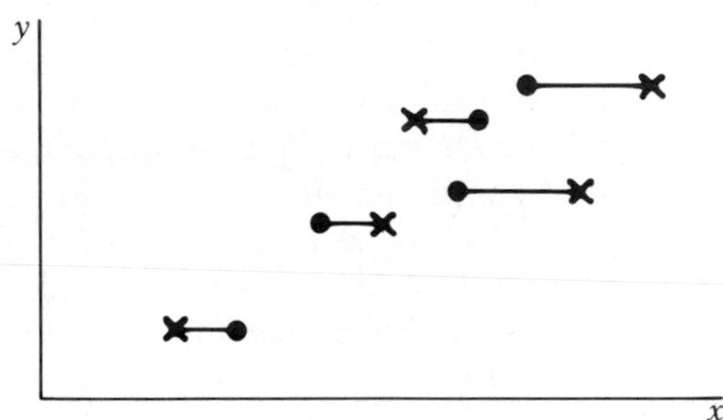

Figure 15.17 Effects of unreliable X measurement

Example 15.25 Refer to Example 15.10 again. Is there any danger of unreliable X measurement?

Solution That depends on how the motor-skills scale was constructed. If the scale was based on well-defined physical measurements or observations, such as time required to crawl a given distance or determination of whether the child can hold a specified object, relatively little measurement error would be expected. On the other hand, if the scale was based on subjective estimates, large error might be a problem. A way to test would be to see if different measurements on a particular baby gave very similar results (reliable measurement) or very different results (unreliable measurement). □

Given the assumptions, several hypothesis tests and confidence intervals are performed in regression. One is an F test for the null hypothesis that *none* of the independent variables have any predictive value at all. Under this null hypothesis, all correlations with Y, and therefore all slopes, will be 0. The test is based on sums of squares (SS) defined previously.

$$\text{SS(Regression)} = \sum (\hat{y}_i - \bar{y})^2$$

$$\text{SS(Residual)} = \sum (y_i - \hat{y}_i)^2$$

Note that SS(Regression) measures the extent to which the predicted values $\hat{y}_i$ deviate from the mean $\bar{y}$. If none of the X's has real predictive value, the predicted values will only deviate randomly from the mean, and SS(Regression) will be small. But if some of the X's have value in predicting Y, then $\hat{y}$ will vary as the X's vary, and SS(Regression) will be larger. SS(Residual) reflects the degree of random variation around the regression equation. As in analysis of variance, the SS values are divided by degrees of freedom (df) to yield mean square (MS) values. The details are shown in the box.

Test That a Regression Equation Has Some Predictive Value

1. H_0: all $\beta_j = 0$
2. H_a: at least one β_j is not 0
3. T.S.: $F = \dfrac{\text{MS(Regression)}}{\text{MS(Residual)}}$

where MS(Regression) = SS(Regression)/k
and MS(Residual) = SS(Residual)/$[n - (k + 1)]$

4. R.R.: reject H_0 if $F > F$-table value with k numerator df and $n - (k + 1)$ denominator df

The SS, df, and MS are calculated by almost all standard regression programs. The names of the SS vary from one program to another. SS(Regression) may be shown as SS(Model) or SS(Explained), and SS(Residual) may be shown as SS(Error) or SS(Unexplained). Whatever names are used, SS(Regression) is shown first, followed by SS(Residual). Many programs also print out the sum of these two SS, which is SS(Total). For example, Minitab computer output from the Bolton study of the predictability of SAT scores from PSAT, IQ, and QPA values included

ANALYSIS OF VARIANCE

DUE TO	DF	SS	MS = SS/DF
REGRESSION	3	1669189	556396
RESIDUAL	78	400256	5131
TOTAL	81	2069445	

Thus $F = 556{,}396/5131 = 108.4$, which is far beyond any F-table value for 3 and 78 df. Thus there is conclusive evidence that the three independent variables have at least some predictive value.

Example 15.26

Rubenstein and Howes (1979) reported on a regression analysis of Y = the developmental level exhibited by children while playing at day care centers. Although they included several independent variables, let us focus on X_1 = level of play of peers, X_2 = extent of adult direction or orders, X_3 = amount of holding/hugging behavior by the adult, and X_4 = amount of adult play with objects.

Note that inferring any cause-and-effect relations from such a study would be very dangerous if the data were obtained in an observational study, without experimental randomization. For instance, if Y and X_2 show a positive correlation, is that because adult direction causes an improvement in development level or because a higher developmental level causes more adult direction?

Assume that the results in Figure 15.18 were obtained from the SAS system. Identify the appropriate SS, MS, and F statistics.

REGRESSION FOR CHILD DEVELOPMENT DATA

GENERAL LINEAR MODELS PROCEDURE

DEPENDENT VARIABLE: DEVELOP   DEVELOPMENT LEVEL OF CHILD (1-99)

SOURCE	DF	SUM OF SQUARES	MEAN SQUARE	F VALUE	PR > F	R-SQUARE	C.V.
MODEL	4	17462.71628666	4365.67907167	20.25	0.0001	0.506235	30.1861
ERROR	79	17032.56942763	215.60214465		ROOT MSE		DEVELOP MEAN
CORRECTED TOTAL	83	34495.28571429			14.68339690		48.64285714

SOURCE	DF	TYPE I SS	F VALUE	PR > F	DF	TYPE III SS	F VALUE	PR > F
PEERPLAY	1	8363.09620872	38.79	0.0001	1	4335.45744815	20.11	0.0001
ADULTDIR	1	5284.73345987	24.51	0.0001	1	1313.46766230	6.09	0.0157
HOLDHUG	1	156.75703522	0.73	0.3964	1	307.96739508	1.43	0.2356
ADULTPLA	1	3658.12958285	16.97	0.0001	1	3658.12958285	16.97	0.0001

SOURCE

PEERPLAY
ADULTDIR
HOLDHUG
ADULTPLA

Figure 15.18 Output for Example 15.26

Solution

There are several tables showing SS. The first one refers to the model as a whole. (Note that the df for MODEL is shown as 4, which is the number of X's in the model.) What SAS calls SS(MODEL) is SS(Regression)and is shown, to a startling number of decimal places, as

17462.71628666. MS(Regression) is shown as 4365.67907167. The residual SS and MS are identified as ERROR, and are 17032.56942763 and 215.60214465. F is shown next to MS(MODEL) as 20.25. □

The F statistic can be written as a function of multiple R^2:

$$F = \frac{R^2/k}{(1 - R^2)/[n - (k + 1)]}$$

For example, in the Bolton example, we have

$$R^2 = \frac{1669189}{2069445} = .80659$$

with $k = 3$ and $n - (k + 1) = 78$. Thus

$$F = \frac{.80659/3}{(1 - .80659)/78} = 108.4$$

once again.

Example 15.27 Refer to Example 15.26. Calculate R^2 and confirm the F value.

Solution Rounded off to two decimal places, we have SS(Regression) = 17462.72 and SS(Total) = 34495.29.

$$R^2 = \frac{17462.72}{34495.29} = .50623$$

The numerator and denominator df were shown in the output as 4 and 79, respectively.

$$F = \frac{.50623/4}{(1 - .50623)/79} = 20.25$$

as found in Example 15.26. □

The null hypothesis of this F test is just about the "nullest" hypothesis there is. It asserts that every one of the carefully chosen explanatory variables is totally worthless as a predictor. Rejecting that null hypothesis isn't exactly a major feat; one has only established that somewhere among the predictors there is some more than random predictability. A more interesting question is Which specific predictors have value? One way to answer that question is with a t test, shown by most computer programs. There is a t test for each coefficient in the regression model. The null hypothesis for, say, coefficient 1 is H_0: $\beta_1 = 0$. This hypothesis does *not* assert that the X_1 variable is worthless as a predictor of Y. It does assert that, once all other predictors have been included, X_1 has no additional predictive value. In other words, the null hypothesis asserts that X_1 has no *additional* predictive value as the *last predictor* in the regression equation. Equivalently, it asserts that the partial correlation of X_1 with Y, controlling for the other X variables, is 0. The test of this null hypothesis requires that an estimated standard error of each regression coefficient be calculated; most standard regression programs perform that calculation automatically. One way to write the estimated standard error of, say, $\hat{\beta}_1$ which is conceptually, but not computationally, useful is

$$\text{est. std. error}(\hat{\beta}_1) = \frac{s_{Y|1,2,\ldots,k}}{\sqrt{\sum (x_{i,1} - \bar{x}_1)^2 (1 - R^2_{1|2,\ldots,k})}}$$

where $R^2_{1|2,\ldots,k}$ is the multiple R^2 when X_1 is treated as the *dependent* variable, with the other X's as independent variables.

Note that the standard error of an estimated partial slope depends on three things:

1. The residual (error) standard deviation. The larger this standard deviation is, the less reliable all estimates are, and the larger is the standard error of a coefficient.

2. The degree of variability of the variable associated with the coefficient. If the X_1 variable doesn't vary much in the data, there's not much information about the effect of varying X_1, and the standard error will be large.

3. The amount of collinearity, or correlation among the independent variables. If the X_1 variable correlates strongly with other predictors, there isn't much information about the effect of varying X_1 while holding other variables constant; thus the estimate of the partial slope of X_1 will be poor, and the standard error will be large.

Example 15.28

Refer to Example 15.26. Further computer output is included in Figure 15.19.

Locate the estimated standard errors of the coefficients.

PARAMETER	ESTIMATE	T FOR H0: PARAMETER=0	PR > |T|	STD ERROR OF ESTIMATE
INTERCEPT	22.46657324	1.38	0.1709	16.25912456
PEERPLAY	5.06792821	4.48	0.0001	1.13015895
ADULTDIR	-3.79924543	-2.47	0.0157	1.53926784
HOLDHUG	-1.80354811	-1.20	0.2356	1.50904399
ADULTPLA	5.50611393	4.12	0.0001	1.33672536

Figure 15.19 Output for Example 15.28

Solution The standard errors are shown under the heading STD ERROR OF ESTIMATE. □

The t test for the null hypothesis that variable X_j has no additional predictive value as the last predictor in a regression model follows the standard t test form, and is shown in the box below.

t Test of an Individual Coefficient in Regression

1. H_0: $\beta_j = 0$
2. H_a: $\beta_j \neq 0$
3. T.S.: $t = \hat{\beta}_j/\text{est. std. error}(\hat{\beta}_j)$
4. R.R.: reject H_0 if $|t| >$ t-table value with $n - (k + 1)$ df

NOTE: It is also possible to perform a one-tailed test.

For example, in the prediction of SAT using PSAT, IQ, and QPA, Minitab output included

	COLUMN	COEFFICIENT	ST. DEV. OF COEF.	T-RATIO = COEF./S.D.
	—	84.28	78.30	1.08
X1	PSAT	7.8289	0.7514	10.42
X2	IQ	−0.2595	0.9744	−0.27
X3	QPA	52.01	17.15	3.03

There were 78 residual df. The t values for PSAT and QPA, shown in the right-most column of the output, were far beyond t-table values with 60 or 120 df, so we may confidently assert that each of these had "last predictor in" predictive value. The t value for IQ is very small. There is no real evidence that IQ is useful in predicting SAT scores for this school, given PSAT and QPA scores.

Example 15.29 Refer to Example 15.26. Locate the values of the t statistic, and the appropriate df. What conclusion can be reached by the t tests?

Solution The column headed T FOR H0: PARAMETER=0 gives the t statistics, and the next column, labeled PR > |T|, gives two-tailed p-values. The p-values for PEERPLAY and ADULTPLA are extremely small, and the p-value for ADULTDIR is fairly small at .0157. Thus we have good evidence that PEERPLAY, ADULTPLA, and ADULTDIR all have incremental predictive value. The p-value for HOLDHUG is .2356, so there is little, if any, evidence that HOLDHUG has incremental predictive value, given the other independent variables. □

There is also a test intermediate between the F test of the null hypothesis that all predictors have no value and the t test of the null hypothesis that one particular predictor has no (additional) value. Naturally, the intermediate test tests the null hypothesis that *some* of the variables have no predictive value; in fact, the null hypothesis is that once one set of x variables is included in the regression model, the remaining variables have no additional value. Some language is needed to do this test. The **complete model** includes all k predictor variables; a **reduced model** includes only a smaller number, say g, of predictors. For example, in a model with quantitative independent variables X_1 and X_2 and dummy variables X_3, X_4, and X_5, the null hypothesis that the dummy variables have no (additional) value is H_0: $\beta_3 = \beta_4 = \beta_5 = 0$. The complete model contains $X_1, \ldots, X_5$; the reduced model contains only X_1 and X_2.

complete model
reduced model

The test of the null hypothesis (that the variables deleted from the complete model to get the reduced model have no additional predictive value) is again an F test. It is summarized in the box.

Testing the Incremental Predictive Value of a Set of X's

1. H_0: $\beta_{g+1} = \cdots = \beta_k = 0$

2. H_a: H_0 is not true

3. T.S.: compute SS(Regression) for both the complete and the reduced model:

$$F = \frac{\dfrac{[\text{SS(Regression, complete)} - \text{SS(Regression, reduced)}]}{k - g}}{\text{MS(Residual, complete)}}$$

4. R.R.: reject H_0 if $F >$ F-table value with $k - g$ numerator and $n - (k + 1)$ denominator df

> *NOTE:* $k - g$ is the number of additional X variables included in the complete model, over and above those included in the reduced model.

For example, Minitab output for the prediction of SAT score given PSAT, IQ, and QPA included

ANALYSIS OF VARIANCE

DUE TO	DF	SS	MS = SS/DF
REGRESSION	3	1669189	556396
RESIDUAL	78	400256	5131
TOTAL	81	2069445	

FURTHER ANALYSIS OF VARIANCE
SS EXPLAINED BY EACH VARIABLE WHEN ENTERED IN THE ORDER GIVEN

DUE TO	DF	SS
REGRESSION	3	1669189
PSAT	1	1621452
IQ	1	556
QPA	1	47182

We may obtain the required SS and MS for testing the null hypothesis that neither IQ nor QPA adds any predictive value, once PSAT is used, from the output. The reduced model includes only the PSAT score, with an SS(Regression) shown as 1,621,452. The complete model includes all three variables. SS(Regression) for the three-variable model is shown as 1,669,189, and MS(Residual, complete) is shown as 5131. Thus the incremental F statistic is

$$F = \frac{(1{,}669{,}189 - 1{,}621{,}452)/2}{5131} = 4.652$$

with $3 - 1 = 2$ and 78 df. Taking $\alpha = .01$, say, we find that the computed F is larger than the table value (roughly 3.6, by interpolation). Thus there is evidence that at least one of the two variables adds predictive value, given PSAT score. Looking at the results of the t test and at the incremental SS, it appears that QPA has some modest additional value, whereas IQ adds little or nothing given PSAT.

Example 15.30 Refer to Example 15.26. Additional SAS output was as shown in Figure 15.20. Test the null hypothesis that X_3 = holding-hugging behavior and X_4 = adult object play have no incremental predictive value.

DF	TYPE I SS	F VALUE	PR > F	DF	TYPE III SS	F VALUE	PR > F
1	8363.09620872	38.79	0.0001	1	4335.45744815	20.11	0.0001
1	5284.73345987	24.51	0.0001	1	1313.46766230	6.09	0.0157
1	156.75703522	0.73	0.3964	1	307.96739508	1.43	0.2356
1	3658.12958285	16.97	0.0001	1	3658.12958285	16.97	0.0001

Figure 15.20 Output for Example 15.30

Solution The column headed (unmemorably) TYPE I SS gives the incremental SS for the independent variables, entered in the order given. The total of the incremental SS for HOLDHUG and ADULTPLA is

$$156.75703522 + 3658.12958285 = 3814.887 \quad \text{(rounded off)}$$

The MS(Residual) value was found in Example 15.26 to be 215.602. Thus

$$F = \frac{3814.887/2}{215.602} = 8.85$$

For 2 and 79 df (the 79 denominator df comes from Example 15.26), the F value is well beyond table values. Thus we can safely conclude that at least one of the two predictors has incremental predictor value. □

The idea of the incremental F test can be used repeatedly. It's possible to begin with a very small model, even one including no predictors at all, and successively include more and more X variables, testing each new inclusion with the F test. Several computer programs compute the change in SS(Regression) as variables are added in a specified sequence. In addition, some perform successive F tests.

stepwise regression

forward selection

backward elimination

The idea of successively adding terms to a regression model, testing each term as it's added, is the basic idea of **stepwise regression.** In **forward selection,** variables are entered into the regression equation one at a time, in order of their incremental contribution to SS(Regression), until the incremental F test fails to exceed a specified table value. In **backward elimination,** variables are removed from the equation, in order of their incremental loss of SS(Regression), until the incremental F test exceeds a specified table value. There are a number of variations of stepwise regression. Unfortunately, two different choices of method can give different results, sometimes greatly different, with the same data.

Example 15.31 Two stepwise regressions (once using forward selection and once using backward elimination) were run using SAS for the data of Example 15.26. The output in Figure 15.21 was obtained.

(a) Identify the sequence of entrance of predictor variables into the forward selection equation.

(b) Why did the backward elimination procedure stop where it did?

(c) Are the results of the two runs consistent?

```
                              FORWARD SELECTION STEPWISE REGRESSION

                       FORWARD SELECTION PROCEDURE FOR DEPENDENT VARIABLE DEVELOP

STEP 1    VARIABLE ADULTDIR ENTERED    R SQUARE = 0.30337879       C(P) =    31.45597638
----------------------------------------------------------------------------------------------------

STEP 2    VARIABLE PEERPLAY ENTERED    R SQUARE = 0.39564333       C(P) =    18.69410329
----------------------------------------------------------------------------------------------------

STEP 3    VARIABLE ADULTPLA ENTERED    R SQUARE = 0.49730705       C(P) =     4.42840599
----------------------------------------------------------------------------------------------------

STEP 4    VARIABLE HOLDHUG ENTERED     R SQUARE = 0.50623486       C(P) =     5.00000000

                                 DF          SUM OF SQUARES        MEAN SQUARE         F      PROB>F

               REGRESSION         4          17462.71628666      4365.67907167     20.25      0.0001
               ERROR             79          17032.56942763       215.60214465
               TOTAL             83          34495.28571429

                                B VALUE          STD ERROR          TYPE II SS         F      PROB>F

               INTERCEPT    22.46657324
               PEERPLAY      5.06792821         1.13015895      4335.45744815     20.11      0.0001
               ADULTDIR     -3.79924543         1.53926784      1313.46766230      6.09      0.0157
               HOLDHUG      -1.80354811         1.50904399       307.96739508      1.43      0.2356
               ADULTPLA      5.50611393         1.33672536      3658.12958285     16.97      0.0001
----------------------------------------------------------------------------------------------------

NO OTHER VARIABLES MET THE 0.5000 SIGNIFICANCE LEVEL FOR ENTRY INTO THE MODEL.
```

(continued)

Figure 15.21 Output for Example 15.31

```
                    BACKWARD ELIMINATION STEPWISE REGRESSION

             BACKWARD ELIMINATION PROCEDURE FOR DEPENDENT VARIABLE DEVELOP

STEP 0   ALL VARIABLES ENTERED      R SQUARE = 0.50623486      C(P) =      5.00000000

---------------------------------------------------------------------------------------------

STEP 1   VARIABLE HOLDHUG REMOVED   R SQUARE = 0.49730705      C(P) =      4.42840599

                               DF          SUM OF SQUARES         MEAN SQUARE          F    PROB>F

           REGRESSION           3          17154.74889158       5718.24963053      26.38    0.0001
           ERROR               80          17340.53682270        216.75671028
           TOTAL               83          34495.28571429

                          B VALUE              STD ERROR         TYPE II SS          F    PROB>F

           INTERCEPT      14.88724666
           PEERPLAY        4.93591848         1.12775606      4152.19910530      19.16    0.0001
           ADULTDIR       -3.50346167         1.52330457      1146.55212873       5.29    0.0241
           ADULTPLA        4.75111663         1.18118816      3506.91922299      16.18    0.0001
---------------------------------------------------------------------------------------------

ALL VARIABLES IN THE MODEL ARE SIGNIFICANT AT THE 0.1000 LEVEL.
```

Figure 15.21 (continued)

Solution

(a) The sequence is indicated as ADULTDIR, PEERPLAY, ADULTPLA, HOLDHUG.

(b) The last line of the output indicates that all remaining variables have statistically significant (at $\alpha = .10$) predictive value. The SAS program stops when such a result occurs.

(c) HOLDHUG was the last entered and the first removed variable. At least in that sense, the results are consistent. We don't know if further removal would be in the opposite order from that of forward selection. □

The effect of adding variables to a model one at a time depends very much on the sequence in which one adds the variables. An X variable may be a statistically significant predictor when added early in a sequence of variables, but not significant when added late. This is particularly true when there are strong correlations among variables. If variables X_2 and X_3 (say) are highly correlated, each can serve as a surrogate for the other. Thus, if X_2 is added to the sequence of predictors before X_3, it may be useful, but if it is added after X_3 is included, the X_3 variable may have picked up all the predictive value, making X_2 worthless as a last predictor in. It is even possible, with some particular correlation patterns, that an X variable may explain *more* Y variability when added late in a sequence than it does when added early!

Exercises for Section 15.4

15.32. In Exercise 15.24, we had $n = 16$, $k = 3$, and found SS(Regression) = 59.00, SS(Residual) = 96.00, and SS(Total) = 155.00.

a. Calculate the value of the F statistic.
b. Is the result statistically significant at $\alpha = .01$?
c. Place bounds on the p-value of the F test.
d. What is the interpretation of the result?

15.33. In Exercise 15.24, the coefficients of X_1, X_2, and X_3 were 3.00, −2.00, and 1.00, respectively, and their standard errors were 2.00, 2.00, and 2.83, respectively. Perform t tests for each coefficient, using $\alpha = .10$. What is the conclusion of each test?

15.34. In the data of Exercise 15.24, SS(Regression) for the model involving only X_1 was 49.00. Test the null hypothesis that the coefficients of X_2 and X_3 are both 0, using $\alpha = .10$. State your conclusion carefully.

15.35. Suppose that in a regression study there were three X variables, all highly correlated with each other. Computer output yielded the following:

SOURCE	DF	SS	MS
REGRESSION	3	286.721	95.574
RESIDUAL	48	53.419	1.113
TOTAL	51	340.140	

VARIABLE	COEFFICIENT	STD. ERROR	T
INTERCEPT	22.2388	0.7740	28.73
X1	0.0071	0.2114	0.03
X2	0.2160	0.1839	1.18
X3	0.0698	0.0697	1.00

a. Calculate the F statistic for the null hypothesis that all slopes are 0. Is F statistically detectable (significant) at $\alpha = .01$? Put bounds on the p-value.

b. Are the t statistics significant at reasonable α levels?

15.36. Additional computer output showed the increments to SS(Regression) for variables entered in the specified sequence:

SEQUENCE:	X1 X2 X3
X1	284.448
X2	1.118
X3	1.116

SEQUENCE:	X2 X3 X1
X2	285.119
X3	1.521
X1	0.001

a. Test the null hypothesis that X_2 and X_3 have no predictive value, given X_1, using $\alpha = .05$. What is your conclusion?

b. Test the null hypothesis that X_3 and X_1 have no predictive value, given X_2, using $\alpha = .05$. What is your conclusion?

15.37. The overall *F* test performed in Exercise 15.35 was highly significant, yet none of the *t* tests in Exercise 15.35 or *F* tests in Exercise 15.36 were significant. What explains the apparent contradiction?

15.38. Additional output for the study of Exercise 15.28 is shown in Figure 15.22.

```
DEP VARIABLE: EXPECT   PREGAME PERSONAL PERFORMANCE EXPECTATION
                        SUM OF        MEAN
SOURCE      DF         SQUARES      SQUARE      F VALUE      PROB>F
MODEL        3         327.829     109.276       50.684      0.0001
ERROR      183         394.556    2.156046
C TOTAL    186         722.385
    ROOT MSE          1.468348    R-SQUARE       0.4538
    DEP MEAN          4.860963    ADJ R-SQ       0.4449
    C.V.              30.20694

                     PARAMETER    STANDARD    T FOR H0:                  VARIABLE
VARIABLE    DF        ESTIMATE       ERROR  PARAMETER=0   PROB > |T|       LABEL

INTERCEP     1        0.979260    0.643614        1.522       0.1299  INTERCEPT
ABILITY      1        0.455259    0.106372        4.280       0.0001  RATING OF OWN ABILITY
ESTEEM       1        0.126483    0.184090        0.687       0.4929  SCORE ON SELF ESTEEM SCALE
ANXIETY      1        0.148642    0.052018        2.858       0.0048  COMPETITIVE TRAIT ANXIETY
```

Figure 15.22 Output for Exercise 15.38

a. Locate the value of the *F* statistic for testing the null hypothesis that all slopes are 0.

b. SAS computes a *p*-value for the *F* statistic, shown as PROB > F. What does the *p*-value indicate about the predictive value of the independent variables?

c. Locate the standard errors and *t* statistics for the partial slopes.

d. The two-tailed *p*-value for each *t* statistic is shown in the column labeled PROB > |T|. Which of the tests are significant at the usual α levels?

15.39. SPSS-X output for the study of Exercise 15.30 included that shown in Figure 15.23.

```
MULTIPLE R            .42543          ANALYSIS OF VARIANCE
R SQUARE              .18099                           DF     SUM OF SQUARES      MEAN SQUARE
ADJUSTED R SQUARE     .16410          REGRESSION        2          197.49064         98.74532
STANDARD ERROR       3.03531          RESIDUAL         97          893.66936          9.21309

                                      F =      10.71794       SIGNIF F =  .0001

------------------ VARIABLES IN THE EQUATION ------------------

VARIABLE              B          SE B        BETA          T   SIG T

AGE         -9.53299E-05       .01949 -5.109E-04       -.005   .9961
EDUCN            -.42614       .10459    -.42567      -4.075   .0001
(CONSTANT)      13.25808      1.64084                  8.080   .0000
```

Figure 15.23 Output for Exercise 15.39

a. Locate the value of the overall F statistic.

b. The p-value is shown as SIGNIF F. What does it indicate about the predictive value of AGE and EDUCN?

15.40. Refer to Figure 15.23. Locate the t statistics and associated p-values. What do these quantities indicate about the predictive value of AGE and of EDUCN?

15.5 Confidence and Prediction Intervals in Regression

The inferences discussed in Section 15.4 were all stated as hypothesis tests. In this section we'll consider confidence-interval estimation for the true values of various coefficients and correlations, and prediction intervals for predicting new values of Y for any specified set of X values.

Confidence intervals for the coefficients of a general linear model follow the standard "estimate plus or minus table value times standard error" format. The standard errors of the regression coefficients are usually calculated by a computer program. An expression for that standard error, depending on the residual standard deviation, the amount of variability in the corresponding X variable, and the collinearity of that X with the other independent variables, was given in Section 15.4. A t table is used, where the df are the residual (error) df $n - (k + 1)$.

$100(1 - \alpha)\%$ Confidence Interval for a Regression Slope

$$\hat{\beta}_j - t_{\alpha/2}[\text{est. std. error}(\hat{\beta}_j)] < \beta_j < \hat{\beta}_j + t_{\alpha/2}[\text{est. std. error}(\hat{\beta}_j)]$$

where $t_{\alpha/2}$ is based on $n - (k + 1)$ df, n = sample size, and k = number of x terms.

For the data from Example 15.3, predicting SAT score from PSAT, IQ, and QPA scores, computer output gave the following coefficients and standard errors, for 78 residual df:

COLUMN	COEFFICIENT	ST. DEV. OF COEF.
—	84.28	78.30
PSAT	7.8289	0.7514
IQ	−0.2595	0.9744
QPA	52.01	17.15

We do not have t-table values for 78 df. Rough interpolation between values for 60 and 120 df indicates that $t_{.025}$ is about 1.99. Thus the 95% confidence intervals for the three slopes are

$$7.8289 - 1.99(0.7514) < \beta_{\text{PSAT}} < 7.8289 + 1.99(0.7514)$$
$$-0.2595 - 1.99(0.9744) < \beta_{\text{IQ}} < -0.2595 + 1.99(0.9744)$$
$$52.01 - 1.99(17.15) < \beta_{\text{QPA}} < 52.01 + 1.99(0.9744)$$

or

$$6.33 < \beta_{\text{PSAT}} < 9.32$$
$$-2.20 < \beta_{\text{IQ}} < 1.68$$
$$17.88 < \beta_{\text{QPA}} < 86.14$$

Example 15.32 Refer to the output shown in Figure 15.19. Calculate 90% confidence intervals for all partial slopes.

Solution The estimated partial slopes were shown in the column labeled ESTIMATE. The standard errors were shown in the column labeled STD ERROR OF ESTIMATE. There were 79 residual df; rough interpolation indicates that $t_{.05}$ is about 1.667. Thus the 90% confidence intervals are

$$5.0679 - 1.667(1.1302) < \beta_{PEERPLAY} < 5.0679 + 1.667(1.1302)$$
$$-3.7992 - 1.667(1.5393) < \beta_{ADULTDIR} < -3.7992 - 1.667(1.5393)$$
$$-1.8035 - 1.667(1.5090) < \beta_{HOLDHUG} < -1.8035 - 1.667(1.5090)$$
$$5.5061 - 1.667(1.3367) < \beta_{ADULTPLA} < 5.5061 - 1.667(1.3367)$$
□

It's also possible to construct confidence intervals for correlation and partial correlation coefficients. Unfortunately, the method is critically dependent on the assumption that the data are sampled from a (multivariate) normal population. In particular, inference about correlation coefficients is dubious when the X values were systematically controlled by the experimenter.

If the data can be assumed to be normally distributed (in particular, if a scatter plot of the two variables has roughly an elliptical shape), a confidence interval for a correlation can be constructed using a transformation called **Fisher's Z.** This transformation, easily done by a calculator, is

Fisher's Z

$$Z = .5 \log_e \frac{1 + r_{YX}}{1 - r_{YX}}$$

The standard error of Fisher's Z is, to a very good approximation for any sample size, $1/\sqrt{n-3}$. A confidence interval is found by first calculating an interval for "population Z," then transforming back to correlation terms.

100(1 − α)% Confidence Interval for a Correlation Coefficient

Confidence interval for "population Z":

$$Z_L < \text{"population } Z\text{"} < Z_U$$

where

$$Z_L = Z - z_{\alpha/2} \frac{1}{\sqrt{n-3}}$$

and

$$Z_U = Z + z_{\alpha/2} \frac{1}{\sqrt{n-3}}$$

Confidence interval for population correlation coefficient ρ_{YX}:

$$\frac{e^{2Z_L} - 1}{e^{2Z_L} + 1} < \rho_{YX} < \frac{e^{2Z_U} - 1}{e^{2Z_U} + 1}$$

Example 15.33 For the data of Example 15.2, we calculated $r_{YX} = .5266$, with $n = 221$. Calculate a 99% confidence interval for ρ_{YX}.

Solution First, a plot of the data (not shown here) indicated an elliptical, "fuzzy football" shape, so we may reasonably use the Fisher's Z method.

$$Z = .5 \log_e \frac{1 + .5266}{1 - .5266} = 0.5854$$

The standard error is $1/\sqrt{221 - 3} = .06773$. The required table value is $z_{.005} = 2.576$. Thus

$$Z_L = 0.5854 - 2.576(.06773) = 0.4109$$

and

$$Z_U = 0.5854 + 2.576\,(.06773) = 0.7599$$

The 99% confidence interval for ρ is

$$\frac{e^{2(0.4109)} - 1}{e^{2(0.4109)} + 1} < \rho < \frac{e^{2(0.7599)} - 1}{e^{2(0.7599)} + 1}$$

or

$$.3892 < \rho < .6410$$

□

Confidence intervals for correlations often are quite wide, even for moderately large sample sizes. For example, suppose that $n = 103$, $r_{YX} = .60$, and the scatter plot appears to be bivariate normal. A 95% confidence interval for the true correlation is

$$.460 < \rho_{YX} < .711$$

In terms of squared correlation, the limits are .212 to .506. There are several popular methods of statistical analysis, notably stepwise regression and factor analysis, that depend on relatively small-sized differences among sample correlations. Unless the sample sizes are quite large, perhaps even over 1000, those small differences among sample correlations may well be the result of sheer random variation. Thus, the use of methods that are sensitive to small differences in correlations may be quite dangerous, even if the sample size is fairly large.

Confidence intervals for a partial correlation coefficient may be calculated, using a minor modification of the correlation confidence interval defined in the preceding box. The only mechanical change is that the standard error $1/\sqrt{n - 3}$ should be replaced by $1/\sqrt{n - 3 - c}$, where c is the number of controlled variables in the partial correlation coefficient. Thus, for $r_{Y1|2,4}$, variables X_2 and X_4 are being controlled, and $c = 2$. The standard error for Fisher's Z would be

$$\frac{1}{\sqrt{n-3-2}} = \frac{1}{\sqrt{n-5}}$$

Again, the confidence interval can be quite wide.

Example 15.34 For a sample of size 86, the partial correlation of Y with X_2, controlling for X_1, X_3, and X_4, was found to be .483. Calculate a 95% confidence interval for the population partial correlation.

Solution We have $n = 86$ and $c = 3$. Fisher's Z is

$$Z = .5 \log_e \frac{1 + .483}{1 - .483} = 0.5269$$

The required table value $z_{.025}$ is the old familiar 1.96. The standard error is

$$\frac{1}{\sqrt{86-3-3}} = 0.1118$$

Thus

$$Z_L = 0.5269 - 1.96(0.1118) = 0.3078$$

and

$$Z_U = 0.5269 + 1.96(0.1118) = 0.7460$$

The confidence interval is

$$\frac{e^{2Z_L} - 1}{e^{2Z_L} + 1} < \rho_{Y2|1,3,4} < \frac{e^{2Z_U} - 1}{e^{2Z_U} + 1}$$

or

$$\frac{e^{2(0.3078)} - 1}{e^{2(0.3078)} + 1} < \rho_{Y2|1,3,4} < \frac{e^{2(0.7460)} - 1}{e^{2(0.7460)} + 1}$$

$$.298 < \rho_{Y2|1,3,4} < .633$$

Here again, the interval is quite wide. □

As noted previously, inferences based on Fisher's Z can be badly incorrect, for any sample size, if the population is not multivariate normal. The reason is that the standard error term is based on purely theoretical calculations, not on data. (Note that the standard error term doesn't involve the data at all.) If the assumption of normality is wrong, the theoretical calculations are wrong, and the supposed confidence level may be far off. This nonrobustness is very similar to the nonrobustness of inferences about variances discussed in Chapter 12—for virtually the same reason. The jackknife method discussed in Chapter 12 is also useful in making inferences about correlations. See Miller (1974). This method requires a special computer program.

Example 15.35 The data underlying Example 15.34 contained some data points far from the bulk of the data. What does this indicate about the validity of the 95% confidence level in Example 15.34? Does the fact that the sample size is a relatively large 86 help any?

Solution If the data contain serious outliers, it is unlikely that the data came from a multivariate normal distribution. Thus the claimed 95% confidence may be seriously wrong. For Fisher's Z, the sensitivity of the confidence to nonnormality holds regardless of sample size, so the relatively large n is no help in making the 95% more nearly correct. Of course, a large n will make the confidence interval narrower, but that's a different issue. □

Now we turn to a slightly different type of interval. We want a prediction interval for the Y value associated with specified X values. For instance, suppose that a study related Y = score on a test of conservation of number to X = age in months, and that the prediction equation was

$$\hat{y} = -141.95 + 2.825x$$

What is the predicted Y value when $X = 66$ months, and how much variation around that value can be expected? The question has two interpretations. We can consider the *mean* Y score of all individuals aged 66 months, or the Y score of *one individual* aged 66 months. In either case, the predicted Y value is obtained by plugging the specified X value into the equation:

$$\hat{y} = -141.95 + 2.825(66) = 44.50$$

The difference between mean prediction and individual prediction lies in the applicable standard error. Because it's easier to predict a mean than to predict an individual score, the standard error for mean prediction should be, and is, smaller than the standard error for individual prediction.

Standard Errors for Mean and Individual Prediction, Simple Regression

For predicting the mean of all Y scores corresponding to $X = x^*$, the estimated standard error is

$$s_{Y|X}\sqrt{\frac{1}{n}+\frac{(x^*-\bar{x})^2}{S_{xx}}}$$

where $s_{Y|X}$ is the residual standard deviation, n is the sample size used to develop the regression equation, and $\bar{x}$ and S_{xx} are calculated from the original data.

For predicting the Y score of an individual having $X = x^*$, the estimated standard error is

$$s_{Y|X}\sqrt{1+\frac{1}{n}+\frac{(x^*-\bar{x})^2}{S_{xx}}}$$

NOTE: Both standard errors are based on the residual df $n - 2$.

In the conservation example, computations (not shown here) indicated that the residual standard deviation turned out to be 15.72, n was 48, $\bar{x}$ was 70.77, and S_{xx} was 992.48. The estimated prediction standard errors when $X = 66$ are

$$\text{Mean:}\quad 15.72\sqrt{\frac{1}{48}+\frac{(66-70.77)^2}{992.48}} = 3.288$$

$$\text{Individual:}\quad 15.72\sqrt{1+\frac{1}{48}+\frac{(66-70.77)^2}{992.48}} = 16.060$$

Note that the standard error for mean prediction is, as expected, the smaller of the two.

Example 15.36 For the data of Example 15.10, we found $\hat{y} = 18.607 + 0.5067x$, $s_{Y|X} = 6.755$, $n = 221$, $\bar{x} = 30.425$, and $S_{xx} = 14936.02$. Find both estimated standard errors for predicting Y when $X = 60$.

Solution

$$\text{Mean:}\quad 6.755\sqrt{\frac{1}{221}+\frac{(60-30.425)^2}{14936.02}} = 1.697$$

$$\text{Individual:}\quad 6.755\sqrt{1+\frac{1}{221}+\frac{(60-30.425)^2}{14936.02}} = 6.965$$ □

Both estimated standard errors depend on the residual standard deviation and the sample size. The last term in both standard errors is an **extrapolation penalty.** The farther x^*, the new X value, is from $\bar{x}$, the mean of the old X values, the more one is extrapolating from the data. As x^* gets farther and farther from $\bar{x}$, the extrapolation penalty increases and the standard errors get larger.

extrapolation penalty

Example 15.37 The last term in the standard errors of Example 15.36 was relatively large. Why?

Solution The chosen value of x^* was 60, far from the old mean, 30.425. Thus a large extrapolation was being made. □

Once the standard error has been calculated, it's easy to obtain prediction intervals. The usual "estimate plus-or-minus standard error" format applies once again. The relevant table is the t table, with df $n - 2$. For the conservation study, 95% prediction intervals for the mean and individual Y scores corresponding to $x^* = 66$, with $48 - 2 = 46$ df, are

$$\text{Mean:}\quad 44.50 \pm 2.01(3.288) = 44.50 \pm 6.61$$
$$\text{Individual:}\quad 44.50 \pm 2.01(16.060) = 44.50 \pm 32.28$$

Example 15.38 Refer to Examples 15.36 and 15.37. An individual has an X score of 60. Would a Y score of 25 be unusual?

Solution From Example 15.36, we find that the predicted score is $\hat{y} = 18.607 + 0.5067(60) = 49.01$. In Example 15.37, we found a standard error for individual prediction of 6.965. A 99% confidence interval would be

$$49.01 - 2.58(6.965) < Y < 49.01 + 2.58(6.965)$$

or

$$31.04 < Y < 66.98$$

A score of 25 falls well outside a 99% confidence interval, and thus would definitely be unusual. □

The ideas of prediction intervals can be extended to the case where there are several X variables. A computer program is required to do the arithmetic. The extrapolation problem becomes more complex. It is possible that each separate X value may be well within the range of data, yet the combination of X values may be a large extrapolation. Consider, for example, predicting teachers' salaries given age and years of teaching experience. One might well have 30-year-old teachers in the sample, and one might well also have teachers with 25 years of experience. Yet I rather doubt that one would have a 30-year-old teacher with 25 years of experience in the sample!

Exercises for Section 15.5

15.41. Refer to the data of Exercise 15.2. In Exercise 15.12, we calculated the least-squares linear prediction equation and the residual standard deviation.

a. Calculate the estimated standard error of the slope.
b. Calculate a 95% confidence interval for the slope.
c. On the basis of this confidence interval, can you say that the slope is detectably different from 0, using $\alpha = .05$?

15.42. For the data of Exercise 15.2, find the correlation coefficient and calculate a 95% confidence interval for it. Does the confidence interval include the possibility of a 0 correlation?

15.43. Refer again to the data of Exercise 15.2.

a. Compute a 95% prediction interval for the mean Y score of all individuals having an X score of 28.
b. Compute a 95% prediction interval for the Y score of one individual having an X score of 28.
c. Is there a serious problem of extrapolation in either of these prediction intervals?

15.44. Refer to the data from Exercise 15.5 and the computations performed in Exercise 15.13.

a. Find the estimated standard error of the slope of X_1.
b. Calculate a 90% confidence interval for the slope.

15.45. Refer to Exercise 15.44. Calculate a 90% confidence interval for the correlation coefficient. Would you judge the interval to be wide or narrow?

15.46. Refer to Exercise 15.44.

a. Calculate a 95% prediction interval for the Y score of an individual with an X score of 10.

b. Why is the interval so wide?

15.47. Refer to Figure 15.22 in Exercise 15.38.

a. Locate the estimated standard errors of the slopes.

b. Calculate 95% confidence intervals for all slopes. Which of these intervals, if any, indicate that the true slope may be 0?

15.48. In the data of Exercise 15.28, the correlation between EXPECT and ANXIETY is .37357 and the sample size is 188. Calculate a 95% confidence interval for the population correlation. In terms of the predictive value of ANXIETY, what is the range of squared correlation values?

15.49. Refer to Figure 15.15. The estimated standard errors of the slopes are indicated as SE B. Calculate 99% confidence intervals for the true slopes.

15.6 Checking Assumptions in Regression Studies

The inference methods that have been discussed in the last two sections rest on some assumptions. Although we have specified the assumptions, we haven't yet discussed ways to check those assumptions; nor have we determined which of the assumptions are most important. In this section, we'll take care of both issues. We begin by restating the assumptions.

Assumptions of Regression Analysis

1. The mathematical model

$$E(Y) = \beta_0 + \beta_1 X_1 + \cdots + \beta_k X_k$$

(where some of the X variables may be nonlinear functions of others, dummy variables, or product terms) is correct.

2. The random error terms ϵ_i are statistically independent.

3. The variance of $\epsilon_i = \sigma_\epsilon^2$ is constant for all i.

4. The population of errors is normally distributed.

5. The X values are reliably measured and contain no measurement error.

The first assumption is that the proper mathematical form has been used in the regression model. In simple regression with a single independent variable, a scatter plot of Y vs. X will usually indicate any severe nonlinearity. In multiple regression with many X's, individual scatter plots of Y vs. each X_j often fail to reveal nonlinearities, because the nonlinearity is masked by the effects of other X variables. A better way to detect nonlinearities is by plotting residuals (from a linear regression model) against each X. If the residual plot shows a pronounced nonlinear pattern, such as that shown in Figure 15.24, a transformation such as log (X_j) or X_j^2 should be used.

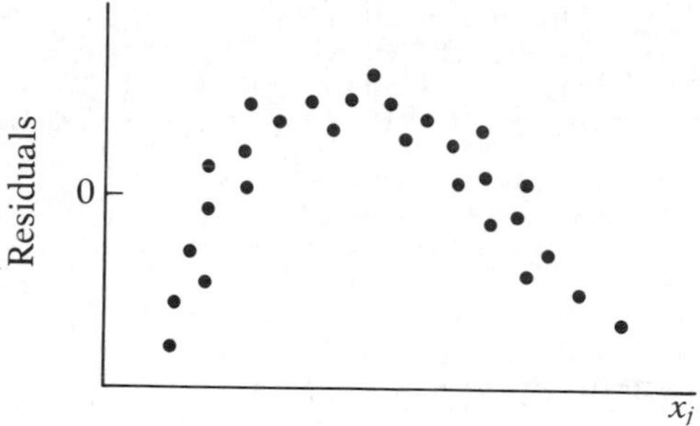

Figure 15.24 Scatter plot showing nonlinear pattern

There are two problems with using linear regression with nonlinear data. First, the linear regression will fail to capture the full degree of predictability that could be had, so the results will be less significant (both statistically and scientifically) than they should be. Second, predictions made by extrapolating the line beyond the actual data can be very seriously in error. See Figure 15.25 for an illustration; if a prediction is made for a new X value as shown, the predicted value (using the straight-line prediction) will be seriously wrong. Extrapolation beyond the data is always hazardous. It is even more so if the mathematical form of the model is wrong.

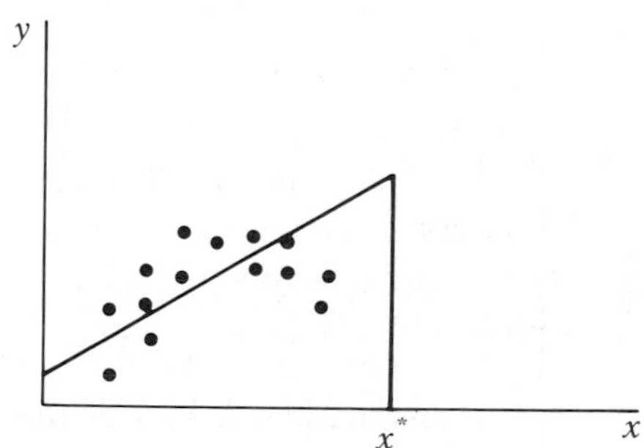

Figure 15.25 Extrapolation with a curvilinear relation

Example 15.39 Residual plots for the data of Example 15.3 are shown in Figure 15.26. Is there evidence of serious nonlinearity?

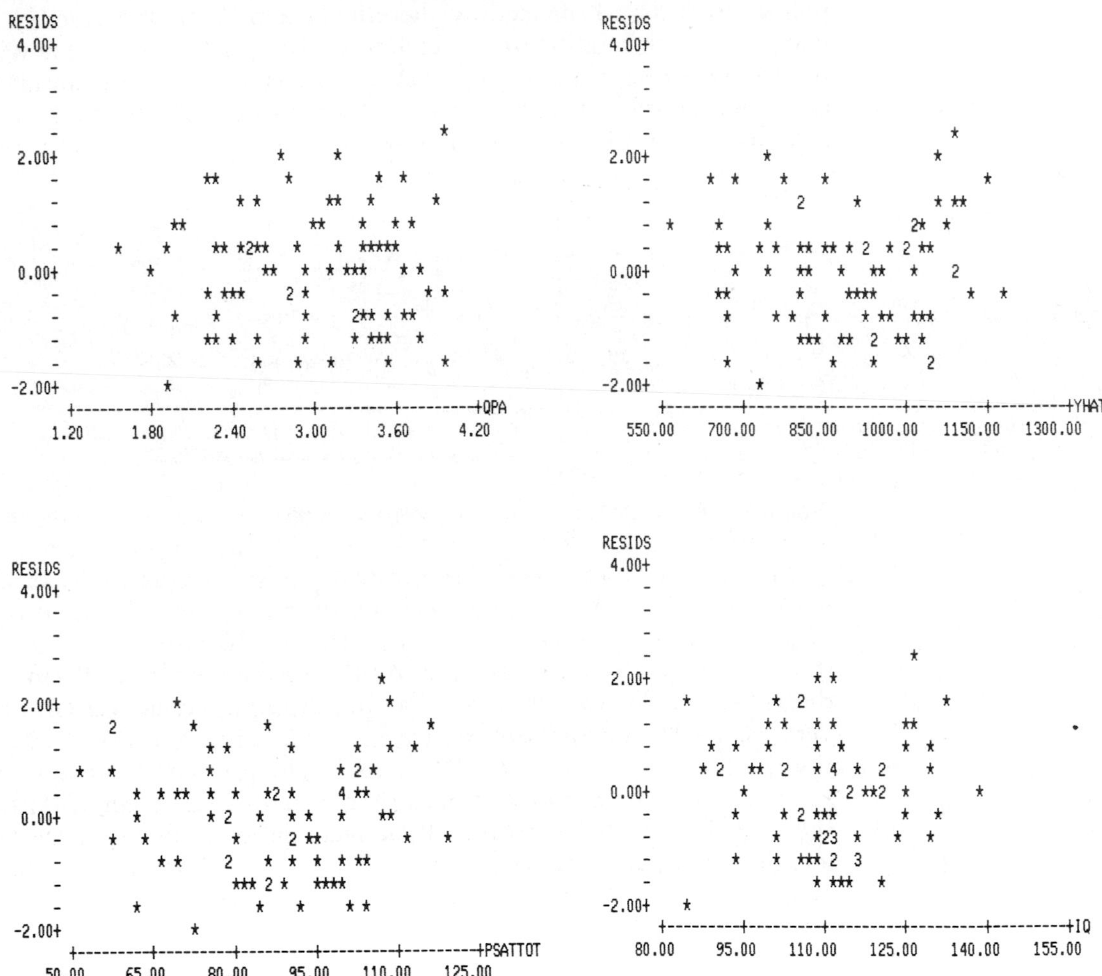

Figure 15.26 Residual plots for Example 15.39

Solution No. There's no indication of a curved pattern in any of the plots. □

The most critical of the assumptions is the assumption that observations are statistically independent, from one observation to the next. The assumption of independence is particularly suspect when the ob-

autocorrelated errors

servations are made one after the other in time. For *time series* data, there is a considerable likelihood of **autocorrelated errors**—errors that are correlated, one with the next. For example, suppose that in a study of infants, one records a developmental measure for a particular baby every month; the independent variable is month number and the dependent variable is the developmental measure. It's very likely that a baby that is developmentally advanced (for a particular age) will still be advanced the next month. Similarly, a baby that is developmentally slow one month is likely to be slow the next month. Thus, there will be a month-to-month correlation of the development scores, violating the assumption of independent errors.

Example 15.40

Refer to Example 15.10, in which motor-skills scores were obtained from boys aged 6 months and 15 months. Is there a danger of autocorrelation in this study?

Solution

It seems unlikely. The data were scores for different children at the same age, rather than the same child at different ages. There's little reason to think that developmental changes in one child will be correlated with developmental changes in some other child. There should be correlation between X and Y scores, but no correlation of the prediction errors. □

residuals

One way to detect autocorrelated errors is to plot the **residuals**—differences between actual Y values and predicted, $\hat{Y}$, values—against time. The ith residual is an estimate of the ith error term. If the errors are independent, a scatter plot of the residuals versus time should be completely random. Autocorrelation (also called serial correlation) is indicated if there is a nonrandom pattern to the residuals. A scatter plot showing autocorrelation appears in Figure 15.27; notice that there is a roughly cyclical pattern to the residuals, over time.

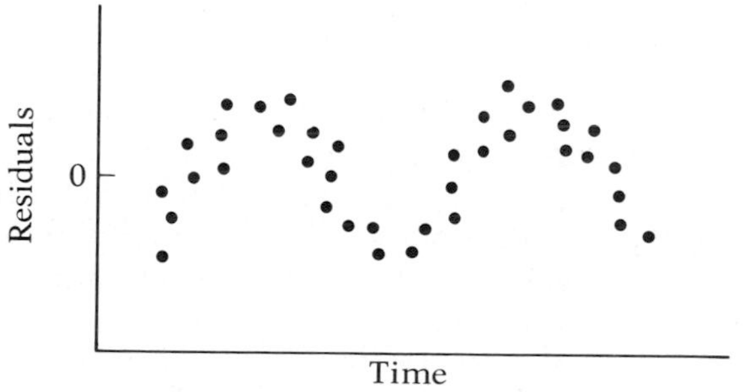

Figure 15.27 Residuals plotted against time, showing autocorrelation

If autocorrelation is present, all variances and standard errors are likely to be seriously wrong, for any sample size. Independence is the most critical assumption in all standard error calculations. If the independence assumption is wrong, the indicated standard errors may be very wrong. In particular, if there is *positive* autocorrelation—positive residuals usually following positive ones, negative residuals usually following negative ones—calculated standard errors will be too small, perhaps much too small. Thus confidence intervals will be narrower than they should be, and hypothesis tests more significant than they should be.

Example 15.41 A computer program was written to simulate regression analysis with autocorrelated error terms. The sample size was 50, the slope of the (only) X variable was 0, and the correlation of successive errors was about .50. Results were as shown in Table 15.5.

	number of times H0:	"slope is 0"	is rejected in favor of
alpha	"slope > 0"	"slope < 0"	total (alpha doubled)
0.100	228	231	459
0.050	161	175	336
0.025	128	142	270
0.010	85	84	169
0.005	66	63	129

	mean	variance
slope	-0.0002	0.000353
resid. variance	1.2420	0.114555
t for the slope	-0.0194	3.188672
Durbin-Watson	1.1532	0.069225
standard error	0.0108	0.000002

Table 15.5

(a) The population variance of the error terms was 1.33333. Does the average residual variance approximate that value?

(b) The theoretical mean and variance of a t statistic, computed on the assumption of 0 autocorrelation, are 0 and (df)/(df − 2), respectively. Do the mean and variance of the 1000 t statistics approximate these values?

(c) Does it appear that the nominal α values are approximately correct?

Solution

(a) The average residual variance is slightly too small at 1.2420.

(b) With $n = 50$ and a single independent variable, df $= 50 - (1 + 1) = 48$. Thus the variance theoretically should be $48/46 = 1.043$. The actual t statistics are far more variable, with a variance of about 3.189.

(c) The numbers of rejections are far higher than the nominal α value would indicate. In the $\alpha = .10$ row, there should be $1000(.10) = 100$ in each of the "slope > 0" and "slope < 0" columns, for a total of 200. In fact, there are 228 and 231, for a total of 459. In the $\alpha = .005$ row, the values should be 5, 5, and 10, and actually are 66, 63, and 129. Autocorrelation leads to far too many t values in the tails of the nominal distribution, as reflected in the large actual variance of the t statistics. ☐

Durbin-Watson statistic

A summary statistic often reported as a test for autocorrelation is the **Durbin-Watson statistic** d. d is based on the squared difference of each residual from the next one; if there is positive autocorrelation, these differences, and therefore the d statistic, will be small. The Durbin-Watson statistic is

$$d = \frac{\sum (e_{t+1} - e_t)^2}{\sum (e_t)^2}$$

where the residuals are denoted as e_t rather than e_i to emphasize that the autocorrelation issue most commonly arises in time-series data. A useful interpretation of the Durbin-Watson d statistic is that d is approximately equal to $2(1 - r_{t,t+1})$, where $r_{t,t+1}$ is the correlation coefficient of residuals at two successive times. If there is independence, this correlation should be near 0. It can be shown [see any econometrics book; e.g., Johnston (1972)] that even modest autocorrelations such as $r_{t,t+1} = .25$ can have a serious effect on inference in regression. Therefore, a Durbin-Watson statistic less than about $1.5 = 2(1 - .25)$ indicates a serious autocorrelation problem.

Example 15.42

Refer to Example 15.40. Computer output showed $d = 1.75$. Is there reason to be suspicious of the standard errors shown in other parts of the output?

Solution

Not really. The value of d is above our rule-of-thumb value 1.5, although not at its most desired value, 2. In addition, the fact that the data aren't time-series data suggests that there should be no autocorrelation. □

homoscedasticity A slightly less critical assumption is the assumption of **homoscedasticity**—constant variability around the regression equation. The most common violation of this assumption occurs when variability increases as the predicted value increases. To detect this problem, plot the residual values against the predicted values. In Figure 15.28, there's a strong indication of nonconstant variability. Notice that "the plot thickens," with the variability of the residuals getting larger as the predicted values get larger.

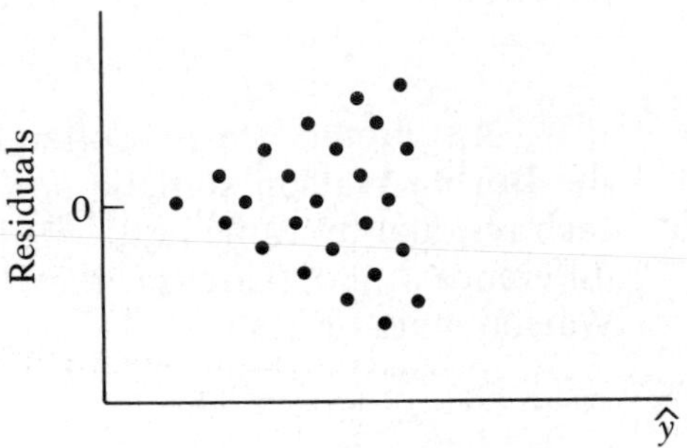

Figure 15.28 Residuals plotted against predicted values—nonconstant variance

One difficulty with using this particular residual plot is that the eye tends to see the range of data, rather than the standard deviation. When there are many points at or near a particular predicted value, the range will naturally tend to be larger than when there are few points at or near the predicted value. If the constant-variance assumption holds, and if most of the predicted values fall in the middle of the $\hat{Y}$ axis, the plot of residuals vs. predicted values should look like Figure 15.29. Ideally, the plot should be shaped like a "fuzzy football."

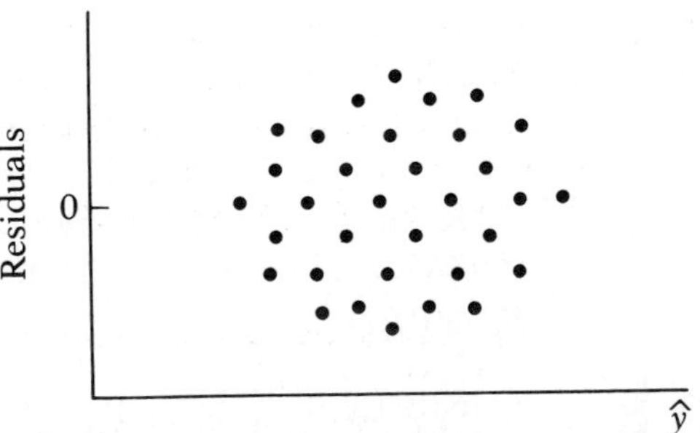

Figure 15.29 Residuals plotted against predicted values—constant variance

Example 15.43

A plot of residuals vs. predicted values for the data of Example 15.3 is shown below in Figure 15.30. Is there evidence of nonconstant variance (heteroscedasticity)?

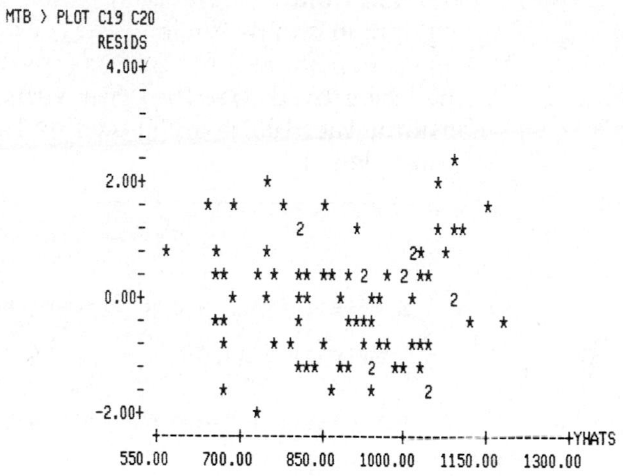

Figure 15.30 Residual plot for Example 15.43

Solution

There's no obvious fan-shaped pattern. Thus, there is no obvious problem of heteroscedasticity. □

When heteroscedasticity occurs, there are several consequences. The nominal α level of tests or confidence intervals may be somewhat in error. The least-squares method may be relatively inefficient, yielding standard errors that are higher than necessary. A particular problem is prediction intervals. If the X values used in the prediction correspond to high variance error, the prediction interval (which is based on the constant-variance assumption) will be too narrow. Conversely, if the X values correspond to low variance error, the interval will be too wide. If nonconstant variance is detected, one useful strategy is to transform the dependent variable Y. Taking the logarithm of Y or defining $Y' = Y/X$ often will eliminate the problem.

Example 15.44

A computer program was written to simulate regression analysis with nonconstant variance. The data were generated in such a way that

$Y = 0 + 0X + \epsilon$, with the standard deviation of the error term proportional to the X value. Two regression models were fitted, the linear model

$$Y = \beta_0 + \beta_1 X + \epsilon$$

and the transformed model

$$\frac{Y}{X} = \beta_0 \frac{1}{X} + \beta_1 + \epsilon$$

It can be proved that the error variance for the transformed model is constant. The results are shown in Table 15.6.

```
                     Regression Analysis with Non-Constant Variance

Simulation of simple linear regression (1000 samples)

Sample size is 50


For y=Beta0+(Beta1)(x)+(x)(error)

one-tail:

          number of times H0:  "beta1 is 0" is rejected in favor of
alpha      "beta1 > 0"     "beta1 < 0"        total (alpha doubled)
0.100          106             137                  243
0.050           59              67                  126
0.025           33              34                   67
0.010           18              13                   31
0.005            8               5                   13


For (y/x)=(Beta0/x)+Beta1+error

one-tail:

          number of times H0:  "beta1 is 0" is rejected in favor of
alpha      "beta1 > 0"     "beta1 < 0"        total (alpha doubled)
0.100          107              99                  206
0.050           54              54                  108
0.025           28              28                   56
0.010            7               7                   14
0.005            5               3                    8
```

Table 15.6

For which model is the nominal α value more accurate?

Solution In the linear model, the number of times that the null hypothesis is rejected is consistently too high. In the transformed model, the number of rejections is quite close to the expected number. □

Another assumption in regression analysis is that the errors are normally distributed. The most serious violation of this assumption occurs when there are outliers. Usually, but not always, outlier values will show up in plots of residuals vs. either independent variable values or predicted values, such as Figure 15.31. In that figure, there appear to be two substantial outliers.

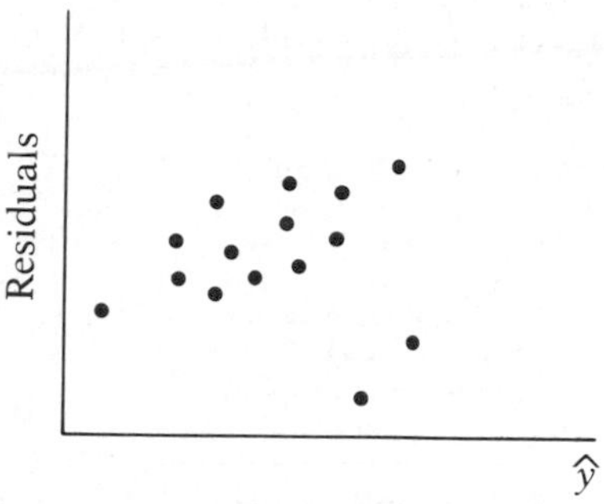

Figure 15.31 Residual plots with outliers

Unfortunately, the most damaging outliers may not show up in residual plots. If the outliers occur at points having great influence or leverage, as discussed in Section 15.2, or if there are several outliers, the regression equation may be twisted so strongly toward the outliers that the residuals from the outliers will be relatively innocuous. Some computer programs will identify high-influence observations. A good practice is to rerun the regression analysis without these points. If any of the coefficients change substantially, the deleted, high-influence points are suspected of being outliers.

When one or more outliers are suspected, there are two possible approaches. One is simply to delete the suspected outliers from the data. This is potentially a misleading approach. By deleting outliers, one may well get a better estimate of the prediction equation, but at the cost of understating the variability around that equation and overstating the amount of correlation and degree of statistical significance of the resulting analysis. If this approach is adopted, basic ethics require that the deletion of data points be reported along with the analysis. An alternative approach is to use various "robust" or "resistant" estimates. See Mosteller and Tukey (1977). These methods are much less sensitive to outliers and don't require that one cavalierly discard parts of the data. Some robust regression methods are already available for standard computer packages (e.g., the RLINE procedure in Minitab); in the future, more of them will be.

Example 15.45 Examine the residual plots for the Bolton data shown in Figure 15.32. Is there any indication of troublesome outliers?

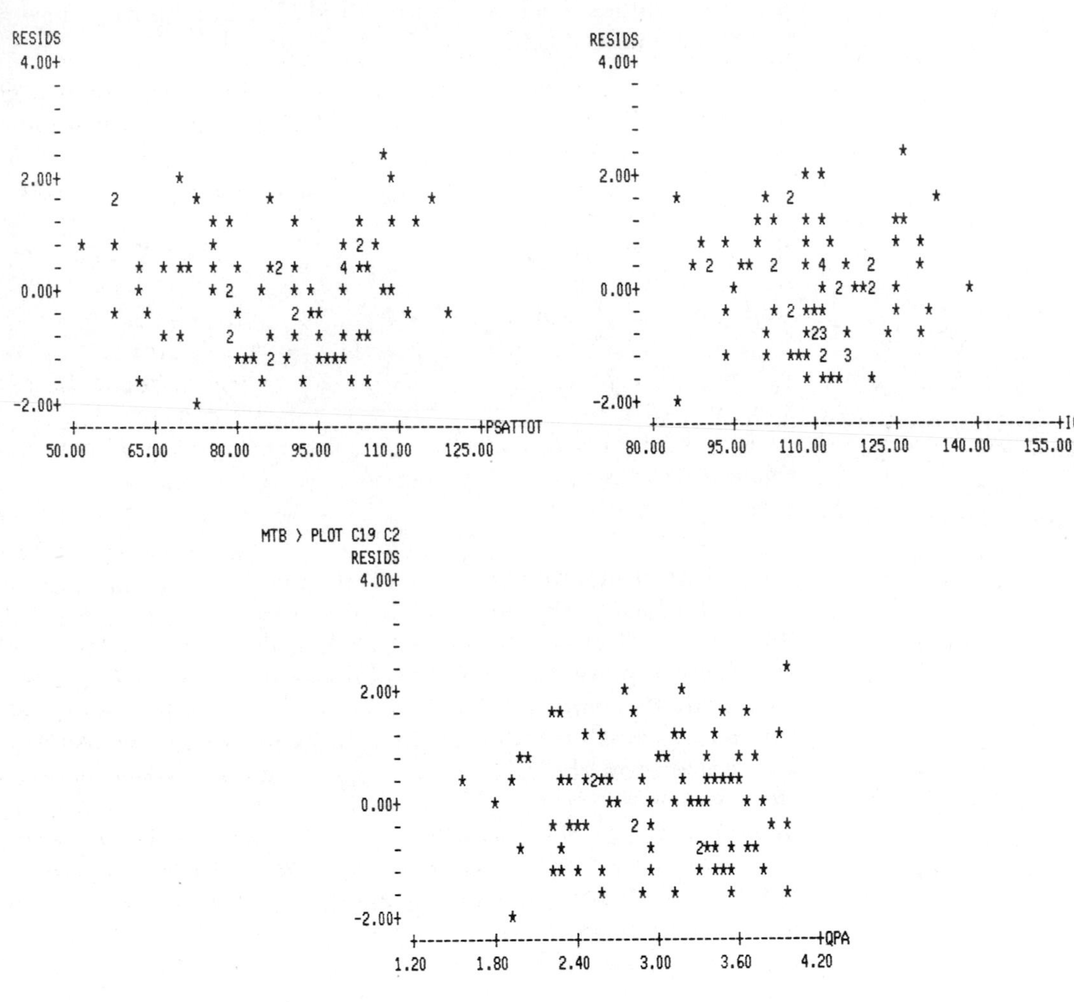

MTB > PLOT C19 C8

Figure 15.32 Output for Example 15.45

Solution There is one rather large positive residual that shows up in all the plots. This residual consistently is near the high end of all X scales and thus may have relatively high influence. The analysis was redone omitting this point. The results (not shown here) showed that the coefficients changed very little, the residual standard deviation decreased slightly, and the R^2 value increased a very small amount. Generally, the results are quite similar, so the potential outlier is not cause for great concern. □

One final potential violation of assumptions can't be detected by residual plots. It's assumed that the values of the independent variable are *reliably measured.* That is, it is assumed that the measured X values are not contaminated by (a substantial amount of) random measurement error. In Section 15.4 we noted that violation of this assumption will bias the coefficient of the associated X variable toward 0, and tend to reduce the magnitude of a correlation. When an unreliably measured X variable is correlated with (collinear with) other X variables, there can be complicated effects on the coefficients of the other X variables, as well. Detecting an unreliably measured X variable can't usually be done by looking at the data. One will only have the observed X values, and will not know for sure if they've been measured reliably. Sometimes one can guess that there's a reliability problem merely from the nature of the variable. For example, suppose that one independent variable in a regression study is "reading readiness" of 5-year-old children. It's unlikely that so slippery a concept as reading readiness can be measured with great accuracy, and therefore it is likely that there is a problem of unreliable measurement.

When one suspects that an X variable will be measured poorly, it's wise to try to minimize the correlation of that variable with other X variables. That way, the bad effects of the unreliable measurement will be limited to the coefficient of the unreliable variable, rather than contaminating the estimates of other coefficients. For example, suppose that we had an unreliable measure of reading readiness, and that children were assigned to two groups—group A for those with low readiness scores receiving some intensive teaching and group B for those with high readiness receiving normal teaching. There will be a strong correlation between the readiness score X_1 and the group dummy variable $X_2 = 0$ for A, 1 for B. Automatically, those with $X_2 = 0$ will have low X_1 scores and those with $X_2 = 1$ will have high X_1 scores. The unreliable measurement of X_1 may cause one to miss the effect of X_2. This phenomenon is shown in Figure 15.33. The solid line shows an assumed true relation between reading achievement (the Y variable) and reading readiness (X_1). In the true relation, the A group does better; the A line is higher. But in the estimated (dotted) line, the unreliable-measurement bias has reduced the slope, and it appears that the B group does better; the dotted B line is higher. A better experimental design would randomly assign children to groups, ignoring X_1. Thus there would be little correlation between X_1 and X_2, and the bias would not contaminate the coefficient of X_2.

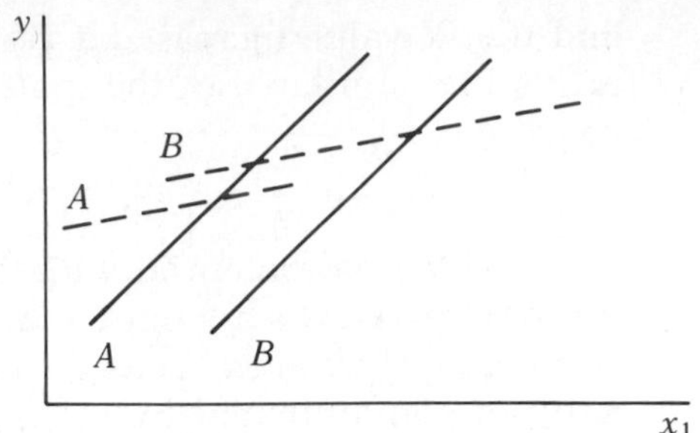

Figure 15.33 Effect of unreliable-X bias: correlated independent variables

Example 15.46

A regression analysis of data taken from young children involved three independent variables: X_1 = prematurity of birth to nearest month (0 indicates born within ±2 weeks of nominal due date, 1 indicates approximately one month premature, 2 indicates approximately two months premature, and −1 indicates one month "postmature"); X_2 = age, in weeks; and X_3 = birth order (1 = first child born to mother, 2 = second child, etc.).

Which variable or variables seem sensitive to unreliability? Is there likely to be an unreliable-X effect on other variables?

Solution

The X_1 variable is severely rounded off; that alone indicates a degree of error of measurement. Further, exact dates of conception are not always known, so there may be further unreliability of this measure. The other two X variables should be measured without any serious error.

Assuming a reasonably random sample of children, there's no obvious reason why X_1 should be strongly correlated with either of the other X's. Unless an unexpected collinearity (X correlation) shows up in the data, there's little reason to be concerned that the unreliability of X_1 will affect the estimates of the effects of X_2 and X_3. □

The list of potential violations of assumptions and of plots needed to check these assumptions may seem forbiddingly long. In this section it's been suggested that you check for

1. correct mathematical form, by plotting residuals against each independent variable;

2. independence, by plotting residuals against time (when the data have been collected as a time series) and by checking the Durbin-Watson statistic;

3. constant variance around the regression equation, by plotting residuals against predicted values;

4. normality, by looking for outliers in the plots and possibly by redoing the regression omitting high-influence points;

5. reliable measurement of the X values, by considering whether in fact these variables are accurately measured.

But in fact, the time required to perform these checks, using any decent computer package, is only a few minutes. When compared to the days and weeks involved in collecting the data for a typical regression-correlation analysis, the time required to check the assumptions is tiny. If the data indicate that one or more of the assumptions are suspect, there are some "tricks of the trade" that can be used to yield a better analysis. See Ott and Hildebrand (1983) for some simple suggestions, or Draper and Smith (1982) for more sophisticated approaches. The need for such methods won't be known unless researchers routinely check the assumptions when doing any regression-correlation analysis. It would be desirable if any statistical analysis proposed for publication in the literature of psychology, education, and the social sciences would involve more than mechanical number-crunching, if referees of such papers would demand that authors check assumptions for reasonableness, and if readers of such papers would be skeptical of statistical analysis unaccompanied by consideration of assumptions. Unfortunately, current practice doesn't really require thoughtful statistical analysis. Until that situation changes, you can be alert to the dangers of mindless statistical analysis by others, and do your own statistical analyses thoughtfully and critically.

Exercises for Section 15.6

15.50. Refer to the data of Exercise 15.2 and the plot made in that exercise. Is there any evidence of violation of assumptions?

15.51. A regression analysis related mother's rating of a child's development level to teacher's rating of the same child. Both mother and teacher rated the child on a 0–20 scale. Output is shown in Figure 15.34.

a. Is there a statistically detectable relationship, according to the output?

RESIDUAL PLOTS FOR DEVELOPMENTAL STUDY

PLOT OF TEACHER*MOTHER   LEGEND: A = 1 OBS, B = 2 OBS, ETC.

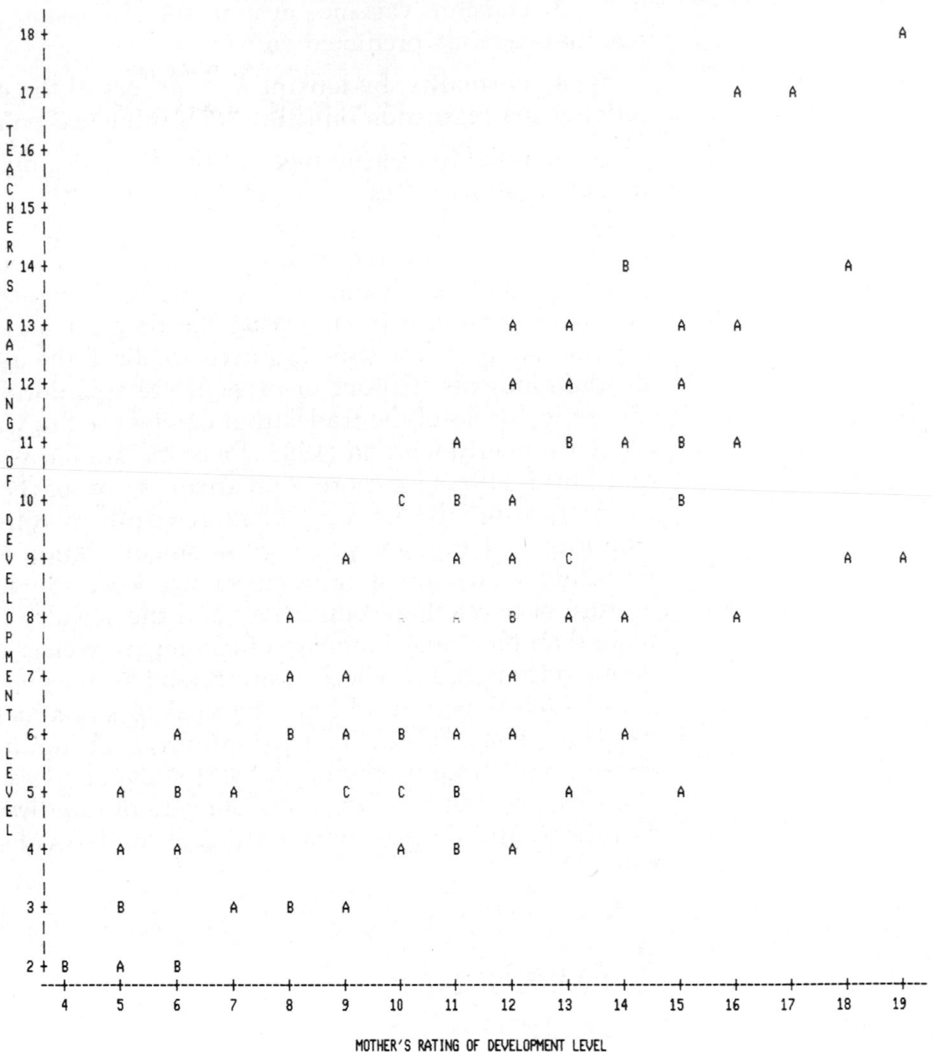

REGRESSION OF DEVELOPMENT RATINGS

GENERAL LINEAR MODELS PROCEDURE

DEPENDENT VARIABLE: TEACHER   TEACHER'S RATING OF DEVELOPMENT LEVEL

SOURCE	DF	SUM OF SQUARES	MEAN SQUARE	F VALUE	PR > F	R-SQUARE	C.V.
MODEL	1	688.50768872	688.50768872	113.23	0.0001	0.571196	31.9251
ERROR	85	516.87162162	6.08084261		ROOT MSE		TEACHER MEAN
CORRECTED TOTAL	86	1205.37931034			2.46593646		7.72413793

Figure 15.34 Output for Exercise 15.51

PARAMETER	ESTIMATE	T FOR H0: PARAMETER=0	PR > |T|	STD ERROR OF ESTIMATE
INTERCEPT	-0.93412162	-1.09	0.2780	0.85556030
MOTHER	0.78547297	10.64	0.0001	0.07381740

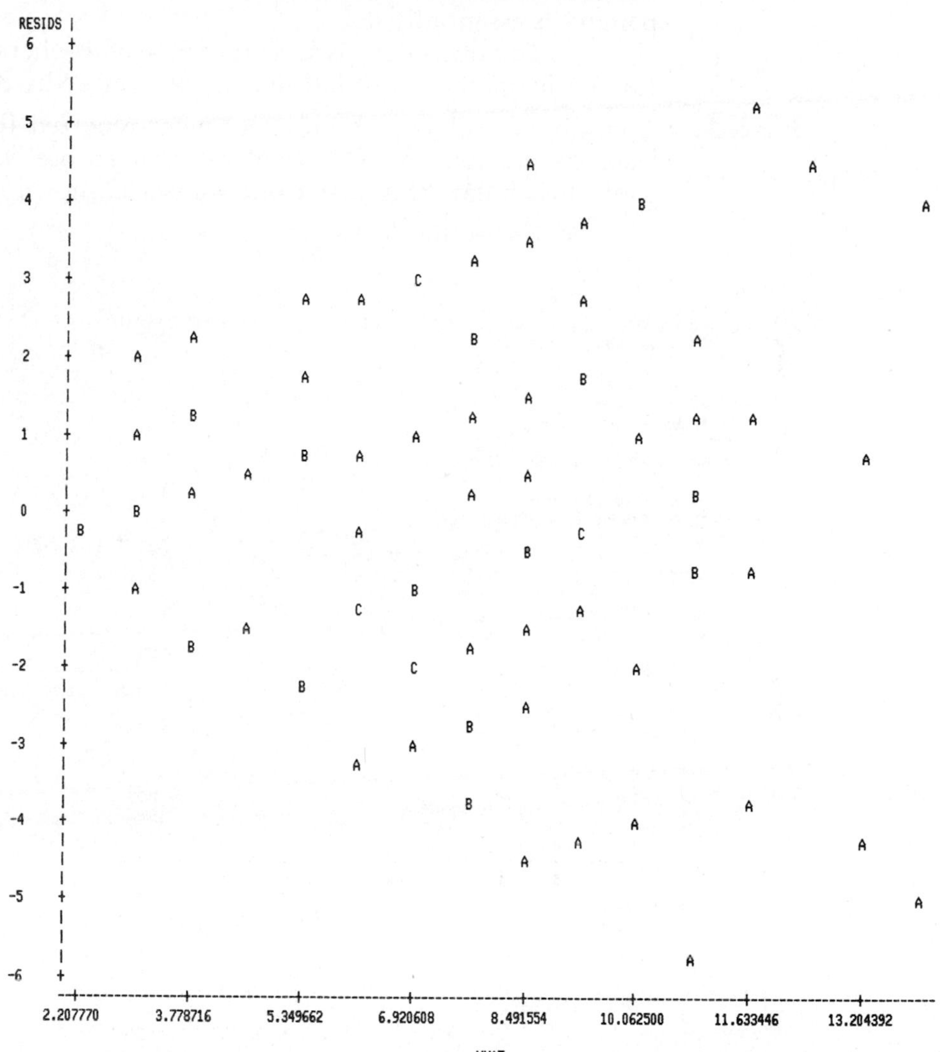

b. What violation of assumptions is most conspicuous?

c. Does the violation found in part b call your answer to part a into question?

15.52. Refer to Exercise 15.51.

a. Calculate a 95% prediction interval for the teacher's rating of a child with a mother's rating equal to 6. Assume that the extrapolation penalty is essentially 0.

b. Can one safely assume that 95% of teachers' ratings of children rated 6 by mothers will fall in this interval? Why or why not?

15.53. A number of subjects in an experiment reported the number of hours slept the previous night. A problem-solving score was reported for each subject each day. Output for one subject is shown in Figure 15.35.

a. Locate the R^2 value.

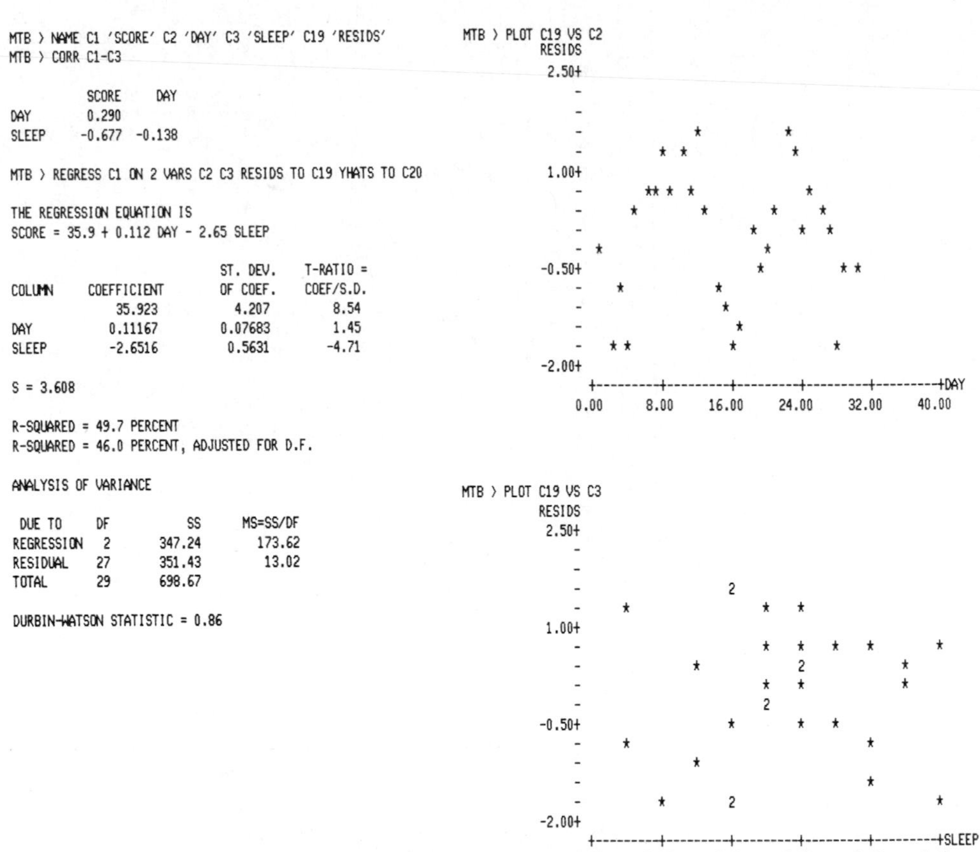

```
MTB > NAME C1 'SCORE' C2 'DAY' C3 'SLEEP' C19 'RESIDS'
MTB > CORR C1-C3

           SCORE      DAY
DAY        0.290
SLEEP     -0.677   -0.138

MTB > REGRESS C1 ON 2 VARS C2 C3 RESIDS TO C19 YHATS TO C20

THE REGRESSION EQUATION IS
SCORE = 35.9 + 0.112 DAY - 2.65 SLEEP

                                 ST. DEV.    T-RATIO =
COLUMN       COEFFICIENT         OF COEF.    COEF/S.D.
                  35.923            4.207         8.54
DAY              0.11167          0.07683         1.45
SLEEP            -2.6516           0.5631        -4.71

S = 3.608

R-SQUARED = 49.7 PERCENT
R-SQUARED = 46.0 PERCENT, ADJUSTED FOR D.F.

ANALYSIS OF VARIANCE

 DUE TO       DF          SS      MS=SS/DF
REGRESSION     2      347.24        173.62
RESIDUAL      27      351.43         13.02
TOTAL         29      698.67

DURBIN-WATSON STATISTIC = 0.86
```


Figure 15.35 Output for Exercise 15.53

b. Calculate an F statistic. Is the prediction statistically significant at $\alpha = .01$?

c. Summarize the results of the t tests.

15.54. Refer to Figure 15.35. What violations of assumptions seem to be present in the study? Do these violations call your answers to Exercise 15.53 into question?

15.55. A normal probability plot of the residuals from the regression study of Exercise 15.39 is shown in Figure 15.36.

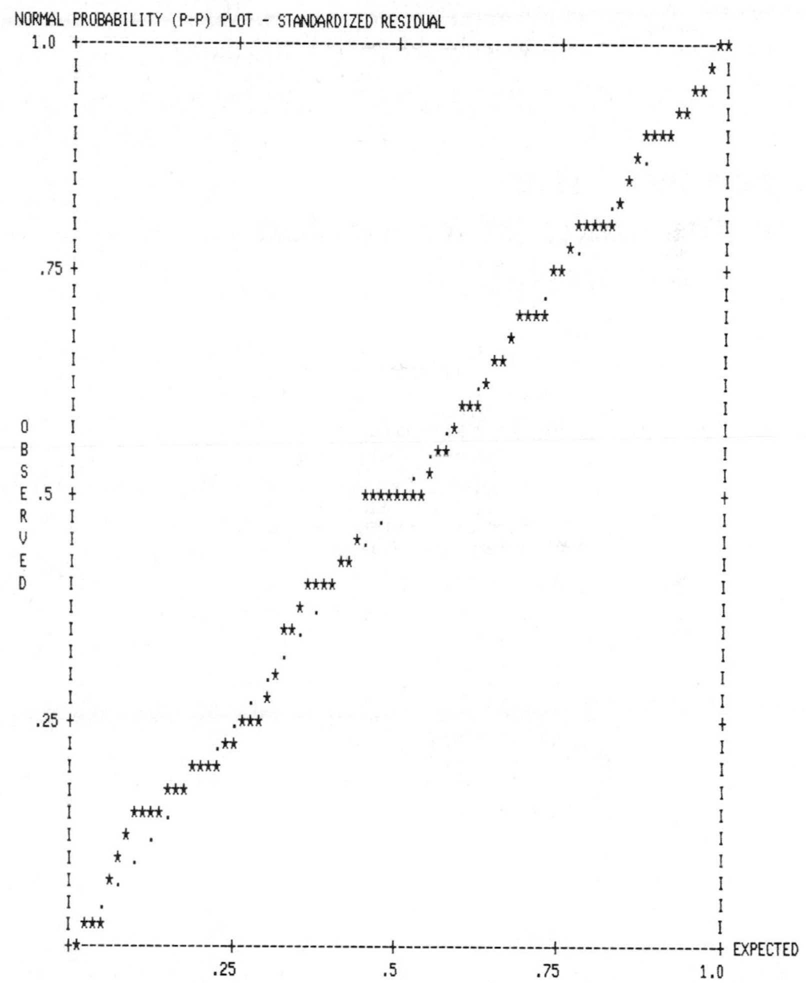

Figure 15.36 Output for Exercise 15.55

a. Is there evidence of outliers or other nonnormal properties of the residual distribution?

b. Should autocorrelation be a problem, given the nature of the study?

15.56. Refer to Exercise 15.39. Is there likely to be a problem of unreliable X measurement? If so, can such a problem be seen in the plots of Exercise 15.55?

15.57. A study was done of the relation between expenditure per pupil and student achievement on a standard test for 24 independent high schools. Scatter plots of achievement against expenditure showed a modest positive relation. However, 3 schools had quite high expenditures, but very low achievement scores. What consequence will this fact have on a regression analysis of the data?

Appendix: The Mathematics of Regression and Correlation

A number of mathematical results about regression and correlation have been used in this chapter. In this appendix, we'll prove some of the results. Other results involve the use of matrix notation and are beyond the scope of this book.

We consider some of the basic results for simple regression, with a single X variable. The first step is to show that the least-squares principle yields the stated $\hat{\beta}_0$ and $\hat{\beta}_1$ coefficients. An algebraic proof using no calculus can be done, but it's rather awkward. Here we use a more direct, but calculus-based, proof.

Result 15.1 The coefficients $\hat{\beta}_0$ and $\hat{\beta}_1$ that minimize SS(Residual) are given by

$$\hat{\beta}_1 = \frac{S_{xy}}{S_{xx}}$$

and

$$\hat{\beta}_0 = \frac{\sum y - \hat{\beta}_1 \sum x}{n}$$

where

$$S_{xy} = \sum xy - \frac{\left(\sum x\right)\left(\sum y\right)}{n}$$

and

$$S_{xx} = \sum x^2 - \frac{\left(\sum x\right)^2}{n}$$

Proof

$$\text{SS(Residual)} = \sum [y - (\hat{\beta}_0 + \hat{\beta}_1 x)]^2$$

To find the values of the coefficients that minimize SS(Residual), we must equate first derivatives to 0. Note that the variables are $\hat{\beta}_0$ and $\hat{\beta}_1$, rather than x and y; x and y are data. The derivative with respect to $\hat{\beta}_0$ is

$$\sum 2[y - (\hat{\beta}_0 + \hat{\beta}_1 x)](-1) = 0$$

Divide through by $2(-1)$ and expand out the sum to get

$$\sum y - \hat{\beta}_0 \sum 1 - \hat{\beta}_1 \sum x = 0$$

or

$$\text{(Eq. 1)} \quad \sum y = \hat{\beta}_0(n) + \hat{\beta}_1 \sum x$$

because $\Sigma\ 1$ is $(1 + 1 + \cdots + 1) = n$. A similar process—taking the derivative with respect to $\hat{\beta}_1$, equating to 0, and solving—yields

$$\text{(Eq. 2)} \quad \sum xy = \hat{\beta}_0 \sum x + \hat{\beta}_1 \sum x^2$$

Thus we have two equations in two unknowns, and must solve the equations. Multiply (Eq. 1) by $(\Sigma\ x)/n$ and subtract the result from (Eq. 2).

$$\sum xy - \left(\sum x\right)\frac{\left(\sum y\right)}{n} = \hat{\beta}_0\left(\sum x - \frac{\sum x}{n}n\right) + \hat{\beta}_1\left(\sum x^2 - \frac{\left(\sum x\right)^2}{n}\right)$$

The left side of this equation is S_{xy} by definition. The factor multiplying $\hat{\beta}_0$ is $\Sigma\ x - \Sigma\ x = 0$. The factor multiplying $\hat{\beta}_1$ is S_{xx}, also by definition. So

$$S_{xy} = \hat{\beta}_1 S_{xx} \quad \text{or} \quad \hat{\beta}_1 = \frac{S_{xy}}{S_{xx}}$$

The equation for $\hat{\beta}_0$ is a rearrangement of (Eq. 1). Subtract $\hat{\beta}_1 \Sigma x$ from both sides of the equation and divide both sides by n to obtain the equation for $\hat{\beta}_0$.

Strictly speaking, we should verify that these equations give a minimum value for SS(Residual), as opposed to a maximum or a saddle point. The calculus that's needed is a bit more difficult, so we won't do a formal proof. Notice that it must be impossible to have a maximum for SS(Residual). That would imply that one could find a "most-squares" line that was the absolutely worst-fitting line to the data. Pretty clearly, no matter how badly fitting one line is to the data, one can always draw another line that's even worse!

The equation for the $\hat{\beta}_1$ coefficient can be rewritten to show the relation between the slope and the correlation coefficient r_{YX}.

Result 15.2

$$\hat{\beta}_1 = r_{YX} \frac{s_y}{s_x}$$

where s_x and s_y are the standard deviations of X and Y, respectively.

Proof

Recall that

$$r_{YX} = \frac{S_{xy}}{\sqrt{S_{xx} S_{yy}}}$$

The short-cut formula for the variance given in Chapter 2 is

$$s_x^2 = \frac{\sum x^2 - \left(\sum x\right)^2 \Big/ n}{n - 1}$$

$$= \frac{S_{xx}}{n - 1}$$

and similarly

$$s_y^2 = \frac{S_{yy}}{n - 1}$$

So

$$r_{YX}\frac{s_y}{s_x} = \frac{S_{xy}}{\sqrt{S_{xx}S_{yy}}} \frac{\sqrt{S_{yy}/(n-1)}}{\sqrt{S_{xx}/(n-1)}} = \frac{S_{xy}}{(\sqrt{S_{xx}})^2}$$

after canceling. So

$$r_{YX}\frac{s_y}{s_x} = \frac{S_{xy}}{S_{xx}} = \hat{\beta}_1$$

as desired.

When a variable is expressed in z-score form, its standard deviation automatically equals 1. Result 15.2 indicates that if both X and Y are stated in z-score form, the slope coefficient of a simple regression equals the correlation coefficient.

The use of z-scores also is helpful in indicating the effects of various correlations in multiple regression. In particular, it indicates the effect of 0 correlations between independent variables.

Result 15.3 If two independent variables X_1 and X_2 have correlation 0, the coefficient $\hat{\beta}_1$ in the multiple regression equation

$$\hat{Y} = \hat{\beta}_0 + \hat{\beta}_1 X_1 + \hat{\beta}_2 X_2$$

equals the coefficient $\hat{\beta}_1$ in the simple regression equation

$$\hat{Y} = \hat{\beta}_0 + \hat{\beta}_1 X_1$$

Proof If Y is expressed in z-score form, $\Sigma\, y = 0$, because the mean of a set of z-scores is 0. Also, $\Sigma\, y^2 = n - 1$, because the variance s_y^2 is 1 in z-score form, so

$$1 = s_y^2 = \sum \frac{(y - \bar{y})^2}{n-1}$$
$$= \sum \frac{y^2}{n-1}$$

because $\bar{y} = 0$ in z-scores. A similar argument shows that when both X and Y are in z-score form,

$$\sum xy = (n-1) r_{YX}$$

The coefficients in multiple regression can be derived using calculus, as in Result 15.1. When there are two independent variables X_1 and X_2, three equations are found by taking derivatives.

$$\text{(Eq. 1)} \quad \sum y = \hat{\beta}_0(n) + \hat{\beta}_1 \sum x_1 + \hat{\beta}_2 \sum x_2$$

$$\text{(Eq. 2)} \quad \sum x_1 y = \hat{\beta}_0 \sum x_1 + \hat{\beta}_1 \sum x_1^2 + \hat{\beta}_2 \sum x_1 x_2$$

$$\text{(Eq. 3)} \quad \sum x_2 y = \hat{\beta}_0 \sum x_2 + \hat{\beta}_1 \sum x_1 x_2 + \hat{\beta}_2 \sum x_2^2$$

When all three variables are in *z*-score form, these equations become

$$\text{(Eq. 1)} \quad 0 = \hat{\beta}_0(n) + \hat{\beta}_1(0) + \hat{\beta}_2(0)$$

$$\text{(Eq. 2)} \quad (n-1)r_{Y1} = \hat{\beta}_0(0) + \hat{\beta}_1(n-1) + \hat{\beta}_2(n-1)r_{12}$$

$$\text{(Eq. 3)} \quad (n-1)r_{Y2} = \hat{\beta}_0(0) + \hat{\beta}_1(n-1)r_{12} + \hat{\beta}_2(n-1)$$

where r_{Y1} is the correlation between Y and X_1, r_{Y2} is the correlation between Y and X_2, and r_{12} is the correlation between X_1 and X_2. If the two X variables are uncorrelated, $r_{12} = 0$ and (Eq. 2) becomes

$$(n-1)r_{Y1} = \hat{\beta}_1(n-1)$$

or

$$\hat{\beta}_1 = r_{y1}$$

exactly as in simple regression.

Result 15.3 has been proved for the *z*-score form of regression. The same result holds for a raw-score regression, because

$$\text{raw-score slope} = (z\text{-score slope})\left(\frac{S_y}{S_x}\right)$$

in general. When X_1 and X_2 are uncorrelated,

$$\text{raw-score slope} = r_{y1}\left(\frac{S_y}{S_x}\right)$$

exactly the expression for $\hat{\beta}_1$ given in Result 15.3.

Now we turn to properties of the residuals $y - \hat{y}$. We will prove that the average residual is 0 and that there is no correlation between the residuals and the X values. Both properties are reasonable ones. If the average residual were, say, positive, there would be a tendency for the regression equation to predict too low; the best-predicting line should not (and does not) have any such tendency. Also, the best-predicting line should (and does) use all the correlation there is between X and Y.

Result 15.4 In simple regression,

1. The sum and average of the residuals is 0.
2. The correlation of the residuals and the X values is 0.

Proof

$$\sum(y - \hat{y}) = \sum(y - \hat{\beta}_0 - \hat{\beta}_1 x)$$

$$= \sum y - \hat{\beta}_0(n) - \hat{\beta}_1 \sum x$$

In the proof of Result 15.1, we found two equations to solve. The first of those equations immediately says that $\Sigma(y - \hat{y}) = 0$. If the sum of the residuals is 0, the average residual must also be 0, proving part 1 of the result.

To prove part 2, we note that the numerator of $r_{\text{residuals},x}$ is

$$\sum (\text{residuals})x - \frac{\left(\sum \text{residuals}\right)\left(\sum x\right)}{n} = \sum (\text{residuals})x$$

because Σ residuals $= 0$. Now

$$\sum (\text{residuals})x = \sum(y - \hat{\beta}_0 - \hat{\beta}_1 x)x$$

$$= \sum xy - \hat{\beta}_0 \sum x - \hat{\beta}_1 x^2$$

The second equation in the proof of Result 15.1 immediately says that this value is 0. Because the numerator of the correlation is 0, the correlation itself must be 0.

Result 15.4 is useful in proving that SS(Total) is the sum of "explained" and "unexplained" parts, namely SS(Regression) and SS(Residual).

Result 15.5

$$\text{SS(Total)} = \text{SS(Residual)} + \text{SS(Regression)}$$

where

$$\text{SS(Total)} = \sum(y - \bar{y})^2$$

$$\text{SS(Residual)} = \sum(y - \hat{y})^2$$

and

$$\text{SS(Regression)} = \sum(\hat{y} - \bar{y})^2$$

Proof The proof is another expand-the-square proof, with the expansion this time around $\hat{y}$.

$$\text{SS(Total)} = \sum(y - \hat{y} + \hat{y} - \bar{y})^2$$
$$= \sum(y - \hat{y})^2 + 2\sum(y - \hat{y})(\hat{y} - \bar{y}) + \sum(\hat{y} - \bar{y})^2$$

The first term in the expansion is SS(Residual), and the last term is SS(Regression). The only thing to show is that the middle term is 0.

$$\sum(y - \hat{y})(\hat{y} - \bar{y}) = \sum(y - \hat{y})(\hat{\beta}_0 + \hat{\beta}_1 x - \bar{y})$$
$$= \hat{\beta}_0 \sum(y - \hat{y}) + \hat{\beta}_1 \sum(y - \hat{y})x - \bar{y}\sum(y - \hat{y})$$
$$= \hat{\beta}_1 \sum(y - \hat{y})x$$

because Result 15.4 says that the sum of the residuals $y - \hat{y}$ is 0.

In the proof of part 2 of Result 15.4, we showed that $\Sigma(y - \hat{y})x = 0$, which completes the proof that the middle term of the expansion of SS(Total) is 0, and therefore the proof of Result 15.5.

Finally, we turn to a proof of the relation between the correlation coefficient and SS(Regression).

Result 15.6 $\text{SS(Regression)} = r_{YX}^2\text{SS(Total)}$

Proof

$$\text{SS(Regression)} = \sum(\hat{\beta}_0 + \hat{\beta}_1 x - \bar{y})^2$$
$$= \sum(\bar{y} - \hat{\beta}_1\bar{x} + \hat{\beta}_1 x - \bar{y})^2$$

by the equation for calculating $\hat{\beta}_0$.

$$\text{SS(Regression)} = (\hat{\beta}_1)^2 \sum (x - \bar{x})^2$$

But $\Sigma\,(x - \bar{x})^2 = (n - 1)s_x^2$ and, from Result 15.2, $\hat{\beta}_1 = r_{YX}(s_y/s_x)$. Therefore,

$$\text{SS(Regression)} = \left[r_{YX}\left(\frac{s_y}{s_x}\right) \right]^2 (n-1)s_x^2$$
$$= (n-1)r_{YX}^2 s_y^2$$
$$= r_{YX}^2 \sum (y-\bar{y})^2 = r_{YX}^2 \text{SS(Total)}$$

Although we won't carry out the details, it's also possible to use an expand-the-square argument to prove that, if X_1 and X_2 are uncorrelated, then

$$\text{SS(Regression using } X_1 \text{ and } X_2)$$
$$= \text{SS(Regression using } X_1) + \text{SS(Regression using } X_2)$$

The key is to show that the cross-product term sums to 0; it turns out that the condition that $r_{12} = 0$ is exactly what's needed to prove that.

Chapter Exercises

15.58. Carne and Kirton (1982) looked at the relation between X = Kirton Adaptation-Innovation Inventory scores and Y = Myers-Briggs Type indicators. Particularly, suppose that the following data are obtained on the Kirton innovation scale and the Myers-Briggs intuition scale.

X	32	36	42	49	51	58	63	66	69	70	70	73	78	82	85	86
Y	65	58	89	79	71	84	82	70	93	90	73	79	103	112	79	81
X	86	90	92	93	93	93	95	95	95	97	98	100	100	100	101	103
Y	120	88	90	106	115	89	100	90	120	81	94	120	11	96	89	110
X	103	106	109	111	115	117	120	122	125	128	129	130	132	137	140	149
Y	110	125	93	130	127	89	106	130	106	143	98	110	146	130	105	121
X	160															
Y	115															

a. Draw a scatter plot of the data. Does there appear to be some degree of linear relation?

b. Is there any evidence of a nonlinear relation?

15.59. Refer to the data of Exercise 15.58.

a. Calculate the coefficients for a linear regression equation.
b. Interpret the coefficients.

15.60. Refer to the data of Exercise 15.58.

a. Calculate r_{YX} and the coefficient of determination.
b. Calculate SS(Residual) and SS(Total).
c. How useful is X in predicting Y?

15.61. Ingram and Zurawski (1981) conducted a study of admissions criteria for a graduate program in psychology. Among the variables studied were Y = rating of the overall quality of the application (1 to 9, 1 = very poor, 5 = average, 9 = very good), X_1 = rated quality of recommendation letters (also on a 1 to 9 scale), and X_2 = undergraduate grade average. Assume that the data were

Y	1	1	1	1	2	2	2	2	2	2	2
X_1	3	2	8	2	4	5	4	5	4	3	5
X_2	2.62	2.81	2.42	3.12	2.47	2.63	2.94	2.26	2.84	2.75	2.98
Y	2	2	3	3	3	3	3	3	3	4	4
X_1	2	3	3	6	7	3	3	4	5	4	7
X_2	2.64	2.91	3.21	3.38	3.00	3.20	2.86	3.20	3.34	2.98	3.28
Y	4	4	4	4	5	5	5	5	5	5	5
X_1	7	6	5	6	3	8	7	6	7	7	5
X_2	2.90	3.16	3.24	3.52	3.31	3.62	3.54	3.38	3.80	3.37	3.20
Y	5	5	6	6	6	6	6	6	6	6	7
X_1	6	7	4	7	6	6	7	6	9	5	6
X_2	3.62	3.67	3.70	3.45	3.48	2.90	3.64	3.45	3.00	3.60	3.52
Y	7	7	7	7	7	7	7	7	7	8	8
X_1	8	7	7	8	9	7	8	5	4	8	7
X_2	3.50	3.49	3.58	3.40	3.29	3.40	3.51	3.89	3.62	3.74	3.68
Y	8	8	8	9	9	9					
X_1	9	9	6	8	9	9					
X_2	3.60	3.55	3.80	3.90	3.72	3.91					

a. Plot Y vs. X_1, possibly using a computer program package. Does there seem to be some degree of linear relation? Is there any evidence of nonlinearity or outliers?

b. Plot Y vs. X_2. Again, is there a linear relation? Any nonlinearity or outliers?

c. From the plots of parts a and b, can you judge which of the two X variables is the better predictor?

15.62. Refer to the data of Exercise 15.61.

a. Calculate the correlations of Y with X_1, and of Y with X_2.
b. Interpret the numerical value of these correlations carefully.
c. Which X variable is the better predictor, in these data?

15.63. Refer again to the data of Exercise 15.61.

a. Calculate the coefficients of the regression equation for predicting Y from X_1.

b. What is the interpretation of the slope coefficient in this situation?

c. What, if anything, is the interpretation of the intercept in this situation?

15.64. Refer to the scatter plots constructed in Exercise 15.61. Is there any evidence of violation of assumptions?

15.65. The following data are artificial, to clarify a point.

X	2	2	3	3	4	4	5	5	6	6
Y	15	11	19	21	24	22	23	21	17	15

a. Plot the data. Will a linear prediction equation be of any use at all?

b. Calculate r_{YX}. Interpret its numerical value.

c. Does the coefficient of determination adequately reflect the value of X for predicting Y in this case? Why?

15.66. A study was designed to explore how ability on a pursuit task could be affected by levels of steady background noise and by levels of intermittent noise. The data were as follows, with Y = pursuit score, X_1 = level of steady noise, and X_2 = level of intermittent noise.

Y	62	51	47	50	52	49	54	47	51	56	46	41
X_1	30	30	30	30	40	40	40	40	50	50	50	50
X_2	15	20	25	30	15	20	25	30	15	20	25	30

Computer output yielded the following:

COLUMN	COEFFICIENT	ST. DEV. OF COEF.	T-RATIO = COEF./S.D.
—	72.000	7.514	9.58
X_1	−0.2000	0.1434	−1.39
X_2	−0.6000	0.2094	−2.87

THE ST. DEV OF Y ABOUT REGRESSION LINE IS
S = 4.055
WITH (12 − 3) = 9 DEGREES OF FREEDOM

R-SQUARED = 53.0 PERCENT
R-SQUARED = 42.6 PERCENT, ADJUSTED FOR D.F.

ANALYSIS OF VARIANCE

DUE TO	DF	SS	MS = SS/DF
REGRESSION	2	167.00	83.50
RESIDUAL	9	148.00	16.44
TOTAL	11	315.00	

FURTHER ANALYSIS OF VARIANCE
SS EXPLAINED BY EACH VARIABLE WHEN ENTERED IN THE ORDER GIVEN

DUE TO	DF	SS
REGRESSION	2	167.00
X_1	1	32.00
X_2	1	135.00

DURBIN-WATSON STATISTIC = 2.24

a. Identify the coefficients of the multiple regression equation.
b. Find the coefficient of determination. What does it mean?

15.67. Refer to Exercise 15.66.

a. Calculate the r^2 values for predicting Y from X_1 alone and from X_2 alone.
b. How does the R^2 value obtained in Exercise 15.66 compare to the sum of the r^2 values found in part a?
c. Plot X_2 vs. X_1. What must the correlation of X_1 and X_2 equal?

15.68. Calculate the residuals for the data of Exercise 15.66. Do any of the values appear to be outliers?

15.69. Plot the residuals calculated in Exercise 15.68 against the predicted values. What potential violation of assumptions is being checked in this plot? Does it appear to be a problem?

15.70. A less than brilliant student claimed that the constant variance assumption was violated because the numbers in the ST. DEV. OF COEF. column in Exercise 15.66 were not all the same. What's wrong with the student's statement?

15.71. In an experiment about the effect of expectations, subjects tried to solve a series of problems. They were told what the average score supposedly was, and also the supposed average time required. The dependent variable, CORRECT, was the number of correct solutions to problems. The supposed average score and average time were, in fact, varied from one subject to the next. Computer output (SAS) is shown in Figure 15.37.

```
                                   RESULTS FOR STRAIGHT LINE MODEL

                                   GENERAL LINEAR MODELS PROCEDURE

DEPENDENT VARIABLE: CORRECT    ACTUAL NUMBER CORRECT

SOURCE              DF    SUM OF SQUARES      MEAN SQUARE     F VALUE      PR > F      R-SQUARE        C.V.

MODEL                2     2267.75694444    1133.87847222       12.09      0.0001      0.187156     34.0160

ERROR              105     9849.15972222      93.80152116                 ROOT MSE              CORRECT MEAN

CORRECTED TOTAL    107    12116.91666667                                9.68511854               28.47222222


                              T FOR H0:     PR > |T|      STD ERROR OF
PARAMETER       ESTIMATE    PARAMETER=0                    ESTIMATE

INTERCEPT    16.59027778           4.17       0.0001        3.98129627
STATEDAV      0.34861111           4.83       0.0001        0.07218861
STATTIME      2.11111111           0.92       0.3572        2.28280433
```

Figure 15.37 Output for Exercise 15.71

a. How useful are the two variables together in predicting CORRECT?

b. Use the R^2 value shown to verify the F value shown. Is there a statistically detectable relation between the dependent and independent variables?

c. What do the t statistics and associated p-values indicate about the predictive value of the two independent variables?

15.72. Residual plots for the regression study of Exercise 15.71 are shown in Figure 15.38.

a. What assumption appears to be violated?
b. How would you modify the analysis of the data?

15.73. A study of children focused on whether a child would define a word by its use. The independent variables of interest were the number of words the child knew and the child's age. Suppose that Minitab output of the data included that shown in Figure 15.39.

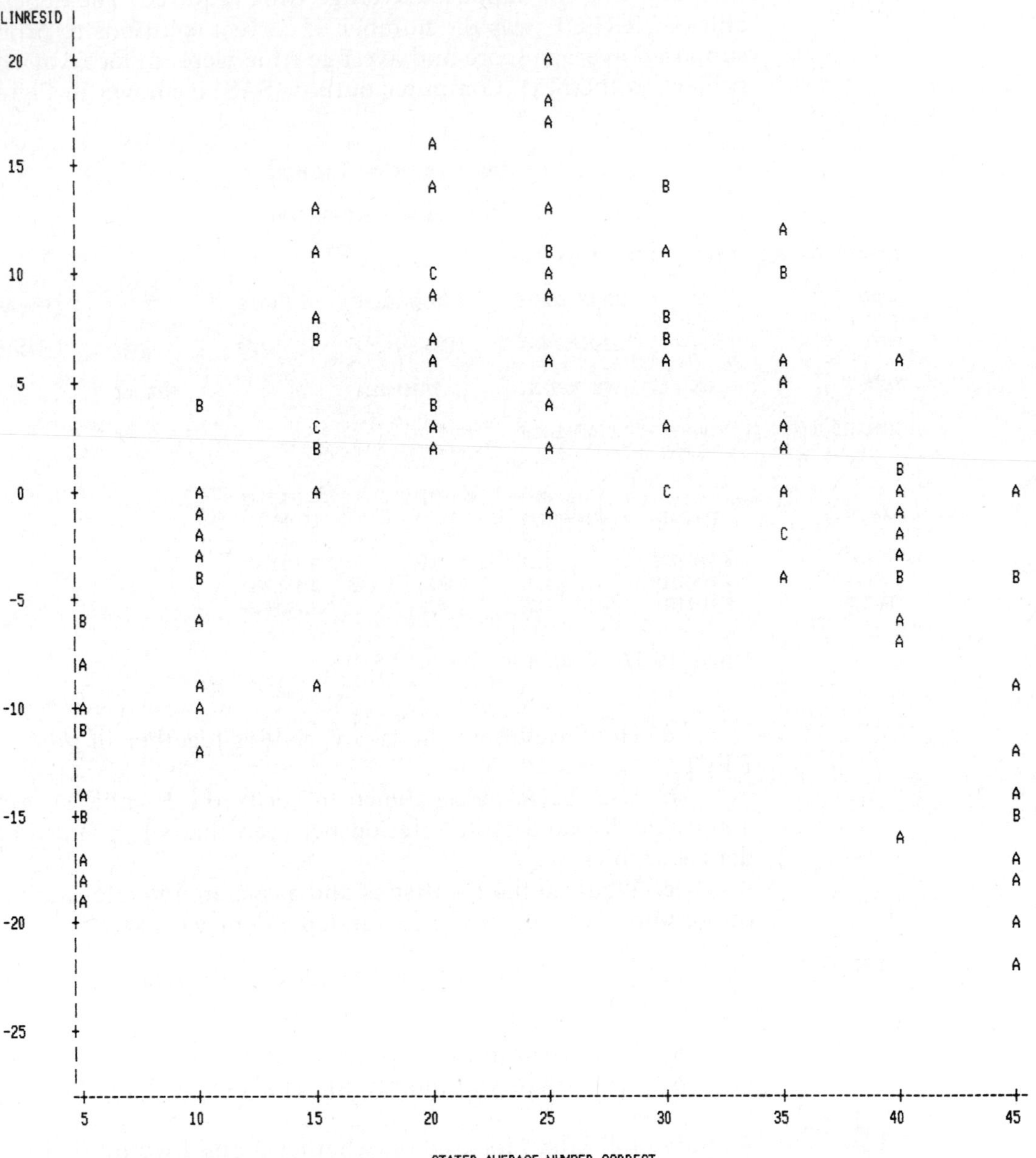

Figure 15.38 Output for Exercise 15.72

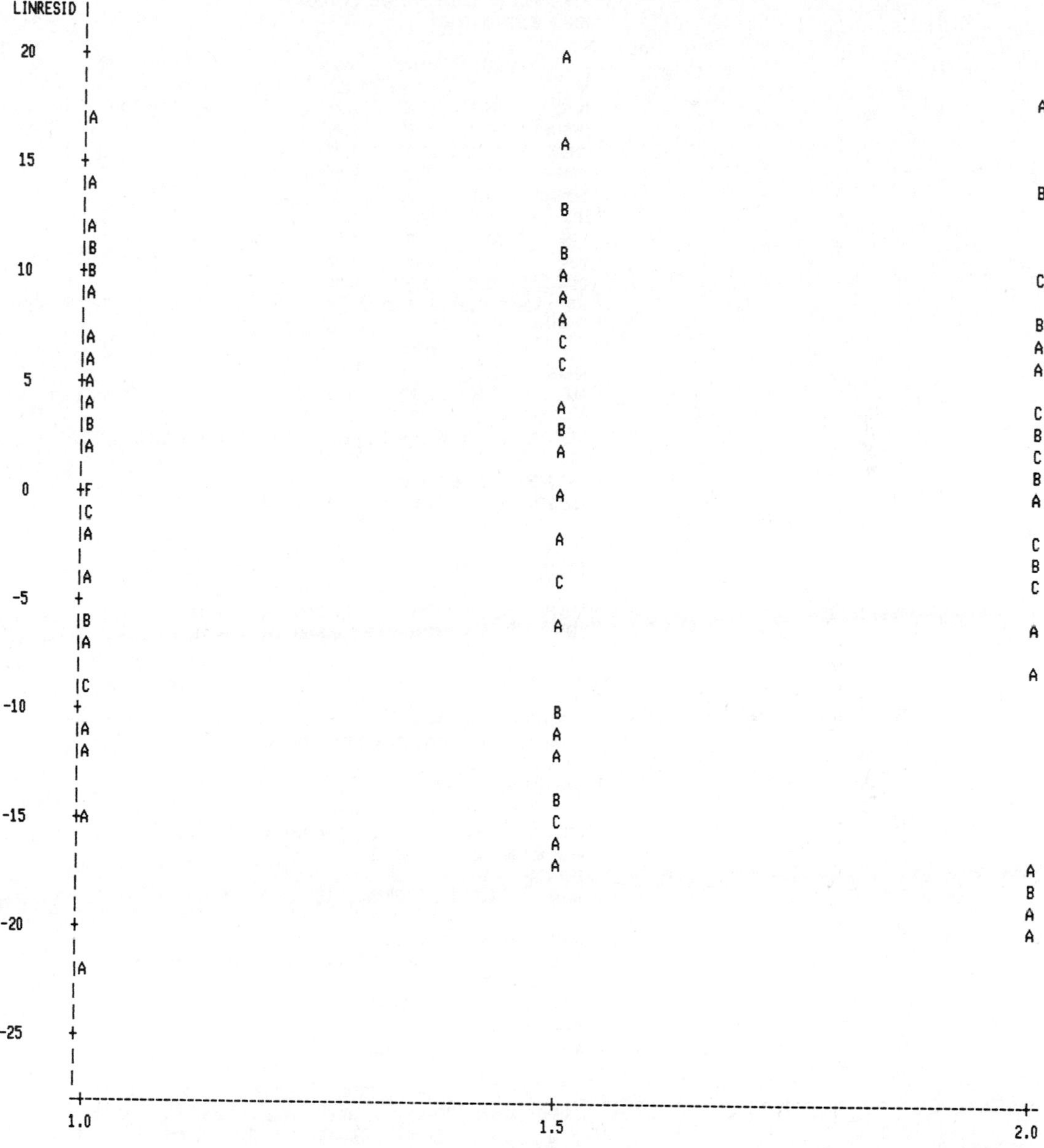
RESIDUAL PLOTS FOR STRAIGHT LINE MODEL
PLOT OF LINRESID*STATTIME   LEGEND: A = 1 OBS, B = 2 OBS, ETC.
LINRESID
20
15
10
5
0
-5
-10
-15
-20
-25
1.0
1.5
2.0
STATED AVERAGE TIME NEEDED

```
MTB > NAME C1 'USEDEFN' C2 'KNOWN' C3 'AGE'
MTB > NOTE C1 IS PERCENTAGE OF WORDS DEFINED BY USE; E.G. "YOU EAT IT"
MTB > NOTE C2 IS NUMBER OF WORDS CHILD CLAIMS TO KNOW, MAX 43
MTB > NOTE C3 IS CHILD'S AGE IN YEARS
MTB > DESCRIBE C1-C3

          USEDEFN     KNOWN       AGE
N             117       117       117
MEAN         56.9     27.47      7.79
MEDIAN       57.7     28.00      7.00
TMEAN        57.1     27.39      7.77
STDEV        18.1      6.30      2.51
SEMEAN        1.7      0.58      0.23
MAX          89.5     41.00     12.00
MIN          10.8     15.00      4.00
Q3           72.0     32.00     10.00
Q1           41.2     22.50      6.00
MTB > CORRELATION OF C1-C3

          USEDEFN    KNOWN
KNOWN      -0.882
AGE        -0.859    0.863

MTB > REGRESS C1 ON 2 VARS IN C2 C3 RESIDS IN C19 PREDICTEDS IN C20

THE REGRESSION EQUATION IS
USEDEFN = 122 - 1.59 KNOWN - 2.74 AGE

                                 ST. DEV.     T-RATIO =
COLUMN        COEFFICIENT        OF COEF.     COEF/S.D.
                 121.890            3.386         36.00
KNOWN            -1.5879           0.2286         -6.95
AGE              -2.7431           0.5739         -4.78

S = 7.824

R-SQUARED = 81.5 PERCENT
R-SQUARED = 81.2 PERCENT, ADJUSTED FOR D.F.

ANALYSIS OF VARIANCE

 DUE TO      DF            SS       MS=SS/DF
REGRESSION    2         30844          15422
RESIDUAL    114          6979             61
TOTAL       116         37823

FURTHER ANALYSIS OF VARIANCE
SS EXPLAINED BY EACH VARIABLE WHEN ENTERED IN THE ORDER GIVEN
 DUE TO      DF            SS
REGRESSION    2         30844
KNOWN         1         29446
AGE           1          1398

                          Y     PRED. Y    ST.DEV.
 ROW    KNOWN     USEDEFN       VALUE    PRED. Y   RESIDUAL    ST.RES.
   2     20.0      55.000      76.416      1.133    -21.417     -2.77R
  12     35.0      22.900      41.625      1.381    -18.725     -2.43R
  36     37.0      10.800      30.220      1.420    -19.420     -2.52R
  53     32.0      31.300      49.132      1.183    -17.832     -2.31R
 100     39.0      51.300      29.788      1.576     21.512      2.81R

R DENOTES AN OBS. WITH A LARGE ST. RES.

DURBIN-WATSON STATISTIC = 1.86
```

Figure 15.39 Output for Exercise 15.73

a. Perform an F test of the null hypothesis that all partial slopes equal 0. What conclusion can be reached?

b. Locate the unadjusted values and adjusted values of $R^2_{\text{USEDEFN}|\text{KNOWN,AGE}}$. How much effect does adjusting for df have in this study?

c. What conclusions can be drawn from the t statistics?

d. Calculate 95% confidence intervals for the true values of the partial slopes.

15.74. Use the FURTHER ANALYSIS OF VARIANCE portion of Figure 15.39 to calculate $r^2_{\text{USEDEFN,AGE}|\text{KNOWN}}$. Interpret the resulting number.

15.75. Residual plots for the data of Exercise 15.73 are shown in Figure 15.40.

a. In the plot of residuals (C19) against predicted values (C20), is there evidence of a serious nonconstant variance problem?

b. In the plots of residuals against independent variables, is there a clear curvature?

c. The Minitab output of Exercise 15.73 labeled five observations as having large residuals. In the residual plots, do there appear to be wild outliers?

15.76. A researcher obtained scores from each subject on scales of stereotypical masculine behavior, stereotypical feminine behavior, and empathy toward others. These variables, as well as the sex of the subject (1 if female, 0 if male), were used as independent variables in a regression model. The dependent variable is WEEPING, a measure of the subject's propensity to cry in various situations. A portion of the SAS output is shown in Figure 15.41.

a. Locate the F statistic and associated p-value. What conclusion can be reached based on them?

b. What conclusions can be reached from the indicated t values?

c. Are the independent variables strongly collinear?

15.77. **a.** Figure 15.41 showed three additional variables, SEX*MASCUL, SEX*FEMININ, and SEX*EMPATHY. (The * symbol stands for multiplication.) What is the purpose of incorporating these terms in the model?

b. Additional SAS output for a regression model excluding these terms is shown in Figure 15.42.

Test the null hypothesis that the true, population coefficients of the three product terms are all 0.

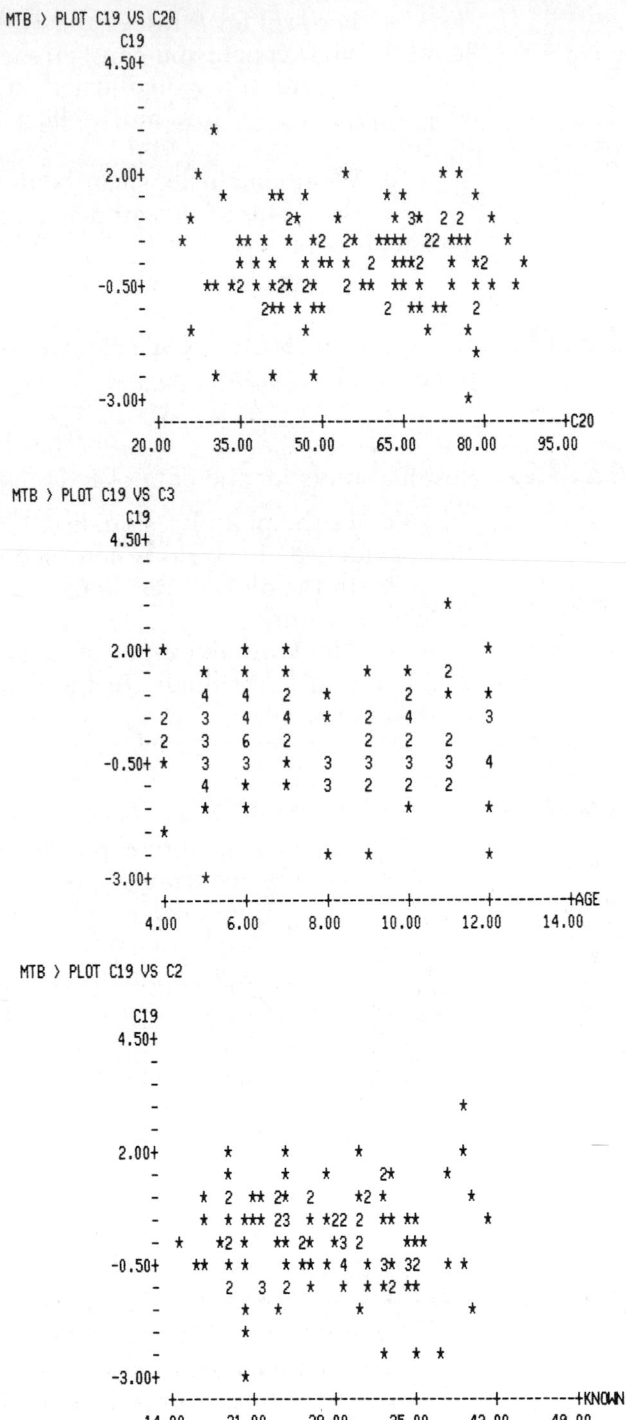

Figure 15.40 Output for Exercise 15.75

CORRELATIONS FOR CRYING DATA

VARIABLE	N	MEAN	STD DEV	SUM	MINIMUM	MAXIMUM
MASCUL	140	46.67142857	9.72976066	6534.00000000	16.00000000	70.00000000
FEMININ	140	63.80000000	10.69471731	8932.00000000	22.00000000	87.00000000
EMPATHY	140	73.67142857	52.09308808	10314.00000000	-101.00000000	220.00000000
WEEPING	140	57.67142857	30.50473318	8074.00000000	-18.00000000	126.00000000

CORRELATION COEFFICIENTS / PROB > |R| UNDER H0:RHO=0 / N = 140

	MASCUL	FEMININ	EMPATHY	WEEPING
MASCUL	1.00000	-0.42887	-0.81492	-0.40368
	0.0000	0.0001	0.0001	0.0001
FEMININ	-0.42887	1.00000	0.85705	0.46066
	0.0001	0.0000	0.0001	0.0001
EMPATHY	-0.81492	0.85705	1.00000	0.52600
	0.0001	0.0001	0.0000	0.0001
WEEPING	-0.40368	0.46066	0.52600	1.00000
	0.0001	0.0001	0.0001	0.0000

FULL MODEL FOR CRYING DATA

GENERAL LINEAR MODELS PROCEDURE

DEPENDENT VARIABLE: WEEPING

SOURCE	DF	SUM OF SQUARES	MEAN SQUARE	F VALUE	PR > F	R-SQUARE	C.V.
MODEL	7	40204.74917489	5743.53559641	8.51	0.0001	0.310834	45.0598
ERROR	132	89140.13653940	675.30406469		ROOT MSE		WEEPING MEAN
CORRECTED TOTAL	139	129344.88571429			25.98661318		57.67142857

PARAMETER	ESTIMATE	T FOR H0: PARAMETER=0	PR > |T|	STD ERROR OF ESTIMATE
INTERCEPT	89.53183112	2.26	0.0257	39.69022387
SEX	-108.04201491	-2.20	0.0296	49.13914874
MASCUL	-0.10456529	-0.06	0.9527	1.75908496
FEMININ	-0.92863142	-0.56	0.5756	1.65456613
EMPATHY	0.43042806	0.77	0.4420	0.55811748
SEX*MASCUL	0.73599698	0.35	0.7281	2.11248650
SEX*FEMININ	1.29765166	0.62	0.5364	2.09350769
SEX*EMPATHY	-0.10382699	-0.15	0.8810	0.69233332

Figure 15.41 Output for Exercise 15.76

15.78. A plot of the residuals and predicted values for the regression model of Exercise 15.77 is shown in Figure 15.43.

a. Is there evidence of considerable heteroscedasticity?
b. Are there extreme outliers in the plot?

FIRST-ORDER MODEL FOR CRYING DATA

GENERAL LINEAR MODELS PROCEDURE

DEPENDENT VARIABLE: WEEPING

SOURCE	DF	SUM OF SQUARES	MEAN SQUARE	F VALUE	PR > F	R-SQUARE	C.V.
MODEL	4	36304.00657279	9076.00164320	13.17	0.0001	0.280676	45.5207
ERROR	135	93040.87914149	689.19169734		ROOT MSE		WEEPING MEAN
CORRECTED TOTAL	139	129344.88571429			26.25246079		57.67142857

PARAMETER	ESTIMATE	T FOR H0: PARAMETER=0	PR > |T|	STD ERROR OF ESTIMATE
INTERCEPT	21.34302340	0.92	0.3588	23.18103672
SEX	0.38231936	0.08	0.9382	4.92203539
MASCUL	0.72886786	0.75	0.4555	0.97390583
FEMININ	-0.56397845	-0.56	0.5740	1.00084912
EMPATHY	0.51718511	1.59	0.1138	0.32489144

Figure 15.42 Output for Exercise 15.77

15.79. A study obtained scores for 129 subjects on three scales, NCT, SAT, and IQ. Minitab output for the data is shown in Figure 15.44.

a. Calculate a 95% confidence interval for the population correlation of NCT and IQ. Does it include 0?

b. Test the null hypothesis that the coefficient of IQ in the regression model is 0. Use a two-sided alternative and $\alpha = .05$.

c. Do the results of parts a and b give conflicting indications of the value of IQ in predicting NCT? If so, why did this occur?

15.80. Residual plots for the regression study of Exercise 15.79 are shown in Figure 15.45.

What assumptions can be checked using these plots? In your judgment, are any of these assumptions violated?

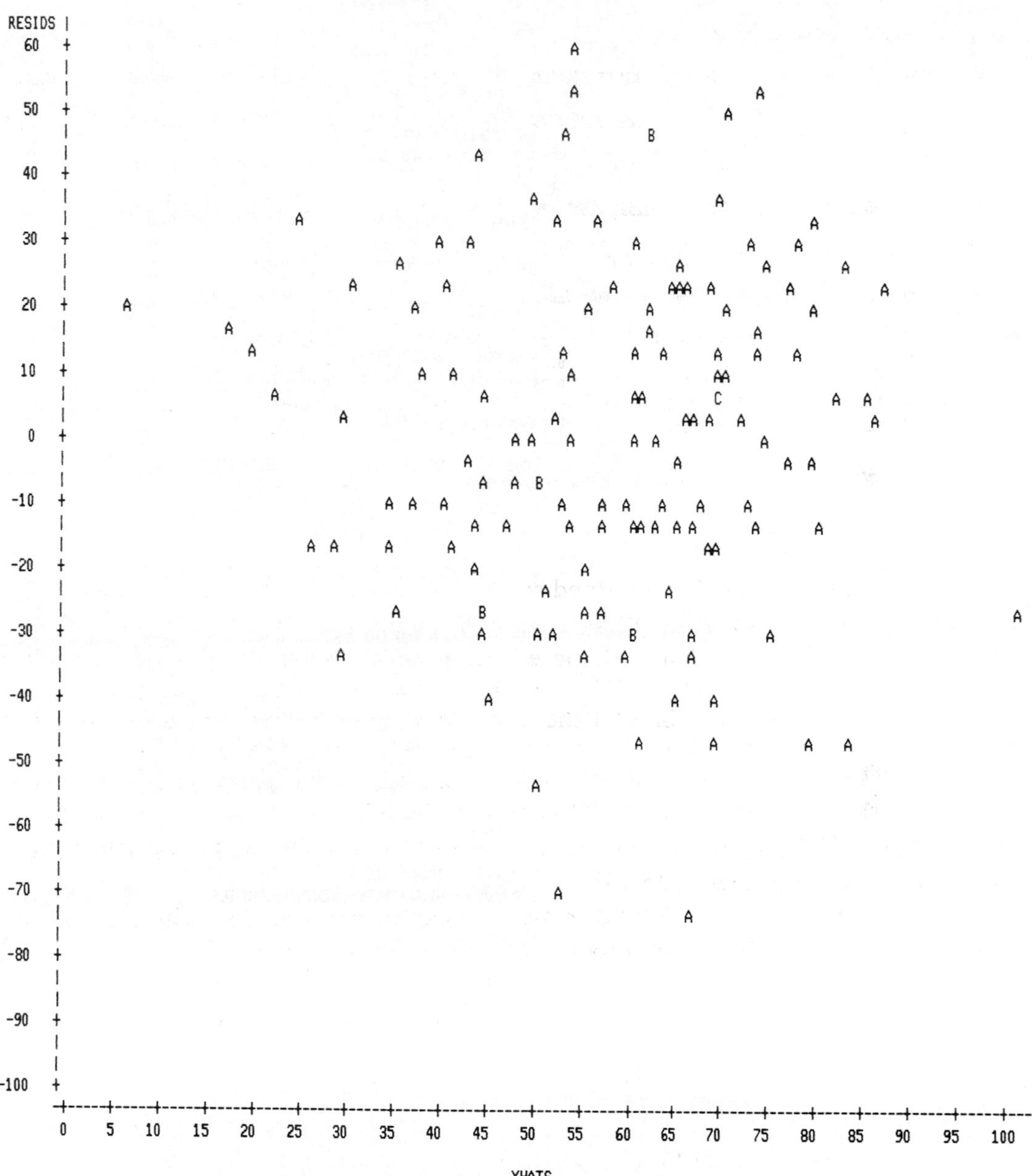

Figure 15.43 Output for Exercise 15.78

```
MTB > NAME C1 'NCT' C2 'SAT' C3 'IQ'
MTB > CORRELATIONS OF C1-C3

              NCT      SAT
SAT          0.785
IQ           0.691    0.819

MTB > REGRESS C1 ON 1 VAR IN C2

THE REGRESSION EQUATION IS
NCT = 35.1 + 0.498 SAT

                                     ST. DEV.    T-RATIO =
COLUMN        COEFFICIENT            OF COEF.    COEF/S.D.
                  35.054               3.505       10.00
SAT              0.49778             0.03480       14.30

S = 5.927

R-SQUARED = 61.7 PERCENT
R-SQUARED = 61.4 PERCENT, ADJUSTED FOR D.F.

ANALYSIS OF VARIANCE

 DUE TO       DF            SS      MS=SS/DF
REGRESSION     1        7186.6        7186.6
RESIDUAL     127        4461.5          35.1
TOTAL        128       11648.1

MTB > REGRESS C1 ON 2 VARS IN C3 C2 RESIDS C19 YHATS C20


THE REGRESSION EQUATION IS
NCT = 31.6 + 0.108 IQ + 0.423 SAT

                                     ST. DEV.    T-RATIO =
COLUMN        COEFFICIENT            OF COEF.    COEF/S.D.
                  31.587               4.173        7.57
IQ               0.10783             0.07126        1.51
SAT              0.42294             0.06037        7.01

S = 5.897

R-SQUARED = 62.4 PERCENT
R-SQUARED = 61.8 PERCENT, ADJUSTED FOR D.F.

ANALYSIS OF VARIANCE

 DUE TO       DF            SS      MS=SS/DF
REGRESSION     2        7266.2        3633.1
RESIDUAL     126        4381.9          34.8
TOTAL        128       11648.1
```

Figure 15.44 Output for Exercise 15.79

```
MTB > NAME C19 'RESIDS' C20 'YHATS'
MTB > PLOT C19 VS C20
```

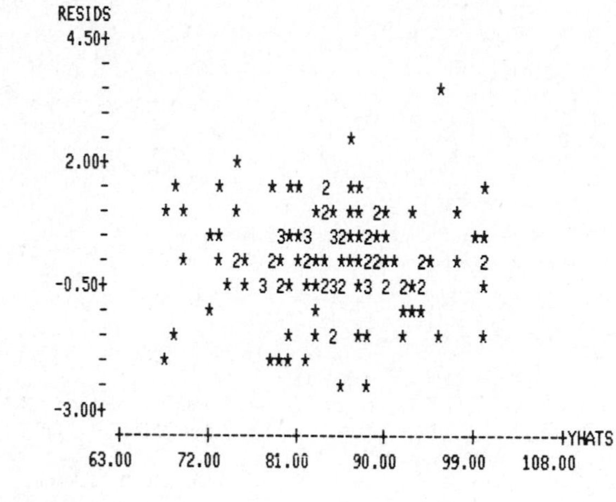


```
MTB > PLOT C19 VS C2
```

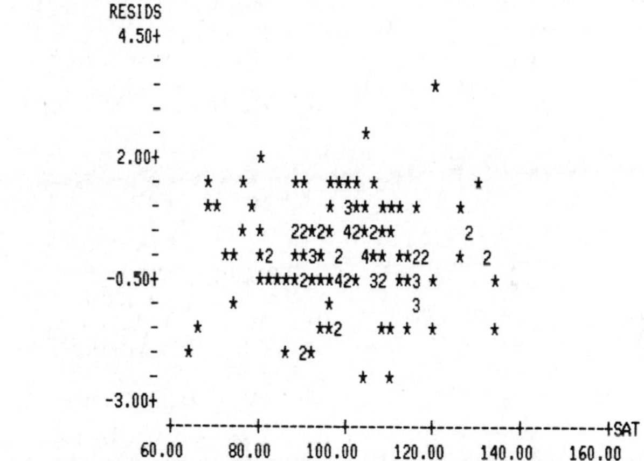

Figure 15.45 Output for Exercise 15.80

References

Ahlgren, A., and Johnson, D. W. Sex differences in cooperative and competitive behavior from the second through the twelfth grades. *Developmental Psychology,* 1979, 15, 45–49.

Anderson, T. W. *An Introduction to Multivariate Statistical Analysis.* New York: John Wiley, 1958.

Beez, W. V. Influence of biased psychological reports on teacher behavior and pupil performance. In Miles, M. B. and Charters, W. W. Jr., *Learning in Social Settings.* Boston: Allyn and Bacon, 1970.

Belsley, D. A., Kuh, E., and Welsch, R. E. *Regression Diagnostics: Identifying Influential Observations and Sources of Collinearity.* New York: John Wiley, 1980.

Bennett, B. W., and Grosser, G. S. Movement toward androgyny in college females through experiential education. *Journal of Psychology,* 1981, 107, 177–183.

Berry, W. Personal communication, 1985.

Bishop, Y. M. M., Fienberg, S. E., and Holland, P. W. *Discrete Multivariate Analysis: Theory and Practice.* Cambridge, Mass.: The MIT Press, 1975.

Bolton, D. Personal communication, 1984.

Boone, S. L., and Montare, A. Aggression and family size. *Journal of Psychology,* 1979, 103, 67–70.

Burnham, J. R. Experimenter bias and lesion labeling. Unpublished manuscript. West Lafayette, Indiana: Purdue University, 1966.

Buss, A. H., Iscoe, I., and Buss, E. H. The development of embarrassment. *Journal of Psychology,* 1979, 103, 227–230.

Carne, G. C., and Kirton, M. M. Styles of creativity—test-score correlations between Kirton adaption-innovation inventory and Myers-Briggs-type indicator. *Psychological Reports,* 1982, 50, 31–36.

Cohen, J. *Statistical Power Analysis for Behavioral Sciences,* revised edition. New York: Academic Press, 1977.

Cohen, J., and Cohen, P. *Applied Multiple Regression/Correlation Analysis for the Behavioral Sciences.* Hillsdale, N.J.: Erlbaum Associates/New York: Halsted Press, 1975.

Croft, A. C. Do spelling tests measure the ability to spell? *Educational and Psychological Measurement,* 1982, 42, 715–723.

Davis, H., and Unruh, W. R. Word memory in non-psychotic depression. *Perceptual and Motor Skills,* 1980, 51, 699–705.

Devore, J. L. *Probability and Statistics for Engineering and the Sciences.* Belmont, Cal.: Brooks-Cole, 1982.

Dey, M. K. Anagram solution speed as a joint function of manifest anxiety and number of category sets. *American Journal of Psychology,* 1978, 91, 81–88.

Dovidio, J. F. Sex, costs, and helping behavior. *Journal of Psychology,* 1982, 112, 231–236.

Draper, N., and Smith, H. *Applied Regression Analysis,* 2nd edition. New York: John Wiley, 1982.

Feldman, K. A., and Newcomb, T. M. *The Impact of College on Students.* San

Francisco: Jossey-Bass, 1969.
Froman, T., and Hubert, L. J. Application of prediction analysis to developmental priority. *Psychological Bulletin,* 1980, 81, 136–146.
Gallistel, C. R. Personal communication, 1983.
Ghatala, E. S. The effect of internal generation of information on memory performance. *American Journal of Psychology,* 1981, 81, 443–450.
Gibbons, J. D., and Pratt, J. W. *Concepts of Nonparametric Theory.* New York: Springer-Verlag, 1983.
Goodman, L. A., and Kruskal, W. H. Measures of association for cross-classifications. *Journal of the American Statistical Association,* 1954, 49, 732–764.
Goodwin, R. S., and Michel, G. F. Head orientation position during birth and in the infant neonatal period, and hand preference at nineteen weeks. *Child Development,* 1981, 52, 819–826.
Grizzle, J. E., Starmer, C. F., and Koch, G. G. Analysis of categorical data by linear models. *Biometrika,* 1969, 25, 489–504.
Groeneveld, R. A., and Meeden, G. Measuring skewness and kurtosis. *The Statistician,* 1984, 33, 391–399.
Guttman, L. An outline of the statistical theory of prediction. In Horst, P., et al., editors, *The Prediction of Personal Adjustment.* New York: Social Science Research Council, Bulletin 48, 1941.
Hanushek, E. A., and Jackson, J. E. *Statistical Methods for Social Scientists.* New York: Academic Press, 1977.
Hildebrand, D. K., Laing, J. D., and Rosenthal, H. *Analysis of Ordinal Data.* Sage University Paper number 8. Beverly Hills, Cal.: Sage Publications, 1977.
Hildebrand, D. K., Laing, J. D., and Rosenthal, H. *Prediction Analysis of Cross Classifications.* New York: John Wiley, 1977.
Hodges, J., and Lehmann, E. Estimates of location based on rank tests. *Annals of Mathematical Statistics,* 1963, 34, 598–611.
Hollander, M., and Wolfe, D. *Nonparametric statistical methods.* New York: John Wiley, 1974.
Howell, D. C. *Statistical Methods for Psychology.* Boston: Duxbury Press, 1982.
Ingram, R. E., and Zurawski, R. Choosing clinical psychologists: An examination of the utilization of admissions criteria. *Professional Psychology,* 1981, 12, 684–689.
Inhelder, B., and Piaget, J. *The Early Growth of Logic in the Child.* New York: Harper and Row, 1964.
Isbitsky, J. R., and White, D. R. Externality and locus of control in obese children. *Journal of Psychology,* 1981, 107, 163–172.
Jacobs, D. Successful empathy training. *Journal of Humanistic Psychology,* 1981, 21, 39–56.
Johnston, J. *Econometric Methods,* 2nd edition. New York: McGraw-Hill, 1972.
Kaufmann, G., and Bengtson, G. Effect of mode of presentation on performance in a familiar task. *Scandinavian Journal of Psychology,* 1980, 21, 61–63.
Keller, H. T., and Tetlow, E. W. Participation in community theater and type of personality. *Psychological Reports,* 1981, 47, 711–714.
Kemeny, J. G., and Snell, J. L. *Finite Markov Chains.* Princeton, N.J.: Van Nostrand, 1960.
Kirk, R. E. *Experimental Design: Procedures for the Behavioral Sciences,* 2nd edition. Belmont, Cal.: Brooks-Cole, 1984.

Kleinbaum, D. G., and Kupper, L. L. *Applied Regression Analysis and Other Multivariable Methods.* North Scituate, Mass.: Duxbury Press, 1978.

Kleinke, C. L., and Nicholson, T. A. Black and white children's awareness of de facto race and sex differences. *Developmental Psychology,* 1979, 15, 84–86.

Larsen, R. J., and Marx, M. L. *An Introduction to Mathematical Statistics and Its Applications.* Englewood Cliffs, N.J.: Prentice-Hall, 1981.

Lasky, R. E., Klein, R. E., Yarbrough, C., and Kallio, K. D. The predictive validity of infant assessments in rural Guatemala. *Child Development,* 1981, 52, 847–856.

Lehmann, E. *Testing Statistical Hypotheses.* New York: John Wiley, 1959.

Lindgren, H. C. *The Psychology of College Success.* New York: John Wiley, 1969.

Lopez, E. M. Increasing motivation with performance-contingent reward. *Journal of Psychology,* 1981, 108, 59–65.

Loucks, S., Kobos, J. C., Stanton, B., Burstein, A. G., and Lawlis, G. F. Sex-related psychological characteristics of medical students. *Journal of Psychology,* 1979, 102, 119–123.

Meyer, W. J., and Thompson, G. G. Sex differences in the distribution of teacher approval and disapproval among sixth- grade children. *Journal of Educational Psychology,* 1956, 47, 385–396.

Miller, R. G. The jackknife—a review. *Biometrika,* 1974, 61, 1–15.

Mood, A. M., Graybill, F. A., and Boes, D. C. *Introduction to the Theory of Statistics.* New York: McGraw-Hill, 1974.

Morelli, G., Andrews, L., and Morelli, R. The relation involving personality variables, problem relevance, rationality, and anxiousness among college men. *Cognitive Therapy and Research,* 1982, 6, 57–62.

Morrison, D. F. *Multivariate Statistical Methods,* 2nd edition. New York: McGraw-Hill, 1976.

Mosteller, F., and Tukey, J. W. *Data Analysis and Regression: A Second Course in Statistics.* Reading, Mass.: Addison-Wesley, 1977.

Mukherji, B. R., Abramson, L. Y., and Martin, D. J. Induced depressive mood and attributional patterns. *Cognitive Therapy and Research,* 1982, 6, 15–21.

Odoroff, C. A comparison of minimum logit chi-square estimation and maximum likelihood estimation in 2×2×2 and 3×2×2 contingency tables: Tests for interaction. *Journal of the American Statistical Association,* 1970, 65, 1617–1631.

Oler, J. *The Power of Sequential Model Selection Techniques.* Unpublished doctoral dissertation, University of Pennsylvania, 1978.

Otis, L. Selective exposure to the film *Close Encounters. Journal of Psychology,* 1979, 101, 293–295.

Ott, L. *An Introduction to Statistical Methods and Data Analysis,* 2nd edition. Boston: Duxbury Press, 1984.

Ott, L., and Hildebrand, D. K. *Statistical Thinking for Managers.* Boston: Duxbury Press, 1983.

Patnaik, P. B. The noncentral χ^2 and F distributions and their approximations. *Biometrika,* 1949, 36, 202–232.

Peters, M., and Durding, B. M. Footedness of left-handers and right-handers. *American Journal of Psychology,* 1979, 92, 133–142.

Pharis, M. E., and Manosevitz, M. Parental models of infancy: A note on gender preference for firstborns. *Psychological Reports,* 1981, 47, 763–768.

Platt, J. E., and Cohen, S. Mental rotation task performance as a function of age and training. *Journal of Psychology,* 1981, 108, 173–178.

Plomin, R., Foch, T. T., and Rowe, D. C. Bobo clown aggression in childhood: Environment, not genes. *Journal of Research in Personality,* 1981, 15, 331–342.

Porac, C., and Coren, S. *Lateral Preferences and Human Behavior.* New York: Springer-Verlag, 1981.

Raaheim, K., Kaufmann, G., and Bengtson, G. Attempts to predict intelligent behavior II: A study of problem solving. *Scandinavian Journal of Psychology,* 1980, 21, 119–121.

Rubenstein, J. L., and Howes, C. Caregiving and infant behavior in day care and in homes. *Developmental Psychology,* 1979, 15, 1–24.

Ryan, T. A., Jr., and Joiner, B. L. Normal probability plots and tests for normality. Unpublished technical report, Department of Statistics, Pennsylvania State University, 1974.

Sarason, I. G. Test anxiety, stress, and social support. *Journal of Personality,* 1981, 49, 101–114.

Scheffé, H. *The Analysis of Variance.* New York: John Wiley, 1959.

Schumm, W. R., Bollman, S. R., and Jurich, A. P. Validity of Edmond's Marital Conventionalization Scale. *Journal of Psychology,* 1981, 109, 65–71.

Segal, J. Coalition formation and players' incentives. *Journal of Psychology,* 1981, 107, 261–266.

Sternlicht, M. Fears of institutionalized mentally retarded adults. *Journal of Psychology,* 1979, 101, 67–72.

Stoline, M. R. The status of multiple comparisons: Simultaneous estimation of all pairwise comparisons in one-way ANOVA designs. *American Statistician,* 1981, 35, 134–141.

Strang, H. R. The effects of challenging goal instructions upon goal setting and performance in a reaction time task. *Journal of Psychology,* 1981, 107, 241–246.

Tellevik, J. M. Language and problem-solving ability: A comparison between deaf and hearing adolescents. *Scandinavian Journal of Psychology,* 1981, 22, 97–100.

Touliatos, J., and Bedeian, A. G. Maternal expectancies for achievement, IQ, and autonomous achievement-motivation in kindergarten children. *Psychological Reports,* 1982, 50, 51–54.

Tukey, J. W. *Exploratory Data Analysis.* Belmont, Mass.: Addison-Wesley, 1977.

Tukey, J. W. The problem of multiple comparisons. Unpublished report, Department of Statistics, Princeton University.

Vandewiele, M. Wolof adolescents' dreams. *Journal of Psychology,* 1981, 109, 3–10.

Waxman, S. Personal communication, 1984.

Weisberg, S., and Cook, R. D. *Residuals and influence in regression.* London: Chapman Hall, 1982.

Welch, B. L. The significance of the difference between two means when the population variances are unequal. *Biometrika,* 1938, 29, 350–362.

Winer, B. J. *Statistical Principles in Experimental Design,* 2nd edition. New York: McGraw-Hill, 1971.

Wold, A. H. Sequence of information and processing strategies in oral language. *Scandinavian Journal of Psychology,* 1982, 23, 267–272.

Tables

Table 1

Binomial probabilities (n between 2 and 6)

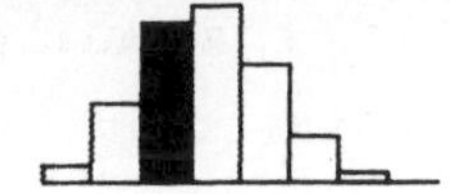

n=2

π

y↓	.05	.10	.15	.20	.25	.30	.35	.40	.45	.50	
0	.9025	.8100	.7225	.6400	.5625	.4900	.4225	.3600	.3025	.2500	2
1	.0950	.1800	.2550	.3200	.3750	.4200	.4550	.4800	.4950	.5000	1
2	.0025	.0100	.0225	.0400	.0625	.0900	.1225	.1600	.2025	.2500	0
	.95	.90	.85	.80	.75	.70	.65	.60	.55	.50	y↑

n=3

π

y↓	.05	.10	.15	.20	.25	.30	.35	.40	.45	.50	
0	.8574	.7290	.6141	.5120	.4219	.3430	.2746	.2160	.1664	.1250	3
1	.1354	.2430	.3251	.3840	.4219	.4410	.4436	.4320	.4084	.3750	2
2	.0071	.0270	.0574	.0960	.1406	.1890	.2389	.2880	.3341	.3750	1
3	.0001	.0010	.0034	.0080	.0156	.0270	.0429	.0640	.0911	.1250	0
	.95	.90	.85	.80	.75	.70	.65	.60	.55	.50	y↑

n=4

π

y↓	.05	.10	.15	.20	.25	.30	.35	.40	.45	.50	
0	.8145	.6561	.5220	.4096	.3164	.2401	.1785	.1296	.0915	.0625	4
1	.1715	.2916	.3685	.4096	.4219	.4116	.3845	.3456	.2995	.2500	3
2	.0135	.0486	.0975	.1536	.2109	.2646	.3105	.3456	.3675	.3750	2
3	.0005	.0036	.0115	.0256	.0469	.0756	.1115	.1536	.2005	.2500	1
4	.0000	.0001	.0005	.0016	.0039	.0081	.0150	.0256	.0410	.0625	0
	.95	.90	.85	.80	.75	.70	.65	.60	.55	.50	y↑

n=5

π

y↓	.05	.10	.15	.20	.25	.30	.35	.40	.45	.50	
0	.7738	.5905	.4437	.3277	.2373	.1681	.1160	.0778	.0503	.0313	5
1	.2036	.3281	.3915	.4096	.3955	.3602	.3124	.2592	.2059	.1563	4
2	.0214	.0729	.1382	.2048	.2637	.3087	.3364	.3456	.3369	.3125	3
3	.0011	.0081	.0244	.0512	.0879	.1323	.1811	.2304	.2757	.3125	2
4	.0000	.0005	.0022	.0064	.0146	.0284	.0488	.0768	.1128	.1563	1
5	.0000	.0000	.0001	.0003	.0010	.0024	.0053	.0102	.0185	.0313	0
	.95	.90	.85	.80	.75	.70	.65	.60	.55	.50	y↑

n=6

π

y↓	.05	.10	.15	.20	.25	.30	.35	.40	.45	.50	
0	.7351	.5314	.3771	.2621	.1780	.1176	.0754	.0467	.0277	.0156	6
1	.2321	.3543	.3993	.3932	.3560	.3025	.2437	.1866	.1359	.0938	5
2	.0305	.0984	.1762	.2458	.2966	.3241	.3280	.3110	.2780	.2344	4
3	.0021	.0146	.0415	.0819	.1318	.1852	.2355	.2765	.3032	.3125	3
4	.0001	.0012	.0055	.0154	.0330	.0595	.0951	.1382	.1861	.2344	2
5	.0000	.0001	.0004	.0015	.0044	.0102	.0205	.0369	.0609	.0938	1
6	.0000	.0000	.0000	.0001	.0002	.0007	.0018	.0041	.0083	.0156	0
	.95	.90	.85	.80	.75	.70	.65	.60	.55	.50	y↑

Table 1 cont'd

Binomial probabilities (n between 7 and 10)

n=7

y↓	π .05	.10	.15	.20	.25	.30	.35	.40	.45	.50	
0	.6983	.4783	.3206	.2097	.1335	.0824	.0490	.0280	.0152	.0078	7
1	.2573	.3720	.3960	.3670	.3115	.2471	.1848	.1306	.0872	.0547	6
2	.0406	.1240	.2097	.2753	.3115	.3177	.2985	.2613	.2140	.1641	5
3	.0036	.0230	.0617	.1147	.1730	.2269	.2679	.2903	.2918	.2734	4
4	.0002	.0026	.0109	.0287	.0577	.0972	.1442	.1935	.2388	.2734	3
5	.0000	.0002	.0012	.0043	.0115	.0250	.0466	.0774	.1172	.1641	2
6	.0000	.0000	.0001	.0004	.0013	.0036	.0084	.0172	.0320	.0547	1
7	.0000	.0000	.0000	.0000	.0001	.0002	.0006	.0016	.0037	.0078	0
	.95	.90	.85	.80	.75	.70	.65	.60	.55	.50	y↑

n=8

y↓	π .05	.10	.15	.20	.25	.30	.35	.40	.45	.50	
0	.6634	.4305	.2725	.1678	.1001	.0576	.0319	.0168	.0084	.0039	8
1	.2793	.3826	.3847	.3355	.2670	.1977	.1373	.0896	.0548	.0313	7
2	.0515	.1488	.2376	.2936	.3115	.2965	.2587	.2090	.1569	.1094	6
3	.0054	.0331	.0839	.1468	.2076	.2541	.2786	.2787	.2568	.2188	5
4	.0004	.0046	.0185	.0459	.0865	.1361	.1875	.2322	.2627	.2734	4
5	.0000	.0004	.0026	.0092	.0231	.0467	.0808	.1239	.1719	.2188	3
6	.0000	.0000	.0002	.0011	.0038	.0100	.0217	.0413	.0703	.1094	2
7	.0000	.0000	.0000	.0001	.0004	.0012	.0033	.0079	.0164	.0313	1
8	.0000	.0000	.0000	.0000	.0000	.0001	.0002	.0007	.0017	.0039	0
	.95	.90	.85	.80	.75	.70	.65	.60	.55	.50	y↑

n=9

y↓	π .05	.10	.15	.20	.25	.30	.35	.40	.45	.50	
0	.6302	.3874	.2316	.1342	.0751	.0404	.0207	.0101	.0046	.0020	9
1	.2985	.3874	.3679	.3020	.2253	.1556	.1004	.0605	.0339	.0176	8
2	.0629	.1722	.2597	.3020	.3003	.2668	.2162	.1612	.1110	.0703	7
3	.0077	.0446	.1069	.1762	.2336	.2668	.2716	.2508	.2119	.1641	6
4	.0006	.0074	.0283	.0661	.1168	.1715	.2194	.2508	.2600	.2461	5
5	.0000	.0008	.0050	.0165	.0389	.0735	.1181	.1672	.2128	.2461	4
6	.0000	.0001	.0006	.0028	.0087	.0210	.0424	.0743	.1160	.1641	3
7	.0000	.0000	.0000	.0003	.0012	.0039	.0098	.0212	.0407	.0703	2
8	.0000	.0000	.0000	.0000	.0001	.0004	.0013	.0035	.0083	.0176	1
9	.0000	.0000	.0000	.0000	.0000	.0000	.0001	.0003	.0008	.0020	0
	.95	.90	.85	.80	.75	.70	.65	.60	.55	.50	y↑

n=10

y↓	π .05	.10	.15	.20	.25	.30	.35	.40	.45	.50	
0	.5987	.3487	.1969	.1074	.0563	.0282	.0135	.0060	.0025	.0010	10
1	.3151	.3874	.3474	.2684	.1877	.1211	.0725	.0403	.0207	.0098	9
2	.0746	.1937	.2759	.3020	.2816	.2335	.1757	.1209	.0763	.0439	8
3	.0105	.0574	.1298	.2013	.2503	.2668	.2522	.2150	.1665	.1172	7
4	.0010	.0112	.0401	.0881	.1460	.2001	.2377	.2508	.2384	.2051	6
5	.0001	.0015	.0085	.0264	.0584	.1029	.1536	.2007	.2340	.2461	5
6	.0000	.0001	.0012	.0055	.0162	.0368	.0689	.1115	.1596	.2051	4
7	.0000	.0000	.0001	.0008	.0031	.0090	.0212	.0425	.0746	.1172	3
8	.0000	.0000	.0000	.0001	.0004	.0014	.0043	.0106	.0229	.0439	2
9	.0000	.0000	.0000	.0000	.0000	.0001	.0005	.0016	.0042	.0098	1
10	.0000	.0000	.0000	.0000	.0000	.0000	.0000	.0001	.0003	.0010	0
	.95	.90	.85	.80	.75	.70	.65	.60	.55	.50	y↑

Binomial probabilities (n between 12 and 16)

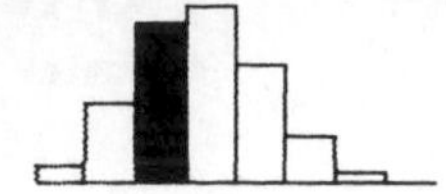

n=12

π

y ↓	.05	.10	.15	.20	.25	.30	.35	.40	.45	.50	
0	.5404	.2824	.1422	.0687	.0317	.0138	.0057	.0022	.0008	.0002	12
1	.3413	.3766	.3012	.2062	.1267	.0712	.0368	.0174	.0075	.0029	11
2	.0988	.2301	.2924	.2835	.2323	.1678	.1088	.0639	.0339	.0161	10
3	.0173	.0852	.1720	.2362	.2581	.2397	.1954	.1419	.0923	.0537	9
4	.0021	.0213	.0683	.1329	.1936	.2311	.2367	.2128	.1700	.1208	8
5	.0002	.0038	.0193	.0532	.1032	.1585	.2039	.2270	.2225	.1934	7
6	.0000	.0005	.0040	.0155	.0401	.0792	.1281	.1766	.2124	.2256	6
7	.0000	.0000	.0006	.0033	.0115	.0291	.0591	.1009	.1489	.1934	5
8	.0000	.0000	.0001	.0005	.0024	.0078	.0199	.0420	.0762	.1208	4
9	.0000	.0000	.0000	.0001	.0004	.0015	.0048	.0125	.0277	.0537	3
10	.0000	.0000	.0000	.0000	.0000	.0002	.0008	.0025	.0068	.0161	2
11	.0000	.0000	.0000	.0000	.0000	.0000	.0001	.0003	.0010	.0029	1
12	.0000	.0000	.0000	.0000	.0000	.0000	.0000	.0000	.0001	.0002	0
	.95	.90	.85	.80	.75	.70	.65	.60	.55	.50	y ↑

n=14

π

y ↓	.05	.10	.15	.20	.25	.30	.35	.40	.45	.50	
0	.4877	.2288	.1028	.0440	.0178	.0068	.0024	.0008	.0002	.0001	14
1	.3593	.3559	.2539	.1539	.0832	.0407	.0181	.0073	.0027	.0009	13
2	.1229	.2570	.2912	.2501	.1802	.1134	.0634	.0317	.0141	.0056	12
3	.0259	.1142	.2056	.2501	.2402	.1943	.1366	.0845	.0462	.0222	11
4	.0037	.0349	.0998	.1720	.2202	.2290	.2022	.1549	.1040	.0611	10
5	.0004	.0078	.0352	.0860	.1468	.1963	.2178	.2066	.1701	.1222	9
6	.0000	.0013	.0093	.0322	.0734	.1262	.1759	.2066	.2088	.1833	8
7	.0000	.0002	.0019	.0092	.0280	.0618	.1082	.1574	.1952	.2095	7
8	.0000	.0000	.0003	.0020	.0082	.0232	.0510	.0918	.1398	.1833	6
9	.0000	.0000	.0000	.0003	.0018	.0066	.0183	.0408	.0762	.1222	5
10	.0000	.0000	.0000	.0000	.0003	.0014	.0049	.0136	.0312	.0611	4
11	.0000	.0000	.0000	.0000	.0000	.0002	.0010	.0033	.0093	.0222	3
12	.0000	.0000	.0000	.0000	.0000	.0000	.0001	.0005	.0019	.0056	2
13	.0000	.0000	.0000	.0000	.0000	.0000	.0000	.0001	.0002	.0009	1
14	.0000	.0000	.0000	.0000	.0000	.0000	.0000	.0000	.0000	.0001	0
	.95	.90	.85	.80	.75	.70	.65	.60	.55	.50	y ↑

n=16

π

y ↓	.05	.10	.15	.20	.25	.30	.35	.40	.45	.50	
0	.4401	.1853	.0743	.0281	.0100	.0033	.0010	.0003	.0001	.0000	16
1	.3706	.3294	.2097	.1126	.0535	.0228	.0087	.0030	.0009	.0002	15
2	.1463	.2745	.2775	.2111	.1336	.0732	.0353	.0150	.0056	.0018	14
3	.0359	.1423	.2285	.2463	.2079	.1465	.0888	.0468	.0215	.0085	13
4	.0061	.0514	.1311	.2001	.2252	.2040	.1553	.1014	.0572	.0278	12
5	.0008	.0137	.0555	.1201	.1802	.2099	.2008	.1623	.1123	.0667	11
6	.0001	.0028	.0180	.0550	.1101	.1649	.1982	.1983	.1684	.1222	10
7	.0000	.0004	.0045	.0197	.0524	.1010	.1524	.1889	.1969	.1746	9
8	.0000	.0001	.0009	.0055	.0197	.0487	.0923	.1417	.1812	.1964	8
9	.0000	.0000	.0001	.0012	.0058	.0185	.0442	.0840	.1318	.1746	7
10	.0000	.0000	.0000	.0002	.0014	.0056	.0167	.0392	.0755	.1222	6
11	.0000	.0000	.0000	.0000	.0002	.0013	.0049	.0142	.0337	.0667	5
	.95	.90	.85	.80	.75	.70	.65	.60	.55	.50	y ↑

n=16 continued on next page

Table 1 cont'd

Binomial probabilities (n between 16 and 20)

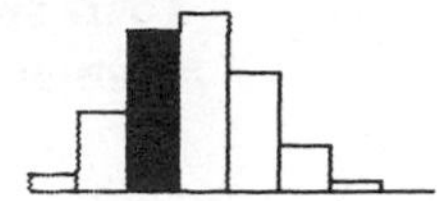

n=16 (continued from previous page)

y ↓	.05	.10	.15	.20	.25	.30	.35	.40	.45	.50	
12	.0000	.0000	.0000	.0000	.0000	.0002	.0011	.0040	.0115	.0278	4
13	.0000	.0000	.0000	.0000	.0000	.0000	.0002	.0008	.0029	.0085	3
14	.0000	.0000	.0000	.0000	.0000	.0000	.0000	.0001	.0005	.0018	2
15	.0000	.0000	.0000	.0000	.0000	.0000	.0000	.0000	.0001	.0002	1
	.95	.90	.85	.80	.75	.70	.65	.60	.55	.50	y ↑

n=18

y ↓	.05	.10	.15	.20	.25	.30	.35	.40	.45	.50	
0	.3972	.1501	.0536	.0180	.0056	.0016	.0004	.0001	.0000	.0000	18
1	.3763	.3002	.1704	.0811	.0338	.0126	.0042	.0012	.0003	.0001	17
2	.1683	.2835	.2556	.1723	.0958	.0458	.0190	.0069	.0022	.0006	16
3	.0473	.1680	.2406	.2297	.1704	.1046	.0547	.0246	.0095	.0031	15
4	.0093	.0700	.1592	.2153	.2130	.1681	.1104	.0614	.0291	.0117	14
5	.0014	.0218	.0787	.1507	.1988	.2017	.1664	.1146	.0666	.0327	13
6	.0002	.0052	.0301	.0816	.1436	.1873	.1941	.1655	.1181	.0708	12
7	.0000	.0010	.0091	.0350	.0820	.1376	.1792	.1892	.1657	.1214	11
8	.0000	.0002	.0022	.0120	.0376	.0811	.1327	.1734	.1864	.1669	10
9	.0000	.0000	.0004	.0033	.0139	.0386	.0794	.1284	.1694	.1855	9
10	.0000	0000	.0001	.0008	.0042	.0149	.0385	.0771	.1248	.1669	8
11	.0000	.0000	.0000	.0001	.0010	.0046	.0151	.0374	.0742	.1214	7
12	.0000	.0000	.0000	.0000	.0002	.0012	.0047	.0145	.0354	.0708	6
13	.0000	.0000	.0000	.0000	.0000	.0002	.0012	.0045	.0134	.0327	5
14	0000	.0000	.0000	.0000	.0000	.0000	.0002	.0011	.0039	.0117	4
15	.0000	.0000	.0000	.0000	.0000	.0000	.0000	.0002	.0009	.0031	3
16	.0000	.0000	.0000	.0000	.0000	.0000	.0000	.0000	.0001	.0006	2
17	.0000	.0000	.0000	.0000	.0000	.0000	.0000	.0000	.0000	.0001	1
	.95	.90	.85	.80	.75	.70	.65	.60	.55	.50	y ↑

n=20

y ↓	.05	.10	.15	.20	.25	.30	.35	.40	.45	.50	
0	.3585	.1216	.0388	.0115	.0032	.0008	.0002	.0000	.0000	.0000	20
1	.3774	.2702	.1368	.0576	.0211	.0068	.0020	.0005	.0001	.0000	19
2	.1887	.2852	.2293	.1369	.0669	.0278	.0100	.0031	.0008	.0002	18
3	.0596	.1901	.2428	.2054	.1339	.0716	.0323	0123	.0040	.0011	17
4	.0133	.0898	.1821	.2182	.1897	.1304	.0738	.0350	.0139	.0046	16
5	.0022	.0319	.1028	.1746	.2023	.1789	.1272	.0746	.0365	0148	15
6	.0003	.0089	.0454	.1091	.1686	.1916	.1712	.1244	.0746	.0370	14
7	.0000	.0020	.0160	.0545	.1124	.1643	.1844	.1659	.1221	.0739	13
8	.0000	.0004	.0046	.0222	.0609	.1144	.1614	.1797	.1623	.1201	12
9	.0000	.0001	.0011	.0074	.0271	.0654	.1158	.1597	.1771	.1602	11
10	.0000	.0000	.0002	.0020	.0099	.0308	.0686	.1171	.1593	.1762	10
11	.0000	.0000	.0000	.0005	.0030	.0120	.0336	.0710	.1185	.1602	9
12	.0000	.0000	.0000	.0001	.0008	.0039	.0136	.0355	.0727	.1201	8
13	.0000	.0000	.0000	.0000	.0002	.0010	.0045	.0146	.0366	.0739	7
14	.0000	.0000	.0000	.0000	.0000	.0002	.0012	.0049	.0150	.0370	6
15	.0000	.0000	.0000	.0000	.0000	.0000	.0003	.0013	.0049	.0148	5
16	.0000	.0000	.0000	.0000	.0000	.0000	.0000	.0003	.0013	.0046	4
17	.0000	.0000	.0000	.0000	.0000	.0000	.0000	.0000	.0002	.0011	3
18	.0000	.0000	.0000	.0000	.0000	.0000	.0000	.0000	.0000	.0002	2
	.95	.90	.85	.80	.75	.70	.65	.60	.55	.50	y ↑

Table 1 cont'd

Binomial probabilities (n=50 and 100)

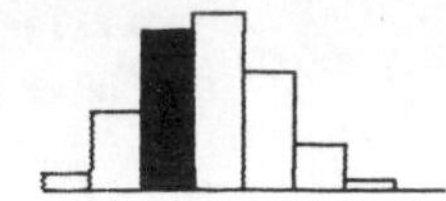

n=50

y↓	.05	.10	.15	.20	.25	.30	.35	.40	.45	.50	
0	.0769	.0052	.0003	.0000	.0000	.0000	.0000	.0000	.0000	.0000	50
1	.2025	.0286	.0026	.0002	.0000	.0000	.0000	.0000	.0000	.0000	49
2	.2611	.0779	.0113	.0011	.0001	.0000	.0000	.0000	.0000	.0000	48
3	.2199	.1386	.0319	.0044	.0004	.0000	.0000	.0000	.0000	.0000	47
4	.1360	.1809	.0661	.0128	.0016	.0001	.0000	.0000	.0000	.0000	46
5	.0658	.1849	.1072	.0295	.0049	.0006	.0000	.0000	.0000	.0000	45
6	.0260	.1541	.1419	.0554	.0123	.0018	.0002	.0000	.0000	.0000	44
7	.0086	.1076	.1575	.0870	.0259	.0048	.0006	.0000	.0000	.0000	43
8	.0024	.0643	.1493	.1169	.0463	.0110	.0017	.0002	.0000	.0000	42
9	.0006	.0333	.1230	.1364	.0721	.0220	.0042	.0005	.0000	.0000	41
10	.0001	.0152	.0890	.1398	.0985	.0386	.0093	.0014	.0001	.0000	40
11	.0000	.0061	.0571	.1271	.1194	.0602	.0182	.0035	.0004	.0000	39
12	.0000	.0022	.0328	.1033	.1294	.0838	.0319	.0076	.0011	.0001	38
13	.0000	.0007	.0169	.0755	.1261	.1050	.0502	.0147	.0027	.0003	37
14	.0000	.0002	.0079	.0499	.1110	.1189	.0714	.0260	.0059	.0008	36
15	.0000	.0001	.0033	.0299	.0888	.1223	.0923	.0415	.0116	.0020	35
16	.0000	.0000	.0013	.0164	.0648	.1147	.1088	.0606	.0207	.0044	34
17	.0000	.0000	.0005	.0082	.0432	.0983	.1171	.0808	.0339	.0087	33
18	.0000	.0000	.0001	.0037	.0264	.0772	.1156	.0987	.0508	.0160	32
19	.0000	.0000	.0000	.0016	.0148	.0558	.1048	.1109	.0700	.0270	31
20	.0000	.0000	.0000	.0006	.0077	.0370	.0875	.1146	.0888	.0419	30
21	.0000	.0000	.0000	.0002	.0036	.0227	.0673	.1091	.1038	.0598	29
22	.0000	.0000	.0000	.0001	.0016	.0128	.0478	.0959	.1119	.0788	28
23	.0000	.0000	.0000	.0000	.0006	.0067	.0313	.0778	.1115	.0960	27
24	.0000	.0000	.0000	.0000	.0002	.0032	.0190	.0584	.1026	.1080	26
25	.0000	.0000	.0000	.0000	.0001	.0014	.0106	.0405	.0873	.1123	25
26	.0000	.0000	.0000	.0000	.0000	.0006	.0055	.0259	.0687	.1080	24
27	.0000	.0000	.0000	.0000	.0000	.0002	.0026	.0154	.0500	.0960	23
28	.0000	.0000	.0000	.0000	.0000	.0001	.0012	.0084	.0336	.0788	22
29	.0000	.0000	.0000	.0000	.0000	.0000	.0005	.0043	.0208	.0598	21
30	.0000	.0000	.0000	.0000	.0000	.0000	.0002	.0020	.0119	.0419	20
31	.0000	.0000	.0000	.0000	.0000	.0000	.0001	.0009	.0063	.0270	19
32	.0000	.0000	.0000	.0000	.0000	.0000	.0000	.0003	.0031	.0160	18
33	.0000	.0000	.0000	.0000	.0000	.0000	.0000	.0001	.0014	.0087	17
34	.0000	.0000	.0000	.0000	.0000	.0000	.0000	.0000	.0006	.0044	16
35	.0000	.0000	.0000	.0000	.0000	.0000	.0000	.0000	.0002	.0020	15
36	.0000	.0000	.0000	.0000	.0000	.0000	.0000	.0000	.0001	.0008	14
37	.0000	.0000	.0000	.0000	.0000	.0000	.0000	.0000	.0000	.0003	13
38	.0000	.0000	.0000	.0000	.0000	.0000	.0000	.0000	.0000	.0001	12
	.95	.90	.85	.80	.75	.70	.65	.60	.55	.50	y↑

(column headings are values of π)

n=100

y↓	.05	.10	.15	.20	.25	.30	.35	.40	.45	.50	
0	.0059	.0000	.0000	.0000	.0000	.0000	.0000	.0000	.0000	.0000	100
1	.0312	.0003	.0000	.0000	.0000	.0000	.0000	.0000	.0000	.0000	99
2	.0812	.0016	.0000	.0000	.0000	.0000	.0000	.0000	.0000	.0000	98
3	.1396	.0059	.0001	.0000	.0000	.0000	.0000	.0000	.0000	.0000	97
4	.1781	.0159	.0003	.0000	.0000	.0000	.0000	.0000	.0000	.0000	96
5	.1800	.0339	.0011	.0000	.0000	.0000	.0000	.0000	.0000	.0000	95
6	.1500	.0596	.0031	.0001	.0000	.0000	.0000	.0000	.0000	.0000	94
7	.1060	.0889	.0075	.0002	.0000	.0000	.0000	.0000	.0000	.0000	93
8	.0649	.1148	.0153	.0006	.0000	.0000	.0000	.0000	.0000	.0000	92
9	.0349	.1304	.0276	.0015	.0000	.0000	.0000	.0000	.0000	.0000	91
10	.0167	.1319	.0444	.0034	.0001	.0000	.0000	.0000	.0000	.0000	90
11	.0072	.1199	.0640	.0069	.0003	.0000	.0000	.0000	.0000	.0000	89
	.95	.90	.85	.80	.75	.70	.65	.60	.55	.50	y↑

(column headings are values of π)

n=100 continued on next page

Table 1 cont'd

Binomial probabilities (n=100)

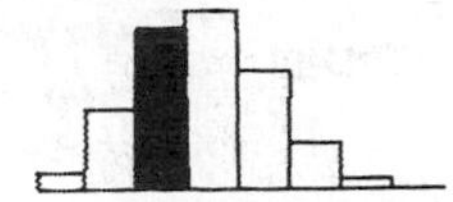

n=100 (continued from previous page)

y ↓	.05	.10	.15	.20	.25	.30	.35	.40	.45	.50	
12	.0028	.0988	.0838	.0128	.0006	.0000	.0000	.0000	.0000	.0000	88
13	.0010	.0743	.1001	.0216	.0014	.0000	.0000	.0000	.0000	.0000	87
14	.0003	.0513	.1098	.0335	.0030	.0001	.0000	.0000	.0000	.0000	86
15	.0001	.0327	.1111	.0481	.0057	.0002	.0000	.0000	.0000	.0000	85
16	.0000	.0193	.1041	.0638	.0100	.0006	.0000	.0000	.0000	.0000	84
17	.0000	.0106	.0908	.0789	.0165	.0012	.0000	.0000	.0000	.0000	83
18	.0000	.0054	.0739	.0909	.0254	.0024	.0001	.0000	.0000	.0000	82
19	.0000	.0026	.0563	.0981	.0365	.0044	.0002	.0000	.0000	.0000	81
20	.0000	.0012	.0402	.0993	.0493	.0076	.0004	.0000	.0000	.0000	80
21	.0000	.0005	.0270	.0946	.0626	.0124	.0009	.0000	.0000	.0000	79
22	.0000	.0002	.0171	.0849	.0749	.0190	.0017	.0001	.0000	.0000	78
23	.0000	.0001	.0103	.0720	.0847	.0277	.0032	.0001	.0000	.0000	77
24	.0000	.0000	.0058	.0577	.0906	.0380	.0055	.0003	.0000	.0000	76
25	.0000	.0000	.0031	.0439	.0918	.0496	.0090	.0006	.0000	.0000	75
26	.0000	.0000	.0016	.0316	.0883	.0613	.0140	.0012	.0000	.0000	74
27	.0000	.0000	.0008	.0217	.0806	.0720	.0207	.0022	.0001	.0000	73
28	.0000	.0000	.0004	.0141	.0701	.0804	.0290	.0038	.0002	.0000	72
29	.0000	.0000	.0002	.0088	.0580	.0856	.0388	.0063	.0004	.0000	71
30	.0000	.0000	.0001	.0052	.0458	.0868	.0494	.0100	.0008	.0000	70
31	.0000	.0000	.0000	.0029	.0344	.0840	.0601	.0151	.0014	.0001	69
32	.0000	.0000	.0000	.0016	.0248	.0776	.0698	.0217	.0025	.0001	68
33	.0000	.0000	.0000	.0008	.0170	.0685	.0774	.0297	.0043	.0002	67
34	.0000	.0000	.0000	.0004	.0112	.0579	.0821	.0391	.0069	.0005	66
35	.0000	.0000	.0000	.0002	.0070	.0468	.0834	.0491	.0106	.0009	65
36	.0000	.0000	.0000	.0001	.0042	.0362	.0811	.0591	.0157	.0016	64
37	.0000	.0000	.0000	.0000	.0024	.0268	.0755	.0682	.0222	.0027	63
38	.0000	.0000	.0000	.0000	.0013	.0191	.0674	.0754	.0301	.0045	62
39	.0000	.0000	.0000	.0000	.0007	.0130	.0577	.0799	.0391	.0071	61
40	.0000	.0000	.0000	.0000	.0004	.0085	.0474	.0812	.0488	.0108	60
41	.0000	.0000	.0000	.0000	.0002	.0053	.0373	.0792	.0584	.0159	59
42	.0000	.0000	.0000	.0000	.0001	.0032	.0282	.0742	.0672	.0223	58
43	.0000	.0000	.0000	.0000	.0000	.0019	.0205	.0667	.0741	.0301	57
44	.0000	.0000	.0000	.0000	.0000	.0010	.0143	.0576	.0786	.0390	56
45	.0000	.0000	.0000	.0000	.0000	.0005	.0096	.0478	.0800	.0485	55
46	.0000	.0000	.0000	.0000	.0000	.0003	.0062	.0381	.0782	.0580	54
47	.0000	.0000	.0000	.0000	.0000	.0001	.0038	.0292	.0736	.0666	53
48	.0000	.0000	.0000	.0000	.0000	.0001	.0023	.0215	.0665	.0735	52
49	.0000	.0000	.0000	.0000	.0000	.0000	.0013	.0152	.0577	.0780	51
50	.0000	.0000	.0000	.0000	.0000	.0000	.0007	.0103	.0482	.0796	50
51	.0000	.0000	.0000	.0000	.0000	.0000	.0004	.0068	.0386	.0780	49
52	.0000	.0000	.0000	.0000	.0000	.0000	.0002	.0042	.0298	.0735	48
53	.0000	.0000	.0000	.0000	.0000	.0000	.0001	.0026	.0221	.0666	47
54	.0000	.0000	.0000	.0000	.0000	.0000	.0000	.0015	.0157	.0580	46
55	.0000	.0000	.0000	.0000	.0000	.0000	.0000	.0008	.0108	.0485	45
56	.0000	.0000	.0000	.0000	.0000	.0000	.0000	.0004	.0071	.0390	44
57	.0000	.0000	.0000	.0000	.0000	.0000	.0000	.0002	.0045	.0301	43
58	.0000	.0000	.0000	.0000	.0000	.0000	.0000	.0001	.0027	.0223	42
59	.0000	.0000	.0000	.0000	.0000	.0000	.0000	.0001	.0016	.0159	41
60	.0000	.0000	.0000	.0000	.0000	.0000	.0000	.0000	.0009	.0108	40
61	.0000	.0000	.0000	.0000	.0000	.0000	.0000	.0000	.0005	.0071	39
62	.0000	.0000	.0000	.0000	.0000	.0000	.0000	.0000	.0002	.0045	38
63	.0000	.0000	.0000	.0000	.0000	.0000	.0000	.0000	.0001	.0027	37
64	.0000	.0000	.0000	.0000	.0000	.0000	.0000	.0000	.0001	.0016	36
65	.0000	.0000	.0000	.0000	.0000	.0000	.0000	.0000	.0000	.0009	35
66	.0000	.0000	.0000	.0000	.0000	.0000	.0000	.0000	.0000	.0005	34
67	.0000	.0000	.0000	.0000	.0000	.0000	.0000	.0000	.0000	.0002	33
68	.0000	.0000	.0000	.0000	.0000	.0000	.0000	.0000	.0000	.0001	32
69	.0000	.0000	.0000	.0000	.0000	.0000	.0000	.0000	.0000	.0001	31
	.95	.90	.85	.80	.75	.70	.65	.60	.55	.50	π ↑

Source: Computed by D. K. Hildebrand.

Table 2

Cumulative Poisson probabilities (mu between .1 and 5.0)

y	.1	.2	.3	.4	.5	.6	.7	.8	.9	1.0
0	.9048	.8187	.7408	.6703	.6065	.5488	.4966	.4493	.4066	.3679
1	.9953	.9825	.9631	.9384	.9098	.8781	.8442	.8088	.7725	.7358
2	.9998	.9989	.9964	.9921	.9856	.9769	.9659	.9526	.9371	.9197
3	1.0000	.9999	.9997	.9992	.9982	.9966	.9942	.9909	.9865	.9810
4	1.0000	1.0000	1.0000	.9999	.9998	.9996	.9992	.9986	.9977	.9963
5	1.0000	1.0000	1.0000	1.0000	1.0000	1.0000	.9999	.9998	.9997	.9994
6	1.0000	1.0000	1.0000	1.0000	1.0000	1.0000	1.0000	1.0000	1.0000	.9999

y	1.1	1.2	1.3	1.4	1.5	1.6	1.7	1.8	1.9	2.0
0	.3329	.3012	.2725	.2466	.2231	.2019	.1827	.1653	.1496	.1353
1	.6990	.6626	.6268	.5918	.5578	.5249	.4932	.4628	.4337	.4060
2	.9004	.8795	.8571	.8335	.8088	.7834	.7572	.7306	.7037	.6767
3	.9743	.9662	.9569	.9463	.9344	.9212	.9068	.8913	.8747	.8571
4	.9946	.9923	.9893	.9857	.9814	.9763	.9704	.9636	.9559	.9473
5	.9990	.9985	.9978	.9968	.9955	.9940	.9920	.9896	.9868	.9834
6	.9999	.9997	.9996	.9994	.9991	.9987	.9981	.9974	.9966	.9955
7	1.0000	1.0000	.9999	.9999	.9998	.9997	.9996	.9994	.9992	.9989
8	1.0000	1.0000	1.0000	1.0000	1.0000	1.0000	.9999	.9999	.9998	.9998

y	2.1	2.2	2.3	2.4	2.5	2.6	2.7	2.8	2.9	3.0
0	.1225	.1108	.1003	.0907	.0821	.0743	.0672	.0608	.0550	.0498
1	.3796	.3546	.3309	.3084	.2873	.2674	.2487	.2311	.2146	.1991
2	.6496	.6227	.5960	.5697	.5438	.5184	.4936	.4695	.4460	.4232
3	.8386	.8194	.7993	.7787	.7576	.7360	.7141	.6919	.6696	.6472
4	.9379	.9275	.9162	.9041	.8912	.8774	.8629	.8477	.8318	.8153
5	.9796	.9751	.9700	.9643	.9580	.9510	.9433	.9349	.9258	.9161
6	.9941	.9925	.9906	.9884	.9858	.9828	.9794	.9756	.9713	.9665
7	.9985	.9980	.9974	.9967	.9958	.9947	.9934	.9919	.9901	.9881
8	.9997	.9995	.9994	.9991	.9989	.9985	.9981	.9976	.9969	.9962
9	.9999	.9999	.9999	.9998	.9997	.9996	.9995	.9993	.9991	.9989
10	1.0000	1.0000	1.0000	1.0000	.9999	.9999	.9999	.9998	.9998	.9997
11	1.0000	1.0000	1.0000	1.0000	1.0000	1.0000	1.0000	1.0000	.9999	.9999

y	3.1	3.2	3.3	3.4	3.5	3.6	3.7	3.8	3.9	4.0
0	.0450	.0408	.0369	.0334	.0302	.0273	.0247	.0224	.0202	.0183
1	.1847	.1712	.1586	.1468	.1359	.1257	.1162	.1074	.0992	.0916
2	.4012	.3799	.3594	.3397	.3208	.3027	.2854	.2689	.2531	.2381
3	.6248	.6025	.5803	.5584	.5366	.5152	.4942	.4735	.4532	.4335
4	.7982	.7806	.7626	.7442	.7254	.7064	.6872	.6678	.6484	.6288
5	.9057	.8946	.8829	.8705	.8576	.8441	.8301	.8156	.8006	.7851
6	.9612	.9554	.9490	.9421	.9347	.9267	.9182	.9091	.8995	.8893
7	.9858	.9832	.9802	.9769	.9733	.9692	.9648	.9599	.9546	.9489
8	.9953	.9943	.9931	.9917	.9901	.9883	.9863	.9840	.9815	.9786
9	.9986	.9982	.9978	.9973	.9967	.9960	.9952	.9942	.9931	.9919
10	.9996	.9995	.9994	.9992	.9990	.9987	.9984	.9981	.9977	.9972
11	.9999	.9999	.9998	.9998	.9997	.9996	.9995	.9994	.9993	.9991
12	1.0000	1.0000	1.0000	.9999	.9999	.9999	.9999	.9998	.9998	.9997
13	1.0000	1.0000	1.0000	1.0000	1.0000	1.0000	1.0000	1.0000	.9999	.9999

y	4.1	4.2	4.3	4.4	4.5	4.6	4.7	4.8	4.9	5.0
0	.0166	.0150	.0136	.0123	.0111	.0101	.0091	.0082	.0074	.0067
1	.0845	.0780	.0719	.0663	.0611	.0563	.0518	.0477	.0439	.0404
2	.2238	.2102	.1974	.1851	.1736	.1626	.1523	.1425	.1333	.1247
3	.4142	.3954	.3772	.3594	.3423	.3257	.3097	.2942	.2793	.2650
4	.6093	.5898	.5704	.5512	.5321	.5132	.4946	.4763	.4582	.4405
5	.7693	.7531	.7367	.7199	.7029	.6858	.6684	.6510	.6335	.6160
6	.8786	.8675	.8558	.8436	.8311	.8180	.8046	.7908	.7767	.7622
7	.9427	.9361	.9290	.9214	.9134	.9049	.8960	.8867	.8769	.8666
8	.9755	.9721	.9683	.9642	.9597	.9549	.9497	.9442	.9382	.9319
9	.9905	.9889	.9871	.9851	.9829	.9805	.9778	.9749	.9717	.9682
10	.9966	.9959	.9952	.9943	.9933	.9922	.9910	.9896	.9880	.9863
11	.9989	.9986	.9983	.9980	.9976	.9971	.9966	.9960	.9953	.9945
12	.9997	.9996	.9995	.9993	.9992	.9990	.9988	.9986	.9983	.9980
13	.9999	.9999	.9998	.9998	.9997	.9997	.9996	.9995	.9994	.9993
14	1.0000	1.0000	1.0000	.9999	.9999	.9999	.9999	.9999	.9998	.9998
15	1.0000	1.0000	1.0000	1.0000	1.0000	1.0000	1.0000	1.0000	.9999	.9999

Table 2 cont'd

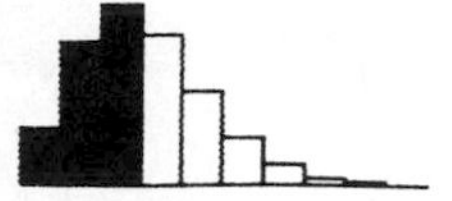

Cumulative Poisson probabilities
(mu between 5.5 and 20.0)

y	μ 5.5	6.0	6.5	7.0	7.5	8.0	8.5	9.0	9.5	10.0
0	.0041	.0025	.0015	.0009	.0006	.0003	.0002	.0001	.0001	.0000
1	.0266	.0174	.0113	.0073	.0047	.0030	.0019	.0012	.0008	.0005
2	.0884	.0620	.0430	.0296	.0203	.0138	.0093	.0062	.0042	.0028
3	.2017	.1512	.1118	.0818	.0591	.0424	.0301	.0212	.0149	.0103
4	.3575	.2851	.2237	.1730	.1321	.0996	.0744	.0550	.0403	.0293
5	.5289	.4457	.3690	.3007	.2414	.1912	.1496	.1157	.0885	.0671
6	.6860	.6063	.5265	.4497	.3782	.3134	.2562	.2068	.1649	.1301
7	.8095	.7440	.6728	.5987	.5246	.4530	.3856	.3239	.2687	.2202
8	.8944	.8472	.7916	.7291	.6620	.5925	.5231	.4557	.3918	.3328
9	.9462	.9161	.8774	.8305	.7764	.7166	.6530	.5874	.5218	.4579
10	.9747	.9574	.9332	.9015	.8622	.8159	.7634	.7060	.6453	.5830
11	.9890	.9799	.9661	.9467	.9208	.8881	.8487	.8030	.7520	.6968
12	.9955	.9912	.9840	.9730	.9573	.9362	.9091	.8758	.8364	.7916
13	.9983	.9964	.9929	.9872	.9784	.9658	.9486	.9261	.8981	.8645
14	.9994	.9986	.9970	.9943	.9897	.9827	.9726	.9585	.9400	.9165
15	.9998	.9995	.9988	.9976	.9954	.9918	.9862	.9780	.9665	.9513
16	.9999	.9998	.9996	.9990	.9980	.9963	.9934	.9889	.9823	.9730
17	1.0000	.9999	.9998	.9996	.9992	.9984	.9970	.9947	.9911	.9857
18	1.0000	1.0000	.9999	.9999	.9997	.9993	.9987	.9976	.9957	.9928
19	1.0000	1.0000	1.0000	1.0000	.9999	.9997	.9995	.9989	.9980	.9965
20	1.0000	1.0000	1.0000	1.0000	1.0000	.9999	.9998	.9996	.9991	.9984
21	1.0000	1.0000	1.0000	1.0000	1.0000	1.0000	.9999	.9998	.9996	.9993
22	1.0000	1.0000	1.0000	1.0000	1.0000	1.0000	1.0000	.9999	.9999	.9997
23	1.0000	1.0000	1.0000	1.0000	1.0000	1.0000	1.0000	1.0000	.9999	.9999

y	μ 11.0	12.0	13.0	14.0	15.0	16.0	17.0	18.0	19.0	20.0
0	.0000	.0000	.0000	.0000	.0000	.0000	.0000	.0000	.0000	.0000
1	.0002	.0001	.0000	.0000	.0000	.0000	.0000	.0000	.0000	.0000
2	.0012	.0005	.0002	.0001	.0000	.0000	.0000	.0000	.0000	.0000
3	.0049	.0023	.0011	.0005	.0002	.0001	.0000	.0000	.0000	.0000
4	.0151	.0076	.0037	.0018	.0009	.0004	.0002	.0001	.0000	.0000
5	.0375	.0203	.0107	.0055	.0028	.0014	.0007	.0003	.0002	.0001
6	.0786	.0458	.0259	.0142	.0076	.0040	.0021	.0010	.0005	.0003
7	.1432	.0895	.0540	.0316	.0180	.0100	.0054	.0029	.0015	.0008
8	.2320	.1550	.0998	.0621	.0374	.0220	.0126	.0071	.0039	.0021
9	.3405	.2424	.1658	.1094	.0699	.0433	.0261	.0154	.0089	.0050
10	.4599	.3472	.2517	.1757	.1185	.0774	.0491	.0304	.0183	.0108
11	.5793	.4616	.3532	.2600	.1848	.1270	.0847	.0549	.0347	.0214
12	.6887	.5760	.4631	.3585	.2676	.1931	.1350	.0917	.0606	.0390
13	.7813	.6815	.5730	.4644	.3632	.2745	.2009	.1426	.0984	.0661
14	.8540	.7720	.6751	.5704	.4657	.3675	.2808	.2081	.1497	.1049
15	.9074	.8444	.7636	.6694	.5681	.4667	.3715	.2867	.2148	.1565
16	.9441	.8987	.8355	.7559	.6641	.5660	.4677	.3751	.2920	.2211
17	.9678	.9370	.8905	.8272	.7489	.6593	.5640	.4686	.3784	.2970
18	.9823	.9626	.9302	.8826	.8195	.7423	.6550	.5622	.4695	.3814
19	.9907	.9787	.9573	.9235	.8752	.8122	.7363	.6509	.5606	.4703
20	.9953	.9884	.9750	.9521	.9170	.8682	.8055	.7307	.6472	.5591
21	.9977	.9939	.9859	.9712	.9469	.9108	.8615	.7991	.7255	.6437
22	.9990	.9970	.9924	.9833	.9673	.9418	.9047	.8551	.7931	.7206
23	.9995	.9985	.9960	.9907	.9805	.9633	.9367	.8989	.8490	.7875
24	.9998	.9993	.9980	.9950	.9888	.9777	.9594	.9317	.8933	.8432
25	.9999	.9997	.9990	.9974	.9938	.9869	.9748	.9554	.9269	.8878
26	1.0000	.9999	.9995	.9987	.9967	.9925	.9848	.9718	.9514	.9221
27	1.0000	.9999	.9998	.9994	.9983	.9959	.9912	.9827	.9687	.9475
28	1.0000	1.0000	.9999	.9997	.9991	.9978	.9950	.9897	.9805	.9657
29	1.0000	1.0000	1.0000	.9999	.9996	.9989	.9973	.9941	.9882	.9782
30	1.0000	1.0000	1.0000	.9999	.9998	.9994	.9986	.9967	.9930	.9865
31	1.0000	1.0000	1.0000	1.0000	.9999	.9997	.9993	.9982	.9960	.9919
32	1.0000	1.0000	1.0000	1.0000	1.0000	.9999	.9996	.9990	.9978	.9953
33	1.0000	1.0000	1.0000	1.0000	1.0000	.9999	.9998	.9995	.9988	.9973

Source: Computed by D. K. Hildebrand.

Table 3
Normal curve areas

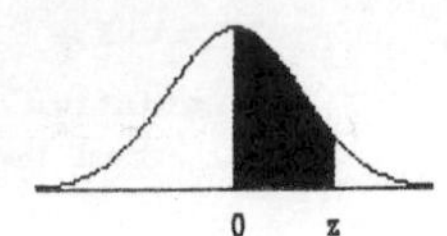

z	.00	.01	.02	.03	.04	.05	.06	.07	.08	.09
0.00	.0000	.0040	.0080	.0120	.0160	.0199	.0239	.0279	.0319	.0359
0.10	.0398	.0438	.0478	.0517	.0557	.0596	.0636	.0675	.0714	.0753
0.20	.0793	.0832	.0871	.0910	.0948	.0987	.1026	.1064	.1103	.1141
0.30	.1179	.1217	.1255	.1293	.1331	.1368	.1406	.1443	.1480	.1517
0.40	.1554	.1591	.1628	.1664	.1700	.1736	.1772	.1808	.1844	.1879
0.50	.1915	.1950	.1985	.2019	.2054	.2088	.2123	.2157	.2190	.2224
0.60	.2257	.2291	.2324	.2357	.2389	.2422	.2454	.2486	.2517	.2549
0.70	.2580	.2611	.2642	.2673	.2704	.2734	.2764	.2794	.2823	.2852
0.80	.2881	.2910	.2939	.2967	.2995	.3023	.3051	.3078	.3106	.3133
0.90	.3159	.3186	.3212	.3238	.3264	.3289	.3315	.3340	.3365	.3389
1.00	.3413	.3438	.3461	.3485	.3508	.3531	.3554	.3577	.3599	.3621
1.10	.3643	.3665	.3686	.3708	.3729	.3749	.3770	.3790	.3810	.3830
1.20	.3849	.3869	.3888	.3907	.3925	.3944	.3962	.3980	.3997	.4015
1.30	.4032	.4049	.4066	.4082	.4099	.4115	.4131	.4147	.4162	.4177
1.40	.4192	.4207	.4222	.4236	.4251	.4265	.4279	.4292	.4306	.4319
1.50	.4332	.4345	.4357	.4370	.4382	.4394	.4406	.4418	.4429	.4441
1.60	.4452	.4463	.4474	.4484	.4495	.4505	.4515	.4525	.4535	.4545
1.70	.4554	.4564	.4573	.4582	.4591	.4599	.4608	.4616	.4625	.4633
1.80	.4641	.4649	.4656	.4664	.4671	.4678	.4686	.4693	.4699	.4706
1.90	.4713	.4719	.4726	.4732	.4738	.4744	.4750	.4756	.4761	.4767
2.00	.4772	.4778	.4783	.4788	.4793	.4798	.4803	.4808	.4812	.4817
2.10	.4821	.4826	.4830	.4834	.4838	.4842	.4846	.4850	.4854	.4857
2.20	.4861	.4864	.4868	.4871	.4875	.4878	.4881	.4884	.4887	.4890
2.30	.4893	.4896	.4898	.4901	.4904	.4906	.4909	.4911	.4913	.4916
2.40	.4918	.4920	.4922	.4925	.4927	.4929	.4931	.4932	.4934	.4936
2.50	.4938	.4940	.4941	.4943	.4945	.4946	.4948	.4949	.4951	.4952
2.60	.4953	.4955	.4956	.4957	.4959	.4960	.4961	.4962	.4963	.4964
2.70	.4965	.4966	.4967	.4968	.4969	.4970	.4971	.4972	.4973	.4974
2.80	.4974	.4975	.4976	.4977	.4977	.4978	.4979	.4979	.4980	.4981
2.90	.4981	.4982	.4982	.4983	.4984	.4984	.4985	.4985	.4986	.4986
3.00	.4987	.4987	.4987	.4988	.4988	.4989	.4989	.4989	.4990	.4990

z	area
3.50	.49976737
4.00	.49996833
4.50	.49999660
5.00	.49999971

Source: Computed by P. J. Hildebrand.

Table 4

Percentage points of the t-distribution

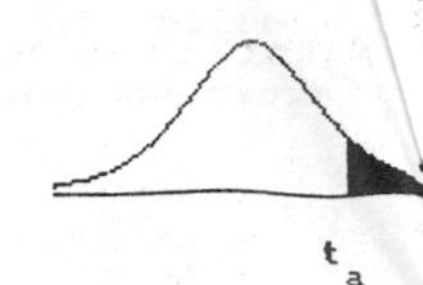

df	a = .1	a = .05	a = .025	a = .01	a = .005	a = .001
1	3.078	6.314	12.706	31.821	63.657	318.309
2	1.886	2.920	4.303	6.965	9.925	22.327
3	1.638	2.353	3.182	4.541	5.841	10.215
4	1.533	2.132	2.776	3.747	4.604	7.173
5	1.476	2.015	2.571	3.365	4.032	5.893
6	1.440	1.943	2.447	3.143	3.707	5.208
7	1.415	1.895	2.365	2.998	3.499	4.785
8	1.397	1.860	2.306	2.896	3.355	4.501
9	1.383	1.833	2.262	2.821	3.250	4.297
10	1.372	1.812	2.228	2.764	3.169	4.144
11	1.363	1.796	2.201	2.718	3.106	4.025
12	1.356	1.782	2.179	2.681	3.055	3.930
13	1.350	1.771	2.160	2.650	3.012	3.852
14	1.345	1.761	2.145	2.624	2.977	3.787
15	1.341	1.753	2.131	2.602	2.947	3.733
16	1.337	1.746	2.120	2.583	2.921	3.686
17	1.333	1.740	2.110	2.567	2.898	3.646
18	1.330	1.734	2.101	2.552	2.878	3.610
19	1.328	1.729	2.093	2.539	2.861	3.579
20	1.325	1.725	2.086	2.528	2.845	3.552
21	1.323	1.721	2.080	2.518	2.831	3.527
22	1.321	1.717	2.074	2.508	2.819	3.505
23	1.319	1.714	2.069	2.500	2.807	3.485
24	1.318	1.711	2.064	2.492	2.797	3.467
25	1.316	1.708	2.060	2.485	2.787	3.450
26	1.315	1.706	2.056	2.479	2.779	3.435
27	1.314	1.703	2.052	2.473	2.771	3.421
28	1.313	1.701	2.048	2.467	2.763	3.408
29	1.311	1.699	2.045	2.462	2.756	3.396
30	1.310	1.697	2.042	2.457	2.750	3.385
40	1.303	1.684	2.021	2.423	2.704	3.307
60	1.296	1.671	2.000	2.390	2.660	3.232
120	1.289	1.658	1.980	2.358	2.617	3.160
240	1.285	1.651	1.970	2.342	2.596	3.125
inf.	1.282	1.645	1.960	2.326	2.576	3.090

Source: Computed by P. J. Hildebrand.

Table 5

Percentage points of the chi-square distribution (a>.5)

df	a=.999	a=.995	a=.99	a=.975	a=.95	a=.9
1	.000002	.000039	.000157	.000982	.003932	.01579
2	.002001	.01003	.02010	.05064	.1026	.2107
3	.02430	.07172	.1148	.2158	.3518	.5844
4	.09080	.2070	.2971	.4844	.7107	1.064
5	.2102	.4117	.5543	.8312	1.145	1.610
6	.3811	.6757	.8721	1.237	1.635	2.204
7	.5985	.9893	1.239	1.690	2.167	2.833
8	.8571	1.344	1.646	2.180	2.733	3.490
9	1.152	1.735	2.088	2.700	3.325	4.168
10	1.479	2.156	2.558	3.247	3.940	4.865
11	1.834	2.603	3.053	3.816	4.575	5.578
12	2.214	3.074	3.571	4.404	5.226	6.304
13	2.617	3.565	4.107	5.009	5.892	7.042
14	3.041	4.075	4.660	5.629	6.571	7.790
15	3.483	4.601	5.229	6.262	7.261	8.547
16	3.942	5.142	5.812	6.908	7.962	9.312
17	4.416	5.697	6.408	7.564	8.672	10.09
18	4.905	6.265	7.015	8.231	9.390	10.86
19	5.407	6.844	7.633	8.907	10.12	11.65
20	5.921	7.434	8.260	9.591	10.85	12.44
21	6.447	8.034	8.897	10.28	11.59	13.24
22	6.983	8.643	9.542	10.98	12.34	14.04
23	7.529	9.260	10.20	11.69	13.09	14.85
24	8.085	9.886	10.86	12.40	13.85	15.66
25	8.649	10.52	11.52	13.12	14.61	16.47
26	9.222	11.16	12.20	13.84	15.38	17.29
27	9.803	11.81	12.88	14.57	16.15	18.11
28	10.39	12.46	13.56	15.31	16.93	18.94
29	10.99	13.12	14.26	16.05	17.71	19.77
30	11.59	13.79	14.95	16.79	18.49	20.60
40	17.92	20.71	22.16	24.43	26.51	29.05
50	24.67	27.99	29.71	32.36	34.76	37.69
60	31.74	35.53	37.48	40.48	43.19	46.46
70	39.04	43.28	45.44	48.76	51.74	55.33
80	46.52	51.17	53.54	57.15	60.39	64.28
90	54.16	59.20	61.75	65.65	69.13	73.29
100	61.92	67.33	70.06	74.22	77.93	82.36
120	77.76	83.85	86.92	91.57	95.70	100.62
240	177.95	187.32	191.99	198.98	205.14	212.39

Table 5 cont'd

Percentage points of the chi-square distribution (a<.5)

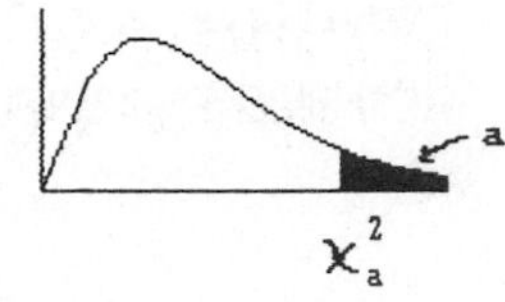

a=.1	a=.05	a=.025	a=.01	a=.005	a=.001	df
2.706	3.841	5.024	6.635	7.879	10.83	1
4.605	5.991	7.378	9.210	10.60	13.82	2
6.251	7.815	9.348	11.34	12.84	16.27	3
7.779	9.488	11.14	13.28	14.86	18.47	4
9.236	11.07	12.83	15.09	16.75	20.52	5
10.64	12.59	14.45	16.81	18.55	22.46	6
12.02	14.07	16.01	18.48	20.28	24.32	7
13.36	15.51	17.53	20.09	21.95	26.12	8
14.68	16.92	19.02	21.67	23.59	27.88	9
15.99	18.31	20.48	23.21	25.19	29.59	10
17.28	19.68	21.92	24.72	26.76	31.27	11
18.55	21.03	23.34	26.22	28.30	32.91	12
19.81	22.36	24.74	27.69	29.82	34.53	13
21.06	23.68	26.12	29.14	31.32	36.12	14
22.31	25.00	27.49	30.58	32.80	37.70	15
23.54	26.30	28.85	32.00	34.27	39.25	16
24.77	27.59	30.19	33.41	35.72	40.79	17
25.99	28.87	31.53	34.81	37.16	42.31	18
27.20	30.14	32.85	36.19	38.58	43.82	19
28.41	31.41	34.17	37.57	40.00	45.31	20
29.62	32.67	35.48	38.93	41.40	46.80	21
30.81	33.92	36.78	40.29	42.80	48.27	22
32.01	35.17	38.08	41.64	44.18	49.73	23
33.20	36.42	39.36	42.98	45.56	51.18	24
34.38	37.65	40.65	44.31	46.93	52.62	25
35.56	38.89	41.92	45.64	48.29	54.05	26
36.74	40.11	43.19	46.96	49.65	55.48	27
37.92	41.34	44.46	48.28	50.99	56.89	28
39.09	42.56	45.72	49.59	52.34	58.30	29
40.26	43.77	46.98	50.89	53.67	59.70	30
51.81	55.76	59.34	63.69	66.77	73.40	40
63.17	67.50	71.42	76.15	79.49	86.66	50
74.40	79.08	83.30	88.38	91.95	99.61	60
85.53	90.53	95.02	100.43	104.21	112.32	70
96.58	101.88	106.63	112.33	116.32	124.84	80
107.57	113.15	118.14	124.12	128.30	137.21	90
118.50	124.34	129.56	135.81	140.17	149.45	100
140.23	146.57	152.21	158.95	163.65	173.62	120
268.47	277.14	284.80	293.89	300.18	313.44	240

Source: Computed by P. J. Hildebrand.

Table 6

Percentage points of the F-distribution (df_2 between 1 and 6)

df_2	a	1	2	3	4	5	6	7	8	9	10
		df_1									
1	.25	5.83	7.50	8.20	8.58	8.82	8.98	9.10	9.19	9.26	9.32
	.10	39.86	49.50	53.59	55.83	57.24	58.20	58.91	59.44	59.86	60.19
	.05	161.4	199.5	215.7	224.6	230.2	234.0	236.8	238.9	240.5	241.9
	.025	647.8	799.5	864.2	899.6	921.8	937.1	948.2	956.7	963.3	968.6
	.01	4052	5000	5403	5625	5764	5859	5928	5981	6022	6056
2	.25	2.57	3.00	3.15	3.23	3.28	3.31	3.34	3.35	3.37	3.38
	.10	8.53	9.00	9.16	9.24	9.29	9.33	9.35	9.37	9.38	9.39
	.05	18.51	19.00	19.16	19.25	19.30	19.33	19.35	19.37	19.38	19.40
	.025	38.51	39.00	39.17	39.25	39.30	39.33	39.36	39.37	39.39	39.40
	.01	98.50	99.00	99.17	99.25	99.30	99.33	99.36	99.37	99.39	99.40
	.005	198.5	199.0	199.2	199.2	199.3	199.3	199.4	199.4	199.4	199.4
	.001	998.5	999.0	999.2	999.2	999.3	999.3	999.4	999.4	999.4	999.4
3	.25	2.02	2.28	2.36	2.39	2.41	2.42	2.43	2.44	2.44	2.44
	.10	5.54	5.46	5.39	5.34	5.31	5.28	5.27	5.25	5.24	5.23
	.05	10.13	9.55	9.28	9.12	9.01	8.94	8.89	8.85	8.81	8.79
	.025	17.44	16.04	15.44	15.10	14.88	14.73	14.62	14.54	14.47	14.42
	.01	34.12	30.82	29.46	28.71	28.24	27.91	27.67	27.49	27.35	27.23
	.005	55.55	49.80	47.47	46.19	45.39	44.84	44.43	44.13	43.88	43.69
	.001	167.0	148.5	141.1	137.1	134.6	132.8	131.6	130.6	129.9	129.2
4	.25	1.81	2.00	2.05	2.06	2.07	2.08	2.08	2.08	2.08	2.08
	.10	4.54	4.32	4.19	4.11	4.05	4.01	3.98	3.95	3.94	3.92
	.05	7.71	6.94	6.59	6.39	6.26	6.16	6.09	6.04	6.00	5.96
	.025	12.22	10.65	9.98	9.60	9.36	9.20	9.07	8.98	8.90	8.84
	.01	21.20	18.00	16.69	15.98	15.52	15.21	14.98	14.80	14.66	14.55
	.005	31.33	26.28	24.26	23.15	22.46	21.97	21.62	21.35	21.14	20.97
	.001	74.14	61.25	56.18	53.44	51.71	50.53	49.66	49.00	48.47	48.05
5	.25	1.69	1.85	1.88	1.89	1.89	1.89	1.89	1.89	1.89	1.89
	.10	4.06	3.78	3.62	3.52	3.45	3.40	3.37	3.34	3.32	3.30
	.05	6.61	5.79	5.41	5.19	5.05	4.95	4.88	4.82	4.77	4.74
	.025	10.01	8.43	7.76	7.39	7.15	6.98	6.85	6.76	6.68	6.62
	.01	16.26	13.27	12.06	11.39	10.97	10.67	10.46	10.29	10.16	10.05
	.005	22.78	18.31	16.53	15.56	14.94	14.51	14.20	13.96	13.77	13.62
	.001	47.18	37.12	33.20	31.09	29.75	28.83	28.16	27.65	27.24	26.92
6	.25	1.62	1.76	1.78	1.79	1.79	1.78	1.78	1.78	1.77	1.77
	.10	3.78	3.46	3.29	3.18	3.11	3.05	3.01	2.98	2.96	2.94
	.05	5.99	5.14	4.76	4.53	4.39	4.28	4.21	4.15	4.10	4.06
	.025	8.81	7.26	6.60	6.23	5.99	5.82	5.70	5.60	5.52	5.46
	.01	13.75	10.92	9.78	9.15	8.75	8.47	8.26	8.10	7.98	7.87
	.005	18.63	14.54	12.92	12.03	11.46	11.07	10.79	10.57	10.39	10.25
	.001	35.51	27.00	23.70	21.92	20.80	20.03	19.46	19.03	18.69	18.41

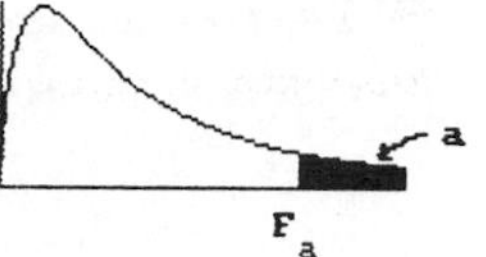

				df_1							
12	**15**	**20**	**24**	**30**	**40**	**60**	**120**	**240**	**inf.**	**a**	**df_2**
9.41	9.49	9.58	9.63	9.67	9.71	9.76	9.80	9.83	9.85	.25	1
60.71	61.22	61.74	62.00	62.26	62.53	62.79	63.06	63.19	63.33	.10	
243.9	245.9	248.0	249.1	250.1	251.1	252.2	253.3	253.8	254.3	.05	
976.7	984.9	993.1	997.2	1001	1006	1010	1014	1016	1018	.025	
6106	6157	6209	6235	6261	6287	6313	6339	6353	6366	.01	
3.39	3.41	3.43	3.43	3.44	3.45	3.46	3.47	3.47	3.48	.25	2
9.41	9.42	9.44	9.45	9.46	9.47	9.47	9.48	9.49	9.49	.10	
19.41	19.43	19.45	19.45	19.46	19.47	19.48	19.49	19.49	19.50	.05	
39.41	39.43	39.45	39.46	39.46	39.47	39.48	39.49	39.49	39.50	.025	
99.42	99.43	99.45	99.46	99.47	99.47	99.48	99.49	99.50	99.50	.01	
199.4	199.4	199.4	199.5	199.5	199.5	199.5	199.5	199.5	199.5	.005	
999.4	999.4	999.4	999.5	999.5	999.5	999.5	999.5	999.5	999.5	.001	
2.45	2.46	2.46	2.46	2.47	2.47	2.47	2.47	2.47	2.47	.25	3
5.22	5.20	5.18	5.18	5.17	5.16	5.15	5.14	5.14	5.13	.10	
8.74	8.70	8.66	8.64	8.62	8.59	8.57	8.55	8.54	8.53	.05	
14.34	14.25	14.17	14.12	14.08	14.04	13.99	13.95	13.92	13.90	.025	
27.05	26.87	26.69	26.60	26.50	26.41	26.32	26.22	26.17	26.13	.01	
43.39	43.08	42.78	42.62	42.47	42.31	42.15	41.99	41.91	41.83	.005	
128.3	127.4	126.4	125.9	125.4	125.0	124.5	124.0	123.7	123.5	.001	
2.08	2.08	2.08	2.08	2.08	2.08	2.08	2.08	2.08	2.08	.25	4
3.90	3.87	3.84	3.83	3.82	3.80	3.79	3.78	3.77	3.76	.10	
5.91	5.86	5.80	5.77	5.75	5.72	5.69	5.66	5.64	5.63	.05	
8.75	8.66	8.56	8.51	8.46	8.41	8.36	8.31	8.28	8.26	.025	
14.37	14.20	14.02	13.93	13.84	13.75	13.65	13.56	13.51	13.46	.01	
20.70	20.44	20.17	20.03	19.89	19.75	19.61	19.47	19.40	19.32	.005	
47.41	46.76	46.10	45.77	45.43	45.09	44.75	44.40	44.23	44.05	.001	
1.89	1.89	1.88	1.88	1.88	1.88	1.87	1.87	1.87	1.87	.25	5
3.27	3.24	3.21	3.19	3.17	3.16	3.14	3.12	3.11	3.10	.10	
4.68	4.62	4.56	4.53	4.50	4.46	4.43	4.40	4.38	4.36	.05	
6.52	6.43	6.33	6.28	6.23	6.18	6.12	6.07	6.04	6.02	.025	
9.89	9.72	9.55	9.47	9.38	9.29	9.20	9.11	9.07	9.02	.01	
13.38	13.15	12.90	12.78	12.66	12.53	12.40	12.27	12.21	12.14	.005	
26.42	25.91	25.39	25.13	24.87	24.60	24.33	24.06	23.92	23.79	.001	
1.77	1.76	1.76	1.75	1.75	1.75	1.74	1.74	1.74	1.74	.25	6
2.90	2.87	2.84	2.82	2.80	2.78	2.76	2.74	2.73	2.72	.10	
4.00	3.94	3.87	3.84	3.81	3.77	3.74	3.70	3.69	3.67	.05	
5.37	5.27	5.17	5.12	5.07	5.01	4.96	4.90	4.88	4.85	.025	
7.72	7.56	7.40	7.31	7.23	7.14	7.06	6.97	6.92	6.88	.01	
10.03	9.81	9.59	9.47	9.36	9.24	9.12	9.00	8.94	8.88	.005	
17.99	17.56	17.12	16.90	16.67	16.44	16.21	15.98	15.86	15.75	.001	

Table 6 cont'd

Percentage points of the F-distribution (df_2 between 7 and 12)

df_2	a	df_1 1	2	3	4	5	6	7	8	9	10
7	.25	1.57	1.70	1.72	1.72	1.71	1.71	1.70	1.70	1.69	1.69
	.10	3.59	3.26	3.07	2.96	2.88	2.83	2.78	2.75	2.72	2.70
	.05	5.59	4.74	4.35	4.12	3.97	3.87	3.79	3.73	3.68	3.64
	.025	8.07	6.54	5.89	5.52	5.29	5.12	4.99	4.90	4.82	4.76
	.01	12.25	9.55	8.45	7.85	7.46	7.19	6.99	6.84	6.72	6.62
	.005	16.24	12.40	10.88	10.05	9.52	9.16	8.89	8.68	8.51	8.38
	.001	29.25	21.69	18.77	17.20	16.21	15.52	15.02	14.63	14.33	14.08
8	.25	1.54	1.66	1.67	1.66	1.66	1.65	1.64	1.64	1.63	1.63
	.10	3.46	3.11	2.92	2.81	2.73	2.67	2.62	2.59	2.56	2.54
	.05	5.32	4.46	4.07	3.84	3.69	3.58	3.50	3.44	3.39	3.35
	.025	7.57	6.06	5.42	5.05	4.82	4.65	4.53	4.43	4.36	4.30
	.01	11.26	8.65	7.59	7.01	6.63	6.37	6.18	6.03	5.91	5.81
	.005	14.69	11.04	9.60	8.81	8.30	7.95	7.69	7.50	7.34	7.21
	.001	25.41	18.49	15.83	14.39	13.48	12.86	12.40	12.05	11.77	11.54
9	.25	1.51	1.62	1.63	1.63	1.62	1.61	1.60	1.60	1.59	1.59
	.10	3.36	3.01	2.81	2.69	2.61	2.55	2.51	2.47	2.44	2.42
	.05	5.12	4.26	3.86	3.63	3.48	3.37	3.29	3.23	3.18	3.14
	.025	7.21	5.71	5.08	4.72	4.48	4.32	4.20	4.10	4.03	3.96
	.01	10.56	8.02	6.99	6.42	6.06	5.80	5.61	5.47	5.35	5.26
	.005	13.61	10.11	8.72	7.96	7.47	7.13	6.88	6.69	6.54	6.42
	.001	22.86	16.39	13.90	12.56	11.71	11.13	10.70	10.37	10.11	9.89
10	.25	1.49	1.60	1.60	1.59	1.59	1.58	1.57	1.56	1.56	1.55
	.10	3.29	2.92	2.73	2.61	2.52	2.46	2.41	2.38	2.35	2.32
	.05	4.96	4.10	3.71	3.48	3.33	3.22	3.14	3.07	3.02	2.98
	.025	6.94	5.46	4.83	4.47	4.24	4.07	3.95	3.85	3.78	3.72
	.01	10.04	7.56	6.55	5.99	5.64	5.39	5.20	5.06	4.94	4.85
	.005	12.83	9.43	8.08	7.34	6.87	6.54	6.30	6.12	5.97	5.85
	.001	21.04	14.91	12.55	11.28	10.48	9.93	9.52	9.20	8.96	8.75
11	.25	1.47	1.58	1.58	1.57	1.56	1.55	1.54	1.53	1.53	1.52
	.10	3.23	2.86	2.66	2.54	2.45	2.39	2.34	2.30	2.27	2.25
	.05	4.84	3.98	3.59	3.36	3.20	3.09	3.01	2.95	2.90	2.85
	.025	6.72	5.26	4.63	4.28	4.04	3.88	3.76	3.66	3.59	3.53
	.01	9.65	7.21	6.22	5.67	5.32	5.07	4.89	4.74	4.63	4.54
	.005	12.23	8.91	7.60	6.88	6.42	6.10	5.86	5.68	5.54	5.42
	.001	19.69	13.81	11.56	10.35	9.58	9.05	8.66	8.35	8.12	7.92
12	.25	1.46	1.56	1.56	1.55	1.54	1.53	1.52	1.51	1.51	1.50
	.10	3.18	2.81	2.61	2.48	2.39	2.33	2.28	2.24	2.21	2.19
	.05	4.75	3.89	3.49	3.26	3.11	3.00	2.91	2.85	2.80	2.75
	.025	6.55	5.10	4.47	4.12	3.89	3.73	3.61	3.51	3.44	3.37
	.01	9.33	6.93	5.95	5.41	5.06	4.82	4.64	4.50	4.39	4.30
	.005	11.75	8.51	7.23	6.52	6.07	5.76	5.52	5.35	5.20	5.09
	.001	18.64	12.97	10.80	9.63	8.89	8.38	8.00	7.71	7.48	7.29

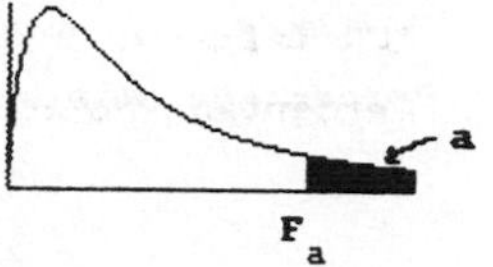

df1: 12	15	20	24	30	40	60	120	240	inf.	a	df_2
1.68	1.68	1.67	1.67	1.66	1.66	1.65	1.65	1.65	1.65	.25	7
2.67	2.63	2.59	2.58	2.56	2.54	2.51	2.49	2.48	2.47	.10	
3.57	3.51	3.44	3.41	3.38	3.34	3.30	3.27	3.25	3.23	.05	
4.67	4.57	4.47	4.41	4.36	4.31	4.25	4.20	4.17	4.14	.025	
6.47	6.31	6.16	6.07	5.99	5.91	5.82	5.74	5.69	5.65	.01	
8.18	7.97	7.75	7.64	7.53	7.42	7.31	7.19	7.13	7.08	.005	
13.71	13.32	12.93	12.73	12.53	12.33	12.12	11.91	11.80	11.70	.001	
1.62	1.62	1.61	1.60	1.60	1.59	1.59	1.58	1.58	1.58	.25	8
2.50	2.46	2.42	2.40	2.38	2.36	2.34	2.32	2.30	2.29	.10	
3.28	3.22	3.15	3.12	3.08	3.04	3.01	2.97	2.95	2.93	.05	
4.20	4.10	4.00	3.95	3.89	3.84	3.78	3.73	3.70	3.67	.025	
5.67	5.52	5.36	5.28	5.20	5.12	5.03	4.95	4.90	4.86	.01	
7.01	6.81	6.61	6.50	6.40	6.29	6.18	6.06	6.01	5.95	.005	
11.19	10.84	10.48	10.30	10.11	9.92	9.73	9.53	9.43	9.33	.001	
1.58	1.57	1.56	1.56	1.55	1.54	1.54	1.53	1.53	1.53	.25	9
2.38	2.34	2.30	2.28	2.25	2.23	2.21	2.18	2.17	2.16	.10	
3.07	3.01	2.94	2.90	2.86	2.83	2.79	2.75	2.73	2.71	.05	
3.87	3.77	3.67	3.61	3.56	3.51	3.45	3.39	3.36	3.33	.025	
5.11	4.96	4.81	4.73	4.65	4.57	4.48	4.40	4.35	4.31	.01	
6.23	6.03	5.83	5.73	5.62	5.52	5.41	5.30	5.24	5.19	.005	
9.57	9.24	8.90	8.72	8.55	8.37	8.19	8.00	7.91	7.81	.001	
1.54	1.53	1.52	1.52	1.51	1.51	1.50	1.49	1.49	1.48	.25	10
2.28	2.24	2.20	2.18	2.16	2.13	2.11	2.08	2.07	2.06	.10	
2.91	2.85	2.77	2.74	2.70	2.66	2.62	2.58	2.56	2.54	.05	
3.62	3.52	3.42	3.37	3.31	3.26	3.20	3.14	3.11	3.08	.025	
4.71	4.56	4.41	4.33	4.25	4.17	4.08	4.00	3.95	3.91	.01	
5.66	5.47	5.27	5.17	5.07	4.97	4.86	4.75	4.69	4.64	.005	
8.45	8.13	7.80	7.64	7.47	7.30	7.12	6.94	6.85	6.76	.001	
1.51	1.50	1.49	1.49	1.48	1.47	1.47	1.46	1.45	1.45	.25	11
2.21	2.17	2.12	2.10	2.08	2.05	2.03	2.00	1.99	1.97	.10	
2.79	2.72	2.65	2.61	2.57	2.53	2.49	2.45	2.43	2.40	.05	
3.43	3.33	3.23	3.17	3.12	3.06	3.00	2.94	2.91	2.88	.025	
4.40	4.25	4.10	4.02	3.94	3.86	3.78	3.69	3.65	3.60	.01	
5.24	5.05	4.86	4.76	4.65	4.55	4.45	4.34	4.28	4.23	.005	
7.63	7.32	7.01	6.85	6.68	6.52	6.35	6.18	6.09	6.00	.001	
1.49	1.48	1.47	1.46	1.45	1.45	1.44	1.43	1.43	1.42	.25	12
2.15	2.10	2.06	2.04	2.01	1.99	1.96	1.93	1.92	1.90	.10	
2.69	2.62	2.54	2.51	2.47	2.43	2.38	2.34	2.32	2.30	.05	
3.28	3.18	3.07	3.02	2.96	2.91	2.85	2.79	2.76	2.72	.025	
4.16	4.01	3.86	3.78	3.70	3.62	3.54	3.45	3.41	3.36	.01	
4.91	4.72	4.53	4.43	4.33	4.23	4.12	4.01	3.96	3.90	.005	
7.00	6.71	6.40	6.25	6.09	5.93	5.76	5.59	5.51	5.42	.001	

Table 6 cont'd

Percentage points of the F-distribution (df_2 between 13 and 18)

df_2	a	df_1 1	2	3	4	5	6	7	8	9	10
13	.25	1.45	1.55	1.55	1.53	1.52	1.51	1.50	1.49	1.49	1.48
	.10	3.14	2.76	2.56	2.43	2.35	2.28	2.23	2.20	2.16	2.14
	.05	4.67	3.81	3.41	3.18	3.03	2.92	2.83	2.77	2.71	2.67
	.025	6.41	4.97	4.35	4.00	3.77	3.60	3.48	3.39	3.31	3.25
	.01	9.07	6.70	5.74	5.21	4.86	4.62	4.44	4.30	4.19	4.10
	.005	11.37	8.19	6.93	6.23	5.79	5.48	5.25	5.08	4.94	4.82
	.001	17.82	12.31	10.21	9.07	8.35	7.86	7.49	7.21	6.98	6.80
14	.25	1.44	1.53	1.53	1.52	1.51	1.50	1.49	1.48	1.47	1.46
	.10	3.10	2.73	2.52	2.39	2.31	2.24	2.19	2.15	2.12	2.10
	.05	4.60	3.74	3.34	3.11	2.96	2.85	2.76	2.70	2.65	2.60
	.025	6.30	4.86	4.24	3.89	3.66	3.50	3.38	3.29	3.21	3.15
	.01	8.86	6.51	5.56	5.04	4.69	4.46	4.28	4.14	4.03	3.94
	.005	11.06	7.92	6.68	6.00	5.56	5.26	5.03	4.86	4.72	4.60
	.001	17.14	11.78	9.73	8.62	7.92	7.44	7.08	6.80	6.58	6.40
15	.25	1.43	1.52	1.52	1.51	1.49	1.48	1.47	1.46	1.46	1.45
	.10	3.07	2.70	2.49	2.36	2.27	2.21	2.16	2.12	2.09	2.06
	.05	4.54	3.68	3.29	3.06	2.90	2.79	2.71	2.64	2.59	2.54
	.025	6.20	4.77	4.15	3.80	3.58	3.41	3.29	3.20	3.12	3.06
	.01	8.68	6.36	5.42	4.89	4.56	4.32	4.14	4.00	3.89	3.80
	.005	10.80	7.70	6.48	5.80	5.37	5.07	4.85	4.67	4.54	4.42
	.001	16.59	11.34	9.34	8.25	7.57	7.09	6.74	6.47	6.26	6.08
16	.25	1.42	1.51	1.51	1.50	1.48	1.47	1.46	1.45	1.44	1.44
	.10	3.05	2.67	2.46	2.33	2.24	2.18	2.13	2.09	2.06	2.03
	.05	4.49	3.63	3.24	3.01	2.85	2.74	2.66	2.59	2.54	2.49
	.025	6.12	4.69	4.08	3.73	3.50	3.34	3.22	3.12	3.05	2.99
	.01	8.53	6.23	5.29	4.77	4.44	4.20	4.03	3.89	3.78	3.69
	.005	10.58	7.51	6.30	5.64	5.21	4.91	4.69	4.52	4.38	4.27
	.001	16.12	10.97	9.01	7.94	7.27	6.80	6.46	6.19	5.98	5.81
17	.25	1.42	1.51	1.50	1.49	1.47	1.46	1.45	1.44	1.43	1.43
	.10	3.03	2.64	2.44	2.31	2.22	2.15	2.10	2.06	2.03	2.00
	.05	4.45	3.59	3.20	2.96	2.81	2.70	2.61	2.55	2.49	2.45
	.025	6.04	4.62	4.01	3.66	3.44	3.28	3.16	3.06	2.98	2.92
	.01	8.40	6.11	5.18	4.67	4.34	4.10	3.93	3.79	3.68	3.59
	.005	10.38	7.35	6.16	5.50	5.07	4.78	4.56	4.39	4.25	4.14
	.001	15.72	10.66	8.73	7.68	7.02	6.56	6.22	5.96	5.75	5.58
18	.25	1.41	1.50	1.49	1.48	1.46	1.45	1.44	1.43	1.42	1.42
	.10	3.01	2.62	2.42	2.29	2.20	2.13	2.08	2.04	2.00	1.98
	.05	4.41	3.55	3.16	2.93	2.77	2.66	2.58	2.51	2.46	2.41
	.025	5.98	4.56	3.95	3.61	3.38	3.22	3.10	3.01	2.93	2.87
	.01	8.29	6.01	5.09	4.58	4.25	4.01	3.84	3.71	3.60	3.51
	.005	10.22	7.21	6.03	5.37	4.96	4.66	4.44	4.28	4.14	4.03
	.001	15.38	10.39	8.49	7.46	6.81	6.35	6.02	5.76	5.56	5.39

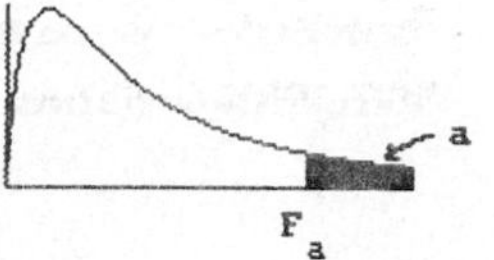

df_1											
12	15	20	24	30	40	60	120	240	inf.	a	df_2
1.47	1.46	1.45	1.44	1.43	1.42	1.42	1.41	1.40	1.40	.25	13
2.10	2.05	2.01	1.98	1.96	1.93	1.90	1.88	1.86	1.85	.10	
2.60	2.53	2.46	2.42	2.38	2.34	2.30	2.25	2.23	2.21	.05	
3.15	3.05	2.95	2.89	2.84	2.78	2.72	2.66	2.63	2.60	.025	
3.96	3.82	3.66	3.59	3.51	3.43	3.34	3.25	3.21	3.17	.01	
4.64	4.46	4.27	4.17	4.07	3.97	3.87	3.76	3.70	3.65	.005	
6.52	6.23	5.93	5.78	5.63	5.47	5.30	5.14	5.05	4.97	.001	
1.45	1.44	1.43	1.42	1.41	1.41	1.40	1.39	1.38	1.38	.25	14
2.05	2.01	1.96	1.94	1.91	1.89	1.86	1.83	1.81	1.80	.10	
2.53	2.46	2.39	2.35	2.31	2.27	2.22	2.18	2.15	2.13	.05	
3.05	2.95	2.84	2.79	2.73	2.67	2.61	2.55	2.52	2.49	.025	
3.80	3.66	3.51	3.43	3.35	3.27	3.18	3.09	3.05	3.00	.01	
4.43	4.25	4.06	3.96	3.86	3.76	3.66	3.55	3.49	3.44	.005	
6.13	5.85	5.56	5.41	5.25	5.10	4.94	4.77	4.69	4.60	.001	
1.44	1.43	1.41	1.41	1.40	1.39	1.38	1.37	1.36	1.36	.25	15
2.02	1.97	1.92	1.90	1.87	1.85	1.82	1.79	1.77	1.76	.10	
2.48	2.40	2.33	2.29	2.25	2.20	2.16	2.11	2.09	2.07	.05	
2.96	2.86	2.76	2.70	2.64	2.59	2.52	2.46	2.43	2.40	.025	
3.67	3.52	3.37	3.29	3.21	3.13	3.05	2.96	2.91	2.87	.01	
4.25	4.07	3.88	3.79	3.69	3.58	3.48	3.37	3.32	3.26	.005	
5.81	5.54	5.25	5.10	4.95	4.80	4.64	4.47	4.39	4.31	.001	
1.43	1.41	1.40	1.39	1.38	1.37	1.36	1.35	1.35	1.34	.25	16
1.99	1.94	1.89	1.87	1.84	1.81	1.78	1.75	1.73	1.72	.10	
2.42	2.35	2.28	2.24	2.19	2.15	2.11	2.06	2.03	2.01	.05	
2.89	2.79	2.68	2.63	2.57	2.51	2.45	2.38	2.35	2.32	.025	
3.55	3.41	3.26	3.18	3.10	3.02	2.93	2.84	2.80	2.75	.01	
4.10	3.92	3.73	3.64	3.54	3.44	3.33	3.22	3.17	3.11	.005	
5.55	5.27	4.99	4.85	4.70	4.54	4.39	4.23	4.14	4.06	.001	
1.41	1.40	1.39	1.38	1.37	1.36	1.35	1.34	1.33	1.33	.25	17
1.96	1.91	1.86	1.84	1.81	1.78	1.75	1.72	1.70	1.69	.10	
2.38	2.31	2.23	2.19	2.15	2.10	2.06	2.01	1.99	1.96	.05	
2.82	2.72	2.62	2.56	2.50	2.44	2.38	2.32	2.28	2.25	.025	
3.46	3.31	3.16	3.08	3.00	2.92	2.83	2.75	2.70	2.65	.01	
3.97	3.79	3.61	3.51	3.41	3.31	3.21	3.10	3.04	2.98	.005	
5.32	5.05	4.78	4.63	4.48	4.33	4.18	4.02	3.93	3.85	.001	
1.40	1.39	1.38	1.37	1.36	1.35	1.34	1.33	1.32	1.32	.25	18
1.93	1.89	1.84	1.81	1.78	1.75	1.72	1.69	1.67	1.66	.10	
2.34	2.27	2.19	2.15	2.11	2.06	2.02	1.97	1.94	1.92	.05	
2.77	2.67	2.56	2.50	2.44	2.38	2.32	2.26	2.22	2.19	.025	
3.37	3.23	3.08	3.00	2.92	2.84	2.75	2.66	2.61	2.57	.01	
3.86	3.68	3.50	3.40	3.30	3.20	3.10	2.99	2.93	2.87	.005	
5.13	4.87	4.59	4.45	4.30	4.15	4.00	3.84	3.75	3.67	.001	

Table 6 cont'd

Percentage points of the F-distribution (df_2 between 19 and 24)

df_2	a	1	2	3	4	5	6	7	8	9	10
		df_1									
19	.25	1.41	1.49	1.49	1.47	1.46	1.44	1.43	1.42	1.41	1.41
	.10	2.99	2.61	2.40	2.27	2.18	2.11	2.06	2.02	1.98	1.96
	.05	4.38	3.52	3.13	2.90	2.74	2.63	2.54	2.48	2.42	2.38
	.025	5.92	4.51	3.90	3.56	3.33	3.17	3.05	2.96	2.88	2.82
	.01	8.18	5.93	5.01	4.50	4.17	3.94	3.77	3.63	3.52	3.43
	.005	10.07	7.09	5.92	5.27	4.85	4.56	4.34	4.18	4.04	3.93
	.001	15.08	10.16	8.28	7.27	6.62	6.18	5.85	5.59	5.39	5.22
20	.25	1.40	1.49	1.48	1.47	1.45	1.44	1.43	1.42	1.41	1.40
	.10	2.97	2.59	2.38	2.25	2.16	2.09	2.04	2.00	1.96	1.94
	.05	4.35	3.49	3.10	2.87	2.71	2.60	2.51	2.45	2.39	2.35
	.025	5.87	4.46	3.86	3.51	3.29	3.13	3.01	2.91	2.84	2.77
	.01	8.10	5.85	4.94	4.43	4.10	3.87	3.70	3.56	3.46	3.37
	.005	9.94	6.99	5.82	5.17	4.76	4.47	4.26	4.09	3.96	3.85
	.001	14.82	9.95	8.10	7.10	6.46	6.02	5.69	5.44	5.24	5.08
21	.25	1.40	1.48	1.48	1.46	1.44	1.43	1.42	1.41	1.40	1.39
	.10	2.96	2.57	2.36	2.23	2.14	2.08	2.02	1.98	1.95	1.92
	.05	4.32	3.47	3.07	2.84	2.68	2.57	2.49	2.42	2.37	2.32
	.025	5.83	4.42	3.82	3.48	3.25	3.09	2.97	2.87	2.80	2.73
	.01	8.02	5.78	4.87	4.37	4.04	3.81	3.64	3.51	3.40	3.31
	.005	9.83	6.89	5.73	5.09	4.68	4.39	4.18	4.01	3.88	3.77
	.001	14.59	9.77	7.94	6.95	6.32	5.88	5.56	5.31	5.11	4.95
22	.25	1.40	1.48	1.47	1.45	1.44	1.42	1.41	1.40	1.39	1.39
	.10	2.95	2.56	2.35	2.22	2.13	2.06	2.01	1.97	1.93	1.90
	.05	4.30	3.44	3.05	2.82	2.66	2.55	2.46	2.40	2.34	2.30
	.025	5.79	4.38	3.78	3.44	3.22	3.05	2.93	2.84	2.76	2.70
	.01	7.95	5.72	4.82	4.31	3.99	3.76	3.59	3.45	3.35	3.26
	.005	9.73	6.81	5.65	5.02	4.61	4.32	4.11	3.94	3.81	3.70
	.001	14.38	9.61	7.80	6.81	6.19	5.76	5.44	5.19	4.99	4.83
23	.25	1.39	1.47	1.47	1.45	1.43	1.42	1.41	1.40	1.39	1.38
	.10	2.94	2.55	2.34	2.21	2.11	2.05	1.99	1.95	1.92	1.89
	.05	4.28	3.42	3.03	2.80	2.64	2.53	2.44	2.37	2.32	2.27
	.025	5.75	4.35	3.75	3.41	3.18	3.02	2.90	2.81	2.73	2.67
	.01	7.88	5.66	4.76	4.26	3.94	3.71	3.54	3.41	3.30	3.21
	.005	9.63	6.73	5.58	4.95	4.54	4.26	4.05	3.88	3.75	3.64
	.001	14.20	9.47	7.67	6.70	6.08	5.65	5.33	5.09	4.89	4.73
24	.25	1.39	1.47	1.46	1.44	1.43	1.41	1.40	1.39	1.38	1.38
	.10	2.93	2.54	2.33	2.19	2.10	2.04	1.98	1.94	1.91	1.88
	.05	4.26	3.40	3.01	2.78	2.62	2.51	2.42	2.36	2.30	2.25
	.025	5.72	4.32	3.72	3.38	3.15	2.99	2.87	2.78	2.70	2.64
	.01	7.82	5.61	4.72	4.22	3.90	3.67	3.50	3.36	3.26	3.17
	.005	9.55	6.66	5.52	4.89	4.49	4.20	3.99	3.83	3.69	3.59
	.001	14.03	9.34	7.55	6.59	5.98	5.55	5.23	4.99	4.80	4.64

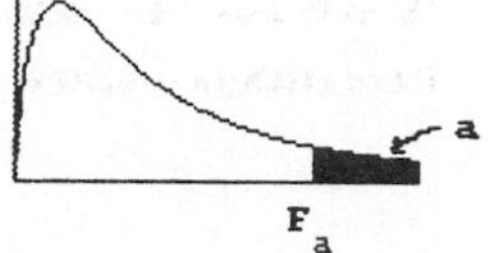

df_1											
12	15	20	24	30	40	60	120	240	inf.	a	df_2
1.40	1.38	1.37	1.36	1.35	1.34	1.33	1.32	1.31	1.30	.25	19
1.91	1.86	1.81	1.79	1.76	1.73	1.70	1.67	1.65	1.63	.10	
2.31	2.23	2.16	2.11	2.07	2.03	1.98	1.93	1.90	1.88	.05	
2.72	2.62	2.51	2.45	2.39	2.33	2.27	2.20	2.17	2.13	.025	
3.30	3.15	3.00	2.92	2.84	2.76	2.67	2.58	2.54	2.49	.01	
3.76	3.59	3.40	3.31	3.21	3.11	3.00	2.89	2.83	2.78	.005	
4.97	4.70	4.43	4.29	4.14	3.99	3.84	3.68	3.60	3.51	.001	
1.39	1.37	1.36	1.35	1.34	1.33	1.32	1.31	1.30	1.29	.25	20
1.89	1.84	1.79	1.77	1.74	1.71	1.68	1.64	1.63	1.61	.10	
2.28	2.20	2.12	2.08	2.04	1.99	1.95	1.90	1.87	1.84	.05	
2.68	2.57	2.46	2.41	2.35	2.29	2.22	2.16	2.12	2.09	.025	
3.23	3.09	2.94	2.86	2.78	2.69	2.61	2.52	2.47	2.42	.01	
3.68	3.50	3.32	3.22	3.12	3.02	2.92	2.81	2.75	2.69	.005	
4.82	4.56	4.29	4.15	4.00	3.86	3.70	3.54	3.46	3.38	.001	
1.38	1.37	1.35	1.34	1.33	1.32	1.31	1.30	1.29	1.28	.25	21
1.87	1.83	1.78	1.75	1.72	1.69	1.66	1.62	1.60	1.59	.10	
2.25	2.18	2.10	2.05	2.01	1.96	1.92	1.87	1.84	1.81	.05	
2.64	2.53	2.42	2.37	2.31	2.25	2.18	2.11	2.08	2.04	.025	
3.17	3.03	2.88	2.80	2.72	2.64	2.55	2.46	2.41	2.36	.01	
3.60	3.43	3.24	3.15	3.05	2.95	2.84	2.73	2.67	2.61	.005	
4.70	4.44	4.17	4.03	3.88	3.74	3.58	3.42	3.34	3.26	.001	
1.37	1.36	1.34	1.33	1.32	1.31	1.30	1.29	1.28	1.28	.25	22
1.86	1.81	1.76	1.73	1.70	1.67	1.64	1.60	1.59	1.57	.10	
2.23	2.15	2.07	2.03	1.98	1.94	1.89	1.84	1.81	1.78	.05	
2.60	2.50	2.39	2.33	2.27	2.21	2.14	2.08	2.04	2.00	.025	
3.12	2.98	2.83	2.75	2.67	2.58	2.50	2.40	2.35	2.31	.01	
3.54	3.36	3.18	3.08	2.98	2.88	2.77	2.66	2.60	2.55	.005	
4.58	4.33	4.06	3.92	3.78	3.63	3.48	3.32	3.23	3.15	.001	
1.37	1.35	1.34	1.33	1.32	1.31	1.30	1.28	1.28	1.27	.25	23
1.84	1.80	1.74	1.72	1.69	1.66	1.62	1.59	1.57	1.55	.10	
2.20	2.13	2.05	2.01	1.96	1.91	1.86	1.81	1.79	1.76	.05	
2.57	2.47	2.36	2.30	2.24	2.18	2.11	2.04	2.01	1.97	.025	
3.07	2.93	2.78	2.70	2.62	2.54	2.45	2.35	2.31	2.26	.01	
3.47	3.30	3.12	3.02	2.92	2.82	2.71	2.60	2.54	2.48	.005	
4.48	4.23	3.96	3.82	3.68	3.53	3.38	3.22	3.14	3.05	.001	
1.36	1.35	1.33	1.32	1.31	1.30	1.29	1.28	1.27	1.26	.25	24
1.83	1.78	1.73	1.70	1.67	1.64	1.61	1.57	1.55	1.53	.10	
2.18	2.11	2.03	1.98	1.94	1.89	1.84	1.79	1.76	1.73	.05	
2.54	2.44	2.33	2.27	2.21	2.15	2.08	2.01	1.97	1.94	.025	
3.03	2.89	2.74	2.66	2.58	2.49	2.40	2.31	2.26	2.21	.01	
3.42	3.25	3.06	2.97	2.87	2.77	2.66	2.55	2.49	2.43	.005	
4.39	4.14	3.87	3.74	3.59	3.45	3.29	3.14	3.05	2.97	.001	

Table 6 cont'd

Percentage points of the F-distribution (df_2 between 25 and 30)

		df_1									
df_2	a	1	2	3	4	5	6	7	8	9	10
25	.25	1.39	1.47	1.46	1.44	1.42	1.41	1.40	1.39	1.38	1.37
	.10	2.92	2.53	2.32	2.18	2.09	2.02	1.97	1.93	1.89	1.87
	.05	4.24	3.39	2.99	2.76	2.60	2.49	2.40	2.34	2.28	2.24
	.025	5.69	4.29	3.69	3.35	3.13	2.97	2.85	2.75	2.68	2.61
	.01	7.77	5.57	4.68	4.18	3.85	3.63	3.46	3.32	3.22	3.13
	.005	9.48	6.60	5.46	4.84	4.43	4.15	3.94	3.78	3.64	3.54
	.001	13.88	9.22	7.45	6.49	5.89	5.46	5.15	4.91	4.71	4.56
26	.25	1.38	1.46	1.45	1.44	1.42	1.41	1.39	1.38	1.37	1.37
	.10	2.91	2.52	2.31	2.17	2.08	2.01	1.96	1.92	1.88	1.86
	.05	4.23	3.37	2.98	2.74	2.59	2.47	2.39	2.32	2.27	2.22
	.025	5.66	4.27	3.67	3.33	3.10	2.94	2.82	2.73	2.65	2.59
	.01	7.72	5.53	4.64	4.14	3.82	3.59	3.42	3.29	3.18	3.09
	.005	9.41	6.54	5.41	4.79	4.38	4.10	3.89	3.73	3.60	3.49
	.001	13.74	9.12	7.36	6.41	5.80	5.38	5.07	4.83	4.64	4.48
27	.25	1.38	1.46	1.45	1.43	1.42	1.40	1.39	1.38	1.37	1.36
	.10	2.90	2.51	2.30	2.17	2.07	2.00	1.95	1.91	1.87	1.85
	.05	4.21	3.35	2.96	2.73	2.57	2.46	2.37	2.31	2.25	2.20
	.025	5.63	4.24	3.65	3.31	3.08	2.92	2.80	2.71	2.63	2.57
	.01	7.68	5.49	4.60	4.11	3.78	3.56	3.39	3.26	3.15	3.06
	.005	9.34	6.49	5.36	4.74	4.34	4.06	3.85	3.69	3.56	3.45
	.001	13.61	9.02	7.27	6.33	5.73	5.31	5.00	4.76	4.57	4.41
28	.25	1.38	1.46	1.45	1.43	1.41	1.40	1.39	1.38	1.37	1.36
	.10	2.89	2.50	2.29	2.16	2.06	2.00	1.94	1.90	1.87	1.84
	.05	4.20	3.34	2.95	2.71	2.56	2.45	2.36	2.29	2.24	2.19
	.025	5.61	4.22	3.63	3.29	3.06	2.90	2.78	2.69	2.61	2.55
	.01	7.64	5.45	4.57	4.07	3.75	3.53	3.36	3.23	3.12	3.03
	.005	9.28	6.44	5.32	4.70	4.30	4.02	3.81	3.65	3.52	3.41
	.001	13.50	8.93	7.19	6.25	5.66	5.24	4.93	4.69	4.50	4.35
29	.25	1.38	1.45	1.45	1.43	1.41	1.40	1.38	1.37	1.36	1.35
	.10	2.89	2.50	2.28	2.15	2.06	1.99	1.93	1.89	1.86	1.83
	.05	4.18	3.33	2.93	2.70	2.55	2.43	2.35	2.28	2.22	2.18
	.025	5.59	4.20	3.61	3.27	3.04	2.88	2.76	2.67	2.59	2.53
	.01	7.60	5.42	4.54	4.04	3.73	3.50	3.33	3.20	3.09	3.00
	.005	9.23	6.40	5.28	4.66	4.26	3.98	3.77	3.61	3.48	3.38
	.001	13.39	8.85	7.12	6.19	5.59	5.18	4.87	4.64	4.45	4.29
30	.25	1.38	1.45	1.44	1.42	1.41	1.39	1.38	1.37	1.36	1.35
	.10	2.88	2.49	2.28	2.14	2.05	1.98	1.93	1.88	1.85	1.82
	.05	4.17	3.32	2.92	2.69	2.53	2.42	2.33	2.27	2.21	2.16
	.025	5.57	4.18	3.59	3.25	3.03	2.87	2.75	2.65	2.57	2.51
	.01	7.56	5.39	4.51	4.02	3.70	3.47	3.30	3.17	3.07	2.98
	.005	9.18	6.35	5.24	4.62	4.23	3.95	3.74	3.58	3.45	3.34
	.001	13.29	8.77	7.05	6.12	5.53	5.12	4.82	4.58	4.39	4.24

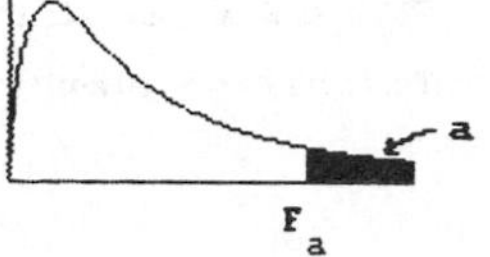

df_1: 12	15	20	24	30	40	60	120	240	inf.	a	df_2
1.36	1.34	1.33	1.32	1.31	1.29	1.28	1.27	1.26	1.25	.25	25
1.82	1.77	1.72	1.69	1.66	1.63	1.59	1.56	1.54	1.52	.10	
2.16	2.09	2.01	1.96	1.92	1.87	1.82	1.77	1.74	1.71	.05	
2.51	2.41	2.30	2.24	2.18	2.12	2.05	1.98	1.94	1.91	.025	
2.99	2.85	2.70	2.62	2.54	2.45	2.36	2.27	2.22	2.17	.01	
3.37	3.20	3.01	2.92	2.82	2.72	2.61	2.50	2.44	2.38	.005	
4.31	4.06	3.79	3.66	3.52	3.37	3.22	3.06	2.98	2.89	.001	
1.35	1.34	1.32	1.31	1.30	1.29	1.28	1.26	1.26	1.25	.25	26
1.81	1.76	1.71	1.68	1.65	1.61	1.58	1.54	1.52	1.50	.10	
2.15	2.07	1.99	1.95	1.90	1.85	1.80	1.75	1.72	1.69	.05	
2.49	2.39	2.28	2.22	2.16	2.09	2.03	1.95	1.92	1.88	.025	
2.96	2.81	2.66	2.58	2.50	2.42	2.33	2.23	2.18	2.13	.01	
3.33	3.15	2.97	2.87	2.77	2.67	2.56	2.45	2.39	2.33	.005	
4.24	3.99	3.72	3.59	3.44	3.30	3.15	2.99	2.90	2.82	.001	
1.35	1.33	1.32	1.31	1.30	1.28	1.27	1.26	1.25	1.24	.25	27
1.80	1.75	1.70	1.67	1.64	1.60	1.57	1.53	1.51	1.49	.10	
2.13	2.06	1.97	1.93	1.88	1.84	1.79	1.73	1.70	1.67	.05	
2.47	2.36	2.25	2.19	2.13	2.07	2.00	1.93	1.89	1.85	.025	
2.93	2.78	2.63	2.55	2.47	2.38	2.29	2.20	2.15	2.10	.01	
3.28	3.11	2.93	2.83	2.73	2.63	2.52	2.41	2.35	2.29	.005	
4.17	3.92	3.66	3.52	3.38	3.23	3.08	2.92	2.84	2.75	.001	
1.34	1.33	1.31	1.30	1.29	1.28	1.27	1.25	1.24	1.24	.25	28
1.79	1.74	1.69	1.66	1.63	1.59	1.56	1.52	1.50	1.48	.10	
2.12	2.04	1.96	1.91	1.87	1.82	1.77	1.71	1.68	1.65	.05	
2.45	2.34	2.23	2.17	2.11	2.05	1.98	1.91	1.87	1.83	.025	
2.90	2.75	2.60	2.52	2.44	2.35	2.26	2.17	2.12	2.06	.01	
3.25	3.07	2.89	2.79	2.69	2.59	2.48	2.37	2.31	2.25	.005	
4.11	3.86	3.60	3.46	3.32	3.18	3.02	2.86	2.78	2.69	.001	
1.34	1.32	1.31	1.30	1.29	1.27	1.26	1.25	1.24	1.23	.25	29
1.78	1.73	1.68	1.65	1.62	1.58	1.55	1.51	1.49	1.47	.10	
2.10	2.03	1.94	1.90	1.85	1.81	1.75	1.70	1.67	1.64	.05	
2.43	2.32	2.21	2.15	2.09	2.03	1.96	1.89	1.85	1.81	.025	
2.87	2.73	2.57	2.49	2.41	2.33	2.23	2.14	2.09	2.03	.01	
3.21	3.04	2.86	2.76	2.66	2.56	2.45	2.33	2.27	2.21	.005	
4.05	3.80	3.54	3.41	3.27	3.12	2.97	2.81	2.73	2.64	.001	
1.34	1.32	1.30	1.29	1.28	1.27	1.26	1.24	1.23	1.23	.25	30
1.77	1.72	1.67	1.64	1.61	1.57	1.54	1.50	1.48	1.46	.10	
2.09	2.01	1.93	1.89	1.84	1.79	1.74	1.68	1.65	1.62	.05	
2.41	2.31	2.20	2.14	2.07	2.01	1.94	1.87	1.83	1.79	.025	
2.84	2.70	2.55	2.47	2.39	2.30	2.21	2.11	2.06	2.01	.01	
3.18	3.01	2.82	2.73	2.63	2.52	2.42	2.30	2.24	2.18	.005	
4.00	3.75	3.49	3.36	3.22	3.07	2.92	2.76	2.68	2.59	.001	

Table 6 cont'd

Percentage points of the F-distribution (df_2 at least 40)

df_2	a	df_1 1	2	3	4	5	6	7	8	9	10
40	.25	1.36	1.44	1.42	1.40	1.39	1.37	1.36	1.35	1.34	1.33
	.10	2.84	2.44	2.23	2.09	2.00	1.93	1.87	1.83	1.79	1.76
	.05	4.08	3.23	2.84	2.61	2.45	2.34	2.25	2.18	2.12	2.08
	.025	5.42	4.05	3.46	3.13	2.90	2.74	2.62	2.53	2.45	2.39
	.01	7.31	5.18	4.31	3.83	3.51	3.29	3.12	2.99	2.89	2.80
	.005	8.83	6.07	4.98	4.37	3.99	3.71	3.51	3.35	3.22	3.12
	.001	12.61	8.25	6.59	5.70	5.13	4.73	4.44	4.21	4.02	3.87
60	.25	1.35	1.42	1.41	1.38	1.37	1.35	1.33	1.32	1.31	1.30
	.10	2.79	2.39	2.18	2.04	1.95	1.87	1.82	1.77	1.74	1.71
	.05	4.00	3.15	2.76	2.53	2.37	2.25	2.17	2.10	2.04	1.99
	.025	5.29	3.93	3.34	3.01	2.79	2.63	2.51	2.41	2.33	2.27
	.01	7.08	4.98	4.13	3.65	3.34	3.12	2.95	2.82	2.72	2.63
	.005	8.49	5.79	4.73	4.14	3.76	3.49	3.29	3.13	3.01	2.90
	.001	11.97	7.77	6.17	5.31	4.76	4.37	4.09	3.86	3.69	3.54
90	.25	1.34	1.41	1.39	1.37	1.35	1.33	1.32	1.31	1.30	1.29
	.10	2.76	2.36	2.15	2.01	1.91	1.84	1.78	1.74	1.70	1.67
	.05	3.95	3.10	2.71	2.47	2.32	2.20	2.11	2.04	1.99	1.94
	.025	5.20	3.84	3.26	2.93	2.71	2.55	2.43	2.34	2.26	2.19
	.01	6.93	4.85	4.01	3.53	3.23	3.01	2.84	2.72	2.61	2.52
	.005	8.28	5.62	4.57	3.99	3.62	3.35	3.15	3.00	2.87	2.77
	.001	11.57	7.47	5.91	5.06	4.53	4.15	3.87	3.65	3.48	3.34
120	.25	1.34	1.40	1.39	1.37	1.35	1.33	1.31	1.30	1.29	1.28
	.10	2.75	2.35	2.13	1.99	1.90	1.82	1.77	1.72	1.68	1.65
	.05	3.92	3.07	2.68	2.45	2.29	2.18	2.09	2.02	1.96	1.91
	.025	5.15	3.80	3.23	2.89	2.67	2.52	2.39	2.30	2.22	2.16
	.01	6.85	4.79	3.95	3.48	3.17	2.96	2.79	2.66	2.56	2.47
	.005	8.18	5.54	4.50	3.92	3.55	3.28	3.09	2.93	2.81	2.71
	.001	11.38	7.32	5.78	4.95	4.42	4.04	3.77	3.55	3.38	3.24
240	.25	1.33	1.39	1.38	1.36	1.34	1.32	1.30	1.29	1.27	1.27
	.10	2.73	2.32	2.10	1.97	1.87	1.80	1.74	1.70	1.65	1.63
	.05	3.88	3.03	2.64	2.41	2.25	2.14	2.04	1.98	1.92	1.87
	.025	5.09	3.75	3.17	2.84	2.62	2.46	2.34	2.25	2.17	2.10
	.01	6.74	4.69	3.86	3.40	3.09	2.88	2.71	2.59	2.48	2.40
	.005	8.03	5.42	4.38	3.82	3.45	3.19	2.99	2.84	2.71	2.61
	.001	11.10	7.11	5.60	4.78	4.25	3.89	3.62	3.41	3.24	3.09
inf.	.25	1.32	1.39	1.37	1.35	1.33	1.31	1.29	1.28	1.27	1.25
	.10	2.71	2.30	2.08	1.94	1.85	1.77	1.72	1.67	1.63	1.60
	.05	3.84	3.00	2.60	2.37	2.21	2.10	2.01	1.94	1.88	1.83
	.025	5.02	3.69	3.12	2.79	2.57	2.41	2.29	2.19	2.11	2.05
	.01	6.63	4.61	3.78	3.32	3.02	2.80	2.64	2.51	2.41	2.32
	.005	7.88	5.30	4.28	3.72	3.35	3.09	2.90	2.74	2.62	2.52
	.001	10.83	6.91	5.42	4.62	4.10	3.74	3.47	3.27	3.10	2.96

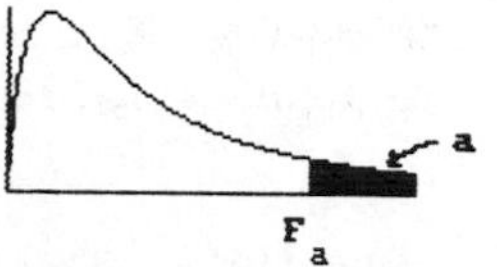

df$_1$											
12	**15**	**20**	**24**	**30**	**40**	**60**	**120**	**240**	**inf.**	**a**	**df$_2$**
1.31	1.30	1.28	1.26	1.25	1.24	1.22	1.21	1.20	1.19	.25	**40**
1.71	1.66	1.61	1.57	1.54	1.51	1.47	1.42	1.40	1.38	.10	
2.00	1.92	1.84	1.79	1.74	1.69	1.64	1.58	1.54	1.51	.05	
2.29	2.18	2.07	2.01	1.94	1.88	1.80	1.72	1.68	1.64	.025	
2.66	2.52	2.37	2.29	2.20	2.11	2.02	1.92	1.86	1.80	.01	
2.95	2.78	2.60	2.50	2.40	2.30	2.18	2.06	2.00	1.93	.005	
3.64	3.40	3.14	3.01	2.87	2.73	2.57	2.41	2.32	2.23	.001	
1.29	1.27	1.25	1.24	1.22	1.21	1.19	1.17	1.16	1.15	.25	**60**
1.66	1.60	1.54	1.51	1.48	1.44	1.40	1.35	1.32	1.29	.10	
1.92	1.84	1.75	1.70	1.65	1.59	1.53	1.47	1.43	1.39	.05	
2.17	2.06	1.94	1.88	1.82	1.74	1.67	1.58	1.53	1.48	.025	
2.50	2.35	2.20	2.12	2.03	1.94	1.84	1.73	1.67	1.60	.01	
2.74	2.57	2.39	2.29	2.19	2.08	1.96	1.83	1.76	1.69	.005	
3.32	3.08	2.83	2.69	2.55	2.41	2.25	2.08	1.99	1.89	.001	
1.27	1.25	1.23	1.22	1.20	1.19	1.17	1.15	1.13	1.12	.25	**90**
1.62	1.56	1.50	1.47	1.43	1.39	1.35	1.29	1.26	1.23	.10	
1.86	1.78	1.69	1.64	1.59	1.53	1.46	1.39	1.35	1.30	.05	
2.09	1.98	1.86	1.80	1.73	1.66	1.58	1.48	1.43	1.37	.025	
2.39	2.24	2.09	2.00	1.92	1.82	1.72	1.60	1.53	1.46	.01	
2.61	2.44	2.25	2.15	2.05	1.94	1.82	1.68	1.61	1.52	.005	
3.11	2.88	2.63	2.50	2.36	2.21	2.05	1.87	1.77	1.66	.001	
1.26	1.24	1.22	1.21	1.19	1.18	1.16	1.13	1.12	1.10	.25	**120**
1.60	1.55	1.48	1.45	1.41	1.37	1.32	1.26	1.23	1.19	.10	
1.83	1.75	1.66	1.61	1.55	1.50	1.43	1.35	1.31	1.25	.05	
2.05	1.94	1.82	1.76	1.69	1.61	1.53	1.43	1.38	1.31	.025	
2.34	2.19	2.03	1.95	1.86	1.76	1.66	1.53	1.46	1.38	.01	
2.54	2.37	2.19	2.09	1.98	1.87	1.75	1.61	1.52	1.43	.005	
3.02	2.78	2.53	2.40	2.26	2.11	1.95	1.77	1.66	1.54	.001	
1.25	1.23	1.21	1.19	1.18	1.16	1.14	1.11	1.09	1.07	.25	**240**
1.57	1.52	1.45	1.42	1.38	1.33	1.28	1.22	1.18	1.13	.10	
1.79	1.71	1.61	1.56	1.51	1.44	1.37	1.29	1.24	1.17	.05	
2.00	1.89	1.77	1.70	1.63	1.55	1.46	1.35	1.29	1.21	.025	
2.26	2.11	1.96	1.87	1.78	1.68	1.57	1.43	1.35	1.25	.01	
2.45	2.28	2.09	1.99	1.89	1.77	1.64	1.49	1.40	1.28	.005	
2.88	2.65	2.40	2.26	2.12	1.97	1.80	1.61	1.49	1.35	.001	
1.24	1.22	1.19	1.18	1.16	1.14	1.12	1.08	1.06	1.00	.25	**inf.**
1.55	1.49	1.42	1.38	1.34	1.30	1.24	1.17	1.12	1.00	.10	
1.75	1.67	1.57	1.52	1.46	1.39	1.32	1.22	1.15	1.00	.05	
1.94	1.83	1.71	1.64	1.57	1.48	1.39	1.27	1.19	1.00	.025	
2.18	2.04	1.88	1.79	1.70	1.59	1.47	1.32	1.22	1.00	.01	
2.36	2.19	2.00	1.90	1.79	1.67	1.53	1.36	1.25	1.00	.005	
2.74	2.51	2.27	2.13	1.99	1.84	1.66	1.45	1.31	1.00	.001	

Source: Computed by P. J. Hildebrand.

Table 7

Critical values for the Wilcoxon signed rank test (n=5(1)54)

One-sided	Two-sided	n=5	n=6	n=7	n=8	n=9	n=10	n=11	n=12	n=13	n=14
p=.1	p=.2	2	3	5	8	10	14	17	21	26	31
p=.05	p=.1	0	2	3	5	8	10	13	17	21	25
p=.025	p=.05		0	2	3	5	8	10	13	17	21
p=.01	p=.02			0	1	3	5	7	9	12	15
p=.005	p=.01				0	1	3	5	7	9	12
p=.0025	p=.005					0	1	3	5	7	9
p=.001	p=.002						0	1	2	4	6

One-sided	Two-sided	n=15	n=16	n=17	n=18	n=19	n=20	n=21	n=22	n=23	n=24
p=.1	p=.2	36	42	48	55	62	69	77	86	94	104
p=.05	p=.1	30	35	41	47	53	60	67	75	83	91
p=.025	p=.05	25	29	34	40	46	52	58	65	73	81
p=.01	p=.02	19	23	27	32	37	43	49	55	62	69
p=.005	p=.01	15	19	23	27	32	37	42	48	54	61
p=.0025	p=.005	12	15	19	23	27	32	37	42	48	54
p=.001	p-.002	8	11	14	18	21	26	30	35	40	45

One-sided	Two-sided	n=25	n=26	n=27	n=28	n=29	n=30	n=31	n=32	n=33	n=34
p=.1	p=.2	113	124	134	145	157	169	181	194	207	221
p=.05	p=.1	100	110	119	130	140	151	163	175	187	200
p=.025	p=.05	89	98	107	116	126	137	147	159	170	182
p=.01	p=.02	76	84	92	101	110	120	130	140	151	162
p=.005	p=.01	68	75	83	91	100	109	118	128	138	148
p=.0025	p=.005	60	67	74	82	90	98	107	116	126	136
p=.001	p=.002	51	58	64	71	79	86	94	103	112	121

One-sided	Two-sided	n=35	n=36	n=37	n=38	n=39	n=40	n=41	n=42	n=43	n=44
p=.1	p=.2	235	250	265	281	297	313	330	348	365	384
p=.05	p=.1	213	227	241	256	271	286	302	319	336	353
p=.025	p=.05	195	208	221	235	249	264	279	294	310	327
p=.01	p=.02	173	185	198	211	224	238	252	266	281	296
p=.005	p=.01	159	171	182	194	207	220	233	247	261	276
p=.0025	p=.005	146	157	168	180	192	204	217	230	244	258
p=.001	p=.002	131	141	151	162	173	185	197	209	222	235

One-sided	Two-sided	n=45	n=46	n=47	n=48	n=49	n=50	n=51	n=52	n=53	n=54
p=.1	p=.2	402	422	441	462	482	503	525	547	569	592
p=.05	p=.1	371	389	407	426	446	466	486	507	529	550
p=.025	p=.05	343	361	378	396	415	434	453	473	494	514
p=.01	p=.02	312	328	345	362	379	397	416	434	454	473
p=.005	p=.01	291	307	322	339	355	373	390	408	427	445
p=.0025	p=.005	272	287	302	318	334	350	367	384	402	420
p=.001	p=.002	249	263	277	292	307	323	339	355	372	389

Source: Computed by P. J. Hildebrand.

Table 8

Percentage points of the studentized range statistic

Error df	alpha	t = number of treatment means: 2	3	4	5	6	7	8	9	10	11
5	.05	3.64	4.60	5.22	5.67	6.03	6.33	6.58	6.80	6.99	7.17
	.01	5.70	6.98	7.80	8.42	8.91	9.32	9.67	9.97	10.24	10.48
6	.05	3.46	4.34	4.90	5.30	5.63	5.90	6.12	6.32	6.49	6.65
	.01	5.24	6.33	7.03	7.56	7.97	8.32	8.61	8.87	9.10	9.30
7	.05	3.34	4.16	4.68	5.06	5.36	5.61	5.82	6.00	6.16	6.30
	.01	4.95	5.92	6.54	7.01	7.37	7.68	7.94	8.17	8.37	8.55
8	.05	3.26	4.04	4.53	4.89	5.17	5.40	5.60	5.77	5.92	6.05
	.01	4.75	5.64	6.20	6.62	6.96	7.24	7.47	7.68	7.86	8.03
9	.05	3.20	3.95	4.41	4.76	5.02	5.24	5.43	5.59	5.74	5.87
	.01	4.60	5.43	5.96	6.35	6.66	6.91	7.13	7.33	7.49	7.65
10	.05	3.15	3.88	4.33	4.65	4.91	5.12	5.20	5.35	5.49	5.61
	.01	4.48	5.27	5.77	6.14	6.43	6.67	6.87	7.05	7.21	7.36
11	.05	3.11	3.82	4.26	4.57	4.82	5.03	5.30	5.35	5.49	5.61
	.01	4.39	5.15	5.62	5.97	6.25	6.48	6.67	6.84	6.99	7.13
12	.05	3.08	3.77	4.20	4.52	4.75	4.95	5.12	5.27	5.39	5.51
	.01	4.32	5.05	5.50	5.84	6.10	6.32	6.51	6.67	6.81	6.94
13	.05	3.06	3.73	4.15	4.45	4.69	4.88	5.05	5.19	5.32	5.43
	.01	4.26	4.96	5.40	5.73	5.98	6.19	6.37	6.53	6.67	6.79
14	.05	3.03	3.70	4.11	4.41	4.64	4.83	4.99	5.13	5.25	5.36
	.01	4.21	4.89	5.32	5.63	5.88	6.08	6.26	6.41	6.54	6.66
15	.05	3.01	3.67	4.08	4.37	4.59	4.78	4.94	5.08	5.20	5.31
	.01	4.17	4.84	5.25	5.56	5.80	5.99	6.16	6.31	6.44	6.55
16	.05	3.00	3.65	4.05	4.33	4.56	4.74	4.90	5.03	5.15	5.26
	.01	4.13	4.79	5.19	5.49	5.72	5.92	6.08	6.22	6.35	6.46
17	.05	2.98	3.63	4.02	4.30	4.52	4.70	4.86	4.99	5.11	5.21
	.01	4.10	4.74	5.14	5.43	5.66	5.85	6.01	6.15	6.27	6.38
18	.05	2.97	3.61	4.00	4.28	4.49	4.67	4.82	4.96	5.07	5.17
	.01	4.07	4.70	5.09	5.38	5.60	5.79	5.94	6.08	6.20	6.31
19	.05	2.96	3.59	3.98	4.25	4.47	4.65	4.79	4.92	5.04	5.14
	.01	4.05	4.67	5.05	5.33	5.55	5.73	5.89	6.02	6.14	6.25
20	.05	2.95	3.58	3.96	4.23	4.45	4.62	4.77	4.90	5.01	5.11
	.01	4.02	4.64	5.02	5.29	5.51	5.69	5.84	5.97	6.09	6.19
24	.05	2.92	3.53	3.90	4.17	4.37	4.54	4.68	4.81	4.92	5.01
	.01	3.96	4.55	4.91	5.17	5.37	5.54	5.69	5.81	5.92	6.02
30	.05	2.89	3.49	3.85	4.10	4.30	4.46	4.60	4.72	4.82	4.92
	.01	3.89	4.45	4.80	5.05	5.24	5.40	5.54	5.65	5.76	5.85
40	.05	2.86	3.44	3.79	4.04	4.23	4.39	4.52	4.63	4.73	4.82
	.01	3.82	4.37	4.70	4.93	5.11	5.26	5.39	5.50	5.60	5.69
60	.05	2.03	3.40	3.74	3.98	4.16	4.31	4.44	4.55	4.65	4.73
	.01	3.76	4.28	4.59	4.82	4.99	5.13	5.25	5.36	5.45	5.53
120	.05	2.80	3.36	3.68	3.92	4.10	4.24	4.36	4.47	4.56	4.64
	.01	3.70	4.20	4.50	4.71	4.87	5.01	5.12	5.21	5.30	5.37
inf.	.05	2.77	3.31	3.63	3.86	4.03	4.17	4.29	4.39	4.47	4.55
	.01	3.64	4.12	4.40	4.60	4.76	4.88	4.99	5.08	5.16	5.23

Source: Abridged from Table 29, Biometrika Tables for Statisticians, Vol. 1, 2nd ed. New York: Cambridge, 1958. Edited by E. S. Pearson and H. O. Hartley.

Table 9

Lower percentage points of the correlation between data and normal scores

n	alpha .10	.05	.01
4	.8951	.8734	.8318
5	.9033	.8804	.8320
10	.9347	.9180	.8804
15	.9506	.9383	.9110
20	.9600	.9503	.9290
25	.9662	.9582	.9408
30	.9707	.9639	.9490
40	.9767	.9715	.9597
50	.9807	.9764	.9664
60	.9835	.9779	.9710
75	.9865	.9835	.9757

Source: Ryan, T., Joiner, B., and Ryan, B., Minitab Reference Manual, Pennsylvania State University.

Answers to Exercises

Chapter 2

2.1 **(a)** Extroverts: 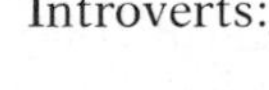 Introverts:

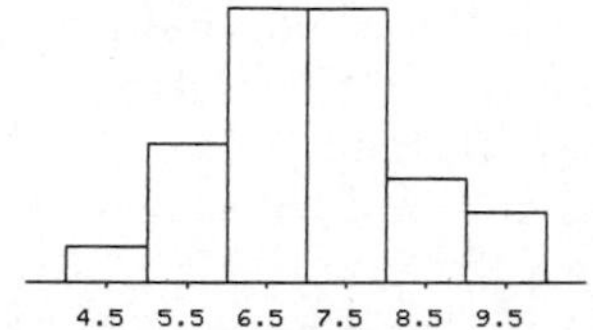

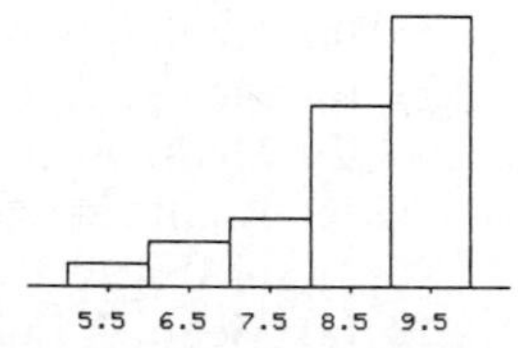

(b) For extroverts, the data are symmetric and unimodal. For introverts, the data are left-skewed. For extroverts, a typical score would be about 7.0; for introverts, it would be about 8.5.

(c) Extroverts:

Stem	Leaves
4	7
5	2 3 5 8
6	0 1 1 4 6 6 7 9
7	0 0 3 4 4 4 6 9
8	2 4 8
9	1 8

Introverts:

Stem	Leaves
5	4
6	7 9
7	0 1 6
8	0 2 2 5 7 9 9 9
9	0 0 0 1 2 3 3 5 5 6 7 8

The stem and leaf display is merely a "sideways histogram."

2.2 **(a)**

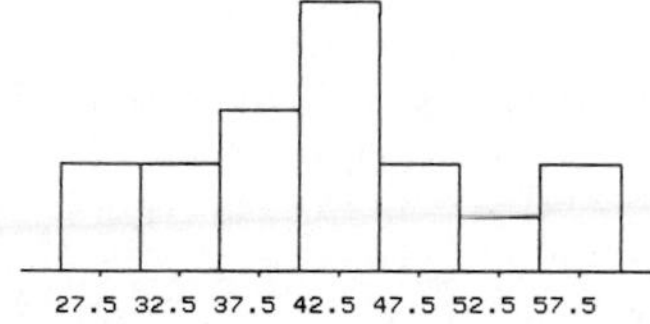

(b)

Stem	Leaves
2	5 8
3	1 3 5 8 9
4	0 1 2 2 3 5 8
5	3 8 9

(c)

Stem	Leaves
2	5 8
3	1 3
3	5 8 9
4	0 1 2 2 3
4	5 8
5	3
5	8 9

Data are unimodal and nearly symmetric.

2.3 **(a)**

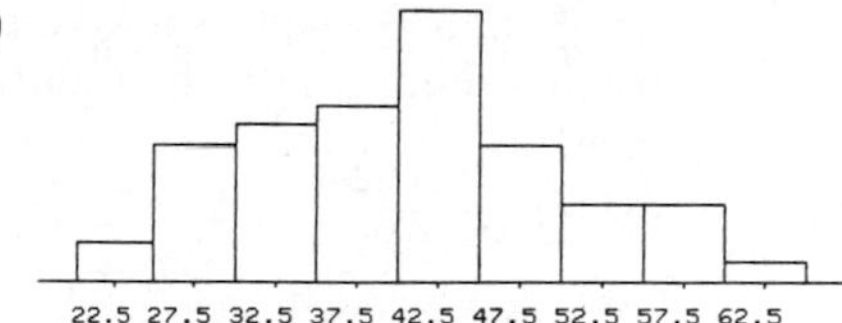

One possible stem and leaf display is

Stem	Leaves
2	0 3
2	5 5 6 7 8 9 9
3	0 0 0 1 1 2 4 4
3	5 5 7 7 8 8 9 9 9
4	0 0 0 0 0 0 0 0 1 1 2 4 4 4
4	5 7 7 8 8 8 9
5	0 0 1 3
5	5 7 7 9
6	2

(b) Both sets of data are roughly symmetric and unimodal.

(c) We can't tell.

2.4 **(a)** Grouped in classes 0.00–0.49, 0.50–0.99, etc., the data appear multimodal.

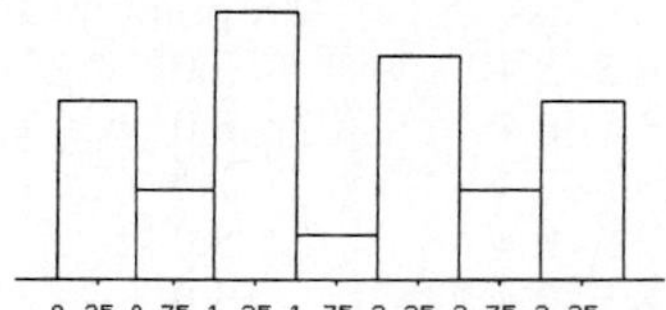

(b) Type A:

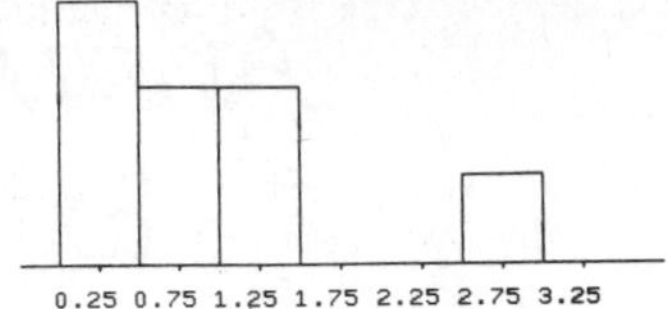

Type B:

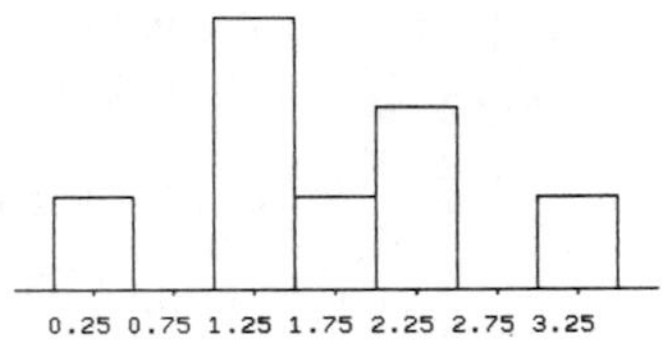

Type C:

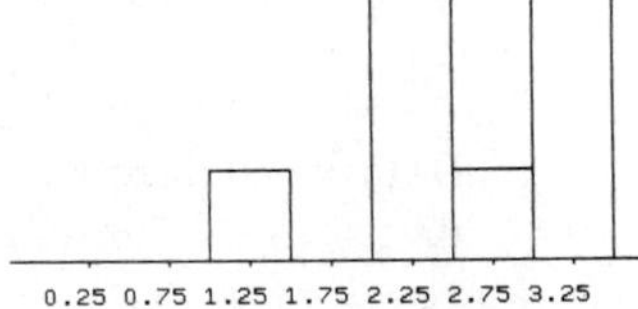

The A data are right-skewed, the B data are symmetric, and the C data are left-skewed.

(c) Type C

2.5 (a)

```
3 | 6 6
4 | 6 3 4 3 9 4 2 1 6 2 0 9
5 | 1 4 1 5 6 7 3 4
6 | 4 8 7 2 8
7 | 6 2 0
```

The data are slightly right-skewed.

(b)

These data are also slightly right-skewed.

2.6 (a) 6.95 (extrovert), 8.9 (introvert); introverts are higher by 1.95.
(b) 6.97 (extrovert), 8.50 (introvert); introverts are higher by 1.53.
(c) The left-skewness of the introvert data

2.7 (a)

Class	Midpoint	Frequency
4.5–5.4	5.0	3
5.5–6.4	6.0	6
6.5–7.4	7.0	10
7.5–8.4	8.0	4
8.5–9.4	9.0	2
9.5–10.4	10.0	1

(b) The mean is approximately 6.96.
(c) The median is approximately 6.9.
(d) Mean = 7.04, median = 7.0, slightly higher but close

2.8 (a) Taking $L = 5.45$, 30th percentile = 6.25
(b) Taking $L = 5.45$, $p = .196$

2.9 (a) Good: mean = 41.18, median = 41; poor: mean = 39.70, median = 40
(b) We couldn't guess!

2.10 (a) Mean = 40.04
(b) Median = 40; results are not useful.

2.11 (a) Means: A, 0.834; B, 1.625; C, 2.416
(b) Medians: A, 0.67; B, 1.50; C, 2.50
(c) Right-skewness of A and left-skewness of C

2.12 (a) Mean = 1.625
(b) Mean = 1.625 again

2.13 (a) No; data are nearly symmetric.
(b) Mean = 52.67, median = 51.43 (using 20–29, 30–39, etc.)
(c) Mean = 52.60, median = 53.50

2.14 (a) 20th percentile = 39.5, 80th percentile = 57.0
(b) $p = .863$

2.15 (a) Good: 34; poor: 42
(b) Not necessarily
(c) Range increases with larger sample size.

2.16 (a) Good: 9.57; poor: 9.68
(b) Virtually equal

2.17 (a) Good: 10; poor: 15.5
(b) Inner fences: good, 20 and 60; poor, 8.25 and 70.25; no outliers
(c)

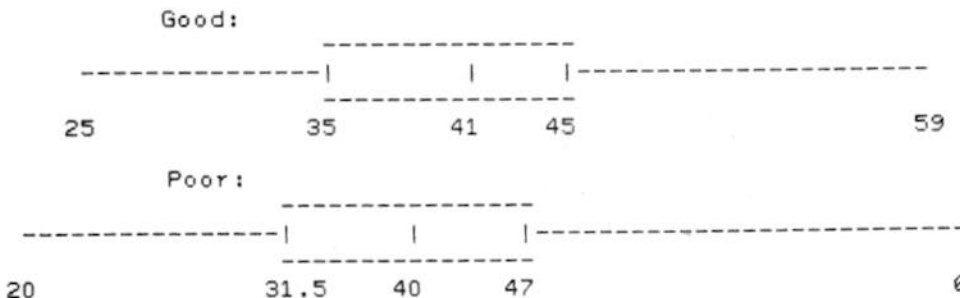

2.18 (a) Extroverts: 1.5; introverts, 1.3
(b) 5.4 (introverts)
(c)

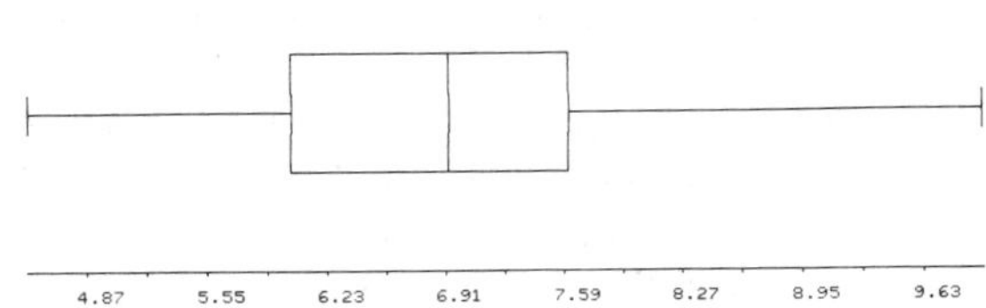

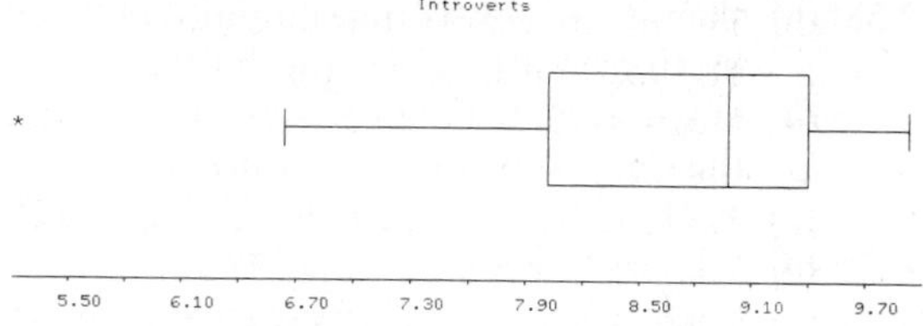

2.19 **(a)** Extroverts: $s = 1.25$; introverts: $s = 1.11$
(b) Extroverts: 69.2%; introverts: 73.1%. Extroverts sample has a better approximation to 68%.

2.20 Satisfaction: $\sigma^2 = 242.5733$ and $\sigma = 15.575$
Conventionalization: $\sigma^2 = 122.8989$ and $\sigma = 11.086$

2.21 **(a)** Use $\Sigma\, y_i^2 = 86{,}795$ and $\Sigma\, y_i = 1579$.
(b) .9667
(c) Quite close; histogram is nearly bell-shaped.
(d) 96.67% is well above 75%.
(e) All scores fall in this range, fitting the Empirical Rule better than Chebyshev's Inequality.

2.22 **(a)** Within 1 standard deviation, .625; within 2, 1.00
(b) No; the histogram isn't bell-shaped.
(c) Far above the Chebyshev bounds
(d) 87.5% for both 1 and 2 standard deviations, way off because the data aren't close to bell-shaped

2.23 **(a)** Trimodal data; the Empirical Rule won't work.

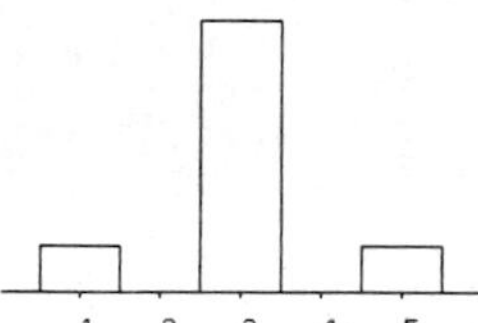

(b) .7503, not close
(c) Very close to .75

2.24 BMDP: MEAN 1.6245833, MEDIAN 1.5000000, ST.DEV. 1.0089425
Minitab: MEAN 1.62, MEDIAN 1.50, STDEV 1.01
SAS: MEAN 1.62458, 50% MED 1.5, STD DEV 1.00894
SPSS-X: MEAN 1.625, MEDIAN 1.500, STD DEV 1.009

2.25 **(a)** Introverts data are more skewed, to the left.
(b) Extroverts: $\text{skew}_1 = 0.015$, $\text{skew}_2 = 0.3157$, $\text{skew}_3 = 0.020$.
Introverts: $\text{skew}_1 = -0.362$, $\text{skew}_2 = -1.103$, $\text{skew}_3 = -0.500$. The introverts data is definitely more skewed.

2.26 **(a)** No; the histogram is roughly bell-shaped.
(b) $\text{skew}_1 = 0.0184$; the mean is only .0184 standard deviation above the median.
(c) $\text{skew}_3 = 0.0248$; someone betting "above the median" would win only 2.48 cents per dollar.

2.27 **(a)** $\text{skew}_1 = 0.1526$, $\text{skew}_3 = 0.2671$; moderately affected.
(b) skew_2 goes from 0.2889 to 2.2005; seriously affected.

2.28 **(a)** Data nearly symmetric, but there may be heavy tails.
(b) ALL DATA SKEW is nearly 0. The magnitudes of the HALF DATA SKEW measures exceed the theoretical value of 0.27 for a bell-shaped distribution. The data are heavy-tailed.

2.29 **(a)** Set A:

```
2 | 6
3 | 2
3 | 7
4 | 0 2 4
4 | 6 7 8 9 9
5 | 1 1 2 3 4
5 | 6 8
6 | 0
```

Set B:

```
2 | 6 6 7 8 9
3 | 1 3
3 | 5 8
4 | 3
4 | 9
5 | 1
5 | 7
6 | 2
6 | 5 7 9
7 | 1 2
```

(b) Set A is slightly left-skewed; set B is bimodal and slightly right-skewed.
(c) Bottom half, −0.333; top half, 0.276; close to normal

(d) Bottom half, 0.3636; top half, −0.3536; bimodality

2.30 **(a)** For A, mean = median = 4.0. For B, mean = 4.0, median = 5.0. The median for B is larger than the median for A.

(b) For A, mean = median = 4.0. For B, mean = 3.5, median = 5.0. The mean for A is bigger than the mean for B, but the median for A is less than the median for B.

(c) The mean is more sensitive.

2.31 The recording should, and does, increase the standard deviation from 1.49 to 2.99.

2.32 **(a)** Black: mean = 2.40, median = 2.00; white: mean = 3.53, median = 4.00

(b) Black: s = 0.968; white: s = 1.04

2.33 **(a)** Black: mean = 2.30, median = 2.00; white: mean = 3.667, median = 4.00. The relation still holds.

2.34 **(a)** The data range from 59 to 145. One plausible set of intervals is 55–64, 65–74, etc.

(b)

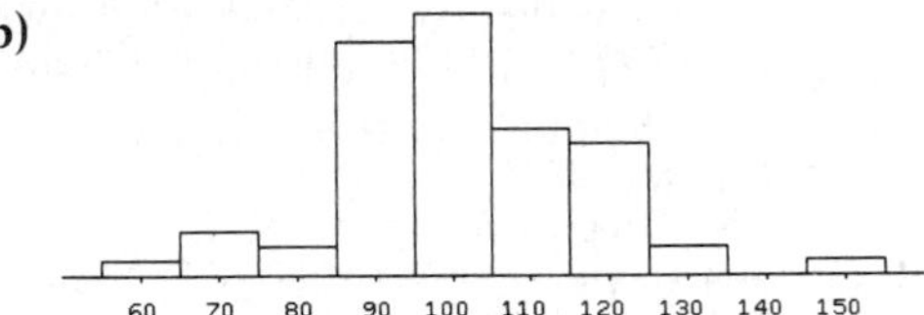

(c) Data appear roughly normal.

2.35

5 | 9
6 | 9
7 | 0 3 7
8 | 4 5 5 6 6 8 8 9
9 | 0 0 1 1 1 2 4 4 4 5 6 6 7 7 8 8 9
10 | 0 0 0 1 1 2 3 3 3 4 5 5 7 9 9
11 | 0 1 1 1 2 5 5 8 9
12 | 2 2 3 3 4 7
13 | 4
14 | 5

Yes, it's roughly normal.

2.36 **(a)** $\bar{y}$ = 100.58, median = 100.0

(b) s = 16.0

(c) 46, or 0.742, slightly above 68%

2.37 **(a)** 95th percentile = 124.4

(b) p = .369

(c) They are unequal; they are different ideas entirely.

2.38 **(a)** $skew_3$ = 0.048 (complete data), −0.370 (bottom half), 0.205 (top half)

(b) Yes; complete-data $skew_3$ is near 0, and the half-data $skew_3$ values are near ±.27.

2.39 **(a)**

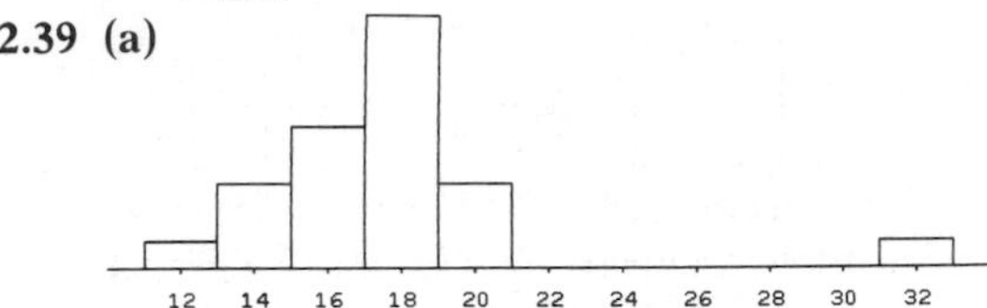

(b) Data are roughly symmetric except for an outlier. There is no upper limit on the times, but there is a lower limit.

2.40

12 | 7
13 | 1 9
14 | 1
15 | 0 9 9
16 | 3 6
17 | 0 4 6 6 7
18 | 0 5 8 8
19 | 5 5
20 | 1

(plus the outlier value 31 | 8)

2.41 $\bar{y}$ = 17.54, median = 17.50. The mean should be larger for right-skewed data.

2.42 $skew_1$ = 0.0095, $skew_3$ = 0.016; very slight right-skewness

2.43 **(a)** s = 3.81

(b) .8635, much larger than .68

2.44 **(a)**

1 | 2 7 7 7 8
2 | 0 2 4 5
3 | 1 3 4 5 5 6 7 7 8
4 | 0 0 1 1 1 2 3 3 3 4 5 6 7 7 8 8
5 | 9
6 | 0 6 7
7 | 2 6
8 | 1
9 | 8

(b) Slightly right-skewed with a heavy right tail

2.45 $\bar{y}$ = 4.20, $\bar{y}$ = 4.10; both slightly above 4.0

2.46 **(a)** $\bar{y}$ = 4.208, s = 1.85

(b) .6905, very close.

2.47 **(a)** 1.5 (extroverts), 1.3 (introverts)

(b) One moderate outlier in the introverts data

2.48 **(a)** IQR = 20.0

(b) Outliers are 59 and 145.

2.49 $skew_3 = -0.370$ and 0.205 (for bottom and top halves), near the values for a normal distribution

2.50 (a)

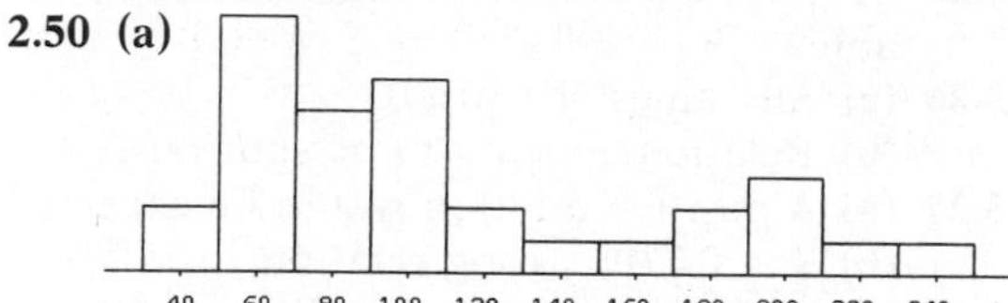

(b) The data are right-skewed and possibly bimodal.

(c) If we use stems 4, 5, ..., 24, there are too many classes.

2.51 (a) $\bar{y} = 108.88$, median = 92.0

(b)

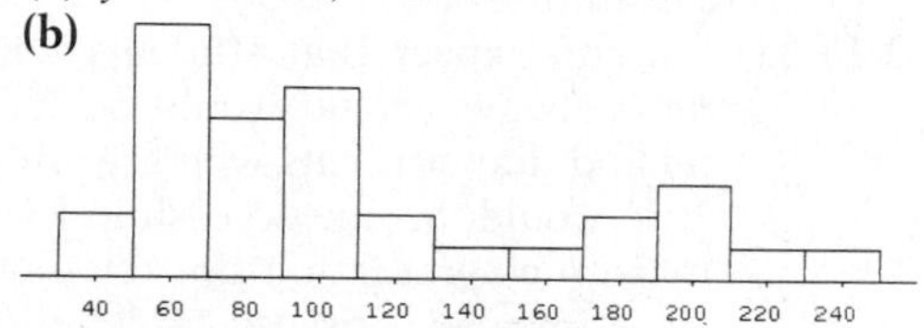

(c) Data are right-skewed.

2.52 IQR = 83.0; no outliers

2.53 (a) $\bar{y} = 108.88$, $s = 57.0$

(b) .9375, close to .95

(c) No; it is far above .75.

2.54 $skew_1 = 0.296$, $skew_2 = 0.878$, $skew_3 = 0.338$; all are positive, indicating right-skewness.

2.55 (a) Roughly 30

(b) median = 31

(c) IQR = 10.5; LO outliers at 6 and 9, HI outlier at 53

2.56 (a) 26

(b) 74.29%, close to 68%

2.57 $skew_1 = 0.016$, $skew_3 = 0.022$; very slight right-skewness

2.58 $skew_3$ (bottom half) $= -0.343$; $skew_3$ (top half) $= 0.485$. The right tail, in particular, is heavier than that of normal data.

2.59 (a) Each boxplot is nearly symmetric.

(b) Near middle of output; nearly symmetric

(c) $skew_1 = -0.112$; slight left-skewness

2.60 (a) MEAN 17.3, STD DEV 6.24717, 50% MED 18

(b) SKEWNESS = −0.148287, also negative

(c) Yes; nearly symmetric

2.61 MEAN 17.3000000, MEDIAN 18.0000000, ST.DEV. 6.2471688

2.62 (a) Right-skewed

(b) 11 values shown as HI outliers

(c) $skew_1 = 0.264$, indicating right-skewness

(d) Note that the program should have used rows 1 to 59 and rows 59 to 117. The result is $skew_3$ and shows right-skewness.

Chapter 3

3.1 (a) $\lambda = 0.133$

(b) Knowing age leads to a 13.3% reduction in error in predicting embarrassment.

3.2 $\lambda = 0.024$, smaller than the value in 3.1

3.3 (a) (yes, 3–4), (no, 5), (no, 6), (no, 7), (no, 8), (no, 9), (no, 10), (no, 11–12)

(b) 111 errors

(c) $\nabla = 0.193$; 19.3% error reduction

(d) precision = 0.390

3.4 (a) $\nabla = 0.201$, precision = 0.414

(b) Yes; the first rule has smaller predictive value and lower precision than this rule.

3.5 (a) $\lambda = 0.103$

(b) 10.3% error reduction

3.6 (a) 271 errors whether or not sex is known

(b) "Dreamer" is always the modal class.

3.7 $\nabla = 0.166$

3.8 precision = 0.480

3.9 $\nabla = 0.136$

3.10 (a) 855 pairs

(b) $\tau = 0.164$

(c) $\gamma = 0.319$

3.11 (a) (4, Small), (4, T/E), (4, Syst), (5, None), (5, Syst), (6, None), (6, Syst), (7, None), (7, Small), (8, None), (8, Small), (8, T/E)

(b) $\nabla = 0.507$, precision = 0.560

3.12 (a) Larger frequencies are on the main diagonal of the table.

(b) $\tau = 0.461$, $\gamma = 0.806$

3.13 (a)

		Aptitude			
		L	M	H	Total
Major	E	53	38	15	106
	O	208	217	148	573
	Total	261	255	163	679

(b) $\lambda = 0$, $\nabla = 0.092$

3.14 (a) $\lambda = 0$, in all regions

(b) In E, $\nabla = -0.073$; in S, $\nabla = 0.264$; in MW, $\nabla = 0.203$; in W, $\nabla = 0.122$.

(c) No. λ shows no relation; the relation found by ∇ does not disappear when region is controlled.

3.15 (a)

		X_1			
		L	M	H	Total
Y	L	108	27	10	145
	M	32	146	33	211
	H	5	27	112	144
	Total	145	200	155	500

(b) Yes; most cases are (L, L), (M, M), or (H, H).

3.16 (a) Yes; the frequencies on the main diagonal are larger.

(b) Yes, in both cases

(c) No; the relation does not disappear.

3.17 (a) Yes; again the main-diagonal frequencies are larger.

(b) To an extent, the relation is spurious; the relations in the subgroups are much less evident.

3.18 (a) $\lambda = 0.133$

(b) $\lambda = 0.260$

(c) Some relation, not extremely strong

3.19 (a)

		Age in Years			
		3&4	5	School Age	Total
Embarrassment	Yes	9	19	197	225
	No	26	13	89	128
	Total	35	32	286	353

(b) $\lambda = 0.133$; no change

3.20 (a) $\nabla = 0.222$

(b) $\nabla = 0.222$; no change

3.21 (a) $\lambda = 0.245$; some relation

(b) $\lambda = 0$; λ did not capture a relation.

3.22 $\gamma = -0.429$; fairly strong negative relation

3.23 (a) $\lambda = 0.158$

(b) $\lambda = 0.161$

(c) There is some relation, but not a strong relation.

3.24 (a) $\nabla = 0.247$

(b) precision = 0.664

(c) λ uses a different prediction rule.

3.25 "Type of classroom" is not an ordinal variable.

3.26 (a) All values of λ are 0.

(b) Relation appears to be spurious.

3.27 (a) A positive relation would be expected.

(b) $\gamma = 0.730$; strong relation

3.28 $\nabla = 0.520$; fairly strong relation

3.29 Arrows between initial reading ability and type of classroom and between initial reading ability and end-of-year ability; no arrow between type of classroom and end-of-year ability

3.30 (a) I would expect that students who almost always attend would be successful and that students who are often absent would be unsuccessful; I would make no prediction for those who sometimes are absent.

(b) $\nabla = 0.405$

(c) precision = 0.4625

3.31 $\lambda = 0.375$, predicting success; $\lambda = 0$, predicting attendance. The values are quite different.

3.32 $\tau = 0.206$; it is meaningful because both variables are ordinal.

3.33 Using λ, there appears to be little relation.

3.34

		Ethnicity			
		B	H	W	Total
Prejudice	Yes	51	27	35	113
	No	34	27	54	115
	Total	85	54	89	228

For predicting prejudice knowing ethnicity, $\lambda = 0.150$, indicating a modest relation.

3.35 (a) GAMMA is shown as 0.876.

(b) If the column variable is the dependent variable, LAMBDA ASYMMETRIC is 0.397; if the row variable is the dependent variable, LAMBDA ASYMMETRIC is 0.623.

3.36 (a) For MOTHERS, the LAMBDA values are 0.15385 and 0.16667. For FATHERS, the LAMBDA values are 0.35714 and 0.33333.

(b) Yes, if one regards the EITHER and DON'T KNOW categories as in between BOY and GIRL.

(c) γ is large for mothers and λ is fairly large for fathers.

Chapter 4

4.1 **(a)** Relative frequency (derived from past data)
(b) Relative frequency
(c) Personal, subjective
(d) Personal

4.2 **(a)** {PNN, NPN, NNP, PPN, PNP, NPP, PPP, NNN}
(b) A = {PNN, NPN, NNP}, B = {PPN, PPP}, C = {PPP}

4.3 A and B, A and C, but not B and C (which both contain PPP).

4.4 **(a)** A typical outcome could be $(n_1, n_2, \ldots, n_6)$, where n_i is the number of errors announced by subject i.
(b) An outcome belongs to E if at least four $n_i = 8$.

4.5 **(a)** $P(A \text{ or } B) = .7$
(b) $P(A \text{ and } B) = 0$
(c) $P(\text{not-}A \text{ and not-}B) = .3$

4.6 **(a)** $P(A \text{ or } B) = .5$
(b) $P(A \text{ and not-}B) = .1$
(c) $P(\text{neither one}) = .5$

4.7 **(a)** For example, $P(D \text{ and not-}E \text{ and not-}F) = .05$
(b) $P(D \text{ and } E \text{ and not-}F) = .15$
(c) $P(\text{none}) = 1 - P(\text{at least 1}) = 1 - .80 = .20$

4.8 **(a)** $P(A \text{ and not-}D) = .30$
(b) $P(A \text{ or } D) = .60$
(c) $P(\text{neither}) = 1 - .60 = .40$

4.9 **(a)** $P(M \text{ and not-}C) = .42$
(b) $P(\text{not-}C|M) = 840/900$
(c) $P(C) = .045$

4.10 No. $P(C|M) = 60/900 = .0666 \neq P(C) = .045$

4.11 **(a)**
$$P(C_1 \text{ and } C_2 \text{ and } C_3) = \frac{90(89)(88)}{2000(1999)(1998)}$$
(b)
$$\text{Probability} = \frac{1160(1159)(1158)}{2000(1999)(1998)}$$

4.12 **(a)** $P(C_1 \text{ and } C_2 \text{ and } C_3) = (90/2000)^3$
(b) Probability $= (1160/2000)^3$

4.13 **(a)** $P(\text{no reward}) = .07776$
(b) $P(\text{reward only on trial 2}) = (.6)(.4)(.6)^3$
(c) Assuming that the question implies reward on *exactly* one of the first two trials, $P(R_1 \text{ or } R_2 \text{ but not both}) = .48$.

4.14 **(a)** $P(F|E) = .375$
(b) $P(E \text{ and } F|D) = .1667$
(c) $P(D \text{ and } E \text{ and } F|D \text{ or } E \text{ or } F) = .0625$

4.15 **(a)** $P(\text{all 6 report no errors}) = (.1)^6$
(b) Probability $= (.6)^3(.3)(.1)^2$

4.16 Assumed independence; dubious if all subjects communicate

4.17 $P(\text{reports 0}|\text{does not report 8 or more}) = .1/.4 = .25$

4.18 **(a)** $P(\text{W and A}) = .12$
(b) $P(\text{W and C or better}) = .24$
(c) $P(\text{W or A}) = .68$

4.19 **(a)** $P(\text{all three are W and A}) = (.12)^3$
(b) $P(\text{no A's}) = (.8)^3$
(c) $P(\text{all A's}|\text{all W's}) = .008$

4.20 **(a)** $P(\text{exactly 2 get C or better grades}) = .384$
(b) By tree, $P(\text{2 A's}|\text{exactly 2 get C or better}) = .024/.384 = .0625$

4.21 **(a)** $P(\text{exactly 1}) = .40$
(b) $P(\text{A}|\text{exactly 1}) = .75$
(c) $P(\text{A}|\text{B}) = .45/.55$

4.22 **(a)** $P(\text{exactly 2}) = .345$
(b) $P(\text{A and B}|\text{exactly 2}) = .045/.345$
(c) $P(\text{A}|\text{exactly 2}) = .285/.345$

4.23 **(a)** $P(\text{high}) = .051$
(b) $P(\text{depressed}|\text{high}) = .48/.51$

4.24 **(a)** $P(\text{C and SD}) = .105$
(b) $P(\text{D}|\text{C and SD}) = .096/.105$
(c) $P(\text{D}|\text{C or SD}) = .304/.555$

4.25 **(a)** $P(\text{correct on all 4}) = .80(.90)^4 + .20(.50)^4 = .53738$
(b) $P(\text{W}|\text{correct on all 4}) = .80(.90)^4/.53738 = .9767$

4.26 **(a)** We assumed $P(\text{second question correct}|\text{W}) = .9$, regardless of the correctness of the first answer; we also assumed $P(\text{second question correct}|\text{P}) = .5$, regardless of the first question.
(b) Unreasonable if successive questions were related

4.27 **(a)** P(at least 2 A's) = .246
(b) Final grade and first midterm grade are conditionally independent given second midterm grade.
(c) It depends on the exams. If the course were strictly cumulative, then only the most recent exam might matter; in that case, the assumption is reasonable.

4.28 **(a)** Not independent; P(E and H) = 15/679 = .022 and P(E)P(H) = (106/679)(163/679) = .037
(b) Not conditionally independent; P(E and H|MW) = 3/200 = .015 and P(E|MW)P(H|MW) = (23/200)(62/200) = .036
(c) Not conditionally independent; P(E and MW|H) = 3/163 = .018 and P(E|H)P(MW|H) = (15/163)(62/163) = .035

4.29 **(a)** P(yes and high) = .131
(b) Not independent
(c) P(yes and high|moderate) = .11
(d) Not conditionally independent

4.30 No. For example, all λ values are 0.

4.31 **(a)** P(all 4 years +) = .136224
(b) By a rather large tree, P(+ after fourth| + after first) = .701144

4.32 Grade 1 level is irrelevant, given grade 2 level; thus the two children will be equally likely to be + after grade 3.

4.33 **(a)** $P(A$ or B or $C)$ = .6
(b) P(at least 1) = $P(A$ or B or $C)$
(c) P(none) = .4
(d) $P(A$ or B|not-$C)$ = .50

4.34 **(a)** $P(A$ and B and C and $D)$ = .0016
(b) P(none) = .4096
(c) P(at least 2) = .1808
(d) P(either C or D or both|at least 2) = .1552/.1808

4.35 **(a)** $P(D$ or E or $F)$ = .88; 88 outcomes are in D or E or F.
(b) D and E and F = {51, 53, ..., 73}; $P(D$ and E and $F)$ = .12
(c) $P(F|D)$ = .50
(d) P(exactly 1) = .38

4.36 D and E are independent, as are D and F. None of the events are mutually exclusive.

4.37 **(a)** It could be the expert's subjective opinion, or a relative frequency based on past data.
(b) Assuming that completely outgrowing autism implies becoming a functional adult, P(functional but not completely outgrowing) = 2/9.

4.38 **(a)** P(first usable subject on trial 3) = .081
(b) P(second usable subject on trial 4) = .0243
(c) P(one usable) = .2916

4.39 **(a)** P(A on both) = .06
(b) Independence; implausible

4.40 **(a)** P(all 3 receive B's) = .064
(b) P(1 A, 1 B, 1 C) = .048

4.41 **(a)** P(D) = .57
(b) P(H|D) = .0526

4.42 **(a)** P(no H) = .59049
(b) P(exactly 2 H) = .0729
(c) P(first H at 4th person) = .0729

4.43 No. For example, P(high maturity and high dependency|6 months) $\neq$ P(high maturity| 6 months)P(high dependency|6 months)

4.44 **(a)** P(high maturity and high dependency) = .11
(b) P(high maturity)P(high dependency) = .3(.4) = .12 $\neq$ .11 so the events are not independent.

Chapter 5

5.1 **(a)** {SSS, SSN, SNS, SNN, NSS, NSN, NNS, NNN}
(b) For example, $P(\text{SNS}) = \dfrac{6(4)(5)}{10(9)(8)}$
(c)

Outcome	SSS	SSN	SNS	SNN
x value	3	2	2	1

Outcome	NSS	NSN	NNS	NNN
x value	2	1	1	0

(d)

x	0	1	2	3
$f_X(x)$	1/30	3/10	1/2	1/6

5.2 **(a)** {S, NS, NNS, NNNS, NNNNS}
(b) For example,

$$P(\text{NNNNS}) = \frac{4(3)(2)(1)(6)}{10(9)(8)(7)(6)}$$

(c)

Outcome	S	NS	NNS	NNNS	NNNNS
y value	1	2	3	4	5

(d)

y	1	2	3	4	5
$f_Y(y)$	3/5	4/15	1/10	1/35	1/210

5.3 **(a), (b), (c)** For example, the top branch could be RRRR, with $w = 0$, and probability $(.2)^4$.

(d)

w	0	1	2	3	4
$f_W(w)$	.0016	.0256	.1536	.4096	.4096

5.4 **(a)** For example, one outcome is CCICII (with C meaning correct and I meaning incorrect); this outcome has $w = 3$.

(b) For example, $P(\text{CCICII}) = (.2)^3(.8)^3$

(c)

w	0	1	2	3
$f_W(w)$	$(.2)^6$	$6(.8)^1(.2)^5$	$15(.8)^2(.2)^4$	$20(.8)^3(.2)^3$

w	4	5	6
$f_W(w)$	$15(.8)^4(.2)^2$	$6(.8)^5(.2)^1$	$(.8)^6$

(d) Probabilities have the form (number of paths)$(.8)^w(.2)^{n-w}$.

5.5 **(a), (b)** For example, the path (right, rewarded, left, rewarded) has $y = 1$ and probability (.5)(.3)(.1)(.2) = .003.

(c)

y	0	1	2
$f_Y(y)$	.275	.325	.400

5.6

x	0	1	2
$f_X(x)$	.1900	.4575	.3525

5.7 **(a)** For example, the path corresponding to (middle digit flashes, left one doesn't, right one does) has $y = 2$ and probability 1(.5)(.5).

(b)

y	1	2	3
$f_Y(y)$	.25	.50	.25

5.8 Draw a large tree indicating both which digits flash and whether a digit that is flashed is in the memory set.

z	0	1
$f_Z(z)$	.57421875	.35546875

z	2	3
$f_Z(z)$	.06640625	.00390625

5.9 $E(X) = 1.80$, $\text{Var}(X) = 0.56$

5.10 $\mu_Y = 1.571$, $\sigma_Y = 0.8206$

5.11 $E(Y) = 1.125$, $E(X) = 1.1625$

5.12 $\mu_X = 7.14$, $\sigma_X = 2.337$

5.13

t	0	1	2	3	4
$f_T(t)$	.0001	.0004	.0010	.0020	.0035

t	5	6	7	8
$f_T(t)$	.0056	.0088	.0142	.0227

t	9	10	11	12
$f_T(t)$	.0352	.0486	.0608	.0721

t	13	14	15	16
$f_T(t)$	.0840	.0985	.1170	.1330

t	17	18	19	20
$f_T(t)$	.1300	.1025	.0500	.0100

5.14 $E(T) = 2E(X)$, $\text{Var}(T) = 2\text{Var}(X)$

5.15 $\mu_Z = E(Z) = 0.50$, $\sigma_Z^2 = \text{Var}(Z) = 0.4062$

5.16 **(a)** $P(U > .7) = .3$

(b) $P(\text{at least 2 are greater than } .7) = .216$

5.17 $E(U) = 0.5$, because the density is symmetric around 0.5.

5.18 $\sigma_U = 0.2887$

5.19 **(a)** $P(.7 < Y < .9) = .5844$

(b) $P(Y < .6) = .0302$

5.20 $\mu_Y = 0.8333$, $\sigma_Y = 0.1034$

5.21 **(a)** The probability is 0.9558

(b) It is very close to the Empirical Rule figure and consistent with Chebyshev's Inequality.

(c) Not in this case

5.22 **(a)** $P(0.02 < T < 0.06) = .3691$

(b) $P(T > 0.1) = .1353$

5.23 **(a)** $\mu_T = 0.05$

(b) $\sigma_T^2 = 0.0025$

5.24 The probability is .8647, well above 68%.

5.25 $F_T(t) = 1 - e^{-20t}$, $P(0.02 < T < 0.06) = F_T(0.06) - F_T(0.02) = .3691$, $P(T > 0.10) = 1 - F_T(0.10) = .1353$

5.26 **(a)**

x	2	3	4	5	6
$f_X(x)$	.1	.2	.4	.2	.1

(b) The probabilities are symmetric around 4; $E(X) = 4.0$.
(c) $\sigma_X = 1.0954$
(d) $P(\mu_X - \sigma_X < X < \mu_X + \sigma_X) = .8$

5.27 **(a)**

y	4	5	6	7	8
$f_Y(y)$	.42	.27	.19	.08	.04

(b) $E(Y) = 5.05$
(c) $\text{Var}(Y) = 1.2875$

5.28 **(a)** $E(X) = 5.0$
(b) Probabilities are symmetric around 5.
(c) $\sigma_X = 1.0954$

5.29 **(a)** $E(Y) = 5.0$, $\sigma_Y = 1.265$
(b) The Y probabilities are more spread out; the standard deviation of Y should be larger than the standard deviation of X.

5.30 $E(U) = 2$, $\text{Var}(U) = 2$

5.31 **(a)**

t	0	1	2	3	4
$f_T(t)$	.04	.08	.12	.16	.20

t	5	6	7	8
$f_T(t)$	.16	.12	.08	.04

(b) $E(T) = 4$, $\text{Var}(T) = 4$
(c) $E(T) = 2E(U)$, $\text{Var}(T) = 2\text{Var}(U)$

5.32 **(a)**

t	0	1	2	3	4
$f_T(t)$	.008	.024	.048	.080	.120

t	5	6	7	8	9
$f_T(t)$	.144	.152	.144	.120	.080

t	10	11	12
$f_T(t)$	.048	.024	.004

(b) $E(T_3) = 6$, $\text{Var}(T_3) = 6$
(c) $E(T_3) = 3E(U)$, $\text{Var}(T_3) = 3\text{Var}(U)$. It appears that $E(T_n) = nE(U)$ and $\text{Var}(T_n) = n\text{Var}(U)$.

5.33

y	1	2	3	4	5
$f_Y(y)$	.5000	.2500	.1250	.0625	.0625

$E(Y) = 1.9375$, $\text{Var}(Y) = 1.4366$

5.34 $E(Y) = 1.9980$, $\text{Var}(Y) = 1.7873$

5.35 $E(Y)$ seems to be approaching 2.0000. It is hard to see what $\text{Var}(Y)$ is approaching.

5.36 **(a)** It is a triangle, with a peak at $x = 1$, $f(x) = 1$.
(b) $P(X < 0.5) = .125$

5.37 **(a)** By symmetry, the mean is 1.0.
(b) $\sigma_X = 0.4082$
(c) The probability is .6498, close to .68 because the density is not far from bell-shaped.

Chapter 6

6.1 **(a)** Yes; they are independent success/failure trials, each with probability .75 of success.
(b) Yes; the number of trials is fixed and order is irrelevant.

6.2 **(a)** $P(Y = 8) = \dfrac{12!}{8!\,4!}(.75)^8(.25)^4$

Similar expressions hold for 9 and 10 successes.
(b) $P(Y = 8) = .1936$, $P(Y = 9) = .2581$, $P(Y = 10) = .2323$
(c) $P(8 \le Y \le 10) = .5549$

6.3 $E(Y) = 9.0$, $\sigma_Y = 1.50$

6.4 Independence is unlikely.

6.5 **(a)** Negative binomial
(b) Probability $= .02435$
(c) Expected value $= 32$, variance $= 96$

6.6 **(a)** Learning by the experimenter may lead to increasing probability of success.
(b) Yes; the number of trials is fixed.

6.7 **(a)** With $n = 100$ and $\pi = .95$, $P(Y \ge 90) = .9885$
(b) $\mu_Y = 95$, $\sigma_Y = 2.179$

6.8 **(a)** Negative binomial
(b) $P(Y = 110) = \dfrac{109!}{99!\,10!}(.95)^{100}(.05)^{10}$
(c) $E(Y) = 105.263$, $\text{Var}(Y) = 5.54$

6.9 The probability of two firings at the same time is negligible and firings are independent of one another.

6.10 **(a)** $P(12 \text{ or fewer}) = .9362$
(b) $P(8) = .1395$
(c) $P(\text{more than } 10) = .1841$

6.11 $\mu = 8$, $\sigma = 2.828$

6.12 **(a)** $P(0.1 < X < 0.4) = .4684$
(b) $E(X) = 0.25$, $\sigma_X = 0.25$

6.13 No answers will change.

6.14 Find $P(10 \text{ or fewer failures})$ with $\mu = 100(.05) = 5.0$ for failures. From Table 2, $P(10 \text{ or fewer}) = F(10) = .9885$, a close approximation.

6.15 **(a)** $P(10 \text{ or fewer}) = .9863$
(b) Quite good; n is large, the probability of success is small, and the mean is 5.

6.16 **(a)** $P(X = 4) = .0902$
(b) No simultaneous signals and no dependence of signals
(c) $\mu_X = 2$, $\sigma_X = 1.414$

6.17 $P(Y < 1) = .8647$

6.18 **(a)** The mechanism depends *only* on the result in the previous minute.
(b) Yes; the probabilities don't change over time.

6.19 **(a)**

		Next Minute	
		On	Off
This Minute	On	.09	.91
	Off	.03	.97

(b)

		Two Minutes Ahead	
		On	Off
This Minute	On	.0354	.9646
	Off	.0318	.9682

6.20 **(a)**

		Four Minutes Ahead	
		On	Off
This Minute	On	.031927	.968073
	Off	.031914	.968086

(b) About .032 and .968

6.21 To 6 decimal places, both matrices have identical rows (.031915 and .968085). These are the long-run probabilities.

6.22 **(a)** Learner status depends only on the preceding tasks, not on any previous tasks.
(b) The transition probability matrix doesn't change from one task to another.

6.23 **(a)** Probability = .82
(b) Probability = .026

6.24 **(a)**

Two steps

.82	.17	.01
.17	.66	.17
.01	.17	.82

Four steps

.7014	.2533	.0453
.2533	.4934	.2533
.0453	.2533	.7014

Eight steps

.558175	.314117	.127708
.314117	.371765	.314117
.127708	.314117	.558175

(b) The probabilities seem to be approaching .333333, as may most easily be seen in the middle row.

6.25 **(a)** 1.40, −0.80, −2.333
(b) 133, 91, 120

6.26 **(a)** $P(80 < \text{IQ} < 120) = .8164$
(b) $P(\text{IQ} > 125) = .0485$
(c) $P(75 < \text{IQ} < 85) = .1102$

6.27 **(a)** $z = 0.84$
(b) IQ = 112.6
(c) $z = \pm 1.645$
(d) IQ = 75.325 and 124.675

6.28 $P(\text{IQ} > 140 | \text{IQ} > 130) = .1667$

6.29 **(a)** 0.75, 0.10, −0.10, 0.75
(b) 97, 49, 133.6

6.30 **(a)** $P(70 < \text{depression} < 90) = .383$
(b) $P(\text{depression} > 130) = .0062$
(c) Depression = 112.9

6.31 **(a)** $\mu = 60$, $\sigma = 3.873$
(b) $P(X \geq 50) = .9951$

6.32 **(a)** $P(X \geq 49.5) = .9966$
(b) Yes; the expected number of successes is 60 and the expected number of failures is 20.

6.33 **(a)** $\mu = 240$, $\sigma^2 = 240$
(b) $P(X < 200) = .0049$

6.34 (a) $P(X < 199.5) = .0045$
(b) Negligible difference; approximation should be good with $\mu = 240$.

6.35 (a) $P(38 < Y < 62) = .7698$
(b) $P(Y > 68) = .0359$, $P(Y < 32) = .0359$
(c) 80th percentile = 58.4, 20th percentile = 41.6

6.36 (a) $P(Y < 68|Y > 50) = .9282$
(b) $P(Y < 0 \text{ or } Y > 100) = .00000058$

6.37 (a) $P(4 \text{ appear}) = \frac{6!}{4!\,2!}(.8)^4(.2)^2$
(b) $P(4 \text{ appear}) = .2458$

6.38 $\mu = 4.80$, $\sigma^2 = 0.96$, $\sigma = 0.9798$

6.39 (a) $P(28 \text{ sessions}) = \frac{27!}{19!\,8!}(.8)^{20}(.2)^8$
(b) $\mu = 25.0$, $\sigma = 2.50$

6.40 The probability of appearing is constant over subjects; whether or not one subject appears doesn't change the probability that another will appear. These seem plausible to us.

6.41 (a) $P(Y \le 3) = .1512$
(b) $P(4 \le Y \le 8) = .6960$
(c) $\mu_Y = 6.0$, $\sigma_Y = 2.449$

6.42 (a) $P(\text{time} > 2) = .2231$
(b) Expected time = 1.333 minutes

6.43 Flashes don't occur together and the occurrence of one flash doesn't change the probability that another will occur; in Exercise 6.42, the expected rate stays constant.

6.44 (a)

		To		
		A	S	N
	A	.2	.1	.7
From	S	.1	.3	.6
	N	.1	.1	.8

(b) Yes; the same probabilities apply over time.
(c) The probabilities of future settings depend only on the current setting, not on past history.

6.45 $P(N \text{ at time } 3|A \text{ at time } 0) = .764$

6.46 (a) No predictability
(b) .111111, .125000, .763889

6.47 No; the independence assumption is violated.

6.48 (a) $\mu = 5.0$
(b) $P(Y = 5) = .2023$
(c) $P(Y \ge 5) = .5852$

6.49 (a) Expected credit = 0
(b) $P(Y \ge 4) = .4787$

6.50 (a) $P(Y \le 20) = .1488$
(b) $\mu_Y = 25$, $\sigma_Y = 4.330$
(c) $P(Y \le 20) = .1251$

6.51 $P(Y \le 20) = .1492$, a much closer approximation

6.52 (a) $P(Y \ge 4) = .3522$
(b) $\mu_Y = 3.0$

6.53 (a) $P(Y \ge 4) = .2643$ (no correction), or .3783 (with correction)
(b) μ is only 3.0, so the approximation isn't great.

6.54 $P(Y \ge 4) = .3528$, a very close approximation

6.55 (a) $P(Y \ge 5) = .2746$
(b) The agency doesn't receive simultaneous applications, and the arrival of an application doesn't change the probability that others will arrive.

6.56 $P(\text{time} < 1/12) = .253$

6.57 (a) $P(V|F) = P(\text{director leaves}) = .04$, $P(F|V) = P(\text{new director}) = .12$
(b) The process has no memory and probabilities don't change.
(c) $P(\text{filled at } 4|\text{vacant at } 0) = .3766$

6.58 $P(F)$ is approximately .75.

6.59 (a) .0668, .3085, .6915, respectively
(b) $P(\text{serious}|Y < 35) = .257$

6.60 (a) $\mu_Y = 24.1722$, $\sigma_Y^2 = 20.9621$
(b) $P(20 \le Y \le 30) = .7166$
(c) Yes; the expected numbers of successes and failures are large.

6.61 $\mu = 148.93$

6.62 (a) $P(55 < Y < 70) = .0265$
(b) $P(Y < 55|Y < 70) = .057$

6.63 (a) $P(X \ge 73) = .0023$
(b) Excellent approximation

6.64 Little effect for large expected numbers; in fact, the corrected probability is .0029.

Chapter 7

7.1 All U. S. children of those ages

7.2 Draw randomly from a numbered list of all children.

7.3 Bias toward day-care children (working mothers); bias toward very young children

7.4 Randomly from a list of all current members

7.5 Bias toward large district; also, members of a single board will tend to think similarly.

7.6 Serious possibility of nonresponse bias

7.7 Against nonmembers

7.8 Gets separate estimates; may reduce variance, if the strata are homogeneous

7.9 Yes; the means are very different.

7.10 There is no explicit randomization.

7.11 The observer is not "blind" to the treatment.

7.12 Randomization and blinding are accomplished. There still may be a bias in selecting the subjects.

7.13 There is no randomization and the observers are not blind to the treatment.

7.14 Either randomly choose a total of 50 patients for the drug from the combined pool or randomly choose 25 from each hospital for the drug. Don't tell the directors who has gotten the drug.

7.15 If it was thought that patients of different ages would have different reactions to the drug

7.16 The parameter is the population fraction having fears of the supernatural. The statistic is the sample fraction.

7.17 If there are differences between men and women in the proportion fearing the supernatural. It would be useless if there was as much variability within men and within women as within the whole population.

7.18 In a large number of samples, 8.53% will have 11 patients reporting fears.

7.19 In a large number of samples, 18.53% will yield a mean of 0 and 19.77% will yield a mean of 0.0625.

7.20 The Monte Carlo fractions are close to the theoretical values.

7.21 No; different statistics have different theoretical distributions.

7.22 If suicide-proneness is indeed related to depression, stratification would be useful. Stratification changes the theoretical distribution, if it's at all useful.

7.23 No; the population distribution will be different.

7.24 In a large number of samples, 53% will have a λ value between .001 and .099.

7.25 The Monte Carlo frequencies are quite close to the theoretical frequencies.

7.26 **(a)** $E(\bar{Y}) = 0.191$
(b) $\sigma_{\bar{Y}} = 0.1905$

7.27 The standard error, 0.1905

7.28 The expected value doesn't change; $\sigma_{\bar{Y}}$ decreases from .1905 all the way to .1904.

7.29 No; the sampling fraction is nearly irrelevant.

7.30 No; there is negligible difference between the strata.

7.31 **(a)** $E(\bar{Y}) = 6.885$
(b) $\sigma_{\bar{Y}} = 0.04835$

7.32 No change

7.33 $P(\bar{Y} > 0.300) = .2843$

7.34 No; n is small, the probability is one-sided, and the population may be skewed.

7.35 **(a)** $E(\bar{Y}) = 7.50$, $\sigma_{\bar{Y}} = 0.25$
(b) $P(7.00 < \bar{Y} < 8.00) = .9544$

7.36 Yes; n is fairly large and the probability is two-sided.

7.37 **(a)** The probability in Exercise 7.35 is close to correct.
(b) The probability in the right tail may be higher than the approximation indicates.

7.38 **(a)** Over many samples, the standard deviation of the statistic will be 0.060.
(b) $P(\text{skew} > .100) = .1423$

7.39 **(a)** The theoretical probabilities for the Groeneveld-Meeden statistic are normal.
(b) No; the rules of thumb for a mean may not apply to a different statistic.

7.40 **(a)** It should equal $E(\bar{Y}) = 50.000$.
(b) It is estimating the standard error, 3.1623; the Monte Carlo estimate is quite close.

7.41 The distribution is, and should be, normal.

7.42 There is obvious bias.

7.43 It is biased toward concert-goers. There's no randomization.

7.44 **(a)** Yes; all are close to 3.00.
(b) The theoretical standard errors are

being estimated; they are 0.5477, 0.3464, 0.2000, and 0.1414, close to the estimates.

7.45 **(a)** There's no obvious curve or S-shape.
(b) The plot goes up in small steps instead of continuously.

7.46 The distribution of MIN may not be normal. The expected value certainly will not equal the population mean; the expected value must be less than that. The standard error formula that applies to a mean will not apply to MIN.

7.47 **(a)** Biased sample
(b) No randomization

7.48 **(a)** Over a large number of samples, the average value will be 0.27 and the standard deviation will be 0.042.
(b) P(top-half skew > 0.312) $= .1587$
(c) A normal theoretical distribution; we need rules of thumb to decide if and when this is true.

7.49 **(a)** Expected value = 50.0
(b) Standard error = 5.7735
(c) None

7.50 The standard error will be wrong, because there will be dependence among the groups of five scores.

7.51 **(a)** The average should be, and is, near 0.00.
(b) The theoretical standard errors are 0.2582 and 0.1826, very close to the Monte Carlo results.

7.52 **(a)** Given a symmetric, continuous population, the sample size should be large enough to make the normal approximation pretty close.
(b) The normal plot is nearly a straight line. Thus heavy-tailness is not critical.

7.53 **(a)** Yes; the normal plot is nearly linear.
(b) The standard errors are being approximated. These standard errors are smaller than those in Exercise 7.51.

7.54 **(a)** Plot curved, indicating skewness
(b) Yes; the mean is much larger than the median.

7.55 First normal, then skewed, then outlier-prone

Chapter 8

8.1 Both are (close to) unbiased.

8.2 $\hat{\mu}_1$ has smaller variance and is more efficient.

8.3 **(a)** No, by the Central Limit Theorem
(b) By plotting the means that were obtained

8.4 The data appear outlier-prone.

8.5 There is a bit of an S-shape in the plot, confirming outlier-proneness.

8.6 Groeneveld-Meeden $\text{skew}_3 = -0.008$ (all data); virtually no skewness. For the bottom half, $\text{skew}_3 = -0.625$; for the top half, $\text{skew}_3 = 0.462$, both larger in magnitude than .27 (the value for a normal population). The data are heavy-tailed (outlier-prone).

8.7 No; use a trimmed mean or median.

8.8 No; the 1.4427 value assumes an exponential population. The estimator is not nominally robust.

8.9 For severe right-skewness, $n = 31$ may not be enough to appeal to the Central Limit Theorem; mean may not be nominally robust.

8.10 Yes; there's a curve in the plot.

8.11 .0271, .0277, .0279, .0277; estimate = 0.90

8.12 $\hat{\mu} = (3.6)/4 = 0.90$

8.13 3334, 4963, 6524, 7800, 8605, 9049, 9005, 8584. The estimate should be about 10, perhaps larger. Try more values.

8.14 Estimate is $-6/\log(0.59) - 1 = 10.37$.

8.15 .0779, .0997, .1030, .0948; estimate is roughly 0.70.

8.16 Estimate $= (4.02)/6 = 0.67$

8.17 **(a)** $21.18 < \mu < 24.74$
(b) In the long run, 90% of such intervals will contain the true value of μ_{pop}.

8.18 $\mu_{pop} > 24.74$

8.19 The sample mean may not be most efficient. A more efficient estimator would yield a narrower confidence interval.

8.20 **(a)** Data appear nearly normal.
(b) Plot is roughly a line, indicating nearly normal data.

8.21 **(a)** $0.24 < \mu_{pop} < 1.56$
(b) $\mu_{pop} < 1.45$
(c) No, because n is only 4 and the population is skewed. One-sided intervals are poorer approximations than two-sided intervals.

8.22 Yes, by the Central Limit Theorem. The two-sided interval will have a slightly closer approximation.

8.23 $20.83 < \mu_{pop} < 25.09$; required $n = 57$

8.24 **(a)** Yes, by the Central Limit Theorem and the near-normality of the data.
(b) No; relative efficiency depends very little on n.

8.25 **(a)** $90.148 < \mu_{pop} < 106.612$
(b) $85.490 < \mu_{pop} < 111.270$, more than 50% wider

8.26 Required $n = 174$

8.27 **(a)** Plot the data to see if they appear roughly normal. Skewness with a small n would make the nominal confidence level incorrect; heavy-tailness would make the interval inefficient.
(b) A large n would take care of skewness concerns but not inefficiency concerns.

8.28 **(a)**

```
 6 | 9
 7 | 6 9 9
 8 | 2 4 6 6 7
 9 | 2 5 6 7 8 9 9
10 | 0 0 2 3 3 4 5 6 6 7 7 9 9
11 | 0 1 1 2 3 6 6 6 8
12 | 5 7
13 | 0 7
```

No severe skewness or obvious outliers
(b) The whole-data measure indicates very mild left-skewness. The half-data measures are larger in magnitude than .27, indicating heavy tails.
(c) The 69 and 137 values are mild outliers.
(d) The outliers indicate that the mean is an inefficient estimator, with an unnecessarily large standard error.

8.29 $98.01 < \mu_{pop} < 107.09$, width 9.08

8.30 $n = 217$

8.31 $-0.025 <$ population $\text{skew}_3 < 0.183$

8.32 The skew_3 statistic has an approximately normal sampling distribution (for this sample size and population).

8.33 By calculus or computation, $\hat{\pi} = .16$

8.34 The estimator is not quite unbiased.

8.35 The boxplots all are nearly symmetric around 0, indicating that all three estimators are unbiased. The boxplot for the median is narrowest, indicating that the median is most efficient.

8.36 The averages are close to 0, indicating unbiased estimators. The standard deviation (which will be close to the standard error) is smallest for the median, indicating that the median is most efficient.

8.37 **(a)** $14.51 < \mu_{pop} < 17.97$
(b) $\mu_{pop} < 17.80$

8.38 **(a)** A single curve (as opposed to an S shape) indicates skewness.
(b) No; the Central Limit Theorem will apply with $n = 142$. If anything, the confidence level for the one-sided interval may be slightly in error.

8.39 $n = 426$

8.40 **(a)** $0.179 <$ population $\gamma < 0.673$
(b) Yes; all reasonable values for population γ are positive numbers, indicating an increasing relation.
(c) $0.219 <$ population γ

8.41 Because the standard error involves $\sqrt{n}$, doubling n does not cut the width in half.

8.42 **(a)** The boxplot is narrowest for the median, indicating that it is the most efficient estimator; the trimmed mean is second best. The mean is a distant third.
(b) The same conclusion holds for both sample sizes.

8.43 Yes; the standard deviation is smallest for the median, almost as small for the trimmed mean, and much larger for the mean.

8.44 **(a)** $n = 666$
(b) No; the mean is most efficient for a normal population.

8.45 **(a)** A smaller n will do.
(b) A smaller n will do.

8.46 Estimate = 0.35

8.47 Choose various values for the parameter. Draw repeated samples using each parameter value, calculate the estimates for each sample, and average the results for each parameter value. See if the average estimate (approximately) equals the parameter value.

Chapter 9

9.1 **(a)** A probability of .5 indicates complete unpredictability.
(b) H_a: $\pi > .5$ might be taken to test if women hope for girl babies.
(c) The possibility of bias in either direction

9.2 Type I error would be to claim that there was a bias when there was none. Type II error would be to claim no bias when there was one.

9.3 **(a)** $\alpha = .0106$
(b) R.R.: $y \leq 7$ or $y \geq 17$

9.4 Support the research hypothesis.

9.5 **(a)** H_0: $\pi = .50$ (same proportion as would lend 10 cents), H_a: $\pi < .50$ (smaller proportion would lend 30 cents)
(b) Claiming that fewer students would lend 30 cents when the same proportion would lend 30 cents as would lend 10 cents

9.6 **(a)** By Table 1, $\alpha = .1012$
(b) R.R.: $y \leq 19$

9.7 Reject H_0; support H_a.

9.8 Power probabilities are .2860, .5610, and .8096.

9.9 **(a)** Power probabilities are .0425, .1560, and .3888. Power decreased.
(b) Power probabilities are .2415, .6223, and .9123. Power increased for $\pi = .40$ and $\pi = .35$.

9.10 Add binomial probabilities for $n = 24$, $\pi = .70$, $y \leq 7$.

9.11 **(a)** R.R.: $y \leq 17$ or $y \geq 33$
(b) Power $= .7741$
(c) Power is larger in (b), because n is larger.

9.12 **(a)** $\mu = 4.0$ indicates no belief either way.
(b) H_a: $\mu > 4.0$

9.13 Data contain outliers; median test would have better power.

9.14 H_0: $\mu = 4$, H_a: $\mu > 4$, T.S.: $z = \dfrac{\bar{y} - 4}{1.00/\sqrt{127}}$, R.R.: $z > z_\alpha$, for any specified α. Conclusion: $z = 12.5$; reject H_0 for any reasonable α.

9.15 **(a)** H_0: $\mu = 427$
(b) H_a: $\mu \neq 427$, to test for any difference

9.16 R.R.: $|z| > 1.645$

9.17 H_0: $\mu = 427$, H_a: $\mu \neq 427$, T.S.: $z = \dfrac{\bar{y} - 427}{223.2/\sqrt{22}}$, R.R.: $|z| > 1.645$. Conclusion: $z = 0.946$; retain H_0.

9.18 Nominal α may be slightly wrong, but the test is two-tailed and the z statistic is not close to the rejection region.

9.19 **(a)** $p = .34$
(b) Not statistically significant ($p = .34$)
(c) The apparent difference *might* be random variation.

9.20 **(a)** Population may be skewed, hence p-value may be inaccurate.
(b) Not crucial; p-value far from usual α values

9.21 **(a)** H_a was one-sided; p-value should be one-tailed.
(b) p is much less than .0001.
(c) Yes, extremely so

9.22 $n = 127$ should yield a good normal approximation; p will be some very small number.

9.23 **(a)** No directional hypothesis is indicated. Use a two-tailed p.
(b) p-value $= 2(.006) = .012$
(c) Effect is small, but conclusively shown.

9.24 **(a)** $z = 1.46$ is correct.
(b) $p = .0721$, using one-tailed test
(c) Not significant at $\alpha = .05$, significant at $\alpha = .10$
(d) He accepted H_0 rather than retaining it.

9.25 **(a)** $378.73 < \mu < 565.27$
(b) Interval includes 427.
(c) Retain H_0; the result is statistically insignificant.

9.26 **(a)** $\mu > 4.90$
(b) 4.0 is not included.
(c) Reject H_0; the result is statistically significant.

9.27 Reject H_0 at $\alpha = .01$; thus $p < .01$.

9.28 **(a)** $0.096 < \mu < 0.786$
(b) $\mu = 0$ is not included.
(b) Only .03 standard deviation

9.29 **(a)** $-6.64 < \mu < 111.06$
(b) Very wide because n is small
(c) Coaching may have no value; it may

have great value.

9.30 **(a)** Power = .4761
(b) Required n = 253

9.31 Power = $P(z > -1.80)$ = .9641, required n is only 24.

9.32 **(a)** Power = .1446
(b) Required n = 515

9.33 Yes, the Central Limit Theorem will apply very well.

9.34 **(a)** Power = .0465, required n = 731
(b) Power = .1003, required n = 915
(c) Power = .1075, required n = 804
(d) Decreasing α, decreasing $|\mu_a - \mu_0|$, and increasing σ all decrease the power for a fixed n.

9.35 **(a)** Yes; there are multiple success/failure trials, the probability of success should be constant, the trials should be independent, and the number of trials should be a fixed number.
(b) H_0: $\pi = .20$
(c) Folklore says H_a: $\pi > .20$.

9.36 **(a)** $\alpha = .0322$
(b) If $\pi = .15$ (a nonboundary value), $P(\text{reject } H_0)$ will decrease.
(c) Power = .3990

9.37 **(a)** H_0: $\pi = .20$, H_a: $\pi > .20$, T.S.: Y = the number of questions for which b is correct, R.R.: $y \geq 8$. Conclusion: $y = 7$; retain H_0.

9.38 **(a)** H_0: $\mu_{\text{diff}} = 0$
(b) H_a: $\mu_{\text{diff}} \neq 0$, because no specific direction has been indicated.

9.39 R.R.: $z > |2.58|$, where $z = \dfrac{\bar{Y} - 0}{12.50/\sqrt{30}}$

9.40 $\beta = .6517$

9.41 **(a)** H_0: $\mu = 0$, H_a: $\mu \neq 0$, T.S.: z (as in Exercise 9.39), R.R.: $|z| > 2.58$. Conclusion: $z = 2.81$; reject H_0. It appears that the population mean is positive.
(b) $p = 2(.0025) = .0050$

9.42 $0.52 < \mu < 12.30$. Reject H_0 because $\mu_0 = 0$ is not included within the interval.

9.43 Data are outlier-prone. A median test might be (even) more conclusive.

9.44 **(a)** Whoever judges success should be blind to the patient's treatment, and patients should be assigned randomly to treatments.
(b) H_0: $\pi = .30$
(c) H_a: $\pi > .30$

9.45 Type I: Claim incorrectly that the new treatment is better.
Type II: Claim incorrectly that the new treatment is not better.

9.46 **(a)** R.R.: $y \geq 8$
(b) $P(Y \geq 8) = .0266$; lower for a nonboundary value in H_0 and much less than .10

9.47 Power = .5982

9.48 R.R.: $y \geq 9$ and power = .4018. Power decreases (and β increases) as α decreases.

9.49 **(a)** H_0: $\pi = .30$, H_a: $\pi > .30$, T.S.: Y = the number of successfully treated patients, R.R.: $y \geq 8$. Conclusion: $y = 12$; reject H_0.
(b) Yes; $y = 12$ is in that R.R.
(c) Part (a) indicates $p < .10$, part (b) indicates $p < .05$.

9.50 **(a)** μ = mean performance of entire fourth-grade population
(b) H_0: $\mu = 3.62$, H_a: $\mu > 3.62$
(c) The new curriculum might have negative effects.

9.51 **(a)** T.S.: $z = \dfrac{\bar{Y} - 3.62}{0.65/\sqrt{94}}$, R.R.: $z > 1.645$
(b) R.R.: $z > 2.33$

9.52 **(a)** $\beta = .0057$
(b) $\beta = .0322$, larger because α is smaller
(c) No; a p-value refers to a single sample; β is long-run. Also, the p-value calculations assume H_0 is true.

9.53 The first four steps have been stated. Conclusion: $z = 2.54$; reject H_0.

9.54 $\mu > 3.680$ doesn't include 3.60; reject H_0.

9.55 $p = .0055 < \alpha$ in either case.

9.56 **(a)** Yes; we rejected H_0.
(b) No; that's not a statistical question.

9.57 **(a)** H_0: $\mu_{\text{pop}} = 500$, H_a: $\mu_{\text{pop}} \neq 500$
(b) R.R.: $|z| > 1.645$

9.58 $\beta = 1 - \text{power} = 1 - P(z > -0.86) = .1949$

9.59 $z = -1.73$; support the research hypothesis.

9.60 $486.54 < \mu_{\text{pop}} < 499.66$ doesn't contain μ_0.

9.61 **(a)** $p = .0836$, which is not less than .05
(b) No; the difference seems very small.

Chapter 10

10.1 $t_{.025} = 2.228$, 2.086, and 2.042 for 10, 20, and 30 df. As df increases, table value decreases.

10.2 Very close, especially for large a; farthest apart at .001.

10.3 Using 240 (close to 244) df, $t_{.025} = 1.97$ vs. $z_a = 1.96$.

10.4 $t_{.01} = 2.384$

10.5 $2.83 < \mu_{pop} < 3.37$

10.6 No; $n = 216$ is large and two-tailed probabilities are used.

10.7 $37.29 < \mu_{pop} < 48.89$

10.8 $\mu_{pop} < 47.81$

10.9 Data are left-skewed and n is small; probabilities suspect. One-tailed probabilities, as in Exercise 10.8, are poorer approximations.

10.10 $|t| = |-2.449| > t_{.10/2} = 1.65$; reject H_0.

10.11 $|t|$ falls between t-table values for one-tailed areas .005 and .01; for a two-tailed test, $.01 < p < .02$.

10.12 Confidence interval excluded μ_0, hence H_0 was rejected. A 90% confidence interval is equivalent to a two-tailed test at $\alpha = .10$.

10.13 $t = -1.885 < -t_{.05,10\text{ df}} = -1.812$; reject H_0 (barely).

10.14 $.025 < p < .05$; small sample size and skewness make the approximation dubious.

10.15 **(a)** Rats are paired by litter.
(b) $-0.827 < \mu_{diff} < 8.661$
(c) No. $\mu_{diff} = 0$ is included in the interval; retain H_0.

10.16 $.10 < p < .20$

10.17 Differences appear to contain outliers at -18 and 23.

10.18 T.S.: $|T_-| = 10 \leq 13$, the Table 7 value with effective $n = 11$; result is significant. Signed-rank test is more efficient for an outlier-prone population.

10.19 **(a)** Before and after data are paired by supervisor.
(b) $\mu_{diff} > 1.413$
(c) 0 not included; reject H_0.

10.20 $t = 5.01$; $p < .001$. H_0 is again rejected at $\alpha = .01$.

10.21 No; data should be independent and appear nearly normal.

10.22 T.S.: $|T_-| = 26.5 \leq 76$; same conclusion.

10.23 T.S.: $Y =$ the number of positive differences $= 21$. Using a normal approximation without continuity correction, we have $z = 3.40$. The p-value is very small.

10.24 $1.9 \leq \text{median} \leq 3.9$. 0 is not included, so H_0 is rejected, as it was in Exercise 10.23.

10.25 Normal approximation without correction: $z = -5.85$. Reject H_0.

10.26 Both tests rejected H_0, but for a skewed population such as there seems to be here, $\mu \neq$ median; different hypotheses are being tested.

10.27 **(a)** $t_{.05} = 1.681$
(b) $\delta = -0.568$

10.28 Power = .9868

10.29 Power = .13; n is not big enough.

10.30 Power = .38

10.31 **(a)** $12.35 < \mu < 13.77$
(b) Retain H_0; the interval includes $\mu_0 = 13.70$.

10.32 H_0: $\mu = 13.70$, H_a: $\mu \neq 13.70$, $t = \dfrac{13.06 - 13.70}{1.52/\sqrt{20}} = -1.88$. Because $|t| < t_{.05/2} = 2.093$, retain H_0.

10.33 $.05 < p < .10$

10.34 The data are discrete and the t distribution is continuous. However, if the population is symmetric, the t probabilities provide a decent approximation.

10.35 **(a)** $\mu < 6.72$
(b) Reject H_0: $\mu = 7.0$.

10.36 $t = -2.571 < t_{.05} = -1.671$; reject H_0 with $.005 < p < .01$.

10.37 Population is very right-skewed. We have $n = 61$ but a one-tailed probability. One-tailed probabilities are poor approximations for skewed populations.

10.38 **(a)** H_0: $\mu = 50$ is true; we approximate α.
(b) Simulation values are very close to nominal probabilities.

10.39 **(a)** H_a is true; we are approximating power $= P(\text{reject } H_0 | H_a \text{ is true})$.
(b) Power = .4168, close to the simulated power .437

10.40 **(a)** IQ
(b) $\bar{d} = 6.667$, $s_{\bar{d}} = 11.007$

(c) $t = 2.098 > t_{.10/2} = 1.796$; reject H_0.
(d) $.05 < p < .10$, two-tailed

10.41 $T = |T_-| = 13.5 \leq 17$, the Table 7 value for $n = 12$, $\alpha = .10$; reject H_0.

10.42 Mildly outlier-prone data suggest use of signed-rank test. Both tests give the same conclusion.

10.43 $y = 3$ negative scores; retain H_0.

10.44 (a) α; H_0 is true.
(b) Yes, for both tests.

10.45 (a) H_a is true.
(b) Signed-rank test

10.46 $1.47 < \mu < 11.35$

10.47 (a) $t_{diff} = 2.65 > t_{.05/2} = 2.045$; reject H_0.
(b) Reject H_0 because the interval excludes 0.
(c) $.01 < p < .02$

10.48 (a) $t = 1.807 > t_{.05} = 1.734$; reject H_0.
(b), (c) Exclude the 0 difference; the first four ranks are $-17, -5, -2, -1$.
(d) $T = |T_-| = 25 \leq 40$; reject H_0.

10.49 (a) Reject H_0 for all α.
(b) $p < .005$

10.50 The data are nearly symmetric with outliers; use signed-rank test.

10.51 $4.3 \leq \text{median}_{pop} \leq 4.7$ excludes 4.0.

10.52 Skewness makes nominal α probabilities grossly wrong.

10.53 (a) Same subjects
(b) T: MEAN = 0 is 3.8591.
(c) PROB > |T| is .000277142, two-tailed.
(d) Conclusively, the means differ; μ_A is higher.

10.54 (a) SGN RANK is 477.5.
(b) PROB > |S| is .0001.
(c) The signed-rank test yields the smaller p-value.

10.55 Data are skewed. As seen in Exercise 10.52, the signed-rank p-value can't be believed. The two-tailed p-value of the t test should be close.

10.56 (a) The TTEST p-value is .22, not significant.
(b) The WTEST p-value is .064, not conclusive but indicative.

10.57 (a) Tests are based on difference data.
(b) Several possible mild outliers; signed-rank is preferred.

Chapter 11

11.1 (a) Expected value = 3.7, standard error = 3.72
(b) Expected value = 3.7, standard error = 3.65
(c) Expected value = 3.7, standard error = 3.74

11.2 Smallest standard error in (b), best approximation in (a)

11.3 Normal approximation will be best, standard error will be smallest.

11.4 (a) $s^2_{pooled} = 189.40$
(b) Estimated standard error = 5.84
(c) $t = -0.98$, $|t| < t_{.10/2} = 1.68$, so result is not statistically significant.
(d) $|t| < t_{.20/2} = 1.31$, so $p > .20$ (two-tailed).

11.5 $-15.52 < \mu_1 - \mu_2 < 4.10$. Retain H_0, as in Exercise 11.4, because 0 is included.

11.6 $t' = -0.66$, with approximate df = 6.77; not significant

11.7 Both samples are severely skewed. Variances and sample sizes are unequal, so t' is "less worse." Neither one is really appropriate.

11.8 (a) $3.05 < \mu_1 - \mu_2 < 30.95$
(b) Yes, because 0 is not included in the 95% confidence interval.

11.9 $t = 2.527$ with 22 df. t is between $t_{.01} = 2.508$ and $t_{.005} = 2.819$, so $2(.005) < p < 2(.01)$.

11.10 $t' = 2.527$ with 21.66 df; same bounds on the p-value

11.11 No outliers or skewness in the data, data should be independent, variances are close, and n's are equal; no violations

11.12 $3.03 < \mu_1 - \mu_2 < 30.97$; very slightly wider

11.13 $z = 2.28$, $p = .0226$

11.14 Same conclusion. Assumptions of t tests appeared to be met, so the rank sum test is no more (or less) believable.

11.15 $z = 0.15$; retain H_0.

11.16 In Exercise 11.15, $p = .8808$, two-tailed. The p-values for the t tests may be distorted by skewness.

11.17 $U = 91$

11.18 Each n should equal 131.

11.19 Yes; power (for a two-tailed test) $= P(z > -6.3)$, nearly 1.

11.20 Respectively, power = .0427, .1335, and .3085.

11.21 **(a)** Power increases.
(b) Power increases.
(c) Power increases.

11.22 For 99% confidence, each n should be 148. For other confidence levels, different n's would be needed.

11.23 $t = 1.14$, two-tailed p-value $> .20$. Retain H_0.

11.24 In Exercise 10.15, p-value $< .20$. Pairing has been somewhat effective.

11.25 **(a)** $1.385 < \mu_{\text{diff}} < 5.155$
(b) $-1.970 < \mu_{\text{diff}} < 8.516$

11.26 Greater width assuming independent samples indicates that pairing was effective.

11.27 $t_{\text{diff}} = 3.54$, $t_{\text{pooled}} = 1.27$; t_{diff} is much more conclusive.

11.28 **(a)** $s^2_{\text{pooled}} = 52.959$
(b) Estimated standard error = 2.554
(c) $-1.92 < \mu_1 - \mu_2 < 11.90$

11.29 $-2.97 < \mu_1 - \mu_2 < 12.95$

11.30 **(a)** $t_{\text{pooled}} = 1.954$, 40 df; $.05 < p < .10$
(b) $t' = 1.709$, 35 df; $.05 < p < .10$
(c) Same conclusion; t' is more reliable here.

11.31 **(a)** No obvious direction; use H_a: $\mu_B - \mu_G \neq 0$.
(b) $t_{\text{pooled}} = 3.016$, 24 df; reject H_0.
(c) Data are right-skewed, not normal; the population variances may well be unequal (and the n's are not quite balanced). There's no dependence problem.

11.32 $t' = 2.823$, 12 df; reject H_0.

11.33 **(a)** $z = 2.47$; reject H_0.
(b) Data are badly skewed; the rank sum test is best.

11.34 **(a)** H_0 true; approximating α
(b) The population variances (and the n's) are unequal.
(c) t' should be, and is, more reliable.

11.35 **(a)** .8888
(b) No; it is a long-run probability, the power.
(c) Each $n = 213$

11.36 **(a)** H_a: $\mu_1 \neq \mu_2$ is true; approximating power
(b) The results are virtually identical.
(c) Yes. The calculated power was .8888; the simulation gives .884 (or .883).

11.37 The data are paired.

11.38 $t_{\text{diff}} = 5.294$, 11 df; significant with tiny p-value. Pairing reduced variability.

11.39 **(a)** $t_{\text{pooled}} = -1.881$, 23 df; $.05 < p < .10$
(b) Data are grossly skewed; also, the larger s^2 is for the sample with the smaller n.

11.40 $z = 1.78$, $p = .0750$; same conclusion

11.41 **(a)** $\mu_1 = 60$ and $\mu_2 = 50$; H_0 is false.
(b) Population distributions are outlier-prone, not normal.
(c) Rank sum test has higher (better) power.

11.42 **(a)** H_0 is true.
(b) The populations aren't normal.
(c) Yes; for both tests, we see about 200, 100, 50, 20, and 10, as we should.
(d) One-tailed results are slightly better for the rank sum test.

11.43 **(a)** $1.16 < \mu_I - \mu_M < 19.50$
(b) $-0.12 < \mu_I - \mu_M < 20.78$
(c) Data are paired by age, so (b) is appropriate.

11.44 Matching was ineffective.

11.45 The second proposal gives a slightly smaller standard error, 3.204.

11.46 **(a)** Power = .8531, two-tailed
(b) Power = .8729

11.47 **(a)** There are far too many rejections of H_0.
(b) There's no real difference.

11.48 **(a)** Again, far too many rejections
(b) It's worse!

11.49 **(a)** t_{pooled} is VARIANCES EQUAL $t = -3.7072$; t' is VARIANCES UNEQUAL t. When $n_1 = n_2$, $t_{\text{pooled}} = t'$, algebraically.

11.50 **(a)** SUM OF SCORES = S = 970.00
(b) $p = .0001$; same conclusion (conclusively support H_a)

Chapter 12

12.1 $\chi^2_{.975} = 6.908$, $\chi^2_{.025} = 28.85$

12.2 $P\left(\dfrac{25s^2}{100} > 34.38\right) = .10$ so $P(s^2 > 137.52) = .10$.

12.3 Yes; χ^2 probabilities are very sensitive to nonnormality.

12.4 Approximately, $\chi^2_{.05} = 249.92$ and $\chi^2_{.95} = 181.78$. $2.226 < \sigma_{pop} < 2.610$

12.5 No; large-sample arguments are valid for sample means, not for sample variances.

12.6 T.S.: $\chi^2 = 5.184 > \chi^2_{.95} = 3.940$. Retain H_0.

12.7 χ^2 probabilities are highly sensitive to nonnormality, so the nominal α may be erroneous.

12.8 $6.482 < \sigma_{pop} < 15.535$

12.9 The data in Exercise 10.15 are outlier-prone. Again, χ^2 probabilities are sensitive to nonnormality.

12.10 **(a)** 3.00
(b) 4.00
(c) 4.28
(d) 2.69

12.11 $P(F > 2.86) = .025$

12.12 $0.61 < \sigma_1/\sigma_2 < 2.12$ includes the possibility that the ratio is 1.00. Retain H_0; the discrepancy is not statistically significant.

12.13 $F = 1.289 < F_{.25} = 1.51$ (or 1.52); the two-tailed p-value must be greater than $2(.25) = .50$.

12.14 Yes; the data were very nonnormal and the method is sensitive.

12.15 $F = 0.244 < 1/F_{.10/2,6,26} = 1/2.47 = 0.405$; reject H_0.

12.16 $1/F_{.01/2,6,26} = 1/4.10 = 0.244 =$ calculated F. Thus $p = .01$.

12.17 The data were very skewed and tests for variances are very sensitive to nonnormality. The results are dubious.

12.18 **(a)** $11.37 < \sigma_{pop} < 20.52$
(b) $\sigma_{pop} > 11.83$

12.19 **(a)** $\chi^2 = 12.31 < \chi^2_{.95,23\ df} = 13.09$; reject H_0.
(b) Neither one can be used; the one-sided interval goes in the wrong direction.

12.20 The data are left-skewed, and variance influence is unreliable with nonnormal populations.

12.21 **(a)** $0.2693 < \dfrac{\sigma_S}{\sigma_B} < 1.4448$
(b) The interval includes a ratio of 1.00. The difference is not statistically detectable; retain H_0.

12.22 Yes. Any nonnormality makes variance-test probabilities suspect.

12.23 **(a)** $\chi^2_{.95} = 372.26$, $\chi^2_{.05} = 467.44$
(b) $173.65 < \sigma^2_{pop} < 218.05$

12.24 No; this interval is much wider.

12.25 No; inferences about variances do not become less sensitive to nonnormality as n increases.

12.26 **(a)** They're absurdly bad.
(b) The jackknife isn't as bad.

12.27 **(a)** Yes; they're somewhat closer to the nominal values.
(b) No; the χ^2 results are even worse for small α.
(c) Not for variance inferences

12.28 **(a)** $F = 2.285$, $.01 < p < .025$
(b) F is sensitive to nonnormality.

12.29 $t' = 4.756$ is "off the table" so $p < .001$.

12.30 The jackknife p-value is more believable for nonnormal data.

12.31 **(a)** Very bad; nominal two-tailed $\alpha = .01$ has actual α approximately .08.

12.32 $F = 1.57$ is not close to significant at $\alpha = .05$; the p-value (actually two-tailed) is .2596.

Chapter 13

13.1 **(a)** Let π denote the population proportion of children who could use the system within one hour. H_0: $\pi = .80$. To test that more than 80% can use the system, we want H_a: $\pi > .80$.
(b) R.R.: $z > 1.645$, or equivalently $Y \geq 45$

13.2 $P(Y \geq 45) = .0480$, so R.R.: $y \geq 45$, as in Exercise 13.1.

13.3 Yes; $y = 45$ falls (barely) in the R.R.

13.4 **(a)** By the hypothesis-test-equivalent method, $.856 < \pi < .998$; by the approximate method, $.830 < \pi < .970$
(b) Either way, reject H_0; also, intervals are similar.

13.5 $\pi > .81$

13.6 **(a)** H_0: $\pi = .50$, with a two-sided H_a
(b) R.R.: Assuming $\alpha = .05$, $y \leq 34$ or $y \geq 54$.
(c) The expected numbers of successes and failures are both 44 under H_0, so the approximation will be good.

13.7 Question asks about the power of the test.
(a) Power = .48
(b) n must be 259.

13.8 $z = -2.77$, two-tailed p-value $= .0056$
13.9 Test-equivalent: $.276 < \pi < .472$; approximate: $.252 < \pi < .452$. The intervals have about the same length, and both exclude the value $\pi = .50$; there's little difference.
13.10 **(a)** $.474 < \pi < .794$
(b) $.407 < \pi < .745$
13.11 The intervals have width greater than .30 and seem useless.
13.12 $n = 385$
13.13 Neither interval contains .40; by either method, reject H_0.
13.14 **(a)** A wide interval indicates poor power.
(b) Power = .65
13.15 $n = 48$
13.16 $\chi^2 = 15.21 < \chi^2_{.01,11\ df} = 24.72$. Retain H_0. The sample frequencies are compatible with the hypothesis.
13.17 None of Exp < 1, but 5 of 12 Exp (41.83%) < 5. The value of χ^2 is not close to significant, even at $\alpha = .01$, so precision is not critical.
13.18 $\chi^2 = 5.80 < \chi^2_{.05,3\ df}$. Retain the hypothesis.
13.19 No; the Exp condition only relates to the correctness of using χ^2 tables. The sample size is not large and the actual frequencies vary greatly. We may only say that the data don't reject the theory.
13.20 **(a)** Power = .13
(b) By trial and error, roughly $n = 400$
13.21 **(a)** Exp are 16.5, 9.9, and 6.6; χ^2 is correct.
(b) χ^2 is not significant, but more than the expected number of subjects became fully functional. There is a suggestion that the new treatment is better, but the result might be a fluke.
13.22 Power = .19, a very poor value, because n is only 33 and the noncentrality parameter is small.
13.23 **(a)** Direct calculation
(b) $\chi^2 > \chi^2_{.005,2\ df} = 10.60$, so the result is significant at $\alpha = .005$ or larger.
(c) No; all Exp are much larger than 5.
13.24 No; the discrepancies are small. Because n is large, the difference is statistically detectable, but the actual proportions seem reasonably close to the hypothesized ones.
13.25 **(a)** 1000(.10) = 100
(b) 1, 5, 5, 5, and 34, respectively. These values just barely meet the rule of thumb for a good χ^2-table approximation.
13.26 Yes; the actual frequencies are close to the nominally expected frequencies.
13.27 **(a)** .1660, .1676, .1664, .1664, .1676, and .1660, respectively
(b) $\chi^2 = 4.017 < \chi^2_{.05,5\ df} = 11.07$; retain H_0.
13.28 **(a)** Yes; there's no obvious curve.
(b) No, using Table 9.
13.29 **(a)**

Embarrassment	*Age* 3 & 4	5	6	7	8
Yes	22.31	20.40	28.05	32.51	36.38
No	12.69	11.60	15.95	18.49	20.67

etc.

(b) $\chi^2 = 27.61$, $p < .001$

13.30 **(a)**

		Age 3–4	5	6–12
Embarrassment	Yes	9	19	197
	No	26	13	89

(b) $\chi^2 = 25.44$, $p < .001$
(c) Both are highly conclusive, part (b) even more so.
13.31 **(a)** $\nabla = 0.222$, modest relation
(b) $\nabla = 0.222$
13.32 $\chi^2 = 8.19$, $.025 < p < .05$
13.33 $\gamma = -0.429$ is moderately strong.
13.34 Yes; more than half the Exp are less than 5. The χ^2 statistic is in a "gray area," so a good approximation is needed.
13.35 $\lambda = 0.026$; weak relation
13.36 **(a)**

	Preference Boy	Girl	Either
Father	8.5	5	11.5
Mother	8.5	5	11.5

(b) The observed frequencies virtually equal the expected; there is no evidence of dependence.

13.37 (a)

	Belief		
	Boy	Girl	Either
Father	14.5	5	5.5
Mother	14.5	5	5.5

(b) Observed frequencies differ quite a bit from expected; dependence is suggested.
(c) $\chi^2 = 27.60$, $p < .001$; definite dependence

13.38 (a) No; $71/988 < .10$, $32/988 < .05$, and $4/988 < .005$
(b) $E(\chi^2) = \text{df} = 2$, in this case. Average χ^2 is less than 2.

13.39 Again, simulated α probabilities are close to nominal.

13.40 (a) Appears to be a positive relation
(b) $X^2 = 22.87 > \chi^2_{.05} = 15.51$; significant
(c) $.001 < p < .005$

13.41 $\chi^2 = 25.06$, also significant with the same p-value bounds

13.42 No; all expected frequencies are much greater than 5.

13.43 (a)

		Week 3		
		P	F	G
Week 1	P	19	21	14
	F	21	34	22
	G	14	25	30

Some relation is indicated.
(b) The relation is spurious (indirect), by way of Week 2.
(c) Retain H_0; X^2 isn't close to significant.

13.44 No; X^2 isn't even close to the table values.

13.45 (a) $X^2 = 13.87$
(b) $X^2 > 4.605$; reject H_0.

13.46 No; all expected frequencies exceed 5 and X^2 is huge.

13.47 No; whichever p-value one uses is very large.

13.48 The X^2 statistic isn't remotely close to significant; any inaccuracy isn't a problem.

13.49 (a)

	Response		
	1	2	3
White	.20	.15	.15
Black	.15	.15	.20

(b) H_0: independence; it is false.
(c) $\delta^2 = 2.8571$
(d) Power = .32, approximately

13.50 Trial and error shows that n must be something close to 900.

13.51 (a) $\delta^2 = 4.7619$
(b) Power = .43

13.52 The power in Exercise 13.51 is higher because δ^2 is larger.

13.53 (a) H_0: The distribution of opinions is unchanged.
(b) Use the definition of χ^2 with Exp = 96.4, 73.4, and 30.2, respectively.
(c) $\chi^2 = 5.492 < \chi^2_{.05,2\text{ df}} = 5.991$; retain H_0.

13.54 No; H_0 is never proved, only retained.

13.55 (a) $.149 < \pi < .261$
(b) $.159 < \pi < .270$
(c) Yes, it matters. The interval in (a) contains .151, but the interval in (b) doesn't. The (b) interval is preferable because it's equivalent to a z test.

13.56 In Exercise 13.53, we retained H_0. By creating a smaller table in Exercise 13.55, we could reject H_0, indicating a more effective test.

13.57 (a) CHI-SQUARE = 8.91671 and SIGNIFICANCE (p−value) = .1124. Not conclusive; retain H_0.
(b) No; all Exp > 1 and only 16.7% of Exp < 5.
(c) No; the λ values are small.

13.58 (a) $X^2 = 27.05108$; could also use $\chi^2 = 25.85210$.
(b) No. The p-value is .000 (or .001); reject the model.
(c) No; all Exp > 1 and only 5 of 24 (20.83%) Exp < 5. Even if the p-value is slightly in error, it will be very small.

Chapter 14

14.1 (a)

Group	$\bar{y}_{.j}$	s_j
Control	0.0000	0.0000
Ordinary	3.3333	1.6583
Special	6.8889	1.2693

(b) $\bar{y}_{..} = 3.4074$, SS(Between) = 213.63, SS(Within) = 34.89

(c) $F = 73.48 > F_{.01} = 5.82$; the result is statistically significant.
(d) $F > F_{.001} = 9.34$; therefore, $p < .001$.

14.2 Possible nonnormality (in Special) and nonconstant variance. The F is far beyond table values; even severe violations won't change the conclusion.

14.3 $H = 22.056 > \chi^2_{.001} = 13.82$; same conclusion

14.4 (a) $\bar{y}_{..} = 4.333$

(b)

Source	SS	df	MS	F
Between	79.0	2	39.5	2.64
Within	854.3	57	15.0	
Total	933.3	59		

(c) $.05 < p < .10$

14.5 (a) The data appear right-skewed, and the population variances may differ.
(b) With an unbalanced design, nonconstant variances could be a serious problem and cast doubt on the F result.

14.6 (a) $H = 7.191$, $.025 < p < .05$
(b) Given the potential violations of F assumptions, the Kruskal-Wallis result is more believable. The conclusions are similar, with H being more conclusive.

14.7 (a) Selection biases, no randomization, unbalanced design
(b) Minimally, randomize with a balanced design.

14.8 (a)

Source	SS	df	MS	F
Between	2777.1	2	1388.6	33.38
Within	1372.8	33	41.6	
Total	4149.9	35		

(b) $F = 33.38$
(c) $p < .001$

14.9 Yes; there is a downward trend over time, which invalidates the F test.

14.10 $H = 24.24$, $p < .001$. No; both tests assume independence.

14.11 (a) $-5.16 < \mu_1 - \mu_2 < -1.51$
$-8.72 < \mu_1 - \mu_3 < -5.06$
$-5.38 < \mu_2 - \mu_3 < -1.73$
(b) All null hypotheses may be rejected; 0 is never included in the confidence interval.

14.12 Yes; the Tukey method is sensitive to nonconstant variance.

14.13 (a) $\hat{L}$ is a comparison; the constants add to 0. The quantity $(18\bar{Y}_2 + 12\bar{Y}_3)/30$ is the mean for all depressed subjects.
(b) $-0.053 < L < 4.319$, not significant

14.14 (a) $|\hat{L}| = 1.222 < W = 3.47$; not significant
(b) $|\hat{L}| = 1.222 < S = 3.62$; not significant
(c) Same conclusion

14.15 (a) $F = 1.954 < 2.92$ (roughly); retain H_0.
(b) $W = 15.08$ (roughly); no significant differences
(c) Same conclusion
(d) $\hat{L} = (0.5)\bar{Y}_{C1} + (0.5)\bar{Y}_{C2} - (0.5)\bar{Y}_{E1} - (0.5)\bar{Y}_{E2} = -9.5$
Use a post hoc comparison to preserve experimentwise α; $|\hat{L}| < S = 11.70$; not significant.

14.16 (a) Nonconstant variance
(b) The Tukey procedure

14.17 $SS(\hat{L}) = 68.26$, 86.4% of SS(Between)

14.18 $|\hat{L}| = 2.133 < S_{\hat{L}} = 2.51$; not significant

14.19 (a) $L_1 = \mu_2 - \mu_3$,
$L_2 = \mu_1 - 0.5(\mu_2 + \mu_3)$
(b) $0(1) + 1(-0.5) + (-1)(-0.5) = 0$
(c) $SS(\hat{L}_1) = 56.903$, $SS(\hat{L}_2) = 156.734$
(d) Yes; we have $3 - 1 = 2$ orthogonal comparisons.

14.20 In both cases, $|\hat{L}| > S_{\hat{L}}$; both comparisons are significant.

14.21 (a) $SS(\hat{L}) = 902.5$
(b) $|\hat{L}| = 9.5 < S_{\hat{L}} = 10.75$; not significant
(c) Not significant

14.22 (a) For $\hat{L}_1$ and $\hat{L}_2$, $0.5(0) + 0.5(0) + (-0.5)(0) + (-0.5)(0) = 0$
For $\hat{L}_1$ and $\hat{L}_3$, $0.5(-1) + 0.5(1) + (-0.5)(0) + (-0.5)(0) = 0$
For $\hat{L}_2$ and $\hat{L}_3$, $0(-1) + 0(1) + (-1)(0) + 1(0) = 0$
(b) $SS(\hat{L}_1) = 902.5$, $SS(\hat{L}_2) = 11.25$, $SS(\hat{L}_3) = 2.45$; the sum is 916.2 = SS(Between).

14.23 (a)

Source	SS	df	MS	F
Handedness	493.92	1	493.92	0.135
Finger	592.48	1	592.48	0.162
Interaction	18064.48	1	18064.48	4.945
Error	394559.91	108	3653.33	

(b) $F = 4.945 < F_{.01} = 6.85$, roughly. It is not significant at $\alpha = .01$; it is significant at .05.
(c) Profiles cross; there appears to be interaction.
(d) Yes; it is more than 30 times the SS for either main effect.

14.24 (a) Yes; the c weights add to 0 across rows and down columns.
(b) $SS(\hat{L}) = 18064.48 = SS(\text{Interaction})$. This $\hat{L}$ uses the only df for interaction in the data.
(c) Significant at .05, not at .01.

14.25 (a)

Source	SS	df	MS	F
Handedness	493.92	1	493.92	0.135
Finger	18064.48	1	18064.48	4.945
Interaction	592.48	1	592.48	0.162
Error	394559.91	108	3653.33	

SS(Error) is based on the same n's and standard deviations.
(b) No. $F = 0.162 < 1$; F can't be significant.
(c) Yes. $F = 4.945 > F_{.05} = 3.92$ (roughly)

14.26 (a)

Source	SS	df	MS	F
Anxiety	3.375	1	3.375	2.71
Number	116.805	3	38.935	31.22
Interaction	35.445	3	11.815	9.47
Error	109.73	88	1.247	

(b) The plot shows crossing, but generally the same slopes. There is modest interaction.
(c) SS(Interaction)/SS(Total) = .133, indicating a moderate explanatory value.
(d) There is reasonable consistency in the patterns, so it seems meaningful.

14.27 (a) $|\hat{L}| = 3.0 > S_{\hat{L}} = 1.64$; it is significant.
(b) Weights add to 0 across rows and down columns.
(c) $SS(\hat{L}) = 27.0$, 76% of SS(Interaction)

14.28 (a) No, it's a comparison among levels of the column factor.
(b) $SS(\hat{L}) = 98.415$
(c) Post hoc test: $|\hat{L}| = 1.247 > S_{\hat{L}} = 0.794$; significant

14.29 (a) Means, with standard deviations in parentheses:

	Abstract	Concrete
PS	327.4 (36.6)	285.1 (40.8)
SS	275.5 (42.5)	313.7 (40.1)
B	258.4 (49.1)	298.8 (31.0)

(b)

Source	SS	df	MS	F
Major	7708.0	2	3854.0	2.36
Method	2196.0	1	2196.0	1.34
Interaction	22207.0	2	11104.0	6.80
Error	88182.0	54	1633.0	

(c) Nothing serious, perhaps a mild outlier

14.30 (a) Yes; there's a strongly nonparallel profile plot.
(b) $F = 6.80 > F_{.005} = 5.9$ (roughly); significant at $\alpha = .005$, though not at the extreme $\alpha = .001$

14.31 (a) $F = 1.34 < F_{.25} = 1.35$; p-value $> .25$.
(b) Not very meaningful, given the serious interaction

14.32 (a) Profiles are nearly parallel.
(b) Little interaction

14.33 (a)

Source	SS	df	MS	F
Neuroticism	149.784	1	149.784	35.66
Relevance	85.162	2	42.581	10.14
Interaction	2.359	2	1.180	0.28
Error	226.800	54	4.200	

(b) SS(Interaction) is small and $F < 1$ (thus $F <$ all table values).
(c) For both factors, $F > F_{.001}$, so $p < .001$.

14.34 (a) M and H means differ significantly, as do L and H means.
(b) $SS(\hat{L}) = 84.681$, account for 99% of SS(Relevance)

14.35 (a) Weights sum to 0 across rows and down columns.
(b) $SS(\hat{L}) = 1.875$
(c) $|\hat{L}| = 0.75 < S_{\hat{L}} = 7.94$; not significant

14.36 (a) No; all interaction SS are relatively small.
(b) None are significant at $\alpha = .05$.

(c) All are significant with $p < .001$.

14.37 (a) All pairs are significantly different, by the Tukey procedure.
(b) $SS(\hat{L}_1) = 153.76$

14.38 (a) $1(-0.5) + 0(1) + (-1)(-0.5) = 0$
(b) $SS(\hat{L}_2) = 0.48$, SS(Anxiety) = 154.24

14.39 (a) Schools and instruction are factors; child and team are not factors.
(b) Schools is a random factor. Instruction is fixed.
(c) No nested factors.

14.40 No, because of the random factor

14.41 (a) Rats, strain, and maze are factors.
(b) Rats should be random, strain probably should be treated as fixed, and maze might be regarded as random if these mazes were only a sample.
(c) Rat is nested within strain.

14.42 (a) $F = 0.90$, not significant
(b) $F = 12.78$, $p < .001$

14.43 The difference between any *P* mean and any *C* mean is significant—that is, larger than 3.848.

14.44 (a) $\hat{L}$ = (average of *P* means) − (average of *C* means) = 7.00
(b) $SS(\hat{L}) = 1881.6$
(c) Using the planned comparison method, $\hat{L} > S_{\hat{L}} = 1.42$.

14.45 (a) Factors are experimenter, lesion, and label. Rat is not a factor.
(b) Lesion and label are fixed; experimenter is a random factor.
(c) All factors are crossed.

14.46 (a) No; *F* is only 1.21, not significant at $\alpha = .10, .05$, or $.01$.
(b) Both main effects are significant (detectable).

14.47 The SS for label is higher than the SS for lesion.

14.48 (a) $\delta^2 = 2.833$
(b) Power = .10, roughly.

14.49 (a) Power increases.
(b) Power decreases.
(c) Power increases.

14.50 (a) $\delta^2 = 2.04$
(b) Power = .18, approximately

14.51 Power increases; the means are farther apart.

14.52 Increasing the number of categories without increasing the overall *n* decreases power.

14.53 Increasing denominator df increases the power.

14.54 (a) $\delta^2 = 14$
(b) Power = .91, roughly

14.55 For one reason, to reduce the effect of nonconstant variance

14.56 (a)

Source	SS	df	MS	F
Between	37.17	2	18.58	7.42
Within	112.75	45	2.51	
Total	149.92	47		

(b) $F > F_{.05} = 3.23$, approximately. Yes; the difference is detectable (significant).
(c) $.001 < p < .005$

14.57 No serious problems

14.58 (a) $-0.609 < \mu_A - \mu_B < 2.109$
$0.766 < \mu_A - \mu_C < 3.484$
$0.016 < \mu_B - \mu_C < 2.734$
(b) Both *A* and *C* and also *B* and *C*

14.59 (a) $SS(\hat{L}) = 32.667$, 87% of SS(Groups)
(b) $|\hat{L}| = 1.75 > S_{\hat{L}} = 1.233$. Yes; it's significant.
(c) $0.5(1) + 0.5(-1) + (-1)(0) = 0$

14.60 Yes; the scores change in opposite directions.

14.61 Use the definitions.

14.62 (a) SS(Interaction) is largest, agreeing with Exercise 14.60.
(b) $F = 5.306 > F_{.10} = 2.30$ (roughly); significant with *p* slightly less than .005

14.63 (a) SS(Between) = 6042.25, SS(Within) = 12761.09, SS(Total) = 18803.34
(b) $\eta^2 = 0.321$, only a fairly good predictor
(c) $F = 6.945 > F_{.01} = 4.31$ (roughly); in fact, $p < .001$.

14.64 (a) Only the means for grades 6 and 10 differ significantly.
(b) The Kramer modification has no effect when the design is balanced.

14.65 (a) Populations not normal; there are outliers.
(b) $H = 19.990$; same conclusion

14.66 No; the profiles are roughly parallel.

14.67 (a) 0.629, 0.252, and 0.003, respectively

(b) No; interaction accounts for very little of SS(Total).

14.68 (a) $F = 0.803 < 1$ and therefore never significant
(b) Yes; both F's are far beyond all F-table values.

14.69 Yes; a repetition would use the same encoding and test conditions.

14.70 (a) The factors are child, group, and angle. Child should be a random factor, but the others should be fixed.
(b) Angle is within-subjects; group is between-subjects.
(c) Child is nested in group and crossed with angle; group and angle are crossed.

14.71 (a) There is some crossing of profiles.
(b) SS(Interaction) is much less than SS(Angle) but is larger than SS(Training); thus there is some apparent interaction.
(c) $F = 1.159 < F_{.25} = 1.41$; not significant

14.72 (a) Training is not significant at $\alpha = .05$, but $.05 < p < .10$. Angle is significant with $p < .001$.

14.73 The differences between $\bar{y}_1$ and $\bar{y}_2$ and between $\bar{y}_3$ and $\bar{y}_4$ are *not* statistically significant; all other differences are.

14.74 (a) $\hat{L} = 2.39$
(b) $SS(\hat{L}) = 11.4242$, .98 of SS(Angle)
(c) Yes; $|\hat{L}| > 0.496$

14.75 (a) No; all SS for these interactions are relatively small.
(b) No; all F's are less than $F_{.10,3,64} = 2.18$ (roughly).

14.76 (a) $F = 19.864 > F_{.05} = 2.76$. Reject H_0.
(b) $p < .001$

14.77 (a) 7.4, 8.2, 10.05, and 8.35, respectively
(b) Means 1 and 2 are *not* significantly different, nor are means 2 and 4. All other pairs are significantly different.

14.78 (a) $\hat{L} = 10.05 - 7.4 = 2.65$
(b) $SS(\hat{L}) = 70.225$
(c) By Tukey method, $2.65 > 0.935$, so significant

14.79 (a) No; F is only 1.18.
(b) None are significant.

14.80 There is severe nonnormality, specifically severe outliers. This will affect both procedures about equally.

14.81 (a) $H = 21.645$, 4 df
(b) $p < .001$, a much different result

14.82 (a) $F = 4.11$, $.005 < p < .01$, a very different conclusion
(b) There still are three outliers.

14.83 (a) $F = 12.38 > F_{.01} = 4.39$; there is a detectable difference.
(b) PR > F is .0001.

14.84 The 70 mean differs from all the other means.

14.85 (a) The interaction SS, shown as SUBTASKS BY DEADENDS, is fairly small. Interaction is not large.
(b) No; the p-value (Sig. of F) is .451.

14.86 Variability increases as the mean increases.

14.87 (a) No; SS(Interaction) is very small.
(b) For both tests, p is shown as .000.
(c) No obvious pattern

14.88 (a) The profiles are nearly parallel; there's little interaction.
(b) $F < 1$; not significant.
(c) For both tests, $p < .001$.

14.89 The standard deviation for stables is slightly but consistently larger than the standard deviation for neurotics.

14.90 Means 3 and 1 are significantly different, as are means 3 and 2.

14.91 (a) There is modest nonparallelism (interaction).
(b) Yes, significant at $\alpha = .05$; PR > F is .0497.
(c) Yes; for both tests, $p = .0001$.

14.92 (a) $\hat{L} = .5(\bar{y}_{4.} + \bar{y}_{5.}) - .5(\bar{y}_{2.} + \bar{y}_{3.})$
(b) Assuming a post-hoc comparison, $|\hat{L}| < S_{\hat{L}} = 0.907$; the comparison is significant.

14.93 (a) Presumably fixed (like treatment factors).
(b) $\eta^2 = 0.146$. $F = 8.03 > F_{.05} = 2.21$; there is significant interaction.

14.94 There is a time trend. Dependence of observations on time is a serious violation of assumptions.

Chapter 15

15.1 (a) The rate of change of Y as X changes is

constant.

(b) Slope = predicted change in aptitude score when reading score changes by one point; intercept = predicted aptitude score when reading score is 0.

15.2 No evidence of nonlinearity

15.3 **(a)** The slope is the rate of increase of speed per unit change in intensity.

(b) The plot is roughly linear; a transformation might also be tried.

15.4 Yes; the plot is very close to linear.

15.5 **(a)** No obvious curve

(b) If X_1 increases by 1 and X_2 remains constant, Y is predicted to increase by 2; if X_2 increases by 1 and X_1 remains constant, Y is predicted to increase by 3. If both X_1 and X_2 are 0, Y is predicted to be 10. The intercept is meaningful because $X_1 = X_2 = 0$ is possible.

(c) There is no relation at all.

(d) Because X_1 and X_2 are uncorrelated

15.6 **(a)** Nothing obvious

(b) There is positive correlation.

(c) Yes; the coefficient of X_1 changed when X_2 was omitted, indicating that X_1 and X_2 were correlated.

15.7 **(a)** The intercept term is the predicted solution time if the solved puzzle is shown for $x_1 = 0$ seconds (that is, not shown at all) and if strategy C was suggested ($x_2 = x_3 = 0$). Because $x_1 = 0$ would be a drastic change in condition, the intercept isn't very meaningful. Given any suggested strategy, the solution time is predicted to decrease 5.1 seconds per second of increased exposure time of the solution; suggesting strategy A predicts a 4.7-second decrease, compared to the baseline, C; suggesting B predicts a 9.3-second increase.

(b) A nonlinear model is needed.

15.8 **(a)** Plug in values to the dummy variables to obtain an equation for each strategy. The 80.1 and −5.0 coefficients are the intercept and slope for strategy C, the baseline. The −4.4 and 9.6 coefficients are intercept differences, and the 0.2 and −0.3 coefficients are slope differences, all relative to strategy C.

(b) No; the slope differences (coefficients of the product terms) are small.

(c) We still need a (different) nonlinear model.

15.9 The slopes represent predicted changes in Y per unit change in the variable, holding all other variables constant. The coefficient of X_3 indicates the predicted effect of special training. The intercept is meaningless.

15.10 Interactions allow for different slopes for training vs. no training.

15.11 Both are linear in the coefficients; both fit the GLM.

15.12 **(a)** $S_{xy} = 703.88$, $S_{xx} = 688.94$, $S_{yy} = 809.76$

(b) $\hat{y} = 9.304 + 1.022x$

(c) SS(Residual) = 90.7

(d) $s_{Y|X} = 2.46$

15.13 **(a)** $\hat{y} = 10 + 2x_1$

(b) Respectively, 1, 2, −3, 6, −1, −5, 4, 0, −4, 2, −1, −1, 2, 0, −2

(c) SS(Residual) $= 1^2 + \cdots + (-2)^2 = 122$; $s_{Y|1} = 3.063$

(d) SS(Residual) = 1742 − 10(150) − 2(60) = 122

15.14 **(a)** $\hat{y} = 10 + 3x_2$

(b) $s_{Y|2} = 3.419$

(c) X_1; smaller $s_{Y|X}$

15.15 **(a)** r_{xy} is larger for X_1 (.704 vs. .610); X_1 is again the better predictor.

(b) The correlation is 0; a slope will not change if the other variable is omitted from the prediction.

15.16 **(a)** $\hat{y} = 10 + 3x_1$

(b) $\hat{y} = 10 + 5.5x_2$

15.17 $s_{Y|1} = 2.935$, $s_{Y|2} = 3.282$; X_1 is better.

15.18 $r_{y1} = .841$, $r_{y2} = .796$; X_1 is better.

15.19 **(a)** $\hat{y} = -9.817 + 48.697x$

(b) $\hat{y} = -21.170 + 10.357x$

(c) $s_{Y|X}$ is 11.807 for $D = 0.4$; it is 5.897 for $D = 4.0$.

15.20 **(a)** No; the units are different.

(b) No; for the same reason.

15.21 The retransformed model has slightly higher SS(Residual) for the $D = 0.4$ data, but a much lower one for $D = 4.0$.

15.22 **(a)** $r_{yx} = .942$

(b) $s_{yy} = 809.76$, SS(Residual) = 90.71

(c) $(.942)^2 = .888 = (809.76 - 90.71)/809.76$

15.23 Yes, as the sum of two r^2 values; the X's are uncorrelated.

15.24 **(a)** Correlation = .562; error reduction = $(.562)^2$ = .316

(b) −.241 and .232, respectively; X_1 is best.

15.25 **(a)** $R^2_{Y|1,2,3}$ = .381

(b) No; the X's are correlated.

15.26 $r^2_{Y3|1,2}$ = .010; adding X_3 explains only 1.0% of the unexplained Y variation.

15.27 Adjusted $R^2_{Y|1,2,3}$ = .226

15.28 **(a)** ABILITY, correlation .65453

(b) R-SQUARE =0.4538 means 45.38% of Y variability has been accounted for.

(c) Very little; ADJ R-SQ is close to R-SQUARE.

15.29 No; the independent variables are correlated.

15.30 **(a)** EDUCN has the larger correlation (in magnitude) with WELL.

(b) .42543, .18099

(c) .16410, only slightly less

15.31 **(a)** −.000095 is tiny in context.

(b) Near 0

15.32 **(a)** F = 2.458

(b) No; $F_{.01}$ = 5.95.

(c) $.10 < p < .25$

(d) There is no conclusive evidence of predictive value.

15.33 The t values are 1.50, −1.00, and 0.35; none of them are larger in magnitude than $t_{.10/2,12\text{ df}}$ = 1.782. There's no conclusive evidence that adding any one X to the others improves predictability.

15.34 $F = 2.94 > F_{.10,2,12} = 2.81$; there is evidence that (at least one of) X_2 and X_3 have additional predictive value when added to X_1 as a predictor.

15.35 **(a)** $F = 85.87$, $p < .001$

(b) None of them are significant.

15.36 **(a)** $F = 1.004 < F_{.05,2,48}$; retain H_0.

(b) $F = 0.684$; retain H_0.

15.37 The X's are highly correlated; any one of them serves about as well as all of them.

15.38 **(a)** F VALUE = 50.684

(b) $p = .0001$; there is conclusive evidence of predictability.

(c) In the STANDARD ERROR and T FOR H0 columns

(d) ABILITY, ANXIETY

15.39 **(a)** F = 10.71794

(b) p is tiny; definite predictability.

15.40 There's no evidence that AGE adds predictive value to EDUCN; EDUCN definitely adds predictive value to AGE.

15.41 **(a)** Estimated standard error = .0937

(b) $0.822 < \beta_1 < 1.222$

(c) Yes; interval excludes 0.

15.42 $.843 < \rho < .979$; 0 is not close to this interval.

15.43 **(a)** $36.635 < \mu < 39.205$

(b) $32.522 < Y < 43.318$

(c) No; $x^* = 28$ is very close to $\bar{x} = 29.41$.

15.44 **(a)** Estimated standard error = 0.5587

(b) $1.004 < \beta_1 < 2.996$

15.45 $.380 < \rho < .874$ seems very wide.

15.46 **(a)** $16.135 < Y < 43.865$

(b) Severe extrapolation

15.47 **(a)** The relevant standard errors (in the STANDARD ERROR column) are 0.106372, 0.184090, and 0.052018.

(b) $0.247 < \beta_{ABILITY} < 0.664$
$-0.234 < \beta_{ESTEEM} < 0.487$
$0.047 < \beta_{ANXIETY} < 0.251$

The true slope of ESTEEM may be 0.

15.48 $0.2431 < \rho < 0.4908$, $0.0591 < \rho^2 < 0.2409$

15.49 $-0.0513 < \beta_{AGE} < 0.0515$
$-0.7018 < \beta_{EDUCN} < -0.1504$

15.50 No nonlinearity in plot, no time-ordered autocorrelation, constant variability around the line, no outliers, should be no measurement error.

15.51 **(a)** Yes; p = .0001

(b) Heteroscedasticity

(c) Not badly; heteroscedasticity has only mild effect on the test, and the test statistic was huge.

15.52 **(a)** With $n = 87$, $-1.16 < Y < 8.71$

(b) No; heteroscedasticity distorts such intervals.

15.53 **(a)** R^2 = .497

(b) $F = 13.33 > F_{.01,2,27} = 5.49$; significant

(c) t for SLEEP is significant, t for DAY is not.

15.54 Autocorrelation; cyclical pattern in plot against DAY and Durbin-Watson only 0.86. Autocorrelation biases R^2, F, and t; the answers to Exercise 15.53 are doubtful.

15.55 **(a)** No; normal plot is linear.

(b) No; not time series data.

15.56 No; age and education can be measured accurately. We could not see this problem, given only the measured data.

15.57 There are three outlier points; these points will diminish the correlation substantially.

15.58 (a) There's a generally increasing relation.
(b) There's no obvious curve in the plot.

15.59 (a) $\hat{\beta}_1 = .505$, $\hat{\beta}_0 = 52.033$
(b) $\hat{\beta}_1$ indicates that a one-point increase in Kirton score predicts a .505 increase in Myers-Briggs score. The intercept term corresponds to $X = 0$, a value not present in the data; thus $\hat{\beta}_0$ is not meaningful.

15.60 (a) $r_{YX} = .719$, $r_{YX}^2 = .517$
(b) SS(Residual) = 9641.525, SS(Total) = 19960
(c) X accounts for 51.7% of variation in Y.

15.61 (a) Modest linear relation in the plot; an outlier at $x_1 = 8$, $y = 1$
(b) Some linear relation, no obvious curve or outliers
(c) X_2 appears better.

15.62 (a) $r_{Y1} = .643$, $r_{Y2} = .749$
(b) X_1 alone accounts for $(.643)^2 = .413$ of the squared error of Y; X_2 alone accounts for $(.749)^2 = .561$ of the squared error of Y.
(c) X_2 is better.

15.63 (a) $\hat{y} = .165 + .806x_1$
(b) An increase of 1 in recommendation predicts an increase of .806 in quality rating.
(c) It's meaningless; a 0 recommendation isn't possible.

15.64 There's at least one outlier. No obvious nonlinearity, nonconstant variance, or autocorrelation can be seen.

15.65 (a) The data are curved; a linear prediction would be a poor fit.
(b) $r_{YX} = .285$, $r^2 = .081$. Linear prediction accounts for only 8.1% of Y variation.
(c) No; we need a nonlinear prediction.

15.66 (a) $\hat{\beta}_0 = 72.000$, $\hat{\beta}_1 = -0.200$, $\hat{\beta}_2 = -0.600$
(b) R-SQUARED = .530 means that X_1 and X_2 together account for 53% of the variation in Y.

15.67 (a) $r_{Y1}^2 = (-.319)^2 = .102$, $r_{Y2}^2 = (.655)^2 = .429$
(b) Multiple R^2 is .530; the sum of the individual r^2 values is .531, equal except for roundoff error.
(c) The correlation must be 0.

15.68 −5, 3, 4, −2, 3, 3, −5, −1, 2, −6, 1, 3. None of them are far from the rest. All are within the "fences" defined in Chapter 2; there are no outliers.

15.69 We are checking for nonconstant variance; it doesn't seem to be a serious problem.

15.70 The constant variance assumption refers to the residuals, not to the standard errors.

15.71 (a) Only 18.7% of variation accounted for
(b) $F = 12.088$; $p = .0001$. There's a detectable relation.
(c) STATEDAV definitely adds predictive value; no evidence that STATTIME adds value.

15.72 (a) Model isn't correct; there's a curve in the STATEDAV plot.
(b) Include a $(\text{STATEDAV})^2$ term.

15.73 (a) $F = 252.8$; definitely some predictive value
(b) $R^2 = .815$ (unadjusted) and .812 (adjusted); negligible effect of adjustment
(c) Both variables add predictive value.
(d) $-2.0405 < \beta_1 < -1.1353$
$-3.8794 < \beta_2 < -1.6068$

15.74 Partial $r^2 = .167$; adding AGE accounts for 16.7% of the residual variability from the KNOWN prediction.

15.75 (a) No fan shape; no problem
(b) No curve; no problem
(c) Nothing appears absurdly wild.

15.76 (a) $F = 8.51$, $p = .0001$; there is some predictive value in the equation.
(b) Only SEX fairly definitely adds predictive value.
(c) Yes, especially EMPATHY and FEMININ

15.77 (a) To allow for interaction
(b) Incremental $F = 1.925$, 3 and 132 df, $.10 < p < .25$; no conclusive evidence that null hypothesis is false

15.78 (a) No fan shape
(b) Nothing that jumps out

15.79 **(a)** $.678 < \rho < .826$; 0 is not included.
(b) $t = 1.51$, not close to significant at $\alpha = .05$
(c) Part (a) indicates that IQ by itself has predictive value for NCT; part (b) indicates that IQ may not add predictive value to SAT in predicting NCT. Note that SAT and IQ are correlated.

15.80 Constant variance (YHAT plot), nonlinearity (SAT plot), and no outliers (both plots). There may be one large positive outlier, with rather high leverage.

Index

For Review: Key Concepts and Definitions (continued)

Test statistic (T.S.) a summary figure derived from the data, used to test the hypotheses

Rejection region values of the T.S. that are deemed to support H_a

Type I (false positive) error rejecting H_0 when it is true

Type II (false negative) error retaining H_0 when H_a is true

α $P(\text{reject } H_0 | H_0 \text{ true})$

β $P(\text{retain } H_0 | H_a \text{ true})$

Power the probability that a statistically significant result will be found; $P(\text{reject } H_0 | H_a \text{ true}) = 1 - \beta$

***p*-value** probability of obtaining a test statistic value at least as extreme as the actual value, given H_0; small values indicate conclusive rejection of H_0

Statistically significant result a result that allows rejection of the null hypothesis and support of the research hypothesis

Universal Rejection Region reject H_0 if and only if the *p*-value is less than α

Rejection region related to confidence interval reject H_0 if and only if the H_0 value is not in the confidence interval

Size of effect hypothetical difference between the true mean and the H_0 mean, often taken as a fraction of σ_{pop}

Chapter 10

Estimated standard error standard error obtained by replacing an unknown population value such as σ_{pop} with a sample estimate such as s

***t* statistic** notation used when true standard error is replaced by estimated standard error

Degrees of freedom (df) the number of deviations that are free to vary; $df = n - 1$ in a single sample

Assumptions of *t* methods observations are independent and drawn from a normal population

Paired samples samples in which each observation in one sample is experimentally linked to a specific observation in another sample

Paired-sample methods methods based on analysis of the single sample of differences

Signed-rank test test based on ranking differences in order of magnitude; less sensitive than a *t* test to outliers

Median (sign) test a binomial test for a hypothetical median

Power of *t* probability that H_a will be supported when it is true; depends on n, α, σ_{pop}, and deviation of true mean from H_0 mean

Chapter 11

Assumptions of two-sample methods independence within and between samples

Balanced design design calling for an equal number of observations in both samples; minimizes the effect of violation of normality and constant variance assumptions

Pooled variance average of sample variances, weighted by df

Heteroscedasticity nonconstant variance; causes t_{pooled} probabilities to be wrong when design is not balanced

***t′* method** an approximate *t* test for use with unbalanced designs and possible nonconstant variance

Rank sum method a rank-based method useful with independent samples from nonnormal populations

Relative efficiency a measure of the effectiveness of one test relative to another

Advantage of pairing if successful, pairing reduces the amount of variability and increases efficiency

Chapter 12

χ^2 distribution useful for making inferences about variance and standard deviation; used only on positive numbers; mean = df, variance = 2(df); sensitive to nonnormality for any n

***F* distribution** useful for comparing variances; used only on positive numbers; mean approximately 1; also sensitive to nonnormality for any n

Jackknife methods useful for making inferences about variances; less sensitive to nonnormality, particularly for large n

Ansari-Bradley method rank method for comparing variances; doesn't assume normality

Chapter 13

Normal approximation to binomial basis for test and confidence interval for a proportion; close if expected numbers of successes and failures are both at least 5

χ^2 goodness-of-fit test a method for testing whether sample frequencies are consistent with a hypothetical set of probabilities

Multinomial assumptions for χ^2 tests independent observations, constant probability of success, fixed n

Danger of goodness-of-fit test the real research hypothesis is formulated as a null hypothesis

Noncentral χ^2 the distribution underlying power calculations for a χ^2 test

χ^2 independence test test to see if there is a relation between two qualitative variables

Strength of relation must be measured by methods of Chapter 3, not by χ^2

χ^2 tables acceptable if all Exp ≥ 1 and at least 80% of Exp are ≥ 5

Log-linear model approach useful in testing more complex hypotheses, including conditional independence

Chapter 14

Factor an experimental independent variable

Level a particular value of a factor